Molecular Biology and Genetics of the Lepidoptera

Marian R. Goldsmith   František Marec

Principal translator   He Ningjia

西南师范大学出版社
国家一级出版社 全国百佳图书出版单位
CRC Press
Taylor & Francis Group

鳞翅目昆虫的分子生物学和遗传学

玛丽安·戈德史密斯　弗兰蒂塞克·马莱克　编著

何宁佳 主译

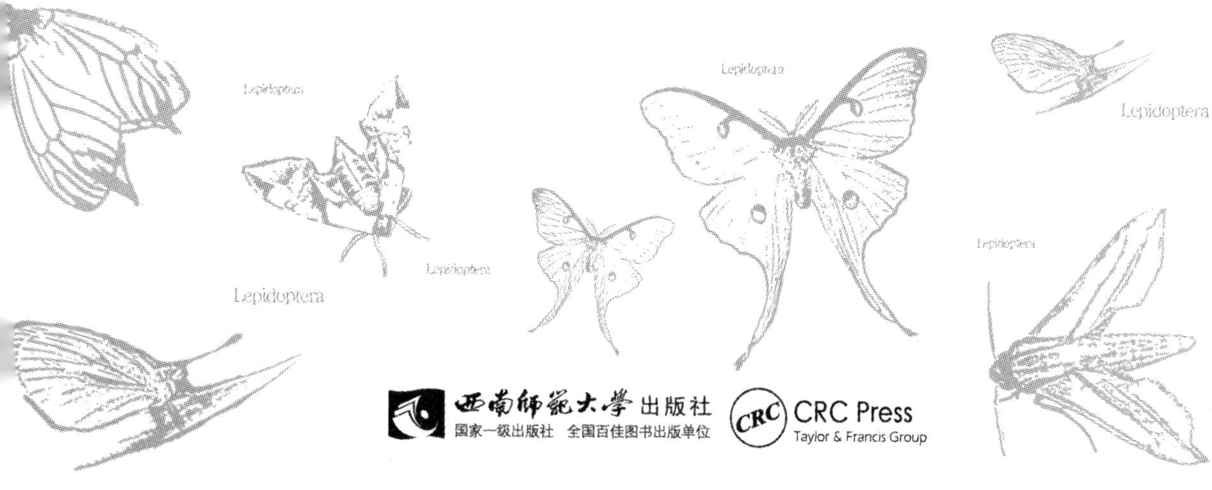

西南师范大学出版社
国家一级出版社　全国百佳图书出版单位
CRC Press
Taylor & Francis Group

Molecular Biology and Genetics of the Lepidoptera / by Marian R. Goldsmith and František Marec /
ISBN:978-1-4200-6014-0

版贸核渝字(2010)第 229 号

图书在版编目(CIP)数据

鳞翅目昆虫的分子生物学和遗传学/(美)戈德史密斯,(捷克)马莱克编著;何宁佳主译.—重庆:西南师范大学出版社,2010.12
ISBN 978-7-5621-5079-4

Ⅰ.①鳞… Ⅱ.①戈… ②马… ③何… Ⅲ.①鳞翅目—分子生物学②鳞翅目—分子遗传学 Ⅳ.①Q969.42

中国版本图书馆 CIP 数据核字(2010)第 214434 号

鳞翅目昆虫的分子生物学和遗传学

Marian R. Goldsmith　František Marec 编著
何宁佳　主译

出 版 人:周安平
责任编辑:杜珍辉　任志林　伯古娟　卢　旭
书籍设计:CASTALY 尚品视觉　周　娟　尹　恒
出版发行:西南师范大学出版社
印　　刷:重庆科情印务有限公司
开　　本:787 mm×1092 mm　1/16
印　　张:25.75
插　　页:8
字　　数:580 千字
版　　次:2011 年 3 月第 1 版
印　　次:2011 年 3 月第 1 次
书　　号:ISBN 978-7-5621-5079-4

定　　价:56.00 元

作者简介

玛丽安·戈德史密斯：美国罗德岛大学生命科学系主任、教授。她与福蒂斯·卡法托斯(Fotis Kafatos)一道在1988年发起成立了国际鳞翅目分子生物学和遗传学研讨会。该研讨会每隔两到三年在希腊的克里姆帕里(Kolympari)召开，届时各国从事鳞翅目昆虫研究的科学家汇聚一堂，参加这个在世界上独一无二的研讨会。作为使用分子生物学工具来进行研究的遗传学家，戈德史密斯早期的研究以家蚕为模式系统。1972年她在哈佛大学跟随福蒂斯·卡法托斯做博士后研究期间，经过了两个月在日本蚕桑实验站(现更名为日本国立农业生物科学研究所NIAS)与伊藤敏雄和位于三岛的日本国立遗传研究所的田岛弥太郎合作研究后(由美国国家科学基金资助)，她首次把家蚕引入福蒂斯·卡法托斯的实验室，从事其卵壳蛋白的表达和调控研究。随后，利用多次的休假和短期访问的机会，玛丽安与日本九州大学的坂口文吾、土井良宏，日本国立健康研究所(东京)的前川秀彰，日本国立蚕丝昆虫研究所(筑波)的田村秀树、原和二郎，东京大学的小林正彦、岛田透建立了研究合作关系。1997～1998年她作为访问教授在东京大学工作。与此同时，她与NIAS的三田和英在家蚕的基因组计划的实施中建立了长期的合作关系。近年来，作为客座教授的她与中国上海植物生理生态研究所的黄永平、苗雪霞一道开发了针对复杂形状的作图和定位克隆工具。她是美国科学促进会成员。因在家蚕卵壳蛋白基因分子遗传学研究中所作出的科学贡献，她获得了2002年度日本蚕丝学会奖。她与亚当·威尔金斯在1995年共同编著了由剑桥大学出版社出版的《鳞翅目昆虫分子模式系统》一书。

弗兰蒂塞克·马莱克：捷克共和国契斯凯巴德杰维契捷克科学院生物中心昆虫研究所的资深研究员，南波黑米亚大学科学学院分子细胞生物学和遗传学教授。从1990年起成为德国洪堡基金会资助的研究员。当他在1991～1992年和1998年以洪堡学者的身份工作时，与德国吕贝克医科大学生物研究所的瓦尔特·特劳特建立了长期的合作关系。在20世纪90年代末，他与北海道大学(日本札幌)应用生物学实验室的佐原健建立了紧密的合作关系。与多数遗传学家一样，他最初研究果蝇，对突变遗传学感兴趣，但不久就迷上了鳞翅目昆虫的遗传学。他早期的研究兴趣包括采用化学和辐射突变的方法开发针对鳞翅目害虫的遗传控制策略、鳞翅目物种抗离子射线的染色体机制，以及减数分裂染色体配对中联会复合体的形成（该复合体是一个特殊的核结构，介导两条同源染色体的紧密联会）。他大部分早期工作是在地中海粉螟中完成的，地中海粉螟在遗传研究中是仅次于家蚕的第二大鳞翅目模式生物。他最近的研究集中在昆虫端粒的分子结构上、昆虫端粒 DNA 重复序列的种系发生，以及用先进的分子细胞遗传学的方法研究鳞翅目性染色体的分子分化和进化。此外，他与联合国粮农组织（FAO）和国际原子能机构（IAEA）成立的核技术粮农应用联合司（Joint FAO/IAEA Division of Nuclear Techniques in Food and Agriculture，奥地利，维也纳）属下的害虫防治部门合作，采用新方法来构建鳞翅目害虫的不育系。借助这种方法，他致力于开发苹果蠹蛾的转基因不育系，旨在产生单雄后代来对这种害虫进行防控。他与美国农业部农业研究局的雅基马农业研究实验室丽萨·列文合作来完成此项工作。

Contributors

Hiroaki Abe
Department of Biological Production
Faculty of Agriculture
Tokyo University of Agriculture and Technology
Tokyo, Japan

Joaquin Baixeras
Cavanilles Institute of Biodiversity and Evolutionary Biology
University of Valencia
Valencia, Spain

Philip Batterham
Department of Genetics
Bio21 Institute
University of Melbourne
Parkville, Victoria, Australia

Simon W. Baxter
Department of Zoology
School of Biology
University of Cambridge
Cambridge, United Kingdom

Patrícia Beldade
Evolutionary Biology Group
Institute of Biology
Leiden University
Leiden, The Netherlands

Bryony C. Bonning
Department of Entomology
Iowa State University
Ames, Iowa

Adriana D. Briscoe
Department of Ecology and Evolutionary Biology
University of California, Irvine
Irvine, California

John Brown
Systematic Entomology Laboratory
USDA
National Museum of Natural History
Smithsonian Institution
Washington, D.C.

Nicola Chamberlain
School of Biological Sciences
University of Exeter in Cornwall
Penryn, United Kingdom

Derek Collinge
CSIRO Entomology
Canberra, Australia
and
School of Biochemistry and Molecular Biology
Australia National University
Canberra, Australia

Michael P. Cummings
Center for Bioinformatics and Computational Biology
University of Maryland
College Park, Maryland

Donald R. Davis
Department of Entomology
National Museum of Natural History
Smithsonian Institution
Washington, D.C.

Richard H. ffrench-Constant
School of Biological Sciences
University of Exeter in Cornwall
Penryn, United Kingdom

Tsuguru Fujii
Graduate School of Agriculture and Life Science
The University of Tokyo
Tokyo, Japan

Marian R. Goldsmith
Biological Sciences Department
University of Rhode Island
Kingston, Rhode Island

Karl Gordon
CSIRO Entomology
Canberra, Australia

Fred Gould
Department of Entomology and the Keck Center for Behavioral Biology
North Carolina State University
Raleigh, North Carolina

Astrid T. Groot
Max Planck Institute for Chemical Ecology
Department of Entomology
Jena, Germany

David G. Heckel
Department of Entomology
Max Planck Institute for Chemical Ecology
Jena, Germany

Keith R. Hopper
United States Department of Agriculture
Agricultural Research Service
Newark, Delaware

Chris D. Jiggins
Department of Zoology
School of Biology
University of Cambridge
Cambridge, United Kingdom

Keiko Kadono-Okuda
National Institute of Agrobiological Sciences
Tsukuba, Japan

Michael R. Kanost
Department of Biochemistry
Kansas State University
Manhattan, Kansas

Akito Y. Kawahara
Department of Entomology
University of Maryland
College Park, Maryland

František Marec
Biology Centre ASCR
Institute of Entomology
and
Faculty of Science
University of South Bohemia
České Budějovice, Czech Republic

Owen McMillan
Department of Genetics
North Carolina State University
Raleigh, North Carolina

Christine Merlin
Department of Neurobiology
University of Massachusetts Medical School
Worcester, Massachusetts

James B. Nardi
Department of Entomology
University of Illinois
Urbana, Illinois

Sara J. Oppenheim
Department of Entomology
North Carolina State University
Raleigh, North Carolina

Cynthia S. Parr
Human-Computer Interaction Lab
University of Maryland
College Park, Maryland

Jerome C. Regier
University of Maryland Biotechnology Institute
University of Maryland
College Park, Maryland

Steven M. Reppert
Department of Neurobiology
University of Massachusetts Medical School
Worcester, Massachusetts

Hugh M. Robertson
Department of Entomology
University of Illinois at Urbana-Champaign
Urbana, Illinois

Amanda D. Roe
Department of Entomology
University of Minnesota
St. Paul, Minnesota

Daniel Rubinoff
Department of Plant and Environmental Protection Sciences
University of Hawaii
Honolulu, Hawaii

Suzanne V. Saenko
Evolutionary Biology Group
Institute of Biology
Leiden University
Leiden, The Netherlands

Ken Sahara
Laboratory of Applied Molecular Entomology
Research Institute of Agriculture
Hokkaido University
Sapporo, Japan

Coby Schal
Department of Entomology and the Keck Center for Behavioral Biology
North Carolina State University
Raleigh, North Carolina

Nina Richtman Schmidt
Department of Entomology
Iowa State University
Ames, Iowa

Toru Shimada
Graduate School of Agriculture and Life Science
The University of Tokyo
Tokyo, Japan

Thomas J. Simonsen
Department of Biological Sciences
University of Alberta
Edmonton, Alberta, Canada

Marilou P. Sison-Mangus
Department of Ecology and Evolutionary Biology
University of California, Irvine
Irvine, California

Michael R. Strand
Department of Entomology
University of Georgia
Athens, Georgia

Wee Tek Tay
CSIRO Entomology
Canberra, Australia

Walther Traut
Universität Lübeck
Zentrum für Medizinische Strukturbiologie
Institut für Biologie
Lübeck, Germany

Gissella M. Vásquez
Department of Entomology and the Keck Center for Behavioral Biology
North Carolina State University
Raleigh, North Carolina

Andreas Vilcinskas
Institute of Phytopathology and Applied Zoology
Interdisciplinary Research Center
Justus-Liebig-University of Giessen
Giessen, Germany

Niklas Wahlberg
Department of Biology
University of Turku
Turku, Finland

Kevin W. Wanner
Department of Plant Sciences and Plant Pathology
Montana State University
Bozeman, Montana

Susan J. Weller
Department of Entomology
University of Minnesota
St. Paul, Minnesota

Adam Williams
Department of Genetics
Bio21 Institute
University of Melbourne
Parkville, Victoria, Australia

Andreas Zwick
Department of Entomology
University of Maryland
College Park, Maryland

译者

主译：何宁佳

参译：查幸福　童晓玲　王　菲　甘　玲　梁九波

付　强　徐云敏　亓希武　刘碧朗　王长春

绪言

鳞翅目昆虫的蛾子和蝴蝶是种类繁多的种群之一，在数量上仅次于鞘翅目的甲虫，由超过150 000的物种组成。它们植食性的幼虫遍布全球，是栖息环境的重要组成部分。不管是作为农业和林业的主要害虫，还是作为传粉媒介和其他动物的食物来源，鳞翅目昆虫的许多物种对人类社会都产生了深远的影响。家蚕(*Bombyx mori*)是其中的特例。为获取蚕丝，家蚕被人工饲养多年，特别是在亚洲(中国、日本、韩国、泰国)，南亚的印度，欧洲(法国、意大利、俄罗斯、罗马利亚、保加利亚)和20世纪的南美巴西，蚕丝业在30多个国家中成为了重要的经济产业。

鳞翅目昆虫可能是研究得最为广泛的无脊椎动物，特别是拥有美丽翅膀的蝴蝶，曾经吸引了职业研究者和业余昆虫爱好者的注意。然而，除了家蚕，只有为数不多的鳞翅目昆虫得到了遗传学家的青睐。例如，在J. 塞勒从几种鳞翅目昆虫中发现性染色体前，在20世纪初，L. 唐克斯特和G. H. 雷纳从胡麻斑尺蠖蛾(*Abraxas grossulariata*)性连锁遗传中推测得到了鳞翅目昆虫典型的异型配子生殖(即雌染色体为ZW)。基于20世纪30年代R. 高施密特的工作，舞毒蛾(*Lymantria dispar*)成为了研究雌雄间性广为人知的材料。在早期的遗传研究模式生物中，粉螟(*Ephestia kuehniella*)作出了突出贡献。A. 库恩和他的同事利用收集的粉螟眼色突变体完成了眼色素合成途径的大量研究，后由库恩研究小组成员之一的E. 卡斯帕利提出了“一个基因，一个酶”的假说。持续至20世纪60年代的这些遗传学研究和包括对几百种蛾子和蝴蝶染色体数目的调查可详见1971年由帕加玛出版社出版，R. 罗宾逊执笔的《鳞翅目昆虫遗传学》(*Lepidoptera Genetics*)。这本书还提及了著名的由工业化引发蛾子黑化的例子，讲述了在1848年英国的曼彻斯特地区最初观察到了赤蛾(*Biston betularia*)的黑化现象。在该书中,相当多的篇幅关注了蛱蝶(*Hypolimnas*)和凤蝶(*Papilio*)警戒拟态的进化和遗传,采用这种警戒拟态的蝴蝶的翅模纹模拟了有毒的乳草斑蝶。今天教科书里有关警戒拟态的理论是基于这群科学家的研究(包括 J. W. Z. 布劳尔,C. A. 克拉克,P. M. 谢泼德),并得到了 D. 查尔斯沃思和 B. 查尔斯沃思进一步的详尽阐述。

在很多方面,本书是 1995 年戈德史密斯和威尔金斯出版的《鳞翅目昆虫分子模式系统》一书的后续。在那本书中,着重强调了选中的和正在形成的模式生物的研究,旨在利用它们的特征和在实验上的优势引起科学界的关注。家蚕在实验室从桑叶喂养发展到人工饲料喂养以及它们较大的个体,在蚕丝业高度发达国家的前沿基础研究中得到了广泛应用;同时,烟草天蛾(*Manduca sexta*)也成为美国基础研究(遗传除外)的首要模式昆虫。这两种昆虫至今仍是鳞翅目的模式昆虫,家蚕在分子遗传学、结构及功能基因组学上保持优势,烟草天蛾活跃在先天性免疫学、嗅觉和神经生物学、内分泌学和生物化学领域的前沿。随着分子生物学和基因组技术的发展,个体的大小和规模化饲养已不再重要,因此本书提及的对鳞翅目模式系统的选择更多地反映了利用蛾子和蝴蝶自身的优点来阐明其独有的研究问题。本书的编写初衷并不想包罗万象,只是再次提供针对基础生物学研究计划的展示平台以及源于开发针对鳞翅目害虫防治新方法的迫切需要。对本书未能涵盖的重要模式系统和关键领域的研究表示歉意,但请相信,这里提及的研究会激励读者进一步去阅读相关文献,去了解那些令人着迷和兴奋的研究。

本书第一章对鳞翅目昆虫的系统发生学现状进行了综述,综述聚焦于能代表模式系统的进化枝,在进化的背景下才能了解其本质。随后的章节对目前拥有和进展中的鳞翅目昆虫基因组学和后基因组学丰富的知识进行了综述,包括家蚕(第二章),引人注目的缪氏拟态,釉蛱蝶(*Heliconius*)(第六章)和主要的作物害虫铃夜蛾(*Helicoverpa*)(第十二章)。作为本书的前半部分的重要主题探讨了性别二型性对鳞翅目昆虫生物学多个方面的影响,包括染色体结构、性染色体系统和性别决定（第三、四章），与交配行为关联的蝴蝶的视觉（第七章）、生物钟（第八章），以及性激素的产生和信号接收（第十章）。同样的，对控制蝴蝶翅发育和翅模式基因的研究在过去十年取得了显著的进步，在进化的背景下正在逐渐地产生新的信息（第五、六章）。较其他主要的昆虫而言，从鳞翅目昆虫化学受体的结构和功能多样性上获取的知识是近年研究的又一大突破（第九章），该方面的研究很大程度上依赖于家蚕基因组测序计划。

本书的许多话题直接和非直接地涵盖了另一个重要的话题——昆虫的防控。这些内容包括利用鳞翅目昆虫研究特异性宿主范围的遗传学和神经生理学（第十一章），杀虫剂抗性机理（第十三章），对已经建立起来的昆虫先天性免疫基础研究的进一步扩展（第十四章），探索携带多分病毒的寄生物与鳞翅目昆虫宿主间的相互关系（第十七章）。害虫防治的实用例子包括用鳞翅目昆虫来源的抗菌肽和毒力因子来尝试作为人类病原物的潜在治疗试剂，或者用于产生抗病植物（第十五章），用不同方法对多种体腔内毒素进行传递（第十六章）。最后一个话题是关于病毒的传递系统和功能研究，我们用首例鳞翅目昆虫图位法克隆来结束本书。家蚕浓核病毒是严重影响蚕丝产业的病原物，该方法成功地克隆了抗家蚕浓核病毒 *Bm*DNV-2 的基因 *nsd*-2（第十八章）。如同本书所提及的许多研究计划，该研究代表了鳞翅目昆虫研究的一个制高点，所用的研究策略、技术和序列信息在 20 世纪 90 年代中期还可望不可及。随着最近发表的家蚕基因组的“第二版”信息（见家蚕基因组专刊 *Insect Biochemistry and Molecular Biology*，38（12），December 2008）以及其他鳞翅目昆虫基因组计划的跟进，更多的研究进展会在不久的将来得以面世。

最后，我们感谢 CRC 出版社《昆虫学系列当代话题丛书》的编辑汤姆·米勒，是他策划了整本书并使我们成为本书的编撰者；没有他的鼓励、不断的促进、施加的恰到好处的压力、创造性地保持我们工作热情的方式以及对我们的信任，我们不可能完成这项工作。感谢 CRC 出版社的资深编辑约翰·苏尔兹科的耐心。Taylor & Francis Group 的协调者帕特·罗宾逊,她的帮助和对我们问题的及时回复使该书的出版得以实现,也许她还未意识到她的帮助对我们的重要性。我们还要感谢盖尔·勒纳尔准备样稿,凯利·彭诺耶承担了使手稿得以成型的大量繁杂的工作。最后,我们感谢本书的诸多作者,为他们的辛勤工作、手稿的按时提交、对所提意见的慷慨回复、对该计划的大力支持以及对紧张的写作计划和偶尔出现的通信不畅(由我们负责)的容忍。他们对该书的贡献是无价的。

玛丽安·戈德史密斯和弗兰蒂塞克·马莱克

写于金斯敦和捷克布杰约维采

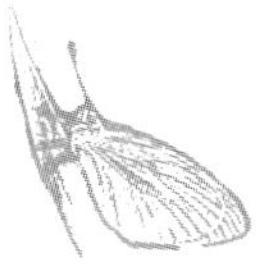

目 录
CONTENTS

第一章 鳞翅目昆虫模型系统的进化框架

Amanda D. Roe, Susan J. Weller, Joaquin Baixeras, John Brown, Michael P. Cummings, Donald R. Davis, Akito Y. Kawahara, Cynthia S. Parr, Jerome C. Regier, Daniel Rubinoff, Thomas J. Simonsen, Niklas Wahlberg, and Andreas Zwick

1. 前言

鳞翅目昆虫是地球上种类最丰富和最容易识别的物种之一，已经有记录与描述的至少有 150000 种（Kristensen 和 Skalski 1998）。它们是具有超级生物多样性的 4 个完全变态昆虫目之一，另外的 3 个目分别是双翅目（蝇）、鞘翅目（甲虫）和膜翅目（黄蜂、蜜蜂和蚂蚁）。仅蝴蝶这一种昆虫的种类数量就超过鸟纲的伯劳属，接近 18000 种（Kristensen 和 Skalski 1998）。通常，我们通过 3 种特征来辨别鳞翅目昆虫，分别是翅膀有鳞毛，可伸长的刺吸式口器（针状吻）以及完全变态的发育特征，其幼虫常被称作“毛虫”。在生物学研究史上，已经在包括发育、遗传、分子生物学、生理学、进化学和生态学多个研究领域使用各个种属的鳞翅目昆虫建立了无数的模型系统（Bates 1861；Müller 1879；Ford 1964；Ehrlich 和 Raven 1967；Kettlewell 1973）。我们之所以使用鳞翅目昆虫作为模型系统，主要是因为该类昆虫具有一系列适合于研究的生物学特征（Bolker 1995）。鳞翅目昆虫的超凡魅力主要归于它们异乎寻常多样的翅膀颜色模式和幼虫形貌，这也是专业人士和业余爱好者贪婪地收集它们的原因（图 1.1；Salmon 2000）。许多常见的鳞翅目昆虫种类的幼虫个体较大，在实验室往往比较容易饲养，这也为早期在发育和疾病相关方面的研究提供了便利。除此之外，鳞翅目昆虫对于人类社会经济发展的影响也促进了鳞翅目模型系统的发展。例如，在亚洲因为有长期的桑蚕养殖的传统，使吐丝昆虫（家蚕）成为被人类饲养的少数几种昆虫之一。然而，鳞翅目昆虫作为一种主要的植食性昆虫，它们中有很多是农业和林业中的重要害虫。我们在本书中所讨论的几个鳞翅目昆虫种类都已被开发成模型系统，它们被开发成模型系统的初衷是为了了解鳞翅目昆虫与它们宿主植物之间的关系，而最终则是要使用研究所得的成果去控制整个害虫群体。

除了实际的应用，鳞翅目昆虫模型系统也为基础研究提供了方便，这方面的研究工作主要包括翅膀斑纹模式的形成，神经系统发育，以及发育相关基因之间的相互作用等（例如 *Bicyclus*，*Manduca*）。另外，我们已经拥有鳞翅目基因组的数据（Mita et al. 2004；Xia et al. 2004；Jiggins et al. 2005）。现在已有 4 个基因组被 GenBank 数据库收录（http://www.ncbi.nlm.nih.gov，accessed April 16，2008），它们是家蚕（蚕蛾科 Bombycidae），热带蝴蝶（蛱蝶科 Nymphalidae），庆网蛱蝶（蛱蝶科 Nymphalidae）和草地贪夜蛾（夜蛾科 Noctuidae）。相比较而言，鳞翅目昆虫的基因组比较大：家蚕（*Bombyx mori*）的基因组约有 475Mbp，烟草天蛾（*Manduca sexta*）的大概有 500Mbp，绿棉铃虫（*Heliothis virescens*）的有 400Mbp 左右，以及釉蛱蝶（*Heliconius*）约有 292Mbp（J. S. Johnston，unpublished；Goldsmith，Shimada，and Abe 2005；Jiggins et al. 2005）。这些基因组几乎是已经完成的黑腹果蝇（*Drosophila melanogaster*）的基因组（175Mbp）的 2.5 倍，是蜜蜂（*Apis mellifera*，236Mbp）或者冈比亚按蚊（*Anopheles gambiae*，280Mbp）基因组的 1.6 倍（Goldsmith，Shimada，Abe 2005；Honeybee Genome Sequencing Consortium 2006）。关于鳞翅目基因组更加深入的阐述请参照第二章和第六章以及其中的参考文献。

本章中我们综述了最初的模型系统的选择是怎样影响随后的深入研究以及系统发

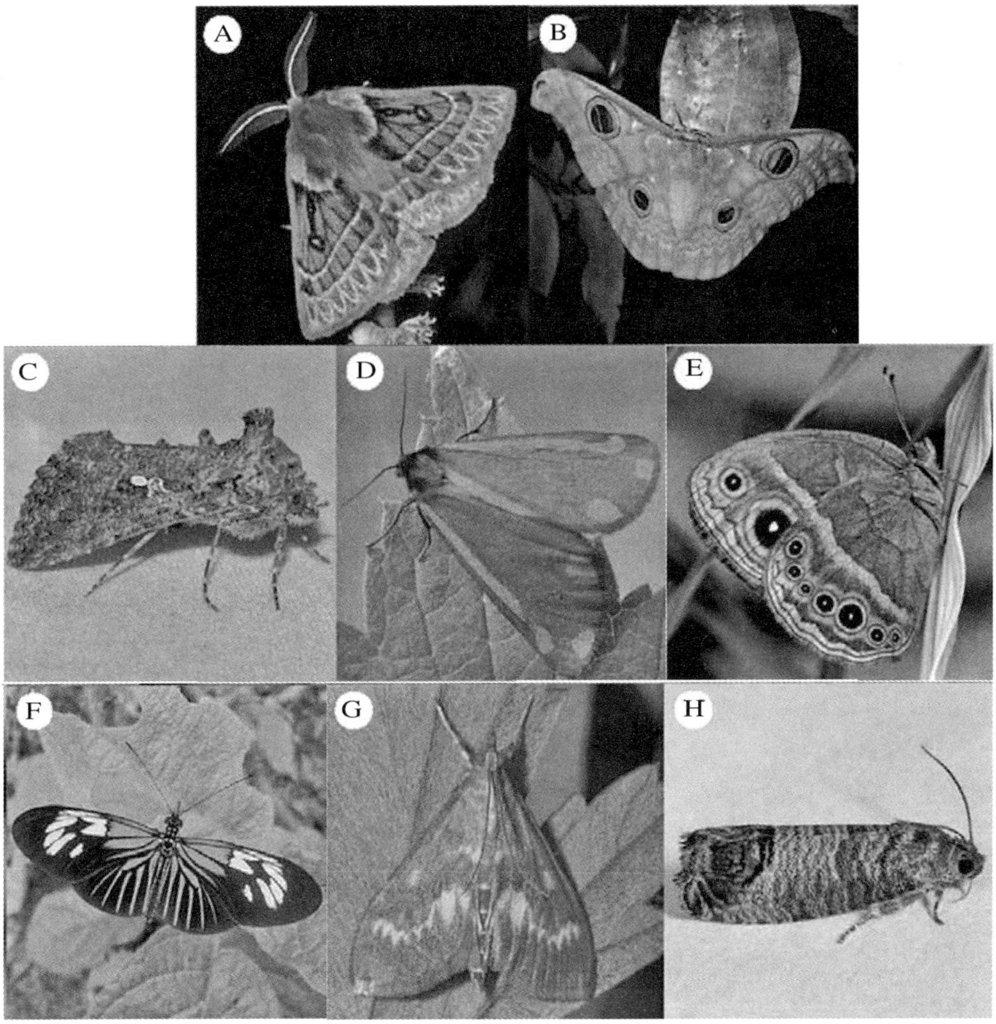

图 1.1　含有模式系统的总科的代表(彩色版本图片见附录图 1.1)。A:蚕蛾总科，*Anthela oressarcha*(A. Zwick);B:蚕蛾总科，*Antheraea larissa*(A. Kawahara);C:夜蛾总科,粉纹夜蛾(*Trichoplusia ni*)(M. Dreiling);D:夜蛾总科,朱砂夜蛾(*Tyria jacobaeae*)(D. Dictchburn);E:凤蝶总科,热带蝴蝶(*Bicyclus anynana*)(A. Monteiro 和 W. Piel); F:凤蝶总科,邮差蝴蝶(*Heliconius erato*)(K. Garwood);G:螟蛾总科,玉米螟(*Ostrinia nubilalis*)(S. Nanz);H:卷蛾总科,苹果蠹蛾(*Cydia pomonella*)(N. Schneider)。

生学的研究在将模型系统定位到一个广义的进化框架下的作用。接下来,我们综述了已知的总科在系统发生学上的关系,这些科含有多个模型系统,也探讨了这些模型系统之间在系统发育上的位置关系。目前已知的总科之间的关系,正面临着分子水平研究的挑战。而且,在匮乏和残缺的化石记录面前,确定分化时间非常复杂。最后,我们发起了关于鳞翅目系统发生学全球性研究的倡导,其目的是将以往那些分离的研究群体联系起来。这些全球性的研究将促进一个鳞翅目系统学研究新时代的到来。

2. 系统发生学与模型系统

模型系统使得研究者们可以集中资源和力量深入细致地研究基本的生物学问题。这种研究资源和人力的集中固然是极其重要的，但是模型系统的真正力量在于将集中研究所得到的成果推广到更大的生物群体中去（Kellogg 和 Shaffer 1993；Bolker 1995）。为了总结这些研究的结果，将模型系统放到它们的进化背景下进行比较是至关重要的。一个进化的框架或者系统发生学能够通过多种研究分析方法进行推论（Swofford et al. 1996；Holder 和 Lewis 2003）。一些可遗传的、独立的特征已经用于系统发生学的分析过程中，它们包括分子水平上的特征（例如 DNA，RNA 和氨基酸序列），表型上的特征（例如形态结构、同工酶）和发育性状（例如个体发育时期或者信号通路），以此用于推论物种之间在进化上的关系。

没有系统发生学，我们从各个模型系统所得到的知识和研究成果仍然是各自分离的。我们将不能概括出吐丝昆虫（如家蚕）、欧洲玉米螟（*Ostrinia nubilalis*）以及果蝇（*Drosophila*）之间的共同特征。通过理解物种之间的进化关系，对于关键性状（例如特征进化）起源的假说能够通过一个群体总结出来（Mabee 2000；Collins et al. 2005）。系统发生学为重建古老的性状提供了方法，另外也为深入理解性状的分化和同源性提供了依据（Mabee 2000；Felsenstein 2004）。例如，灯蛾科（Arctiidae）的系统发生学分析显示吡咯里西啶生物碱隔离发生在灯蛾科最大的亚科灯蛾亚科（Arctiinae）分化之前。对于灯蛾亚科在后来的进化过程中的丢失和所获得性状的重建将能在一个更精细的尺度上进行研究（例如，Weller，Jacobson 和 Conner 1999；DaCosta et al. 2006）。

系统发生学重建的方法同样可以用于研究基因表达与形态性状之间的相互关系，通过研究可以推测该基因在对应形态性状进化上的作用。存在已久的关于节肢动物头部分节和脑部进化的争论在很大程度上是通过研究 *Hox* 基因的表达情况来解决的（例如 Cook et al. 2001）。因此，置于系统发生学的背景下，关于模型系统的研究能够为以前未能解决的许多关于形态进化的难题提供重要的启示。

除了为遗传、发育以及形态性状等在进化上提供研究启示和线索之外，系统发生学也能够用于为模型系统的使用确定其潜在优势和侧重点（Bolker 1995）。模型系统的特征（例如是否容易饲养、身体大小以及经济重要性）能够在无意间影响建立在这个模型系统上的其他研究所得出的结论。如果不放在合适的系统发生学的背景下，模型系统可能导致误导性的结论，这些结论会影响更大的群体。通过将所研究的模型系统与其他的模式和非模型系统放到同一个进化的背景下去分析，可以在鉴定出潜在的优势资源的同时避免产生错误的推论。

3. 鳞翅目系统发生的概述

根据当前的研究，鳞翅目分为 126 个科和 46 个亚科（Kristensen 和 Skalski 1998）。当前各个总科（见图 1.2）之间的进化关系如同一张由已解决的系统发育问题组成的拼图。鳞翅目之间的最基本的关系是基于形态学的研究建立起来的（Kristensen 1984；Davis 1986; Nielsen 和 Kristensen 1996; Kristensen 和 Skalski 1998)，并已经在分子水平上得以确认(Wiegmann et al. 2000; Wiegmann, Regier 和 Mitter 2002)。这些早期分离出的世系也被称为"非双孔亚目"(命名的群体在第 7 分支下)，它们仅仅占鳞翅目总数的一小部分，而且其中包含了很少的模型系统(见图 1.2)，因此在本书中没有进行深入的讨论。相反，双孔亚目(Ditrysia)分支则包括了几乎 99%有描述的种类，但其在总科中的相互关系却知之甚少。尽管如此，这一分支还是包含了当前所有的模型系统。这一世系中的成员基于一系列共同产生的形态性状(称为"共同衍征")而被划归为一个群体，其中包括特化的雌性生殖器(Kristensen 和 Skalski 1998)。Minet(1991)在一篇综述中基于所有生命时期的形态学研究和假定总科的基本分布提出了几个嵌套在双孔亚目内的高度相似群体。尽管直到现在这种组织方法还没有经过系统发生学理论的验证，为了便于展示，我们暂时按目前的惯例采用 Minet 的组织方法(见"目前的研究和将来的方向"章节)。

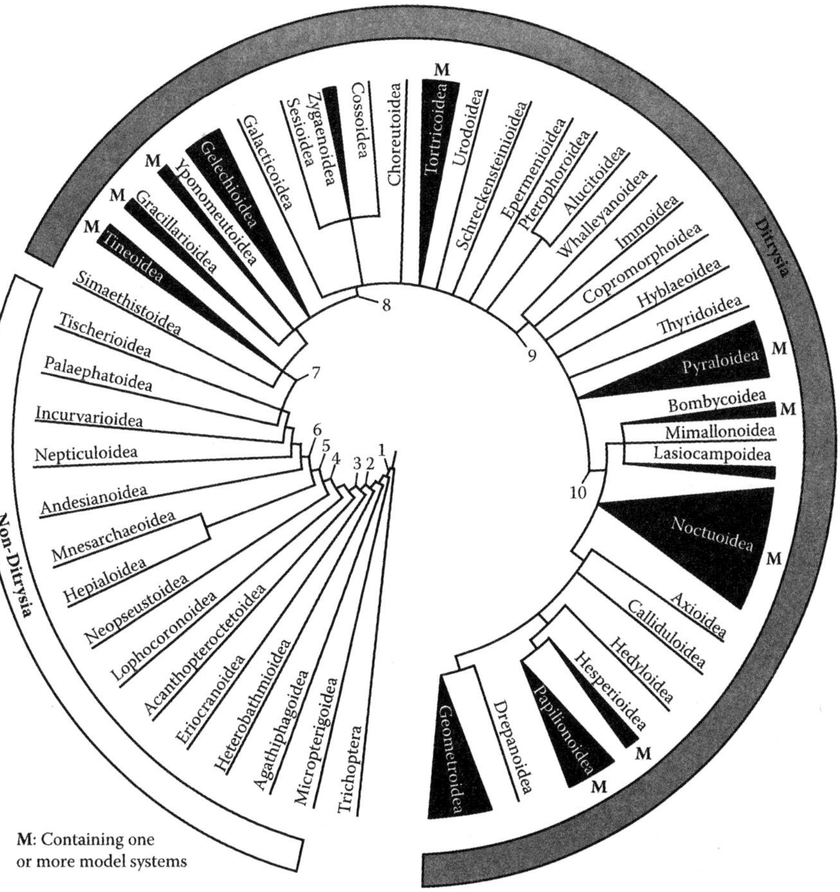

图 1.2　根据形态学特征构建的目前理论上的鳞翅目总科系统发育关系，其中包括了鳞翅目的一个姊妹群毛翅目(改编自图 2.2，Kristensen 和 Skalski 1998)。代表高水平的分类的节点在图中做了标记，例如大鳞翅目；1＝鳞翅目，2＝有喙亚目，3＝Coelolepida，4＝Myoglossata，5＝新鳞翅目，6＝异脉次目，7＝双孔亚目，8＝Apoditrysia，9＝Obtectomera，10＝大鳞翅目。图中给出了含有模式系统的总科。图中分支的宽度代表了当前描述种类的大致比例。

我们对拥有多个模型系统的总科做了一个系统发生的概述。可能的话对这些模型系统的分布是否足够阐明总科的系统发生分化，是否能够毫无怀疑地应用到其他成员的研究中去进行了评述。因为一个单一的模型系统，在系统发生上只能算做一个样本，尽管一个模型系统有了广泛的研究，但我们不能以此类推其他的总科的关系。我们将根据各科在图 1.2 中系统发生的位置对它们进行讨论。

4. 选择的双孔亚目总科概述

在双孔亚目中，33 个总科中的 6 个含有多个模型系统，这 6 个总科分别是：蚕蛾总科（例如，家蚕）、夜蛾总科（例如，虎蛾、舞毒蛾、地老虎蛾），凤蝶总科和弄蝶总科（例如，蝴蝶和弄蝶），螟蛾总科（象鼻虫）以及卷蛾总科（卷叶虫）（见图 1.1）。卷蛾总科，与鳞翅目的另外几个世系由于它们的身体较小，有时也被称作“小鳞翅目”。然而，这一点比较容易混淆，因为身体较小的性状也出现在所谓的“大鳞翅目”中（例如，小夜蛾；Fibiger 和 Lafontaine 2005），同时“小鳞翅目”中也存在个体较大的世系（例如，刺蛾科，刺蛾幼虫；蝙蝠蛾科，鬼蛾）。尽管鳞翅目的模型系统主要集中在双孔亚目，但是还有很多总科并未被模型系统所代表（见图 1.2）。三个种类丰富却并没有被用作模型系统的总科包括尺蛾总科（例如，尺蠖）、麦蛾总科（例如，鞘蛾、遮瑕蛾和卷蛾）和斑蛾总科（例如，斑蛾和刺蛾幼虫），应该是将来模型系统的重要候选者。

4.1 卷蛾总科（Tortricoidea）

卷蛾总科，包括卷蛾科，是小鳞翅目中种类数量仅次于麦蛾总科的第二大科，其中已有描述的种类有 9100 个（Brown 2005）。卷蛾科这一单系的识别是通过一系列的成虫、幼虫和蛹的特征来完成的。虽然在过去的一个世纪中，卷蛾科的几个群体已经被一个或者更多的研究者认为应该属于其他的科，但是现在普遍认为这些类群属于卷蛾科的下属类群（Horak 1998）。本科通常包括三个亚科：卷蛾亚科、细狭云卷蛾亚科和小卷蛾亚科，又可分为 22 族和 957 属（Horak 1998；Brown 2005）。小卷蛾亚科和细狭云卷蛾亚科是通过令人信服的形态学共同衍征来分类辨别的，然而卷蛾亚科几乎全部是确定的并系或者多系的种类。卷蛾科的昆虫在除南极洲之外的所有大陆上都有发现，其中卷蛾科最丰富的地域是热带地区。热带地区的物种极为丰富，绝大多数的物种还不为人所知（Horak 1998；Brown 2005）。

“卷叶蛾”这一名字已经被用于整个卷蛾科，该名字是根据其幼虫的生活习性而来的，幼虫通过折叠或者卷曲食物（植物的叶子）来构建它们的巢穴，然而卷蛾总科的幼虫具有更广泛的食性，包括形成虫瘿、以茎根为食、以果实为食、以种子为食以及以花为食的种类。除此之外，还有极少部分的卷蛾科是捕食者或者寄居者。在卷蛾科的其他特征中，许多卷蛾科的成虫往往可以通过在静息时其翅膀上钟形的花纹来辨别。

许多卷蛾科的昆虫是重要的农业、林业和观赏植物的害虫，其中世界上 164 个属和 687 个种的卷蛾科昆虫被认为是重要的害虫（Zhang 1994）。在所有被研究的卷蛾科的昆虫中，研究得最广泛的是云杉卷蛾（云杉卷叶蛾，卷蛾亚科：黄卷蛾族），它是北美的一种重要的森林害虫，专食松科植物。当前已经从多个角度对该昆虫进行了广泛的研究，包括形态学(例如,Walters,Albert 和 Zacharuk 1998),信息素化学(例如,Delisle,Picimbon 和 Simard 1999),宿主植物偏好性(例如,Albert 1991,2003),生理学(例如,Hock,Albert 和 Sandoval 2007),行为学(例如,Wallace,Albert 和 McNeil 2004)以及寄生虫和病原微生物学(例如,Quayle et al. 2003)。由于云杉卷蛾、云杉黄卷蛾、红带卷蛾、褐卷蛾、棉褐带卷蛾、双斜卷蛾以及云杉卷蛾亚科的几个属比较模糊的关系,通过许多形态学、生物学和生态学的特征,在一定程度上云杉卷蛾可以划分为一个独立的相对比较大的分支。

内蠹蛾(苹果蠹蛾,小卷蛾亚科:小卷蛾族)代表了卷蛾总科的另外一个模型系统。最早的记录见于欧洲,然而实际上当前该种在世界上广泛存在。虽然最初该昆虫只被认为是苹果和梨子(海棠属和梨属;蔷薇科)栽培中的害虫,事实上它们拥有更加广泛的宿主群,双子叶植物的 6 个科的植物都是它们的宿主,包括:壳斗科、芸香科、蔷薇科、桑科、胡桃科以及山龙眼科。在过去的 30 多年中,无数的研究集中在区分它的 6 种信息素,触角受体、信息素分泌腺和交配行为等领域,目的是通过以上的研究来阻断其繁殖过程(Ahmad 和 Al-Gharbawi 1986; Arn 1991; McDonough et al. 1993; Backman 1997; El-Sayed et al 1991; Addison 2005; Trematerra 和 Sciarretta 2005)。对该类昆虫的检疫在国际贸易协定关于特殊农产品的检疫规定中具有重要的意义(例如,Wearing et al. 2001)。

卷蛾科的其他种类正逐渐受到重视,其中包括梨小食心虫(梨小食心虫,小卷蛾亚科:小卷蛾族)。就像苹果蠹蛾一样,梨小食心虫已经变成了坚果类的世界性害虫;红带卷蛾(红带卷蛾,卷蛾亚科:黄卷蛾属),是一种高度杂食性的北美果树害虫;假苹果蠹蛾(假苹果蠹蛾,小卷蛾亚科,小卷蛾族),在非洲是茄科(辣椒属和茄属)和芸香科(柑橘属)植物重要的害虫;斜带卷叶蛾(斜带卷叶蛾,卷蛾亚科:黄卷蛾属),是另外一种广食性的北美卷叶蛾。尽管卷蛾科昆虫具有经济重要性,但是目前我们却没有对这个科在系统发生学上进行研究。另外,我们也无法评论这些模型系统的当前种系发生的分布,因为目前在这个科的系统发生学上的研究仅仅集中在卷叶蛾和小食心虫这两个科。

4.2 螟蛾总科(Pyraloidea)

螟蛾总科,或者称作螟蛾,是一个非常大的总科,包括大约 16000 个已经有记录的种类(Heppner 1991; Munroe 和 Solis 1998),另外至少还有大量的物种还不为人所知。通过一系列的形态学上的特征,我们可以将螟蛾总科划分为一个单系,这些特征包括在第二腹节上存在配对的鼓膜器官(膜样听觉结构),一个基部鳞毛管状长嘴(有时存在)以及典型的翅脉模式(Munroe 和 Solis 1998; Nuss 2006)。

螟蛾总科的幼虫具有高度多样性的生活习性。大多数幼虫是隐蔽的进食者,躲在组织内部或者蹼状叶,穗丝或者蛀屑里进食。主要进食植物的组织(Neunzig 1987; Munroe 和 Solis 1998)。一些种类的幼虫是捕食性的或者寄生在膜翅目的巢穴或者介壳科昆虫上(例如,Neunzig 1997),其他的一些种类则附着在无生命的植物材料上(例如,储存的货物)。许多草螟蛾科的幼虫营水生生活方式,以水下的植物为食(例如,water veneer,淡水蛾,草螟蛾科:草螟蛾亚科)。许多螟蛾科的幼虫也是重要的害虫。欧洲玉米螟(*Ostrinia numbilalis*, Crambidae: Pyraustinae)和水稻螟虫(*Chilo* spp. 和 *Scirpophaga* spp. ,Crambidae：Crambidae 和 Schoenobiinae)破坏多种农田作物。另外，印度谷蛾(Plodia interpunctella，螟蛾科：斑螟亚科)、地中海粉蛾(Ephestia kuehniella，螟蛾科：斑螟亚科)以及粉斑螟蛾(*Cadra cautella*，螟蛾科：斑螟亚科)是重要的危害干货害虫(Neunzig 1987)。蜡螟(*Galleria mellonella* 和 *Achroia grisella*，螟蛾科：Galleriinae)会破坏蜜蜂的巢穴(Neunzig 1987)。反过来，尽管这样做存在一定的风险，一些螟蛾科的昆虫还是被用于入侵植物的生物控制。仙人掌螟蛾(螟蛾科：斑螟亚科)曾被澳大利亚和其他一些地区引入用作控制仙人掌的过度繁殖(Common 1990；Zimmermann，Moran，and Hoffmann 2000)，然而现在它们已经成为了北美地区本地仙人掌的入侵昆虫(Solis，Hight 和 Gordon 2004)。

现在，根据一系列形态学上的特征，螟蛾总科被分为两个科：螟蛾科(Pyralidae)和草蛾科(Crambidae)(Minet 1983；Munroe 和 Solis 1998；Goater，Nuss 和 Speidel 2005)。螟蛾科包含 5 个亚科，草蛾科含有 16 到 17 个亚科(Munroe 和 Solis 1998；Solis 和 Maes 2002；Goater，Nuss 和 Speidel 2005)。从复合树可以看出，现存的基于形态的假说，对理解各个亚科之间的相互关系作用不大(Solis 和 Mitter 1992；Solis 和 Maes 2002)。然而，正在进行的分子水平上的研究为辨析螟蛾总科之间内在的关系提供了强有力的保证。

至少 5 个螟蛾科的种类(它们全是害虫)已经变成了主要的模型系统(见图 1.3)，其中包括欧洲玉米螟和水稻螟。这两个模型系统已经被用于研究信息素合成遗传学和神经生物学(Roelofs 和 Rooney 2003；Jurenka 2004)、生理学(Hodkova 和 Hodek 2004；Srinivasan，Giri 和 Gupta 2006)以及杀虫剂抗性(Coates，Hellmich 和 Lewis 2006)。大蜡螟(*Galleria mellonella*)已经广泛地用于研究免疫应激的遗传和生理学机制，以及用于人类疾病的病原体模型(见第十五章)。除了以上所涉及的领域之外，印度谷蛾和地中海谷蛾被用于鳞翅目中肠生理学的研究、信息素的检测、丝的合成与结构以及杀虫剂抗性的研究(Beckemeyer 和 Shirk 2004；Srinivasan，Giri 和 Gupta 2006；Siaussat et al. 2007)。在早期，地中海谷蛾作为一种研究色素合成和翅模式发育的遗传模型系统，它甚至比用作实验室标准模型的黑腹果蝇还要早(Robinson 1971；Leibenguth 1986)。

虽然这类昆虫未被选择为模式昆虫，但是螟蛾总科的模型系统在系统发生学上的影响是显著的，这也反映了螟蛾总科的这两个科所包含亚科的多样性。尽管这些结果只是基于害虫种类，有一定的片面性，但是这样的分布模式足以使得这些从模型系统得出的结论能够被推广到其他的亚科。

4.3 凤蝶总科和弄蝶总科(Papilionoidea 和 Hesperioidea)

对科学家和普通大众来说，凤蝶总科(真正的蝴蝶)比起弄蝶总科(弄蝶)无疑更为人熟知(Grimaldi 和 Engel 2005)。它们的昼夜节律、美轮美奂的外观以及容易着手研究等特点使得它们成为人们最乐于收集和观察的昆虫。有记录的凤蝶总科大约有 18000 个种类。与蛾类相比凤蝶总科的阿尔法分类研究非常充分，其物种的总数可能与它们实际的物种数接近，而且已经记录下了许多物种的自然历史。蝴蝶已经在生命科学研究的多个领域被用作模式生物，包括保护生物学、生态学、生理学、进化生物学、进化发育生物学以及分子生物学。已有若干的研究成果可以证明对于这类昆虫的科学研究兴趣的广泛性(Vane-Wright 和 Ackery 1984; Nijhout 1991; Boggs,Watt 和 Ehrlich 2003; Ehrlich 和 Hanski 2004)。

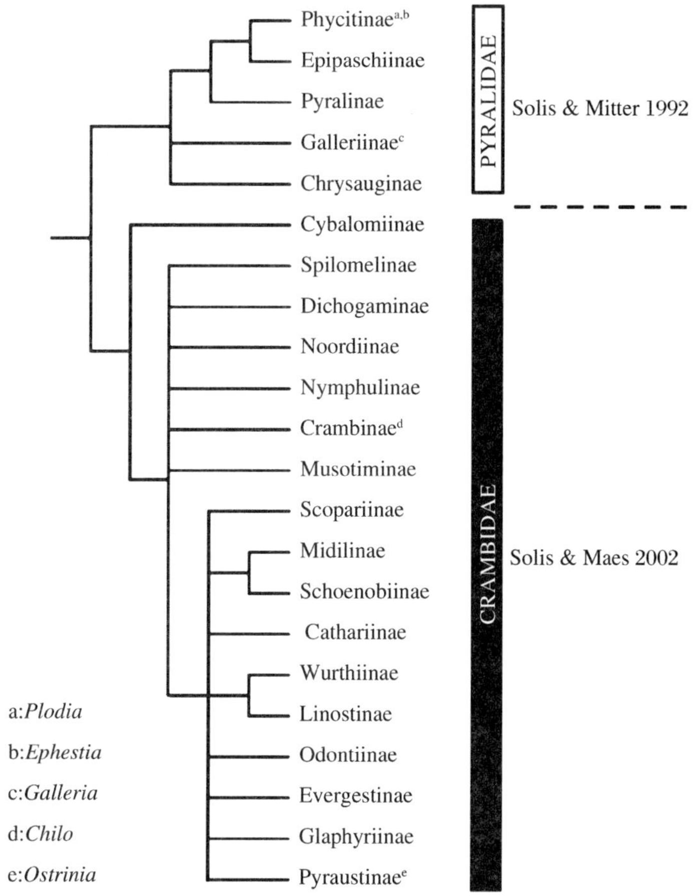

图 1.3 基于成虫形态学特征构建的复合系统发生树,阐明了螟蛾总科各亚科之间推论的系统发生关系。该进化树是根据 Solis,Mitter (1992; Crambidae) 以及 Solis,Maes (2002; Pyralidae)的研究结果构建的。字符上标指出了模式系统的位置。

最早要追溯到 Reuter 在 1896 年的研究工作,从那时起蝴蝶主要世系之间的进化关系一直是许多研究的主题。Ehrlich 1958 年发表的文章为以后的研究工作铺平了道路,以后的科学家可以基于性状特征的严格细致的分析去推论这个群体中主要世系之间的进化关系。凤蝶总科和弄蝶总科一直被认为是分类学上的姐妹群,这一点的主要根据是两者节律性的习性,除此之外 6 个潜在的形态学特征也是它们成为姐妹群的证据(de Jong,Vane-Wright 和 Ackery 1996)。现代分子生物学的证据同样支持两个总科作为姐妹群的假说(Wahlberg et al. 2005)。

弄蝶总科仅仅包含弄蝶科,也就是弄蝶(skippers),大概有 4000 种。凤蝶总科包括所谓的真蝴蝶,传统上将它们划分为 4~5 个科(例如,Ehrlich 1958; Kristensen 1976;de Jong,Vane-Wright 和 Ackery 1996; Ackery,de Jong 和 Vane-Wright 1998; Wahlberg et al. 2005)。凤蝶科和粉蝶科的区分很清楚，它们之间的区分标准自从19世纪提出之后就一直没有什么变动。最近对这些科的研究已经解决了这些亚科和其他下属群体之间的关系(Caterino et al. 2001; Braby, Vila 和 Pierce 2006; Nazari, Zakharov 和 Sperling 2007)。相反，蚬蝶科的分类关系还不确定，有学者把它划为一个独立的科(例如，Eliot 1973; Lamas 2004; Wahlberg et al. 2005)，也有学者把其归入灰蝶科的一个亚科(例如，Ehrlich 1958; Kristensen 1976; de Jong, Vane-Wright 和 Ackery 1996; Ackery, de Jong 和 Vane-Wright 1998)。现在的分子水平的数据显示蚬蝶形成了一个独立的世系，与灰蝶科是姐妹世系的关系，因此应该把蚬蝶科看成是一个独立的科(Wahlberg et al. 2005)。

蛱蝶科的分类一直比较混乱，一些研究者将它们划分为9个不同的科，包括斑蝶科和绡蝶科(Smart 1975)。尽管这些科在系统进化上嵌套在蛱蝶科(Ehrlich 1958; de Jong, Vane-Wright 和 Ackery 1996; Brower 2000; Wahlberg, Weingartner 和 Nylin 2003; Freitas 和 Brown 2004)，但是这个科的昆虫形态上唯一的共同衍征就是分布在触角上的三条纵向的脊(Kristensen 1976; Ackery, de Jong 和 Vane-Wright 1998)。最新的分子生物学研究(Wahlberg et al. 2005)确定这些世系形成了一个单系群体，这也支持了将它们作为蛱蝶科的亚科或者族的观点。

凤蝶总科的5个科(凤蝶科、粉蝶科、蚬蝶科、灰蝶科和蛱蝶科)和弄蝶科在系统发生学上的关系已经被研究人员利用形态学和分子生物学的方法进行了研究(Ehrlich 1958; Kristensen 1976; de Jong, Vane-Wright 和 Ackery 1996; Weller, Pashley 和 Martin 1996; Wahlberg et al. 2005)。图1.4中所示的就是我们现在所理解的以上6个科在系统发生学上的关系，但是粉蝶科在系统发育树上的位置是不稳定的。在一些分析中，粉蛾科被认为是凤蝶科的姐妹世系。要解决这个问题，还需要更多的数据支持。

在分子生物学和遗传学研究中，几个种类的蝴蝶已经被用作了模型系统。凤蝶(凤蝶总科:凤蝶科)一直是化学生态学和物种形成方面研究的中心(Scriber, Tsubaki 和 Lederhouse 1995)。在蚬蝶科中，"checkerspot"蝴蝶在群体生物学研究中占据着显著的位置(例如，Ehrlich 和 Hanski 2004)。釉蛱蝶(釉蛱蝶亚科)一直占据着拟态、物种形成和拟态系统遗传学的中心地位(Mallet, McMillan 和 Jiggins 1998; 见第六章釉蛱蝶的翅颜色模式遗传学)。热带蝴蝶(*Bicyclus anynana*)是蛱蝶科的另一重要的模型系统，是研究翅模式形成的首选模型系统(Beldade 和 Brakefield 2002;见第五章热带蝴蝶翅模式的进化和发育遗传学)。

4.4 蚕蛾总科(Bombycoidea)

蚕蛾总科(Minet 1994)包括大约5000个已有记录的种类，蛾子的个体中等或者较大(图1.1 A,B)，分属于650个属和12个科:澳洲蚕蛾科、枯叶带蛾科、家蚕科、水蜡蛾科(蚬蛾科)、茂蛾科、桦蛾科、带蛾科、枯叶蛾科、栎蛾科、忍冬蛾科、大蛾科以及天蛾科。蚕蛾具有一个胸静脉(Minet 1991)和一个前翅静脉的共同衍征(共同获得的特征)(A. Zwick,未发表)，这一事实支持了蚕蛾总科的单系学说，另外幼虫前腿表皮修饰也可以看作是这一学说的潜在证据(Hasenfuss 1999)。缺少共同衍征很大程度上源于该科昆虫的成虫生命期的急剧缩短，因此导致本总科中各科的关系在系统发生分析方面使用的共同

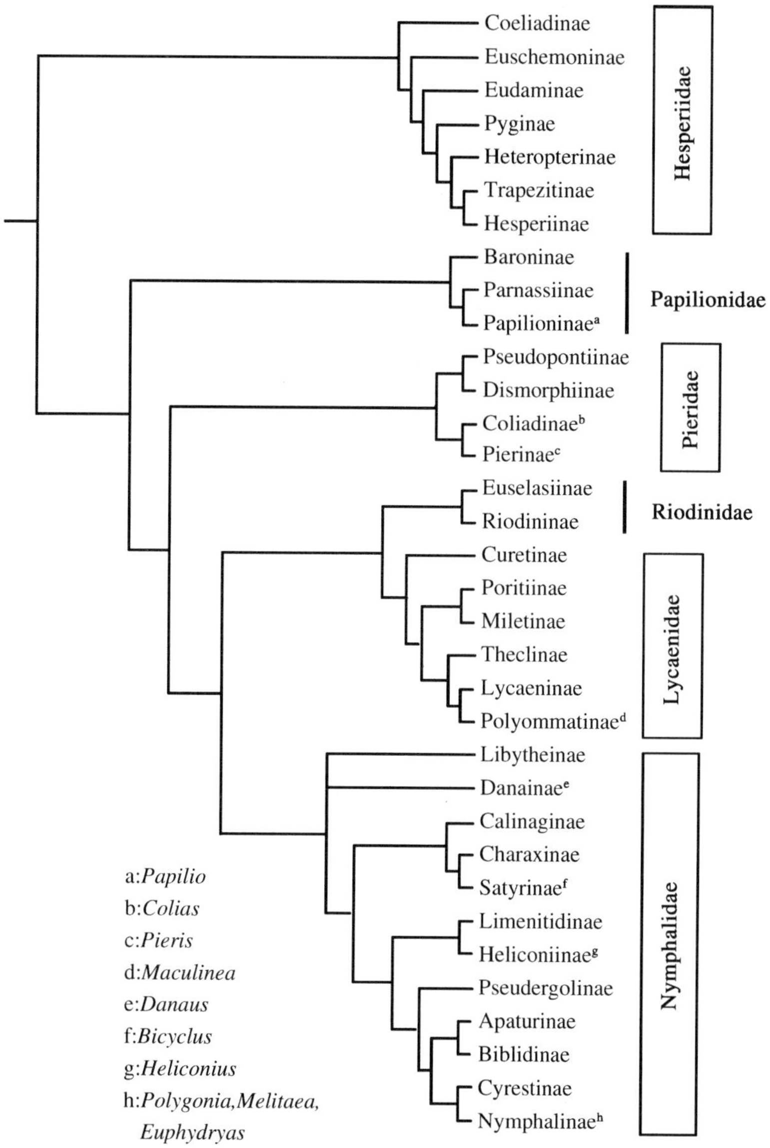

图 1.4　复合系统发生树表明了蝴蝶的各科和各亚科之间的推论的系统发生关系。各科之间的关系摘自 Wahlberg et al. (2005)，**各科内的亚科之间的关系摘自多个不同的研究，包括：**Hesperiidae (Warren 2006)，Papilionidae (Caterino et al. 2001)，Pieridae (Braby，Vila **和** Pierce 2006)，Riodinidae **和** Lycaenidae (Wahlberg et al. 2005)**和** Nymphalidae (Wahlberg **和** Wheat 2008).

结构很少(例如，口器和翅耦合的机制；Minet 1991，1994)。因此，总的说来我们对蚕蛾总科中各科之间的关系知之甚少(图 1.5)。

在蚕蛾总科内，只有三个科(家蚕科、天蛾科和大蛾科)的物种被广泛用作模型系统。已有两个独立的分子水平上的证据显示，以上这三个科之间可能形成了一个单系群(Regier et al. 2008; Zwick 2008)。以上两个研究有部分与 Brock(1971)和 Minet(1991，

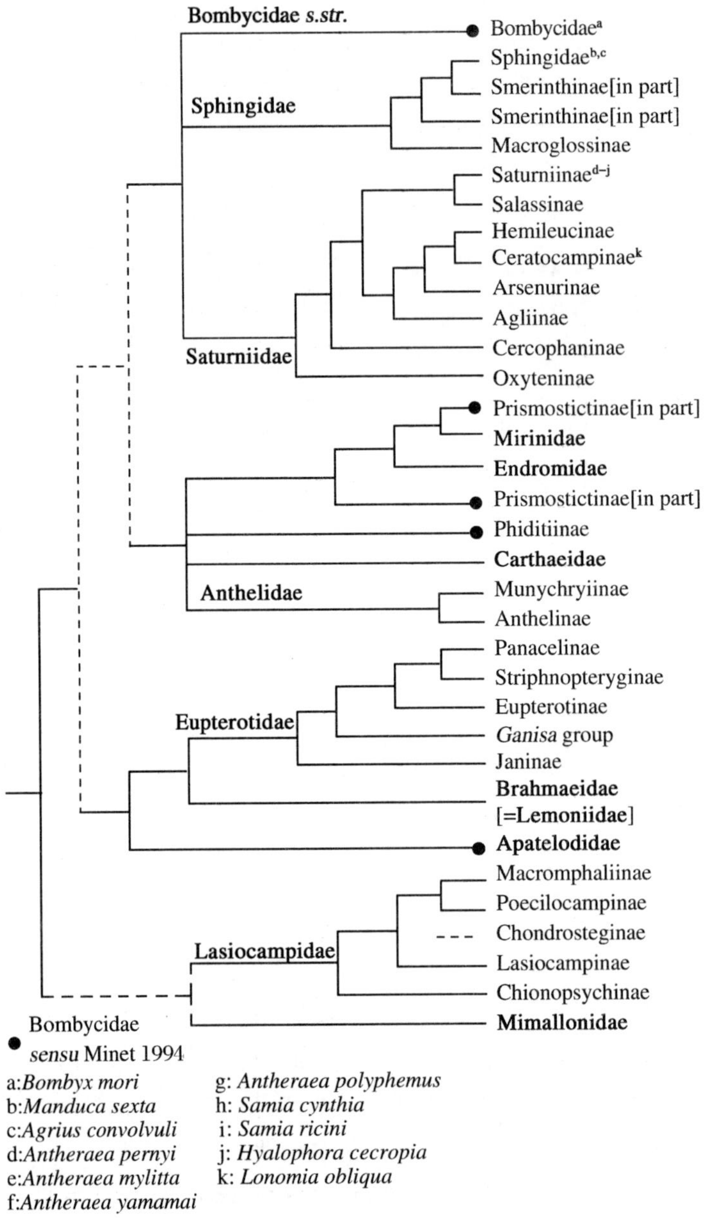

图 1.5　基于 Kawahara et al. (2009), Regier et al. (2001, 2002, 2008) **和** Zwick (2008) **等的分子水平的研究构建的复合系统发生树。用黑色实心圆点标记了** Bombycidae(*sensu* Minet 1994)。

1994)的基于单系学说得到的研究结果相矛盾。

蚕蛾总科以家蚕(家蚕科)最为人所知,家蚕是鳞翅目最著名的模型系统。当前分子生物学的研究(Regier et al. 2008; Zwick 2008; 见图 1.5)表明家蚕科(*sensu* Minet 1994)为非自然存在,应该被定义为野桑蚕的一个亚科。尽管已经对家蚕有了广泛的认识,但是对于家蚕科的其他品种依然知之甚少。这个科迫切需要进行分类学上的全面修订,另外这个科的40个属的350个已有记录的种缺少系统发生学上的研究。作为一个模型系统,家蚕一直被广泛地用于基础研究、生物技术和蚕学研究中(见 Goldsmith,Shimada 和 Abe 2005;一个关于分子生物学和遗传学的综述,见第二章和第四章)。当前,大量分子水平的数据已经被公开,包括两个全基因组测序(Mita et al. 2004; Xia et al. 2004);细菌人工染色体(BAC)文库;表达序列标签(ESTs)以及分子连锁、遗传和物理图谱(Goldsmith,Shimada 和 Abe 2005)等数据。现在,在野桑蚕中开展了部分遗传学的研究,野桑蚕被认为是家蚕的野生祖先(Arunkumar,Metta 和 Nagaraju 2006)。然而,关于家蚕和野桑蚕在遗传学上的丰富信息与蚕蛾科其他昆虫完全空白的研究现状形成了鲜明的对比。

天蛾科是另一个含有模式物种的科,包含了三个亚科:天蛾亚科、短吻天蛾亚科和长喙天蛾亚科。最近的分子生物学研究(Regier et al. 2001; Kawahara et al. 2009)揭示了这些亚科之间的进化关系,部分研究结果与当前亚科/族的分类相矛盾,却与天蛾科种类关系的形态学研究结果相一致(Rothschild 和 Jordan 1903; Nakamura 1976; Kitching 和 Cadiou 2000,Kitching 2002,2003)。虽然我们仍对许多天蛾科的种类知之甚少,但对于其在分类学,未成熟时期,以及对于部分种类的生物学的研究已经相当充分(Kitching 和 Cadiou 2000)。

一个熟知的模式昆虫是烟草天蛾。该昆虫已经被广泛用于生物化学、生理学、形态学以及营养生态学方面的研究(Slansky 1993; Willis,Wilkins 和 Goldsmith 1995)。3800个 GenBank 登录号(包括 EST 序列)和两个 BAC 文库反映了烟草天蛾作为模型系统的普遍性(Sahara et al. 2007)。另外一个正在逐渐受到重视的天蛾科的模式昆虫是分布广泛的甘薯天蛾(*Agrius convolvuli*),甘薯天蛾已经被用于许多生理学和免疫学研究,还包括一些分子遗传学方面的研究。

蚕蛾总科的第三个广为人知的科是大蛾科。当前这个科的系统发育假说(图 1.5)在很大程度上是以分子生物学研究为基础的(Regier et al. 2002,2008; Zwick 2008),唯一基于形态学上的部分是由 Michener 于1952年提出来的。在分子生物学研究方面,大蛾科的8个亚科之间的系统发生关系在充足数据的支持下已经被全面解决。而其他水平的系统发生学假说却受到了限制(Friedlander et al. 1998; Rubinoff 和 Sperling 2002,2004; Regier et al. 2005)。与天蛾科一样,大蛾科的分类学一直以来在全球范围就有充分的研究,并且获取了许多种类的幼虫期和生命周期方面的信息。

大蛾科中的9个种已经用作分子生物学的模型系统,这9个种来自大蛾科的8个亚科中的两个,天蛾科亚科和农神蛾亚科。模型系统中的4个同属于一个属,包括中国柞蚕、印度柞蚕、天蚕和多音天蚕。柞蚕主要分布在北半球,这个属包括大约70个已经有记录的种和多个亚种(Paukstadt,Brosch 和 Paukstadt 2000)。Regier 等人(2005)提出了一个基于70个种中的16个种(其中包括4个模型系统)的系统发生学假说。他们在分子水平上研究了形态上的进化和卵壳气洞冠的发育。柞蚕用于丝的生产,并且和烟草天蛾、家蚕一样,广泛地用于基础研究(Goldsmith 和 Wilkins 1995)。中国柞蚕(*Antheraea pernyi*)一直是研究生物钟分子机制的重要模型系统(见第八章和其中的文献)。然而,与家蚕和烟草天蛾相比,在 GenBank 中登陆的这4个柞蚕种的序列数目可以忽略不计,仅有印度柞蚕(大约4000条)和多音天蚕(大约700条)序列,而且这些序列大多数来自

EST 测序。

另外三个模型系统来自眉纹大蚕蛾科，这是大蚕蛾族的姐妹族(图 1.5)，包括樗蚕、蓖麻蚕和惜古比天蚕。因为同物异名和命名的不统一，亚洲属的樗蚕的分类学非常的复杂和混乱。幸运的是，Peigler 和 Naumann 在 2003 年发表了一个关于这个属的分类学的综合修订。在非分类学文献中，"蓖麻蚕"一直没有统一地被认作是有效物种，或为樗蚕的一个亚种，或为另一种形式的樗蚕。然而，可能以上所有的说法都是错误的。现在，蓖麻蚕被认为是 *S. canningi* 樗蚕的家养形式，与野生的樗蚕没有任何的关系(Peigler 和 Naumann 2003)。因此，我们需谨慎对待过去研究对樗蚕种的鉴定，这些研究为证据标本的常规累计，甚至也包括对模型系统的研究提供了大量的数据。相似地，新北极区的惜古比天蚕的命名也一直很混乱，其中含有三个种和许多个亚种。所有的类群在圈养中很容易杂交，自然杂交就出现在接触地带(Tuskes，Tuttle 和 Collins 1996)。在作为模型系统而用于研究领域的广泛性方面，樗蚕和惜古比天蚕是相类似的，但都比不上柞蚕(Goldsmith 和 Wilkins 1995)。这些种在 GenBank 中的登录号的数目比柞蚕要少；然而，国际上正在着手建立一个包含这些属的 EST 的数据库(www. cdfd. org. in/wildsilkbase/home. php)，这些工作可以大大增加我们对它们基因组的认识。

不同于其他天蚕蛾科的模式生物，南美的 Lonomia 毛虫(*Lonomia obliqua*)(ceratocampinae)的应用范围小于前者，被特别用于研究其高毒性幼虫和这种毒素的抗凝性质(Veiga et al. 2005)。尽管研究范围较窄，Lonomia 毛虫是天蚕蛾科在 GenBank 中拥有第二多信息量的物种，其 EST 库为蚕蛾总科为数不多的 EST 库中的一员。Lemaire (1972，2002)的研究表明 Lonomia 毛虫包括 12 个以上的种。

剩余的蚕蛾总科中不含有分子模型系统，并且大部分是研究相对较少，在经济上没有价值，而且种群数比较小的物种。枯叶蛾科(例如帐篷毛虫)是一个重要的例外，其在 150 个属中大约有 1500 个种，并且在几个属中，有几个种是重要的害虫(例如天幕毛虫、马尾松毛虫和青枯叶蛾属)。枯叶蛾科是单系的结论得到了分子和形态学数据的有力支持(Zwick 2008)。尽管如此，还是缺乏对这个科的种系发生关系的研究(Regier et al. 2001，2008；Zwick 2008)。

根据目前有限的知识，相较于其他科而言，这三个包含分子模型系统的蚕蛾总科之间的关系比较近(图 1.5)(Regier et al. 2001，2008；Zwick 2008)。为了最大限度地利用和使蚕蛾总科现存的庞大数据产生最大的价值，需要在许多分类水平上研究其种系发生关系。在枯叶蛾科中，需要发展模型系统，对早期分化出去的种进行研究来增加目前蚕蛾总科的模型系统的种系发生宽度。目前，关于蚕蛾总科的概括被局限在它的模型系统的种系发生位置上。

4.5 夜蛾总科(Noctuoidea)

夜蛾总科是鳞翅目中最大的总科，大约在 7200 个属中有描述的种就有 70000 个(Kitching 和 Rawlins 1998)。这个总科包括一些我们非常了解的鳞翅目昆虫：灯蛾科(ermines，footman 和虎蛾)，毒蛾科(舞毒蛾和丛蛾)，夜蛾科(地老虎，deltoids，owlets 和 underwings)和舟蛾科(prominents 和 processionary 蛾)。夜蛾被定义为：在成虫期具有一个胸膜(听觉器官)和相关腹部结构，并且在幼虫的后胸出现一对刚毛——澳舟蛾科

oenosandrid 的幼虫是一个例外，它只有一根刚毛（Miller 1991；Kitching 和 Rawlins 1998）。

夜蛾总科有许多生态上比较重要的种，包括一些对森林和农业产生最大破坏的害虫（Kitching 和 Rawlins 1998）。另外，特别是灯蛾，它的化学生态学和交配行为一直是研究的焦点（由 Conner 和 Weller 在 2004 综述）。在北美，众所周知的种包括舞毒蛾（舞毒蛾，毒蛾科：orgyiinae 亚科），玉米耳虫（谷实夜蛾，夜蛾科：夜蛾亚科），烟草青虫（棉铃虫），甘蓝尺蠖（粉纹夜蛾，夜蛾科：金翅夜蛾亚科）和秋黏虫（苜蓿夜蛾，夜蛾科：Xyleninae 亚科）。一些种的毛虫具有成团的（刺人的）要脱落的刚毛，对人类是危险的（例如舞毒蛾，Thaumetopoeia processionaria，夜蛾科：Thaumetopoeinae；见 Kitching 和 Rawlins 1998 综述）。

在这个总科中，由澳舟蛾科 Oenosandridae，舟蛾科和墨西哥舟蛾科 Doidae 组成的基本谱系缺乏模型系统（Miller 1991；Kitching 和 Rawlins 1998）。然而，模型系统主要集中在三个巨大的科，夜蛾（大约 50000 个种；例如棉铃虫、石榴棉铃虫、斜纹夜蛾），灯蛾科（大约 11000 个种；例如雄响盒蛾 *Utetheisa ornatrix*，豹灯蛾 *Arctia caja*，黑条灰灯蛾 *Creatonotus gangis*）和毒蛾（大约 500 个种；例如舞毒蛾）。根据权威分析，这些科形成一个有着部分疑问的谱系进化枝，这些谱系可被视为总科或科（Kitching 和 Rawlins 1998）。依据易于观察的性状，例如它们的后翅脉、腿的锯齿性状，以及幼虫形态的各个方面，这些科和有疑问的谱系可进行不同的划分（Kitching 和 Rawlins 1998）。

夜蛾总科的分类，特别是夜蛾科，近来经历了较大的改动和重分类。传统上灯蛾科被视为毒蛾的姐妹谱系；并且四个分类群——拟灯夜蛾亚科 Aganainae（Hypsidae），瘤蛾亚科 Nolinae，Hermiinae 亚科和 Pantheinae 亚科——被看作灯蛾科的亚科，夜蛾科的亚科，或者与灯蛾科同源的独立科（见 Kitching 和 Rawlins 1998 的综述；Jacobson 和 Weller 2002；Fibiger 和 Lafontaine 2005）。开始于十多年前的分子研究结果一直暗示要根据它们的后翅脉表面来判断，剩下的夜蛾谱系可以被划分成两个主要的谱系或进化枝（图 1.6）。这些中的一个，是具有 trifine 的后翅脉的瘤蛾亚科 Noctuids（例如，M_3，Cu1A，对应的 Cu1B；Mitchell，Mitter 和 Regier 2006）。其他的进化枝大部分由具有四个细微后翅脉的瘤蛾亚科 noctuid 亚科组成（Weller et al. 1994；Mitchell et al. 1997；Mitchell，Mitter 和 Regier 2000）。在图 1.6A 中，除了毒蛾和灯蛾以外的所有分类群都曾经被分到夜蛾中。从而研究表明传统的夜蛾不是单系统发生的。

对这个的分类的修改层出不穷（参见 Lafontaine 和 Fibiger 2006）。最受到关注的研究，开始于 2005 年 6 月，三个标志性的刊物（虽然这些刊物各自有其局限性和优势）展示了系统发生的细节并且三次重写了夜蛾的分类（Fibiger 和 Lafontaine 2005；Mitchell，Mitter 和 Regier 2006；Lafontaine 和 Fibiger 2006）。分子水平上的研究（Mitchell，Mitter 和 Regier 2006）对比较模糊的谱系不具有完整的取样，但是对两个编码蛋白的核基因，1 α 延伸因子、多巴脱羧酶提供了严格的数据分析（*Ddc*；图 1.6A），这给被形态假说所反对的进化枝提供了坚实的统计学支持。形态学上的结论不是基于正式的系统发生的数据分析，但是对形态学领域的平面图提供了一个权威的综述（图 1.6B）。因而在没有

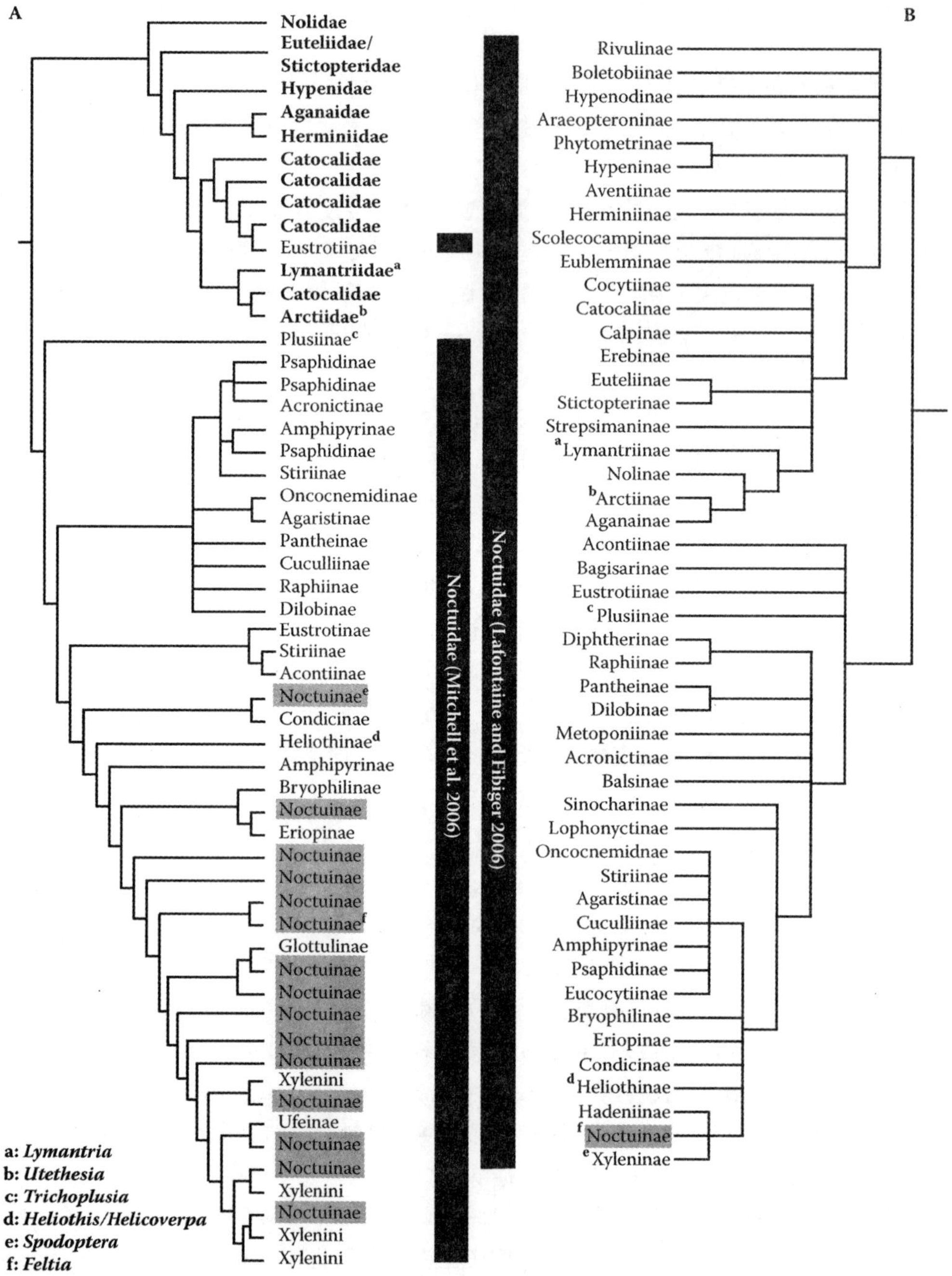

图 1.6　在夜蛾总科分支内，quadrafid 前翅的假设的系统发生关系。没有柱状标示的是 quadrifine 后翅，垂直的柱状标示的是 trifine 后翅亚科；上标表示模式系统中的位置。A，基于 *EF-1α* 和 *Ddc* 基因的序列(见图 2 和 3)构建的最大简约分子进化树，改编自 Mitchell，Mitter 和 Regier (2006)。减少的分支代表了科与亚科之间的关系。B，Lafontaine 和 Fibiger (2006)提出的选择性的分类和进化关系，是基于微夜蛾提出的，因为微夜蛾在进化树中位置不确定，所以图中没有显示。

后继调查研究的情况下是不能解决结果的不一致性的。所有的这些研究结果都证实了传统的夜蛾是非单系发生的。

夜蛾科的模型系统都集中在“夜蛾”(图 1.6A),包括大多数的农业害虫。这些模式在这个进化枝中跨过了种系发生的差异性,从 *T. ni*(金翅夜蛾亚科)到斜纹夜蛾(*Noctuinae*;阴影部分,图 1.6)。近来的系统发生研究有助于阐明一些亚科(例如,夜蛾亚科 Heliothinae,Fang et al. 1997;Cho et al. 2008)与属的关系(例如,斜纹夜蛾;Pogue 2002)。剩余的模型系统只出现在灯蛾科和毒蛾科(图 1.6B;灯蛾科和毒蛾;Lafontaine 和 Fibiger 2006)。而在舟蛾科、墨西哥舟蛾科、澳舟蛾科和早期分化的谱系不存在模型系统。模型系统分布的残缺破坏了由横跨整个总科所得到的观察结果的可信性。

5. 目前的研究和将来的方向

5.1 分子水平上双孔次目的种系发生学与化石的作用

所有先前关于种系发生的讨论都展示在图 1.2 上。这个种系进化关系作为一个假说是有用的,但是在双孔次目内这个总科关系是临时的。双孔次目中总科的关系极少被研究并且从没有进行过明确的种系发生分析。直到近期,研究的焦点才集中在理解很少的一些总科的相互关系上(例如:蚕蛾总科、凤蝶总科、螟蛾总科和夜蛾总科)。因此,对鳞翅目种系发生框架的全面理解仍然没有完成。

幸运的是,目前鳞翅目的种系发生研究主要集中在对双孔次目中总科相互关系的理解上。两项正在进行和互补的新计划始于 2006。第一个是汇集鳞翅目昆虫的进化树计划(LepAToL,总部设在美国),其目的在于用来自于数百个种的核基因序列(达到 26 个基因座位)调查所有鳞翅目中总科的相互关系。第二个是鳞翅目双孔次目的系统分类学计划(总部设在芬兰),其目标在于调查相当数目的种中的核和线粒体的 8 个基因座(与 LepAToL 研究的 26 个核基因基本没有重叠)和 200 个形态特征。这两个计划的进展定期在 http://www.leptree.net 网站上更新。

由 LepAToL 小组主持的对总科相互关系研究的最初报道于 2007 年 12 月在网上发布。这个报道包括了集中在 obtectomera 这个进化枝上的 123 个种(26 个总科)(节 9,图 1.2)。测序了 5 个编码蛋白的核基因,测序长度达 6.7kb,这些基因编码的蛋白质包括多巴脱羧酶、周期基因、无翅基因、烯醇化酶和 CAD(氨甲酰磷酸合成酶 2,天冬氨酸氨甲酰基转移酶和氨甲酰天冬氨酸脱氢酶)。

这个研究的主要结果总结如下:大多数总科可被复原为单系,并且包括了几个具有代表性的总科,总科之间的相互关系得到了解决和数据的支持(例如卷叶蛾总科、螟蛾总科、夜蛾总科)。在较广泛的规模上,分析复原了几个图 1.2 所描述的进化枝;尽管如此;存在两个有意义的特例:蝴蝶(凤蝶总科)和 allies(弄蝶总科丝角蝶科)超出了宏观鳞翅目的进化枝,并且这些总科一直与一个或多个微观鳞翅目总科更近;螟蛾总科一直与剩余的宏观鳞翅目形成一种新的姐妹谱系关系。尽管这个结论在多个总科的单个节点上得不到强有力的支持,但是把全面的分子水平上的数据与先前的种系发生假说进行比较发现,测试结果与以前定义的宏观鳞翅目的单系发生假说是相矛盾的。因而,迄今的发现暗示对鳞翅目种系发生的框架在多个水平上的改进在不久的未来将会得到实现。

在多个水平上很好地解决种系发生关系对鳞翅目的全面研究而言是重要的。但是要完全理解许多有意思的性状的进化关系则需要关于进化事件重建的明确时间点的知识。在大多数情况下,这种知识最可靠的来源是化石记录。不幸的是,鳞翅目的化石记录比其他较大的昆虫目(Labandeira 和 Sepkoski Jr. 1993)都稀少得多,特别是那种在岩石里形成的压缩化石(Grimaldi 和 Engel 2005)。Grimaldi 和 Engel (2005)指出,因为鳞翅目昆虫的鳞毛与水相排斥,导致它们很少会沉入水中形成沉积物。

从广义来看,鳞翅目进化时间的化石证据可以总结如下:最早的鳞翅目化石可以追溯到大约 1.9 亿年前的侏罗纪早期(mya;Whalley 1986;Grimaldi 和 Engel 2005)。现存最老的鳞翅目昆虫化石在黎巴嫩琥珀中被发现,它源于距今 1.25 亿年前的白垩纪早期,琥珀中有翅蛾(*Parasabatinca aftimacrai*)(Whalley 1978),属于现存最老的科,即小翅蛾科。同样研究人员从黎巴嫩琥珀中发现的一个幼虫化石,被归为喙亚目这个进化枝。这个进化枝的特点是成虫首次出现喙。尽管对这些化石的解释是很迫切的,但是双孔次目第一个以树叶化石记录的形式出现在距今 9700 万年前的白垩纪中期,这个说法是根据现存的 Phyllocnistinae(潜叶蛾科)(Labandeira et al. 1994)而来的。尽管缺少来自于白垩纪其他的双次亚目化石,但是大鳞翅目(macrolepidoptera)的古新纪(距今 5600 万年)化石的出现,暗示着许多其他的双次亚目谱系都是在白垩纪期间产生的,白垩纪在距今 6550 万年前结束(Grimaldi 和 Engel 2005)。许多现存的双次亚目科首先出现在始新世波罗的海的琥珀中(440 万年前),并且其他的许多化石是在始新世、渐新世弗洛里森特页岩中发现的(3800 万年前;见 Grimaldi 和 Engel 2005 综述)。

用目前的系统发生研究整合这些化石的信息,必须解决两个问题。第一,化石必须要放置到目前的分子系统发生学中去。起码化石与总科的对应(或者更准确地与科和属对应)需要根据形态学的特征进行判断(共同衍征)。目前,多数现存的结论都是仅仅依据对现存物种主观印象相似性的基础上得出的,而且大多数化石还没有被相关专家检测研究(de Jong 2007)。第二是方法问题。分子进化与速率恒定时常相违背,即突变不是以类似于时钟的方式累积的(例如,Kumar 2005)。为了根据分子分析得到更精确的起源点,正在开发用来解释突变速率不均匀的复杂分析方法(例如,Welch 和 Bromham2005;Rutschmann 2006)。最好的情况是,未来的分析不光包括分子水平上的数据,而且应该包括形态学上的基础。尽管化石不完整,也可尝试着与建立的系统发生相融合。带有讽刺意味的是,只有当形态学在种间的研究和在分子间一样广泛时,才有可能实现利用分子系统学揭示鳞翅目进化历史的希望。

5.2 鳞翅目系统学的虚拟群落构建

在过去的二十年间,形态分类学家肩负起把鳞翅目模型系统放置到进化背景中的工作重任。历史上,分类学家都是相对独立地工作,文章一般都是一个作者,偶尔有两个。这种典型的慢进度研究使得依靠它们的系统发生学去建立模型系统的研究者一直都有挫折感。与此相反的是,研究模型系统的许多研究者以较大的合作形式发表文章(Cronin 和 Franks 2006)。分类学研究的缓慢传播一直是建立团体的一个历史性障碍。

系统性的工作常常是以博物馆或地域性专题文章的形式公布，这样的文章只有在最大的图书馆才能找到。只有最近的成果可以以数码形式提供，但通常都超出了许多世界性组织的经济承受范围。历史上研究者依靠维持个人的关系网与这个领域的发展保持同步，通过文章和样本的交换、不定期的拜访，加上召开会议来促使科学上的互换，来扩大和维持研究的网络。因而，相当多的工作是孤立地或者是以一些小的研究团队的形式来完成的。如同在其他昆虫中的研究一样，这样的状况导致鳞翅目昆虫形态学上的命名法的独特发展，常常对一个特定的种类使用高度特殊的词汇（例如对蝴蝶的生殖器术语的特殊命名）。由于支离破碎的词汇的影响，对研究鳞翅目多个种类感兴趣的研究者不得不去学习来源于特别的科和总科的特殊术语。这样的状况阻碍了鳞翅目系统学研究的快速发展。

为了消除这些障碍和形成一个庞大的、全球性的、相互合作的虚拟组织，几个网站已经采取了一些措施。它们中的一些是根据类群建立的，集中在特别的总科。例如，GlobIZ 是螟蛾总科数据库（http://www. pyraloidea. org），尺蠖网站论坛（http://www. herbulot. de），蛱蝶科系统学小组（http://www. nymphalidae. utu. fi），Tortricid. net（http://www. tortricid. com）和 Gelechioid 工作小组（Gelechioidea 框架，http://www. msstate. edu/org/mississippientmuseum/Lepidoptera/GelechioidFramework. htm）。

几个基于网络的计划开始致力于把所有的分类学家或研究鳞翅目的专家在一个虚拟的社区中链接起来。欧洲分类学研究所（EDIT；http://www. e-taxonomy. eu）旨在把欧洲的分类学研究联接起来。EDIT 试图在数字化上使得欧洲的研究成果与全球的分类学研究的努力相适应，特别是与全球分类学倡议相适应。鳞翅目是 EDIT 和 Lepsys 优先研究的目标种类，Lepsys 是欧洲鳞翅目系统学家的协会（http://www. lepsys. eu），协调欧洲的鳞翅目研究。

LepTree（http://www. leptree. net）与 EDIT 一样，支持一个虚拟社区，在那里，分子水平和形态学水平的计划都得到支持。为了最大化的开放和交流，开发出有着开放资源内容的操作系统 Drupal（http://www. drupal. org）的网站，主要是利用它的讨论和协同功能。这个网站被进一步定制来以开放资源的形式存储生物和社会数据（http://www. openrdf. org），语义上丰富的 OWL（网站本体语言）和 RDF（资源描述框架）形式用于数据的存储和分享（例如，Mabee et al. 2007）。作为这些技术的一个例子，一些亚社区一起工作，建立了一个形态学术语的词汇表，大家能相互参照和分享这个词汇表用于注释图片的描述。社区的数据将会自动发送到类似生命百科的门户网站并且使得全世界的分类学家都能以数字数据的形式获得。

这些全球性网站的努力（EDIT，Lepsys，LepTree）代表了介于随意讨论和“大科学”这种丰富数据的合作之间的一个中间状态。它们促进了知识、术语学和实验方案的交流，由此建立了一个更大、更全球性的社区（Preece 2000）。有这样工具的网站能够建立鼓励合作和远程培训的虚拟社区，而且还可以迅速简单地发表结果。因此，这个鳞翅目系统学家的虚拟社区正极力促成大家一起有效地工作来回答一系列比较的和进化的问题。

5.3 总结与未来的模式系统

模型系统分散在双孔次目中，并且为广泛的生物过程提供了见解。通过对这些模型系统进化发生位置的总结，发现几个主要的鳞翅目模型系统的代表性明显不足，这种不足在非双孔次目以及尺蛾总科和斑蛾总科中表现最明显。这种研究的不足需要通过从谱系中选择和研究新模型来解决。

进化框架对选择新模型系统或生物进行比较分析提供了指导。几个作者已经提出了在发展新模型系统时需要考虑的因素(Mabee 2000；Santini 和 Stellwag 2002；Collins et al. 2005；Jenner 和 Willis 2007)，谨慎地选择分类群并进行比较将会使机械性和比较结果达到最大程度。首先，一个模型应该被选来有利于进一步的研究主题，鉴定形态相似的(古代或者原始的)性状，寻求解决新性状的起源问题。例如，理解鳞翅目的翅膀模式(Ramos 和 Monteiro 2007；参见第五、六章)，基因组大小(Gregory 和 Hebert 2003；Goldsmith，Shimada 和 Abe 2005)，信息素合成(参见第十章)将不仅需要在鳞翅目中而且要在毛刺目中的模型中进行慎重的选择。其次，在实验室中获得和饲养模型系统应该是可行的。例如，家蚕被证明是容易喂养的，在几千年前就已经家养来获得丝织品，这使得它在科学研究中处于引人注目的地位。另一方面，特别是目前缺乏模型系统的基底谱系的鳞翅目昆虫，其幼虫食谱以及成虫的繁殖行为是未知的。在我们成功培育这些分类群和发展它们成为能存活的模型系统之前，我们需要加强对这些谱系基本生物学的理解。最后，系统发生的抽样与包含模型系统的谱系之内和之间的关系必须经过检查来确保与被检查的形状相对应的分类群抽样的合适范围。目前欧洲和北美将努力提供更强劲的系统发生框架以用于精确估计模式和非模式分类群的关系和选择未来研究的新模型系统。

6. 致谢

我们对 LepTree 学界对这章的修改所提供的交流和建议，特别是 M. Horak 和 C. Mitter 两人提出的详细而富有洞察力的建议和评述表示感谢。对匿名的 SEL 评议者提出的有帮助的建议表示感谢。这项研究工作受到国家科学基金 No. 0531639（提供给：Mitter，Cummings，Parr，Regier，Roe，Weller)NSF DEB-0212910 基金（提供给：Regier)，明尼苏达实验站基金 No. MN-17-022（提供给：Roe，Weller)，嘉士伯基金(提供给：Simonsen)，巴伦西亚大学基金 UV-AE-20070204（提供给：Baixeras)和加拿大国家科学和技术研究委员会(提供给：Sperling，Simonsen)的资助。

参考文献

[1]Ackery，P. R.，R. de Jong，and R. I. Vane-Wright. 1998. The butterflies：Hedyloidea，Hesperoidea and Papilionoidea. In *Lepidoptera，moths and butterflies. 1 Evolution，systematics and biogeography*，ed. N. P. Kristensen，263－300. *Handbook of Zoology*，Vol. 4，Part 35，*Arthropoda*：*Insecta*. Berlin and New York：Walter de Gruyter.

[2]Addison，M. F. 2005. Suppression of codling moth *Cydia pomonella* L.（Lepidoptera：Tortricidae) populations in South African apple and pear orchards using sterile insect release. *Acta Hort*. 671：255－57.

[3]Ahmad，T. R.，and Z. A. Al－Gharbawi. 1986. Effects of pheromone trap design and placement on catches of codling moth *Cydia pomonella* males. *J. Appl. Entomol*. 102：52－57.

[4]Albert, P. J. 1991. A review of some host-plant chemicals affecting the feeding and oviposition behaviours of the eastern spruce budworm, *Choristoneura fumi ferana* Clem. (Lepidoptera: Tortricidae). *Mem. Entomol. Soc. Can.* 159:13—18.

[5]Albert, P. J. 2003. Electrophysiological responses to sucrose from a gustatory sensillum of the larval maxillary palp of the spruce budworm, *Choristoneura fumiferana* (Clem.) (Lepidoptera: Tortricidae). *J. Insect Physiol.* 49:733—38.

[6]Arn, H. 1991. Sex pheromones. In *Tortricid pests: Their biology, natural enemies and control*, ed. L. P. S. Van der Gees and H. H. Evenhuis, 187—207. Amsterdam: Elsevier.

[7]Arunkumar, K. P., M. Metta, and J. Nagaraju. 2006. Molecular phylogeny of silkmoths reveals the origin of domesticated silkmoth, *Bombyx mori* from Chinese *Bombyx mandarina* and paternal inheritance of *Antheraea proylei* mitochondrial DNA. *Mol. Phylogenet. Evol.* 40:419—27.

[8]Arzberger, P., and T. Finholt. 2002. *Data and collaboratories in the biomedical community*. Ballston, VA: Nat. Inst. Health.

[9]Backman, A.-C. 1997. Pheromones released by codling moth females and mating disruption dispensers. *IOBC West Palearc. Reg. Sec. Bull.* 20:175—80.

[10]Bates, H. W. 1861. Contributions to an insect fauna of the Amazon valley. Lepidoptera: Heliconidae. *Trans. Linn. Soc.* 23:495—566.

[11]Beckemeyer, E. F., and P. D. Shirk. 2004. Development of the larval ovary in the moth, *Plodia interpunctella*. *J. Insect Physiol.* 50:1045—51.

[12]Beldade, P., and P. M. Brakefield. 2002. The genetics and evo-devo of butterfly wing patterns. *Nat. Rev. Genet.* 3:442—52.

[13]Boggs, C. L., W. B. Watt, and P. R. Ehrlich. 2003. *Butterflies: Ecology and evolution taking flight*. Chicago: Univ. Chicago Press.

[14]Bolker, J. A. 1995. Model systems in developmental biology. *Bioessays* 17:451—55.

[15]Braby, M. F., R. Vila, and N. E. Pierce. 2006. Molecular phylogeny and systematics of the Pieridae (Lepidoptera: Papilionoidea): higher classification and biogeography. *Zool. J. Linn.* Soc. 147:239—75.

[16]Brock, J. P. 1971. A contribution towards an understanding of the morphology and phylogeny of the ditrysian Lepidoptera. *J. Nat. Hist.* 5:29—102.

[17]Brower, A. V. Z. 2000. Phylogenetic relationships among the Nymphalidae (Lepidoptera), inferred from partial sequences of the *wingless* gene. *Proc. Biol. Sci.* 267:1201—11.

[18]Brown, J. W. 2005. *World catalogue of insects, Volume* 5: *Tortricidae (Lepidoptera)*. Stenstrup, Denmark: Apollo Books.

[19]Caterino, M. S., R. D. Reed, M. M. Kuo, and F. A. H. Sperling. 2001. A partitioned likelihood analysis of swallowtail butterfly phylogeny (Lepidoptera: Papilionidae). *Syst. Biol.* 50:106—27.

[20]Cho, S., A. Mitchell, C. Mitter, J. C. Regier, M. Matthews, and R. Robertson. 2008. Molecular phylogenetics of heliothine moths (Lepidoptera: Noctuidae: Heliothinae), with comments on the evolution of host range and pest status. *Syst. Entomol.* 33:581—94.

[21]Coates, B. S., R. L. Hellmich, and L. C. Lewis. 2006. Sequence variation in trypsin-and chymotrypsin-like cDNAs from the midgut of *Ostrinia nubilalis*: Methods for allelic differentiation of candidate *Baccillus thuringiensis* resistance genes. *Insect Mol. Biol.* 15:13—24.

[22]Collins, A. G., P. Cartwright, C. S. McFadden, and B. Schierwater. 2005. Phylogenetic context and basal metazoan model systems. *Integr. Comp. Biol.* 45:585—94.

[23]Common, I. F. B. 1990. *Moths of Australia*. Carlton: Melbourne Univ. Press.

[24]Conner, W. E., and S. J. Weller. 2004. A quest for alkaloids. The curious relationships between tiger moths and plants containing pyrrolizidine alkaloids. In *Advances in insect chemical ecology*, ed. R. T. Cardé and J. Millar, 248—82. Cambridge: Cambridge Univ. Press.

[25]Cook, C., M. L. Smith, M. J. Telford, A. Bastianello, and M. Akam. 2001. *Hox* genes and the phylogeny of the arthropods. *Curr. Biol.* 11:759—63.

[26]Cronin, B., and S. Franks. 2006. Trading cultures: Resource mobilization and service rendering in the life sciences as revealed in the journal article's paratext. *J. Am. Soc. Inf. Sci. Technol.* 57:1909—18.

[27]DaCosta, M. A., P. Larson, J. P. Donahue, and S. J. Weller. 2006. Phylogeny of milkweed tussocks (Arctiidae: Arctiinae, Phaegopterini) and its implications for evolution of ultrasound communication. *Ann. Entomol. Soc. Am.* 99:723—42.

[28]Davis, D. R. 1986. A new family of monotrysian moths from austral South America (Lepidoptera: Palaephatidae), with a phylogenetic review of the Monotrysia. *Smithson. Cont. Zool.* 434:1—202.

[29]de Jong, R. 2007. Estimating time and space in the evolution of the Lepidoptera. *Tijdschr. Entomol.* 150:319—46.

[30]de Jong, R., R. I. Vane-Wright, and P. R. Ackery. 1996. The higher classification of butterflies (Lepidoptera): Problems and prospects. *Entomol. Scand.* 27:65—101.

[31]Delisle, J., J.-F. Picimbon, and J. Simard. 1999. Physiological control of pheromone production in *Choristoneura fumiferana* and *C. rosaceana*. *Arch. Insect Biochem. Physiol.* 42:253—68.

[32]Ehrlich, P. R. 1958. The comparative morphology, phylogeny and higher classification of the butterflies (Lepidoptera: Papilionoidea). *Univ. Kan. Sci. Bull.* 39:305—70.

[33]Ehrlich, P. R., and I. Hanski. 2004. *On the wings of checkerspots: A model system for population biology*. Oxford: Oxford Univ. Press.

[34]Ehrlich, P. R., and P. H. Raven. 1967. Butterflies and plants: A study in coevolution. *Evolution* 18:586—608.

[35]El-Sayed, A., M. Bengtsson, S. Rauscher, J. Lofqvist, and P. Witzgall. 1999. Multicomponent sex pheromone in codling moth (Lepidoptera: Tortricidae). *Environ. Entomol.* 28:775—79.

[36]Eliot, J. N. 1973. The higher classification of the Lycaenidae (Lepidoptera): A tentative arrangement. *Bull. Br. Mus. Nat. Hist. Entomol.* 28:373—505.

[37]Fang, Q. Q., C. Soowon, J. C. Regier, et al. 1997. A new nuclear gene for insect phylogenetics: Dopa decarboxylase is informative of relationships within Heliothinae (Lepidoptera: Noctuidae). *Syst. Biol.* 46:269—83.

[38]Felsenstein, J. 2004. *Inferring phylogenies*. Sunderland, MA: Sinauer Associates.

[39]Fibiger, M., and J. D. Lafontaine. 2005. A review of the higher classification of the Noctuoidea (Lepidoptera) with special reference to the Holarctic fauna. *Esperiana* 11:7—92.

[40]Ford, E. B. 1964. *Ecological genetics*. London: Chapman & Hall.

[41]Freitas, A. V. L., and K. S. J. Brown. 2004. Phylogeny of the Nymphalidae (Lepidoptera: Papilionoidea). *Syst. Biol.* 53:363—83.

[42]Friedlander, T. P., K. R. Horst, J. C. Regier, C. Mitter, R. S. Peigler, and Q. Q. Fang. 1998. Two nuclear genes yield concordant relationships within Attacini (Lepidoptera: Saturniidae). *Mol. Phylogenet. Evol.* 9:131—40.

[43]Goater, B., M. Nuss, and W. Speidel. 2005. Pyraloidea I (Crambidae: Acentropinae, Evergesti-

nae, Heliothelinae, Schoenobiinae, Scopariinae). In *Microlepidoptera of Europe* 4, ed. P. Huemer and O. Karsholt, 1—304. Stenstrup, Denmark: Apollo Books.

[44]Goldsmith, M. R., T. Shimada, and H. Abe. 2005. The genetics and genomics of the silkworm, *Bombyx mori*. *Annu. Rev. Entomol.* 50:71—100.

[45]Goldsmith, M. R., and A. S. Wilkins. 1995. *Molecular model systems in the Lepidoptera*. New York: Cambridge Univ. Press.

[46]Gregory, T. R., and P. D. N. Hebert. 2003. Genome size and variation in lepidopteran insects. *Can. J. Zool.* 81:1399—1405.

[47]Grimaldi, D. A. 1996. *Amber: Window to the past*. New York: Abrams/Am. Mus. Nat. Hist.

[48]Grimaldi, D. A. 1999. The co-radiations of pollinating insects and angiosperms in the Cretaceous. *Ann. Mo. Bot. Gard.* 86:373—406.

[49]Grimaldi, D. A., and M. S. Engel. 2005. *Evolution of the insects*. New York: Cambridge Univ. Press.

[50]Hasenfuss, I. 1999. The adhesive devices in larvae of Lepidoptera (Insecta, Pterygota). *Zoomorphology* 119:143—62.

[51]Heppner, J. B. 1991. Faunal regions and the diversity of Lepidoptera. *Trop. Lepid.* 2, Suppl. 1:1—85.

[52]Hinton, H. E. 1946. On the homology and nomenclature of the setae of lepidopterous larvae, with some notes on the phylogeny of the Lepidoptera. *Trans. Entomol. Soc. Lond.* 97:1—37.

[53]Hock, V., P. J. Albert, and M. Sandoval. 2007. Physiological differences between two sugar-sensitive neurons in the galea and the maxillary palp of the spruce budworm larva *Choristoneura fumiferana* (Clem.) (Lepidoptera: Tortricidae). *J. Insect Physiol.* 53:59—66.

[54]Hodkova, M., and I. Hodek. 2004. Photoperiod, diapause, and cold-hardiness. *Eur. J. Entomol.* 101:445—58.

[55]Holder, M., and P. O. Lewis. 2003. Phylogeny estimation: traditional and Bayesian approaches. *Nat. Rev. Genet.* 4:275—84.

[56]Honeybee Genome Sequencing Consortium. 2006. Insights into social insects from the genome of the honeybee *Apis mellifera*. *Nature* 443:931—49.

[57]Horak, M. 1998. The Tortricoidea. In *Lepidoptera, moths and butterflies. 1 Evolution, systematics and biogeography*, ed. N. P. Kristensen, 199—215. *Handbook of Zoology*, Vol. 4, Part 35, *Arthropoda: Insecta*. Berlin and New York: Walter de Gruyter.

[58]Jacobson, N. L., and S. J. Weller. 2002. *A cladistic study of the tiger moth family Arctiidae (Noctuoidea) based on larval and adult morphology*. Thomas Say Monograph Series, Entomol. Soc. Amer.

[59]Jenner, R. A., and M. A. Willis. 2007. The choice of model organisms in evo-devo. *Nat. Rev. Genet.* 8:311—19.

[60]Jiggins, C. D., J. Mavarez, M. Beltrán, J. S. Johnston, and E. Bermingham. 2005. A genetic linkage map of the mimetic butterfly *Heliconius melpomene*. *Genetics* 171:557—70.

[61]Jurenka, R. 2004. Insect pheromone biosynthesis. *Top. Curr. Chem.* 239:97—132.

[62]Kawahara, A. Y., A. A. Mignault, J. C. Regier, I. J. Kitching, and C. Mitter. 2009. Phylogeny and biogeography of hawkmoths (Lepidoptera: Sphingidae): Evidence from five nuclear genes. /PLoS ONE/ 4:e5719.

[63]Kellogg, E. A., and H. B. Shaffer. 1993. Model organisms in evolutionary studies. *Syst. Biol.* 42:

409—14.

[64]Kettlewell,H. B. D. 1973. *The evolution of melanism, the study of a recurring necessity; with special reference to industrial melanism in Lepidoptera*. Oxford: Clarendon Press.

[65]Kitching,I. J. 2002. The phylogenetic relationships of Morgan's Sphinx, *Xanthopan morganii* (Walker), the tribe Acherontiini, and allied long-tongued hawkmoths (Lepidoptera: Sphingidae, Sphinginae). *Zool. J. Linn. Soc.* 135:471—527.

[66]Kitching,I. J. 2003. Phylogeny of the death's head hawkmoths, *Acherontia* [Laspeyres], and related genera (Lepidoptera: Sphingidae: Sphinginae: Acherontiini). *Syst. Entomol.* 28:71—88.

[67]Kitching,I. J., and J. -M. Cadiou. 2000. *Hawkmoths of the world: An annotated and illustrated revisionary checklist (Lepidoptera: Sphingidae)*. Ithaca: Nat. Hist. Mus. London & Cornell Univ. Press.

[68]Kitching,I. J., and J. E. Rawlins. 1998. The Noctuoidea. In *Lepidoptera, moths and butterflies. 1. Evolution, systematics, and biogeography*, ed. N. P. Kristensen, 351—401. *Handbook of Zoologie*, Vol. 4, Part 35, *Arthropoda: Insecta*. Berlin and New York: Walter de Gruyter.

[69]Kristensen,N. P. 1976. Remarks on the family-level phylogeny of butterflies (Insecta, Lepidoptera, Rhopalocera). *Z. Zool. Syst. Evol.* 14:25—33.

[70]Kristensen, N. P. 1984. Studies on the morphology and systematics of primitive Lepidoptera. *Steenstrupia* 10:141—91.

[71]Kristensen,N. P., and A. W. Skalski. 1998. Phylogeny and palaeontology. In *Lepidoptera, moths and butterflies. 1. Evolution, systematics, and biogeography*, ed. N. P. Kristensen, 7—25. *Handbook of Zoologie*, Vol. 4, Part 35, *Arthropoda: Insecta*. Berlin and New York: Walter de Gruyter.

[72]Kumar,S. 2005. Molecular clocks: Four decades of evolution. *Nat. Rev. Genet.* 6:654—62.

[73]Labandeira,C. C., D. L. Dilcher, D. R. Davis, and D. L. Wagner. 1994. Ninety-seven million years of angiosperm-insect association: Palaeobiological insights into the meaning of coevolution. *Proc. Natl. Acad. Sci. U. S. A.* 9:12278—82.

[74]Labandeira,C. C., and J. J. Sepkoski, Jr. 1993. Insect diversity in the fossil record. *Science* 261: 310—15.

[75]Lafontaine,J. D., and M. Fibiger. 2006. Revised higher classification of the Noctuoidea (Lepidoptera). *Can. Entomol.* 138:610—35.

[76]Lamas,G. 2004. *Atlas of Neotropical Lepidoptera. Checklist: Part 4A, Hesperioidea — Papilionoidea*. Gainesville: Scientific Publishers.

[77]Leibenguth, F. 1986. Genetics of the flour moth, *Ephestia kühniella*. *Agric. Zool. Rev.* 1: 39—72.

[78]Lemaire,C. 1972. Révision du genre *Lonomia* (Walker) [Lep. Attacidae]. *Ann. Soc. Entomol. Fr.* 8:767—861.

[79]Lemaire,C. 2002. *The Saturniidae of America (=Attacidae). Hemileucinae*. Keltern: Goecke & Evers.

[80]Mabee,P. M. 2000. Development data and phylogenetic systematics: Evolution of the vertebrate limb. *Amer. Zool.* 40:789—800.

[81]Mabee,P. M., M. Ashburner, Q. Cronk, et al. 2007. Phenotype ontologies: the bridge between genomics and evolution. *Trends Ecol. Evol.* 22:345—50.

[82]Mallet,J., W. O. McMillan, and C. D. Jiggins. 1998. Mimicry and warning color at the boundary between races and species. In *Endless forms: Species and speciation*, ed. D. J. Howard and S. H. Berloch-

er,390—403. New York: Oxford Univ. Press.

[83]McDonough,L. M. ,H. G. Davis,P. S. Chapman,and C. L. Smithhisler. 1993. Response of male codling moths (*Cydia pomonella*) to components of conspecific female sex pheromone glands in flight tunnel tests. *J. Chem. Ecol.* 19:1737—48.

[84]Michener,C. D. 1952. The Saturniidae (Lepidoptera) of the Western hemisphere. *Bull. Am. Mus. Nat. Hist.* 98:335—502.

[85]Miller,J. S. 1991. Cladistics and classification of the Notodontidae (Lepidoptera: Noctuoidae) based on larval and adult morphology. *Bull. Am. Mus. Nat. Hist.* 204:1—230.

[86]Minet,J. 1983. Etude morphologique et phylogénétique des organs tympaniques des Pyraloidea. 1-généralities et homologies. (Lep. Glossata). *Ann. Soc. Entomol. Fr.* 19:175—207.

[87]Minet,J. 1991. Tentative reconstruction of the ditrysian phylogeny (Lepidoptera: Glossata). *Entomol. Scand.* 22:69—95.

[88]Minet,J. 1994. The Bombycoidea: Phylogeny and higher classification (Lepidoptera: Glossata). *Entomol. Scand.* 25:63—88.

[89]Mita,K. ,M. Kasahara,S. Sasaki,et al. 2004. The genome sequence of silkworm,*Bombyx mori*. *DNA Res.* 11:27—35.

[90]Mitchell,A. ,S. Cho,J. C. Regier,C. Mitter,R. W. Poole,and M. Matthews. 1997. Phylogenetic utility of elongation factor—1α in Noctuoidea (Insecta: Lepidoptera): The limits of synonymous substitution. *Mol. Biol. Evol.* 14:381—90.

[91]Mitchell,A. ,C. Mitter,and J. C. Regier. 2000. More taxa or more characters revisited: Combining data from nuclear protein-encoding genes for phylogenetic analyses of Noctuoidea (Insecta: Lepidoptera). *Syst. Biol.* 49:202—24.

[92]Mitchell,A. ,C. Mitter,and J. C. Regier. 2006. Systematics and evolution of the cutworm moths (Lepidoptera: Noctuidae): Evidence from two protein-coding nuclear genes. *Syst. Entomol.* 31:21—46.

[93]Müller,F. 1879. *Ituna and Thyridia*; a remarkable case of mimicry in butterflies. *Trans. Entomol. Soc. Lond.* 20—29.

[94]Munroe,E. ,and M. A. Solis. 1998. The Pyraloidea. In *Lepidoptera,moths and butterflies. 1 Evolution,systematics and biogeography*,ed. N. P. Kristensen,233—56. *Handbook of Zoology*,Vol. 4,Part 35,*Arthropoda:Insecta*. Berlin and New York: Walter de Gruyter.

[95]Nakamura,M. 1976. An inference on the phylogeny of Sphingidae in relation to habits and the structures of their immature stages. *Yugatô* 63:19—28.

[96]Nazari,V. ,E. V. Zakharov,and F. A. H. Sperling. 2007. Phylogeny,historical biogeography,and taxonomic ranking of Parnassiinae (Lepidoptera,Papilionidae) based on morphology and seven genes. *Mol. Phylogenet. Evol.* 42:131—56.

[97]Neunzig,H. H. 1987. Pyralidae (Pyraloidea). In *Immature insects*,ed. F. W. Stehr,462—94. Dubuque: Kendall Hunt Publishing Co.

[98]Neunzig,H. H. 1997. Pyraloidea,Pyralidae (part). In *The moths of America north of Mexico*,ed. R. B. Dominick,D. C. Ferguson,J. G. Franclemont,R. W. Hodges,and E. G. Munroe,fascicle 15. 4:1—157. Washington,DC: The Wedge Entomological Research Foundation.

[99]Nielsen,E. S. ,and N. P. Kristensen. 1996. The Australian moth family Lophocoronidae and the basal phylogeny of the Lepidoptera (Glossata). *Invertebr. Taxon.* 10:1199—1302.

[100]Nijhout,H. F. 1991. *The development and evolution of butterfly wing patterns*. Washington,

DC: Smithsonian Institution Press.

[101]Nuss, M. 2006. Global information system of Pyraloidea (GlobIZ). http://www.pyraloidea.org (accessed July 31, 2007).

[102]Paukstadt, U., U. Brosch, and L. H. Paukstadt. 2000. Preliminary checklist of the names of the worldwide genus *Antheraea* Hübner, 1819 ("1816") (Lepidoptera: Saturniidae). Part I. *Galathea. Ber. Kreis. Nürnber. Entomol. e. V.*, Suppl. 9:1—59.

[103]Peigler, R. S., and S. Naumann. 2003. *A revision of the silkmoth genus* Samia. San Antonio: Univ. Incarnate Word.

[104]Pogue, M., 2002. *A world revision of the genus Spodoptera Guenée (Lepidoptera: Noctuidae)*. Washington, DC: Mem. Am. Entomol. Soc. no. 43.

[105]Preece, J. 2000. *Online communities: Designing usability, supporting sociability*. Chichester: John Wiley & Sons.

[106]Quayle, D., J. Régnière, N. Cappuccino, and A. Dupont. 2003. Forest composition, host-population density, and parasitism of spruce budworm *Choristoneura fumiferana* eggs by *Trichogramma minutum*. *Entomol. Exp. Appl.* 107:215—27.

[107]Ramos, D. M., and A. Monteiro. 2007. Transgenic approaches to study wing color pattern development in Lepidoptera. *Mol. Biosyst.* 3:530—35.

[108]Regier, J. C., C. P. Cook, C. Mitter, and A. Hussey. 2008. A phylogenetic study of the "bombycoid complex" (Lepidoptera) using five protein-coding nuclear genes, with comments on the problem of macrolepidopteran phylogeny. *Syst. Entomol.* 33:175—89.

[109]Regier, J. C., C. Mitter, T. P. Friedlander, and R. S. Peigler. 2001. Phylogenetic relationships in Sphingidae (Insecta: Lepidoptera): Initial evidence from two nuclear genes. *Mol. Phylogenet. Evol.* 20:311—25.

[110]Regier, J. C., C. Mitter, R. S. Peigler, and T. P. Friedlander. 2002. Monophyly, composition, and relationships within Saturniinae (Lepidoptera: Saturniidae): Evidence from two nuclear genes. *Insect Syst. Evol.* 33:9—21.

[111]Regier, J. C., U. Paukstadt, L. H. Paukstadt, C. Mitter, and RS. Peigler. 2005. Phylogenetics of eggshell morphogenesis in *Antheraea* (Lepidoptera: Saturniidae): Unique origin and repeated reduction of the aeropyle crown. *Syst. Biol.* 54:254—67.

[112]Reuter, E. 1896. Über die Palpen der Rhopaloceren. *Acta Soc. Sci. Fennicae* 22:1—578.

[113]Robinson, R. 1971. *Lepidoptera genetics*. Oxford: Pergamon Press.

[114]Roelofs, W. L., and A. P. Rooney. 2003. Molecular genetics and evolution of pheromone biosynthesis in Lepidoptera. *Proc. Natl. Acad. Sci. U. S. A.* 100:9179—84.

[115]Rothschild, W. V., and K. Jordan. 1903. A revision of the lepidopterous family Sphingidae. *Novit. Zool.* 9(Suppl.):1—972.

[116]Rubinoff, D., and F. A. H. Sperling. 2002. Evolution of ecological traits and wing morphology in *Hemileuca* (Saturniidae) based on a two-gene phylogeny. *Mol. Phylogenet. Evol.* 25:70—86.

[117]Rubinoff, D., and F. A. H. Sperling. 2004. Mitochondrial DNA sequence, morphology and ecology yield contrasting conservation implications for two threatened buckmoths (*Hemileuca*: Saturniidae). *Biol. Conserv.* 118:341—51.

[118]Rutschmann, F. 2006. Molecular dating of phylogenetic trees: A brief review of current methods that estimate divergence times. *Divers. Distrib.* 12:35—48.

[119]Sahara,K. ,A. Yoshido,F. Marec,et al. 2007. Conserved synteny of genes between chromosome 15 of *Bombyx mori* and a chromosome of *Manduca sexta* shown by five-color BAC-FISH. *Genome* 50:1061—65.

[120]Salmon,M. A. 2000. *The Aurelian legacy: British butterflies and their collectors*. Berkeley: Univ. California Press.

[121]Santini,F. ,and E. J. Stellwag. 2002. Phylogeny,fossils and model systems in the study of evolutionary developmental biology. *Mol. Phylogenet. Evol.* 24:379—83.

[122]Scriber,J. M. ,Y. Tsubaki,and R. C. Lederhouse. 1995. *Swallowtail butterflies: Their ecology and evolutionary biology*. Gainesville: Scientific Publishers.

[123]Shimodaira,H. 2002. An approximately unbiased test of phylogenetic tree selection. *Syst. Biol.* 51:492—508.

[124]Siaussat,D. ,F. Bozzolan,P. Porcheron,and S. Debernard. 2007. Identification of steroid hormone signaling pathway in insect cell differentiation. *Cell. Mol. Life Sci.* 64:365—76.

[125]Slansky Jr. ,F. ,1993. Nutritional ecology: The fundamental quest for nutrients. In *Caterpillars: Ecological and evolutionary constraints on foraging*,ed. N. E. Stamp and T. M. Casey,29—91. New York: Chapman and Hall.

[126]Smart,P. 1975. *The illustrated encyclopedia of the butterfly world*. London: Salamander Books.

[127]Solis,M. A,S. D. Hight,and D. R. Gordon. 2004. Alert: Tracking the cactus moths as it flies and eats it way westward in the U. S. *News Lepid. Soc.* 46:3—4.

[128]Solis,M. A. ,and K. V. N. Maes. 2002. Preliminary phylogenetic analysis of the subfamilies of Crambidae (Pyraloidea Lepidoptera). *Belg. J. Entomol.* 4:53—95.

[129]Solis,M. A. ,and C. Mitter. 1992. Review and preliminary phylogenetic analysis of the subfamilies of the Pyralidae (sensu stricto) (Lepidoptera: Pyraloidea). *Syst. Entomol.* 17:79—90.

[130]Srinivasan,A. ,A. P. Giri,and V. S. Gupta. 2006. Structural and functional diversities in Lepidoptera serine proteases. *Cell. Mol. Biol. Lett.* 11:132—54.

[131]Swofford,D. L. ,G. J. Olsen,P. J. Waddell,and D. M. Hillis. 1996. Phylogenetic inference. In *Molecular Systematics*,2nd edition,ed. D. M. Hillis,C. Moritz,and B. K. Mable,407—514. Sunderland: Sinauer Associates.

[132]Trematerra,P. ,and A. Sciarretta. 2005. Activity of the kairomone ethyl (E,Z)-2,4-decadienoate in the monitoring of *Cydia pomonella* (L.) during the second annual flight. *Redia* (Italy) 88:57—67.

[133]Tuskes,P. M. ,J. P. Tuttle,and M. M. Collins. 1996. *The wild silk moths of North America. A natural history of the Saturniidae of the United States and Canada*. Ithaca: Cornell Univ. Press.

[134]Vane-Wright,R. I. ,and P. R. Ackery. 1984. *The Biology of Butterflies*. Princeton: Princeton Univ. Press.

[135]Veiga,A. B. G. ,J. M. C. Ribeiro,J. A. Guimarães,and I. M. B. Francischetti. 2005. A catalog for the transcripts from the venomous structures of the caterpillar *Lonomia obliqua*:Identification of the proteins potentially involved in the coagulation disorder and hemorrhagic syndrome. *Gene* 355:11—27.

[136]Wahlberg,N. ,M. F. Braby,A. V. Brower,et al. 2005. Synergistic effects of combining morphological and molecular data in resolving the phylogeny of butterflies and skippers. *Proc. Biol. Sci.* 272:1577—86.

[137]Wahlberg,N. ,E. Weingartner,and S. Nylin. 2003. Towards a better understanding of the higher systematics of Nymphalidae (Lepidoptera: Papilionoidea). *Mol. Phylogenet. Evol.* 28:473—84.

[138]Wahlberg,N. ,and C. W. Wheat. 2008. Genomic outposts serve the phylogenomic pioneers: De-

signing novel nuclear markers for genomic DNA extractions of Lepidoptera. *Syst. Biol.* 57:231—42.

[139]Wallace, E. K., P. J. Albert, and J. N. McNeil. 2004. Oviposition behavior of the eastern spruce budworm *Choristoneura fumiferana* (Clemens) (Lepidoptera: Tortricidae). *J. Insect Behav.* 17:145—54.

[140]Walters, B. D., P. J. Albert, and R. Y. Zacharuk. 1998. Morphology and ultrastructure of chemosensilla on the proboscis of the adult spruce budworm *Choristoneura fumiferana* (Clem.) (Lepidoptera: Tortricidae). *Can. J. Zool.* 76:466—79.

[141]Warren, A. 2006. The higher classification of the Hesperiidae (Lepidoptera: Hesperioidea). PhD diss., Oregon State Univ.

[142]Wearing, C. H., J. Hansen, C. Whyte, C. E. Miller, and J. Brown. 2001. The potential for spread of codling moth via commercial sweet cherry: A critical review and risk assessment. *Crop Prot.* 20:465—88.

[143]Welch, J. J., and L. Bromham, 2005. Molecular dating when rates vary. *Trends Ecol. Evol.* 20: 320—27.

[144]Weller, S. J., N. L. Jacobson, and W. E. Conner. 1999. The evolution of chemical defences and mating systems in tiger moths (Lepidoptera: Arctiidae). *Biol. J. Linn. Soc.* 68:557—78.

[145]Weller, S. J., D. P. Pashley, and J. A. Martin. 1996. Reassessment of butterfly family relationships using independent genes and morphology. *Ann. Entomol. Soc. Am.* 89:184—92.

[146]Weller, S. J., D. P. Pashley, J. A. Martin, and J. L. Constable. 1994. Phylogeny of noctuoid moths and the utility of combining independent nuclear and mitochondrial genes. *Syst. Biol.* 43:194—211.

[147]Whalley, P. E. S. 1978. New taxa of fossil and recent Micropterigidae with a discussion of their evolution and a comment on the origin of the Lepidoptera. *Ann. Tvl. Mus.* 31:71—86.

[148]Whalley, P. E. S. 1986. A review of the current fossil evidence of Lepidoptera in the Mesozoic. *Biol. J. Linn. Soc.* 28:253—71.

[149]Wiegmann, B. M., C. Mitter, J. C. Regier, T. P. Friedlander, D. M. Wagner, and E. S. Nielsen. 2000. Nuclear genes resolve Mesozoic-aged divergences in the insect order Lepidoptera. *Mol. Phylogenet. Evol.* 15:242—59.

[150]Wiegmann, B. M., J. C. Regier, and C. Mitter. 2002. Combined molecular and morphological evidence on phylogeny of the earliest lepidopteran lineages. *Zool. Scripta* 31:67—81.

[151]Willis, J. H., A. S. Wilkins, and M. R. Goldsmith. 1995. A brief history of Lepidoptera as model systems. In *Molecular model systems in the Lepidoptera*, ed. M. R. Goldsmith and A. S. Wilkins, 1—20. Cambridge: Cambridge Univ. Press.

[152]Xia, Q., Z. Zhou, C. Lu, et al. 2004. A draft sequence for the genome of the domesticated silkworm (*Bombyx mori*). *Science* 306:1937—40.

[153]Zhang, B.-C. 1994. *Index of economically important Lepidoptera*. Oxon: CAB International.

[154]Zimmermann, H. G., V. C. Moran, and J. H. Hoffmann. 2000. The renowned cactus moth, *Cactoblastis cactorum*: Its natural history and threat to native *Opuntia* floras in Mexico and United States of America. *Divers. Distrib.* 6:259—69.

[155]Zwick, A. 2008. Molecular phylogeny of Anthelidae and other bombycoid taxa (Lepidoptera: Bombycoidea). *Syst. Entomol.* 33:190—209.

第二章 家蚕遗传学和基因组学的研究进展

Marian R. Goldsmith

1. 前言:家蚕模型

家蚕,很久以来就被作为鳞翅目昆虫生物学的模式生物。家蚕具有相对短的、可预测的生活周期,体型大,易于大规模和实验室饲养等优点,加上在养蚕业方面积累的丰富的基础知识,使家蚕成为鳞翅目昆虫基础研究的典型代表。早在 20 世纪就开始对家蚕进行遗传学研究(家蚕的遗传研究和养殖历史,参见 Eickbush 1995; Yasukochi,Fujii,Goldsmith 2008),因此收集了许多在饲养过程中或由于野桑蚕的基因渗入,产生的自然突变种。野桑蚕与家蚕最为接近,存在于中国、日本和韩国的桑园中,并且能和家蚕杂交产生一些有生育能力的品种。中国、日本、韩国和印度的主要种质库保存着数百的"地理的"或本地的品系和遗传上改良的变种,像其他农业上选育的品种一样,根据它们的经济特征来分类,例如饲养特征和强壮性、茧和丝的质量、生育性和产卵力等(Sohn 2003;表 2.1)。这些研究中心同时保存着许多影响整个变态时期形态、生化和行为上的突变种。日本有大约 430 个突变种(Fujii 1998; 表 2.1),中国保存有超过 600 个突变种(Lu,Dai 和 Xiang 2003)。在 Müller 于 1927 年第一次利用 X 射线诱变果蝇的实验报道后不久,家蚕即被用于照射研究(Tazima 1964)。这些早期的工作,产生和选择了很多形态和行为上的突变体以及染色体畸变的资源。之后是从常染色体到雌性 W 染色体的性别连锁的易位,这些易位产生一些可以看得见的,影响卵、幼虫和茧的突变,并被用于培育自显性别的品种,从而用于蚕丝业生产(Tazima 1964; Nagaraju 1996; Fujii et al. 2006)。因为 W 染色体携带着可能的雌性决定 *Fem* 基因,这些畸变对于受影响的常染色体(Fujiwara 和 Maekawa 1994)和 W 染色体的精细结构定位也是有用的(见第四章 W 染色体易位的讨论和在性别决定研究中的应用)。

表 2.1 家蚕品种和遗传资源库

Table 2.1 Silkworm Stocks and Mutant Resources

Database	Contents	Host	Web Site
SilkwormBase	Silkworm strains and mutant resources; information on mutations	Institute of Genetic Resources Center, Kyushu University	http://www.shigen.nig.ac.jp/silkwormbase/ViewStrainGroup.do
Silkworm Genome Database	Descriptions and photographs of mutations	University of Tokyo Department of Agricultural And Environmental Biology	http://www.ab.a.u-tokyo.ac.jp/bioresource/shimada/database.html
Silkworm "Aburako" (translucent) mutation database	Descriptions, images, and linkage assignments of translucent mutations	National Institute of Agrobiological Sciences, Tsukuba, Japan	http://cse.nias.affrc.go.jp/natuo/en/aburako_top_en.htm
Database of Sericultural Genebank	Silkworm strain resources (453 stocks; mutant and commercial races); mulberry stocks; insect cell lines	National Institute of Agrobiological Sciences, Kobuchizawa, Japan	http://ss.nises.affrc.go.jp/nises/db-eng.html
Silkworm Germplasm	Practical breeding strains (430 stocks); economic characters and morphological traits	Central Sericultural Germplasm Resources Centre, Central Silk Board, Tamil Nadu, India	http://www.silkgermplasm.com

20 世纪 90 年代,家蚕继续作为鳞翅目昆虫的模型用于分子遗传学和基因组学研究,家蚕所具有的丰富的、分化良好的、稳定的实验室品种和大量关于家蚕基础生物学的知

识使其很早就用于经典的或孟德尔遗传学的研究。这些优点和其他一些技术特征，比如在高度近交的情况下，提供低程度 DNA 多态性的起始研究材料，加上其作为模式生物的历史角色，使家蚕成为第一个用于大规模 cDNA 测序(Mita et al. 2003)，并最终完成全基因组测序的鳞翅目昆虫(International Silkworm Genome Consortium 2008)。

在这一章中，将集中对家蚕的分子遗传学和基因组学最近的发展进行阐述；之前工作的综述可以参见 Nagaraju 和 Goldsmith 于 2002 年，Goldsmith，Shimada 和 Abe 于 2005 年的文献，家蚕基因组学最近发展的概述可以参见 Zhou，Yang 和 Zhong 于 2008 年的文献。

2. 家蚕细胞遗传学

家蚕单倍染色体数目是 28，接近于大多数鳞翅目昆虫的染色体数目 31(De Prins 和 Saitoh 2003)，由 WZ：ZZ 型的雌雄性染色体组成(见第三章鳞翅目昆虫性染色体的讨论；第四章家蚕性染色体和性别决定的讨论)。20 世纪 70 年代中期，W. Traut 在粗线期的卵母细胞中第一次在细胞遗传学上鉴定了家蚕的染色体组型(Traut 1976)。尽管这为家蚕染色体和一些染色体区域的相对大小提供了一些资料，但是 WZ 二价体仍然没有被证明。此外，由于家蚕染色体缺乏明显的初级缢痕(着丝粒)，并且具有接近于鳞翅目昆虫典型的全动力学结构(见第三章)，所以通过不同的染色体模式只能鉴定到 6 个染色体。

20 世纪 90 年代以来，伴随着用于染色体原位杂交的放射性和荧光核苷酸衍生物直接标记 DNA 探针技术的发展，家蚕细胞遗传学得到迅速的发展。关于家蚕研究的早期报道是用放射性探针标记 BMC1，BMC1 大小约 5kb，属于长散布核原件 1(LINE-1)家族，BMC1 在家蚕基因组中大约有 3500 个拷贝并且在染色体上分布广泛(Ogura et al. 1994)。Sahara 和他的同事们应用更加精确和灵敏的荧光原位杂交技术(FISH)，鉴定了家蚕染色体末端的端粒上含有标准的昆虫序列$(TTAGG)_n$(Sahara，Marec 和 Traut 1999)。这些探针由长伸展的重复所组成，这对于在相对较小的鳞翅目昆虫染色体上检测定位信号是必需的，并且具有可预测的模式和位点(二价染色体末端形成双联体)，这些优点使它们成为对基本方法进行研究的好靶标。Traut 等(1999)应用比较基因组杂交技术(CGH)研究性染色体的分子分化，比较基因组杂交最初是用于检测癌症组织的染色体异常的。Traut 等应用比较基因组杂交技术第一次在雌蚕中鉴定了野生型的 W 染色体，结果显示，这个雌性决定染色体含有大量的普遍存在的 DNA 重复(CGH 原理见第三章)。

单拷贝序列可视化研究的突破是构建大片段的 BAC 文库(Wu et al. 1999)，从中分离克隆感兴趣基因并可直接用于荧光素标记。通过这种方法，使得在家蚕中很容易地可视化鉴定 W 染色体。在之前的研究中，K. Sahara 和他的同事们利用 BAC 文库得到了完整的 W 染色体，分离所用的引物是从和 W 染色体连锁的非长末端重复反转录转座子中得来的(Sahara et al. 2003)。这些研究第一次为包含 W 染色体在内的重复元件的类型提供了直接的证据(见第四章家蚕 W 染色体结构)，这也预示了 BAC-FISH 技术在鳞翅目昆虫细胞遗传学研究上的分析能力。在以后的改进中，Yasukochi 和他的同事们利用两种荧光标记显示了一个大的 BAC 重叠群的 5′和 3′末端，这个重叠群包含位于第六连锁群上的 *BmHox* 基因群，包括 *Bmzen*、*Dfd*、*Scr*、*Antp*、*Ubx*、*abd-A* 和 *Abd-B* 基因，有 2～3Mb 长(Yasukochi et al. 2004)。通过加入第三种探针，用相同的荧光素标记远端的连锁基因 *labial*，能够直接显示这个单拷贝基因的定位，为大约十分之一染色体长度的距离，与遗传连锁图谱一致。这个标志性的研究提供了利用两种颜色的荧光原位杂交技术确定完整的家蚕染色体组型的方法，这种方法使锚定标记的 BACs 定位到每个连锁群上(Yoshido et al. 2005)，并用 BAC-FISH 技术定位单个基因(例如，Niimi et al. 2006)。这种技术发展到曾用五种颜色来比较家蚕的第十五连锁群和烟草天蛾与之相对应的连锁群上保守标记的定位(Sahara et al. 2007)，表明这种方法不仅能用于鳞翅目昆虫的连锁图谱的研究，也能用于比较基因组学的研究。

3. "经典"连锁图谱

经典连锁图谱包括大约 240 个基因座，通过形态、行为和生化突变来进行标记(Fujii 1998)。利用相似表型的形态突变作为对照基因座来标记特定的连锁群可以发现一些问题；最近，连锁群 26 和连锁群 28 是通过两个独立的黄色幼虫突变 *Sel* 和 *Xan* 来定义的，现在通过序列标记位点(STS; Yasukochi et al. 2005)表明这两个突变在同一个连锁群上，并且用微卫星和简单序列重复(SSRs; Miao et al. 2007)表明它们可能是等位的或很近的连锁。这表明尽管简单的可视的突变仍将被用于检测品种保持中的污染和迅速筛选具有感兴趣的重组体的染色体区域，然而通过聚合酶链式反应(PCR)、单核苷酸多态性(SNPs)或高通量测序技术可以快速精确地评估可用的分子标记，这将使可视化标记的品种在很多应用中显得陈旧，它们最大的价值将是作为基因功能分析的资源。

4. 分子连锁图谱

分子连锁图谱是现代遗传学的基础。它为许多应用提供了一个参考的框架，如决定自发的或诱导的突变的染色体定位，包括具有孟德尔类型遗传的(比如，起主要效应)和复杂遗传的基因，或具有相对小的累加效应或没有累加效应的数量性状；提供支架来拼接大规模物理图谱和全基因组鸟枪法(WGS)测序数据；定位克隆只知道表型效应的基因；反映出全基因组的组织和进化模式。遗传标记可以是随机的，例如随机扩增多态 DNA 序列(RAPDs)或扩增片段长度多态性(AFLPs)，这些仅限于单个的实验室品种或野生种群，或来自于物种特异性的单拷贝染色体位点，或者是通过限制性酶切片段多态性(RFLPs)、剪切扩增多态性序列(CAPS)、STS、SSRs 和 SNPs 得到的基因。所有这些类型的标记现在都用于研究家蚕，并且自从 13 年前首次报道用 RAPDs (Promboon et al. 1995)和 RFLPs 来检测当时并不知道序列的 cDNAs 以来(Shi，Heckel 和 Goldsmith 1995)，在构建家蚕连锁图谱方面取得明显的进步。之前的详细连锁图谱见综述(Nagaraju 和 Goldsmith 2002; Goldsmith，Shimada 和 Abe 2005; Zhou，Yang 和 Zhong 2008)。在随后的一节中，将描述当前的连锁图谱的突出特征，着重强调将它们作为研究家蚕和其他鳞翅目昆虫遗传学资源的策略。

4.1 参考品系

可能是长期驯化的结果，家蚕可以耐受相对高水平的近交。这和许多其他的“野生”鳞翅目昆虫的实验室品系不同。野生鳞翅目昆虫品系由于近交抑制，即有害的隐性突变纯合，它们需要有规律的和野生品种异型杂交来避免种群衰落。这个特点使家蚕成为研究分子遗传学和基因组学特别好的模型，因为可以用一些特征清楚的、广泛散布的、稳定近交的品系作为通用的参考。因此，两个种系已被用来做连锁图谱，一个是在日本资源中心叫 p50(Daizo)，而在中国资源中心叫大造的品系，还有一个是遗传上相对古老的改良品系，中国 108(C108)。p50/大造起源于中国，被认为是一种地理品系，已被用来构建许多用于 EST 计划的 cDNA 文库(Mita et al. 2003)，用于大规模物理图谱计划的 BAC 文库(Yamamoto et al. 2006，2008；Suetsugu et al. 2007)，染色体步移(Koike et al. 2003)，BAC-FISH(Yoshido et al. 2005；Sahara et al. 2007)，全基因组鸟枪测序(Mita et al. 2004；Xia et al. 2004)。尽管这些品系中仍存在一些多态性(Mills 和 Goldsmith 2000；Cheng et al. 2004)，但是如预想的那样，它们的纯合程度已经足够使不同实验室保存的同一品系的 RAPDs 和 RFLPs 位点一致(M. R. Goldsmith，个人意见；Miao et al. 2005)，足够促进 BAC 重叠群的构建(Suetsugu et al. 2007；Yamamoto et al. 2008)，足够从各自独立的计划中比较和组装全基因组鸟枪测序序列(Mita et al. 2004；Xia et al. 2004)。这种相对高的纯合性是选择家蚕作为第一个鳞翅目昆虫进行全基因组测序的主要原因之一(见下文)。

4.2 野桑蚕和家蚕

通过染色体组型研究发现，野桑蚕包括不同的种群。中国野桑蚕(ChBm)的单倍体染色体数目和家蚕相同，为 28 条，但是日本野桑蚕(JaBm)的单倍体染色体数目是 27 条，这些不同可能是导致家蚕在中国得到驯化的部分原因，尽管这种驯化在中国的不同地方是发生了一次还是两次仍在讨论中(Li et al. 2005)。

染色体组型的不同可能是由于家蚕的染色体 14 和染色体 27 或 28 发生融合导致的(Banno et al. 2004)。在韩国的不同区域的不同的野桑蚕种群具有 27 或 28 条染色体，表明它们是在 200 万年前因为朝鲜半岛的形成而发生分离的(Kawanishi et al. 2008)。应用从家蚕中得到的探针进行 BAC-FISH 表明这两个物种的 W 染色体保持着相当高程度的相似性，都含有大量的相同类型的重复序列；然而，利用 BAC 产生的不同的杂交谱表明染色体可能发生过一些重排(Yoshido et al. 2007；见第三章鳞翅目昆虫 W 染色体的构成)。以后更详细的细胞遗传学分析可能揭示在野桑蚕驯化过程中曾发生过其他一些大规模的染色体组型变化，为描绘不同亚群的进化提供里程碑式的研究。

代表性野桑蚕的完整线粒体 DNA 序列是已知的。尽管对家蚕和野桑蚕基于不同的序列的分化估计从 100 万～200 万年(Pan et al. 2008)到最多 700 万年(Yukuhiro et al. 2002)，但是通过比较家蚕和野桑蚕单个(Li et al. 2005)或多个(Arunkumar，Metta 和 Nagaraju 2006；Pan et al. 2008)线粒体基因和区域表明家蚕更近似于中国野桑蚕，这和细胞遗传学证据和历史记录都是相一致的。从 5.8S～28S 核糖体 DNA 的差异中得到的证据表明中国野桑蚕和日本野桑蚕大约有 300 万年的分歧，这与它们的地理分离是一致的(Maekawa et al. 1988)。中国、日本和韩国的野桑蚕个体的核糖体 DNA 中存在杂合型，这些品系仍然和其他野桑蚕或家蚕杂交。因此，研究与驯化有关的核基因必须说明一个给定的染色体的亲本起源，这个染色体可能是从其他物种中引入的，研究者必须在这些限制的条件下在头脑中整合这些数据(Kawanishi et al. 2008)。对不同地理种群的野桑蚕的 mariner-like 元件(MLEs)进行系统发生分析也表明中国是家蚕的起源地(Kawanishi et al. 2007)；有趣的是，许多 MLEs 具有全长度的开放阅读框的特征表明它们可能仍具有活性(Kawanishi et al. 2008)。

野桑蚕的幼虫和成虫具有灰暗的色素和良好的伪装，与此不同的是，家蚕的幼虫都是浅色或白色的，只有第二("眼点")、第五("月牙")、第八("星")体节上有少许斑点，蛾主要是浅褐色，翅上有微弱的条带。许多从野桑蚕引入家蚕的色素模式基因，如幼虫斑纹、翅的模式和茧色等，具有孟德尔遗传规律来影响相应的特征(Fujii 1998)。物种间其他的更细微的性状差别，如行为差异(单个或成群产卵、小枝拟态)、免疫应答等，在与家蚕杂交后应当作为数量性状来研究(见下文)。

野桑蚕到家蚕的驯化长期被人们关注，特别是家蚕是唯一完全依赖于人的饲养而存活的昆虫。家蚕和野桑蚕在行为上、形态上、生理上和生化上的不同可被用于研究直接选择和昆虫生物学之间的影响。哪些遗传的改变改良了家蚕的品质从而使其能够应用于农业？包括：大量饲养条件下的高存活率；高饲料效率；高产丝量，并且蚕丝具有良好的纺织特性；缺乏争斗和行为的改变从而保证了简单的交配，高生育率和产卵力；卵长时期的滞育；肥大可食用的蛹。哪些被采集者简单地选择保留下来，就是说，没受到选择的基因也没有商业的应用价值？利用新的基因组数据和现有的研究方法似乎能容易地回答这些与蚕丝业长期相关的问题，但是选择特征和基因系统仍需要深思熟虑，这将为了解物种形成和人工、自然选择提供新的见解。

4.3 图谱构建策略

大多数的家蚕连锁图谱分两步来构建，第一步是利用雌蚕缺乏重组交换(Maeda 1939)，通过用 F_1 代雌蚕回交亲本的雄蚕，产生 10～15 个基因型后代小种群而建立起连锁群。随后是扩大这个种群，使个体达到 100～200 头，用 F_1 代的雄蚕回交亲本的雌蚕，从而计算在这种低水平的交换率下的重组频率(详情见第六章对蝴蝶及第八章对家蚕的实例操作)。

利用这种策略，Miao 等(2005)构建了一个由 500 多个 SSR 标记锚定到大约 80 个 CAPS 标记上的重组图谱，这些标记对应的基因序列已经被登录到 NCBI 上；连锁群和分布已经用独立的杂交进行检验，杂交的品种来自另一个实验室。相似地，Kadono-Okuda 等(2002)和 Nguu 等(2005)利用 RF02 和 RF50 品系建立了独立的 RFLP 图谱，选用这两个品系是为了使多态性最大化(W. Hara，私人交流)。整合独立构建的图谱需要用相同的标记来完成；由于每个连锁群都建立了各自的参考基因，使得这项工作变得相对简单。只用 F_2 代杂交也是一种可用的方法，这种方法的优点是减少了需要保持的品系的数目和每代需要饲养的数目，缺点是需要在后代中为每个连锁群分出非重组的雌性相关的染色体。这增加了分析的复杂性，特别针对诸如 RAPDs 和 AFLPs 等显性特征，对亲本各自产生的连锁图谱必须用通用的、可参考的、共显性的标记来整合(Yasukochi 1998；详情见第六章)。尽管如此，借助于 RFLPs(Shi，Heckel 和 Goldsmith 1995)和 RAPDs(Promboon et al. 1995；Yasukochi 1998)计划成功地利用 p50 和 C108 的 F_2 代杂交(Promboon et al. 1995；Yasukochi 1998)建立了最初的图谱。其中之一的 RAPD 图谱含有 1000 多个标记(Yasukochi 1998)，其上锚定了 523 个 BAC 重叠群(contigs)，427 个 STS 代表已克隆的基因和表达序列标签(ESTs)(Yasukochi et al. 2006)。

家蚕的遗传图谱建立的方法是，用两步定位策略，通过对从雌回交的 15 个个体亚群的 Southern 杂交将 1688 个核基因定位到连锁群上(Yamamoto et al. 2008)。使用雄回交构建的重组图谱最初包含了超过 500 个 SNP 位点(Yamamoto et al. 2006)，后来通过 BAC 末端序列(BESs)扩展到 1755 个 SNP 位点(Yamamoto et al. 2008)。通过限制性酶切片段指纹图谱和高密度 BAC 与 ESTs 杂交，将 BESs 锚定到 BAC 重叠群上(Yamamoto et al. 2008)。遗传和物理图谱的整合为基因组序列的组装提供了很好的框架(见下文)。

4.4 图谱特征

迄今为止，已经建立了许多独立的参考图谱，如利用 RAPDs(Promboon et al. 1995)和 RFLPs(Hara et al. 2002；Nguu et al. 2005)方法得到的含有大约 200 个标记的中低密度图谱；利用 AFLPs(Tan et al. 2001)、SSRs(Miao et al. 2005)或 SNPs (Yamamoto et al. 2006)方法得到的含有大约 500 个标记的中密度图谱；利用 RAPDs (Yasukochi 1998；Yasukochi et al. 2006)或 SNPs(Yamamoto et al. 2008；及 Yasukochi，Fujii 和 Goldsmith 2008)方法得到的含有超过 1000 个标记的高密度图谱。RFLP 图谱以克隆的 cDNAs 为基础，因此，它们为整合这些计划与其他物种相关的保守序列或基因进行比较提供了普遍的参考。尽管大多数的计划都是使用同一参考品种，但是独立构建的图谱并没有被整合；尽管现在利用组装的全基因组鸟枪序列数据(见下文)能在一定程度上整合这些图谱，但是仍需要提高分辨率和整合度解决这些图谱中的分歧，特别是需要填平用不同标记系统而留下的缺口。因此，基于不同类型的标记的图谱长度仍没有确定。

除基因组范围的连锁图谱以外，利用 RAPDs、SSRs 和 FISSRs(Nagaraja et al. 2005)的组合及 SSRs (Miao et al. 2008)发展了 Z 染色体的分子图谱。这对性别连锁在物种形成中的进化角色提出假设有很重要的作用 (Prowell 1998；Qvarnström 和 Bailey 2009)。家蚕 Z 染色体图谱将为揭示和比较其他鳞翅目昆虫的性别连锁特征提供参考。

4.5 经典连锁图谱和分子连锁图谱的整合

分子图谱和经典的形态生化突变图谱的整合是促进定位克隆在其他鳞翅目昆虫中没有的独特突变基因的一个重要的战略目标。它们为解决许多基础性的问题提供了良好的基础，这些问题包括：胚胎和幼虫的发育(例如 *ki*、肾脏形卵、母性致死、只形成外胚层衍生物、大多数幼虫只活到第一次蜕皮)，突变对整个幼虫形状的影响(例如 *gn*、细节蚕、细长体、体节缢缩；*e*、长节、第四和第五体节之间的节间膜拉伸；*nb*、狭胸、短胸、腹部肥胖)，幼虫的标记(例如具有至少 15 个等位基因的 *p* 基因座)，翅的发育和模式(见第六章家蚕翅突变),幼虫的表皮(例如 *K*、龙角、独特的新月形和星形斑纹突起)等,有超过 35 个品系具有半透明皮肤的突变从而影响氮代谢产物的形成和沉积(例如:*od*,性别连锁的油蚕、Nagaraja et al. 2005;w^{3oe}、白卵、Kômoto et al. 2009)。确定其他一些突变如食性、滞育、蜕皮和对真菌、病毒(见第十八章病毒抗性)的抗性等的生物学基础将为研究鳞翅目害虫的防治开辟新的途径。目前,家蚕的突变和连锁图谱的分布目录参见 SilkwormBase (表 2.1)。

迄今,大多数整合经典图谱和分子图谱的方法是直接地将一两个标准的、简单的形态突变和一个特异的分子连锁群联系起来(例如,Miao et al. 2005; Yasukochi et al. 2005)。现在遗传图谱和物理图谱上可以利用的高密度的分子标记和锚定的基因座,还有用 AFLPs、RAPDs 和 SNPs 等方法产生的无名标记仍然没有在 DNA 水平上被鉴定,因此迫切需要整合整个图谱。

4.6 家蚕突变的分子分析

迄今为止,只有少数的家蚕突变体在 DNA 水平上被研究。包括油蚕或半透明的幼虫突变体,*oq* (Yasukochi,Kanda 和 Tamura 1998; Kômoto 2002)和 *og* (Kômoto et al. 2003),这两种突变体不能将尿酸晶体沉积到表皮上,因为它们编码黄嘌呤脱氢酶和钼辅因子硫化酶的基因分别有缺陷;两种白卵的突变体,编码犬尿氨酸-3-单氧化酶的 *w-1* (white egg 1,Quan et al. 2002)和编码腺苷三磷酸结合转运蛋白的 *w-3* (white egg 3, Abraham et al. 2000; Kômoto et al. 2009),*w-3* 是果蝇 white 基因的同源基因(综述 Goldsmith,Shimada 和 Abe 2005);编码类胡萝卜素结合蛋白的 *Y* 突变体(黄色的血和茧)(Tabunoki et al. 2004; Tsuchida et al. 2004; Sakudoh et al. 2007)。这些突变基因已经通过精巧的猜想被克隆,利用大致的图谱定位和可能的候选基因之间的对应关系,随后通过缺失(Sakudoh et al. 2007)或提前终止(Kômoto et al. 2009)等方法实现功能阻断来寻求分子水平上的证据。

利用大规模的 EST 库、BAC 文库和整合的物理遗传图谱等可以通过以图谱为基础的或定位克隆等方法来分离突变的基因和其对应的野生型的基因。这样可以使突变基因可能存在的染色体区域变小,随后进行测序和功能分析,理想情况下,突变可以通过转基因来回复,从而证明这个靶标基因(定位克隆见第六和十八章)。最近,用这种方法鉴定了三种突变体。第一种是 *nsd-2*(浓病毒 2 抗性基因),这是一种隐性的突变使家蚕能够产生对浓核病毒的抗性,浓核病毒是养蚕业的一种主要病毒,能够编码一个有缺陷的十二次跨膜的蛋白,即可能的 *Bm*dnv-2 受体(见第十八章 *nsd-2* 的克隆和分析)。第二个鉴定的突变基因是通过定位克隆鉴定的无翅突变 *fl*(*flugellos*,蛹期和蛾期无翅)。有趣的是,用蜕皮激素处理培养的翅原基后(Matsuoka 和 Fujiwara 2000),用 RNA 差显技术(Matsunaga 和 Fujiwara 2002)研究表达谱时发现了一些过表达或低表达的候选基因。但是仍然需要对这四个 *fl* 的等位基因进行定位克隆和分子分析,从而鉴定真正的 *fl* 基因座 *fringe*(*fng*),*fringe* 编码 *Notch* 信号途径中的糖基转移酶,*Notch* 信号途径是翅原基形态发生所必需的(Sato et al. 2008)。第三种鉴定的突变是 Ek^{p},这是一个和 *Hox* 基因家族同源的能扰乱幼虫腹足发育的突变。在利用 9 个分开的回交种群中的 2000 多个后代第一次建立了良好的结构遗传图谱之后,对野生型和突变型进行了详细的序列比较并随后在单个 BAC 覆盖的区域内鉴定了这个突变基因(Xiang et al. 2008)。

关于用于图位克隆的分子标记间的突变的报道也开始出现，这些计划的目标是利用这些靶标基因进行品系改良(例如 *pph*、杂食性、促进人工饲料的喂养；Mase et al. 2007)，标记辅助选择(例如 *sch*、伴性赤蚁、用于雄性特异选择；Miao et al. 2008)，或者是其他一些蚕丝业上的实际应用(例如 *sli*、成虫期短，羽化后几天就死掉；Li et al. 2008；见第十八章基因对病毒的抗性和 *nsd-2* 的定位克隆)。其他一些正在进行的工作旨在回答关于基因功能的一些基本问题，例如外层黄茧基因 *C* 影响从血液中吸收类胡萝卜素进入丝腺中(Tabunoki et al. 2004)，并且最近通过 SSRs 对其进行了定位(Zhao et al. 2008)。尽管和色彩斑斓的蝴蝶和其他一些有装饰色彩的蛾子相比，大部分没有色素的家蚕成虫显得相对单调很多，然而，一些身体和翅模式的突变种(例如 *Bm*、黑蛾；*Ws*，野桑蚕翅斑；*Wm*、野桑蚕黑蛾；*wb*、白带黑翅)——这些突变可能是从野桑蚕得来的——是翅模式的拟态和黑化的候选基因，而这些性状一直是鳞翅目昆虫生物学的关键问题(蝴蝶翅模式，见第五、六章)。

4.7 数量性状位点(QTL)

家蚕一鳞翅目昆虫遗传学和基因组学的第二类突变是数量性状或复杂性状，这些性状在野生型品系间差异很大，并且可以应用于育种中。例如，不同品种的季节繁殖率和存活率不同，所以不同杂交种要配合一定的饲养条件，例如温度、湿度、桑叶质量等，一些杂交种最好在春季饲养而另一些最好在夏季或秋季饲养。鉴定这些特征的遗传学基础不仅有利于蚕丝业中的标记辅助的选择，而且可以在鳞翅目害虫防治方面发现新的靶标。尽管用微小的改变很难证明多基因控制的性状的遗传学基础，但是已有报道利用简单重复序列(Chatterjee 和 Mohandas 2003)为桥梁和 AFLPs (Lu et al. 2004)来确定 QTL 图谱的研究。此外，一些实际繁殖中的特征是和一些孟德尔类型的基因相关联的，这使它们成为定位克隆的良好靶标。最初的例子是晚熟基因 *Lm*，它是和 Z 连锁的基因，影响幼虫龄期的持续时间、总的茧重和茧壳的重量等。影响生殖力、产卵力和繁殖能力的其他方面的一些性状可能是研究家蚕繁殖的有效资源。此外，利用这些候选基因可以验证在其他鳞翅目昆虫中建立的物种形成和生殖隔离假说(例如，Sperling 1994；Prowell 1998)。

4.8 家蚕和其他鳞翅目昆虫间保守的同线性

同线性在鳞翅目昆虫中的暗示——具有共同祖先(orthologs)的基因存在于相同连锁群上——首先是在研究性别或 Z 连锁性状时提出，特别是在不同的分类群中，等位基因酶和表型性状影响生殖隔离(Sperling 1994)。家蚕分子遗传图谱的建立为检测常染色体的同线性提供了基础。第一个利用这种方法检验同线性的鳞翅目昆虫是热带蝴蝶(*Heliconius melpomene*)(Yasukochi et al. 2006；Pringle et al. 2007)和烟草天蛾(Sahara et al. 2007)。利用重组图谱(Pringle et al. 2007)和 BAC-FISH (Yasukochi et al. 2006)技术发现，一些已知基因的一系列参考标记存在于三个物种的单个染色体上，并且具有保守的基因顺序。关于热带蝴蝶、邮差蝴蝶和家蚕相关连锁群的更详细的遗传分析和物理图谱可以预测这些物种间的染色体等量性(Papa et al. 2008)。这些研究也显示出相当多的微同线性，表明这些保守的基因种群具有潜在的功能重要性(Papa et al. 2008)。最近的一个基于大于 460 个同源基因，对家蚕和热带蝴蝶之间的核型范围的比较分析发现两者之间具有明显的染色体保守性以及由于染色体的或大或小的易位和反转引起的同线性阻断(Beldade et al. 2009)。在如此广泛分化的分类群范围内的同线性,证明了家蚕提供的参考染色体图谱和其他基因组资源将促进从缺少图谱信息的物种中鉴定新的基因和候选基因。关于同线性的深入讨论见第六章。

5. 基因组资源

家蚕现在具有大规模的基因组资源,包括之前提到的大规模的 EST 库(大约有 238000 条序列,截至 2008 年 10 月),BAC 文库,整合的物理遗传图谱和基因组框架图。在线的数据库提供了其中的部分资源(表 2.2);此外,正在建立芯片数据库和蛋白质组学数据库,这使家蚕成为和果蝇、蚊子、赤拟谷盗及非洲蜂等一样的领先模式生物。随后是关于这些家蚕资源的简要介绍。

表 2.2 基因组及功能基因组资源

Table 2.2 Genomics and Functional Genomics Resources

Database	Contents	Host	Web Site
SilkDB	Silkworm genome database	The Institute of Sericulture and Systems Biology, Southwest University, Chongqing, China	http://silkworm.genomics.org.cn/
KAIKObase	Silkworm genome database	National Institute of Agrobiological Sciences, Tsukuba, Japan	http://sgp.dna.affrc.go.jp/KAIKObase/
SilkBase	ESTs from *Bombyx mori*, *Samia cynthia ricini*	University of Tokyo, Laboratory of Insect Genetics and BioSciences	http://morus.ab.a.u-tokyo.ac.jp/cgi-bin/index.cgi
Full-Length cDNA Database	ESTs from *Bombyx mori* full-length cDNA libraries	University of Tokyo, Laboratory of Insect Genetics and Biosciences	http://papilio.ab.a.u-tokyo.ac.jp/Bombyx_EST/
ButterflyBase	Genomics database for Lepidoptera	Max Planck Society, Jena, Germany; Consortium for Comparative Genomics of Lepidoptera	http://butterflybase.ice.mpg.de/
WildSilkBase	ESTs, *Antheraea assama*, *A. mylitta*, *Samia cynthia ricini*	Center for DNA Fingerprinting and Diagnostics, Hyderabad, India	http://www.cdfd.org.in/wildsilkbase/home.php
SilkSatDB	Silkworm microsatellites	Center for DNA Fingerprinting and Diagnostics, Hyderabad, India	http://210.212.212.7:9999/PHP/SILKSAT/index.php
BmMDB: *Bombyx mori* Microarray Database	Microarray profiles In process: Web site established; one set of published data currently available	The Institute of Sericulture and Systems Biology, Southwest University, Chongqing, China	http://silkworm.swu.edu.cn/microarray/
Silkworm Proteome Database	Two-dimensional gel electrophoresis patterns and spot identification	Center for Genetic Resource Information, National Institute of Genetics, Misima, Japan	http://www.shigen.nig.ac.jp/ISPD/index.jsp
KAIKO 2DDB	Silkworm proteome database: two-dimensional gel electrophoresis patterns and spot identification	National Institute of Agrobiological Sciences, Tsukuba, Japan	http://kaiko2ddb.dna.affrc.go.jp/cgi-bin/search_2DDB.cgi
Bombyx Trap Database	Enhancer trap, GAL4/UAS, and other transgenic lines	Transgenic Silkworm Research Center, National Institute of Agrobiological Sciences, Tsukuba, Japan	http://sgp.dna.affrc.go.jp/ETDB/

5.1 EST库

第一个大规模的EST库包括大约35000条序列，是从36个标准的cDNA文库中每个随机测序1000条而得到的，这些cDNA文库是从不同的组织和发育时期构建得来的，其目标是鉴定大量的不同的家蚕基因产物（现增加到48个文库；见SilkBase，表2.2）。大约报道了11000条独立的序列（Mita et al. 2003），平均每2.7条EST产生一个重叠群，说明这是一种发现基因的可行方法。构建第二个大规模的EST库的目的是为连锁图谱和在基因相关测试中发现SNP，从73235条高质量的EST中产生了12980个重叠群，这些序列是从12个组织或时期特异性的文库中得来的（Cheng et al. 2004）。尽管独立的基因产物的产出率较低（每5.6条EST产生一个独立聚类），但正如预料的，这个计划可以有效地鉴定SNP和小的插入或缺失，显著地扩展了EST库。总的SNP的产出率很低（101个SNPs，27个插入/缺失，1.4/1000碱基对），表明用于全基因组测序的近交品系大造在编码区域上相对纯合度高。之后的计划集中在从单个来源中获得一系列的EST，例如翅原基（Kawasaki et al. 2004）、后部丝腺（Zhong et al. 2005）、表皮（Okamoto et al. 2008）、脂肪体（Cheng et al. 2006）或者某一两个发育时期的全蚕（Oh et al. 2006），目的是为了得到基因的表达模式（见下文）和从中获得感兴趣的序列。从EST库中得到的重叠群的组装（Zhang et al. 2007b）和全cDNAs的测序都正在进行中（K. Mita和T. Shimada，个人交流；全长cDNA文库，见表2.2），这将为注释家蚕基因组序列提供更完整的资源。

野桑蚕数据库包括EST库和全长的cDNAs（表2.2），这是由三种野桑蚕的序列构建的，这些野桑蚕可产生商用的蚕丝，如印度的阿萨姆蚕（*Antheraea assama*），中国柞蚕（*A. mylitta*）和蓖麻蚕（*Samia cynthia ricini*）（Arunkumar et al. 2008）。GenBank中最近登录的这些物种的EST数目迅速增加，从最初的57000条（4000个重叠群和10000 singletons）增加到94500条（*A. assama*有大约35200条；*A. mylitta*有大约39000条；*S. c. ricini*有大约20300条；2008年10月）；为发现有经济价值的基因和基础及进化研究提供了重要资源。

5.2 BAC文库

家蚕现在有四个可用的BAC文库，其中有三个是用p50的雌雄混合的后部丝腺构建的，另一个是用C108品系，并且用*Eco*RI（Koike et al. 2003），*Hind*III（Wu et al. 1999）或*Bam*HI（Yamamoto et al. 2006）中的一种进行部分消化构建的。这些文库中的配对的末端BESs用于构建BAC重叠群、组装全基因组序列、发现SNP和EST用于构建整合的物理遗传图谱（Yamamoto et al. 2006，2008）。用不同的限制性内切酶消化的BAC文库能够详细地比较*Eco*RI和*Bam*HI消化得到的BESs文库。在代表超过家蚕10%基因组的95000条BESs（55Mb）的基础上，Suetsugu等（2007）发现*Bam*HI消化的BESs中长散布元件（LINEs）比*Eco*RI消化的文库要多（*Bam*HI消化的BESs为16.40%，*Eco*RI消化的BESs为5.14%）；另外，短转座元件（TEs）的相对含量正好相反（*Bam*HI消化的BESs为2.07%，*Eco*RI消化的BESs为5.40%）。这两个酶切的BESs中总的鸟嘌呤-胞嘧啶（GC）含量也不相同（*Bam*HI消化的BESs为40.30%，*Eco*RI消化的BESs为

37.45%),这可能是由于酶的识别位点的 GC 含量不同导致的。用 BLASTx 对这两个文库聚类的 BESs 进行分析表明 *Bam*HI 消化的 BESs 对应的蛋白编码序列的频率更高(*Bam*HI 消化的 BESs 为 28.2%,*Eco*RI 消化的 BESs 为 20.2%)。这和家蚕蛋白质编码序列与基因间的 DNA 的相对 GC 含量一致。

这些结果表明了利用互补的消化获得完整基因组是非常重要的,特别是对于仍没有全基因组序列可用的物种。从这些计划获得的 BAC 文库被广泛地用于基因种群的克隆和详细的分析,例如 *Hox* 和相邻的基因(Yasukochi et al. 2004)、*Broad* 复合体(Ijiro et al. 2004)、定位克隆(见上面及第十八章)、利用 BAC-FISH 构建基因的物理图谱(Yoshido et al. 2005; Sahara et al. 2007)、染色体结构的详细分析等(比如,W; Sahara et al. 2003; Yoshido et al. 2007; 第三、四章)。

5.3 全基因组序列草图

完整的基因组序列已成为研究现代分子生物学和进化生物学必需的条件。尽管在对 cDNA 文库中的 EST 进行深入测序时发现了很多表达基因产物,但是表达的基因产物中有很大一部分因为含量太少,如果不用合适的方法则在 cDNA 文库中检测不到,其中包括许多重要的成分,如编码化学感应蛋白的信使 RNAs(mRNA)、小的核 RNA(snRNAs)和小 RNA(miRNAs)。这些基因可以在全基因组的范围内利用它们的独特性质通过计算方法高效地获得。尽管一些结构特征例如内含子和外显子可以通过测序选择的 BAC 克隆来确定,但是利用全基因组序列可以发现许多其他重要的或没有预测到的基因组特征,例如不表达的假基因,距离近的或远的相关的家族成员,基因类群和重复染色体片段等,这都需要一个全局的分析。有一个像家蚕这样的具有参考的基因组序列的模式生物有利于在其他相关物种中发现基因和进行基因组的组装及注释,这些物种的基因组测序可以在低覆盖度的情况下仍有较高的信息产出。最后,具有一个代表性的鳞翅目昆虫的全基因组序列能够迅速地为比较基因组学提供新的机会,为研究不同进化种群的进化历史和回答一些问题提供线索,例如哪些基因和特征是家蚕特有的,哪些是和其他鳞翅目昆虫或其他分类学上的昆虫种群共有的,或与其他节肢动物共有的,或与其他后生动物共有的。

2004 年发布了两个用鸟枪法获得的家蚕基因组序列草图,一个大约是六倍的覆盖度(Xia et al. 2004),另一个是三倍(Mita et al. 2004)。这些最初的数据库的低覆盖度使基因组分成相对小的片段(测序 DNA 的重叠),六倍库和三倍库对应的重叠群的平均大小分别是 12.5kb 和 17.9kb,scaffold(拼接的序列)的平均大小分别是 26.9kb 和 7.8kb。选用共同的 p50/大造进行这两个计划促进了数据的整合,并通过增加基因组的覆盖度来产生一个更完整的组装数据,现在这项工作正在进行中,并且将要包括 fosmids 文库来源的序列和全长的 cDNAs 序列,还将与整合的 SNP-EST-遗传物理图谱进行比对(国际家蚕基因组协会,2008)。在线数据库提供了扩展的 scaffold、预测的基因模式、EST 比对和其他功能(表 2.2;Wang et al. 2005)。

尽管存在断裂,但是这两个基因组序列的广度是相对较高的,通过 BLAST 搜索和序列比对分析表明,约 91%的已知基因和 EST 可以在六倍库的全基因组序列中发现(Xia et al. 2004)。同样的，约 97%基因组序列存在于三倍库的全基因组序列中。此外，一系列完全测序并组装的 BAC 克隆与全基因组数据进行比对发现有 78%～87%的覆盖度(Mita et al. 2004)。重叠群的长度和 scaffolds 在接头组装后将更长，将具有 8～9 倍的覆盖度；这些数据与附加的序列信息整合后，再加上整合的物理遗传图谱，将为准确基因注释提供更有力的平台。

备注：中国和日本的研究团队最近报道了包括 fosmid 和 BAC 克隆序列数据的整合全基因组序列，得到一个 8.5 倍覆盖度的基因组数据(432Mb)(国际家蚕基因组协会，2008)。新的数据在连续性上有明显的改良，重叠群或 scaffold 上的大部分组装的序列至少是 3.7Mb(最长的为 14.3Mb)，87%的序列是通过 SNP 图谱定位到 28 条染色体上的。见 Couble，Xia 和 Mita(2008)关于家蚕基因组结构、基因家族进化和功能的相关文章。

5.4 基因组特征

单倍体的家蚕基因组大小范围是从六倍库的 428.7Mb (Xia et al. 2004)到三倍库的 514Mb(Mita et al. 2004)。这些数据与之前用 DNA-DNA 退火动力学得到的每 1C DNA 0.53pg (Gage 1974)和用细胞光度测定法测量染色的精子和血细胞核酸得到的每 1C DNA0.52pg 的结果相一致(Rasch 1974)。注意这些数据转化值中的错误后得到基因组的大小约为 530Mb，但以 978×10^{6} bp/pg DNA 的标准估计，更接近 518Mb (Doležel et al. 2003)。最近利用更精确的流式细胞仪测量碘化丙啶染色的成虫脑的有丝分裂染色体并以果蝇作为标准得到一个稍微低的结果 450～493Mb(J. S. Johnston，个人交流)，这和六倍库得到的结果相接近，这表明两个全基因组数据整合后的这个值将减少(国际家蚕基因组协会，2008)。

利用从头开始发现基因的算法并校正可能的错误、重复序列、转座元件和假基因等，六倍库预测家蚕基因组包含 18510 个基因，通过对齐的相互最佳匹配得出这些基因的平均大小比果蝇大 2.3 倍(Xia et al. 2004)。和果蝇相比每个基因的外显子略多(平均比值为 1.15)，说明基因的大小不同主要是由于家蚕基因有更大的内含子，至少部分是由于插入的转座元件数目的增加造成的(Xia et al. 2004)。Cot 曲线测量表明家蚕基因组包含 45%的重复元件(Gage 1974)，这和家蚕三倍全基因组库中不同的重复家族含有超过 180000 个拷贝的报道相一致(Mita et al. 2004)。其中的主要部分包括超丰度的短散布重复元件(SINE)*Bm1*(121000 个拷贝)，长散布元件(LINE)或 non-LTR 反转座子 *BMC*1(37000 个拷贝)，*Bm5886*(28000 个拷贝)，*HOPE Bm2*(3380 个拷贝)，三种类型的 Class Ⅱ *mariner* 转座子(大约 10000 个拷贝)，不同的 LTR 反转座子(大约 400 个拷贝)。除了 *Bm1*，*BMC1* 和 LINEs 以外，大部分是在 5′末端被截短(Mita et al. 2004)。在全基因组数据中对 LTR 反转座子进行相关搜索检测到 29 个家族，占全基因组的近 12%，其中有许多是之前没被发现的。但是，除少数外，大部分都是相对小的数目(20～700 之间；Xu et al. 2005)。其中的大部分属于三个主要的群体，*Gypsy*-like，*Copia*-like 和 *Pao-Bel* 群，它们合起来接近占了全基因组的 12%，但 W 染色体除外，它基本上全是由反转座子元件组成的(W 染色体的分子特征见第四章，家蚕转座子的组成和结构综述见 Eickbush 1995；Abe et al. 2005；Goldsmith，Shimada 和 Abe 2005)。

然而大部分元件在染色体上广泛分布，non-LTR 反转座子的两个家族，TRAS1 和 SART1，在单倍体基因组中大约有 1000 个拷贝，与端粒的 TTAGG 重复序列有关（Fujiwara et al. 2005）。它们是有转录活性的，此外，有证据表明它们能通过一些机制转移到端粒重复中并对端粒的保持起一定作用（Fujiwara et al. 2005）。

另一个感兴趣的类群是 Class Ⅱ 元件中的 *piggyBac* 元件，*piggyBac* 转座元件最早是在粉纹夜蛾（*Trichoplusia ni*）的细胞系中发现并转到杆状病毒中的（Fraser et al. 1995），并随后被改造成一个在昆虫转基因中通用的、广泛的载体（例如，，Sarkar et al. 2006），包括家蚕（见下文）。家蚕基因组中鉴定了 98 个 *piggyBac* 样的元件（PBLEs），其中的大部分是截短形式的；然而，在不同组织和发育时期的 cDNA 中观察到一些元件的转录，有五个具有开放阅读框（ORFs），编码全长的转座酶蛋白，表明其中的一些可能是有活性的（Xu et al. 2006）。随后在其他一些鳞翅目昆虫中也发现了 *PBLEs*（Wang et al. 2006；Sun et al. 2008）。

家蚕或其他鳞翅目昆虫基因组中最难明白的一个可能的重复序列类群是和全运动型染色体的特异性的着丝粒结构相联系的特异 DNA 元件，称为弥散着丝粒。查明它们是否像预测的那样具有很多数目和广泛的分布（Ogura et al. 1994），并组成了一个重复序列的类群，或是与六倍基因组序列中发现的一样（Xia et al. 2004），对应于其他物种的全着丝粒蛋白将会很有意思。关于鳞翅目昆虫染色体结构的讨论见第三章。

第一个家蚕全基因组范围的调查得到的成果包括与已知的果蝇基因的同源性比较得到的和胚胎发育相关的基因（Xia et al. 2004），以及转录水平低因而在 EST 库中很难发现的例如化学感受器（Zhou et al. 2006；Gong et al. 2007；Wanner et al. 2007；关于鳞翅目昆虫化学受体的讨论，见第九、十、十一章），snRNAs（Sierra-Montes et al. 2005；Smail et al. 2006）和 miRNAs（Tong，Jin 和 Zhang 2006；He et al. 2008；Yu et al. 2008）等，期待在不久的将来用新的基因组数据和更高级的分析方法来鉴定分散的基因家族成员。

6. 功能基因组学

为了确定家蚕基因和基因组序列的功能或作用而发展出很多研究后基因组学的工具，称为功能基因组学。这些研究工具包括检测特定组织和发育时期基因表达模式的方法（例如 EST 库、微阵列和蛋白质组学），产生转基因昆虫的方法，干扰基因功能的方法，如 RNA 干扰技术等。下面是这些领域中近来研究工作的总结。早先的研究由 Goldsmith，Shimada 和 Abe（2005），Zhou，Yang 和 Zhong（2008）分别做过概述。

6.1 转录谱和蛋白质谱

转录谱是用来表明基因表达的模式，常用于研究昆虫的变态过程，对实验处理的反应，或是病原物或突变造成的影响。迄今为止，对家蚕大部分基因表达的研究都是基于 cDNA 文库的 EST 序列，这些 cDNA 文库是由感兴趣的组织和发育时期构建而来的。这种方法可以知道某个特定 mRNA 的频率从而粗略估计基因的表达，若想更直接和灵敏地在转录水平上检测，可通过 Northern blots，RT-PCR 或微阵列等方法。EST 谱已被用于检测变态过程中的早期胚胎（Hong et al. 2006），胚胎期与幼虫期（Oh et al. 2006），脂肪体(Cheng et al. 2006),丝腺(Zhong et al. 2005),翅原基(Kawasaki et al. 2004)和幼虫表皮(Okamoto et al. 2008)的基因表达的变化。

虽然通过调查EST谱可以粗略估计某个特定的组织或时期的转录活性,但是微阵列(基因芯片)提供了一个更灵敏和综合的基因表达情况。在全基因组序列完成之前,家蚕的微阵列是由单个组织收集的几千条EST发展而来的,例如分化的翅原基(Kawasaki et al. 2004; Ote et al. 2004),或是特定的发育时期例如胚胎期(Hong et al. 2006)。这些数据为以后的分析提供上调或下调的基因(例如,Ote et al. 2005),但是范围不全面。随后,一个全基因组范围的微阵列包括23000条寡核苷酸探针用于调查五龄中期十个组织和器官的表达谱,这些探针是根据已知的基因和全基因组草图预测的基因设计的(Xia et al. 2007)。这项研究将作为研究幼虫期基因表达变化的标准。同样重要的是,这是注释家蚕基因组的一个资源,通过比较一些特定序列在不同组织、不同发育时期或不同昆虫中的表达谱就可以推断其普遍的功能。一个根据超过50000条EST设计的,代表16000个基因的微阵列用于检测培养的血球细胞暴露于细菌后的表达情况的变化(Ha Lee et al. 2007)。尽管只有很少的序列在表达水平上有改变,但是表达模式明显不同,暗示了该技术在揭示协同表达模式和未知基因中的潜力。

蛋白质组学方法也开始得到利用,即用双向电泳加上MS/MS或MALDI-TOF(matrix-assisted laser desorption/ionization－time of flight)来鉴定蛋白。尽管可以利用EST数据和比较的方法鉴定一些蛋白质,但是这需要一个良好注释的鳞翅目昆虫基因组,而这方面的进程却很慢。与早期的转录数据一样,初步的蛋白质组学谱大部分集中在幼虫的组织例如中肠(Kajiwara et al. 2005; Yao et al. 2008)、脂肪体(Kajiwara et al. 2006)、血液(Li et al. 2006)、幼虫到蛹变态时期的骨骼肌(Zhang et al. 2007a)和中后部丝腺(Hou et al. 2007; Liu et al. 2008)等。家蚕蛋白质组学数据现在可在线使用(表2.2),提供了有用的公共资源。关于家蚕蛋白质组学技术及应用见Zhou,Yang和Zhong(2008)的综述。

6.2 转基因

将转基因技术运用到家蚕上获得转基因家蚕在2000年(Tamura et al. 2000)被首次报道并且现在已经得到广泛应用;关于发展有效的基因表达系统和表达载体的早期工作已由Goldsmith,Shimada,Abe(2005)进行综述。前囊胚层胚胎注射仍然是基因导入的最常用方法。在*piggyBac*的基础上发展了许多载体,如之前提到的,通常含有一个家蚕肌动蛋白*A*3(*BmactinA*3)启动子或眼和神经元中特异表达的*3x3P*启动子,接在它们后面的报告基因是一个增强型绿色荧光蛋白(EGFP)或红色荧光蛋白(DsRed)。特定条件下的表达可以用果蝇的热激蛋白70(*hsp*70)启动子来实现(Yamamoto et al. 2004; Dai et al. 2007),已经建立的GAL4/UAS系统可用于在特定组织中利用其他启动子实现靶标基因的表达(Imamura et al. 2003; Tan et al. 2005; Uchino et al. 2006)。最近有报道称在家蚕中能够成功地利用另一转座原件*Minos*中的转座酶和其他成分(Uchino et al. 2007),*Minos*最早是在大型种果蝇(*D. hydei*)中发现的(Franz和Savakis 1991)。尽管用*Minos*的转化水平比以*piggyBac*为基础的载体水平要低,但是它为发展增强子捕获系统提供了一种可选择的转座酶资源。这为揭示组织和时期特异性的调控元件提供了一种有效的方法(Uchino et al. 2008)。最近,利用杆状病毒 AcNPV 设计了以家蚕长散布元件 SART1(可优先整合到端粒中)和 R1(可整合到 28S 核糖体 DNA 中)为基础的载体,用其感染五龄期的幼虫以及在不损伤功能基因的情况下产生大量的重组蛋白(Kawashima et al. 2007)。关于转基因实验的结果已经建立了一个新的数据库(家蚕陷阱数据库;表 2.2)。

6.3 应用

家蚕的遗传工程可被用来研究基因的功能和调控,从蚕茧和血液中获得大量的、有价值的产物,可用于品系的改良,特别是增加家蚕对家蚕核型多角体病毒(BmNPV)的抗性,尽管目前没有达到完全的抗性(Isobe et al. 2004; Kanginakudru et al. 2007)。下面给出的是关于前两个应用的例子。

利用转基因家蚕研究基因的功能已有多种方法,例如引入野生型基因通过突变恢复来证明候选基因的功能(例如,*BmY*,Sakudoh et al. 2007;*Nd-sD*,Inoue et al. 2005;*Bmnsd-2*,Ito et al. 2008);利用野生型基因的异位表达来确定它们在发育和分化中的作用(例如,*Juvenile Hormone Esterase*,Tan et al. 2005;*B. mori doublesex*,*Bmdsx*,Suzuki et al. 2005);利用转基因使启动子分开(*fibroin-H*,Shimizu et al. 2007)和调查 RNA 拼接元件(*Bmdsx*,Funaguma et al. 2005),结合 RNAi 进行基因敲除和沉默等(参见下章)。利用加入附加的转录调控元件的载体,或许可以发现新的启动子捕获系统。利用转基因系统对家蚕和其他的鳞翅目昆虫进行基本的基因功能研究是可行的。

利用家蚕作为生物反应器,利用基因特异性的启动子可以在丝腺的不同部分实现基因的表达,如后部丝腺的丝素轻链(*fibroin-L*)(Inoue et al. 2005; Adachi et al. 2006; Hino,Tomita 和 Yoshizato 2006),*fibrohexamerin*(Royer et al. 2005)和丝素重链(*fibroin-H*)(Kojima et al. 2007; Kurihara et al. 2007),中部丝腺的丝胶(*sericin*)启动子(Ogawa et al. 2007)。裸蛹突变体 *Nd-sD*,因为缺乏丝纤维中的丝素轻链而不能泌丝,可用于在蚕茧中增加重组蛋白的产量(Inoue et al. 2005; Yanagisawa et al. 2007),此外,从家蚕杆状病毒早期(immediate-early 1)(Tomita et al. 2007) 基因中引入一个转录增强子或从多角体蛋白 5′UTR(非翻译区域)中引入一个翻译增强子可以帮助实现高水平的表达(Iizuka et al. 2008)。利用家蚕生产的增值产物包括人的 β 成纤维细胞生长因子(Hino,Tomita 和 Yoshizato 2006),用于细胞培养的掺入到胶原或纤维结合蛋白中的有生物活性的膜(Adachi et al. 2006; Yanagisawa et al. 2007),制药模型如生产猫干扰素(Kurihara et al. 2007)。对这个系统优化后可以实现胶原脯氨酸的羟基化作用,这个在通常情况下表达水平低,不能产生必需的翻译后修饰的酶的亚基,通过家蚕实现了对它的表达 (Adachi et al. 2006) 。

6.4 RNAi

RNA 干扰技术(RNAi)是一个多用途的方法,可用于多种生物中来调查基因的功能和证明定位克隆中的候选基因。RNAi 是通过内源性的核 RNA 诱导的沉默复合体

(RISC)，通过 RISC 暴露于对应于靶标 mRNA 的双链 RNA(dsRNA)而被激活的，来实现 mRNA 的选择性降解。dsRNA 可通过多种途径引入有机体内(关于 RNAi 的更多信息，见第十二章)。家蚕上第一个成功的 RNAi 的实验是注射以 *Bmwh3* 为靶标的 dsRNA 到卵中，*Bmwh3* 是果蝇 *white* 基因的同源物，产生了白卵和半透明的胚胎表型(Quan，Kanda 和 Tamura 2002)。通过注射 dsRNA 到五龄幼虫或蛹中成功地实现基因下调，观察到了很多功能，包括程序性细胞死亡(家蚕组织蛋白酶 *D*，Gui et al. 2006)，免疫应答(家蚕溶菌酶类似蛋白 *I*，Gandhe，Janardhan 和 Nagaraju 2007)，幼虫一蛹的变态(*Bmftz*，Cheng et al. 2008；*cathepsin B*，Wang et al. 2008)，翅的伸展(Huang et al. 2007)，以及目前为止最广泛的研究之一，即一系列关于信息素产生的因子(Matsumoto et al. 2007)。此外，利用转基因技术表达一个在组成型启动子(*period*，Sandrelli et al. 2007；*Bmrelish*，Tanaka et al. 2007)或热激型启动子(*eclosion hormone*，Dai et al. 2007)控制下的反向重复的靶标 RNA 的研究也已经发展起来；后者将实现在特定发育时期进行基因敲除。RNAi 也被用于转染家蚕培养细胞来检测 RNA 甲基转移酶在程序性细胞死亡中的作用(Nie et al. 2008)，并且更有意思的是，这牵涉家蚕 *AGO* 基因(*BmAGO2*)，*BmAGO2* 是 RISC 复合体的关键成分，能抑制双链损伤的修复(Tsukioka et al. 2006)。这种方法在家蚕中是有效的，尽管在其他鳞翅目昆虫中报道很少，表明在广泛应用于其他物种之前仍有一些技术难点和生物障碍需要解决。

7. 总结：丝绸之路的下一步在哪？

现在我们已经到达了丝绸之路的一个新的时代，家蚕长期作为鳞翅目昆虫的模式生物，在发展了新的遗传学和基因组学之后其作用明显加强，这将促进其他物种在很多领域的迅速发展。新的现代分子细胞遗传学，特别是 BAC-FISH，是物理连锁图谱和精细结构连锁图谱的一个重要手段，能促进在其他缺少遗传学背景的物种中研究染色体和基因的结构。新整合的分子遗传图谱表现出广泛的同线性，全基因组序列将促进所有鳞翅目昆虫中的基因发现，并将为研究许多重要的发育功能和有意思的进化问题提供工具。其中需要在 Z 染色体上鉴定尽可能多的基因，因为 Z 染色体被认为在物种形成中起着很重要的作用。鉴定性别决定途径中的基因包括 *Fem* 因子将有助于揭示雌性家蚕基因组中 W 染色体的神秘作用。同样地，现在仍需要应用这些工具，与应用功能基因组学的工具一样，来研究大量的家蚕突变体，它们是鳞翅目昆虫生物学唯一的资源。另一个前沿的问题是探索包括野桑蚕在内不同的地理品种的更全面的遗传学基础。毫无疑问，许多整合到近交系里的复杂性状都能在它们的祖先中找到起源，有些性状受到驯化的强选择压力，会重新分化形成。两者都是很有意思的，并可能用于阐明其他鳞翅目昆虫的进化问题。

参考文献

[1]Abe, H., K. Mita, Y. Yasukochi, T. Oshiki, and T. Shimada. 2005. Retrotransposable elements on the W chromosome of the silkworm, *Bombyx mori*. *Cytogenet. Genome Res*. 110:144—51.

[2]Abraham, E. G., H. Sezutsu, T. Kanda, T. Sugasaki, T. Shimada, and T. Tamura. 2000. Identification and characterisation of a silkworm ABC transporter gene homologous to *Drosophila* white. *Mol. Gen. Genet*. 264:11—19.

[3]Adachi, T., M. Tomita, K. Shimizu, S. Ogawa, and K. Yoshizato. 2006. Generation of hybrid transgenic silkworms that express *Bombyx mori* prolyl-hydroxylase alpha-subunits and human collagens in posterior silk glands: Production of cocoons that contained collagens with hydroxylated proline residues. *J. Biotechnol*. 126:205—19.

[4]Arunkumar, K. P., M. Metta, and J. Nagaraju. 2006. Molecular phylogeny of silkmoths reveals the origin of domesticated silkmoth, *Bombyx mori* from Chinese *Bombyx mandarina* and paternal inheritance of *Antheraea proylei* mitochondrial DNA. *Mol. Phylogenet. Evol*. 40:419—27.

[5]Arunkumar, K. P., A. Tomar, T. Daimon, T. Shimada, and J. Nagaraju. 2008. WildSilkbase: An EST database of wild silkmoths. *BMC Genomics* 9:338.

[6]Banno, Y., T. Nakamura, E. Nagashima, H. Fujii, and H. Doira. 2004. M chromosome of the wild silkworm, *Bombyx mandarina* ($n=27$), corresponds to two chromosomes in the domesticated silkworm, *Bombyx mori* ($n=28$). *Genome* 47:96—101.

[7]Beldade, P., S. V. Saenko, N. Pul, and A. D. Long. 2009. A gene-based linkage map for *Bicyclus anynana* butterflies allows for a comprehensive analysis of synteny with the lepidopteran reference genome. *PLoS Genet*. 5:e1000366.

[8]Chatterjee, S. N., and T. P. Mohandas. 2003. Identification of ISSR markers associated with productivity traits in silkworm, *Bombyx mori* L. *Genome* 46:438—47.

[9]Cheng, D., Q. Y. Xia, J. Duan, et al. 2008. Nuclear receptors in *Bombyx mori*: Insights into genomic structure and developmental expression. *Insect Biochem. Mol. Biol*. 38:1130—37.

[10]Cheng, D. J., Q. Y. Xia, P. Zhao, et al. 2006. EST-based profiling and comparison of gene expression in the silkworm fat body during metamorphosis. *Arch. Insect Biochem. Physiol*. 61:10—23.

[11]Cheng, T. C., Q. Y. Xia, J. F. Qian, et al. 2004. Mining single nucleotide polymorphisms from EST data of silkworm, *Bombyx mori*, inbred strain Dazao. *Insect Biochem. Mol. Biol*. 34:523—30.

[12]Couble, P., K. Mita, and Q. Xia. 2008. Editorial: Silkworm genome. *Insect Biochem. Mol. Biol*. 38:1035.

[13]Dai, H., R. Jiang, J. Wang, et al. 2007. Development of a heat shock inducible and inheritable RNAi system in silkworm. *Biomol. Eng*. 24:625—30.

[14]De Prins, J., and K. Saitoh. 2003. Karyology and sex determination. In *Lepidoptera, moths and butterflies: Morphology, physiology, and development*, ed. N. P. Kristensen, 449—68. Berlin: Walter de Gruyter.

[15]Doležel, J., J. Bartoš, H. Voglmayr, and J. Greilhuber. 2003. Nuclear DNA content and genome size of trout and human. *Cytometry A* 51:127—28.

[16]Eickbush, T. H. 1995. Mobile elements in lepidopteran genomes. In *Molecular model systems in the Lepidoptera*, ed. M. R. Goldsmith and A. S. Wilkins, 77—105. New York: Cambridge University Press.

[17]Franz, G., and C. Savakis. 1991. *Minos*, a new transposable element from *Drosophila hydei*, is a member of the Tc1-like family of transposons. *Nucleic Acids Res*. 19:6646.

[18]Fraser, M. J., L. Cary, K. Boonvisudhi, and H. G. Wang. 1995. Assay for movement of lepidopteran transposon IFP2 in insect cells using a baculovirus genome as a target DNA. *Virology* 211:397—407.

[19]Fujii, H. 1998. *Genetical stocks and mutations of Bombyx mori: Important genetic resources.* Fukuoka, Japan: Kyushu University.

[20]Fujii, T., N. Tanaka, T. Yokoyama, et al. 2006. The female-killing chromosome of the silkworm, *Bombyx mori*, was generated by translocation between the Z and W chromosomes. *Genetica* 127:253—65.

[21]Fujiwara, H., and H. Maekawa. 1994. RFLP analysis of chromosomal fragments in genetic mosaic strains of *Bombyx mori*. *Chromosoma* 103:468—74.

[22]Fujiwara, H., M. Osanai, T. Matsumoto, and K. K. Kojima. 2005. Telomere-specific non-LTR retrotransposons and telomere maintenance in the silkworm, *Bombyx mori*. *Chromosome Res*. 13:455—67.

[23]Funaguma, S., M. G. Suzuki, T. Tamura, and T. Shimada. 2005. The *Bmdsx* transgene including trimmed introns is sex-specifically spliced in tissues of the silkworm, *Bombyx mori*. *J. Insect Sci*. 5:17.

[24]Gage, L. P. 1974. The *Bombyx mori* genome: Analysis by DNA reassociation kinetics. *Chromosoma* 45:27—42.

[25]Gandhe, A. S., G. Janardhan, and J. Nagaraju. 2007. Immune upregulation of novel antibacterial proteins from silkmoths (Lepidoptera) that resemble lysozymes but lack muramidase activity. *Insect Biochem. Mol. Biol*. 37:655—66.

[26]Goldsmith, M. R., T. Shimada, and H. Abe. 2005. The genetics and genomics of the silkworm, *Bombyx mori*. *Annu. Rev. Entomol*. 50:71—100.

[27]Gong, D. P., H. J. Zhang, P. Zhao, Y. Lin, Q. Y. Xia, and Z. H. Xiang. 2007. Identification and expression pattern of the chemosensory protein gene family in the silkworm, *Bombyx mori*. *Insect Biochem. Mol. Biol*. 37:266—77.

[28]Gui, Z. Z., K. S. Lee, B. Y. Kim, et al. 2006. Functional role of aspartic proteinase cathepsin D in insect metamorphosis. *BMC Dev. Biol*. 6:49.

[29]Ha Lee, J., I. Hee Lee, H. Noda, K. Mita, and K. Taniai. 2007. Verification of elicitor efficacy of lipopolysaccharides and peptidoglycans on antibacterial peptide gene expression in *Bombyx mori*. *Insect Biochem. Mol. Biol*. 37:1338—47.

[30]Hara, W., E. Kosegawa, K. Mase, and K. Kadono-Okuda. 2002. Improvement of linkage analysis in the silkworm, *Bombyx mori*, by using cDNA clones' RFLP. *J. Sericol. Sci. Jpn*. 71:95—100.

[31]He, P. A., Z. Nie, J. Chen, et al. 2008. Identification and characteristics of microRNAs from *Bombyx mori*. *BMC Genomics* 9:248.

[32]Hino, R., M. Tomita, and K. Yoshizato. 2006. The generation of germline transgenic silkworms for the production of biologically active recombinant fusion proteins of fibroin and human basic fibroblast growth factor. *Biomaterials* 27:5715—24.

[33]Hong, S. M., S. K. Nho, N. S. Kim, J. S. Lee, and S. W. Kang. 2006. Gene expression profiling in the silkworm, *Bombyx mori*, during early embryonic development. *Zool. Sci*. 23:517—28.

[34]Hou, Y., Q. Xia, P. Zhao, et al. 2007. Studies on middle and posterior silk glands of silkworm (*Bombyx mori*) using two-dimensional electrophoresis and mass spectrometry. *Insect Biochem. Mol. Biol*. 37:486—96.

[35]Huang, J., Y. Zhang, M. Li, et al. 2007. RNA interference-mediated silencing of the *bursicon* gene

induces defects in wing expansion of silkworm. *FEBS Lett*. 581:697—701.

[36]Iizuka, M. , M. Tomita, K. Shimizu, Y. Kikuchi, and K. Yoshizato. 2008. Translational enhancement of recombinant protein synthesis in transgenic silkworms by a 5'-untranslated region of polyhedrin gene of *Bombyx mori* nucleopolyhedrovirus. *J. Biosci. Bioeng*. 105:595—603.

[37]Ijiro, T. , H. Urakawa, Y. Yasukochi, M. Takeda, and Y. Fujiwara. 2004. cDNA cloning, gene structure, and expression of Broad-Complex (BR-C) genes in the silkworm, *Bombyx mori*. *Insect Biochem. Mol. Biol*. 34:963—69.

[38]Imamura, M. , J. Nakai, S. Inoue, G. X. Quan, T. Kanda, and T. Tamura. 2003. Targeted gene expression using the GAL4/UAS system in the silkworm *Bombyx mori*. *Genetics* 165:1329—40.

[39]Inoue, S. , T. Kanda, M. Imamura, et al. 2005. A fibroin secretion-deficient silkworm mutant, *Nd-sD*, provides an efficient system for producing recombinant proteins. *Insect Biochem. Mol. Biol*. 35:51—59.

[40]The International Silkworm Genome Consortium. 2008. The genome of a lepidopteran model insect, the silkworm, *Bombyx mori*. *Insect Biochem. Mol. Biol*. 38:1036—45.

[41]Isobe, R. , K. Kojima, T. Matsuyama, et al. 2004. Use of RNAi technology to confer enhanced resistance to BmNPV on transgenic silkworms. *Arch. Virol*. 149:1931—40.

[42]Ito, K. , K. Kidokoro, H. Sezutsu, et al. 2008. Deletion of a gene encoding an amino acid transporter in the midgut membrane causes resistance to a *Bombyx* parvo-like virus. *Proc. Natl. Acad. Sci. U. S. A*. 105:7523—27.

[43]Kadono-Okuda, K. , E. Kosegawa, K. Mase, and W. Hara. 2002. Linkage analysis of maternal EST cDNA clones covering all twenty-eight chromosomes in the silkworm, *Bombyx mori*. *Insect Mol. Biol*. 11:443—51.

[44]Kajiwara, H. , Y. Ito, A. Imamaki, M. Nakamura, K. Mita, and M. Ishizaka. 2005. Protein profile of silkworm midgut of fifth-instar day-3 larvae. *J. Electrophoresis* 49:61—69.

[45]Kajiwara, H. , Y. Itou, A. Imamaki, M. Nakamura, K. Mita, and M. Ishizaka. 2006. Proteomic analysis of silkworm fat body. *J. Insect Biotech. Sericol*. 75:47—56.

[46]Kanginakudru, S. , C. Royer, S. V. Edupalli, et al. 2007. Targeting *ie-1* gene by RNAi induces baculoviral resistance in lepidopteran cell lines and in transgenic silkworms. *Insect Mol. Biol*. 16:635—44.

[47]Kawanishi, Y. , R. Takaishi, Y. Banno, et al. 2007. Sequence comparison of *Mariner*-like elements among the populations of *Bombyx mandarina* inhabiting China, Korea and Japan. *J. Insect Biotech. Sericol*. 76:79—87.

[48]Kawanishi, Y. , R. Takaishi, M. Morimoto, et al. 2008. A novel maT-type transposable element, BmamaT1, in *Bombyx mandarina*, homologous to the *B. mori* mariner-like element Bmmar6. *J. Insect Biotech. Sericol*. 77:45—52.

[49]Kawasaki, H. , M. Ote, K. Okano, T. Shimada, Q. Guo-Xing, and K. Mita. 2004. Change in the expressed gene patterns of the wing disc during the metamorphosis of *Bombyx mori*. *Gene* 343:133—42.

[50]Kawashima, T. , M. Osanai, R. Futahashi, T. Kojima, and H. Fujiwara. 2007. A novel target-specific gene delivery system combining baculovirus and sequence-specific long interspersed nuclear elements. *Virus Res*. 127:49—60.

[51]Koike, Y. , K. Mita, M. G. Suzuki, et al. 2003. Genomic sequence of a 320-kb segment of the Z chromosome of *Bombyx mori* containing a *kettin* ortholog. *Mol. Genet. Genomics* 269:137—49.

[52]Kojima, K. , Y. Kuwana, H. Sezutsu, et al. 2007. A new method for the modification of fibroin

heavy chain protein in the transgenic silkworm. *Biosci. Biotechnol. Biochem.* 71:2943—51.

[53]Kômoto, N. 2002. A deleted portion of one of the two xanthine dehydrogenase genes causes translucent larval skin in the oq mutant of the silkworm (*Bombyx mori*). *Insect Biochem. Mol. Biol.* 32: 591—97.

[54]Kômoto, N., G. X. Quan, H. Sezutsu, and T. Tamura. 2009. A single-base deletion in an ABC transporter gene causes white eyes, white eggs, and translucent larval skin in the silkworm *w-3(oe)* mutant. *Insect Biochem. Mol. Biol.* 39:152—56.

[55]Kômoto, N., H. Sezutsu, K. Yukuhiro, Y. Banno, and H. Fujii. 2003. Mutations of the silkworm molybdenum cofactor sulfurase gene, *og*, cause translucent larval skin. *Insect Biochem. Mol. Biol.* 33:417—27.

[56]Kurihara, H., H. Sezutsu, T. Tamura, and K. Yamada. 2007. Production of an active feline interferon in the cocoon of transgenic silkworms using the fibroin H-chain expression system. *Biochem. Biophys. Res. Commun.* 355:976—80.

[57]Li, A., Q. Zhao, S. Tang, Z. Zhang, S. Pan, and G. Shen. 2005. Molecular phylogeny of the domesticated silkworm, *Bombyx mori*, based on the sequences of mitochondrial cytochrome b genes. *J. Genet.* 84: 137—42.

[58]Li, O., W. Hara, T. Yokoyama, and O. Ninagi. 2008. Gene mapping of *short lifespan in imago* by cDNA linkage in the silkworm, *Bombyx mori*. *Sanshi-Konchu Biotec* 77:47—52 (in Japanese).

[59]Li, X., X. Wu, W. Yue, J. Liu, G. Li, and Y. Miao. 2006. Proteomic analysis of the silkworm (*Bombyx mori*) hemolymph during developmental stage. *J. Proteome Res.* 5:2809—14.

[60]Liu, W., F. Yang, S. Jia, X. Miao, and Y. Huang. 2008. Cloning and characterization of Bmrunt from the silkworm *Bombyx mori* during embryonic development. *Arch. Insect Biochem. Physiol.* 69:47—59.

[61]Lu, C., F. Y. Dai, Z. H. Xiang. 2003. Studies on the mutation strains of the *Bombyx mori* gene bank. *Scientia Agricultura Sinica* 36:968—75. (In Chinese)

[62]Lu, C., B. Li, A. Zhao, and Z. Xiang. 2004. QTL mapping of economically important traits in silkworm (*Bombyx mori*). *Sci. China C Life Sci.* 47:477—84.

[63]Maeda, T. 1939. Chiasma studies in the silkworm *Bombyx mori*. *Jpn. J. Genet.* 15:118—27.

[64]Maekawa, H., N. Takada, K. Mikitani, et al. 1988. Nucleolus organizers in the wild silkworm *Bombyx mandarina* and the domesticated silkworm *Bombyx mori*. *Chromosoma* 96:263—69.

[65]Mase, K., T. Iizuka, T. Yamamoto, E. Okada, and W. Hara. 2007. Genetic mapping of a food preference gene in the silkworm, *Bombyx mori*, using restriction fragment length polymorphisms (RFLPs). *Genes Genet. Syst.* 82:249—56.

[66]Matsumoto, S., J. J. Hull, A. Ohnishi, K. Moto, and A. Fonagy. 2007. Molecular mechanisms underlying sex pheromone production in the silkmoth, *Bombyx mori*: Characterization of the molecular components involved in bombykol biosynthesis. *J. Insect Physiol.* 53:752—59.

[67]Matsunaga, T. M., and H. Fujiwara. 2002. Identification and characterization of genes abnormally expressed in wing-deficient mutant(flugellos) of the silkworm, *Bombyx mori*. *Insect Biochem. Mol. Biol.* 32:691—99.

[68]Matsuoka, T., and H. Fujiwara. 2000. Expression of ecdysteroid-regulated genes is reduced specifically in the wing discs of the wing-deficient mutant(fl) of *Bombyx mori*. *Dev. Genes Evol.* 210:120—28.

[69]Miao, X., M. Li, F. Dai, C. Lu, M. R. Goldsmith, and Y. Huang. 2007. Linkage analysis of the vis-

ible mutations *Sel* and *Xan* of *Bombyx mori* (Lepidoptera: Bombycidae) using SSR markers. *Eur J. Entomol.* 107:647—52.

[70]Miao, X. X., W. H. Li, M. W. Li, Y. P. Zhao, X. R. Guo, and Y. P. Huang. 2008. Inheritance and linkage analysis of co-dominant SSR markers on the Z chromosome of the silkworm (*Bombyx mori* L.). *Genet. Res.* 90:151—56.

[71]Miao, X. X., S. J. Xub, M. H. Li, et al. 2005. Simple sequence repeat-based consensus linkage map of *Bombyx mori*. *Proc. Natl. Acad. Sci. U. S. A.* 102:16303—08.

[72]Mills, D. R., and M. R. Goldsmith. 2000. Characterization of early follicular cDNA library suggests evidence for genetic polymorphisms in the inbred strain C108 of *Bombyx mori*. *Genes Genet. Syst.* 75: 105—13.

[73]Mita, K., M. Kasahara, S. Sasaki, et al. 2004. The genome sequence of silkworm, *Bombyx mori*. *DNA Res.* 11:27—35.

[74]Mita, K., M. Morimyo, K. Okano, et al. 2003. The construction of an EST database for *Bombyx mori* and its application. *Proc. Natl. Acad. Sci. U. S. A.* 100:14121—26.

[75]Nagaraja, G. M., G. Mahesh, V. Satish, M. Madhu, M. Muthulakshmi, and J. Nagaraju. 2005. Genetic mapping of Z chromosome and identification of W chromosome-specific markers in the silkworm, *Bombyx mori*. *Heredity* 95:148—57.

[76]Nagaraju, J. 1996. Sex determination and sex limited traits in the silkworm, *Bombyx mori*; their application in sericulture. *Indian J. Sericol.* 35:83—89.

[77]Nagaraju, J., and M. R. Goldsmith. 2002. Silkworm genomics—progress and prospects. *Curr. Sci.* 83:415—25.

[78]Nguu, E. K., K. Kadono-Okuda, K. Mase, E. Kosegawa, and W. Hara. 2005. Molecular linkage map for the silkworm, *Bombyx mori*, based on restriction fragment length polymorphisms of cDNA clones. *J. Insect Biotechnol. Sericol.* 74:5—13.

[79]Nie, Z., R. Zhou, J. Chen, et al. 2008. Subcellular localization and RNA interference of an RNA methyltransferase gene from silkworm, *Bombyx mori*. *Comp. Funct. Genomics* 2008:571023.

[80]Niimi, T., K. Sahara, H. Oshima, Y. Yasukochi, K. Ikeo, and W. Traut. 2006. Molecular cloning and chromosomal localization of the *Bombyx Sex-lethal* gene. *Genome* 49:263—68.

[81]Ogawa, S., M. Tomita, K. Shimizu, and K. Yoshizato. 2007. Generation of a transgenic silkworm that secretes recombinant proteins in the sericin layer of cocoon: Production of recombinant human serum albumin. *J. Biotechnol.* 128:531—44.

[82]Ogura, T., K. Okano, K. Tsuchida, et al. 1994. A defective non-LTR retrotransposon is dispersed throughout the genome of the silkworm, *Bombyx mori*. *Chromosoma* 103:311—23.

[83]Oh, J. H., Y. J. Jeon, S. Y. Jeong, et al. 2006. Gene expression profiling between embryonic and larval stages of the silkworm, *Bombyx mori*. *Biochem. Biophys. Res. Commun.* 343:864—72.

[84]Okamoto, S., R. Futahashi, T. Kojima, K. Mita, and H. Fujiwara. 2008. A catalogue of epidermal genes: Genes expressed in the epidermis during larval molt of the silkworm *Bombyx mori*. *BMC Genomics* 9:396.

[85]Ote, M., K. Mita, H. Kawasaki, M. Kobayashi, and T. Shimada. 2005. Characteristics of two genes encoding proteins with an ADAM-type metalloprotease domain, which are induced during the molting periods in *Bombyx mori*. *Arch.* Insect Biochem. *Physiol.* 59:91—98.

[86]Ote, M., K. Mita, H. Kawasaki, et al. 2004. Microarray analysis of gene expression profiles in

wing discs of *Bombyx mori* during pupal ecdysis. Insect Biochem. *Mol. Biol.* 34:775—84.

[87]Pan, M. , Q. Yu, Y. Xia, et al. 2008. Characterization of mitochondrial genome of Chinese wild mulberry silkworm, *Bombyx mandarina* (Lepidoptera: Bombycidae). *Sci. China C Life Sci.* 51:693—701.

[88]Papa, R. , C. M. Morrison, J. R. Walters, et al. 2008. Highly conserved gene order and numerous novel repetitive elements in genomic regions linked to wing pattern variation in *Heliconius* butterflies. *BMC Genomics* 9:345.

[89]Pringle, E. G. , S. W. Baxter, C. L. Webster, A. Papanicolaou, S. F. Lee, and C. D. Jiggins. 2007. Synteny and chromosome evolution in the lepidoptera: Evidence from mapping in *Heliconius melpomene*. *Genetics* 177:417—26.

[90]Promboon, A. , T. Shimada, H. Fujiwara, and M. Kobayashi. 1995. Linkage map of random amplified DNAs (RAPDs) in the silkworm, *Bombyx mori*. *Genet. Res.* 66:1—7.

[91]Prowell, D. P. 1998. Sex linkage and speciation in Lepidoptera. In *Endless forms: Species and speciation*, ed. D. J. Howard and S. H. Berlocher, 309—19. New York: Oxford University Press.

[92]Quan, G. X. , T. Kanda, and T. Tamura. 2002. Induction of the *white egg 3* mutant phenotype by injection of the double-stranded RNA of the silkworm *white* gene. *Insect Mol. Biol.* 11:217—22.

[93]Quan, G. X. , I. Kim, N. Kômoto, et al. 2002. Characterization of the kynurenine 3-monooxygenase gene corresponding to the *white egg 1* mutant in the silkworm *Bombyx mori*. *Mol. Genet. Genomics* 267:1—9.

[94]Qvarnström, A. , and R. I. Bailey. 2009. Speciation through evolution of sex-linked genes. *Heredity* 102:4—15.

[95]Rasch, E. M. 1974. The DNA content of sperm and hemocyte nuclei of the silkworm, *Bombyx mori* L. *Chromosoma* 45:1—26.

[96]Royer, C. , A. Jalabert, M. Da Rocha, et al. 2005. Biosynthesis and cocoon-export of a recombinant globular protein in transgenic silkworms. *Transgenic Res.* 14:463—72.

[97]Sahara, K. , F. Marec, and W. Traut. 1999. TTAGG telomeric repeats in chromosomes of some insects and other arthropods. *Chromosome Res.* 7:449—60.

[98]Sahara, K. , A. Yoshido, N. Kawamura, et al. 2003. W-derived BAC probes as a new tool for identification of the W chromosome and its aberrations in *Bombyx mori*. *Chromosoma* 112:48—55.

[99]Sahara, K. , A. Yoshido, F. Marec, et al. 2007. Conserved synteny of genes between chromosome 15 of *Bombyx mori* and a chromosome of *Manduca sexta* shown by five-color BAC-FISH. *Genome* 50:1061—65.

[100]Sakudoh, T. , H. Sezutsu, T. Nakashima, et al. 2007. Carotenoid silk coloration is controlled by a carotenoidbinding protein, a product of the *Yellow* blood gene. *Proc. Natl. Acad. Sci. U. S. A.* 104:8941—46.

[101]Sandrelli, F. , S. Cappellozza, C. Benna, et al. 2007. Phenotypic effects induced by knock-down of the *period* clock gene in *Bombyx mori*. *Genet. Res.* 89:73—84.

[102]Sarkar, A. , A. Atapattu, E. J. Belikoff, et al. 2006. Insulated *piggyBac* vectors for insect transgenesis. *BMC Biotechnol.* 6:27.

[103]Sato, K. , T. M. Matsunaga, R. Futahashi, et al. 2008. Positional cloning of a *Bombyx* wingless *locus flugellos* (*fl*) reveals a crucial role for fringe that is specific for wing morphogenesis. *Genetics* 179:875—85.

[104]Shi, J. , D. G. Heckel, and M. R. Goldsmith. 1995. A genetic linkage map for the domesticated silkworm, *Bombyx mori*, based on restriction fragment length polymorphisms. *Genet. Res.* 66:109—26.

[105]Shimizu, K., S. Ogawa, R. Hino, T. Adachi, M. Tomita, and K. Yoshizato. 2007. Structure and function of 5′-flanking regions of *Bombyx mori* fibroin heavy chain gene: Identification of a novel transcription enhancing element with a homeodomain protein-binding motif. *Insect Biochem. Mol. Biol.* 37:713—25.

[106]Sierra-Montes, J. M., S. Pereira-Simon, S. S. Smail, and R. J. Herrera. 2005. The silk moth *Bombyx mori* U1 and U2 snRNA variants are differentially expressed. *Gene* 352:127—36.

[107]Smail, S. S., K. Ayesh, J. M. Sierra-Montes, and R. J. Herrera. 2006. U6 snRNA variants isolated from the posterior silk gland of the silk moth *Bombyx mori*. *Insect Biochem. Mol. Biol.* 36:454—65.

[108]Sohn, K-W. 2003. Conservation status of sericulture germplasm resources in the world—II. Conservation status of silkworm (*Bombyx mori*) genetic resources in the world. Papers contributed to expert consultation on promotion of global exchange of sericulture germplasm, Bangkok, Thailand; September 2002. http://www.fao.org/DOCREP/005/AD108E/AD108E00. HTM Contents.

[109]Sperling, F. 1994. Sex-linked genes and species differences in Lepidoptera. *Can. Entomol.* 126: 807—18.

[110]Suetsugu, Y., H. Minami, M. Shimomura, et al. 2007. End-sequencing and characterization of silkworm (*Bombyx mori*) bacterial artificial chromosome libraries. *BMC Genomics* 8:314.

[111]Sun, Z. C., M. Wu, T. A. Miller, and Z. J. Han. 2008. *piggyBac*-like elements in cotton bollworm, *Helicoverpa armigera* (Hubner). *Insect Mol. Biol.* 17:9—18.

[112]Suzuki, M. G., S. Funaguma, T. Kanda, T. Tamura, and T. Shimada. 2005. Role of the male BmDSX protein in the sexual differentiation of *Bombyx mori*. *Evol. Dev.* 7:58—68.

[113]Tabunoki, H., S. Higurashi, O. Ninagi, et al. 2004. A carotenoid-binding protein (CBP) plays a crucial role in cocoon pigmentation of silkworm (*Bombyx mori*) larvae. *FEBS Lett.* 567:175—78.

[114]Tamura, T., C. Thibert, C. Royer, et al. 2000. Germline transformation of the silkworm *Bombyx mori* L. using a *piggyBac* transposon-derived vector. *Nat. Biotechnol.* 18:81—84.

[115]Tan, A., H. Tanaka, T. Tamura, and T. Shiotsuki. 2005. Precocious metamorphosis in transgenic silkworms overexpressing juvenile hormone esterase. *Proc. Natl. Acad. Sci. U. S. A.* 102:11751—56.

[116]Tan, Y. D., C. Wan, Y. Zhu, C. Lu, Z. Xiang, and H. W. Deng. 2001. An amplified fragment length polymorphism map of the silkworm. *Genetics* 157:1277—84.

[117]Tanaka, H., H. Matsuki, S. Furukawa, et al. 2007. Identification and functional analysis of *Relish* homologs in the silkworm, *Bombyx mori*. *Biochim. Biophys. Acta* 1769:559—68.

[118]Tazima, Y. 1964. *The genetics of the silkworm. London*: Academic Press.

[119]Tomita, M., R. Hino, S. Ogawa, et al. 2007. A germline transgenic silkworm that secretes recombinant proteins in the sericin layer of cocoon. *Transgenic Res.* 16:449—65.

[120]Tong, C. Z., Y. F. Jin, and Y. Z. Zhang. 2006. Computational prediction of microRNA genes in silkworm genome. J. *Zhejiang Univ. Sci. B* 7:806—16.

[121]Traut, W. 1976. Pachytene mapping in the female silkworm, *Bombyx mori* L. (Lepidoptera). *Chromosoma* 58:275—84.

[122]Traut, W., K. Sahara, T. D. Otto, and F. Marec. 1999. Molecular differentiation of sex chromosomes probed by comparative genomic hybridization. *Chromosoma* 108:173—80.

[123]Tsuchida, K., C. Katagiri, Y. Tanaka, et al. 2004. The basis for colorless hemolymph and cocoons in the *Y*-gene recessive *Bombyx mori* mutants: A defect in the cellular uptake of carotenoids. *J. Insect Physiol.* 50:975—83.

[124]Tsukioka, H. , M. Takahashi, H. Mon, et al. 2006. Role of the silkworm *argonaute2* homolog gene in doublestrand break repair of extrachromosomal DNA. *Nucleic Acids Res*. 34:1092—101.

[125]Uchino, K. , M. Imamura, H. Sezutsu, et al. 2006. Evaluating promoter sequences for trapping an enhancer activity in the silkworm *Bombyx mori*. *J. Insect Biotech. Sericol*. 75:89—97.

[126]Uchino, K. , M. Imamura, K. Shimizu, T. Kanda, and T. Tamura. 2007. Germ line transformation of the silkworm, *Bombyx mori*, using the transposable element Minos. *Mol. Genet. Genomics* 277:213—20.

[127]Uchino, K. , H. Sezutsu, M. Imamura, et al. 2008. Construction of a *piggyBac*-based enhancer trap system for the analysis of gene function in silkworm *Bombyx mori*. *Insect Biochem. Mol. Biol*. 38: 1165—73.

[128]Wang, G. H. , C. Liu, Q. Y. Xia, Y. F. Zha, J. Chen, and L. Jiang. 2008. Cathepsin B protease is required for metamorphism in silkworm, *Bombyx mori*. *Insect Sci*. 15:201—08.

[129]Wang, J. , X. Ren, T. Miller, and Y. Park. 2006. *piggyBac*-like elements in the tobacco budworm, *Heliothis virescens* (Fabricius). *Insect Mol. Biol*. 15: 435—43.

[130]Wang, J. , Q. Xia, X. He, et al. 2005. SilkDB: a knowledgebase for silkworm biology and genomics. Nucleic Acids Res. 33:D399—402.

[131]Wanner, K. W. , A. R. Anderson, S. C. Trowell, D. A. Theilmann, H. M. Robertson, and R. D. Newcomb. 2007. Female-biased expression of odourant receptor genes in the adult antennae of the silkworm, *Bombyx mori*. *Insect Mol. Biol*. 16:107—19.

[132]Wu, C. , S. Asakawa, N. Shimizu, S. Kawasaki, and Y. Yasukochi. 1999. Construction and characterization of bacterial artificial chromosome libraries from the silkworm, *Bombyx mori*. *Mol. Gen. Genet*. 261:698—706.

[133]Xia, Q. , D. Cheng, J. Duan, et al. 2007. Microarray-based gene expression proifiles in multiple tissues of the domesticated silkworm, *Bombyx mori*. *Genome Biol*. 8:R162.

[134]Xia, Q. , Z. Zhou, C. Lu, et al. 2004. A draft sequence for the genome of the domesticated silkworm (*Bombyx mori*). *Science* 306:1937—40.

[135]Xiang, H. , M. Li, F. Yang, et al. 2008. Fine mapping of E(kp)-1, a locus associated with silkworm (*Bombyx mori*) proleg development. *Heredity* 100:533—40.

[136]Xu, H. F. , Q. Y. Xia, C. Liu, et al. 2006. Identification and characterization of *piggyBac*-like elements in the genome of domesticated silkworm, *Bombyx mori*. *Mol. Genet. Genomics* 276:31—40.

[137]Xu, J. -S. , Q. -Y. Xia, J. Li, G. -Q. Pan, and Z. -Y. Zhou. 2005. Survey of long terminal repeat retrotransposons of domesticated silkworm (*Bombyx mori*). *Insect Biochem. Mol. Biol*. 35:921—29.

[138]Yamamoto, K. , J. Narukawa, K. Kadono-Okuda, et al. 2006. Construction of a single nucleotide polymorphism linkage map for the silkworm, *Bombyx mori*, based on bacterial artificial chromosome end sequences. *Genetics* 173:151—61.

[139]Yamamoto, K. , J. Nohata, K. Kadono-Okuda, et al. 2008. A BAC-based integrated linkage map of the silkworm *Bombyx mori*. *Genome Biol*. 9:R21.

[140]Yamamoto, M. , M. Yamao, H. Nishiyama, et al. 2004. New and highly efficient method for silkworm transgenesis using *Autographa californica* nucleopolyhedrovirus and *piggyBac* transposable elements. *Biotechnol. Bioeng*. 88:849—53.

[141]Yanagisawa, S. , Z. Zhu, I. Kobayashi, et al. 2007. Improving cell-adhesive properties of recombinant *Bombyx mori* silk by incorporation of collagen or fibronectin derived peptides produced by transgenic silkworms. *Biomacromolecules* 8:3487—92.

[142]Yao, H. P. , X. W. Xiang, L. Chen, et al. 2008. Identification of the proteome of the midgut of silkworm, *Bombyx mori* L. by multidimensional liquid chromatography LTQ-Orbitrap mass spectrometry. *Biosci. Rep.* doi:10. 1042/BSR20080144.

[143]Yasukochi, Y. 1998. A dense genetic map of the silkworm, *Bombyx mori*, covering all chromosomes based on 1018 molecular markers. *Genetics* 150:1513—25.

[144]Yasukochi, Y. , L. A. Ashakumary, K. Baba, A. Yoshido, and K. Sahara. 2006. A second-generation integrated map of the silkworm reveals synteny and conserved gene order between lepidopteran insects. *Genetics* 173:1319—28.

[145]Yasukochi, Y. , L. A. Ashakumary, C. Wu, et al. 2004. Organization of the *Hox* gene cluster of the silkworm, *Bombyx mori*: A split of the *Hox* cluster in a non-*Drosophila* insect. *Dev. Genes Evol.* 214: 606—14.

[146]Yasukochi, Y. , Y. Banno, K. Yamamoto, M. R. Goldsmith, and H. Fujii. 2005. Integration of molecular and classical linkage groups of the silkworm, *Bombyx mori* (n = 28). *Genome* 48:626—29.

[147]Yasukochi, Y. , H. Fujii, and M. R. Goldsmith. 2008. Silkworm. In *Genome mapping and genomics in arthropods*, ed. C. Kole and W. Hunter, 43—57. Berlin: Springer-Verlag.

[148]Yasukochi, Y. , T. Kanda, and T. Tamura. 1998. Cloning of two *Bombyx* homologues of the *Drosophila* rosy gene and their relationship to larval translucent skin colour mutants. *Genet. Res.* 71:11—19.

[149]Yoshido, A. , H. Bando, Y. Yasukochi, and K. Sahara. 2005. The *Bombyx mor*i karyotype and the assignment of linkage groups. *Genetics* 170:675—85.

[150]Yoshido, A. , Y. Yasukochi, F. Marec, H. Abe, and K. Sahara. 2007. FISH analysis of the W chromosome in *Bombyx mandarina* and several other species of Lepidoptera by means of *B. mori* W-BAC probes. *J. Insect Biotech. Sericol.* 76:1—7.

[151]Yu, X. , Q. Zhou, S. C. Li, et al. 2008. The silkworm (*Bombyx mori*) microRNAs and their expressions in multiple developmental stages. *PLoS ONE* 3:e2997.

[152]Yukuhiro, K. , H. Sezutsu, M. Itoh, K. Shimizu, and Y. Banno. 2002. Significant levels of sequence divergence and gene rearrangements have occurred between the mitochondrial genomes of the wild mulberry silkmoth, *Bombyx mandarina*, and its close relative, the domesticated silkmoth, *Bombyx mori*. *Mol. Biol. Evol.* 19:1385—89.

[153]Zhao, Y. —P. , M. —W. Li, A. —Y. Xu, et al. 2008. SSR based linkage and mapping analysis of *C*, a yellow cocoon gene in the silkworm, *Bombyx mori*. *Insect Sci.* 15:399—404.

[154]Zhang, P. , Y. Aso, H. Jikuya, et al. 2007a. Proteomic profiling of the silkworm skeletal muscle proteins during larval-pupal metamorphosis. *J. Proteome Res.* 6:2295—303.

[155]Zhang, Y. Z. , J. Chen, Z. M. Nie, et al. 2007b. Expression of open reading frames in silkworm pupal cDNA library. *Appl. Biochem. Biotechnol.* 136:327—43.

[156]Zhong, B. , Y. Yu, Y. Xu, et al. 2005. Analysis of ESTs and gene expression patterns of the posterior silkgland in the fifth instar larvae of silkworm, *Bombyx mori* L. *Sci. China C Life Sci.* 48:25—33.

[157]Zhou, J. J. , Y. Kan, J. Antoniw, J. A. Pickett, and L. M. Field. 2006. Genome and EST analyses and expression of a gene family with putative functions in insect chemoreception. *Chem. Senses* 31:453—65.

[158]Zhou, Z. , H. Yang, and B. Zhong. 2008. From genome to proteome: Great progress in the domesticated silkworm (*Bombyx mori* L.). *Acta Biochim. Biophys. Sin. (Shanghai)* 40:601—11.

第三章 鳞翅目昆虫W染色体的起源和发展

František Marec, Ken Sahara, and Walther Traut

1. 前言

WZ 性染色体系统是鳞翅目昆虫细胞遗传中最明显的特性。大多数动物和一些雌雄异株的植物属于 XY 或者 X0 系统，少部分如鸟类、蛇类、毛翅蝇（毛翅目）和鳞翅目一样，属于 WZ 或者 Z0 系统。在最近一些文献中讲述了鳞翅目昆虫性染色体和性别决定（Traut，Sahara，and Marec 2007）。我们这里主要讲的是 W 染色体的命运。大多数现存的鳞翅目昆虫都具有 W 染色体，令人惊奇的是鳞翅目分支起源于 Z0 染色体系统，W 染色体在进化上出现相对较晚。随后我们将给大家展示 W 染色体的进化与 Y 染色体进化的相似性：它的衰变、与其他染色体的融合、有时甚至发生再次丢失。因此，W 染色体展示了在一个群体中典型的性染色体全面的进化生命周期。

为了更好地了解染色体系统，我们简单地总结了鳞翅目昆虫细胞遗传的基本特性，染色体鉴定的新技术，然后讨论性染色体，尤其是更详细地介绍了 W 染色体。我们将综述关于 W 染色体的组成，分子水平上的分化与进化以及它们在性别决定中的作用。

2. 性染色体外的细胞遗传特性

2.1 染色体数目

鳞翅目昆虫的染色体通常较小、数目多、形状不规则。据报道大多数物种单倍体染色体数目接近 30（Robinson 1971）。在一个可能很古老的模型中，鳞翅目昆虫的染色体数目是 $n=31$，并且典型的核型显示了染色体在大小上逐渐减小（参见 De Prins 和 Saitoh 2003 的综述）。在鳞翅目进化树的某一分支上，具有较少数目核型的物种通过多重染色体融合独立地发生进化；相对来说，它们的染色体较大。古毒蛾属的昆虫是一个明显的例子：古毒蛾（*O. antiqua*）的染色体数目是 $n=14$，*O. thyellina* 的染色体数目是 $n=11$（Traut 和 Clarke 1997）；如大的白蝴蝶，樱桃白纹毒蛾 *Pieris brassicae*（pieridae）的染色体数目是 $n=15$（Doncaster 1912），蓖麻蚕 *Samia cynthia* 的染色体数目变化较大，与所调查的不同群体有关，从 12～14。具有较少染色体数目的物种在凤蝶科中也相当普遍（Brown et al. 2007），在亚热带蝴蝶 *Hypothyris thea*（Nymphalidae；Brown，von Schoultz 和 Suomalainen 2004）和美国亚利桑那巨蝶 *Agathymus aryxna*（Hesperiidae；De Prins 和 Saitoh 2003）这两个物种中发现的染色体数目最少为 5。另一方面，核型研究发现可能是通过染色体分裂导致染色体数目较多。例如，中国柞蚕 *Antheraea pernyi* 的染色体数目为 49，而与其亲缘关系很近的日本柞蚕 *A. yamamai* 只有 31 条染色体，为鳞翅目昆虫典型的数目（Kawaguchi 1934）。多重染色体分裂的典型例子是 *Agrodiaetus dolus*，染色体数目高达 90～125（Lukhtanov，Vila 和 Kandul 2006）。的确，*Agrodiaetus* 属染色体数目显示异常的多样性，单倍体染色体数目从 10～134（Lukhtanov 和 Dantchenko 2002；Lukhtanov et al. 2005）。另外，lycaenid 蝴蝶，Atlas blue，*Polyommatus atlantica* 具有最多的染色体数目为 223，不仅仅是在鳞翅目昆虫中，也是在所有动物中最多的（de Lesse 1970）。

2.2 染色体结构

鳞翅目和它的姊妹群——毛翅目，都是具有全着丝点染色体的生物。与典型的单着丝粒染色体生物(例如双翅目和鞘翅目)相比，全着丝点染色体缺乏明显的初级缢痕(着丝粒)，结果导致在有丝分裂中期姊妹染色单体被平行分开(Suomalainen 1966；Murakami 和 Imai 1974)。正常的分离点是连接着纺锤体微管的大着丝粒。事实上，染色体并不是真正的全着丝点的，但是结点覆盖了染色单体长度的重要部分(Gassner 和 Klemetson 1974；Wolf 1996)。

鳞翅目物种对粒子射线抵抗力很强，尽管这种现象的机制并不是很清楚。高度的抵抗粒子射线的一个明显作用应该归功于其特别的染色体结构(见 Carpenter，Bloem 和 Marec 2005 的综述)。较大的动粒板确保大多数射线诱导的断裂片段不会导致染色体片段的缺失，这种缺失在单着丝粒染色体物种中是很典型的。反而，片段能够持续存在于一定数目的有丝分裂细胞中，甚至能够从生殖细胞传给下一代(Rathjens 1974；Marec et al. 2001)。延长的动粒板降低了由双着丝粒染色体和无着丝粒片段导致的死亡率(Tothová 和 Marec 2001)。这种特性来源于近着丝粒染色体结构，使得鳞翅目具有独特的特性——所谓的遗传不育性(Carpenter，Bloem 和 Marec 2005)和如前所述的在一些鳞翅目昆虫类群的核型进化中有利于保有染色体融合和断裂的产物。

2.3 雌减数分裂过程中交换和交叉的缺失

在鳞翅目昆虫雌个体的减数分裂 I 期不发生染色体交换和交叉，然而，在雄个体中染色体表现出正常顺序的减数分裂，在双线期和终变期具有典型的交叉过程(例如 Traut 1977；Nokkala 1987)。在这个方面，鳞翅目与果蝇和其他高等双翅目相比处在相反的位置，这些昆虫在雄个体中是不发生减数分裂的染色体重组的。在这两种情况下，非交叉减数分裂与异配性别相关。

然而，与果蝇雄性相比，在鳞翅目昆虫雌性减数第一次分裂前期是与雄性相类似的，直至粗线期，染色体进行配对并形成规则的二价体。该过程由蛋白质联会复合体(SCs)介导，该复合体在果蝇雄性中是不存在的(见 Marec 1996 的综述)。鳞翅目雄性和雌性的 SCs 在这个时期唯一显著的不同是在雄性的 SCs 中存在重组结，在雌性中却没有。重组结是由多酶组成的纳米机器(nanomachines)来介导重组的。雌性减数分裂过程没有减数分裂的重组，因此，没有形成交叉，从而不好分辨双线期和终变期。在分裂中期 I，以修饰形式存在的、与二价体联合的 SCs 转换为“消除染色质”，将同源染色体连接在一起，停留在赤道板，在分裂后期 I 染色体分开。染色质消除的过程被认为是代替了交叉和交换，使得同源染色体正常分开(Rasmussen 1977)。

2.4 体细胞组织的多倍型

在鳞翅目昆虫幼虫和成虫中大多数的体细胞组织是多倍型的。多倍体化水平的变化比较大。在幼虫成虫盘、神经中枢和血淋巴中是二倍体核(例如：Trant 和 Scholz 1978；F. Marec，未发表数据)。在成虫盘中也经常看到四倍体。它们中的一部分至少是发生了有丝分裂，很容易从染色体数目上确定四倍体。在四倍体中，核内多倍化是一般规则，所以多倍体水平可以通过细胞色素光度法决定。例如，我们在肠细胞发现多倍化的中间状态，在马氏管中发现高度状态，在丝腺和下颌腺中也是具有高度多倍化程度(Buntrock，Marec 和 Traut，未发表)。家蚕丝腺多倍化记录至少达到 4×10^5C(Gage 1974)。高度多倍化细胞的核通常形状不规则；有别于平常细胞的球状，它们是树枝状的(例如 Traut 和 Schdz 1978)。

3. 染色体的鉴定

鳞翅目昆虫染色体形态较小且形状不规则，尤其是缺乏作为形态学标志的着丝粒，并且缺乏便利的染色体显带技术，这妨碍了用标准的细胞遗传的方法来鉴定染色体。很长时间以来这种状况限制了鳞翅目昆虫细胞染色体的计数。粗线期绘图技术提高了在减数分裂粗线期鉴定染色体的水平(Traut 1976)。在粗线期，染色体为体细胞的两倍，且长度比有丝分裂的时期长，另外，能形成特有的形态。粗线期绘图还能够鉴定某些物种中 WZ 染色体二价体并且能够鉴定少部分的常染色体二价体(Traut 1976；Schulz 和 Traut 1979)。但对于一般的染色体，还未找到其鉴定方法。

所有的现代技术利用了粗线期良好的形态特征。最近，来自细菌人工染色体(BACs)的探针通过荧光原位杂交(FISH)被成功地运用到粗线期来补充建立完整的家蚕染色体核型(图 3.1A，B；Yoshido et al. 2005)。BAC- FISH 技术提供特异的识别位点和锚位点，因此不仅用于鉴定个体的染色体，还是构建基因物理图谱的有力工具(Yasukochi et al. 2004；Niimi et al. 2006)。已经应用该技术研究了家蚕和其他鳞翅目物种保守的同线性基因(Yasukochi et al. 2006；Sahara et al. 2007)。BAC-FISH 是目前鳞翅目核型研究和基因物理图谱构建中的最尖端技术。

在一些物种中，单纯的粗线期绘图技术足以识别已经分化的性染色体。虽然常染色体和 Z 染色体具有明显的染色体间差异，W 染色体通常形成部分或者全部染色均一的线条(例如，Schulz 和 Traut 1979；Traut 和 Marec 1997)。这个方法行不通的时候，一些分子生物学技术对于鉴定 W 染色体将是很有用的，包括比较基因组杂交(CGH)，基因组原位杂交(GISH)，W-BAC-FISH 和 W-染色体染色。

CGH 利用两个不同标记探针共杂交，所用的基因组 DNA 分别来自雌个体和雄个体。CGH 通过颜色或信号强度来区分 W 染色体，也就是说，分别通过结合雌特异探针组分和积累的重复元件的杂交来区分(Traut et al. 1999；Traut，Eickhoff 和 Schorch 2001；Sahara et al. 2003a；Vítková et al. 2007)。利用 CGH 通过杂交雌和雄 DNA 来比较信号，因此，不仅可以区分性染色体而且还可提供序列组成的其他信息(图 3.1C—F)。

GISH 是在很多未标记的竞争性 DNA(这里的全基因组 DNA 来自同型配子个体)存在的条件下，以标记的探针杂交为基础(这里的全基因组 DNA 来自异型配子个体)。GISH 通过强烈结合荧光标记雌基因组的 DNA 来可视化 W 染色体，结合到常染色体和 Z 染色体的将被竞争性 DNA 抑制(图 3.1G；Sahara et al. 2003b；Mediouni et al. 2004；Fuková，Nguyen 和 Marec 2005；Yoshido，Yamada 和 Sahara 2006)。昆虫末端着丝粒探针$(TTAGG)_n$ FISH 结合物和 GISH 对研究东方草丛蛾(*Artaxa subflava*)的特异性染色体对的性染色体组分和具有多重性染色体的物种的粗线期补体是相当有用的(Yoshido，Marec 和 Sahara 2005)。

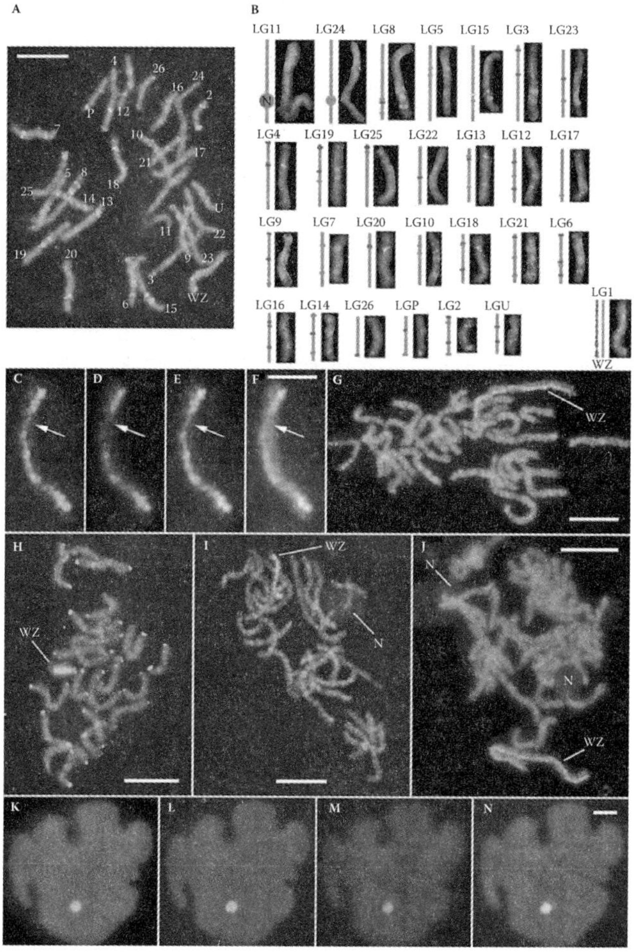

图 3.1　来自鳞翅目昆虫雌性的粗线期(A～J)和多倍体化的分裂间期的核(K～N)的 FISH 杂交图片(彩色版本图片见附录图 3.1)。DAPI(蓝色)复染色的染色体和多倍体核。(A～B)家蚕 BAC-FISH。(A)62 BAC 探针与卵母细胞粗线期核杂交,每一个探针是 Cy-3(红色)和/或荧光(绿色)标记。(B)来自同一个细胞核的 28 个二价体 BAC-FISH 核型组装;每一个 BAC 信号颜色特征可以用于判断区分每一个二价体。二价体用数字和字母(P 和 U)与 LGs 一起标记;WZ(=LG1),是性染色体二价体。(C～F)用 CGH 染色的 *Galleria mellonella* 的 WZ 二价体;(C)Alexa Fluor 488 染料标记的雌基因组探针与全 W 染色体的杂交信号;(D)Cy3(红色)标记雄基因组探针杂交信号;(E)两个探针归并影像;(F)包括复染(蓝色)两个探针归并影像;用两个探针杂交的 W 染色体有均匀的亮点,显示了 W 染色体随机重复片段的存在,除了仅被雌探针杂交的由箭头指向的间隙片段,表明了 W 染色体特异的 DNA 片段。(G)用 Cy-3 标记的雌基因组 DNA(红色)GISH 杂交的 *Cydia pomonella* (n=28)粗线期卵母细胞核,在 WZ 二价体中高亮标记的 W 染色体。(H)用 CY-3 标记的雌基因组 DNA(红色)GISH 杂交的 *Artaxa subflava* (n=22; A. Yoshido 和 K. Sahara, unpublished)粗线期卵母细胞核和$(TTAGG)_n$荧光标记成绿色的探针杂交的可视化端粒;GISH 鉴定了被长 Z 染色体包裹的小的 W 染色体。(I)W-BAC-FISH 染色家蚕粗线期卵母细胞核;通过 W 衍生的 BAC 探针标注了 WZ 二价体中的高亮的 W 染色体,绿色的 FluorX-标记的 5H4C 克隆。单个核仁(N)伴随有常染色体二价体。(J)利用光谱桔红标记的 W-painting 探针 FISH 杂交 *Cadra cautella* (n=30)粗线期补体图;表明了探针与 WZ 二价体中 W 染色体和两个核仁较强的结合能力,每一个都与不同常染色体二价体有关。(K～N)CGH 染色的 *Plodia interpunctella* 多倍体化的马氏管细胞分裂间期细胞核,展示了 DAPI 染色用雌雄基因组 DNA 探针标记的显著高亮的性染色质体;(K)DAPI 复染(蓝色);(I)Alexa Fluor 488 染料标记的雌基因组探针;(M)Cy-3 标记的雄探针(红色);(N)两个探针归并影像;显示了雌探针比雄探针与性染色质较强的结合能力,表明了 W 特异的 DNA 重要的片段。A、B、H 来源于 Atsuo Yoshido (Sapporo);C 至 F、J、K 到 N 来源于 Magda Vítková (České Budějovice)。刻度=10 μm。

Sahara 等人通过区分 W 染色体重复元件来鉴定染色体。他们利用单一的 W- BAC-FISH 作为探针进行 FISH 杂交(W-BAC-FISH),尽管它来自 W-DNA 分子相对较小的部分(大约 170kb),但是对于杂交家蚕的 W 染色体来说是足够的。他们证明了 W 染色体的同质性(家蚕 W 染色体的序列组成见第四章)。

最近,一个新的广泛应用的方法能够可视化和分析鳞翅目 W 染色体。基于来自于高度多倍体化核的 W 染色质的激光显微解剖,例如来自马氏管。因为 W 染色质体由许多重复的 W 染色体形成,所以提供了更多样本材料。通过简并寡核苷酸引物 PCR 扩增(DOP-PCR)和荧光染料标记 DNA,在 FISH 中直接应用该探针给 W 染色体着色(图 3.1J; Fuková et al. 2007; Vítková et al. 2007)。

4. 性染色体系统

具有 XY 或 X0 性染色体的物种能够产生雌决定性和雄决定性配子(雄异型配子),然而 WZ 或 Z0 物种产生雌决定性和雄决定性卵(雌异型配子)。从大量性连锁材料、更多特异的 Z 连锁材料、细胞遗传证据来看,鳞翅目昆虫雌性属于异型配子。

雌个体异配型的染色体系通常是 WZ/ZZ 和 Z/ZZ(或者成为 WZ 或 Z0),这个取决于 W 染色体的有无。在鳞翅目昆虫中 WZ/ZZ 是很普遍的(图 3.2a)。模式昆虫家蚕和实验室常用的物种如地中海粉螟和蜡螟,还有害虫如苹果蠹蛾、舞毒蛾、石榴螟、粉斑螟和谷螟等都是 WZ/ZZ 型(Traut 和 Marec 1997; Mediouni et al. 2004; Fuková,Nguyen 和 Marec 2005; Vítková et al. 2007)。相比之下,Z/ZZ 染色体系是不常见的(图 3.2b)。如在鳞翅目低等物种小翅蛾科(Traut 和 Marec 1997)中,偶尔还在较多的高等物种中如夜蛾科的单梦尼夜蛾 *Orthosia gracilis* (Traut 和 Mosbacher 1968),天蚕蛾科的蓖麻蚕中发现(Yoshido,Marec 和 Sahara 2005)。

多重染色体系统也在鳞翅目中发现,具有两个 W 染色体(W_1W_2Z/ZZ;图 3.2c,d)的系统,如在螟蛾 *Witlesia murana*、卷叶蛾 *Bactra lacteana*(Suomalainen 1969)、红宝石虎蛾 *Phragmatobia fuliginosa*(Traut 和 Marec 1997)和草丛蛾 *O. thyellina*(Yoshido,Marec 和 Sahara 2005)中都有所发现。具有两个不同 Z 染色体($WZ_1Z_2/Z_1Z_1Z_2Z_2$)的系统在小巢蛾属的 6 个物种中曾被发现和报道(图 3.2f; Nilsson,Löfstedt 和 Dävring 1988)。相似的类型在青枯叶蛾(图 3.2e; Rishi,Sahni 和 Rishi 1999)和天蚕 *S. cynthia* 亚种 Shinju 蚕中(Yoshido,Marec 和 Sahara 2005)也有所发现。

多重性染色体系统可能来源于 6 条染色体的分裂和融合。因为它们的全着丝粒染色体的结构,不像在其他物种中,在鳞翅目中不容易发生分裂。*W. murana* 和 *B. lacteana* 的 W_1 和 W_2 染色体是通过分裂起源的,产生 W_1W_2Z/ZZ 类型的性染色体系统。然而,在大多数情况下,常染色体和性染色体的融合似乎是多样性染色体最可能的来源。常染色体与 Z 染色体的融合产生一个新的 Z 染色体(可以用 Z^A 和 A^Z 表示),有两个可以配对的染色体,一个自由的常染色体和初始的 W 染色体,它们的行为与 W 染色体在减数分裂时的行为是一致的。最终性染色体系统形式可以用 W_1W_2Z/ZZ 表示(图 3.2d)。一个常染色体与 W 染色体的融合产生一个新的 W 染色体(可以用 W^A 和 A^w 表示)。在减数分裂过程中它们有两个可以配对的染色体,一个自由的常染色体和初始的 Z 染色体,

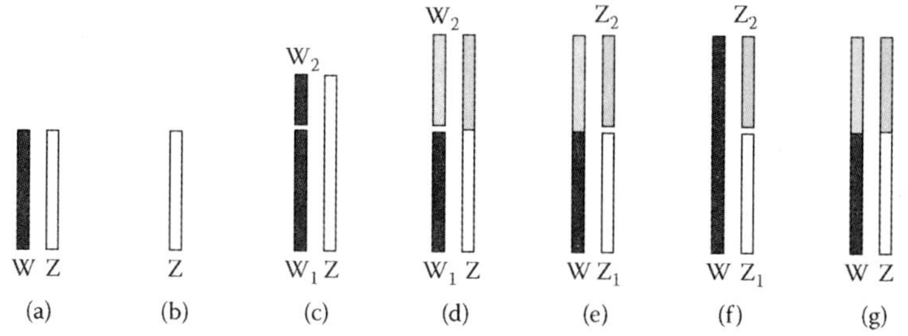

图 3.2　雌性鳞翅目昆虫性染色体组成的框架图。W 和 Z,性染色体;黑色代表 W 染色体异染色质化的部分。而白色则代表起初的 Z 染色体;灰色则代表常染色体或是常染色体起源的片段。(a)代表 WZ,如地中海粉螟、蜡螟和家蚕。(b)Z0 型,如 *Micropterix calthella* 和蓖麻蚕。(c)W_1W_2Z 系统,由一个初始的 Z 染色体和两个异染色质化的 W 染色体,较长的 W_1 和较短的 W_2 组成。它意味着初始 W 染色体发生分裂,如螟蛾(*Witlesia murana*),卷叶蛾科(*Bactra lacteana*)。(d)W_1W_2Z 系统,它是由一个初始的 W(W_1)染色体,一个由初始 Z 染色体与常染色体融合形成的新的 Z 染色体,和一个融合的常染色体同源染色体的 W_2 染色体组成,如红宝石虎蛾(*Phragmatobia fuliginosa*)。(e)W Z_1Z_2 它是由起初的 Z(Z_1)染色体,一个由初始 W 染色体与常染色体(注意这里的常染色体片段是没有异染色质化的)融合形成的新的 W 染色体,及一个与融合常染色体同源染色体的 Z_2 染色体组成,如青枯叶蛾(*Trabala vishnu*)。(f)W Z_1Z_2 它与(e)相似,但是它有一个完全异染色质化的 W 染色体,这就意味着起初的 W 染色体和常染色体有着很长时间的融合,如小巢蛾属(*Yponomeuta cagnagellus*)和小白鼬毛皮飞蛾(*Y. padellus*)。(g)WZ 它是由新产生的 W 和 Z 两个染色体组成的,而它们都是由初始的 W 和 Z 染色体与一对常染色体中的同源染色体融合产生的。如古毒蛾 *Orgyia antiqua* 和树天蚕蛾 *Samia cynthia walkeri*。

它们的行为和 Z 染色体一样。因此一个 $WZ_1Z_2/Z_1Z_1Z_2Z_2$ 系统就产生了。但是这种系统也可能是原来 Z 染色体融合产生的。一个关于原始状态和染色体之间同源性的调查在决定鳞翅目昆虫多重性染色体系统起源的解释方面是很有必要的。

5. W 染色体的异染色质化

在一些种类中,如地中海粉螟、舞毒蛾和粉斑螟的 W 染色体可以在粗线期用 4′,6-氨基-2-苯基(DAPI)(图 3.3a,b)或是用常规的吉姆莎或苔红素(Schulc 和 Traut1979;Traut 和 Marec 1997;Vítková et al. 2007)来识别。W 染色体没有常染色质以及 Z 染色体中存在的典型的染色粒一染色质间区相互间隔形式,而是形成一条看起来更加稠密的或是浓缩更深的染色质片段。CGH 和 GICH 能够使这条染色质更加明显(从图 3.1C—F,G 可以看出)。在一些种类,如家蚕 *B. mori*,蜡螟 *G. mellonella* 和石榴螟 *E. ceratoniae* 中,它们的 W 染色体很难通过传统的显微镜观察区分出来,可以用 CGH 和 GISH 方法将它们区分开来(Traut et al. 1999; Mediouni et al. 2004; Vítková et al. 2007)。这个突出区的产生被认为归因于 W 染色体中存在像反转录转座子这样的大量聚集的重复序列(家蚕 W 染色体上的反转座子见第四章)。我们将这些染色体看作是异染色质的。这种异染色质性质在分裂间期的细胞中更加明显,此时 W 染色体会形成一个密集的染色质体,或者是性染色质或 W 染色质(见下节)。

这条异染色质 W 染色体和常染色质 Z 染色体在很大程度上是非同源的,但是它们在减数分裂过程中保持配对并直至减数第一次分裂中期(Weith 和 Traut 1986; Marec et al. 2001)。但是与常染色体二价体相比，它们出现得迟一些。在地中海粉螟(*E. kuehniella*)中，W 和 Z 染色体之间的长度差异在配对过程中被调整以至于这些性染色体在后面的粗线期中是完全联会的(*Marec* 和 *Traut* 1994)。即使是多条性染色体在减数分裂时也会形成看起来长度已经得到调整的多价体(例如，Seiler 1914；Traut 和 Marec 1997；Yoshido，Marec 和 Sahara 2005)。在配对期间调整染色体的长度的例外是很少的。例如，在 *A. subflava* 中，在粗线期阶段，较短的 W 染色体被较长的 Z 染色体所包围(见图 3.1H)。相似地，热带蝴蝶 *Bicyclus anynana* 的微小 W 染色体形成一个被长的 Z 染色体所环绕的半球体(图 3.3c；A. E. Van't Hof et al. 2008)。

在一些种类中，W 染色体只有部分发生异染色质化，剩下的部分表现为常染色质(Traut 和 Marec 1997)。这与 vapourer moth，古毒蛾 *O. antiqua* (Lymantriidae)中的调查很接近。古毒蛾 *O. antiqua* 含有减少的染色体数目，为 14。这就意味着这些种类的进化历史中多个染色体发生了融合(Trant 和 Clarke 1997)。在粗线期，W 与 Z 形成很长的二价体，这个二价体中 W 染色体发生了部分异染色质化，而剩余的部分以 W 和 Z 染色体的同源染色粒形式存在(Traut 和 Clarke 1997)。只有发生异质化的部分才会被 CGH 和 GISH 显现出来(Traut，Eickhoff 和 Schorch 2001；Yoshido，Marec 和 Sahara 2005)。这个 W-Z 对可能是由一对常染色体和原始的 W(就是现在的异质区域)和原始的 Z 染色体(图 3.2g；Yoshido，Marec 和 Sahara 2005)融合产生的一个 neo-W，neo-Z 对。由于有较近的亲缘关系，染色体数目为 30 的 *O. recens* 仍然含有一个小的并且完全异染色质化的 W 染色体(Robinson 1971；Yoshido，Yamada 和 Sahara 2006)。染色体数为 26 且 W 染色体的部分异质化的树天蚕蛾 *S. Cynthia walkeri* 也被推断有着相似的起源(Yoshido，Marec 和 Sahara 2005)。

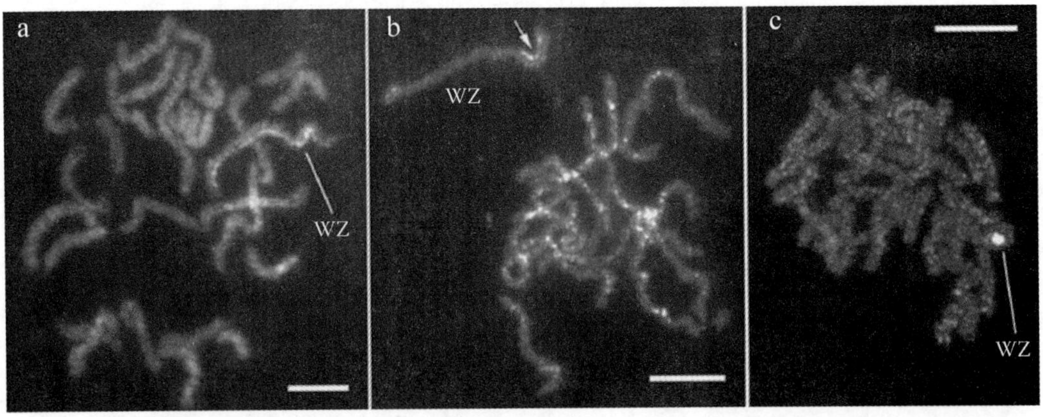

图 3.3　卵母细胞粗线期的荧光图像，(a，b)是用 DAPI 染色或 YOYO-1(c)。(a)苹果蠹蛾 *Cydia pomonella* (n=28)；注意在 WZ 二价体中被 DAPI 染色显现出来的是 W 的异染色质化的片段。(b)古毒蛾 *Orgyia antique* (n=14)，在这些种类中，WZ 二价体是与 1 或 2 染色体二价体融合产生的；注意 W 染色体遗传的那部分成为 DAPI 染色的异染色质(箭头)。(c)热带蝴蝶 *Bicyclus anynana* (n=28)，注意被 YOYO-1 染色的小 W 染色体被较长的 Z 染色体所包围。c 源自 Arjen E. Van't Hof (Liverpool)。刻度=10 μm。

6. W 染色质

W 染色体可导致鳞翅目昆虫分裂间期存在这样一个特性:像哺乳动物一样,鳞翅目昆虫含有一个被称之为雌特异性的性染色质。在大多数种类中,雌体细胞的分裂间期显示有一个或更多个异染色质体,而雄性中不显示。这个雌特异性的性染色质起源于 W 染色体并因此命名为 W 染色体(见 Traut 和 Marec 1996 的综述)。恰恰相反,哺乳动物的性染色质则由存在于雌性躯体组织中的两个 X 染色体中的一条演化而来。

W 染色质以一种可被强烈染色的异染色体存在。这个 W 染色质体在一些鳞翅目昆虫高度多倍化的组织如马氏管和丝腺中更加显著(图 3.1K～N)。在一些拥有完全异染色质 W 的种类中,如在地中海粉螟 *E. kuehniella* 中,它在大部分组织的多倍化过程中仍是个单一的,但却一直增加的个体。只有在营养细胞和卵泡细胞中含有多个 W 染色质体。含有部分异质化 W 的种类中,例如古毒蛾 *O. antiqua* 或是人工诱导的 W-Z 染色体融合体,它们所有的多倍体组织都含有多性染色质体(Traut,Weith 和 Traut 1986; Marec 和 Traut 1994; Traut 和 Marec 1997)。在地中海粉螟 *E. kuehniella* 中,辐射诱导将 Z 染色体上的片段转座到 W 染色体上的转座—T(W;Z)转座,根据转座片段的大小,显示畸形的或是片段化的性染色体(Marec 和 Traut 1994)。畸形或是多倍的性染色质体被认为是由异染色质化的 W 染色体相互聚集并与常染色质染色体在多倍体细胞核中的空间中分散之间相反的趋势引起的(Traut,Weith 和 Traut 1986; Marec 和 Traut 1994)。

7. 鳞翅目昆虫 W 染色体的起源

关系密切的石蛾蝇、毛翅目与鳞翅目一样为雌异配型。这种特性在进化祖先中普遍存在,甚至在毛翅目和鳞翅目分化之前。换句话说,也就是在大约 1.9 亿万年前(Kristensen 和 Skalski 1999; Grimaldi 和 Engel 2005)。在毛翅目中,目前为止被调查的所有种类均为 Z/ZZ 性染色体系统(Klingstedt 1931; Marec 和 Novák 1998; Lukhtanov 2000)。一个最近的关于鳞翅目的 Z/ZZ 与 ZW/ZZ 系统分化的调查显示 Z/ZZ 系统在鳞翅目基底谱系中是一个普遍的性质。同时它在一些更高等的种群中也零星地存在(图 3.4)。这就意味着鳞翅目作为一个单独的进化群也是以 Z/ZZ 性染色体决定机制开始进化的,W 染色体是在后来获得的。图 3.4 中的进化树揭示 W 染色体在"高等的"鳞翅目、双孔目中是一种普遍性质,而这占现有鳞翅目和它姐妹进化分支,小号潜蛾 Tischeriina 中种类的 98%(Traut 和 Marec 1996; Lukhtanov 2000)。因此,W 染色体在双孔亚目 Ditrysia 和小号潜蛾 Tischeriina 的共同祖先中就存在。在时间上,最早的证据是毛翅目和鳞翅目在 1.8 亿万～1.9 亿万年之间,而最早的一个双孔亚目 Ditrysia 种类的化石记录大约在 9700 万年前(Kristensen 和 Skalski 1999; 见 Grimaldi 和 Engel 2005 综述),因为球蚜科 Adeliidae 和曲蛾科 Incruvariidae 还没有 W 染色体,所以更可能接近后期(Lukhtanov 2000)。

在高等的鳞翅目不同种群中,W 染色体再一次随机且独立地消失。这些种类再一次回到 Z/ZZ 系统。在另一些种群中,ZW/ZZ 系统被多性染色体系统,W_1W_2Z/ZZ 和 $WZ_1Z_2/Z_1Z_1Z_2Z_2$ 所取代(图 3.4)。

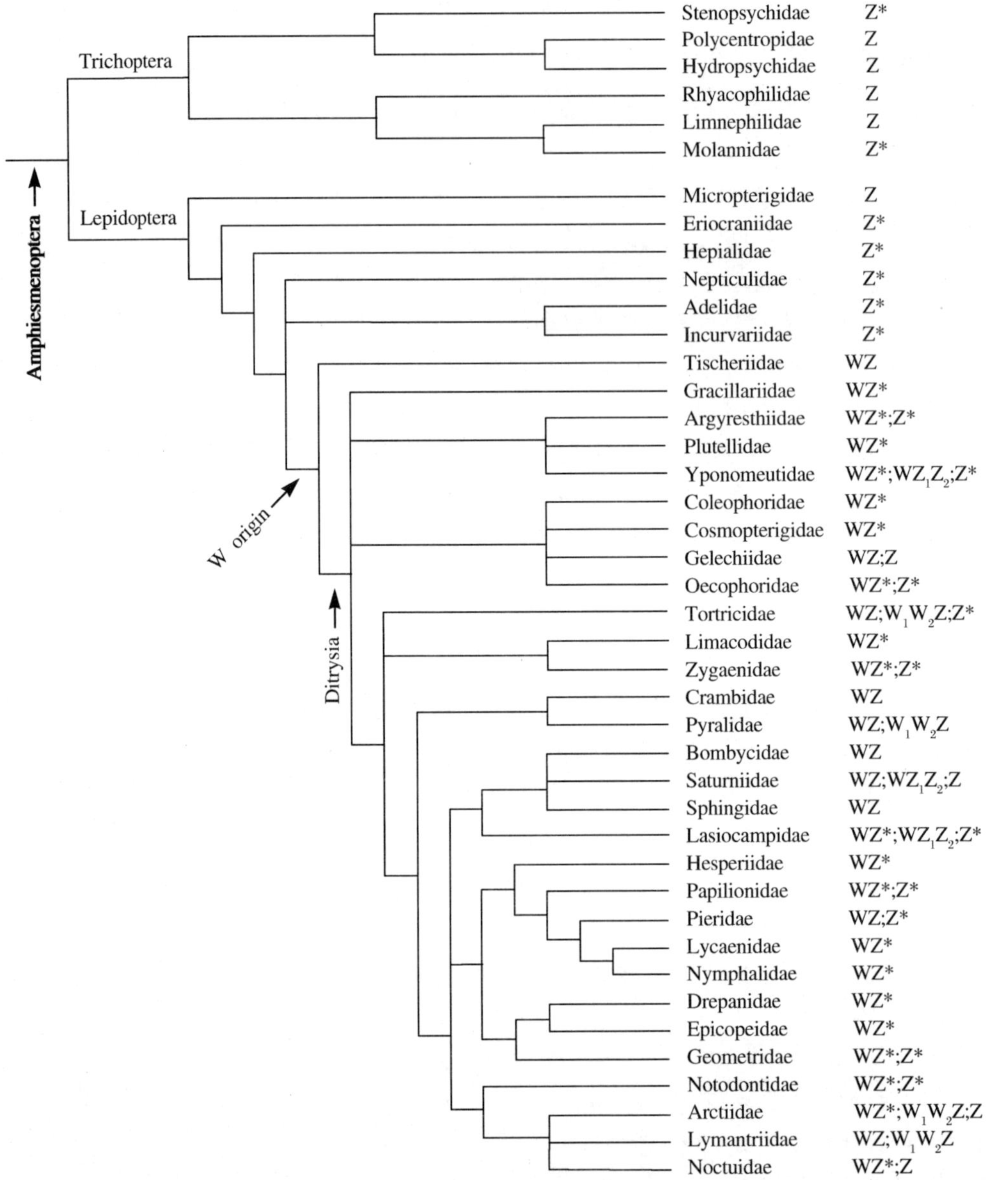

图 3.4　鳞翅目昆虫性染色体系统发生。每个分支右末端都列出了各自家族中雌性染色体组成。由 W 染色体的存在与缺失演变而来的性染色体系统都已经用星号(＊)表示出来。性染色体和 W 染色质数据来自 Traut 和 Marec(1996)以及其中参考资料，Marec 和 Novák(1998)，Lukhtanov(2000)，Traut，Sahara 和 Marec(2007) 以及其中参考资料。这个进化树是基于 Morse(1997)提出的毛翅目及 Kristensen 和 Skalski(1999)提出的鳞翅目。

关于在 Ditrysia 和 Tischeriina 的共同祖先中的 W 染色体是由什么演变而来的这一问题存在很多的假说。在遗传的 Z/ZZ 系统中，Z 染色体与常染色体的融合可能是 ZW/ZZ 系统的起源。根据这个假说，高等的鳞翅目中新的 Z 染色体一部分来自于遗传的 Z，一部分来自于前面的常染色体。自由的常染色体是 W 染色体的起源(Traut 和 Marec 1996)。另一个假说提出一条多余的染色体(B染色体)作为W染色体的起源,这是由于在ditrysian和nonditrysian家族中的染色体数目不稳定暗示了一个常染色体缺失(Lukhtanov 2000)。当Z染色体上足够的分子数据可以应用于与ditrysian和nonditrysian种类中的Z染色体上遗传内容进行比较时,这个问题就得以解决了。

8. 性别决定的功能

W和Z这对性染色体在性别决定中发挥很明显的作用。它在胚胎发育时期提供一个初始信号去选择性地开始向雌或向雄的途径。但是这不能说明是否有初始的性别决定基因定位在W染色体或是Z染色体上。从细胞遗传学上讲,W染色体的存在与否或是Z染色体的数目是可以决定向雌或向雄的发育。

考虑到果蝇中现在的性别染色体系统起源于遗传的Z/ZZ机制,我们期望Z染色体上的雄启动基因与Z的数目机制一起作为鳞翅目昆虫性别决定的基础。这种期望肯定与拥有Z/ZZ性别决定系统的种类相符合,不仅存在于低等谱系中,而且也存在于那些更高等的且丢失W染色体的鳞翅目的种类中。但是,在任何鳞翅目种类中,还未揭示雄启动基因的性质或是它们的准确定位。

但是,在家蚕中,W染色体的存在与否决定了性别。我们可以通过各种各样的多倍体或是非整倍体得知,单一W染色体的存在可导致它们的雌性发育,而与Z染色体的数目没有关系(Tazima 1964)。假定的雌性基因(*Fem*)还未被分离出来或是在W染色体上得到精确定位。但是一系列的对W染色体的转座和缺失研究都证明这些W染色体片段不包含*Fem*基因(Abe et al. 2005b;参见第四章)。除了这种明显的例子,W染色体在性别决定中的角色是不确定的,因为它在除家蚕外的其他多倍体和非整倍体重组体物种中没有被准确地检测到。

性染色体的初始信号会触发一连串的性别决定基因。这个级联反应是从果蝇的研究中了解的,它除了初始信号还需要*Sxl*,*tra*-*tra2*和*dsx*基因的参与。但是这种级联反应在其他昆虫中并不完全是保守的,即使是在非果蝇蝇类中也不存在(见Schütt和Nöthiger 2002的综述)。在家蚕中,*Sxl*,*tra*和*dsx*基因是保守的。但是,只有*dsx*的同源基因在性别决定中发挥作用,而*Sxl*和*tra2*不起作用(Ohbayashi,Suzuki和Shimada 2002; Niimi et al. 2006; Fujii和Shimada 2007;家蚕性别决定在第四章讲述)。在除家蚕之外的鳞翅目种类的性别决定基因还没有进行研究。

9. W染色体的遗传侵蚀

在任何鳞翅目昆虫中,除了家蚕中的*Fem*基因,很少基因被认为是W染色体连锁的候选基因。在家蚕中,一个控制产生大卵的基因,*Esd*,被推测与W染色体连锁(Kawamura 1988,1990),但是它的位置仍未被确定。在地中海粉螟*E. kuehniella*中,雌性中存在两条性染色体之间的非典型传递重排,这说明在W染色体上存在雄性致死因子(Marec et al. 2001)。在东方虎凤蝶*Papilio glaucus*中,黑色形态性状被认为是W连锁的,但对此已经提出了一些质疑(Scriber, Hagen和Lederhouse 1996; Andolfatto, Scriber和Charlesworth

2003)。

寻找遗传性状并将它们结合到特异的连锁群的经典方法已经揭示了只存在很少的W连锁基因。从分子水平上寻找这些基因一直没有成功。在中国的栎树蚕蛾(silk moth),中国柞蚕 *A. pernyi* 中,化性(*period*)基因是Z连锁的,但是在W染色体上也检测到了这个基因的一些拷贝,其中有两个可能有以下两种功能:其中一个被命名为 *perW*,会编码截断蛋白;另一个会产生一个反义RNA (Gotter, Levine 和 Reppert 1999)。

在家蚕中,W染色体的RAPDs(随机扩增DNA多态性),Lambda克隆和BACs都已经分离出来了,结果显示它们主要是由LTR和非LTR反转录转座子组成,这就意味着它们就是家蚕W染色体的主要的结构成分(Abe et al. 2005a; 家蚕W染色体的分子组成详见第四章)。利用一种直接的方法——显微解剖法,Fuková 等(2007)分离出了苹果蠹蛾 *C. pomonella* 的W染色体上的DNA。除了反转录转座子之外,他们还检测了非编码序列,只有少数的序列是单拷贝的和W染色体特有的。大多数都是多拷贝的序列。它们主要的定位位置可以分别在常染色体、W染色体,甚至在Z染色体上。

虽然对于得出这种答案还不是很满意,但是以上的研究告诉我们:至少鳞翅目昆虫W染色体上基因数目是很少的,取而代之的是大量的重复序列。与它起源的常染色体相比,显然W染色体遭受过遗传侵蚀。如果它们对性别决定不起作用或是缺少一些必要的功能,那么它们就是不必要的染色体。很明显从鳞翅目高等类群拥有Z/ZZ性染色体系统来看这点是肯定的(图3.4):它们确实丢失了W染色体,这证明了它是不必要的。但是大多数高等鳞翅目(Ditrysia)中,W染色体被保留下来。所以我们推测它可能含有与这些种类不可缺少的功能的一个或多个基因。

W染色体的最基本的特征与Y染色体相似。与许多Y染色体一样,鳞翅目W染色体是部分或是完全异染色质化的,含有极端的重复序列,并且几乎没有基因。就像是现在的XX/X系统中的Y染色体,W染色体在进化的进程中会偶然丢失。保守的杂合性和它不能够与之对应的染色体重组被认为是这种遗传侵蚀的原因。这个背景选择、选择性的清除的过程和 Muller's ratchet 成为了一些文章和综述的主题(例如 Charlesworth, Charlesworth 和 Marais 2005)。在这里我们可以看到:一个典型的性染色体的全部的进化过程,从它的起源经过结构和分子的分化,与常染色体发生重排到最终丢失,都可以在鳞翅目的昆虫中进行研究,并对性染色体的进化和性别决定的机制带来新的认识。

10. 致谢

F. M. 由捷克共和国的资助机构的206/06/1860专利和昆虫研究所的Z50070508. K. S. 项目支持,感谢推进生物科学基础研究的创新活动(PROBRAIN)和日本学术振兴协会(JSPS)的18380037项目的支持。

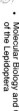

参考文献

[1]Abe, H. , K. Mita, T. Oshiki, and T. Shimada. 2005a. Retrotransposable elements on the W chromosome of the silkworm, *Bombyx mori*. *Cytogenet. Genome Res.* 110:144—51.

[2]Abe, H. , M. Seki, F. Ohbayashi, et al. 2005b. Partial deletions of the W chromosome due to reciprocal translocation in the silkworm *Bombyx mori*. *Insect Mol. Biol.* 14:339—52.

[3]Andolfatto, P. , J. M. Scriber, and B. Charlesworth. 2003. No association between mitochondrial DNA haplotypes and a female-limited mimicry phenotype in *Papilio glaucus*. *Evolution* 57:305—16.

[4]Brown, K. S. J. , A. V. L. Freitas, N. Wahlberg, B. von Schoultz, A. O. Saura, and A. Saura. 2007. Chromosomal evolution in the South American Nymphalidae. *Hereditas* 144:137—48.

[5]Brown, K. S. J. , B. von Schoultz, and E. Suomalainen. 2004. Chromosome evolution in Neotropical Danainae and Ithomiinae (Lepidoptera). *Hereditas* 141:216—36.

[6]Carpenter, J. E. , S. Bloem, and F. Marec. 2005. Inherited sterility in insects. In *Principles and practice in area-wide integrated pest management*, ed. V. A. Dyck, J. Hendrichs, and A. S. Robinson, 115—46. Dordrecht: Springer.

[7]Charlesworth, D. , B. Charlesworth, and G. Marais. 2005. Steps in the evolution of heteromorphic sex chromosomes. *Heredity* 95:118—28.

[8]de Lesse, H. 1970. Les nombres de chromosomes dans le groupe de *Lysandra argester* et leur incidence sur la taxonomie. *Bull. Soc. Entomol. Fr.* 75:64—68.

[9]De Prins, J. , and K. Saitoh. 2003. Karyology and sex determination. In *Lepidoptera, moths and butterflies: Morphology, physiology, and development*, ed. N. P. Kristensen, 449—68. Berlin: Walter de Gruyter.

[10]Doncaster, L. 1912. The chromosomes in the oogenesis and spermatogenesis of *Pieris brassicae*, and in the oogenesis of *Abraxas grossulariata*. *J. Genet.* 2:189—200.

[11]Fujii, T. , and T. Shimada. 2007. Sex determination in the silkworm, *Bombyx mori*: A female determinant on the W chromosome and the sex-determining gene cascade. *Sem. Cell Dev. Biol.* 18:379—88.

[12]Fuková, I. , P. Nguyen, and F. Marec. 2005. Codling moth cytogenetics: Karyotype, chromosomal location of rDNA, and molecular differentiation of sex chromosomes. *Genome* 48:1083—92.

[13]Fuková, I. , W. Traut, M. Vítková, P. Nguyen, S. Kubíčková, and F. Marec. 2007. Probing the W chromosome of the codling moth, *Cydia pomonella*, with sequences from microdissected sex chromatin. *Chromosoma* 116:135—45.

[14]Gage, L. P. 1974. Polyploidization of the silk gland. *J. Mol. Biol.* 86:97—108.

[15]Gassner, G. , and D. J. Klemetson. 1974. A transmission electron microscope examination of hemipteran and lepidopteran gonial centromeres. *Can. J. Genet. Cytol.* 16:457—64.

[16]Gotter, A. L. , J. D. Levine, and S. M. Reppert. 1999. Sex-linked *period* genes in the silkmoth, *Antheraea pernyi*: Implications for circadian clock regulation and the evolution of sex chromosomes. *Neuron* 24:953—65.

[17]Grimaldi, D. A. , and M. S. Engel. 2005. *Evolution of the insects*. New York: Cambridge Univ. Press. Kawaguchi, E. 1934. Zytologische Untersuchungen am Seidenspinner und seinen Verwandten. II. Spermatogenesese bei *Antheraea yamamai* Guérin, *Antheraea pernyi* Guérin, und ihrem Bastard. *Jpn. J. Genet.* 10:135—51.

[18]Kawamura, N. 1988. The egg size determining gene, *Esd*, is a unique morphological marker on

the W chromosome of *Bombyx mori*. *Genetica* 76:195—201.

[19]Kawamura, N. 1990. Is the egg size determining gene, *Esd*, on the W chromosome identical with the sex-linked giant egg gene, *Ge*, in the silkworm? *Genetica* 81:205—10.

[20]Klingstedt, H. 1931. Digametie bei Weibchen der Trichoptere *Limnophilus decipiens* Kol. *Acta Zool. Fennica* 10:1—69.

[21]Kristensen, N. P., and A. W. Skalski. 1999. Phylogeny and palaeontology. In *Lepidoptera, moths and butterflies. 1. Evolution, systematics, and biogeography*, ed. N. P. Kristensen, 7—25. *Handbook of Zoologie*, Vol. 4, Part 35, *Arthropoda: Insecta*. Berlin and New York: Walter de Gruyter.

[22]Lukhtanov, V. A. 2000. Sex chromatin and sex chromosome systems in non-ditrysian Lepidoptera (Insecta). *J. Zool. Syst. Evol. Res.* 38:73—79.

[23]Lukhtanov, V. A., and A. D. Dantchenko. 2002. Principles of the highly ordered arrangement of metaphase I bivalents in spermatocytes of *Agrodiaetus* (Insecta, Lepidoptera). *Chromosome Res.* 10:5—20.

[24]Lukhtanov, V. A., N. P. Kandul, J. B. Plotkin, A. V. Dantchenko, D. Haig, and N. E. Pierce. 2005. Reinforcement of pre-zygotic isolation and karyotype evolution in *Agrodiaetus* butterflies. *Nature* 436:385—89.

[25]Lukhtanov, V. A., R. Vila, and N. P. Kandul. 2006. Rearrangement of the *Agrodiaetus dolus* species group (Lepidoptera, Lycaenidae) using a new cytological approach and molecular data. *Insect Syst. Evol.* 37:325—34.

[26]Marec, F. 1996. Synaptonemal complexes in insects. *Int. J. Insect Morphol. Embryol.* 25:205—33.

[27]Marec, F., and K. Novák. 1998. Absence of sex chromatin correponds with a sex-chromosome univalent in females of Trichoptera. *Eur. J. Entomol.* 95:197—209.

[28]Marec, F., A. Tothová, K. Sahara, and W. Traut. 2001. Meiotic pairing of sex chromosome fragments and its relation to atypical transmission of a sex-linked marker in *Ephestia kuehniella* (Insecta: Lepidoptera). *Heredity* 87:659—71.

[29]Marec, F., and W. Traut. 1994. Sex chromosome pairing and sex chromatin bodies in W-Z translocation strains of *Ephestia kuehniella* (Lepidoptera). *Genome* 37:426—35.

[30]Mediouni, J., I. Fuková, R. Frydrychová, M. H. Dhouibi, and F. Marec. 2004. Karyotype, sex chromatin and sex chromosome differentiation in the carob moth, *Ectomyelois ceratoniae* (Lepidoptera: Pyralidae). *Caryologia* 57:184—94.

[31]Morse, J. C. 1997. Phylogeny of Trichoptera. *Annu. Rev. Entomol.* 42:427—50.

[32]Murakami, A., and H. T. Imai. 1974. Cytological evidence for holocentric chromosomes of the silkworms, *Bombyx mori* and B. *mandarina* (Bombycidae, Lepidoptera). *Chromosoma* 47:167—78.

[33]Niimi, T., K. Sahara, H. Oshima, Y. Yasukochi, K. Ikeo, and W. Traut. 2006. Molecular cloning and chromosomal localization of the *Bombyx Sex-lethal* gene. *Genome* 49:263—68.

[34]Nilsson, N.-O., C. Löfstedt, and L. Dävring. 1988. Unusual sex chromosome inheritance in six species of small ermine moths (*Yponomeuta*, Yponomeutidae, Lepidoptera). *Hereditas* 108:259—65.

[35]Nokkala, S. 1987. Cytological characteristics of chromosome behaviour during female meiosis in *Sphinx ligustri* L. (Sphingidae, Lepidoptera). *Hereditas* 106:169—79.

[36]Ohbayashi, F., M. G. Suzuki, and T. Shimada. 2002. Sex determination in *Bombyx mori*. *Curr. Sci.* 83:466—71.

[37]Rasmussen,S. W. 1977. The transformation of the synaptonemal complex into the "elimination chromatin" in *Bombyx mori* oocytes. *Chromosoma* 60:205—21.

[38]Rathjens,B. 1974. Zur funktion des W-chromatins bei *Ephestia kuehniella* (Lepidoptera): Isolierung und charakterisierung von W-chromatin-mutanten. *Chromosoma* 47:21—44.

[39]Rishi,S. ,G. Sahni,and K. K. Rishi. 1999. Inheritance of unusual sex chromosome evidenced by A^W Z sex trivalent in *Trabala vishnu* (Lasiocampidae,Lepidoptera). *Cytobios* 100:85—94.

[40]Robinson,R. 1971. *Lepidoptera genetics*. Oxford: Pergamon Press.

[41]Sahara,K. ,F. Marec,U. Eickhoff,and W. Traut. 2003a. Moth sex chromatin probed by comparative genomic hybridization (CGH). *Genome* 46:339—42.

[42]Sahara,K. ,A. Yoshido,N. Kawamura,et al. 2003b. W-derived BAC probes as a new tool for identification of the W chromosome and its aberrations in *Bombyx mori*. *Chromosoma* 112:48—55.

[43]Sahara,K. ,A. Yoshido,F. Marec,et al. 2007. Conserved synteny of genes between chromosome 15 of *Bombyx mori* and a chromosome of *Manduca sexta* shown by five-color BAC-FISH. *Genome* 50:1061—65.

[44]Schütt,C. ,and R. Nöthiger. 2002. Structure,function and evolution of sex-determining systems in dipteran insects. *Development* 127:667—77.

[45]Schulz,H.-J. ,and W. Traut. 1979. The pachytene complement of the wildtype and a chromosome mutant strain of the flour moth,*Ephestia kuehniella* (Lepidoptera). *Genetica* 50:61—66.

[46]Scriber,J. M. ,R. H. Hagen,and R. C. Lederhouse. 1996. Genetics of mimicry in the tiger swallowtail butterflies,*Papilio glaucus* and *P. canadensis*(Lepidoptera: Papilionidae). *Evolution* 50:222—36.

[47]Seiler,J. 1914. Das Verhalten der Geschlechtschromosomen bei Lepidopteren. Nebst einem Beitrag zur Kenntnis der Eireifung,Samenreifung und Befruchtung. *Arch. Zellforsch.* 13:159—269.

[48]Suomalainen,E. 1966. Achiasmatische Oogenese bei Trichopteren. *Chromosoma* 18:201—07.

[49]Suomalainen, E. 1969. On the sex chromosome trivalent in some Lepidoptera females. Chromosoma 28:298—308.

[50]Tazima,Y. 1964. *The genetics of the silkworm*. London: Academic Press.

[51]Tothová,A. ,and F. Marec. 2001. Chromosomal principle of radiation-induced F_1 sterility in *Ephestia kuehn*l*iella*(Lepidoptera: Pyralidae). *Genome* 44:172—84.

[52]Traut,W. 1976. Pachytene mapping in the female silkworm *Bombyx mori* L. (Lepidoptera). *Chromosoma* 58:275—84.

[53]Traut,W. 1977. A study of recombination,formation of chiasmata and synaptonemal complexes in female and male meiosis of *Ephestia kuehniella*(Lepidoptera). *Genetica* 47:135—42.

[54]Traut,W. ,and C. A. Clarke. 1997. Karyotype evolution by chromosome fusion in the moth genus *Orgyia*. Hereditas 126:77—84.

[55]Traut,W. ,and F. Marec. 1996. Sex chromatin in Lepidoptera. *Q. Rev. Biol.* 71:239—56.

[56]Traut,W. ,and F. Marec. 1997. Sex chromosome differentiation in some species of Lepidoptera (Insecta). *Chromosome Res.* 5:283—91.

[57] Traut, W. , and C. Mosbacher. 1968. Geschlechtschromatin bei Lepidopteren. *Chromosoma* 25: 343—56.

[58]Traut,W. ,and D. Scholz. 1978. Structure,replication and transcriptional activity of the sex-specific heterochromatin in a moth. *Exp. Cell Res.* 113:85—94.

[59]Traut,W. ,U. Eickhoff,and J.-C. Schorch. 2001. Identification and analysis of sex chromosomes

by comparative genomic hybridization (CGH). *Methods Cell Sci*. 23:157—63.

[60]Traut, W., K. Sahara, and F. Marec. 2007. Sex chromosomes and sex determination in Lepidoptera. *Sex. Dev.* 1:332—46.

[61]Traut, W., K. Sahara, T. D. Otto, and F. Marec. 1999. Molecular differentiation of sex chromosomes probed by comparative genomic hybridization. *Chromosoma* 108:173—80.

[62]Traut, W., A. Weith, G. Traut. 1986. Structural mutants of the W chromosome in *Ephestia* (Insecta, Lepidoptera). *Genetica* 70:69—79.

[63]Van't Hof, A. E., F. Marec, I. J. Saccheri, P. M. Brakefield, B. J. Zwaan. 2008. Cytogenetic characterization and AFLP-based genetic linkage mapping for the butterfly *Bicyclus anynana*, covering all 28 karyotyped chromosomes. *PLoS ONE* 3:e3882.

[64]Vítková, M., I. Fuková, S. Kubíčková, and F. Marec. 2007. Molecular divergence of the W chromosomes in pyralid moths (Lepidoptera). *Chromosome Res*. 15:917—30.

[65]Weith, A., and W. Traut. 1986. Synaptic adjustment, non-homologous pairing, and non-pairing of homologous segments in sex chromosome mutants of *Ephestia kuehniella* (Insecta, Lepidoptera). *Chromosoma* 94:125—31.

[66]Wolf, K. W. 1996. The structure of condensed chromosomes in mitosis and meiosis of insects. *Int. J. Insect Morphol. Embryol.* 25:37—62.

[67]Yasukochi, Y., L. A. Ashakumary, K. Baba, A. Yoshido, and K. Sahara. 2006. A second-generation integrated map of the silkworm reveals synteny and conserved gene order between lepidopteran insects. *Genetics* 173:1319—28.

[68]Yasukochi, Y., L. A. Ashakumary, C. C. Wu, et al. 2004. Organization of the *Hox* gene cluster of the silkworm, *Bombyx mori*: A split of the *Hox* cluster in a non-*Drosophila insect*. *Dev. Genes Evol*. 214:606—14.

[69]Yoshido, A., H. Bando, Y. Yasukochi, and K. Sahara. 2005. The *Bombyx mori* karyotype and the assignment of linkage groups. *Genetics* 170:675—85.

[70]Yoshido, A., F. Marec, and K. Sahara. 2005. Resolution of sex chromosome constitution by genomic in situ hybridization and fluorescence in situ hybridization with $(TTAGG)_n$ telomeric probe in some species of Lepidoptera. *Chromosoma* 114:193—202.

[71]Yoshido, A., Y. Yamada, and K. Sahara. 2006. The W chromosome detection in several lepidopteran species by genomic in situ hybridization (GISH). *J. Insect Biotech. Sericol.* 75:147—51.

第四章 家蚕性染色体和性别决定

Hiroaki Abe, Tsuguru Fujii and Toru Shimada

1. 前言

所有昆虫都是雌雄异体的，它们的性别通常是由基因决定的。已经详细地探讨过果蝇的性别决定的分子和遗传机制(Schütt 和 Nöthiger 2000)，最近已得到蜜蜂 *Apis mellifera* 的性别决定基因(Hasselmann et al. 2008)。

很长时间以前，很多研究者已经做了大量的研究去解决家蚕的性别决定机制问题。性别对产茧的效率有很大的影响，因此，很值得去探索人工控制性别的方法。但是，性别决定机制仍然不清楚。性别决定的最上游元件位于雌性特异的 W 染色体，通过在组织常染色体上的一些基因决定着雌性的自主行为。在这章中，我们首先回顾了 W 染色体的研究历史。然后，我们介绍了 Z 染色体连锁基因的功能，还有常染色体性别决定基因，并简单讨论了在生殖系细胞中性别决定的路径。

2. W 染色体和常染色体间的易位

蚕丝使家蚕成为了重要的经济昆虫。性别决定的研究已经和蚕丝产业需求紧密地联系在一起。在 1906 年，家蚕遗传学的奠基人之一，外山亀太郎(Kametaro Toyama)，认识到了日本和泰国杂交种的杂种优势(Toyama 1906; cited in Tazima in 1964)。在外山的建议下，日本政府决定在养蚕场里强制推行杂交品种并在 1914 年开始分发杂交卵。从那时起，几乎所有日本蚕种场养的蚕都是 F_1 杂交系。为了产生 F_1 杂交代，雌性和雄性必须在蛹期或蛹期之前分开；因此，对卵生产者来说辨别家蚕的性别就变得非常重要了。性别的确定一般是由专业的“辨别者”来确定的，他们的方法是：从末龄幼虫的后腹部上，对雌性来说是检查 Ishiwata 胚盘，对雄性来说是检查 Herald 腺体。但是，这种方法需要技术熟练的专业人员，也不适用于大规模的推广。因此，我们需要一种更加简便实用的性别鉴定方法。

田中(1916)发现家蚕的性别是由性染色体上性连锁基因决定的(Z 连锁)。雌性是 WZ 型(相当于 XY)，雄性是 ZZ 型的(XX；鳞翅目昆虫中拥有异型胚子的性染色体系统的雌性和它的变体请参见第三章)。但是，在性别决定中是 W 染色体还是 Z 染色体起主要的作用仍是不清楚的。后来，通过使用多倍体做的一些实验，桥本发现不管常染色体和 Z 染色体有多少数目，W 染色体对雌性的正向性别决定作用都是很强烈的。多倍体个体，例如，WZZ 和 WWZZ，表现型是正常的雌性。因此，桥本(1933)推断只要有一个 W 染色体存在就可以决定是雌性；W 染色体的存在和功能几乎是确定的。尽管如此，但是在那时，却没有方法从遗传学的角度证明 W 染色体的存在，原因是：虽然已经在 Z 染色体上发现了一些有形态学特征的基因位点，但是没有在 W 染色体上发现。此外，重组(互换)严格地限制在雄性(Maeda 1939)，并且也没有细胞学上的方法去识别 W 染色体。

田岛弥太郎相信，“如果具有显著特征的常染色体基因被人工转移到 W 染色体，那么我们将很容易辨别家蚕的性别。”(Tazima 2001)因此，人们开始了在染色体和常染色体之间进行基因转移的努力。

在研究由 X 射线引起的染色体畸变中，田岛弥太郎获得了一个品系，它们的第二号染色体的两个拷贝发生了融合(分别用 $+^{p}$ 和 p^{Sa} 来作为正常的标记和黑色的标记)。在 1941 年，田岛发现了在 W 染色体和染色体 2 的复合形式($+^{p}p^{Sa}$)之间的基因位移。他把这批蚕分为三类：所有的雌性显示（$+^{p}p^{Sa}$）的特征，所有的雄性显示（p）的特征，没有显示（$+^{p}p^{Sa}$）特征的雄性。他认为这批蚕的复合染色体（$+^{p}p^{Sa}$）已经融合到了 W 染色体上，田岛弥太郎用这批雌性和 p 的雄性测交，然后观察在下一代中的分离模式，从而检测了他的想法。他把有 W 常染色体基因转移的这一品系命名为 W-PSA 品系(技术上称作 T(W;2) $p^{Sa}+^{p}$；Tazima 1941，1944)。作为一个结果，田岛弥太郎证实了桥本的观点，得出了家蚕的雌性是由 W 染色体的存在或缺失所决定的结论。

在田岛实验成功的鼓舞下，一些研究者接受了新的挑战，他们想再诱导出一些性状的（卵色，幼虫标记，还有茧色和血色）等位基因的转移。像在下面探讨的一样，获得了几种 W-基因转移品系，例如，T(W；10) $+^{w-2}$，T(W；2) p^{Sa}，T(W；2) p^{B}，T(W；2)Y。对 W 染色的修饰和分析主要集中于上述结果品系的实际应用。因此，W 染色体的分子生物学研究很多年来一直没有开展。尽管如此，W-基因转移品系对 W 染色体的分子生物学研究也起到了意外的帮助。

性别决定研究中的一个丰硕成果就是源于对遗传嵌合体品系的研究(Goldschmidt 和 Katsuki 1931)。嵌合体品系现在常常作为发育生物学和病理学的研究主题(Mine et al. 1983； Shimada，Ebinuma 和 Kobayashi 1986； Abe，Kobayashi 和 Watanabe 1990)。使用 *mo* 基因和 T(W；2) p^{Sa}，T(W；2) p^{B} 及 T(W；10) $+^{w-2}$，性嵌合体在雌性和雄性之间含有明显界限可以在胚胎、幼虫、成虫阶段得到区分。使用性嵌合体的形态学和生化系研究很强烈地暗示家蚕中缺失如性激素一样可散播的物质(Mine et al. 1983； Fujii 和 Shimada 2007)。

家蚕性染色体一个有趣的特点是：在 Z 染色体中有很多有功能性的基因(Koike et al. 2003)，相反 W 染色体上除了预测的 *Fem* 基因外，很少有功能性基因，这个 *Fem* 基因决定着雌性。W 染色体的另一特征是它不和 Z 染色体或常染色体之间发生重组。由于没有重组，W 染色体处于静止状态，在母女之间传递时不会有什么变化。因此，W 染色体的结构应该是异常稳定的。

3. W 染色体的分子和基因组学分析

对 W 染色体的研究一直都是基于对这些 W 染色体修饰品系的实际应用。尽管对这些品系的基因组 DNA 进行分析已经变得很容易，但多年来一直没有开展对 W 染色体的分子生物学研究。家蚕基因组的研究始于 20 世纪 90 年代。Promboon 等人(1995)构建了 RAPDs 的连锁图，安河内(1998)构建了基于 1018 个分子标记的高密度遗传图谱。但是以上两个连锁图中都没有得到 W 染色体的 RAPD 标记。为了对 W 染色体进行分子生物学分析，阿部等人(1995)试图使用一大组随机的 10-mer 引物去获得 W 染色体的 RAPD 标记，最后获得了 12 组雌特异的 RAPD 标记(W-Kabuki，W-Kamikaze，W-Samurai，BMC1-Kabuki，W-Rikishi，W-Musashi，W-Sasuke，W-Sakura，W-Yukemuri-L，W-Yukemuri-S，W-Bonsai 和 W-Mikan；Abe et al. 1998a；2005b)。随后，他们得到了含有 W-Kabuki 和 W-Samurai RAPD 序列的 λ 噬菌体克隆。令人惊奇的是，W 特异的 RAPD 标记和噬菌体克隆的序列由一个很多反转录原件的嵌套结构构成(Abe et al. 1998a，1998b，2005b；Ohbayashi et al. 1998)。从这些结果中，可以预见 W 染色体完全是由重复序列构成的。另外，可以预期，由于 WZ 染色体对的存在，如果雌性的基因组 DNA 被用于 WGS 分析的话，那么序列数据的组装将会变得非常复杂和困难。因此，不管中国还是日本，都用雄性去进行 WGS 序列分析(Mita et al. 2004；Xia et al. 2004)。出于这些原因，家蚕的 W 染色体的基因组研究变成了一个薄弱的领域。

从 12 个 W 特异的 RAPD 标记中推导出的氨基酸序列中，有 11 个序列和从其他的几种生物中的许多反转座原件序列推出的氨基酸序列显示了相似性(Abe et al. 1998a；2005b)。而且，几乎所有的氨基酸序列都含有 2 个或 3 个反转录原件的边界(Abe et al. 1998a；2005b)。

4. 使用分子标记对畸变 W 染色体的比较

4.1 T(W;3)*Ze* 染色体型(限性 *Zebra*)

一些研究者接受了在 W 染色体和带有可见特征显性等位基因的常染色体之间诱导出新的染色体片段移位的挑战。桥本(1948)在 X 光的诱导下后，分离了一个带有 *Ze* 标记(*Zebra*)的原 3 号染色体的一个片段，现在易位到了 W 染色体。在带有 T(W;3)*Ze* 的品系中，雌性幼虫带有斑纹标记，相反雄性幼虫只有发白的体表。研究发现在 T(W;3) *Ze* 品系的 W 染色体区，缺少了 12 个 W 特异性 RAPD 标记(W-Mikan 和 W-Samurai；Abe et al. 2005b；图 4.1)中的 2 个。

4.2 T(W;10)$+^{w-2}$染色体型(限性黑卵)

多样化的卵色(绒毛膜色)是家蚕所特有的特征。卵色来源于绒毛膜细胞所产生的色素，绒毛膜是覆盖于胚胎上的一层单层膜。绒毛膜细胞起源于卵裂核，所以它的基因型和胚胎的是完全一致的。正常的卵色是茶褐色(黑色)着色，相反，有一种 *w*-2(*white* 2 gene)突变体，卵呈淡黄色-白色着色。田岛、原田、和太田(1951)在用 X 射线照射后，分离到一个易位到 W 染色体的带有$+^{w2}$基因的 10 号染色体片段。在这个品系中，白卵是雄性且基因型是 Z/Z，*w*-2/ *w*-2；黑卵是雌性且基因型是 Z/ T(W；10)$+^{w2}$, *w*-2/*w*-2。T(W；10)$+^{w2}$染色体品系的 W 染色体区域缺少 12 个 W 特异性 RAPD 标记(Abe et al. 2005b；图 4.2)中的一个(W-Mikan)。

4.3 T(W;2)Y 染色体型(限性黄茧)

日本得到了许多茧色的突变体。在日本工业用饲养的蚕吐的丝几乎都是白色的，所以人们习惯上认为白色的茧(无色茧)为正常的茧。在漫长的养蚕史上，曾经观察和保藏了拥有其他茧色的品系，包括黄、金黄、桃红和绿色茧。黄色和桃红色茧源于桑叶中的类胡萝卜素、胡萝卜素、叶黄素的存在。绿色茧源于类黄酮的存在，黄血色的幼虫一般说来也是黄色茧(Doira 1978)。Y 是一个叫黄血基因 *yellow blood* gene(*Y*，2－25.6)的等位基因。有 Y 等位基因的幼虫，由于类胡萝卜素从桑叶经消化器官进入血液，所以血淋巴呈深黄色。

为了构建一个限性黄茧品系，木村、原田和青木(1971)用了 9 年的时间(1961～1969)，使用 γ 射线或 X 射线照射 707 个拥有正常的 W 染色体和正常的 Y 等位基因的雌蛹系。通过这项实验，他们不仅得到了 2 号染色体上带有 Y 等位基因的片段易位到 W

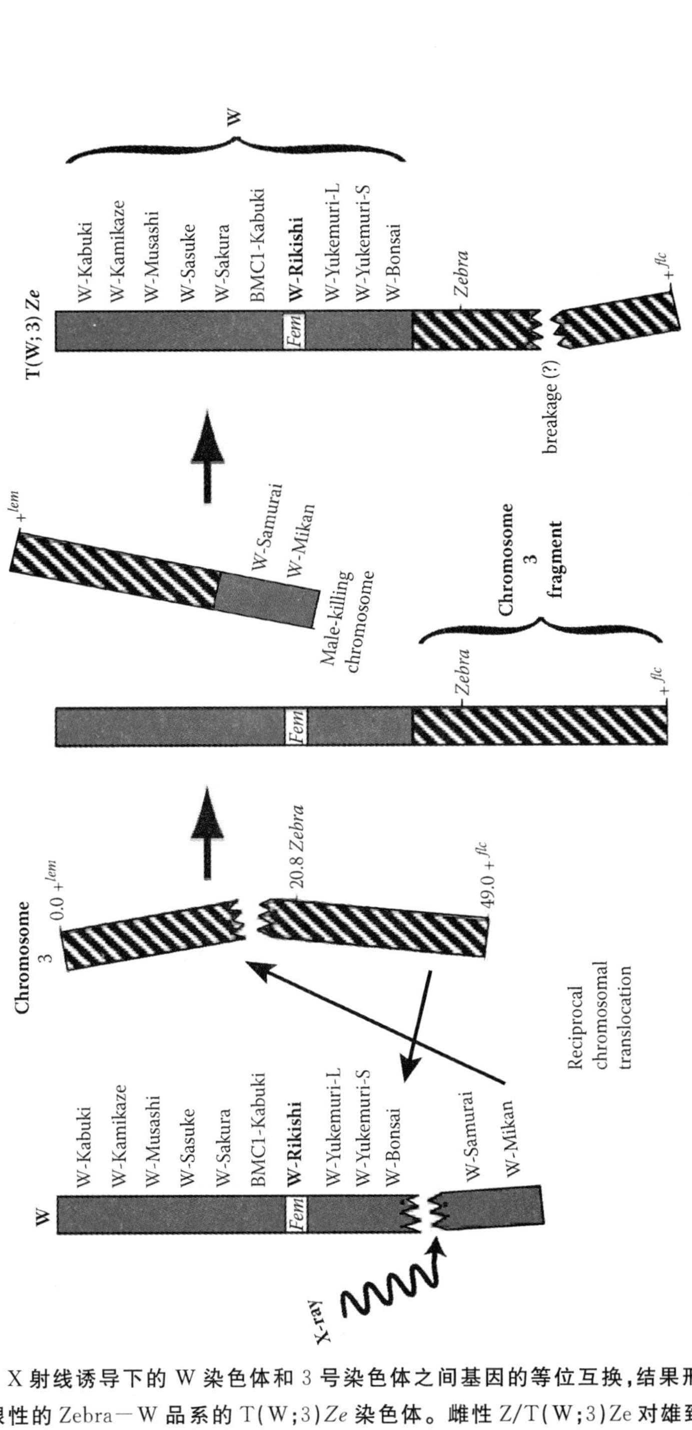

图 4.1　显示了 X 射线诱导下的 W 染色体和 3 号染色体之间基因的等位互换，结果形成了一个雄致死的染色体和限性的 Zebra—W 品系的 T(W;3)*Ze* 染色体。雌性 Z/T(W;3)Ze 对雄致死染色体有耐受性，但是对雄性 Z/Z 是致死的。这暗示了 W 染色体交换到 3 号染色体的那部分片段含有对雄性发育有害的因子(Hashimoto 1953)。遗憾的是，雄性致死染色体在人们能够详细研究它们前就丢失了(Y. Tazima. pers. comm)。

染色体上的品系，而且得到了一个限性的黄茧品系(T(W;2)Y)，这个品系的雌性是黄茧(Z/ T(W;2)Y ，$+^{Y}/+^{Y}$)，雄性是白茧(Z/ Z，$+^{Y}/+^{Y}$)。但是，接下来在对这种品系的商业化饲养过程中，雌性品种出现了几种生理缺陷(体弱，茧壳轻，直肠突出的幼虫)(Niino et al. 1987)。这些缺陷被认为是由 T(W;2)Y 品系 2 号染色体片段的额外部分引起的。为了去除这些额外部分,Niino 等人(1988)用 γ 射线照射这个品系,得到了一个衍生品系,发现 2 号染色体易位片段的非 Y 基因部分已被删除。因此 T(W;2)Y 染色体品系重新被射线所修正。研究者们可以得到几种修正的 T(W;2)Y 染色体品系。

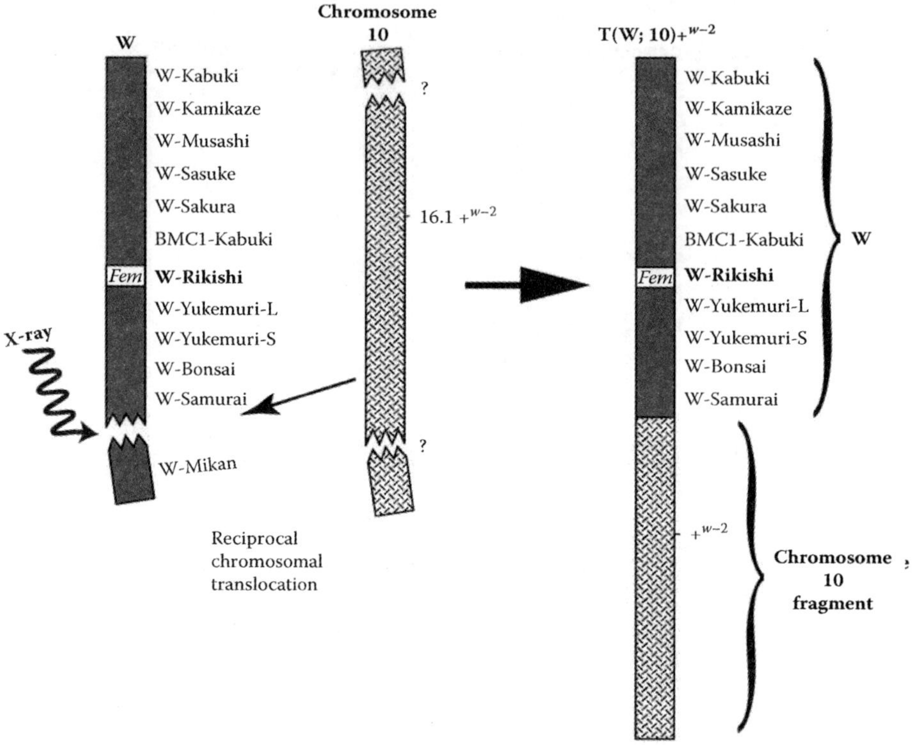

图 4.2　显示 X 射线引导下在 W($+^{W-2}$)染色体和 10 号染色体之间的互换,结果是产生了限性黑卵—W 品系的 T(W;10)$+^{W-2}$染色体。

所有的品系含有三种 T(W;2)Y 染色体中的一种——T(W;2)Y-Abe,T(W;2)Y-Chu 和 T(W;2)Y-Ban——分别由三个独立的研究小组获得。Abe 等(2008)确定了这三种 T(W;2)Y 染色体的 W 染色体特异性 RAPD 标记的数目。T(W;2)Y-Chu 含有这 12 个标记中的 6 个(W-Rikishi,W-Yukemuri-L,W-Yukemuri-S,W-Bomsai,W-Samurai 和 W-Mikan)。但是,另外两个(T(W;2)Y-Abe,T(W;2)Y-Ban)都只含有 12 个标记中的一个(W-Rikishi;图 4.3)。

4.4 $p^{Sa}+^{P}$ W $+^{od}$和 Df($p^{Sa}+^{p}$ W $+^{od}$)*Fem*(雌致死)染色体型

田岛(1948)得到一个复杂的易位片段,$p^{Sa}+^{P}$W $+^{od}$,推测是含有$+^{od}$基因(*od*:清晰半透明,1(Z 染色体)—49.6)的 Z 染色体小片段和 $p^{Sa}+^{P}$W 染色体连接的产物。在一系列使用 $p^{Sa}+^{P}$W $+^{od}$染色体系做的分离实验中,他发现了一种例外的个体,它的表现型是雄性,在用 X 射线照射 Z/$p^{Sa}+^{P}$W $+^{od}$ 雌性后代中用$+^{P}$,p^{Sa},$+^{od}$来标记。田岛(1952)推测这种例外的雄性有一个突变的 W 染色体是由 $p^{Sa}+^{P}$W $+^{od}$染色体品系的雌性决定区域的删除所产生的。为了研究方便,突变 W 染色体被藤井命名为 Df($p^{Sa}+^{P}$ $+^{od}$)*Fem*(2006),然后他在雄性的减数分裂时期用表现型和分子标记对 Df($p^{Sa}+^{P}$W

$+^{od}$)*Fem* 染色体系做了遗传学行为的研究。这些研究揭示了 Df($p^{Sa}+^{P}W+^{od}$)*Fem* 染色体和“删除的 Z 染色体”的一部分连接起来了，这种融合的染色体在雄性的减数分裂中表现为 Z 染色体的性质。藤井等(2006) 把这种非常复杂的染色体命名为 DfZ-DfW，确定了它不仅含有 W 染色体的左边，而且含有 12 个 W 特异性 RAPD 标记的中的 3 个(W-Mikan，W-Samurai 和 W-Bonsai)。这些结果揭示了 W 特异性 RAPD 标记是按照 W-Mikan，W-Samurai 和 W-Bonsai 的顺序排列的(从 W 染色体的左侧末端开始；图 4.4)。

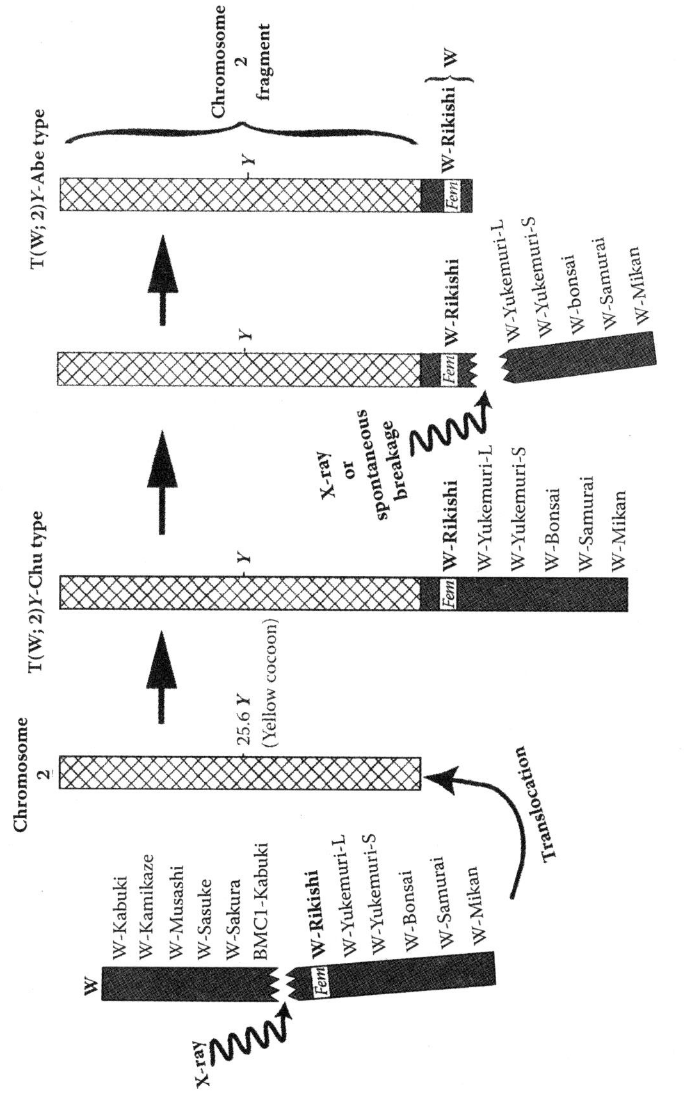

图 4.3 **图示了限性黄茧 W 品系的产生。X 射线的诱导产生** T(W;2)*Y*-Chu **和** T(W;2)*Y*-Abe **染色体品系。*Y*，2 号染色体上的黄血基因。**

4.5 W(B-YL-YS)*Ze* 染色体型

阿部等(2008)使用 X 射线照射限性的 Zebra-W 品系的雌性蛹，它们的染色体型是(T(W;3)*Ze*)，最后分离出了一个例外的带斑纹的雄性幼虫。在雄蛾和雌蛾交配之后，从

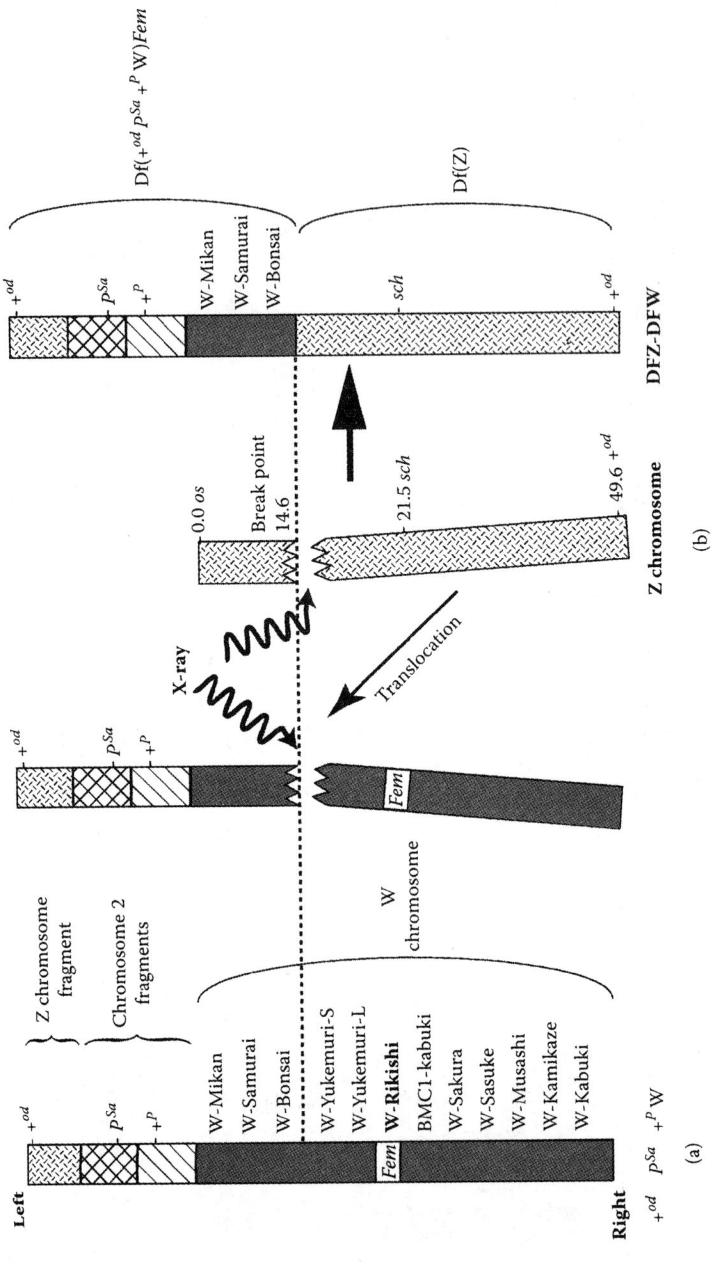

图 4.4　图示的是 X 射线诱发的在$+^{od}\ p^{Sa}+^{P}W$ 染色体和 Z 染色体之间的片段易位。(a)图示的是$+^{od}\ p^{sa}+^{p}W$ 染色体的起源。W 特异性 RAPD 标记的位置在 W 染色体右边标出。*Fem* 是推测的雌决定基因。(b)图示的是 X 射线诱导的$+^{od}\ p^{sa}+^{p}W$ 染色体的破坏(左图),Z 染色体的$+^{od}$,*sch* 基因(中图),结果产生了 DfZ-DfW 染色体(右图)。

雄蛾的腿上提取了基因组 DNA,作为 PCR 的模板。在雄蛾中检测到三个 W 染色体特异的 RAPD 标记(W-Bonsai,W-Yukemuri-L,W-Yukemuri-S)。这些结果暗示了在 T(W;3)*Ze* 染色体中出现过断裂,产生了如图 4.5 所示的含有 W-Bonsai,W-Yukemuri-L 和 W-

Yukemuri-S 的派生 W;3 的易位片段。阿部等(2008)把它命名为 W(B-YL-YS)*Ze* 染色体型。

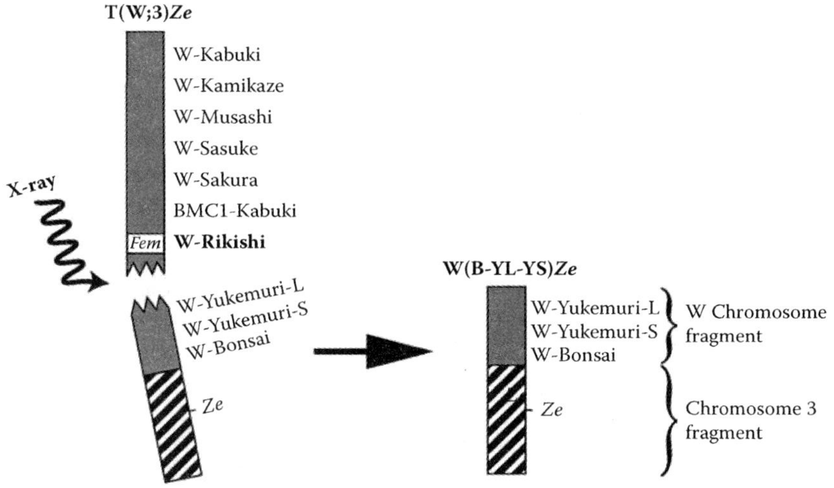

图 4.5　图示的是在 X 射线的诱导下由 T(W;3)*Ze* 染色体型产生 W(B-YL-YS)*Ze* 染色体型的过程。(具体细节见 Abe et al. (2008))。

4.6 Z_1 和 $Ze^W Z_2$ 染色体型

当我们用 X 射线照射断裂的 T(W;3)*Ze* 的染色体型品系,也得到了命名为 Ze^W 的 W 染色体片段,我们确定 Ze^W 片段含有 3 个 W 染色体特异型 RAPD 标记(W-Bonsai,W-Yukemuri-L 和 W-Yukemuri-S)。意料之外的是,发现 Z 染色体也断裂成一个带有 $+^{sch}$(1－21.5)的大片段(Z_1)和带有 $+^{od}$(1－46.9)的小片段(Z_2)。而且,Ze^W 片段和 Z_2 片段连接起来了。把这个连接起来的染色体片段命名为 $Ze^W Z_2$(图 4.6;Fujii et al. 2007)。

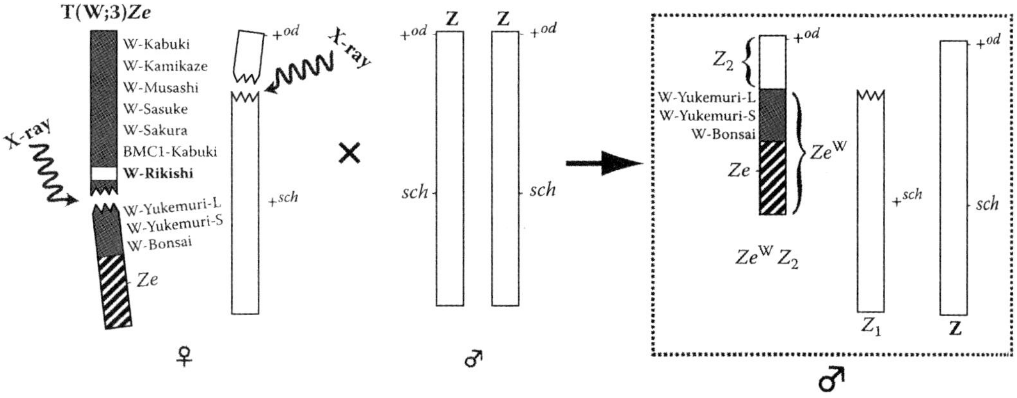

图 4.6　图示的是 X 射线诱导的 T(W;3)*Ze* 染色体的变化的隔离。基因型为 $sch+^{od}/sch+^{od}$ 的雄性和用 X 射线照射的雌性杂交。一个从 T(W;3)*Ze* 染色体上分离到的片段被命名为 Ze^W 片段。我们确信 Ze^W 片段含有 12 个 W 特异 RAPD 标记中的 3 个(W-Bonsai,W-Yukemuri-L 和 W-Yukemuri-S)。Z 染色体分离的长片段带有 $+^{sch}$ 基因位被称为 Z_1(1－21.5),带有 $+^{od}$ 基因位被命名为 Z_2(1－49.6)。有关细节见 Fujii et al. (2007)。

5. 雌决定基因 *Fem* 在 W 染色体的定位

W 染色体特异性 RAPD 标记在 W 染色体变体中的存在或缺失，使推测 *Fem* 基因相对标记位置成为可能。T（W；10）$+^{w-2}$ 染色体型不含有 W-Mikan 标记（Abe et al. 2005b）。T(W;3)*Ze* 染色体型不含有 W-Samurai 或 W-Mikan（Abe et al. 2005b）。雌性致死 DfZ-DfW 染色体型含有 W-Mikan，W-Samurai 和 W-Bonsai（Fujii et al. 2006）。而且，W(B-YL-YS)*Ze* 染色体型含有 W-Bonsai，W-Yukemuri-L 和 W-Yukemuri-S（Abe et al. 2008）。从这些结果中，除了另外两个的相对位置，可以看出有 3 个 W 染色体特异型 RAPD 标记的顺序是可以确定的。从 W 染色体左端开始顺序依次是：W-Mikan，W-Samurai，W-Bonsai 和 W-Yukemuri-L 或 Yukemuri-S（图 4.7）。而且，有证据表明包含这 5 个标记的区域不含有假设的 *Fem* 基因，因为它们的丢失对性别的决定没有影响。

T（W；2）*Y*-Abe 和-Chu 染色体型不包括 W-Kabuki，W-Kamikaze，W-Musashi，W-Sasuke，W-Sakura，或者 BMC1-Kabuki。但是，T（W；2）*Y*-Abe 染色体型含有 W-Rikishi。从这些结果中，W-Kabuki，W-Kamikaze，W-Musashi，W-Sasuke，W-Sakura 和 BMC1-Kabuki 的顺序是不能确定的，但是我们能得出 W-Rikishi 是位于 W 染色体一端的末梢（右端）（图 4.7；Abe et al. 2008）。这些信息可以很好地帮助我们克隆出假设的 *Fem* 基因。

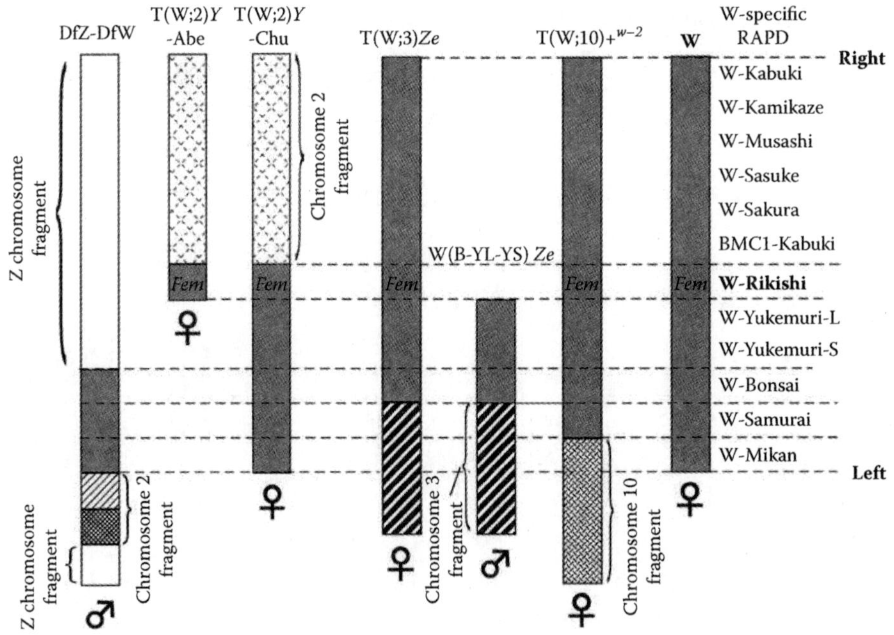

图 4.7 使用 W 染色体的变体，获得的 W 特异性的 RAPD 标记和推测的 *Fem* 基因的高分辨率的定位。6 个（W-Kabuki，W-Kamikaze，W-Musashi，W-Sasuke，W-Sakura 和 BMC1-Kabuki）和 2 个（W-Yukemuri-L，W-Yukemuri-S）W 特异的 RAPD 标记的顺序不能确定。其他的标记定位如图中所示。关于 T（W；3）*Ze*，T（W；10）$+^{W-2}$，DfZ-DfW 染色体型的细节，见阿部等人（2005b）和藤井等（2006）。

6. W 染色体核苷酸序列的特点

为了详细地分析 W 染色体，我们构建了基于 λ 噬菌体的家蚕基因组 DNA 文库。从文库中我们得到了含有 W-Kabuki 和 W-Samurai 的 RAPD 序列的两个 λ 噬菌体的克隆。这两个克隆的 DNA 序列是由许多反转录原件的嵌套结构构成的（Ohbayashi et al. 1998; Abe et al. 2000, 2005b）。

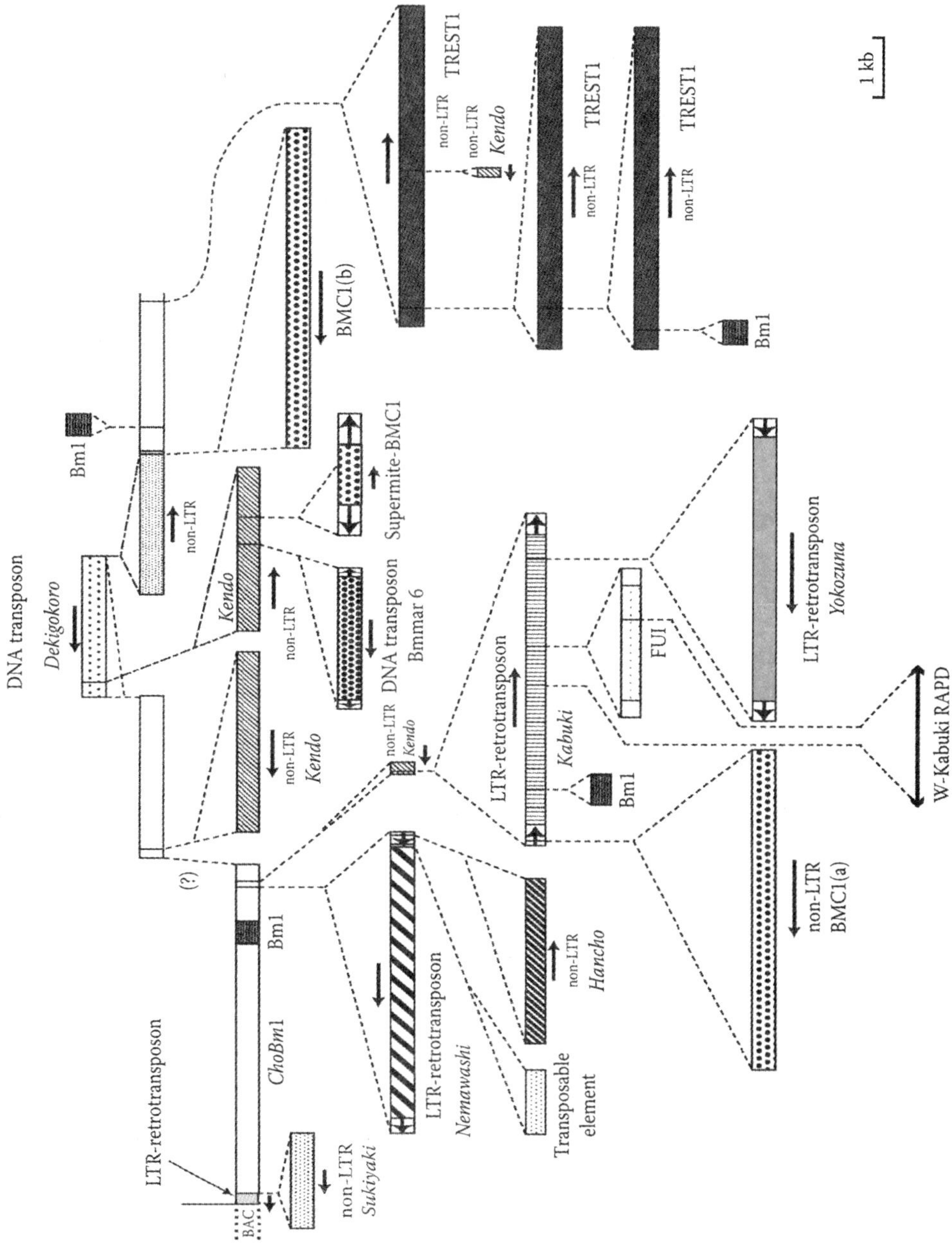

图 4.8　在 W 染色体的序列上存在可转座的原件。在含有 W-Kabuki 标记序列的 BAC 克隆里发现了嵌套的转座原件。这幅图是基于最近的 DNA 序列信息。详见阿部等（2005a）。

在日本，用从家蚕 p50 种系雌雄混合群提取的基因组 DNA 构建了家蚕 BAC 文库，命名为 RPCI-96 文库(Koike et al. 2003)。另外构建了两个家蚕的 BAC 文库，分别使用的是 p50 品系和 C108 品系的基因组 DNA(Wu et al. 1999)。我们用 W 特异的 RAPD 标记鉴定出 W 特异的 BAC 文库克隆，然后用鸟枪法对它们测序。但是，由于存在许多重复的 DNA 元件，特别是 non-LTR 反转录转座子 BMC1(Abe et al. 1998b)和 Kendo(Abe er al. 2002)的存在。发现鸟枪法测序序列的组装是很困难的，并没有得到一个单一的叠连群。然而，克隆序列的小部分被组装成了一个叠连群，并对其进行了生物信息学的数据分析(Abe et al. 2005a)。如图 4.8 中所示，W 染色体的结构特征和其他的染色体的典型结构是不相同的。许多的 LTR 和 non-LTR 反转录转座子，反转录子(Bm1；Adams et al. 1986)，DNA 转座子，以及它们的衍生物都积聚在了 W 染色体上。BAC 克隆和 W-特异的 RAPD 标记的 DNA 序列数据(Abe et al. 1998a，2005a，2005b)，结合由阿部等(2008)得出的结果，认为除了含有推测的 *Fem* 基因的区域外，W 染色体可以被看做一个巨大的无功能的垃圾场。

7. 使用 PCR 标记法对畸变 Z 染色体的结构分析

2004 年，中国的研究小组和日本的研究小组分别独立发表了他们的家蚕 WGS 序列数据(Xia et al. 2004；Mita et al)。通过整合 2 个 WGS 的数据，长 20.4-Mb 的 5 个 scaffold 被分配给 Z 染色体(国际家蚕基因组协会 2008)。这些序列数据(build2)对畸变 Z 染色体的结构分析是非常有用的。

正如上边所提到的，在产生 W 染色体畸变的试验中，我们得到了 2 个 Z 染色体的片段，Z_1 和 Z_2。所以，分析 Z_1 和 Z_2 断裂点附近的区域是可能的。因此，我们同时分析了 Z^{Vg} 和 $+^{od}$ p^{Sa} $+^{p}$ W 染色体，还有以前获得的畸变的 Z 染色体(Fujii et al. 2007)。

7.1 Z_1 和 Z_2($Ze^{W}Z_2$)染色体型

早先，我们描述了 X 射线诱导的 Z 染色体的删除，命名为 Z_1 和 Z_2($Ze^{W}Z_2$)，是由 *sch* 基因位点(1－21.5)和 $+^{od}$ 基因位点(1－49.6)的断裂形成的(图 4.6)。然后我们得到了胚胎致死型卵(Z_1/W)。基于 build2 序列数据，通过使用多种 Z 染色体标记的 PCR 分析，我们发现 Z_1 染色体少了包括近源端的 6-Mb 的部分(Fujii et al. 2007)。

在雄性($Z_1 Ze^{W}Z_2$/Z)和雌性($Z_1 Ze^{W}Z_2$/W)的减数分裂期间我们分析了 Z_1 和 $Ze^{W}Z_2$ 片段的遗传学行为。Z_1 片段和 Z 和 W 染色体的分离是正常的，但是观察到在 $Ze^{W}Z_2$ 片段与 Z 或 W 染色体片段间都是不分离的。而且，由不分离所产生的 2A：Z/ Z_1 雄性中，观察到表现型的缺陷。即 Z/Z_1 雄蛾能爬行，与雌蛾交配，但是不能振翅。为了分析它为什么不能振翅，我们观察到了它的飞行肌发生了曲折，遭到破坏或退化(Fujii et al. 2007)。这种不振翅现象和飞行肌退化的表型不是由 Z_1 引起的，因为雄性 Z/$Z_1 Ze^{W}Z_2$ 能很活跃地振翅。这个结果暗示一种肌肉相关的剂量敏感性基因位于 Z 染色体的删除区域(Fujii et al. 2007)。

7.2 Z^{Vg} 染色体型

Vg 是 X 射线诱导产生的显性基因，结果是引起了成体翅的退化。$+^{Vg}$ *od*/ *Vg* 个体总是显示油蚕 *od* 的表现型，因为 Z 染色体含有 *Vg* 基因（Z^{Vg}染色体），缺少 *od* 基因位点（Tazima 1944）。Z/Z_1 雄性不能振翅，但是 Z/Z^{Vg}能振翅的事实表明 Z^{Vg}染色体的删除区域比 Z_1 删除区域更小。通过 PCR 分析，发现 Z^{Vg}染色体缺少一个 1.5Mb 的空隙区域。而且，成功地得到一个含有断裂点的片段（Fujii et al. 2008）。对片段进行测序使我们确定了精确的近端和远端断裂点。*Vg* 是一个显性突变。因此，如果空隙区被删除的话，那么和翅发育有关的基因的功能将会被阻挠。但是，没有得到预测 build2 基因序列横跨近端和远端的连接点。另一方面，我们在距离远端断裂点很近的地方发现了一个预测基因，与 *apterous*(*ap*)基因相同源，*ap* 基因是果蝇 LIM-HD 家族的成员，在翅的发育过程中起着重要作用。*ap* 基因表达严格限制在翅盘的背部间隔，并对背细胞来说是必需的（Cohen et al. 1992；Diaz-Benjumea 和 Cohen 1993）。*ap* 的异常表达会引起严重的翅功能缺陷（Ricón-Limas et al. 2000）。我们推测家蚕中 *ap* 同源基因的表达被位于这个基因上游的空隙删除所阻挠。

8. Z 染色体在鳞翅目物种中的共线性

安河内等(2006)揭示 *triosephosphate isomerase* 和 *ap* 同源基因定位在 Z 染色体的现象不仅存在于家蚕，也存在于墨尔波墨蝴蝶 *Heliconius melpomene* 中，暗示染色体范围内的共线性存在于大型鳞翅目保守物种中。

众所周知，z 染色体控制化性（滞育）和光周期性（Morohoshi 1980），性连锁 *Lm*（晚熟）位点和化性强烈地联系在了一起。Z 染色体控制滞育也已经在其他的物种得到了报告，包括天蚕蛾 *Samia cynthia*（Saito 1994）和中国柞蚕 *Antheraea pernyi*（Shimada, Yanmauchi 和 Kobayashi 1988）。有趣的是，三个生物钟基因，*BmClk*（果蝇生物钟的同源基因），*Bmper*(*period*)和 *BmCyc*(*cycle*)也都位于 Z 染色体上（http://sgp.dna.affrc.go.jp/KAIKObase/； T. Shimada et al.，unpublished）。只有一个中央时钟系统，*Bmtim*(timeless)位于常染色体。中国柞蚕 *A. pernyi*，*period* 基因也位于 Z 染色体，但是在 W 染色体上有几个不完整的拷贝。正如 Gotter，Levine 和 Reppert（1999)所讨论的那样，*period* 基因的 Z 连锁性可能解释性别出现的不同时间和成虫行为的不同。目前基因组的组装可能会进一步解释鳞翅目昆虫生物钟的性别依赖功能（鳞翅目生理周期的详细机制，见第八章）。

9. 剂量补偿

雄性异型胚子（XY）或半合子（XO）的动物，例如果蝇和线虫，有一个剂量补偿机制去平衡雄性（1 个 X）和雌性（2 个 X）X 染色体的转录数量（Straub 和 Becker 2007）。众所周知，所有影响剂量补偿的突变在果蝇里都是致死的，暗示了剂量补偿机制的重要性(Bhadra et al. 2005)。另一方面，Z染色体的剂量补偿没有在家蚕中观察到。例如，发现来自Z连锁的*T15.180a*和*Bmkettin*基因的mRNA数量，雄性是雌性的2倍(Suzuki，Shimada和Kobayashi 1998，1999)。而且，Koike等(2003)定量了13个靠近*Bmkettin*位点的Z连锁基因的mRNA的表达量，发现有10个在雄性中比雌性中的表达更丰富。这些结果暗示尽管可能有一小部分的基因是剂量补偿的，但是整体上剂量补偿是缺失的。

在果蝇里，至少有5个蛋白质的编码基因(*mof*，*mle*，*msl-1*，*msl-2*和*msl-3*)是必需的(Dahlsveen et al. 2006；Gilfillan et al. 2006)。甄选家蚕EST和基因组序列揭示了预测的*mof*，*mle*和*msl*-3的同源基因。*mof*基因编码一个果蝇染色质重构所必需的组蛋白乙酰转移酶，*mle*编码一个解螺旋酶家族蛋白。相比*mof*，*mle*和*msl-3*基因，*msl-1*和*msl-2*在家蚕基因组中不存在。果蝇*msl-2* mRNA前体在TRA(Transformer)蛋白的帮助下，经历了一个雌性特异性的剪接，它仅仅在雄性里存在功能。果蝇的Msl蛋白形成一个带有2个非编码RNA(*roX1*和*roX2*)的MSL复合体。家蚕缺少*msl-1*，*msl-2*，*roX1*和*roX2*的同源基因的事实可能用于解释家蚕Z连锁基因的染色体范围内剂量补偿的缺失。

为什么家蚕缺少染色体范围内的剂量补偿机制？家蚕Z染色体含有差不多800个基因。众所周知，Z染色体有几个基因对控制雄性化的功能非常重要。一个典型的例子就是编码雌性信息素受体的基因，它位于Z染色体(Sakurai et al. 2004)上并且仅仅在成体雄性的触角中表达。正如Koike等(2003)所讨论的那样，Z染色体含有很多基因是间接飞行肌的组成部件，包括*Bmkettin*和*Bmtitin*。因为雄蛾在飞行和移动方面更加活跃，可以推测肌肉蛋白基因在雄性个体中有偏向性的表达。

10. 果蝇*Doublesex*同源基因*Bmdsx*在性别决定中的角色

正如上边所描述的，家蚕的性别决定首先取决于W染色体的存在或是缺失。W染色体起始了一系列级联反应，能够导致体细胞系和生殖细胞系的雌性模式发育。尽管如此，直到最近几年，我们还不清楚性别决定级联反应的下游机制。其中一个下游基因，家蚕双性基因(*Bmdsx*)，在性别决定中起了一个重要的作用。*Bmdsx*是果蝇*doublesex*(*dsx*)基因的直系同源基因，是果蝇性别分化级联反应下游的一个关键基因(Saccone et al. 2002)。

果蝇中，性别决定的原始信号是X/A的比率，即X染色体和常染色体的数目平衡。X染色体的数目决定了*Sex lethal*(*Sxl*)的表达，*Sxl*是早期胚胎基因中的性别决定的掌控基因。在后胚胎期和胚后时期，依赖于SXL蛋白，*Sxl*被雄特异的剪接阻抑所调控。因此，SXL蛋白仅在雌性中合成。果蝇的SXL蛋白调节了*Sxl*和*transformer*(*tra*)的剪接。功能性的TRA蛋白，仅在雌性里产生，导致了雌特异性的*dsx*剪接，而雄性功能性TRA的缺失产生了雄特异(默认)的*dsx*的剪切。来源于性别特异剪接的雌雄DSX蛋白分别调节雌雄性别分化所必需的基因转录(Yamamoto et al. 1998)。在蜜蜂 *Apis mellifera* 中,染色体的性别决定区域含有性别决定补偿基因(*csd*)和另一个基因,这两个基因都编码 TRA 样 SR 型蛋白(Hasselmann et al. 2008)。这暗示蜜蜂的 *dsx* 也被 TRA 样剪切活化体所调节,尽管 SR 型蛋白和蜜蜂 *dsx* 的相互作用还未见报道。

在家蚕中,*Bmdsx* 是位于 25 号染色体上的单拷贝基因,它类似于果蝇的 *dsx* 被性特异剪接所调控。在这些物种中不同形式的 *Bmdsx* 的 mRNA 分别在雌性和雄性(Ohbayashi et al. 2001)中表达。使用 Hela 细胞提取物,Suzuki 等(2001)发现,*Bmdsx* 剪接的默认模式产生了雌性形式的产物(图 4.9)。这暗示了控制 *Bmdsx* 性别特异剪接的因子不是一个像 TRA 的剪接活化体,而是一些剪接沉默体。很明显,调节 *dsx* 剪接的机制在果蝇和家蚕中是不一样的。

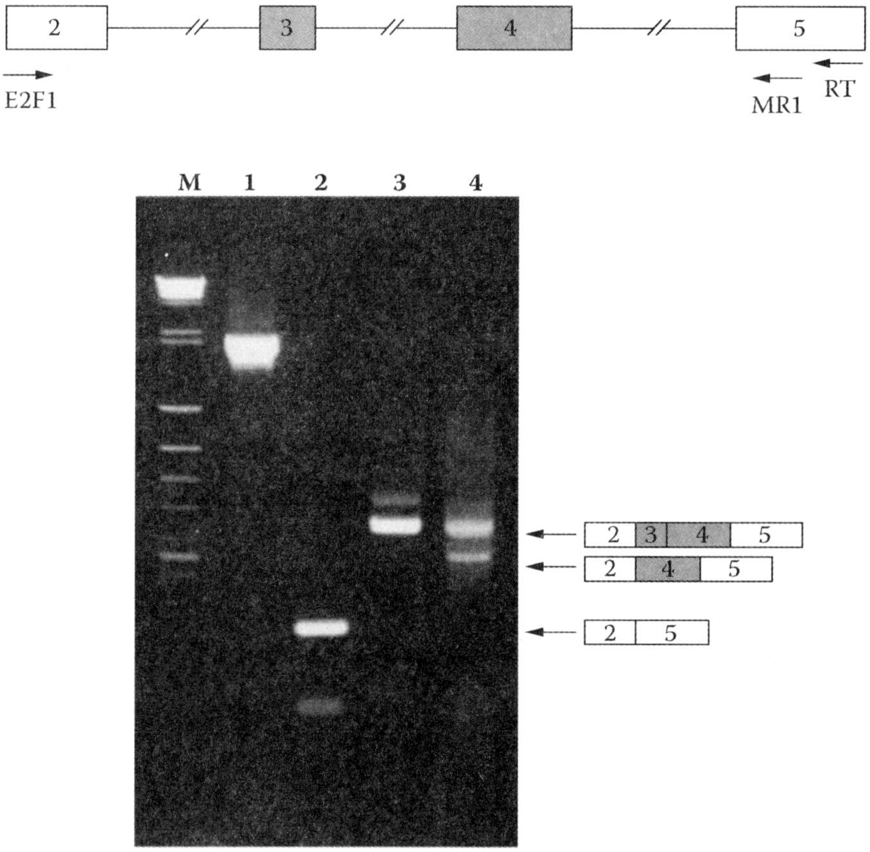

图 4.9 *Bmdsx* **前体** mRNA **的体外剪接产物。前体** mRNA **来源于含有** *Bmdsx*2-5 **外显子的小基因,它在一个含有**(lane 4)**或者不含有**(lane 3) HeLa **细胞核提取物的反应中进行剪接。转录产物用** RT primer **反转录**,cDNAs **用** E2F1 **和** MR1 **引物进行** PCR **扩增。用从精巢**(lane 2)**和卵巢**(lane3)**里提取的** poly(A)+RNA **也同样用相同的** RT-PCR **体系。箭头表示使用** RT-PCT **的引物的位置和方向。**M **代表** DNA **的大小标记。引用自** Suzuki **等人**(2001)。

BmDSX 蛋白含有一个类似锌指结构的 DNA 结合结构域。为了理解 *Bmdsx* 的功能,需知道 BmDSX 蛋白的作用靶标。Suzuki 等(2003)发现在有雌性类型的 *Bmdsx* mRNA 表达的转基因雄蚕中,表达了少量卵黄原蛋白的 mRNA,这在正常的雄性中是检测不到的。相反,在有雄类型的 *Bmdsx* mRNA 表达的转基因雌性中,卵黄原蛋白 mRNA 的表达相对得到了抑制,信息素结合蛋白 mRNA 的表达增加了,并且观察到了一些生殖器官的形态发生了异常(Suzuki et al. 2003,2005)。这些结果强烈地暗示 *Bmdsx* 影响的不仅仅是卵黄原蛋白基因,还有信息素结合蛋白基因的表达。Suzuki 等(2003)发现在位于卵黄原蛋白启动子区域的一个元件,推测它可能是 BmDSX 蛋白的结合位点,实验数据展示了这一元件和雌雄形式的 BmDSX 蛋白之间的分子相互作用。他们也观察到了在转基因蛾中生殖器形态的不正常,这暗示 *Bmdsx* 控制了决定生殖器形态的基因。另外,为了全面理解 *Bmdsx* 的功能,必须在基因组范围中去调查 *Bmdsx* 的靶标。

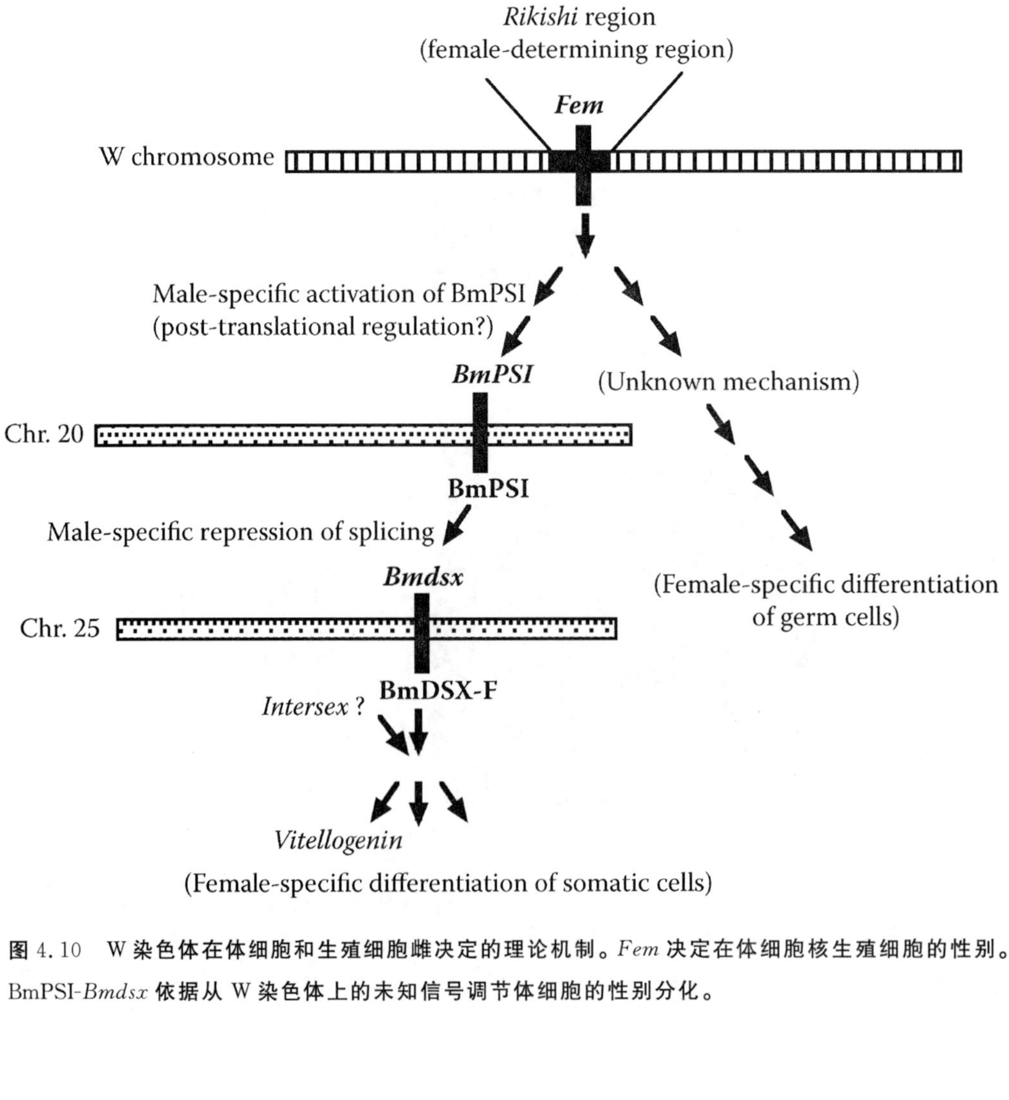

图 4.10 W 染色体在体细胞和生殖细胞雌决定的理论机制。*Fem* 决定在体细胞核生殖细胞的性别。BmPSI-*Bmdsx* 依据从 W 染色体上的未知信号调节体细胞的性别分化。

最近，Suzuki 等(2008)发现了一个位于 *Bmdsx* 第 4 外显子的顺式元件(CE1)，它对雄特异的剪接是必需的。而且，他们发现与 CE1 互相作用的蛋白，BmPSI，果蝇 P 元件体细胞抑制子(PSI)的同源体，含有一个 KH 结构域。用 *Bmpsi* 的 DsRNA 对培养来自雄性的细胞进行 RNAi 试验，诱导产生了一个雌类型的 *Bmdsx* 的 mRNA 的上调表达。他们认为 BmPSI 是 *Bmdsx* 的雄特异的剪接体抑制因子，换言之，是直接的上游因子。*Bmpsi* 基因没有位于性染色体，而是位于 20 号常染色体上。因此，W 染色体及 *Bmpsi* 在性别决定级联中的机制仍然是未知的(图 4.10)。

11. 其他常染色体性别决定基因

Sxl 的序列在包括家蚕的鳞翅目昆虫和双翅目昆虫之间是保守的(Niimi 2006; Traut et al. 2006)。尽管如此，并没有在家蚕的 *Sxl* 同源基因中观察到依赖性别的表达。正如下边所描述的，家蚕基因组中没有 SXL 目标基因 *tra* 的同源基因。而且，*Bmdsx* 的性别特异的剪接体不是被 TRA 所调节，而是被上边所提到的 BmPSI 蛋白所调节。考虑这些证据，我们推测 *Sxl* 可能不是家蚕的性别决定因子。

Suzuki 等(2005)发现利用转基因进行 *Bmdsx* 雌特异形式的异位表达仅诱导家蚕部分的性反转。尤其是试验中仅诱导少量的卵黄原基因 mRNA (Suzuki et al. 2005)的表达。因此，我们可以推测另一个分子与 BmDSX 协同作用产生雌表达基因如卵黄原基因的表达。在果蝇中，已报道雌类型的 DSX 蛋白与雌雄间体(*intersex*，*ix*)基因编码的配偶体蛋白相互作用，从而激活卵黄蛋白基因的表达。家蚕基因组 WGS 组装发现含有 *ix* 基因的同系物(Mita et al. 2004; Siegal 和 Baker 2005)。在家蚕中，尽管目前没有这方面的证据，但这个基因可能与 BmDSX 相互作用来调控卵黄原蛋白和其他基因雌特异性的表达。

家蚕基因组(Mita et al. 2004; Xia et al. 2004)没有 *tra* 基因的同系物，*tra* 在果蝇中编码一个 *dsx* 的调控因子，而 *tra*-2(*tra* 的辅助因子)在家蚕基因组中具有同系物。这与家蚕 *Bmdsx* 不受剪接激活因子调控，而是通过剪接抑制因子 BmPSI 来调控的假说相一致(Suzuki et al. 2001; Suzuki et al. 2008)。

转基因实验(Suzuki et al. 2005)显示在卵巢中生殖细胞发育基本上不受雌特异 Bmdsx 形式转基因的影响。这结果表明家蚕生殖细胞性别决定与体细胞性别决定采用不同的方式，这与果蝇比较类似。在果蝇中，好几个基因包括 *Sxl*、*bam*、*otu*、*ovo* 和 *snf* 调控生殖细胞的性别决定。然而，尽管家蚕基因组中具有 *Sxl*、*otu*、*ovo* 和 *snf* 同源物，但决定果蝇生殖细胞雌方向的重要基因 *bam* 在家蚕基因组中没有。这表明家蚕生殖细胞性别分化途径与果蝇的不同。由于 W 染色体的多倍体和缺失/转座决定体细胞和生殖细胞性别发生的同时进行，因此，一些生殖细胞性别决定未知基因也可能在 W 染色体的控制之下，正如体细胞的性别决定基因一样。

12. 性染色体功能分析和其他鳞翅目昆虫 *Wolbachia* 感染的性别决定

正如上面叙述的，采用性染色体转座来揭示性染色体不同区段或片段的编码功能是必要的。然而，由于在鳞翅目昆虫中染色体较小、数目较多，并且相互间不好区分，所以一般很难分离出染色体转座。而且，当转座发生时，也缺乏遗传标记来鉴定。但是，在螟蛾（*Ephestia kuehniella*）中，大量的性染色体变异包括转座得以分离，并用于随后的研究（Traut 和 Rathjens 1973; Marec 和 Mirchi 1990; 如下）。与家蚕类似，螟蛾性染色体 ZW 为雌，ZZ 为雄。Z 和 W 染色体在性别决定中的作用不清楚。使用光学显微镜，W-Z 二价体在粗线期能通过异染色质线而被识别，这是螟蛾 W 染色体的一个特征（Traut 和 Rathjens 1973）。我们推测一个性染色质体（SB），包含许多浓缩的 W 染色体，可在鳞翅目多倍体雌个体间期中观察到（Ennis 1976; Traut 和 Marec 1996）。在螟蛾中，Traut 和 Rathjens（1973）发现 W 染色体与常染色体的融合伴随着黏性 SB 的断裂（异染色质的组成）。因此，SB 形态成功地被作为一个细胞发生标记来分离由 Y 射线诱导的 T（W;A）染色体（Traut 和 Rathjens 1973; Rathjens 1974）。研究者分析了在卵母细胞粗线期每个 T（W;A）染色体的结构（Traut，Weith 和 Traut 1986）。而且，使用定位于转座常染色体上的常染色体标记基因解析了一些 T（W;A）染色体的特征（Traut，Weith 和 Traut 1986）。

为了使螟蛾 Z 染色体转座到 W 染色体上，即获得 T（W;Z）染色体，Marec 和 Mirchi（1990）将 γ-射线诱导得到的 Z^{dz+}/W 雌个体与 *dz*/*dz*（前翅中心呈灰色）雄个体交配，以筛选 Z^{dz+} 雌个体，这与该物种中性别连锁遗传的正常模式不同。从 Z^{dz+} 雌个体，他们分离到 12 个性染色体突变系，这些突变系 Z 染色体含有其他额外序列，包括一个 $+^{dz}$ 基因（Z^{dz+} 片段），并且其中四个系统中 Z^{dz+} 片段被转座到 W 染色体上。这四个 T（W;Z）转座已被 Marec 和 Traut（1994）实验验证。在接下来的 12 个系统中的 4 个，其 Z^{dz+} 片段与 W 染色体片段的相融合情况见 Marec 等所述（2001）。有趣的是，2 个系统中 Z^{dz+} 片段含有一段 W 染色体的雄个体不能完成胚胎发育，而其他系统中 Z^{dz+} 片段不含或含有一段 W 染色体的雄个体存活。基于这些结果，Marec 等（2001）认为螟蛾 W 染色体的一部分携带或含有杀雄因子，这个因子与正常性别决定途径相互作用。

众所周知，共生（或寄生）细菌 *Wolbachia* 影响许多昆虫和节肢动物的性别决定和生殖。在鳞翅目昆虫中，*Wolbachia* 感染草螟（*Ostrinia furnacalis* 和 *O. scapulalis*）会引起雄性致死（Kageyama 和 Traut 2004; Sakamoto et al. 2007），而在宽边黄粉蝶 sulfur butterfly，*Eurema hecabe* 中，它会引起遗传雄个体（ZZ）转变成雌性（Hiroki et al. 2002）。这意味着 *Wolbachia* 通过一个未知机制与性别决定途径进行相互交流。*Wolbachia* 诱导生殖操控的分子研究可能是研究鳞翅目昆虫性别决定及性染色体在性决定中作用的有用方法。

13. 结论和展望

根据物种和品系的不同，鳞翅目昆虫雌/雄性染色体构成基本上是 WZ/ZZ 或者 Z/

ZZ(Traut,Sahara 和 Marec 2007)。在家蚕和野桑蚕的所有品系中都是 WZ/ZZ 型。在鳞翅目昆虫 W 染色体中,蚕 W 染色体是被解析得最清楚的。但是,还是没人能成功鉴定出雌决定因子。家蚕基因组草图已经完成,基因组的第二次组装数据可在 http://sgp.dna.affrc.go.jp/KAIKObase/获得。但是,还没有进行雄基因组 DNA 的分析以及 W 染色体的全测序。迄今,只知道 W 染色体含有很多转座元件,并且还没有发现蛋白质编码基因(见第三章鳞翅目昆虫 W 染色体的结构、组成及发生)。通过 W 染色体和性别决定机制的深度分析将有可能解答雌决定因子是否为编码一个蛋白质的基因。在家蚕和鳞翅目基因组学中,W 染色体的特征仍是待解答的最大领域之一。

参考文献

[1]Abe,H.,T. Fujii,N. Tanaka,et al. 2008. Identification of the female-determining region of the W chromosome in *Bombyx mori*. *Genetica*. 133:269—82.

[2]Abe,H.,M. Kanehara,T. Terada,et al. 1998a. Identification of novel random amplified polymorphic DNAs (RAPDs) on the W chromosome of the domesticated silkworm,*Bombyx mori*,and the wild silkworm,*B. mandarina*,and their retrotransposable element-related nucleotide sequences. *Genes Genet. Syst*. 73:243—54.

[3]Abe,H.,M. Kobayashi,and H. Watanabe. 1990. Mosaic infection with a densonucleosis virus in the midgut epithelium of the silkworm,*Bombyx mori*. *J. Invertebr. Pathol*. 55:112—17.

[4]Abe,H.,K. Mita,Y. Yasukochi,T. Oshiki,and T. Shimada. 2005a. Retrotransposable elements on the W chromosome of the silkworm *Bombyx mori*. *Cytogenet. Genome Res*. 110:144—51.

[5]Abe,H.,F. Ohbayashi,T. Shimada,T. Sugasaki,S. Kawai,and T. Oshiki. 1998b. A complete full-length non-LTR retrotransposon,BMC1,on the W chromosome of the silkworm,*Bombyx mori*. *Genes Genet. Syst*. 73:353—58.

[6]Abe,H.,F. Ohbayashi,T. Shimada,et al. 2000. Molecular structure of a novel *gypsy*-Ty3-like retrotransposon (*Kabuki*) and nested retrotransposable elements on the W chromosome of the silkworm *Bombyx mori*. *Mol. Gen. Genet*. 263:916—24.

[7]Abe,H.,M. Seki,F. Ohbayashi,et al. 2005b. Partial deletions of the W chromosome due to reciprocal translocation in the silkworm *Bombyx mori*. *Insect Mol. Biol*. 14:339—52.

[8]Abe,H.,T. Shimada,T. Yokoyama,T. Oshiki,and M. Kobayashi. 1995. Identification of random amplified polymorphic DNA on the W chromosome of the Chinese 137 strain of the silkworm,*Bombyx mori* (in Japanese with English summary). *J. Seric. Sci. Jpn*. 64:19—22.

[9]Abe,H.,T. Sugasaki,T. Terada,et al. 2002. Nested retrotransposons on the W chromosome of the wild silkworm Bombyx mandarina. *Insect Mol. Biol*. 11:307—14.

[10]Adams,D. S.,T. H. Eickbush,R. J. Herrera,and P. M. Lizardi. 1986. A highly reiterated family of transcribed oligo(A)-terminated,interspersed DNA elements in the genome of *Bombyx mori*. *J. Mol. Biol*. 187:465—78.

[11]Bhadra,M. P.,U. Bhadra,J. Kundu,and J. A. Birchler. 2005. Gene expression analysis of the

function of the male-specific lethal complex in *Drosophila*. *Genetics* 169:2061—74.

[12]Casper, A., and M. Van Doren. 2006. The control of sexual identity in the *Drosophila* germline. *Development* 133:2783—91.

[13]Cohen, B., M. E. McGufin, C. Pfeile, D. Segal, and S. M. Cohen. 1992. *apterous*, a gene required for imaginal disc development in *Drosophila* encodes a member of the LIM family of developmental regulatory proteins. *Genes Dev*. 6:715—29.

[14]Dahlsveen, I. K., G. D. Gilfillan, V. I. Shelest, et al. 2006. Targeting determinants of dosage compensation in *Drosophila*. *PLoS Genet*. 2:e5.

[15]Diaz-Benjumea, F. J., and S. M. Cohen. 1993. Interaction between dorsal and ventral cells in the imaginal disc directs wing development in *Drosophila*. *Cell* 75:741—52.

[16]Doira, H. 1978. Genetic stocks of the silkworm. In *The silkworm, an important laboratory tool*, ed. Y. Tazima, 53—81. Tokyo: Kodansha Ltd.

[17]Ennis, T. J. 1976. Sex chromatin and chromosome numbers in Lepidoptera. *Can. J. Cytol.* 18:119—30.

[18]Fujii, T., H. Abe, S. Katsuma, et al. 2008. Mapping of sex-linked genes onto the genome sequence using various aberrations of the Z chromosome in *Bombyx mori*. *Insect Biochem. Mol. Biol.* 38: 1072—79.

[19]Fujii, T., and T. Shimada. 2007. Sex determination in the silkworm, *Bombyx mori*: A female determinant on the W chromosome and the sex-determining gene cascade. *Semin. Cell Dev. Biol.* 18:379—88.

[20]Fujii, T., N. Tanaka, T. Yokoyama, et al. 2006. The female-killing chromosome of the silkworm, Bombyx mori, was generated by translocation between the Z and W chromosomes. *Genetica* 127:253—65.

[21]Fujii, T., T. Yokoyama, O. Ninagi, et al. 2007. Isolation and characterization of sex chromosome rearrangements generating male muscle dystrophy and female abnormal oogenesis in the silkworm, *Bombyx mori*. *Genetica* 130:267—80.

[22]Garrett-Engele, C. M., M. L. Siegal, D. S. Manoli, et al. 2002. *intersex*, a gene required for female sexual development in *Drosophila*, is expressed in both sexes and functions together with *doublesex* to regulate terminal differentiation. *Development* 129:4661—75.

[23]Gilfillan, G. D., T. Straub, E. de Wit, F. Greil, et al. 2006. Chromosome-wide gene-specific targeting of the *Drosophila* dosage compensation complex. *Genes Dev*. 290:858—70.

[24]Goldschmidt, R., and K. Katsuki. 1931. Vierte Mitteilung über erblichen Gynandromorphism von *Bombyx mori* L. *Biol. Zbl.* 48:685—99.

[25]Goldsmith, M. R. 1995. Genetics of the silkworm: Revisiting an ancient model system. In *Molecular model systems in the lepidoptera*, ed. M. R. Goldsmith and A. S. Wilkins, 21—76. New York: Cambridge Univ. Press.

[26]Goldsmith, M. R., T. Shimada, and H. Abe. 2005. The genetics and genomics of the silkworm, *Bombyx mori*. *Annu. Rev. Entomol.* 50:71—100.

[27]Gotter, A. L., J. D. Levine, and S. M. Reppert. 1999. Sex-linked *period* genes in the silkmoth, *Antheraea pernyi*: Implications for circadian clock regulation and the evolution of sex chromosomes. *Neuron* 24:953—65.

[28]Hashimoto, H. 1933. The role of the W-chromosome in the sex determination of *Bombyx mori* (in Japanese). *Jpn. J. Genet.* 8:245—47.

[29]Hashimoto, H. 1948. Sex-limited zebra, an X-ray mutation in the silkworm (in Japanese with English summary). *J. Seric. Sci. Jpn.* 16:62—64.

[30]Hashimoto, H. 1953. Genetical studies of *Bombyx mori* L. on the lethal gene which affects the male (in Japanese with English summary). *J. Seric. Sci. Jpn.* 22:200—204.

[31]Hasselmann, M., T. Gempe, M. Schiøtt, C. G. Nunes-Silva, M. Otte, and M. Beye. 2008. Evidence for the evolutionary nascence of a novel sex determination pathway in honeybees. *Nature* 454(7203):519—22.

[32]Hiroki. M., Y. Kato, T. Kamito, and K. Miura. 2002. Feminization of genetic males by a symbiotic bacterium in a butterfly, *Eurema hecabe* (Lepidoptera: Pieridae). *Naturwissenschaften* 89:167—170.

[33]International Silkworm Genome Consortium. 2008. The genome of a lepidopteran model insect, the silkworm *Bombyx mori*. *Insect Biochem. Mol. Biol.* 38:1036—45.

[34]Kageyama, D., and W. Traut. 2004. Opposite sex-specfic effects of *Wolbachia* and interference with the sex determination of its host *Ostrinia scapulalis*. *Proc. Biol. Sci.* 271:251—58.

[35]Kimura, K., C. Harada, and H. Aoki. 1971. Studies on the W-translocation of yellow blood gene in the silkworm (*Bombyx mori*) (in Japanese with English summary). *Jpn. J. Breed.* 21:199—203.

[36]Koike, Y., K. Mita, M. G. Suzuki, et al. 2003. Genomic sequence of a 320-kb segment of the Z chromosome of *Bombyx mori* containing a *kettin* ortholog. *Mol. Genet. Genomics* 269:137—49.

[37]Maeda, T. 1939. Chiasma studies in the silkworm *Bombyx mori*. Jap. J. Genet. 15:118—27.

[38]Marec, F., and R. Mirchi. 1990. Genetic control of the pest Lepidoptera: Gamma-ray induction of translocations between sex chromosomes of *Ephestia kuehniella* Zeller (Lepidoptera: Pyralidae). *J. Stored Prod. Res.* 26:109—16.

[39]Marec, F., A. Tothová, K. Sahara, and W. Traut. 2001. Meiotic pairing of sex chromosome fragments and its relation to atypical transmission of a sex-linked marker in *Ephestia kuehniella* (Insecta: Lepidoptera). *Heredity* 87:659—71.

[40]Marec, F., and W. Traut. 1994. Sex chromosome pairing and sex chromatin bodies in W-Z translocation strains of *Ephestia kuehniella* (Lepidoptera). *Genome* 37:426—35.

[41]Mine, E., S. Izumi, M. Katsuki, and S. Tomino. 1983. Developmental and sex-dependent regulation of storage protein synthesis in the silkworm, *Bombyx mori*. Dev. Biol. 97:329—37.

[42]Mita, K., M. Kasahara, S. Sasaki, et al. 2004. The genome sequence of silkworm, *Bombyx mori*. *DNA Res.* 11:27—35.

[43]Morohoshi, S. 1980. The control of growth and development in *Bombyx mori*. XLI: Control of hormonal antagonistic balance regarding insect development by brain hormone. *Proc. Jpn. Acad. Ser. B*: 56:200—05.

[44]Niimi, T., K. Sahara, H. Oshima, et al. 2006. Molecular cloning and chromosomal localization of the *Bombyx Sex-lethal* gene. *Genome* 49:263—68.

[45]Niino, T., R. Eguchi, A. Shimazaki, and A. Shibukawa. 1988. Breakage by γ-rays of the +*i-lem* locus on the translocated 2nd chromosome in the sex-limited yellow cocoon silkworm. *J. Seric. Sci. Jpn.*

57:75—76.

[46]Niino,T. ,T. Kanda,R. Eguchi,et al. 1987. Defects and structure of translocated chromosome in the sex-limited yellow cocoon strain of the silkworm, *Bombyx mori* (in Japanese with English summary). *J. Seric. Sci. Jpn.* 56:240—46.

[47]Ohbayashi,F. ,T. Shimada,T. Sugasaki,et al. 1998. Molecular structure of the *copia*-like retrotransposable element *Yokozuna* on the W chromosome of the silkworm, *Bombyx mori*. *Genes Genet. Syst.* 73:345—52.

[48]Ohbayashi,F. M. ,G. Suzuki,K. Mita,et al. 2001. A homologue of the *Drosophila doublesex* gene is transcribed into sex-specific mRNA isoforms in the silkworm, *Bombyx mori*. *Comp Biochem. Physiol. B Biochem. Mol. Biol.* 128:145—58.

[49]Osanai,M. ,H. Takahashi,K. K. Kojima,et al. 2004. Essential motifs in the 3′ untranslated region required for retrotransposition and the precise start of reverse transcription in non-long-terminal-repeat retrotransposon SART1. *Mol. Cell. Biol.* 24:7902—13.

[50]Promboon,A. ,T. Shimada,H. Fujiwara,and M. Kobayashi. 1995. Linkage map of random amplified polymorphic DNAs (RAPDs) in the silkworm, *Bombyx mori*. *Genet. Res.* 66:1—7.

[51]Rathjens,B. 1974. Zur Funktion des W-Chromatins bei *Ephestia kuehniella* (Lepidoptera): Isolierung und Charaktesisierung von W-Chromatin-Mutanten. *Chromosoma* 47:21—44.

[52]Rincón-Limas,D. E. ,C. -H. Lu,I. Canal,and J. Botas. 2000. The level of DLDB/CHIP controls the activity of the LIM-homeodomain protein Apterous: Evidence for a functional tetramer complex *in vivo*. *EMBO J.* 19:2602—14.

[53]Saccone,G. ,A. Pane,and L. C. Polito. 2002. Sex determination in flies,fruitflies and butterflies. *Genetica* 116:15—23.

[54]Sakamoto,H. ,D. Kageyama,S. Hoshizaki,and Y. Ishikawa. 2007. Sex-specific death in the Asian corn borer moth (*Ostrinia furnacalis*) infected with *Wolbachia* occours across larval development. *Genome* 50:645—52.

[55]Sakurai,T. ,T. Nakagawa,H. Mitsuno,et al. 2004. Identification and functional characterization of a sex pheromone receptor in the silkmoth *Bombyx mori*. *Proc. Natl. Acad. Sci. U. S. A.* 101:16653—58.

[56]Saito,H. 1994. Diapause response in *Samia cynthia* subspecies and their hybrids (Lepidoptera: Saturniidae). *Appl. Entomol. Zool.* 29:296—98.

[57]Schütt,C. ,and R. Nöthiger. 2000. Structure,function and evolution of sex-determining systems in Dipteran insects. *Development* 127:667—77.

[58]Shimada,T. ,H. Ebinuma,and M. Kobayashi. 1986. Expression of homeotic genes in *Bombyx mori* estimated from asymmetry of dorsal closure in mutant/normal mosaics. *J. Exp. Zool.* 240:335—42.

[59]Shimada,T. ,H. Yamauchi,and M. Kobayashi. 1988. Diapause of the inter-specific F_1 hybrids between *Antheraea yamamai* (Guerin-Meneville) and *A. pernyi* (G. -M.) (Lepidoptera: Saturniidae). *Jpn. J. Appl.* Entomol. Zool. 32:120—25 (In Japanese with English summary).

[60]Siegal,M. L. ,and B. S. Baker. 2005. Functional conservation and divergence of *intersex*,a gene required for female differentiation in *Drosophila melanogaster*. *Dev. Genes Evol.* 215:1—12.

[61]Straub, T. , and P. B. Becker. 2007. Dosage compensation: the beginning and end of generalization. *Nat. Rev. Genet.* 8:47—57.

[62]Suzuki, M. G. , S. Funaguma, T. Kanda, et al. 2003. Analysis of the biological functions of a *doublesex* homologue in *Bombyx mori*. *Dev. Genes Evol*. 213:345—54.

[63]Suzuki, M. G. , S. Funaguma, T. Kanda, et al. 2005. Role of the male BmDSX protein in the sexual differentiation of *Bombyx mori*. *Evol. Dev.* 7:58—68.

[64]Suzuki, M. G. , S. Imanishi, N. Dohmae, et al. 2008. Establishment of a novel in vivo sex-specific splicing assay system to identify a *trans*-acting factor that negatively regulates splicing of *Bombyx* female exons. *Mol. Cell. Biol.* 28:333—43.

[65]Suzuki, M. G. , F. Ohbayashi, K. Mita, and T. Shimada. 2001. The mechanism of sex-specific splicing at the *doublesex* gene is different between *Drosophila melanogaster* and *Bombyx mori*. *Insect Biochem. Mol. Biol.* 31:1201—11.

[66]Suzuki, M. G. , T. Shimada, and M. Kobayashi. 1998. Absence of dosage compensation at the transcription level of a sex-linked gene in a female heterogametic insect, *Bombyx mori*. *Heredity* 81:275—83.

[67]Suzuki, M. G. , T. Shimada, and M. Kobayashi. 1999. *Bm kettin*, homologue of the *Drosophila kettin* gene, is located on the Z chromosome in *Bombyx mori* and is not dosage compensated. *Heredity* 82: 170—79.

[68]Tanaka, Y. 1916. Genetic studies in the silkworm. *J. Coll. Agric. Sapporo* 6:1—33.

[69]Tazima, Y. 1941. A simple method of sex discrimination by means of larval markings in *Bombyx mori* (in Japanese). *J. Seric. Sci. Jpn.* 12:184—88.

[70]Tazima, Y. 1944. Studies on chromosome aberrations in the silkworm. II. Translocation involving second and W-chromosomes (in Japanese with English summary). *Bull. Seric. Exp. Stn.* 12:109—81.

[71]Tazima, Y. 1948. Translocation of the Z chromosome to the W chromosome of the silkworm, *Bombyx mori*. In *Oguma commemoration volume on cytology and genetics* (in Japanese), 88—99.

[72]Tazima, Y. 1952. Inheritance of sex (in Japanese). In *Silkworm genetics*, ed. Y. Tanaka, 351—72. Tokyo: Shokabo.

[73]Tazima, Y. 1964. *The genetics of the silkworm*. London: Academic Press.

[74]Tazima, Y. 2001. *Improvement of biological functions in the silkworm*. Enfield, NH: Science Publishers, Inc. (Translated from Japanese).

[75]Tazima, Y. , C. Harada, and N. Ohta. 1951. On the sex discriminating method by colouring genes of silkworm eggs. I. Induction of translocation between the W and the tenth chromosomes (in Japanese with English summary). *Jpn. J. Breed.* 1:47—50.

[76]Traut, W. , and F. Marec. 1996. Sex chromatin in Lepidoptera. *Q. Rev. Biol.* 71:239—56.

[77]Traut, W. , T. Niimi, K. Ikeo, and K. Sahara. 2006. Phylogeny of the sex-determining gene *Sex-lethal* in insects. *Genome* 49:254—62.

[78]Traut, W. , and B. Rathjens. 1973. Das W-Chromosom von *Ephestia kuehniella* (Lepidoptera) und die Arbeitung des Geschlechtschromatins. *Chromosoma* 41:437—46.

[79]Traut, W. , K. Sahara, and F. Marec. 2007. Sex chromosomes and sex determination in Lepidop-

tera. *Sex. Dev.* 1:332—46.

[80]Traut, W., A. Weith, and G. Traut. 1986. Structural mutants of the W chromosome in *Ephestia* (Insecta, Lepidoptera). *Genetica* 70:69—79.

[81]Wu, C., S. Asakawa, N. Shimizu, et al. 1999. Construction and characterization of bacterial artificial chromosome libraries from the silkworm, *Bombyx mori*. *Mol. Gen. Genet.* 261:698—706.

[82]Xia, Q., Z. Zhou, C. Lu, et al. 2004. A draft sequence for the genome of the domesticated silkworm (*Bombyx mori*). *Science* 306:1937—40.

[83]Yamamoto, D., K. Fujitani, K. Usui, et al. 1998. From behavior to development: Genes for sexual behavior define the neuronal sexual switch in *Drosophila*. *Mech. Dev.* 73:135—46.

[84]Yasukochi, Y. 1998. A dense genetic map of the silkworm, *Bombyx mori*, covering all chromosomes based on 1018 molecular markers. *Genetics* 150:1513—25.

[85]Yasukochi, Y., L. A. Ashakumary, K. Baba, et al. 2006. A second-generation integrated map of the silkworm reveals synteny and conserved gene order between lepidopteran insects. *Genetics* 173:1319—28.

第五章 蝴蝶翅模式的进化和发育遗传学——热带蝴蝶眼斑

Patrícia Beldade and Suzanne V. Saenko

1. 前言

蝴蝶翅膀色彩绚丽的斑纹常常为人们研究包括从系统分类学和进化到发育遗传学和合成色素的生物化学等各种生物问题带来许多启发和灵感。蝴蝶翅模式惊人的多样性及其在适应性上的价值和发育基础上积累的潜在知识，使其成为进化发育生物学这个较新研究领域中的一个最受欢迎的模式系统(Nijhout 1991; Beldade 和 Brakefield 2002; McMillan，Monteiro 和 Kapan 2002; Joron et al. 2006; Parchem，Perry 和 Patel 2007; Wittkopp 和 Beldade 2009)。

蝴蝶翅模式的多样性令人惊奇，在 17000 多个蝴蝶品种中，许多品种都可以通过翅模式识别。即使同一品种，在不同的地理环境、不同季节，雌雄之间和同一翅膀的背腹表面的色彩也会存在巨大差异(例见 Nijhout 1991)。尽管蝴蝶有令人眼花缭乱的斑点、条带和条纹，但大多数蝴蝶的翅模式可以看作是由一个基本的蛱蝶科平面图(nymphalid ground plan) 衍生出来的，这一特点可用来鉴定种内和种间模式元件的同源性(Nijhout 1991)。在这平面图里，不同类型的模式元件，如眼斑、山形饰条和条带平行连续地排列，且各个元件在由翅脉界定的翅腔里沿前后轴重复分布。各个模式元件的独立发育和进化被认为是促进蝴蝶翅模式分化的重要原因。

翅模式对适应能力的重要性及其内在的遗传学基础历来都是生态学和进化研究的焦点，这些研究提出了自然和性别选择是怎样引起群体变异的重要观点，在过去 20 年里，人们详细分析了翅模式形成和分化中发育的进程和路径。这将是本章的重点。

1.1 蝴蝶翅模式的生态学和进化发育生物学

鳞翅目昆虫中有许多例子已纳入教科书，如白桦尺蛾(*Biston betularia*)(Majerus 1998)的工业黑化、柑橘凤蝶(*Papilio*，凤蝶科)和西番莲蝶(*Heliconius*，釉蛱蝶科)(Joron 和 Mallet 1998)的拟态为例的适应色多态型和许多物种因季节性而产生的色彩模式多态型(Brakefield 和 French 1999)。翅颜色模式还提供了一个特殊的适应性进化的例子，即蝴蝶眼斑，其有颜色的同心环类似于脊椎动物的眼睛，有恐吓捕食者或转移目标的功能。蝴蝶翅模式自然变异不仅在防御天敌(Stevens 2005)，还在种内的识别、性别选择和温度调节等方面具有重要的意义。

近年来，与翅颜色模式的生态学和基础遗传学的研究相比，人们更关注于翅模式形成的发育和分子遗传学的研究。蝴蝶的进化发育生物学研究主要以少数的几个品种为研究对象，通过不同的方法去阐明不同翅模式形成的发育和遗传机制。例如，对凤蝶(*Papilio*)的研究就详细分析了色素形成的生化路径及其与整体着色多态型的关系(Koch，Behnecke 和 ffrench-Constant 2002)。对釉蛱蝶(*Heliconius*)的连锁分析，包括传统的低丰度的杂交和近年来高丰度的遗传图谱都明确了许多翅模式(由不同颜色的大斑纹组成)变异的遗传学特征(Joron et al. 2006)。在第六章将详述釉蛱蝶(*Heliconius*)翅模式的遗传学。对鹿眼蛱蝶(*Junonia* (*Precis*) *coenia*)和热带蝴蝶(*Bicyclus anynana*)翅发育的研究揭示了许多参与眼斑形成的发育进程和遗传路径(Nijhout 1991; Beldade 和 Brakefield 2002; McMillan，Monteiro 和 Kapan 2002)。本章将详细叙述在实验群体热带蝴蝶(*B. anynana*)中取得的相关研究进展。

1.2 热带蝴蝶(*B. anynana*)眼斑模式的进化发育生物学

通过保罗·布雷克费尔德(Paul Brakefield)20 年的努力,建立了热带蝴蝶(*B. anynana*)的实验系统(Brakefield,Beldade 和 Zwaan 2009)。这些蝴蝶个体既小到可以饲育和分析较大的群体,同时个体又大到可以对单个个体和特殊组织进行手工操作时不会遇到技术上的难题。这样就能使群体、生物有机体和分子水平的分析进行有机结合。更重要的是逐渐发展的基因组测序和转基因等实验方法都能较易地在这些品系中操作(Ramos 和 Monteiro 2007; Beldade, McMillan 和 Papanicolaou 2008; Beldade et al. 2009),并可与生态学和自然变异的理论知识相结合。

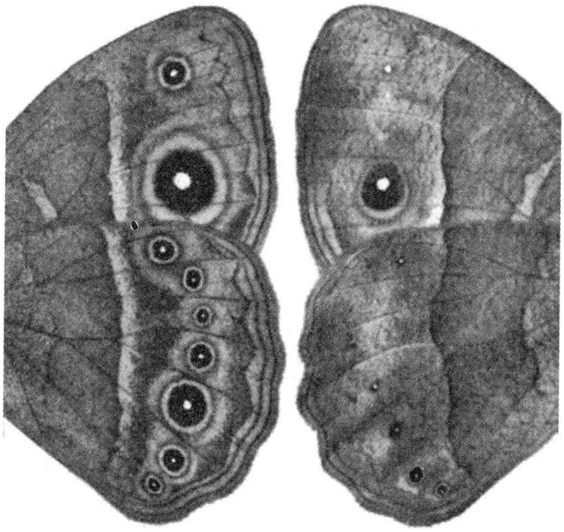

图 5.1 热带蝴蝶(*Bicyclus anynana*)翅模式的表型(彩色版本图片见附录图 5.1)。在野外发生的两种季节型斑纹模式亦可在实验室内模拟,饲育条件为幼虫期 27℃或 20℃,成虫将分别呈现类似于自然界的潮湿季节(左)和干燥季节(右)的两种表型。图片展示的是前后翅的腹面,其背面由于饲育温度并未显示出差异。文章中将详细讨论这种交替选择表型的适应性意义和潜在的生理学基础。

热带蝴蝶(*B. anynana*)翅的眼斑已被证明是一个能引起形态变异的进化和发育结合起来研究的特殊有用系统。其眼斑能起到天敌防御(Lyytinen et al. 2004; Brakefield 和 Frankino 2006)、配偶选择(Robertson 和 Monteiro 2005; Costanzo 和 Monteiro 2007)和区分种内种间的形态变异等作用(Beldade, Brakefield 和 Long 2005; Brakefield 和 Roskam 2006)。在实验室保存了热带蝴蝶(*B. anynana*)不同眼斑表型的遗传多态性群体,为研究在发育过程中基因型变异是如何影响表型变异提供了理想的材料。实验室里也保存了那些在自然群体中因不同季节表现出不同表型的遗传材料,用来研究环境因素是怎样决定这种交替选择的发育轨迹的(图 5.1)。此外,还可以在这约 80 个尾蛱蝶属(*Bicyclus*)的不同的眼斑品种中做形态(Condamin 1973)和分子系统发生(Monteiro 和 Pierce 2001)的研究。就发育机制的研究而言,眼斑在从模式的遗传路径和细胞信号到色素合成的生物化学等方面亦可进行详细的研究(图 5.2)。

在本章,我们重点讲述热带蝴蝶(*B. anynana*)眼斑模式形成和多样化的遗传和发育基础。首先关注的是眼斑形成的机制,包括在蛹翅里细胞的相互作用模式和眼斑不同发育时期中的遗传路径。接下来的部分将讲述眼斑模式在种间变异的机制,包括在眼斑形成中牵涉的不同元件的发育进程及其作用,以及利用基因组信息在全基因组水平上的分析和研究。第三部分内容重点关注眼斑模式的进化多样性,重点讲述新特征的起源和进化以及这些连续重复结构的多样性。最后,我们将讨论该研究的新动向。

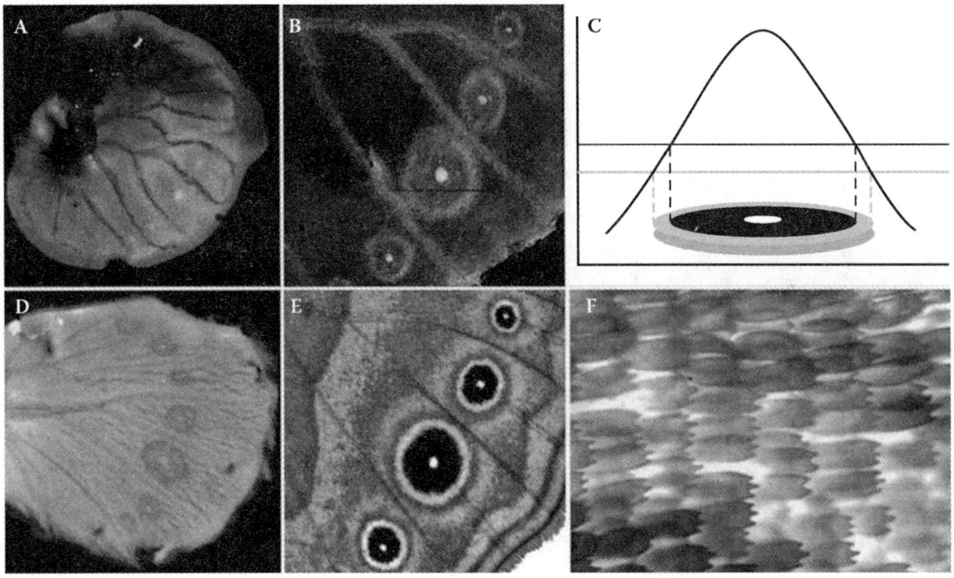

图 5.2 热带蝴蝶(*Bicyclus anynana*)的眼斑发育(彩色版本图片见附录图 5.2)。(A)眼斑的形成始于幼虫的末龄,推测是由于 *Distal-less*(红)和 *engrailed*(绿)基因的作用来确定眼斑的中心(用免疫荧光技术检测蛋白在后翅中的表达部位)。同时,在翅缘部位也能检到 *Distal-less* 基因的表达,*engrailed* 基因在后区室各处的表达是昆虫翅发育的基本特征。(B)这些基因也在蛹翅的初期阶段的眼斑焦点中共表达。在眼斑的内环细胞表达 *Distal-less* 基因,外环细胞表达 *engrailed* 基因。(C) 眼斑色素环的形成可能依赖于一个信号应答机制,通过该机制眼斑中心的细胞产生一个信号分子并发散开来形成一个浓度梯度(曲线所示),临近的细胞则以类似于阈值的方式做出应答(水平线)。(D)表皮细胞在成虫羽化前短时间内产生特殊的颜色。在蛹后期,后翅呈现金色的环和白色的环心成熟鳞毛,随后即会变成棕色背景下的黑色环。(E)成虫翅模式由连续的重复眼斑组成。图中显示的是有 4 个眼斑的成虫后翅的部分区域。(F)由鳞毛排列形成的色彩模式,每一方向都是一个特殊的颜色(黑、黄、棕)。

2. 眼斑形成的发育遗传学

蝴蝶的翅膀可以区分出所有类型的不同色彩的模式元件,不可否认眼斑的发育是研究得最清楚的。在 20 世纪 80 年代早期,Fred Nijhout 开创性地对鹿眼蛱蝶(*J. coenia*)的蛹翅进行外科手术,从而建立了眼斑形成的发育模式。这些分析蝴蝶翅模式进化和发育的实验在他的书里有详细叙述(Nijhout 1991)。在 90 年代中期,Sean Carroll 的实验室关于鹿眼蛱蝶(*J. coenia*)和热带蝴蝶(*B. anynana*)翅发育过程中候选基因表达模式方面的研究就开始探索参与眼斑形成的遗传途径(Carroll et al. 1994; Brakefield et al. 1996; 图 5.2A,B)。

蝴蝶翅由两层表皮膜组成,由其间的翅脉支撑。翅从一龄幼虫期就开始发育,那时只是中胸和后胸两侧表皮细胞的轻微扩大。翅脉的发育和翅原基大小的巨增出现在幼虫末龄期。蛹翅的鳞毛成熟和色素沉积发生在成虫羽化前的短时间内。翅的色彩模式是由单细胞层、单色的鳞毛部分重叠组成的。在发育相互作用中该二维性质的简化模式发生在幼虫的晚期和蛹的早期,那时就决定了鳞毛的成熟和色素沉着,这个时期比出现可见色彩的时期早很多。在此,我们综述眼斑发育过程中不同时期的分子和遗传基础,从幼虫翅原基到蛹早期色彩环的决定再到蛹后期形成真正的颜色。

2.1 眼斑形成的模式:信号传输和应答机制

对即将形成蛹翅眼斑的部位进行表皮标记是非常容易的,因此可以在前翅的背面定位眼斑的中心。这种操作已经确立了这些中心,命名为焦点(foci),具有形成眼斑的性质。将这些眼斑的焦点移植到没有眼斑的部位会在宿主位置出现异位的眼斑,同时若在蛹的早期破坏焦点,成虫翅相应位置的眼斑则会减弱或者完全消失(Nijhout 1980;French 和 Brakefield 1995)。这些结果说明是通过以眼斑焦点作为组织者产生可发散的形成素,再向临近的细胞发出信号来完成该发育模式的规划(图 5.2C)。焦点外传输的信号在翅的表皮形成一个浓度梯度,周围的细胞再对信号浓度以类似于阈值的方式作出应答,随即产生特殊的颜色。也是通过焦点的分解,而不是焦点的产生来形成交替选择眼斑模式(季节型),在眼斑的同心环出现组织信号(French 和 Brakefield 1992)或额外的形成素(Dilão 和 Sainhas 2004)。

尽管眼斑形成的信号应答机制得到实验结果强有力的支持,但是对这个信号传输和应答过程中的基因及其作用的了解是很有限的。相反,对眼斑发育中所涉及的遗传路径的了解则更多,有大量的基因已被证明在不同时期参与了眼斑的发育,下面将详细讲述。

2.2 确立幼虫晚期翅原基眼斑焦点的遗传路径

对热带蝴蝶(*B. anynana*)和鹿眼蛱蝶(*J. coenia*)幼虫和蛹翅的基因表达模式研究表明大量的遗传路径参与眼斑的发育。这些候选的目标基因都包括在昆虫翅发育的路径中,且在黑腹果蝇(*Drosophila melanogaster*)里已有大量的研究报道。果蝇和蝴蝶翅发育中参与区室化的这些基因(例如 *apterous* 分布在背区室,*engrailed* 和 *cubitus interruptus* 分别在后区室和前区室,*Distal-less* 和 *wingless* 基因分布在翅缘)执行相似的功能,只是在蝴蝶里有些基因还同时作用于眼斑发育的不同时期(Carroll et al. 1994;图 5.3 和表 5.1)。

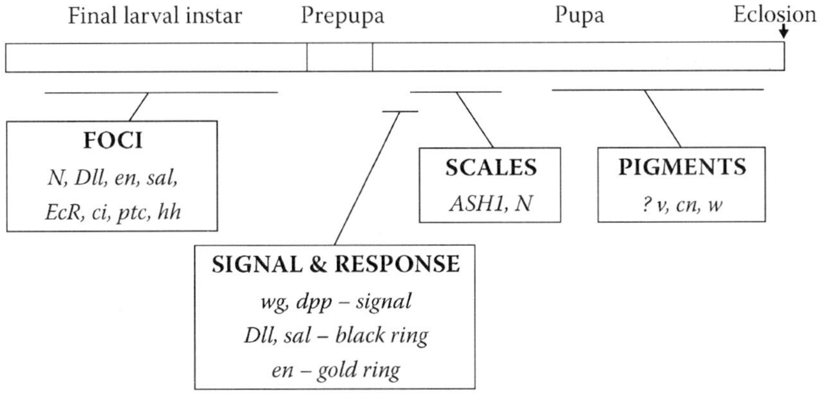

图 5.3 眼斑不同时期发育的相关基因。横线表示热带蝴蝶(*B. anynana*)发育的连续时期,下面的线表示眼斑形成的阶段,方框内罗列的是在翅模式发育过程中检测到表达的相关基因(详见表 5.1)。眼斑焦点的确定、信号应答的相互作用和鳞毛成熟的相关数据主要来至于对热带蝴蝶(*B. anynana*)和鹿眼蛱蝶(*J. coenia*)的研究结果,而色素合成基因的数据则主要来自于那些没有眼斑的品种。这些基因在眼斑形成中的功能仍不清楚。

幼虫末龄期在翅脉分隔的区室中按眼斑的轴向确定了眼斑焦点位置。到目前为止，人们了解的在该过程中最早上调表达的基因是 *Notch*，它最初在脉络的中线表达，然后在未来的焦点细胞中表达(表 5.1)。*Notch* 表达后随即激活 *Distal-less*，然后 *engrailed* 和 *spalt* 的表达上调。同时，Hedgehog 信号途径的基因也在眼斑焦点(*patched* 和 *cubitus interruptus*)及其周围(*hedgehog*)的细胞中表达。然后，蜕皮激素受体蛋白(推测是通过 *Distal-less* 的作用而上调表达)亦在眼斑焦点细胞中表达。这些基因在决定眼斑焦点中的作用是通过在不同眼斑形态的热带蝴蝶(*B. anynana*)中调查这些基因的表达模式来证实的。在后面的部分将会详细叙述相关内容。这些基因是如何精确地相互调节以及如何调控产生焦点信号和形成眼斑的下游遗传路径的过程都还不得而知。尽管如此，我们还是根据黑腹果蝇中已知的相关路径描述这个过程中的遗传调节级联(Evans 和 Marcus 2006; Marcus 和 Evans 2008)。

表 5.1　蝴蝶翅模式形成的基因

Table 5.1 Genes Implicated in Butterfly Wing Pattern Formation

Gene Name	Abbreviation	Molecular Function*	References
Achaete-scute homologue	*ASH1*	Transcription factor	Galant et al. 1998
Cinnabar	*cn*	Kynurenine 3-monooxygenase	Reed and Nagy 2005
Cubitus interruptus	*ci*	Transcription factor	Keys et al. 1999
Decapentaplegic	*dpp*	Morphogen	Monteiro et al. 2006
Distal-less	*Dll*	Transcription factor	Carroll et al. 1994; Brakefield et al. 1996
Ecdysone Receptor	*EcR*	Nuclear receptor with transcription factor activity	Koch et al. 2003
Engrailed	*en*	Transcription factor	Keys et al. 1999; Brunetti et al. 2001
Hedgehog	*hh*	Morphogen	Keys et al. 1999
Notch	*N*	Transmembrane receptor with transcription factor activity	Reed and Serfas 2004
Patched	*ptc*	Transmembrane receptor for *Hedgehog*	Keys et al. 1999
Spalt	*sal*	Transcription factor	Brunetti et al. 2001; Reed, Chen, and Nijhout 2007
Ultrabithorax	*Ubx*	Transcription factor	Lewis et al. 1999
Vermilion	*v*	Tryptophan 2,3-dioxygenase	Reed and Nagy 2005; Beldade, Brakefield. and Long 2005
White	*w*	Pigment precursor transporter	Reed and Nagy 2005
Wingless	*wg*	Morphogen	Monteiro et al. 2006

* Relevant molecular function as per Gene Ontology classification. available also in FlyBase

2.3 眼斑形成中负责信号传输和应答的候选基因

尽管在几十年前人们就认识到眼斑的形成包括焦点的信号传输和表皮的应答，但对其分子机制的了解却很少。近年来，借鉴果蝇翅模式中形成素的研究成果使得我们在这方面有了进一步的认识。安东尼娅(Antónia Monteiro)及其同事的研究表明在热带蝴蝶

(*B. anynana*)早期蛹翅的眼斑区域的细胞(见表 5.1References)中能检测到 Wingless 和 Smad 的磷酸化形式，以及 Decapentaplegic 途径的信号转导蛋白的表达。这与人们推测 Wingless 和 Decapentaplegic 都是候选眼斑信号的结论一致，但仍需要进一步的功能分析来证实它们在眼斑发育中的作用。

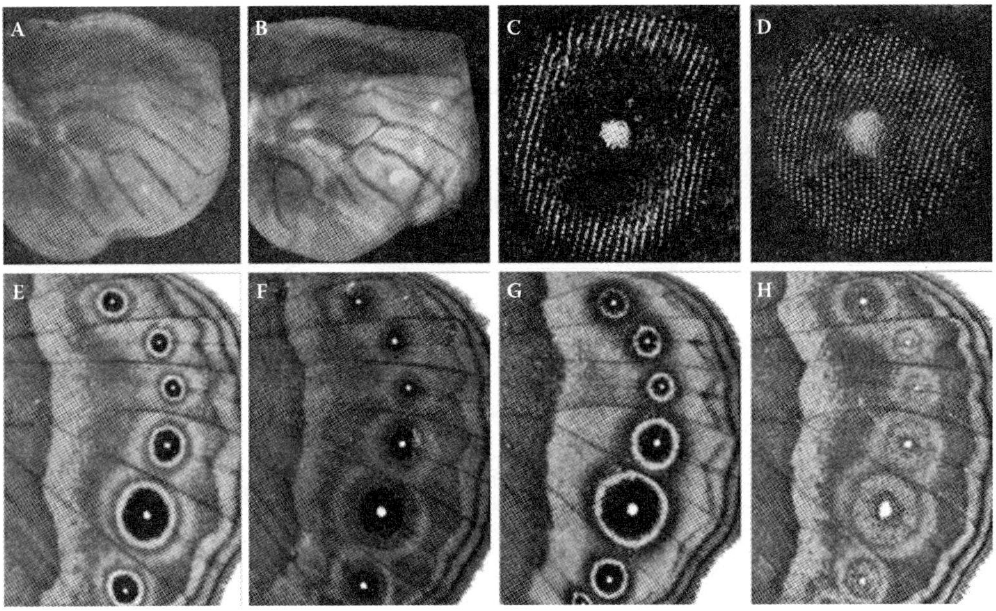

图 5.4 热带蝴蝶(*Bicyclus anynana*)翅模式突变表型(彩色版本图片见附录图 5.4)。(A—D)Engrailed (绿)和 Distal-less(红)蛋白在幼虫翅原基(A,B)和蛹前翅(C,D)中表达。(E—H)热带蝴蝶(*B. anynana*)不同实验群体的成虫后翅的详细腹面形态。(A)在热带蝴蝶(*B. anynana*)野生型的成虫前翅上有两个眼斑，且眼斑的中心在幼虫晚期的翅原基就已经出现。(B)在 *Spotty* 突变体中，幼虫翅原基上有两个额外的眼斑焦点，其位置和成虫前翅额外的两个眼斑的位置一致。(C)在蛹的早期，眼斑的外环就已经形成，engrailed 蛋白也在外部金色的环上表达。(D)在 *Goldeneye* 突变体中，engrailed 的表达遍布整个眼斑，其表达区域与成虫眼斑颜色的构成一致(见 H)。(E)野生型热带蝴蝶(*B. anynana*)的着色模式。(F)*melanine* 突变表现为全黑的表型。(G)*Band* 突变体翅的远端(近半个翅)的区域颜色较浅。(H)*Goldeneye* 突变的眼斑颜色组成发生变化。

虽然眼斑焦点信号的作用模式及其作用的靶基因还不清楚，但目前已鉴定了一些表皮应答级联的基因。对热带蝴蝶(*B. anynana*)基因的表达模式分析表明有许多基因在幼虫翅原基的眼斑焦点表达，并且这些基因的表达可以持续到蛹的早期，另外一些基因在将来发育为眼斑环的细胞内上调表达。值得注意的是，*engrailed* 的表达区域与成虫眼斑外部环(金色)的位置一致，*Distal-less* 和 *spalt* 则在眼斑的内环(黑色)表达(图 5.2B 和 5.4C)。这些基因编码的转录因子可能直接作用于焦点信号途径中，并激活色素生物合成路径，因为无论是在野生型的热带蝴蝶(*B. anynana*)还是颜色模式变异的突变体(例如：金色眼斑 *Goldeneye*)中，这些基因的表达部位都和成虫眼斑的色素环的分布完全吻合(Brunetti et al. 2001；图 5.4D)。在其他蝴蝶品种中，这些转录因子也在眼斑的同心环表达，虽然分布于不同的空间区域，但与成虫眼斑色彩模式仍是相关的(Brunetti et al. 2001)。

2.4 蛹翅晚期的色素合成和鳞毛成熟

大量的研究集中于眼斑发育早期的细胞相互作用和遗传路径。而对后面的发育进程了解较少，在蛹后期，鳞毛形成细胞根据预定模式信息来产生特殊的色素。

在蝴蝶翅鳞毛里有4种主要的色素：黑色素(melanins，从红棕色到黑色)，眼色素(ommochromes)和蝶呤类(pterins，从黄到红)以及黄酮类化合物(flavonoids，白到红或蓝)。鳞毛发育成熟时就同时合成了色素，在大多数蝴蝶品种里，最后基本上形成的都是黑色的鳞毛(Nijhout 1991；Reed 和 Nagy 2005)。就热带蝴蝶(*B. anynana*)的眼斑而言，首先是眼斑中心(白色部分)的鳞毛细胞最先成熟，然后是外环的黄色鳞毛，最后成熟的是中环的黑色鳞毛(Koch et al. 2000；图5.2D—F)。

最近在小红蛱蝶(*Vanessa cardui*)和瓦氏袖蝶(*Heliconius erato*)的研究表明色素合成候选基因的表达与翅不同颜色的分布是相关的(Reed 和 Nagy 2005；Reed，McMillan 和 Nagy 2008)。其中的一些基因(例如：*vermillion*)在热带蝴蝶(*B. anynana*)的翅中也有表达(Beldade，Brakefield 和 Long 2005)，其作用为标记眼斑的不同色素环。我们可以用色素合成突变的实验品种来研究这些或者其他候选基因的表达，从翅的整体、部分到眼斑的环等不同层次研究基因的表达与不同颜色表型的关系(图5.4E—H)。

3. 眼斑模式变异的机制

前面部分介绍的实验主要鉴定了眼斑形成中的大量遗传路径和发育进程。这些遗传路径(比如遗传路径中的基因)和进程(比如眼斑焦点信号强度和表皮的应答阈值)中的那些特殊因素能造成眼斑表型的变异，这是进化发育生物学研究的重要问题。人们用具有不同翅模式表型的热带蝴蝶(*B. anynana*)实验群体来分析眼斑模式变异的遗传、生理和发育基础。这些研究结果将在下面讲述。

3.1 表型可塑性的遗传和生理基础

热带蝴蝶(*B. anynana*)的表型有明显的季节性差异，包括翅的色彩模式也会改变。蝴蝶的这种表型的交替选择是为了更适应干湿季节交替的自然环境。在干燥季节颜色不鲜明的棕色蝴蝶利用干枯的叶子作为背景来隐藏自己，从而躲避天敌。而在潮湿的季节，当蝴蝶在绿色植物上休息的时候，其色彩艳丽的眼斑会把捕食者的注意力吸引到翅缘上，远离脆弱的身体从而保护自己(Brakefield 和 Frankino 2009)。

在实验室里，将蝴蝶在高温和低温下饲养，则会分别出现类似与自然界潮湿和干燥状态下的两种表型(图5.1)。这种材料已经被用来研究环境的差异性是如何影响发育进程，并导致成虫表型的差异(Brakefield，Kesbeke 和 Koch 1998)。而且，还可以通过人为调节蝴蝶在各个发育阶段的饲养温度，使蝴蝶产生干（小眼斑）、湿（大而明显的眼斑）两种翅模式的组成性表达（Brakefield et al. 1996；Koch，Brakefield 和 Kesbeke 1996）。众所周知，内分泌信号，又称蜕皮甾体，在昆虫发育和非遗传多型性的控制上起着至关重要的作用（Riddiford 1995；Nijhout 2003a）。在上面提到的系统中测量蛹期蜕皮激素的滴度，发现激素释放的时间明显不同，腹面眼斑更大的品种在化蛹后蜕皮激素的滴度更高（Koch，Brakefield 和 Kesbeke 1996；Brakefield，Kesbeke 和 Koch 1998）。而且，在小眼斑品种的嫩蛹里注射激素，可以使其成虫的腹面眼斑变大（Brakefield，Kesbeke 和 Koch 1998）。

尽管我们了解很多蜕皮激素在鳞翅目昆虫发育中的功能，但该激素的鉴定以及在眼斑相关遗传路径的上下游的作用模式还不清楚。我们已知 *Distal-less* 转录因子在干、湿季节型的表达水平不同（Brakefield et al. 1996），且该基因已被证明与背面眼斑的大小有关（Beldade，Brakefield 和 Long 2002）。但蛹期血淋巴释放蜕皮激素的时间和 *Distal-less* 基因在眼斑区域表达水平之间的精确关系还不清楚（可参阅 Koch et al. 2003）。大量蝴蝶品种（包括热带蝴蝶，*B. anynana*）（Beldade，McMillan 和 Papanicolaou 2008）的基因组测序将有助于我们了解导致交替选择翅模式表型的环境、基因、激素和发育之间的关系。

3.2 热带蝴蝶（*B. anynana*）实验群体的遗传变异

热带蝴蝶（*B. anynana*）的实验群体不仅可以模拟自然的季节多态型，还隐藏了许多与眼斑模式相关的遗传变异类型（Brakefield，Beldade 和 Zwaan 2009）。在对大规模的实验群体进行人工选择时，已经分离出许多可遗传的变异体，并且眼斑的大小、颜色组成、形状和位置这样的性状产生了循序渐进的变化（McMillan，Monteiro 和 Kapan 2002；Beldade，Brakefield 和 Long 2005）。大量不同翅模式表型的自发突变体都已经被分离并建立了稳定的品系。这些品系不止涉及眼斑的大小（例：*Bigeye*，表型为整体放大的眼斑；Brakefield et al. 1996）、颜色组成（例：*Goldeneye*，表型为没有黑色的眼斑环，全为金色；图 5.4H；Brunetti et al. 2001）、形状（例：*Cyclops*，表型为狭长且融合的眼斑；Brakefield et al. 1996），还有翅模式的其他方面的特征（例：*melanine* 和 *Band*；图 5.4F—G）。

具有引起极端效应的等位基因和引起微小效应的等位基因的位点的相似性程度不得而知，这些效应影响实验和自然群体中数量变异的分离。突变品系和选育品种可以用来研究翅模式变异的遗传组成，还可鉴定色素模式形成和变异的其他发育路径和进程。

3.3 发育进程中的变异

对热带蝴蝶（*B. anynana*）的研究已经从幼虫晚期翅原基的前模式基因表达、蛹早期翅的信号应答、晚期鳞毛成熟和色素沉积等方面详细了解了眼斑形成各个阶段的变异特征。

眼斑发育相关基因在大量突变品种和选育系的幼虫晚期翅原基里的表达模式都有

所不同。例如，在成虫表现为过剩眼斑的突变体（例：*Spotty*，图 5.4B）中，幼虫翅原基里就有眼斑焦点标记基因 *Notch*，*Distal-less* 和 *engrailed* 的表达区，反之，在成虫表现为眼斑缺失的突变品种中（例：*3＋4* 和 *Missing*），就检测不到这些基因的表达（Brakefield et al. 1996；Monteiro et al. 2003；Reed 和 Serfas 2004；Monteiro et al. 2007）。而且，这些基因表达的位置与眼斑的形状（例：在 *Cyclops* 突变中，*Distal-less* 就只在一个狭长的区域表达，这与该突变狭长的眼斑完全吻合；Brakefield et al. 1996）和大小（例：在人工培育的翅膀背面眼斑大小不同的品系中 *Distal-less* 和 *engrailed* 的表达量也有差异；Beldade，Brakefield 和 Long 2005）都比较一致。其他的一些眼斑表型变异的品种，就与幼虫翅原基中眼斑焦点标志基因的表达无关，而与眼斑发育后期的基因表达模式有关。例如：*Goldeneye* 突变体中色素结构的变化与 *engrailed* 基因在蛹翅即将发育为眼斑的位置的表达模式是相关的（Brunetti et al. 2001；Saenko et al. 2008；图 5.4D，H）。以上结果说明这些眼斑表型变异的等位突变品种包含了大量候选的遗传路径。

通过在眼斑表型不同的培育品种之间移植蛹的眼斑诱导中心（眼斑焦点），即可以研究眼斑焦点信号强度和表皮应答阈值水平的变化。例如：眼斑大小的数量变化就很可能与眼斑焦点信号的强弱有关（Monteiro，Brakefield 和 French 1994），而眼斑形状（Monteiro，Brakefield 和 French 1997b）和色素结构（Monteiro，Brakefield 和 French 1997a）的不同就可能源于表皮应答的差异。在热带蝴蝶（*B. anynana*）和美洲虎凤蝶（*Papilio glaucus*）的突变体里研究发现发育过程中鳞毛成熟与鳞毛颜色、鳞毛异时改变的相关性已经被视为引起蝴蝶翅模式变异的一个机制（Koch et al. 2000），但目前还不清楚这些信号应答的差异是怎样影响热带蝴蝶（*B. anynana*）眼斑中鳞毛成熟和色素合成的。

从本质上来说，那些决定成虫表型变异的基因座并不受其他基因表达模式的影响。但事实上，任何一个基因的表达变化都是源于该位点的等位突变（顺式作用元件）或负责调节这个基因的其他基因（反式作用因子）发生变异。哪些顺式作用元件和反式作用因子决定了这些基因的表达变化是表型多样化研究的重要内容。在此，我们将讨论那些决定蝴蝶翅模式变异的基因。

3.4 鉴定决定翅模式变异的基因

候选的发育基因对我们研究究竟是哪些基因座导致了蝴蝶翅表型（包括翅的颜色模式）的各种变异尤为重要。全基因组定位技术已经越来越成熟，随着基因分型成本的降低，我们可以在一个较大的定位系统中检测更多的标记，但从遗传区域的定位到真正鉴定一个决定表型的基因位点仍不是件容易的事。尤其对于非传统的模式生物蝴蝶来说，在没有全基因组测序，也没有足够的各种 DNA 序列多态性信息的情况下就显得更为困难。

对热带蝴蝶（*B. anynana*）实验群体的研究已经表明眼斑发育路径中的候选基因可

以引起个体内眼斑表型的变异。在不同表型的选育系之间杂交，眼斑发育关键基因 *Distal-less* 的 DNA 序列多态性与眼斑大小的数量性状之间在重组后代中呈共分离 (Beldade, Brakefield 和 Long 2002)。尽管该基因与影响眼斑数量的 *Missing* 位点并不连锁(Monteiro et al. 2007)，但它的表达模式变化同样与眼斑的数量变化有相关性。这就说明是 *Distal-less* 的上游基因(可能是 *Notch*)决定了眼斑缺失 (*Missing*)这种表型。鉴定这些决定眼斑表型各种变异的候选基因是非常有必要的。在其他蝴蝶系的翅模式元件中也报道了一些发育候选基因与翅色彩模式变异性状紧密连锁(Kronforst et al. 2006; Clark et al. 2008)。

参与眼斑发育的基因是导致眼斑模式变异的重要候选基因。同时还可以通过分析那些既影响眼斑表型，又影响其他更保守的发育进程(指其遗传基础已经在模式生物中研究很清楚的发育进程)的多效突变体来获得其他的候选基因(Saenko et al. 2008)。然而要直接检测候选基因在翅模式变异中的作用还需要序列和多态型信息，就目前而言，这些信息还非常有限，EST 计划会加快蝴蝶少数实验品种(Beldade, McMillan 和 Papanicolaou 2008; Papanicolaou et al. 2008)(包括热带蝴蝶，*B. anynana*)(Beldade et al. 2006; Long, Beldade 和 Macdonald 2007; Beldade et al. 2009)基因鉴定的进程，新的基因组手段也会有利于阐释发育的理论基础。通过详细分析候选基因和已鉴定的遗传路径，或用全基因组的方法，或者二者相结合，总而言之通过这些方法可以更好地对翅模式变异进行遗传研究。令人兴奋的是，最近的一些比较分析结果表明很多直系同源的基因在家蚕(*Bombyx mori*)与墨尔波墨蝶(*H. melpomene*)之间(Pringle et al. 2007)、家蚕(*B. mori*)与热带蝴蝶(*B. anynana*)之间(Beldade et al. 2009)，以及烟草天蛾(*Manduca sexta*)、家蚕(*B. mori*)和墨尔波墨蝶(*H. melpomene*)之间(Sahara et al. 2007)都是线性保守的，这说明以后对蝴蝶翅模式变异的定位可以借鉴其他鳞翅目昆虫的相关连锁信息，从而有利于我们加快从基因组区域的遗传定位到大量候选基因分析鉴定的进程(Beldade et al. 2009)。

4. 眼斑模式的多样性

目前我们还不知道热带蝴蝶(*B. anynana*)变异相关的机制在多大程度上可以解释亲缘关系较远的蝴蝶品种中眼斑形态的差异；也不清楚有多大可能找到眼斑和其他翅模式元件多样化的分子机制的相似之处。在此，我们将讲述蝴蝶翅模式(特别是眼斑)在研究形态多样性的分子机制中起到的重要作用。

动物形态的多样性可以通过两种方面去描述：一个是进化事件的起源和修饰(例如龟壳和鸟的羽毛)；另一个为连续重复结构的多样性(例如脊椎动物的牙齿和昆虫的体节)。蝴蝶翅膀拥有这两方面的特征：它们的色彩模式是一个新的进化事件，并且是由不同类型的连续同源模式元件组成，在不同重复和不同品种之间表现出多样性。下面我们将从这两个方面讨论蝴蝶眼斑模式多样性的遗传基础。

4.1 形态学上新性状的进化

进化发育生物学的一个标准就是无论是在物种内还是物种间这些保守的遗传路径在有关的不同发育时期和组织都适用。新性状的产生往往是利用那些在不同系统中共有的基因和遗传路径的调换。蝴蝶翅提供了许多有趣的共选择的例子。黑腹果蝇(*Drosophila melanogaster*)中已研究清楚的许多遗传级联与蝴蝶翅发育中的非常相似,已经研究证明的有:与鳞毛形成(*achaete-scute* 和 *Notch* 基因在果蝇中是刚毛发育的基因,而在蝴蝶的鳞毛发育中起重要作用;Galant et al. 1998;Reed 2004)、色素合成(例:果蝇眼和蝴蝶翅中的眼色素路径;Beldade,Brakefield 和 Long 2005;Reed 和 Nagy 2005)和特殊的模式元件(例:*hedgehog* 路径在果蝇翅和蝴蝶的眼斑中也相同;Keys et al. 1999)相关的遗传级联。

胚胎发育和伤口愈合在模式生物中已有大量的研究,且认为是非常保守的。而眼斑形成和胚胎发育(通过分析影响眼斑形成和胚胎形成的多态突变体;Saenko et al. 2008)与创伤的应答(通过分析受伤引起的眼斑形成;Brakefield 和 French 1995;Monteiro et al. 2006)之间的分子机制又具有共通性,这有助于我们借鉴模式系统的有关信息去增进对眼斑进化发育生物学的认识。在眼斑发育背景下,分析这些已经研究清楚的路径不但使我们能把模式系统的知识转化为系统特异性状的理解,而且还为我们对一些出现新功能的重要发育路径进行进化研究提供机会(Beldade 和 Saenko 2009)。

4.2 个体化连续重复的机制

蛱蝶科平面图(Nymphalid ground plan)是蝴蝶翅模式构造特征的最好实例。不同类型模式元件之间的独立性(例如:眼斑和山形饰条)和连续重复元件的关联(例如:同一翅表面的两个眼斑)这一点在大量不同品种的蝴蝶中已经得到证实(Paulsen 和 Nijhout 1993;Nijhout 2003b;Allen 2008)。而且,在所有眼斑区域中基因的表达模式都几乎完全一致,但这些基因并不在其他模式元件中表达(图 5.2B,C)。这些连续重复元件是同时出现然后再分化,还是每次重复发生在不同时间,还是个未解决的问题(Monteiro 2008)。到目前为止,在连续同源眼斑分化的遗传和发育基础方面的研究已经取得一些进展。

尽管连续重复特征之间有很强的关联性,但在热带蝴蝶(*B. anynana*)的人工选育中发现眼斑表型的个别变化存在巨大的可伸缩性(Beldade,Koops 和 Brakefield 2002)。而且,这种数量变化所涉及的候选基因(Beldade,Brakefield 和 Long 2002)和大量表型明显变化的自发突变(Beldade,French 和 Brakefield 2008;Monteiro et al. 2003)都表明只影响连续重复眼斑的亚群。区分热带蝴蝶(*B. anynana*)突变体的各个眼斑的发育和进化程度决定了这种区分的遗传模式,在这个模式里,那些控制重要的眼斑基因表达的特异眼斑调控区域与每个未来的眼斑位置是密切相关的(McMillan,Monteiro 和 Kapan 2002; Monteiro et al. 2003)。对这些候选眼斑基因的调控区域和功能的详细研究将会验证这个遗传模式。

研究的另一个重要方面是确定眼斑形成中究竟是哪些信号传输和应答分子引起了翅表面特征的差异。在选育品系间移植蛹期的眼斑焦点,发现各个眼斑大小的改变主要受焦点信号强度的影响,但是眼斑复合体的大小则主要受表皮应答灵敏度的影响(Beldade,French 和 Brakefield 2008)。这个结果反映眼斑形成中的信号传输和应答分子在翅模式分化中起到的作用有差异,说明眼斑形态的不同方面中所涉及的发育机制可以解释连续重复的进化多样性(Allen et al. 2008)。

5. 展望

以上对热带蝴蝶(*B. anynana*)实验群体的研究在眼斑发育机制和所涉及的复杂的基因及信号路径方面作出了重要贡献。但对于眼斑发育,我们仍有许多不清楚的地方。眼斑发育的一些环节还没有引起足够的重视(见图 4.3),还有很多基因和发育过程都不清楚,也不了解它们之间的相互关系以及是如何受到环境因素的影响而造成眼斑模式的变化的。

在热带蝴蝶(*B. anynana*)和其他一些蝴蝶品系中都已经建立了转基因技术,因此可以更详细地对眼斑基因(包括从形态发生素到色素合成的相关基因)进行功能研究(Ramos 和 Monteiro 2007)。在不同的蝴蝶品种中,已有报道可以用 DNA 电穿孔技术(Golden et al. 2007)或病毒载体(Lewis et al. 1999; Lewis 和 Brunetti 2006)转化翅表皮细胞,并在热带蝴蝶(*B. anynana*)中已成功地运用转座元件载体进行转化(Marcus, Ramos 和 Monteiro 2004)。后者可以与最近发展的精确控制转基因的时空表达的新技术(即激光诱导的热激蛋白)相结合(Ramos et al. 2006),还能促进用可诱导的基因敲除或异位突变等手段去研究基因功能(Ramos 和 Monteiro 2007)。

这些实验技术和热带蝴蝶(*B. anynana*)逐渐丰富的基因序列和多态性信息,再加上其他鳞翅目昆虫的相关信息将帮助我们认识翅模式发育,研究不同品种间翅模式发育和变异的共同点和差异。当在大量的品种中研究清楚这些遗传和发育机制后,再回到自然群体,结合生态学压力的知识来解释这些机制是怎样影响翅模式进化多样性的。

参考文献

[1]Allen,C. E. 2008. The "eyespot module" and eyespots as modules: Development, evolution, and integration of a complex phenotype. *J. Exp. Zoolog. B Mol. Dev. Evol.* 310:179—90.

[2]Allen,C. E. ,P. Beldade,B. J. Zwaan,P. M. Brakefield. 2008. Differences in the selection response of serially repeated color pattern characters: standing variation, development, and evolution. *BMC Evol. Biol.* 8:94.

[3]Beldade, P. , and P. M. Brakefield. 2002. The genetics and evo-devo of butterfly wing patterns. *Nat. Rev. Genet.* 3:442—52.

[4]Beldade, P. , P. M. Brakefield, and A. D. Long. 2002. Contribution of *Distal-less* to quantitative variation in butterfly eyespots. *Nature* 415:315—18.

[5]Beldade, P. , P. M. Brakefield, and A. D. Long. 2005. Generating phenotypic variation: prospects from "evodevo" research on *Bicyclus anynana* wing patterns. *Evol. Dev.* 7:101—07.

[6]Beldade, P. , V. French, and P. M. Brakefield. 2008. Developmental and genetic mechanisms for evolutionary diversification of serial repeats: Eyespot size in *Bicyclus anynana* butterflies. *J. Exp. Zoolog. B Mol. Dev. Evol.* 310:191—201.

[7]Beldade, P. , K. Koops, and P. M. Brakefield. 2002. Developmental constraints versus flexibility in morphological evolution. *Nature* 416:844—47.

[8]Beldade, P. , W. O. McMillan, and A. Papanicolaou. 2008. Butterfly genomics eclosing. *Heredity* 100:150—57.

[9]Beldade, P. , S. Rudd, J. D. Gruber, and A. D. Long. 2006. A wing expressed sequence tag resource for *Bicyclus anynana* butterflies, an evo-devo model. *BMC Genomics* 7:130.

[10]Beldade, P. , S. V. Saenko, N. Pul, and A. D. Long. 2009. A gene-based linkage map for *Bicyclus anynana* butterflies allows for a comprehensive analysis of synteny with the lepidopteran reference genome. *PLoS Genet.* 5: e1000366.

[11]Beldade, P. , and S. V. Saenko. 2009. Conserved developmental processes and the evolution of novel traits: wounds, embryos, veins, and butterfly eyespots. In *Animal Evolution*, M Telford & T Littlewood (eds). Novartis Foundation. (in press)

[12]Brakefield, P. M. , P. Beldade, and B. J. Zwaan. 2009. The African butterfly *Bicyclus anynana*: Evolutionary genetics and evo-devo. In *Emerging model organisms: A laboratory manual*, Vol. 1, ed. R. R. Behringer, A. D. Johnson, and R. E. Krumlauf. Cold Spring Harbor: Cold Spring Harbor Laboratory. pp. 291—330.

[13]Brakefield, P. M. , and W. A. Frankino. 2009. Polyphenisms in Lepidoptera: Multidisciplinary approaches to studies of evolution. In *Phenotypic plasticity in insects. Mechanisms and consequences*, ed. D. W. Whitman and T. N. Ananthakrishnan. Oxford: Oxford Univ. Press. pp. 121—152.

[14]Brakefield, P. M. , and V. French. 1995. Eyespot development on butterfly wings: The epidermal response to damage. *Dev. Biol.* 168:98—111.

[15]Brakefield, P. M. , and V. French. 1999. Butterfly wings: The evolution of development of colour patterns. *Bioessays* 21:391—401.

[16]Brakefield, P. M. , J. Gates, D. Keys, et al. 1996. Development, plasticity and evolution of butterfly wing patterns. *Nature* 384:236—42.

[17]Brakefield, P. M. , F. Kesbeke, and P. B. Koch. 1998. The regulation of phenotypic plasticity of eyespots in the butterfly *Bicyclus anynana*. *Am. Nat.* 152:853—60.

[18]Brakefield, P. M. , and J. C. Roskam. 2006. Exploring evolutionary constraints is a task for an integrative evolutionary biology. Am. Nat. 168, Suppl. 6:S4—13.

[19]Brunetti,C. R. ,J E. Selegue,A. Monteiro,V. French,P. M. Brakefield,and S. B. Carroll. 2001. The generation and diversification of butterfly eyespot color patterns. *Curr. Biol.* 11:1578—85.

[20]Carroll,S. B. ,J. Gates,D. N. Keys,et al. 1994. Pattern formation and eyespot determination in butterfly wings. *Science* 265:109—14.

[21]Clark,R. ,S. M. Brown,S. C. Collins,C. D. Jiggins,D. G. Heckel,and A. P. Vogler. 2008. Colour pattern specification in the Mocker swallowtail *Papilio dardanus*: The transcription factor *invected* is a candidate for the mimicry locus H. *Proc. Biol. Sci.* 275:1181—88.

[22]Condamin, M. 1973. *Monographie du Genre Bicyclus* (Lepidoptera, Satyridae). Dakar: Inst. Fond. Afr. Noire.

[23]Costanzo,K. ,and A. Monteiro. 2007. The use of chemical and visual cues in female choice in the butterfly *Bicyclus anynana*. *Proc. Biol. Sci.* 274:845—51.

[24]Dilão,R. ,and J. Sainhas. 2004. Modelling butterfly wing eyespot patterns. *Proc. Biol. Sci.* 271:1565—69.

[25]Evans,T. M. ,and J. M. Marcus. 2006. A simulation study of the genetic regulatory hierarchy for butterfly eyespot focus determination. *Evol. Dev.* 8:273—83.

[26]French,V. ,and P. M. Brakefield. 1992. The development of eyespot patterns on butterfly wings: Morphogen sources or sinks? *Development* 116:103—09.

[27]French,V. ,and P. M. Brakefield. 1995. Eyespot development on butterfly wings: The focal signal. *Dev. Biol.* 168:112—23.

[28]Galant,R. ,J. B. Skeath,S. Paddock,D. L. Lewis,and S. B. Carroll. 1998. Expression pattern of a butterfly *achaete-scute* homolog reveals the homology of butterfly wing scales and insect sensory bristles. *Curr. Biol.* 8:807—13.

[29]Golden,K. ,V. Saji,N. Markwarth,B. Chen,and A Monteiro. 2007. In vivo electroporation of DNA into the wing epidermis of a butterfly. *J. Insect Sci.* 7:53.

[30]Joron,M. ,C. D. Jiggins,A. Papanicolaou,and W. O. McMillan. 2006. *Heliconius* wing patterns: An evo-devo model for understanding phenotypic diversity. *Heredity* 97:157—67.

[31]Joron, M. , and J. L. B. Mallet. 1998. Diversity in mimicry: Paradox or paradigm? *Trends Ecol. Evol.* 13:461—66.

[32]Keys,D. N. ,D. L. Lewis,J. E. Selegue,et al. 1999. Recruitment of a *hedgehog* regulatory circuit in butterfly eyespot evolution. *Science* 283:532—34.

[33]Koch,P. B. ,B. Behnecke,and R. H. ffrench-Constant. 2000. The molecular basis of melanism and mimicry in a swallowtail butterfly. *Curr. Biol.* 10:591—94.

[34]Koch,P. B. ,P. M. Brakefield,and F. Kesbeke. 1996. Ecdysteroids control eyespot size and wing color pattern in the polyphenic butterfly *Bicyclus anynana* (Lepidoptera: Satyridae). *J. Insect Physiol.* 42:223—30.

[35]Koch,P. B. ,U. Lorenz,P. M. Brakefield,and R. H. ffrench-Constant. 2000. Butterfly wing pattern mutants: Developmental heterochrony and co-ordinately regulated phenotypes. *Dev. Genes Evol.* 210:536—44.

[36]Koch, P. B., R. Merk, R. Reinhardt, and P. Weber. 2003. Localization of ecdysone receptor protein during colour pattern formation in wings of the butterfly *Precis coenia* (Lepidoptera: Nymphalidae) and co-expression with Distal-less protein. *Dev. Genes Evol.* 212:571—84.

[37]Kronforst, M. R., L. G. Young, D. D. Kapan, C. McNeely, R. J. O'Neill, and L. E. Gilbert. 2006. Linkage of butterfly mate preference and wing color preference cue at the genomic location of *wingless*. *Proc. Natl. Acad. Sci. U. S. A.* 103:6575—80.

[38]Lewis, D. L., and C. R. Brunetti. 2006. Ectopic transgene expression in butterfly imaginal wing discs using vaccinia virus. *Biotechniques* 40:48, 50, 52.

[39]Lewis, D. L., M. A. DeCamillis, C. R. Brunetti, et al. 1999. Ectopic gene expression and homeotic transformations in arthropods using recombinant Sindbis viruses. *Curr. Biol.* 9:1279—87.

[40]Long, A. D., P. Beldade, and S. J. Macdonald. 2007. Estimation of population heterozygosity and library con-struction-induced mutation rate from expressed sequence tag collections. *Genetics* 176: 711—14.

[41]Lyytinen, A., P. M. Brakefield, L. Lindstrom, and J. Mappes. 2004. Does predation maintain eyespot plasticity in *Bicyclus anynana*? *Proc. Biol. Sci.* 271:279—83.

[42]Marcus, J. M., and T. M. Evans. 2008. A simulation study of mutations in the genetic regulatory hierarchy for butterfly eyespot focus determination. *Biosystems.* 93:250—255.

[43]Marcus, J. M., D. M. Ramos, and A. Monteiro. 2004. Germline transformation of the butterfly *Bicyclus anynana*. *Proc. Biol. Sci.* 271, Suppl. 5:S263—65.

[44]Majerus, M. E. N. 1998. *Melanism: Evolution in action*. Oxford: Oxford Univ. Press.

[45]McMillan, W. O., A. Monteiro, and D. D. Kapan. 2002. Development and evolution on the wing. *Trends Ecol. Evol.* 17:125—33.

[46]Monteiro A. 2008. Alternative models for the evolution of eyespots and of serial homology on lepidopteran wings. *Bioessays* 30:358—66.

[47]Monteiro, A. F., P. M. Brakefield, and V. French. 1994. The evolutionary genetics and developmental basis of wing pattern variation in the butterfly *Bicyclus anynana*. *Evolution* 48:1147—57.

[48]Monteiro, A., P. M. Brakefield, and V. French. 1997a. Butterfly eyespots: The genetics and development of the color rings. *Evolution* 51:1207—16.

[49]Monteiro, A., P. M. Brakefield, and V. French. 1997b. The genetics and development of an eyespot pattern in the butterfly *Bicyclus anynana*: Response to selection for eyespot shape. *Genetics* 146:287—94.

[50]Monteiro, A., B. Chen, L. C. Scott, et al. 2007. The combined effect of two mutations that alter serially homologous color pattern elements on the fore and hindwings of a butterfly. *BMC Genet.* 8:22.

[51]Monteiro, A., G. Glaser, S. Stockslager, N. Glansdorp, and D. Ramos. 2006. Comparative insights into questions of lepidopteran wing pattern homology. *BMC Dev. Biol.* 6:52.

[52]Monteiro, A., and N. E. Pierce. 2001. Phylogeny of *Bicyclus* (Lepidoptera: Nymphalidae) inferred from COI, COII, and *EF-1* alpha gene sequences. *Mol. Phylogenet. Evol.* 18:264—81.

[53]Monteiro, A., J. Prijs, M. Bax, T. Hakkaart, and P. M. Brakefield. 2003. Mutants highlight the

modular control of butterfly eyespot patterns. *Evol. Dev.* 5:180—87.

[54]Nijhout, H. F. 1980. Pattern formation on lepidopteran wings: Determination of an eyespot. *Dev. Biol.* 80:267—74.

[55]Nijhout, H. F. 1991. *The development and evolution of butterfly wing patterns*. Washington, DC: Smithsonian Inst. Press.

[56]Nijhout, H. F. 2003a. The development and evolution of adaptive polyphenisms. *Evol. Dev.* 5:9—18.

[57]Nijhout, H. F. 2003b. Polymorphic mimicry in *Papilio dardanus*: Mosaic dominance, big effects, and origins. *Evol. Dev.* 5:579—92.

[58]Papanicolaou, A., S. Gebauer-Jung, M. L. Blaxter, W. O. McMillan, and C. D. Jiggins. 2008. Butterfly Base: A platform for lepidopteran genomics. *Nucleic Acids Res.* 36:D582—87.

[59]Parchem, R. J., M. W. Perry, and N. H. Patel. 2007. Patterns on the insect wing. *Curr. Opin. Genet. Dev.* 17:300—08.

[60]Paulsen, S. M., and H. F. Nijhout. 1993. Phenotypic correlation structure among elements of the color pattern in *Precis coenia* (Lepidoptera, Nymphalidae). *Evolution* 47:593—618.

[61]Pringle, E. G., S. W. Baxter, C. L. Webster, A. Papanicolaou, S. F. Lee, and C. D. Jiggins. 2007. Synteny and chromosome evolution in the lepidoptera: Evidence from mapping in *Heliconius melpomene*. *Genetics* 177:417—26.

[62]Ramos, D. M., F. Kamal, E. A. Wimmer, A. N. Cartwright, and A. Monteiro. 2006. Temporal and spatial control of transgene expression using laser induction of the hsp70 promoter. *BMC Dev. Biol.* 6:55.

[63]Ramos, D. M., and A. Monteiro. 2007. Transgenic approaches to study wing color pattern development in Lepidoptera. *Mol. Biosyst.* 3:530—35.

[64]Reed, R. D. 2004. Evidence for Notch-mediated lateral inhibition in organizing butterfly wing scales. *Dev. Genes Evol.* 214:43—46.

[65]Reed, R. D., P. H. Chen, and H. F. Nijhout. 2007. Cryptic variation in butterfly eyespot development: The importance of sample size in gene expression studies. *Evol Dev* 9:2—9.

[66]Reed, R. D., W. O. McMillan, and L. M. Nagy. 2008. Gene expression underlying adaptive variation in Heliconius wing patterns: Non-modular regulation of overlapping cinnabar and vermilion prepatterns. *Proc. Biol. Sci.* 275:37—45.

[67]Reed, R. D., and L. M. Nagy. 2005. Evolutionary redeployment of a biosynthetic module: expression of eye pigment genes vermilion, cinnabar, and white in butterfly wing development. *Evol. Dev.* 7:301—11.

[68]Reed, R. D., and M. S. Serfas. 2004. Butterfly wing pattern evolution is associated with changes in a Notch/ Distal-less temporal pattern formation process. *Curr. Biol.* 14:1159—66.

[69]Riddiford, L. M. 1995. Hormonal regulation of gene expression during lepidopteran development. In *Molecular model systems in the Lepidoptera*, ed. M. R. Goldsmith and A. S. Wilkins, 1st ed., 293—322. Cambridge: Cambridge Univ. Press.

[70]Robertson, K. A., and A. Monteiro. 2005. Female *Bicyclus anynana* butterflies choose males on the basis of their dorsal UV-reflective eyespot pupils. *Proc. Biol. Sci.* 272:1541—46.

[71] Saenko, S. V., V. French, P. M. Brakefield, and P. Beldade. 2008. Conserved developmental

processes and the formation of evolutionary novelties: Examples from butterfly wings. *Philos. Trans. R. Soc. Lond. B Biol. Sci.* 363:1549—55.

[72]Sahara, K., A. Yoshido, F. Marec, et al. 2007. Conserved synteny of genes between chromosome 15 of *Bombyx mori* and a chromosome of *Manduca sexta* shown by five-color BAC—FISH. *Genome* 50: 1061—65.

[73]Stevens, M. 2005. The role of eyespots as anti-predator mechanisms, principally demonstrated in the Lepidoptera. *Biol. Rev. Camb. Philos. Soc.* 80:573—88.

[74]Wittkopp, P. J., and P. Beldade. 2009. Development and evolution of insect pigmentation: Genetic mechanisms and the potential consequences of pleiotropy. *Sem. Cell. Dev. Biol.* 20: 65—71.

第六章 鳞翅目昆虫基因组中鉴定适应性基因的前景：一个蝴蝶色彩模式研究的案例

Simon W. Baxter, Owen McMillan, Nicola Chamberlain, Richard H. ffrench-Constant, and Chris D. Jiggins

1. 背景

鳞翅目是一个变化多样的进化枝，其生态和进化过程一直备受生物学家的关注。这可以追溯到对自然群体的经典进化遗传研究，它为我们认识大自然界作出了重要贡献(Clarke 和 Sheppard 1960；Kettlewell 1973；Ford 1975)。虽然如此，基因型与表型关系的研究依然是棘手的“黑盒子”。与之相反，我们现在所处的时代即使对于研究背景不清楚的基因组，仍能较容易地克隆其主要表型效应基因，这就为我们在分子水平上研究表型进化提供了空前的机会(Feder 和 Mitchell-Olds 2003)。因此，现在对鳞翅目昆虫中控制主要表型性状基因的鉴定是相当重要的，最近几项研究已经在解释表型特征的分子基础方面取得了重大进展，包括从杀虫药抗药性(通过扩散能力)到形态性状(如色彩模式)(Gahan，Gould 和 Heckel 2001；Daborn et al. 2002；Hanski 和 Saccheri 2006；Joron et al. 2007)。

釉蛱蝶 *Heliconius* 翅模式作为性状进化适应辐射的例子引起进化生物学家极大的兴趣，在生态水平上得到了很好的研究(Joron et al. 2007)。墨尔波墨蝶 *Heliconius melpomene* 和瓦氏袖蝶 *H. erato* 属于新热带蝴蝶，同地方种群有相同的翅模式，然而在不同的地理区域则显示出不同模式。虽然墨尔波墨蝶 *Heliconius melpomene* 和瓦氏袖蝶 *H. erato* 不能种间杂交，但是在每个种内不同翅模式表型的品种可以进行杂交。过去的研究表明，表现出釉蛱蝶 *Heliconius* 翅模式元件性状出现与否仅由极少数位点控制(Sheppard et al. 1985)。因此，一个看似简单的遗传系统是快速适应辐射和许多进化聚集事件的基础。狐眼袖蝶(*Heliconius numata*)有一个不同寻常的“超基因”系统，在这个系统中单独的孟德尔位点即可控制整个翅模式的多态性(Joron et al. 2007)。翅模式形成的分子基础为解释不同基因型与表型的联系提供了令人兴奋的机会。我们尤其想知道等位基因的单一位点变化是如何来控制模式发育及这些变异中的分子变化的。另外，这将为我们提供一个很好的机会来研究位点周围的序列变化，这些位点的变化是经过很强的自然选择而重复发生的。更重要的是，所有的这些进化改变都是多次发生的，包括分散与聚集，这就为研究那些没有被覆盖的模式的重复性提供了机会。

本章描述了现在对于 *Heliconius* 可利用的分子资源及如何利用这些资源鉴定控制翅模式变化的基因。这些方法可以用来精确定位其他鳞翅目物种中分离的孟德尔或数量遗传位点。另外，我们还讨论了鳞翅目昆虫中染色体共线性的意义及鉴定控制主要表型的候选基因。因此，本章的主要目标是概括至今在我们工作中所获得的关于 *Heliconius* 的研究结果，将有助于在鳞翅目昆虫中寻找基因。

2. 釉蛱蝶(*Heliconius*)的可利用资源

在过去的五年中，在研究釉蛱蝶 *Heliconius* 蝴蝶的过程中产生了一些新的分子资源，包括细菌人工染色体文库(BAC)、基因组综述序列(GSSs)、表达序列标签(EST)文库、微卫星标记和蝴蝶数据库 ButterflyBase(一个在线数据库)。这些资源在墨尔波墨蝶 *H. melpomene*、瓦氏袖蝶 *H. erato* 和狐眼袖蝶(*H. numata*)3 个目标物种中平行发展，目标就是促进这些基因组进化的比较分析。

2.1 微卫星标记(Microsatellite Markers)

与其他鳞翅类一样,已证明在釉蛱蝶 *Heliconius* 中发展大量的微卫星标记是很困难的(Mavárez 和 González 2006)。虽然如此,在瓦氏袖蝶 *H. erato* 和墨尔波墨蝶 *H. melpomene* 两个物种分别定位了大约 20 个标记,用于在新的定位实验中鉴定染色体连锁群(LGs)(Jiggins et al. 2005; Kapan et al. 2006)。在很多情况下由墨尔波墨蝶 *H. melpomene* 发展的标记在狐眼袖蝶(*H. numata*)中也适用。这两个物种相同位点的数目很少,所以这些标记在比较连锁分析时并不是特别有用。虽然如此,仍有 13 个位点在墨尔波墨蝶 *H. melpomene* 和蓝白相间长翼蛱蝶 *H. cydno* 的群体遗传实验中是有用的,且已经利用杂交研究对隐藏物种的鉴定和杂交物种的形成(Mavárez et al. 2006)。

2.2 基因组文库(Genomic Libraries)

目前插入基因组大片段的 BAC 文库已经可以用于三种釉蛱蝶:墨尔波墨蝶(*H. melpomene*)、狐眼袖蝶(*H. numata*)和瓦氏袖蝶(*H. erato*)。插入片段平均大小和基因组覆盖度为 120kb,8 倍(*H. melpomene*)和 115kb, 7.5 倍(*H. numata*)(available from Amplicon Express)。对于瓦氏袖蝶 *H. erato* 有两个文库可用,是利用不同的选择性核酸内切酶构建而成的,平均插入片段为 153kb 和 175kb(C. Wu, pers. comm.),这样就覆盖了整个基因组 16 倍(Papa et al. 2008)。这些文库均可用于高密度尼龙膜筛选。墨尔波墨蝶 *H. melpomene* 的整个文库已经被组装成 2091 个重叠群(plus singletons; Baxter et al. 2008)。

2.3 基因组综述序列(Genomic Sequence Surveys)

墨尔波墨蝶 *H. melpomene* BAC 文库的 18000 个克隆均已完成测序,一共得到 32528 个高质量单遍基因组综述序列(GSSs)。这些序列随机提供占整个基因组 7%的信息,并且包括了大量的重复元件和推测的基因序列。利用 PartiGene 软件(Parkinson et al. 2004)对这些读长进行聚类,得到 23420 个序列聚类群和单态,最大的聚类群包括了 161 个重复序列。对釉蛱蝶 *Heliconius* 基因组重复元件的初步调查结果见 Papa et al. (2008)。共有 2133 个聚类群翻译后能与 UniProt 蛋白数据库中至少一个目标比对上,其相似性得分值都在 80 以上,有 4121 个聚类群通过与家蚕基因组 tBLASTx 比对得到相同的分值(Ramen contigs from KAIKObase http://sgp.dna.affrc.go.jp/pubdata/genomicsequences.html),因此,可以确信它们一定包含了一些基因的片段或者保守元件。所以,这些序列对基因和标记的鉴定、釉蛱蝶 *Heliconius* 基因组中重复序列的鉴定、通过基于序列的标记锚定 BAC 重叠群均是一种很有用的资源。

2.4 cDNA 文库

构建好的互补 DNA(cDNA)文库是发展基因资源很重要的第一步,是鉴定重要基因功能的所有研究设计的起点。文库应该由全长 cDNA 组成,有方向性(方向与 3′端 5′端 UTRs 引物有关),且必须是归一化的。归一化就是用转录组来补偿转录本丰度的过程。

由于转录本在表达量上存在几个数量级的变化，少数高表达基因常常占一个细胞内 mRNA 的大部分(Bonaldo，Lennon 和 Soares 1996；Hillier et al. 1996)。因此，如果没有归一化，cDNA 文库的重组克隆中插入高表达的基因大部分是“看家基因”(housekeeping genes)，如肌动蛋白、微管蛋白和核糖体蛋白。所以，归一化有助于发现稀有的转录本，这些转录本可能会在生态和发育过程中扮演重要角色(Bouck 和 Vision 2007)。

文库的构建比起过去十年已经变得相对容易了，现在有很多的商品化的试剂盒来分离提取双链 cDNA、归一化和文库构建。我们最近成功运用了 Trimmer-Direct cDNA 标准化试剂盒(Evrogen，Moscow，Russia)。虽然如此，构建一个好的 cDNA 文库仍需要专业技术，并且这个过程可以交给商业公司来做，虽然前期花费较高，但当专业技术与经验受限时，这是最有效率的方法，也是通向成功的捷径。

2.5 转录组测序分析

转录组测序是寻找基因、注释基因组序列和研究发育的有利方法，随着测序成本逐渐降低(仍然在降)，通过适度的 cDNA 文库测序可以很快获得特异组织的转录本组的第一手资料。另外，通过对 cDNA 文库克隆随机测序获得的大量 ESTs 数据包括了各种分类的基因组信息(Bouck 和 Vision 2007)。因此，ESTs 是连接主要模式生物基因组信息的桥梁，也是适应性变化研究、比较作图(Pringle et al. 2007)和表达研究(Reed，McMillan 和 Nagy 2007；Bouck 和 Vision 2007)的重要基础。瓦氏袖蝶(*H. erato*)和墨尔波墨蝶(*H. melpomene*)不同发育时期翅原基的 EST 测序都已初步经完成(Papanicolaou et al. 2005)。就瓦氏袖蝶(*H. erato*)而言，目前已经从一个由不同发育时期的翅组成的 cDNA 文库中测得约 18000 条 EST 序列。由于这个文库并不是一个标准化的文库，有很多基因常被多次测序，其中有 20 个基因已经测过 75 次以上。尽管冗余性增加了花费和减慢基因的发现，但是它也有两个重要的好处。一方面，这些重复意味着我们测的转录本都比较长，有助于下游基因的鉴定和描述；另一方面，由于我们的文库包括了很多个体的材料，所以这些重叠的序列中包含了大量的 SNPs 信息，可用于将来的群体基因组学研究。这 18000 条 EST 序列可以分成 6564 个亚族，即代表了 3000～5000 个不同的基因，包括了昆虫翅发育、信号传输和色素形成中的基因(Papanicolaou et al. 2007)。另外，还有一部分编码区域，预测为分子量相对较大的肽，它们与黑腹果蝇(*Drosophila melanogaster*)和家蚕(唯一的一个完成全基因组测序的鳞翅目昆虫)的基因都没有同源性，推测可能是蝴蝶中的新基因或品种间分化很大的基因。

第二代测序技术为研究转录本差异提供了一个更快更完美的策略，并促进了鳞翅目昆虫功能基因组研究方法的快速发展(Toth et al. 2007；Hudson 2008；Vera et al. 2008)。更重要的是，用该技术研究基因表达特征不再需要构建文库，一个焦磷酸测序反应只需要 1～4μg 双链 cDNA 作为模板。如有必要，cDNA 还可以标准化，然后会很快获得 EST 数据。例如：一个 454 高通量焦磷酸测序仪(Roche Diagonostics，Burgess Hill，West Sussex，UK)可以在 8h 内产生约 100Mb 个碱基序列，包含了约 400000 个序列读长，序列的平均读长可达 200～250bp(bp；Hudson 2008)。除了同聚物外，单一读长的准确性非常高(Margulies et al. 2005；Huse et al. 2007)。而且还有很多重叠的读长，因此精确性就更高了(Wicker et al. 2006)。第一次在墨尔波墨蝶(*H. melpomene*)中对标准化的 cDNA 样本进行的单次测序实验就获得了 6500 个平均长度为 1050bp 的序列。虽然因一些种内的多态性或选择性拼接造成这些短序列读长的拼接组装时具有一定困难(Trombetti et al. 2007),但测得的基因数量与在瓦氏袖蝶(*H. erato*)中大规模的 EST 测序获得的基因数量非常相似。将传统的 Sanger 法测序与第二代 454 测序平台获得的数据结合起来,就可以在短期内通过高质量的比对获得一个高精度的袖蝶转录组数据,在狐眼袖蝶 *H. numata* 和瓦氏袖蝶(*H. erato*)中也已开展了同样的测序工作。

3. 蝴蝶数据库 ButterflyBase

各种 EST 测序产生了大量的数据,这为数据的管理提出挑战。必须去除载体序列和一些低质量的序列,还需对序列之间进行比对,把那些可能是来自于同一位点的序列聚集在一起,将这些重叠的序列组装拼接成一个大的片段(例如:contigs)。而且这些数据应该放在一个有注释信息的数据库里。至少这个数据库应该包括如下信息:(1)与其他生物基因的序列相似性;(2)准确的蛋白质翻译;(3)推测的生物功能和保守的蛋白质功能域。此外,还应考虑 SNPs、表达模式和推测的基因相互作用等信息。现在已有许多可利用的各种自动化的原始数据分析平台和大量鳞翅目特异物种的数据库,例如:SpodoBase(http://bioweb. ensam. inra. fr/spodobase/)是一个苜蓿夜蛾(*Spodoptera frugiperda*)的整合 EST 数据库。

SpodoBase、KAIKObase 和 SilkDB 这些特异物种的数据库都是非常有价值的资源,但分析和进入方式都有不同程度的差异。因此,研究者不能完全利用这些资源把鳞翅目昆虫作为独特的模型来研究功能变异的起源和保持。ButterflyBase 的建立第一次试图把这些鳞翅目昆虫的基因组信息整合在一起(Papanicolaou et al. 2007)。这个数据库把所有鳞翅目的 EST 数据聚在一起,并加以注释。到 2007 年 8 月,该数据库已经把保存的 273077 条 mRNA 序列聚类成了 70867 个基因,其中大部分都已预测了蛋白质序列。GenBank 中保存的所有 6907 条鳞翅目的蛋白质序列与该数据库相比,很多序列都只是一部分或者一些片段。该数据库中的所有序列都用同样的分析平台处理(Parkinson et al. 2004),因此可以在物种间直接作数据的比较分析。这些数据可以用文本检索相似注释的数据,也可以用 BLAST 工具进行检索。传统的 blastall 选项 (BLASTn, tBLASTx 等)和新算法 psi-BLAST 和 msBLAST(用蛋白组研究中获得的短的蛋白序列检索数据库的工具)等各种 BLAST 工具都可以用于检索。

当然,目前的数据分析水平有局限性,特别是用一个自动的分析平台去分析这些原始的序列信息确实有缺点。主要的问题是 EST 序列的聚类可能会有错:同一个基因的不同序列可能会因为等位突变或测序错误而聚为不同的群,相反,亲缘关系较近的旁系同源基因也可能会错误地聚集在一起。例如,在瓦氏袖蝶(*H. erato*)的数据里,6500 多个群(cluster) 可以组成 2000 个超集群(superclusters),每个超集群都由许多具有序列相似性的单个群组成。瓦氏袖蝶(*H. erato*)种内变异很大(平均每 25bp 就有一个多态性位点),最近的分析表明至少有一些超集群是由同一位点的等位突变组成的。任何自动的分析平台都会出现这些错误,就要求我们在利用 ButterflyBase 的数据时要加以考虑。接下来为了克服这些问题，我们将像 FlyBase 数据库那样添加一个人为校正的原件，那样用户就可以系统地校正这些错误。在 ButterflyBase 数据库里，用户可以通过一个注释 wiki 对目标基因添加注释，并且我们鼓励用户利用该功能。

4. 利用杂交的遗传分析

基因组序列提供了一系列有用的工具，但釉蛱蝶(*Heleconius*)系统真正的价值在可以做遗传杂交实验，从而鉴定控制翅模式的自发等位突变。用不同瓦氏袖蝶(*H. erato*)和墨尔波墨蝶(*H. melpomene*)品种杂交，在后代可对性状分离进行遗传分析。用同一物种中翅模式不同的个体进行杂交，通过观察后代分离，分析表型特征的显隐性。

鳞翅目昆虫单倍体的全着丝点染色体数目都比较大，一般是 28～32 条(Suomalainen 1969)。鳞翅目昆虫的染色体系统对物种特征的遗传分析非常有利(第三章对鳞翅目昆虫的染色体的性质特征进行了综述)。在减数分裂中，雄性同源染色体之间发生交换和形成配子，而在雌性中不会发生交换(Maeda 1939; Suomalainen, Cook 和 Turner 1973; Turner 和 Sheppard 1975; Traut 1977)。每个来源于雌性的染色体都完整地保留在后代中，同一染色体上的基因和标记完全连锁。在卵发生时，同源染色体分离，后代平等地遗传其中一条，因此没有来自两个亲本的染色体之间的复杂交换，可以直接判断连锁关系；而在雄性鳞翅目昆虫中，在精子发生时，会发生染色体的交换(Maeda 1939; Turner 和 Sheppard 1975; Traut 1977)。遗传杂交实验中，用带雌性信息的标记把表型特征定位到具体的染色体上，然后再用带雄性信息的标记来定位。

在不同表型的品种间做杂交实验可以分离感兴趣的性状。这对后面在连锁图谱上定位这些位点非常重要。尤其值得注意的是当发生多种交换的时候，很难用单对交配(一对雌雄交配)实验来确定分离模式。例如：通过大量的杂交实验，Eberle 和 Jehle (2006)研究发现苹果小卷蛾(*Cydia pomonella*)的杆状病毒抗性是常染色体控制的非完全显性的性状。然而如果用一个单对交配的杂交系统，那么该抗性特征就会被误认为是性连锁的遗传(Asser-Kaiser et al. 2007)。

根据感兴趣性状的分离特征(是显性还是隐性)，可以实施很多杂交策略。Heckel et al. (1999)根据控制小菜蛾(*Plutella xylostella*)对苏云金杆菌(*Bacillus thuringiensis*, Bt)抗性的单隐性位点的分离特征，而采取回交策略，用一个纯合的隐性抗性个体与一个纯合的敏感系个体杂交，F_1 代的雌性个体再与抗性品种回交，这样就建立了一个拥有雌标记信息的定位群体，用来将该形状定位到连锁群上。同时用 F_1 代的雄性个体与抗性品种回交，建立一个拥有雄标记信息的定位群体来为特定的连锁群产生重组连锁图(Heckel et al. 1999; Baxter et al. 2005;图 6.1A)。

我们用从南美不同地方收集的色彩模式不同的釉蛱蝶(*Heleconius*)杂交获得 F_2 代群体(Jiggins et al. 2005; Tobler et al. 2005; Kapan et al. 2006)。使用 F_2 杂交群体而不用回交群体的主要原因是来自双亲的显隐性性状都可以用这一群体定位(图 6.1B)。而在二选一的回交设计中，研究一个位点需要 4 种不同的定位群体(雌雄)。而且，在系统内和系统间都有丰富的多态性，利用雌完全连锁的特征将标记定位在连锁群上的优点，再用一个 F_2 杂交设计就可以构建一个在亲本间有多态性标记的连锁图（Yasukochi 1998；Jiggins et al. 2005）。

每个单对交配的杂交实验可获得的 F_2 代和回交后代的个体数是非常重要的。在我们的研究中，用一窝 120～180 个个体的群体来初步确定具有翅模式基因位点的染色体和定位其大致区域。Jiggins 还用一个大约只有 70 个个体的 F_2 代小群体做了一个整合的遗传图谱（Jiggins et al. 2005）。接着可以在同样表型分离的其他窝的群体里定位一些紧密连锁的标记，以得到一个更精确的连锁定位图。

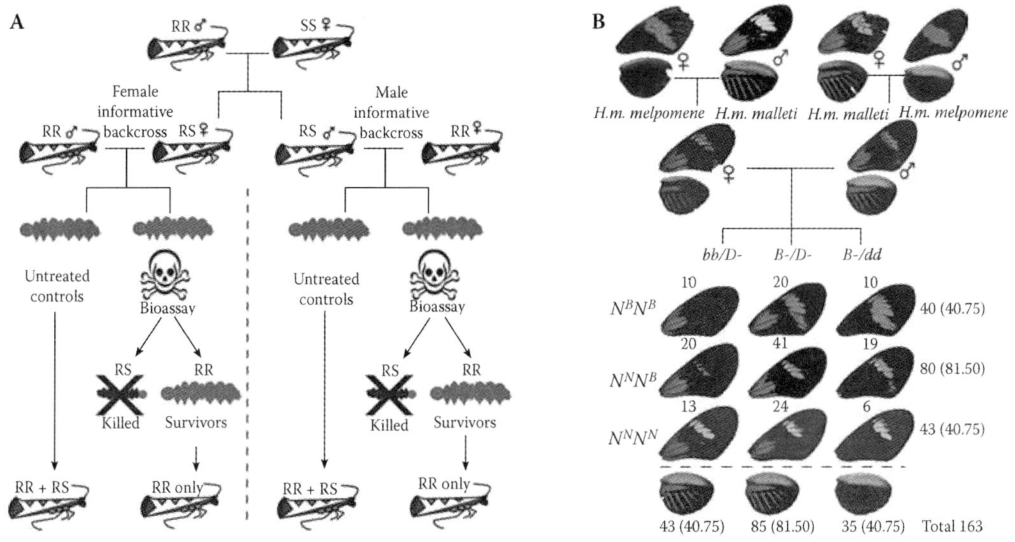

图 6.1　遗传定位的杂交组合（彩色版本图片见附录图 6.1）。（A）一个使小菜蛾（*Plutella xylostella*）对苏云金杆菌（Bt）CrylA 毒素的隐性抗性基因分离的回交组合。抗性（RR）和敏感系（SS）个体单对交配，获得对 Bt 敏感的 F_1 代杂合体（RS）。再用 F_1 雌和 F_1 雄分别和抗性品种回交。F_2 代幼虫一部分用 BT 毒素做生物检测杀死仍为敏感性的杂合子，另一部分不做任何处理直接饲养至化蛾。由于雌完全连锁，可用 AFLP 标记鉴定抗性因子的连锁群。F_1 代雌性个体遗传下来的与抗性基因所在的连锁群会在未处理的群体中按 1R∶1S 分离，在药物检测的群体中表现出固有的分离率。而用 F_1 雄回交的组合可以用于抗性基因连锁的 AFLP 标记来定位该基因的位点，因为在精子发生时会出现交换。在作者同意的前提下，图片改编自 Heckel et al.（1999）。（B）不同翅模式表型的墨尔波墨蝶（*H. melpomene*）品种之间的杂交（44 窝）。*H. m. melpomene* 来自法属圭亚那（French Guiana，位于南美洲北部），表型为 *HmB*（*B*，显性，前翅红条纹），*H. m. malleti* 来自厄瓜多尔（Ecuador），表型为 *HmD*（*D*，显性，后翅和前翅基部呈红色射线状）和 *HmN*（*N*，不完全显性，前翅有黄条纹）。F_1 代表现出三种表型的杂合子（$BbDdN^NN^B$）。163 个 F_2 代个体表现出 9 种表型，各自的比例（如括号中所示）与预期比例值相差无几，为 1∶2∶1，N^NN^N，N^NN^B，N^BN^B 分别为纯合子，杂合子和无 *HmN* 的纯合子，bb/*D*- 无 *HmB* 表型，B-/D-是有红色条纹表型的杂合子，和 B-/*dd* 无 *HmD* 表型。“－”表示由显性基因控制表型。也可参阅 Sheppard et al.（1985），449 页。

5. 用 AFLPs 标记构建遗传连锁图

扩增片段长度多态性（AFLP）标记最初用来建立 DNA 指纹图谱，并不用于研究基因组起源和基因组中的结构复杂性问题(Vos et al. 1995)。十多年后,AFLP 才作为一种重要的分子工具用于连锁定位研究、数量性状基因座(QTL)的鉴定、系统发生树的重建、群体遗传结构的分析和 cDNA 转录组分析(参阅 Bensch 和 Akesson 2005; Meudt 和 Clarke 2007 的综述)。通过对釉蛱蝶遗传标记的分析(例如在回交群体中分离的 AFLP 遗传标记),可以利用具有雌性标记的回交群体鉴定连锁群。而在 F_1 雄为轮回亲本的回交群体中,可以通过重组率来确定定位于同一连锁群上标记之间的距离。

在一定的后代群体中可以获得大量的 AFLP 标记,瓦氏袖蝶(*H. erato*)和墨尔波墨蝶 (*H. melpomene*)的全基因组分子连锁图谱都已完成(Jiggins et al. 2005; Kapan et al. 2006)。在墨尔波墨蝶 (*H. melpomene*)(基因组大小为 292Mb)中,200 多个 AFLP 标记定位到 21 条所有的染色体上,另外还定位了大约 20 个微卫星标记。整个遗传图谱的大小为 1600cM,1cM 约代表 180kb。瓦氏袖蝶(*H. erato*)的基因组更大,为 400Mb,已定位了 300 多个 AFLP 标记,约 2000cM,1cM 约代表 200kb。由于不知道 AFLP 标记的序列,所以只有 AFLP 标记对比较定位研究是不够的,但它们为遗传图谱提供了一个主链,一些基于基因的标记(包括候选基因的标记)可以因此而定位。

6. 候选基因方法

模式生物的分子和细胞路径研究为我们提供了大量的基因,可以作为我们对鳞翅目昆虫感兴趣的表型进行研究的候选基因。一旦连锁图谱建立,就可以系统地把候选基因定位到染色体上,从而检测它是否与我们的目标位点有关。因为基因功能具有保守性,尽管这种方法有一定的局限性,但它已经被成功地运用于鉴定目标基因(Haag 和 True 2000)。

在杀虫剂抗性的案例中,其编码蛋白与杀虫剂相互作用(例如解毒酶或者受体)的基因可能是抗性的候选基因。那些不能遗传定位到与抗性相关的染色体或者位点的候选基因就可以否定它与抗性有直接作用(Baxter et al. 2008)。用这样的方法在烟芽夜蛾 *Heliothis virescens* 中,成功地将一个抗性候选基因(一个钙黏蛋白类蛋白)定位到 *BtR*-4 位点,该位点对杀虫剂 Bt 毒素 Cry1Ac 有 40%～80%的抗性(Gahan,Gould 和 Heckel 2001)。

即使如此,这种方法还是有明显的局限性,只能用于研究那些已有研究的基因或者在这个物种中容易克隆的基因。所以还需要一些技术可以对那些表型的本质不做任何先验假设,以便于鉴定直接控制那些重要表型的基因,而不是把基因简单地预测归类到下游路径里。在釉蛱蝶里,我们试图鉴定控制模式多态性的基因。已经有很多在昆虫翅发育中表达的基因(包括色素合成或信号传导路径的基因)完成了遗传定位,包括 *wingless*,*Dopa-decarboxylase*,*Distal-less*,*invected*,*patched*,*vermilion*,*scalloped*,*cubitus interruptus*,*scarlet*,*white* 和 *apterous* 等基因(Joron et al. 2007)。这些基因中没有一个基因是直接控制颜色模式开关的基因。因此,对翅模式调控的机理在没有预测的情况下,我们就利用定位克隆技术来研究。

7. 鉴定与目标性状连锁的 AFLP 标记

有两种方法去定位一个与目标位点距离较近的 AFLP 标记;一种是在连锁图谱构建中随机产生,第二种是定点 AFLP 混合分组分析法(BSA 法,它是将目标性状在 F_2 代的两组极端表型个体的 DNA 分别混合成两个 DNA 池，然后用相关标记在两池中进行标记与性状间的共分离分析，确定是否连锁及彼此间的遗传距离从而进行基因定位）。Jiggins 等(2005)在墨尔波墨蝶(*H. melpomene*)里鉴定了一个与不完全的隐性色彩模式位点，*HmYb*，紧密连锁的 AFLP 标记"CA_CTAa41"（即 a41），*HmYb* 的表型为后翅上有一个黄色条纹。然后通过对该 AFLP 标记测序的方法在很多类似的窝(蛾区)里做基因型分析，最后，a41 这个标记被定位到离 *HmYb* 位点 1cM 的距离，所以随即产生的标记偶尔也会与目标性状连锁。

然而我们不可能对每个标记都用大量的个体去检测它们在基因组的特定区域。改用 AFLP 混合分组分析法使得我们可以利用不同的 AFLP 引物组合去扫描很少的样本以鉴定在特定区域的标记(Baxter et al. 2008)。这个技术根据性状的两个方面把 AFLP 技术预扩增反应池混合成两个基因池。由于定位群体的个体拥有同样的等位基因，那么这两个混合的基因池之间应该只在与目标区域紧密连锁的标记才表现出差异。只需用定位群体的两个亲本和这两个混合的基因池这 4 个样来筛选 AFLP 引物组合。

用这种方法，我们已经定位了控制墨尔波墨蝶(*H. melpomene*)红色翅模式的位点。用墨尔波墨蝶(*H. melpomene*)中具有两种完全不同的翅表型，红色和黑色的品种进行杂交。*HmB* 为前翅红色条纹，*HmD* 为后翅和前翅基部呈红色射线状，这两个位点都是显性，且都已定位到 18 连锁群。F_2 代表型的分离比为：1 *HmB-/dd* ：2 *HmB-/ D-* ：1 *Hmbb/D-*。因为这些标记都是显性的，"—"表示无论是纯合还是杂合，其表型由显性等位基因控制。AFLP 的模板分别用 F_2 代无前翅斑(*Hmbb*)表型和无放射状斑纹表型(*Hmdd*)的个体组成。一旦 AFLP 产物成功地在个体中扩增出来，就将隐性表型的样本根据表型汇集成两个基因池，并用 256 个 AFLP 引物组合进行筛选。由于这些样本都是同胞个体，所以在多数情况下这两个基因池都表现出相同的 AFLP 带型。而那些只在其中一个样本中出现的带就很可能与控制翅颜色的位点紧密连锁。一共检测到 19 个这样的带，它们全都用于在所有的个体中检测其分离模式。其中 10 个带表现出阳性结果：5 个与 *HmB* 表型缺失有关，另外 5 个与 *HmD* 表型缺失有关。因为在 DNA 池中进行 AFLP PCR 扩增会导致差异，因此高比例的假阳性(19 中有 9 个)是不难想象的，9 个阴性结果可能是由 AFLP PCR 的非特异扩增造成的，10 个阳性带通过丙烯酰胺凝胶电泳和测序来验证，并用几个类似的窝(蛾区)来做基因型检测，从而构建重组连锁图谱。这些标记定位在 7cM 内，说明 *HmB* 和 *HmD* 这两个位点都定位在同样的染色体区域(图 6.2A—C)。

7.1 从 AFLPs 到 BAC Tile Paths

对与目标性状连锁的 AFLP 标记可以测序，然后用于筛选 BAC 文库(通过筛查排列在纤维膜上的克隆)，为了保证对所有的候选基因进行测序，必须对包含这个位点的全部 BAC 克隆进行筛选。尽管大的定位群体可以让我们得到更高精度的遗传图谱，但为了保证这个位点周围整个区域都能被检测到，每一次鉴定步移两端的 BAC 探针之间的重组是十分重要的。根据 AFLP 遗传图谱估计 1cM 在墨尔波墨蝶（*H. melpomene*)约代表 180kb，而在瓦氏袖蝶(*H. erato*) 中 1cM 约代表 280kb(Jiggins et al. 2005; Kapan et al. 2006)。因此目标位点与距离 1cM 的 AFLP 标记之间仍然包括了几个 BAC 克隆长。

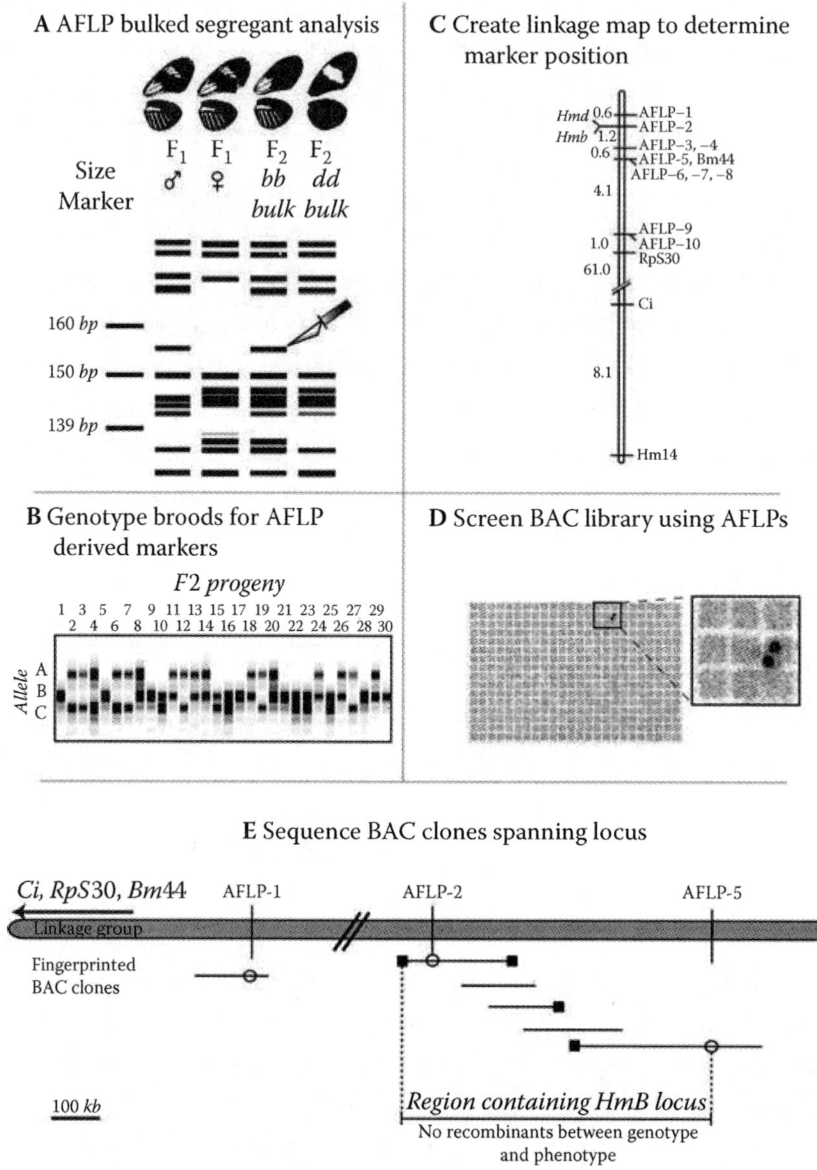

图 6.2 用 44 窝墨尔波墨蝶（*H. melpomene*）鉴定鳞翅目昆虫的主效基因。(A)用 AFLP 分析在 F_1 亲本和 F_2 混合分离基因池(基因池分别由 *Hmbb*/ *D*-和 *HmB*-/*dd* 两种表型的纯合子组成)，图上的带表示只在父本和其中一个基因池出现的带，并通过在个体中进行 AFLP 分析来验证它们确实与表型位点连锁。与表型分离的 AFLP 带进行丙烯酰胺电泳和测序。(B)在许多相同表型分离的群体中对测序后的 AFLP 带进行基因型分析。(C)用 AFLP 标记的基因型构建连锁图谱确定标记之间的顺序。图中显示的有 10 个 AFLP 标记、3 个基因(*cubitus interruptus*，*Ribosomal Protein S30* 和 *Bm44*)和一个微卫星标记 *Hm14*。标记 AFLP-2 与 B 和 D 位点的距离是 0cM. (D) 通过扫描基因组 BAC 文库去分离含有 AFLP 标记的克隆。放大图展示了含有阳性克隆的 BAC 过滤区域(E)对杂交检测到的覆盖了目标基因所在的基因组区域的最小 BAC 进行测序。在此，检测到 AFLP-2 和-5 两个标记分布在非重叠的 BAC contig 里(空心圆圈)。然后用 BAC 的末端序列(黑色方框)发展为新的探针去重新扫描 BAC 文库，以填补 AFLP-2 和-5 两个标记之间的空缺。利用 BAC 末端序列标记和 AFLP 标记得到的重组连锁图谱确定了包含 *HmB* 位点的区域。

为了鉴定控制 *HmYb* 表型的基因，用 AFLP 标记 *a*41 作为探针扫描墨尔波墨蝶(*H. melpomene*)的 BAC 基因组文库。克隆测序和随后的 BAC 扫描共检测了 500kb 以上的区间，包括 6 个 BAC 克隆。为了鉴定 *HmB* 和 *HmD* 位点，同样也已检测了大约 600kb 的区域，且在 *HmB* 尾端的两端还有重组事件发生(Baxter et al. 2008；图 6.2D—E)。

即使如此，从连锁的 AFLP 标记到能完全覆盖目标基因的探针扫描仍是很困难的，就目前来看这可能是最难的一步。最大的问题就是 BAC 文库里的遗传异质性，这就意味着指纹重叠群(contig)不具备应有的广泛性(Baxter et al. 2008)。我们的确也在基因组的某些区域检测到两个平行的 contig 同时定位在相同的位点上，看起来就像是文库中的两个不同的等位基因样本。当检测这个指纹数据的时候，在这些克隆里又并不存在共同的带。简而言之，在墨尔波墨蝶(*H. melpomene*)的 contig 还不足以大到覆盖 2～3 个 BAC 长度，也就意味着为了扫描到更宽的区域我们必须要对文库重复扫描。当然，最好是能用更多纯合的材料来做 BAC 文库。不管怎么说，用特异 contig 的末端序列作为探针来杂交文库的方法已在很大程度上加快了染色体步移的速度。

7.2 对辐射(Radiation)的比较性作图

在确定这些基因的控制模式之前，可以用这些在作图研究中发展起来的分子标记对不同系统的翅模式基因的位置进行精细的比较分析。对于那些亲缘关系太远而不能进行杂交的物种(例如瓦氏袖蝶(*H. erato*)和墨尔波墨蝶 (*H. melpomene*))来说，人们曾推测它们的相似图案是否由同源基因控制的(Mallet 1989；Nijhout 1991)。现在，可以利用连锁的分子标记直接比较物种间的遗传结构。实际上，不同物种间遗传结构的共同之处比人们想象的还多。不仅在瓦氏袖蝶(*H. erato*)和墨尔波墨蝶 (*H. melpomene*)中控制相似表型的主位点具有同源性，而且在另一物种狐眼袖蝶(*Heliconius numata*)中，该区域还控制了一个非常不同的翅模式(Joron et al. 2007；Baxter et al. 2008)。不过我们现在的作图分辨度无法达到比较单基因的程度。但是，这些数据明确显示至少这三种袖蝶品种的翅模式调控存在一定程度的共同性。此外，它还意味着在不同物种中的定位克隆研究具有补充性，并且为研究其他物种提供了帮助。

8. 基因组序列数据的比较

我们已经得到覆盖墨尔波墨蝶 (*H. melpomene*) *yb* 位点和同源的瓦氏袖蝶(*H. erato*) *Cr* 位点的几百 kb 的序列。直接序列比较显示在墨尔波墨蝶 *H. melpomene* 与瓦氏袖蝶 *H. erato* 中基因与基因间的顺序是保守的(图 6.3)。显然，这只是两个釉蛱蝶品种一部分基因组的结果，但是值得注意的是，即使在釉蛱蝶的两个完全不同分支的物种中，基因顺序都表现出保守性。因此目前尚无证据表明基因组的该区域受到很强的适应性选择压力作用而导致了染色体结构的重组。对釉蛱蝶的进一步测序应该可以回答以上的这个例子在釉蛱蝶基因组中基因顺序的保守性和整体的共线性问题上是否具有代表性。

综合这些数据表明墨尔波墨蝶(*H. melpomene*)的基因组大小只相当于瓦氏袖蝶

(*H. erato*)基因组大小 74%,其原因在于内含子与间隔区域的大小的差异(Tobler et al. 2005)。瓦氏袖蝶(*H. erato*)的 *Cr* 区域序列不完整就限制了我们检测这种统计差异的能力。

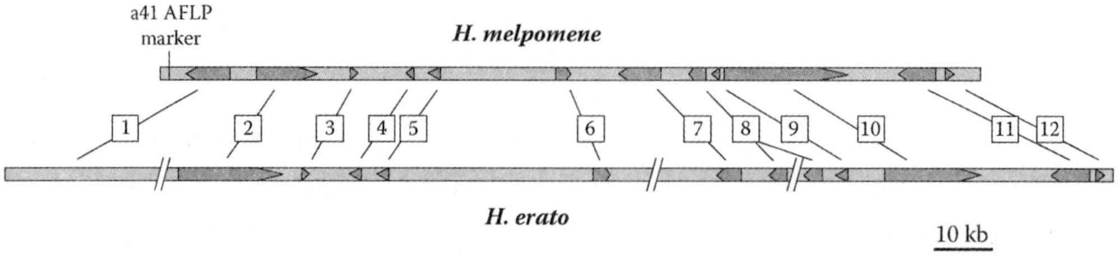

图 6.3 比较临近墨尔波墨蝶(*H. melpomene*)*yb* 位点和瓦氏袖蝶(*H. erato*)*Cr* 位点的基因组序列。用距 *yb* 位点(控制后翅的黄色带纹)1cM 的一个 AFLP 标记 a41 扫描 *H. melpomene* 的 BAC 文库。从该区域分离的一个基因(Rab 蛋白香叶烯基转移酶 β 亚基)作为探针来鉴定 *H. erato* 的同源区域。基因顺序与方向都显示了保守性;但是在 *H. erato* 中有许多 *H. melpomene* 所没有的编码内切酶与反转录酶的基因。为了确定 *H. erato* 中基因的顺序需要对 BAC 序列中的 gap 区域进行测序。1. 酰胺酸释放酶(赤拟谷盗)(*Tribolium castaneum*)。2. 酰胺酸释放酶(赤拟谷盗)(*Tribolium castaneum*)。3. 推测的与 CG10949,CG3838,CG3919 类似的蛋白质。4. 海藻糖酶 1 (*Omphisa fuscidentalis*)。5. 海藻糖酶 1 (*O. fuscidentalis*)。6. B9 蛋白质(非洲爪蟾)(*Xenopus laevis*)。7. 类似于 CG5098-PA。8. 类似于 CG3184-PA(赤拟谷盗)(*T. castaneum*)(在 *H. erato* BAC 序列中该基因跨过了一个间隔)。9. 类似于 CG18292-PA(黑腹果蝇)(*D. melanogaster*)。10. 类似于 CG2519-PB(黑腹果蝇)(*D. melanogaster*)。11. Unkempt(意大利蜜蜂)(*Apis mellifera*)。12. 组蛋白 H3(小家鼠)(*Mus musculus*)。海藻糖酶基因(4 和 5)与酰胺酸释放酶基因(1 和 2)在两个物种中都有基因重复,这些基因重复可能发生在共同的祖先中。虽然基因顺序保守,在 *H. erato* 中内含子和基因间区序列比 *H. melpomene* 的要大,所以在 *H. erato* 中基因密度相对较低。

9. 鉴定翅模式基因——多重证据

对在控制 *HmYb* 翅模式表型基因的功能研究中,如果要做得令人信服的话,就需要用转基因的方法来敲除控制该翅表型的基因,然后再用基因回复的方法加以验证。在这些研究技术还未在鳞翅目昆虫中实现时,在鉴定控制翅模式的基因或影响因子时要寻找三种类型的证据:(1)种群遗传数据。(2)表达数据。(3)抗体染色。我们在杂交区域的任一边中对墨尔波墨蝶 *H. melpomene* 和瓦氏袖蝶 *H. erato* 的 *HmYb* 和 *HmB*/*HmD* BAC walk 中的基因进行测序。我们希望可以找到最终控制翅模式的关键区域和与它连锁的区域之间的明显区别。模式基因大多在幼虫或蛹的翅原基里表达(翅模式的发育也将在本章讨论)。在不同品种里定量分析 BAC walks 鉴定出的基因的表达量可以为我们提供基因表达的时间信息。当然,有可能关键基因本身是由一种 RNA 分子的表达量来调控的,用 BAC 表达芯片实验来研究究竟基因组中哪些区域在翅组织中表达。最后,根据群体遗传学和表达数据筛选出最佳候选基因,我们用其特异的抗体来检测其蛋白在不同发育阶段翅中的表达部位。

10. 大鳞翅目昆虫中的共线性模式

尽管鳞翅目昆虫的染色体数目可以从 5 对到 223 对不等，但最常见的是 28 到 32 对染色体(Suomalainen 1969；Robinson 1971；De Prins 和 Saitoh 2003；见第三章讨论染色体数目部分)。最近的一项重要研究表明，尽管墨尔波墨蝶 *H. melpomene* 与家蚕的染色体数目有很大不同，但在这两个物种中保存着大尺度的染色体共线性(Pringle et al. 2007)。鉴于这两种昆虫在系统发生关系上有一定距离，说明这种模式可能在整个大鳞翅类中都是成立的。最近，一张由家蚕 cDNA 派生标记构建的精细连锁群发表后使得比较分析成为可能(Yasukochi et al. 2006；蚕属的连锁图谱请参阅第二章)。

以前的基于 AFLP 或微卫星标记的图谱并没有提供可以在远缘生物中检测的保守位点。釉蛱蝶有大量的 EST 数据，我们可以对推测与家蚕中定位的标记有同源性的大量基因设计引物。我们主要关注核糖体蛋白，因为它们通常为单拷贝并且在基因组中有着广泛的分布(Heckel 1993)。并且它们在 EST 文库中有很多拷贝，进而作为标记来说相对容易(Papanicolaou et al. 2005；Beldade, McMillan 和 Papanicolaou 2008)。现在这些标记为蝴蝶标记染色体锚定位点和比较连锁作图研究提供了工具。

如上所述，由于在鳞翅目昆虫中雌性个体完全连锁为连锁作图带来很多便利，尤其是我们可以利用非重组组合很快地把标记定位到连锁群上从而很方便地研究共线性(Heckel 1993)。在同一蛾区后代中，我们用 RLFPs 或通过琼脂糖分离出的长度多态性将标记定位到连锁群上。总计，73 个直系同源标记已在墨尔波墨蝶 *H. melpomene* 和家蚕中定位，并且共线性的模式完全保守(Pringle et al. 2007)。在家蚕中连锁的所有标记在釉蛱蝶中都呈类似的共线性(图 6.4)。此外，还有 6 个染色体的融合，位于家蚕两个不同连锁群上的标记在墨尔波墨蝶 *H. melpomene* 中则定位于同一个连锁群上。这并不令人感到意外，因为墨尔波墨蝶 *H. melpomene* 有 21 对染色体，家蚕有 28 对染色体(Brown et al. 1992)。这两种生物可能从具有 31 对染色体的同一祖先进化而来(De Prins 和 Saitoh 2003)，所以额外的染色体融合一定会在两个系统中都存在，只是目前还没有被检测到。因为到目前为止标记的数量是如此之少，产生这样的结果并不在意料之外。

同线性惊人的保守性作为一种进化现象来说非常有趣。鳞翅目昆虫中染色体数目有较大差异，但是很意外的是在釉蛱蝶 *Heliconius* 与家蚕之间并没有太多基因的染色体间重组。对于鉴定受选择的基因来说，这一结果令人鼓舞，因为一个完整的家蚕基因组将成为寻找预测基因的具有重大价值的资源(请参阅第二章中蚕属基因组测序的情况)，即使对于远缘的蝴蝶来说也是如此。当然，对于这些价值不应过高估计，微线性将不会太保守，而且我们推测在差异的物种中将有高频率的倒位与染色体内重组。这种情况在线虫(*Caenorhabditis elegans*)与 *Caenorhabditis briggsae* 的全测序比较基因组中就得到了证实，这两种生物在大共线性上显示了高度的保守性，但在更细微的水平上有高频率

H.melpomene linkage group	*B.mori* linkage group	Number of genes in common
1	4	4
2	16	2
3	6	2
4	21	3
5	3	2
6	9	6
7	2, 11	7
8	25	1
9	7	3
10	5, P	5
11	15	6
12	8, 20	4
13	14, 22	5
14	19	1
15	17	3
16	18	1
17	13, 24	7
18	23, U	5
19	12	2
20	10	2
21	1	2
N/A	26	0
21 Chromosomes	28 Chromosomes	73

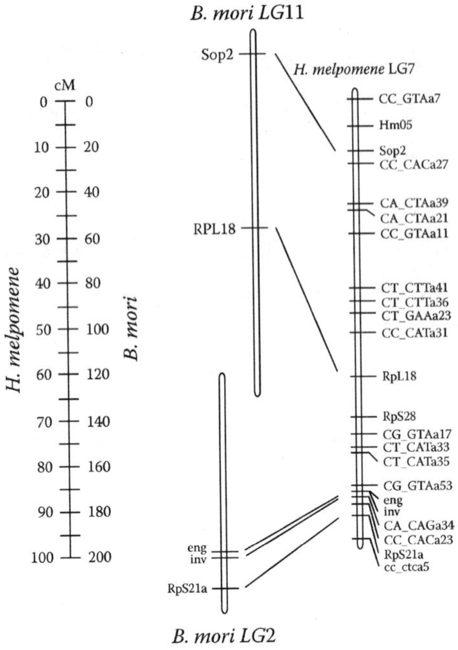

图 6.4　墨尔波墨蝶 *Heliconius melpomene* 与家蚕共线性的保守性。一共有 73 个两个物种共有的标记显示了完全的共线性。共有 6 个可能的染色体融合，正如预料的那样，因为墨尔波墨蝶 *H. melpomene* 有 21 对染色体，家蚕有 28 对染色体。众多染色体的一个连锁图谱——墨尔波墨蝶 *H. melpomene* LG7 在这里展示。注意两物种所用的比例尺不同。图片改编于 Pringle 等(2007)，蚕属的数据来自于 Yasukochi 等(2006)。

的基因组重排(Blaxter 2003)。

因为可以从家蚕每一条染色体中定位有代表性的标记，所以在任何一个可以进行杂交的鳞翅目昆虫中，都可将一个表型标记定位到推测具有与家蚕染色体同源性的连锁群上。通过与家蚕基因组的比对可以从同一染色体中提供额外的候选基因以供作图和进一步定位感兴趣的基因。随着家蚕基因组组装的改良，这将成为研究鳞翅目昆虫的一项有用工具。当下的基因数据只能用于大鳞翅类昆虫，将来它对更基部类群间的线性比较也很重要。

11. 结论

釉蛱蝶提供了一个研究最新的多种适应辐射的好机会。在过去的几年中我们建立了这些蝴蝶的基因组资源，包括应该对比较基因学有帮助的在大鳞翅类昆虫中保守基因在主链上的定位。另外，我们还证明了采用混合分组 AFLP 分析法(bulk-segregant AFLP)去定位特定表型相关基因标记方法的有效性。结合插入基因组大片段的 BAC 文库，该方法可以克隆测序任一基因组中的目标区。为在鳞翅目昆虫基因组中定位感兴趣的基因提供了令人振奋的前景。

12. 致谢

我们对釉蛱蝶基因组协会的 Mathieu Joron, Sean Humphray, Alexie Papanicolaou, Riccardo Papa 和 Jim Mallet 的工作表示感谢。向生物科学与生物技术研究委员会，国家科学基金，国家环境研究委员会(NERC)，莱弗休姆信托基金会，史密森尼热带研究所，英国皇家学会基金表示感谢。对 Autoridad Nacional del Medio Ambiente (ANAM) 允许我们在巴拿马和 Ministerio del Ambiente 允许我们在厄瓜多尔收集蝴蝶表示感谢。

参考文献

[1]Asser-Kaiser, S., E. Fritsch, K. Undorf-Spahn, et al. 2007. Rapid emergence of baculovirus resistance in codling moth due to dominant, sex-linked inheritance. *Science* 317:1916—18.

[2]Baxter, S., R. Papa, N. Chamberlain, et al. 2008. Convergent evolution in the genetic basis of Müllerian mimicry in *Heliconius* butterflies. *Genetics* 180:1567—77.

[3]Baxter, S. W., J.-Z. Zhao, L. J. Gahan, A. M. Shelton, B. E. Tabashnik, and D. G. Heckel. 2005. Novel genetic basis of filed-evolved resistance to Bt toxins in *Plutella xylostella*. *Insect Mol. Biol.* 14:327—34.

[4]Beldade, P., W. O. McMillan, and A. Papanicolaou. 2008. Butterfly genomics eclosing. *Heredity* 100:150—57.

[5]Bensch, S., and M. Akesson. 2005. Ten years of AFLP in ecology and evolution: Why so few animals? *Mol. Ecol.* 14:2899—2914.

[6]Blaxter, M. 2003. Comparative genomics: Two worms are better than one. *Nature* 426:395—96.

[7]Bonaldo, M. F., G. Lennon, and M. B. Soares. 1996. Normalization and subtraction: Two approaches to facilitate gene discovery. *Genome Res.* 6:791—806.

[8]Bouck, A., and T. Vision. 2007. The molecular ecologist's guide to expressed sequence tags. *Mol. Ecol.* 16:907—24.

[9]Brown, K. S., T. C. Emmel, P. J. Eliazar, and E. Suomalainen. 1992. Evolutionary patterns in chromosome numbers in neotropical Lepidoptera. I. Chromosomes of the Heliconiini (Family Nymphalidae: Subfamily Nymphalinae). *Hereditas* 117:109—25.

[10]Clarke, C. A., P. M. Sheppard. 1960. The evolution of mimicry in the butterfly *Papilio dardanus*. *Heredity* 14:163—73.

[11]Daborn, P. J., J. L. Yen, M. Bogwitz, et al. 2002. A single P450 allele associated with insecticide resistance in global populations of *Drosophila*. *Science* 297:2253—56.

[12]De Prins, J., and K. Saitoh. 2003. Karyology and sex determination. In *Lepidoptera, moths and butterflies: Morphology, physiology, and development*, ed. N. P. Kristensen, 449—68. Berlin: Walter de Gruyter.

[13]Eberle, K. E., and J. A. Jehle. 2006. Field resistance of codling moth against *Cydia pomonella* granulovirus (CpGV) is autosomal and incompletely dominant inherited. *J. Invertebr. Pathol.* 93:201

—06.

[14]Feder, M. E. , and T. Mitchell-Olds. 2003. Evolutionary and ecological functional genomics. *Nat. Rev. Genet.* 4:651—57.

[15]Ford, E. B. 1975. *Ecological genetics*. London: Chapman and Hall.

[16]Gahan, L. J. , F. Gould, and D. G. Heckel. 2001. Identification of a gene associated with Bt resistance in *Heliothis virescens*. *Science* 293:857—60.

[17]Haag, E. S. , and J. R. True. 2000. From mutants to mechanisms? Assessing the candidate gene paradigm in evolutionary biology. *Evolution* 55:1077—84.

[18]Hanski, I. , and I. Saccheri. 2006. Molecular-level variation affects population growth in a butterfly metapopulation. *PLoS Biol*. 4:e129.

[19]Heckel, D. G. 1993. Comparative linkage mapping in insects. *Annu. Rev. Entomol*. 38:381—408.

[20]Heckel, D. A. , L. J. Gahan, Y. Liu, and B. E. Tabashnik. 1999. Genetic mapping of resistance to *Bacillus thuringiensis* toxins in diamondback moth using biphasic linkage analysis. *Proc. Natl. Acad. Sci. U. S. A*. 96:8373—77.

[21]Hillier, L. D. , G. Lennon, M. Becker, et al. 1996. Generation and analysis of 280,000 human expressed sequence tags. *Genome Res*. 6:807—28.

[22]Hudson, M. E. 2008. Sequencing breakthroughs for genomic ecology and evolutionary biology. *Mol. Ecol. Res*. 8:3—17.

[23]Huse, S. M. , J. A. Huber, H. G. Morrison, M. L. Sogin, and D. M. Welch. 2007. Accuracy and quality of massively-parallel DNA pyrosequencing. *Genome Biol*. 8:R143.

[24]Jiggins, C. D. , J. Mavarez, M. Beltrán, J. S. Johnston, and E. Bermingham. 2005. A genetic map of the mimetic butterfly, Heliconius melpomene. *Genetics* 171:557—70.

[25]Joron, M. , R. Papa, M. Beltrán, et al. 2007. A conserved supergene locus controls colour pattern diversity in *Heliconius* butterflies. *PLoS Biol*. 4:e303.

[26]Kapan, D. D. , N. Flanagan, A. Tobler, et al. 2006. Localization of Müllerian mimicry genes on a dense linkage map of *Heliconius erato*. *Genetics* 173:735—57.

[27]Kettlewell, H. B. D. 1973. *The evolution of melanism. The study of a recurring necessity*. Oxford: Blackwell.

[28]Maeda, T. 1939. Chiasma studies in the silkworm *Bombyx mori*. *Jpn. J. Genet*. 15:118—27.

[29]Mallet, J. 1989. The genetics of warning colour in Peruvian hybrid zones of *Heliconius erato* and *H. melpomene*. *Proc. R. Soc. Lond. B Biol. Sci*. 236:163—85.

[30]Margulies, M. , M. Egholm, W. E. Altman, et al. 2005. Genome sequencing in microfabricated high-density picolitre reactors. *Nature* 437:376—80.

[31]Mavárez, J. , and M. González. 2006. A set of microsatellite markers for *Heliconius melpomene* and closely related species. *Molecular Ecology Notes* 6:20—23.

[32]Mavárez, J. , C. Salazar, E. Bermingham, C. Salcedo, C. D. Jiggins, and M. Linares. 2006. Speciation by hybridization in Heliconius butterflies. *Nature* 411:868—71.

[33]Meudt, H. M. , and A. C. Clarke. 2007. Almost forgotten or latest practice? AFLP applications,

analyses and advances. *Trends Plant Sci*. 12:106—17.

[34]Nijhout, H. F. 1991 *The development and evolution of butterfly wing patterns*. Washington, DC: Smithsonian Institution Press.

[35]Papa, R., C. M. Morrison, J. R. Walters, et al. 2008. Highly conserved gene order and numerous novel repetitive elements in genomic regions linked to wing pattern variation in *Heliconius butterflies*. *BMC Genomics* 9:345.

[36]Papanicolaou, A., S. Gebauer-Jung, M. L. Blaxter, W. O. McMillan, and C. D. Jiggins. 2007. ButterflyBase: A platform for lepidopteran genomics. *Nucleic Acids Res*. 36:D582—87.

[37]Papanicolaou, A., M. Joron, M. L. Blaxter, W. O. McMillan, and C. D. Jiggins. 2005. Genomic tools and cDNA derived markers for butterflies. *Mol. Ecol*. 14:2883—97.

[38]Parkinson, J., A. Anthony, J. Wasmuth, R. Schmid, A. Hedley, and M. Blaxter. 2004. PartiGene—constructing partial genomes. *Bioinformatics* 20:1398—1404.

[39]Pringle, E. G., S. Baxter, C. L. Webster, A. Papanicolaou, S. F. Lee, and C. D. Jiggins. 2007. Synteny and chromosome evolution in the Lepidoptera: Evidence from mapping in *Heliconius melpomene*. *Genetics* 177:417—26.

[40]Reed, R. D., W. O. McMillan, and L. M. Nagy. 2007. Gene expression underlying adaptive variation in *Heliconius* wing patterns: Non-modular regulation of overlapping cinnabar and vermilion patterns. *Proc. Biol. Sci*. 275:37—45.

[41]Robinson, R. 1971. *Lepidoptera genetics*. Oxford: Pergamon Press.

[42]Sheppard, P. M., J. R. G. Turner, K. S. Brown, W. W. Benson, and M. C. Singer. 1985. Genetics and the evolution of Müllerian mimicry in *Heliconius* butterflies. *Philos. Trans. R. Soc. Lond. B Biol. Sci*. 308:433—613.

[43]Suomalainen, E. 1969. Chromosome evolution in the Lepidoptera. *Chromosomes Today* 2: 132—38.

[44]Suomalainen, E., L. M. Cook, and J. R. G. Turner. 1973. Achiasmatic oogenesis in the heliconiine butterflies. *Hereditas* 74:302—04.

[45]Tobler, A., D. D. Kapan, N. S. Flanagan, et al. 2005. First generation linkage map of the warningly-colored butterfly *Heliconius erato*. *Heredity* 94:408—17.

[46]Toth, A. L., K. Varala, T. C. Newman, et al. 2007. Wasp gene expression supports an evolutionary link between maternal behavior and eusociality. *Science* 318:441—44.

[47]Traut, W. 1977. A study of recombination, formation of chiasmata and synaptonemal complexes in female and male meiosis of *Ephestia kuehniella* (Lepidoptera). *Genetica* 47:135—42.

[48]Trombetti, G. A., R. J. P. Bonnal, E. Rizzi, G. De Bellis, and L. Milanesi. 2007. Data handling strategies for high throughput pyrosequencers. *BMC Bioinformatics* 8, Suppl. 1:S22.

[49]Turner, J. R. G., and P. M. Shepard. 1975. Absence of crossing-over in female butterflies (*Heliconius*). *Heredity* 34:265—69.

[50]Vera, J. C., C. W. Wheat, H. W. Fescemyer, et al. 2008. Rapid transcriptome characterization for a non-model organism using 454 pyrosequencing. *Mol. Ecol*. 17:1636—47.

[51]Vos, P., R. Hogers, M. Bleeker, et al. 1995. AFLP: A new technique for DNA fingerprinting. *Nucleic Acids Res*. 23:4407—14.

[52]Wicker, T., E. Schlagenhauf, A. Graner, T. J. Close, B. Keller, and N. Stein. 2006. 454 sequencing put to the test using the complex genome of barley. *BMC Genomics* 7:275.

[53]Yasukochi, Y. 1998. A dense genetic map of the silkworm, *Bombyx mori*, covering all chromosomes based on 1018 molecular markers. *Genetics* 150:1513—25.

[54]Yasukochi, Y., L. A. Ashakumary, K. Baba, A. Yoshido, and K. Sahara. 2006. A second-generation integrated map of the silkworm reveals synteny and conserved gene order between lepidopteran insects. *Genetics* 173:1319—28.

第七章 蝴蝶眼研究在分子和生理学上的新进展

Marilou P. Sison-Mangus and Adriana D. Briscoe

1. 前言

色觉是一种复杂的性状，能够影响像鳞翅目昆虫这种短命昆虫的生存。在鳞翅目中，色觉系统是各不相同的，蝴蝶的研究最为大家所熟知，这些蝴蝶属于 5 个科。近来几篇综述报道了普通的凤蝶（例如柑桔凤蝶）和粉蝶（例如菜粉蝶）的研究（Stavenga 和 Arikawa 2006；Wakakuwa，Stavenga 和 Arikawa 2007）。在这两个种群中具有互不相同的眼睛，并且在编码视觉色素的视蛋白基因的拷贝数、基因的空间表达模式以及侧面过滤色素的分布上有别于其他蝴蝶家族。迄今为止仅仅只有一项着眼于蚬蝶科（*Apodemia mormo*）的视觉色素的研究（Frentiu et al. 2007）。在这章中我们侧重于介绍灰蝶独特的视觉系统的最新进展，特别是 *Lycaena rubidus* 灰蝶视网膜的性别二态性（灰蝶亚科）和伊眼灰蝶（蓝灰蝶亚科）的色觉行为。视蛋白基因的定位和它们在系统发育中的表达模式清楚地表明所有的蝴蝶眼睛都起源于像蛱蝶的那种比较简单的眼睛（Briscoe 2008）。因而，为了把灰蝶视觉系统的新进展纳入进化框架，我们首先描述蛱蝶比较简单的视觉系统。接着，我们追踪视蛋白基因的分子变化，表达模式以及它们编码的视觉受体在生理学上的变化。最后，我们讨论了灰蝶独特的眼睛设计所产生的潜在行为结果。在这整个过程中，我们提到了一些未来有潜力的研究领域。

2. 蝴蝶小眼的解剖结构

蝴蝶的复眼由成千的小眼组成（Yagi 和 Koyama 1963）。每一个小眼包含一个角膜、一个晶锥和九个感光细胞（R1－9；图 7.1A）以及初级和次级的色素细胞。每个光受体的微绒毛形成感杆小体，其融合在一起形成圆柱形的光学结构——感杆束。感杆束扮演着光波导入器的作用（Nilsson，Land 和 Howard 1988），它从晶锥延伸到基底膜。构成感杆束的微绒毛的排列方式变化取决于物种。在最简单的排列方式中，R1－8 的感杆小体的长度大致相同并且对构成感杆束的贡献差别不大，R9 细胞在感杆束的基部提供一些微绒毛以形成一个分层的结构（Briscoe et al. 2003）。在感杆束的近端，有着成堆的充满着空气的气管泡细胞，这些细胞构成绒毡层，起着干涉镜的作用，能够反射没有被吸收的光，使它透过感杆束，让视觉色素重新吸收返回的光。

绒毡层、视觉色素和过滤色素（图 7.1B）一起存在的情况下使得在大多数蝴蝶眼中都可以看到彩色发光（点睛；图 7.1C）（Stavenga et al. 2001）。点睛的反射光谱在生理学上是很有趣的，在一只完整的蝴蝶中，特别是在光谱的长波部分，它可以作为在小眼中发现的视觉色素吸收光谱的体内非侵袭性探针（Bernard 和 Miller 1970；Vanhoutte 和 Stavenga 2005）。它也可以被用来推断异质性表达的黄色、橙色和红色过滤色素的存在，这些色素修饰了用于激发视觉色素的光波长（Arikawa 和 Stavenga 1997；Arikawa et al. 1999；图 7.1B，C）。尽管如此，蝴蝶的绒毡层结构不是普遍存在的；在凤蝶中不存在（Miller 1979），并且在粉蝶中的分布也是有差异的。粉蝶属的蝴蝶具有绒毡层结构（例子，详见 Briscoe 和 Bernad 2005 的图 2），而襟粉蝶属的蝴蝶则没有这种结构（Takemura，Stavenga 和 Arikawa 2007）。对于那些没有绒毡层结构的蝴蝶，还可以应用细胞内记录的方法测量感光细胞的光谱敏感性。并且对于那些有或者没有绒毡层结构的蝴蝶，组织切片提供了过滤色素分布最直接的证据。

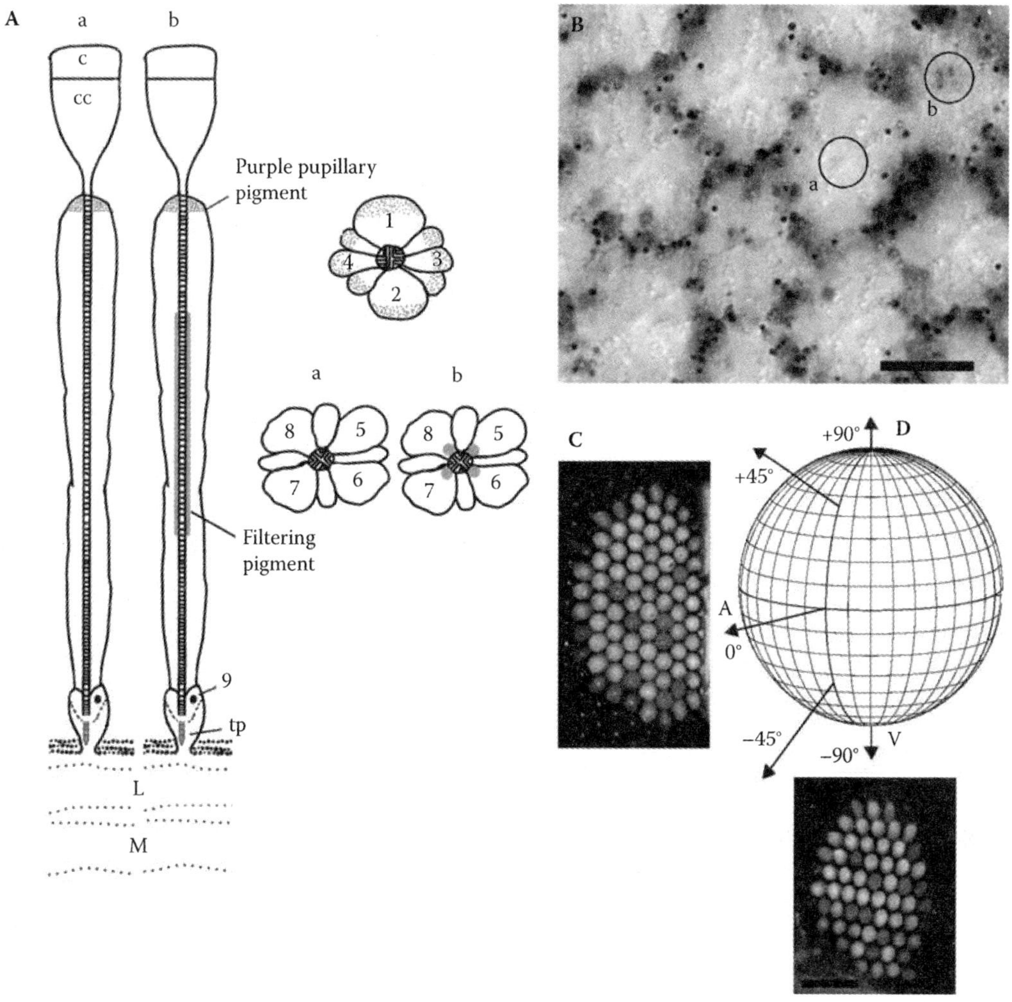

图 7.1　小眼，过滤色素分布和伊眼灰蝶的点睛。(A)腹眼的两种类型小眼，左边是纵向视图，右边是切面视图；非着色(a)红色(b)。紫色瞳孔色素存在于所有 R1—R8 感光细胞的远端，起着调控进入每个小眼光量的作用。(B)红色过滤色素(a)在一些小眼的侧眼中没有而存在于其他侧眼(b)。比例尺，10μm。(C)雌蝶的点睛。小眼的前侧(A)是黄色的，用浅灰色表示，而腹(V)侧是红色的，用深灰色表示。背(D)方向。比例尺，50μm。c，角膜；cc，晶锥；9，第九个感光体；tp，绒毡层；L，层状体；M，髓质。修改自 Sison-Mangus 等 (2008)。

视觉色素是位于各个感光细胞的感杆束微绒毛上吸收光的分子。蝴蝶的视觉色素类似于脊椎动物黑视素，由一个感杆束视蛋白(Briscoe 1998；Kitamoto et al. 1998)和 G 蛋白偶联蛋白亚家族中的成员组成(Provencio et al. 1998)。这些黑视素与对光敏感的发光团(11-顺-3-烃基视黄醛)共价连接(Smith 和 Goldsmith 1990)。因为蝴蝶只用一种发光基团，所以视觉色素的最大吸收光谱取决于视蛋白的氨基酸残基。因而，正是视蛋白使得动物看得见不同波长的光而且由它来产生光感觉。视蛋白遍布在所有的动物中，并且在后生动物进化之前介导光传导级联反应(Terakita 2005; Plachetzki,Degnan 和 Oakley 2007)。动物的视蛋白的分类不仅依据感光细胞的来源(例如，感杆束类型或纤毛类型;Arendt 2003)而且也与形成正确的 G 蛋白偶联受体的特异 G 蛋白亚型有关(Santillo et al. 2006)。

蝴蝶感光细胞也可以根据它们对紫外光(UV;300～400nm)，蓝光(B;400～500nm)和长波光(LW;500～600nm)的敏感性进行粗略分类。利用视网膜电图对光谱敏感性测量的调查，细胞内或外延微光谱分光光度记录器记录结果表明大多数蛾子和蛱蝶的眼睛至少具有一种对紫外光、蓝光和长波光敏感的感光细胞(Briscoe 和 Chittka 2001)。小红蛱蝶(蛱蝶科)*Vanessa cardui* (Nymphalinae)在 360nm，470nm 和 530nm 波长时感光细胞具有最高的敏感性;帝王斑蝶 (斑蝶科)在 340nm，435nm 和 545nm 波长时感光细胞具有最高的敏感性(图 7.2A; Stalleicken, Labhart 和 Mouritsen 2006; Frentiu et al. 2007); 而烟草天蛾，在 357nm，450nm 和 520nm 波长时感光细胞具有最高的敏感性(Bennett 和 Brown 1985)。尽管灰蝶眼睛中感光细胞的生理学数据较少，但是所获数据与这个模式大体是一致的。例如，紫小灰蝶(*Narathura japonica*)成虫的眼睛至少具有三类在 380nm，460nm 和 560nm 波长时有最高敏感性的光受体(Imafuku et al. 2007)。琉璃灰蝶(*Celastrina argiolus*) 至少具有三类在 380nm，440nm 和 560nm 波长时有最高敏感性的光受体(Eguchi et al. 1982); 酢浆灰蝶(*Pseudozizeeria maho*)至少具有三类在 400nm，520nm 和 560nm 波长时有最高敏感性的光受体(Eguchi et al. 1982)。尽管如此，借助外延微光谱分光光度器，四种光受体光谱类型在灰蝶属蝴蝶(包括如下物种:*L. rubidus*，*L. heteronea*，*L. dorcas* 和 *L. nivalis*)的视网膜上被鉴定出来，分别在 360nm，437nm，500nm 和 568nm 波长(*L. nivalis* 是在 575nm)时具有最高敏感性的视觉色素(图 7.2B;Bernard 和 Remington 1991)。早期生理学的研究似乎暗示在一些蛱蝶和灰蝶中具有相似数目(3 个)的视蛋白，然而现在从分子研究的结果来看有一点是明确的，即空间表达谱推断出眼睛中视蛋白基因和光谱受体的数目在各蝴蝶属中有着惊人的差异(见下文)。

3. 蓝色视蛋白在灰蝶中发生基因加倍

蛱蝶科的小红蛱蝶和帝王斑蝶具有最少数目的视蛋白，它们眼睛中仅有紫外光、蓝光和长波光视蛋白基因的一个单拷贝(Briscoe et al. 2003;Sauman et al. 2005)。尽管这些互补 DNA 最初是从眼特异的 cDNA 库中克隆出来的，但是对帝王斑蝶脑的 EST 文库(包含 9484 条特异的 cDNA 序列)进行 BLAST 比对，产生的结果是相同的(Zhu, Casselman 和 Reppert 2008)。除开蛱蝶，其他两个被认作是蛱蝶科的姐妹分类单元的 *L. rubidus*，(灰蝶科)和 *Apodemia mormo*(蚬蝶科)，各自具有四个视蛋白(Campbell, Brower 和 Pierce 2000)。另外，*L. rubidus* 具有两个拷贝的蓝色视蛋白基因(*BRh1*，编码波长在 437nm 最高吸收峰的色素和 *BRh2*，编码波长在 500nm 最高吸收峰的色素)，一个单拷贝的紫外光视蛋白基因(*UVRh*，编码波长在 360nm 最高吸收峰的色素)和一个单拷贝的长波光视蛋白基因(*LWRh1*，编码波长在 568nm 最高吸收峰的色素); 然而蚬蝶 *A. mormo*

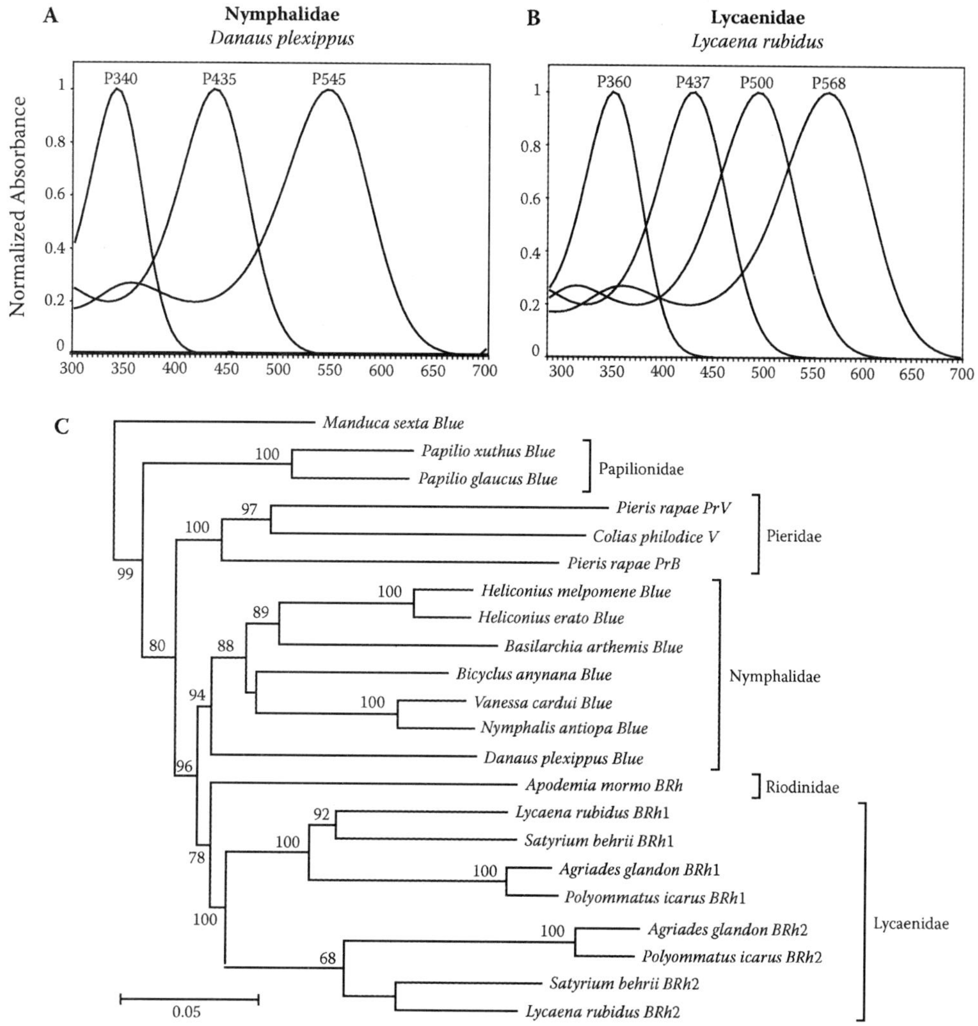

图 7.2 蛱蝶科和灰蝶科典型的视觉色素吸收光谱和鳞翅目蓝视蛋白的进化树。(A)帝王斑蝶的眼睛包含三种视觉色素。P 表示视觉色素的最大吸收峰(λ_{max})。(B) 红翅的 *Lycaena rubidus* 的眼睛包含四种视觉色素。(C)根据对 1077 个核苷酸位点的 neighbor-joining 分析，利用 Tamura-Nei 距离和种间核苷酸替换的异质模式创建蓝视蛋白基因的进化树。这里给出的 Bootstrap 值是基于 500 个最大似然的 Bootstrap 重复计算而来的，而这些重复是利用伽马型参数和不变位点的比例分别为 0.574 和 0.1474 的 GTR＋Γ＋I 模型确定的。GenBank 登录号如下：烟草天蛾(天蛾科；*Manop3*，AD001674)；柑桔凤蝶 (*PxRh4*，AB028217)；东方虎凤蝶(*PglRh6*，AF077192)；菜粉蝶(*PrB*，AB208675；*PrV*，AB208674)；菲罗豆粉蝶(*V*，AY918899)；帝王斑蝶(Blue，AY605544)；*Bicyclus anynana*(*BlueRh*，AY918894)；基红毒蝶(*BlueRh*，AY918906)；墨尔波墨蝶 (*BlueRh*，AY918897)；*Basilarchia arthemis astyanax* (*BlueRh*，AY918902)；丧服蛱蝶(*BlueRh*，AY918893)；小红蛱蝶(*BRh*，AY613987)；*Apodemia mormo* (*BRh*，AY587906)；蓝灰蝶(*BRh1*，DQ402500；*BRh2*，DQ402501)；*Agriades glandon* (*BRh1*，DQ402502；*BRh2*，DQ402503)；*Satyrium behrii* (*BRh1*，DQ402498；*BRh2*，DQ402499)；*Lycaena rubidus*(*BRh1*，AY587902；*BRh2*，AY587903)。进化树修改自 Sison-Mangus 等 (2006)。

具有两个拷贝的长波光视蛋白基因(*LWRh1*,编码波长在 505nm 最高吸收峰的色素;*LWRh2*,编码波长在 600nm 最高吸收峰的色素)并且只有一个单拷贝的紫外光视蛋白基因(*BRh*,编码波长在 450nm 最高吸收峰的色素)和紫外光视蛋白基因(*UVRh*,编码波长在 340nm 最高吸收峰的色素)(Sison-Mangus et al. 2006;Frentiu et al. 2007; Briscoe 2008)。令人吃惊的是,两种蝴蝶已经获得在蓝绿范围(500～505nm)内具有相同最高吸收峰波长的视觉色素,由两种不同的视蛋白基因家族成员所编码,在 *L. rubidus* 中在 500nm 波长有最高吸收峰的视蛋白,而蚬蝶 *A. mormo* 是在 505nm 波长有最高吸收峰的视蛋白。这暗示着这些蝴蝶在超过 7000 万年以前就开始分化(Wahlberg 2007),亲缘较近的蝴蝶不约而同地选择了不同祖先基因来获得相同的视觉色素生理学性质(例如,最高吸收峰的波长)。

菜粉蝶中吸收蓝光的视觉色素在这方面也是值得关注的,因为它们都在 425nm 和 453nm 波长处有最大吸收峰,它们是由加倍的蓝色视蛋白基因 *PrV* 和 *PrB* 编码(Arikawa et al. 2005)。由于迄今被研究的少数的凤蝶和蛱蝶的眼组织只含有一个编码蓝色视蛋白的互补 DNA,所以我们决定通过调查灰蝶和粉蝶基因复制的进化起源来了解它们之间是不是相互独立的。为了达到这个目的,我们调查了来自于 10 个另外的蝴蝶分类属的眼组织特异的 cDNA 文库,包括 7 个灰蝶亚家族中最大的 3 个(灰蝶、蓝灰蝶亚科(Polyommatinae)和线灰蝶亚科(Theclinae)),以及蝴蝶的另外 3 个科(粉蝶科、蛱蝶科和蚬蝶科)。我们总共从这 10 个分类群中克隆了 14 个编码蓝色视蛋白的全长互补 DNA,包括所有被调查的灰蝶亚家族中的 *BRh1* 和 *BRh2* 的同源基因(图 7.2C)。我们在 7 个被调查的蛱蝶亚家族中检测到每个亚家族只有一个蓝色视蛋白互补 DNA。系统发生分析明确地表明 *L. rubidus* 的蓝色视蛋白的进化独立于白粉蝶(*P. rapae crucivora*),这个结果与蛱蝶不同于粉蝶的蓝色视觉色素具有非常不同的最高吸收峰波长这一现象是一致的。我们的结果也表明 *L. rubidus* 蓝色视蛋白基因加倍事件的发生要早于 coppers, hairstreaks 和 blues 的辐射分化之前(灰蝶亚科+线灰蝶亚科+蓝灰蝶亚科; Sison-Mangus et al. 2006)。我们随后调查了伊眼灰蝶亚科亚家族的灰蝶和粉蝶中余下的视蛋白,发现像 *L. rubidus* 一样,除了加倍的蓝色视蛋白,它眼组织包括一个紫外光视蛋白 mRNA 和一个长波光视蛋白的 mRNA (Sison-Mangus et al. 2008)。尽管以前提到的线灰蝶亚科和蓝灰蝶亚科中的各个品种的视网膜电图研究在眼组织中只检测到三个主要的光谱峰,但是我们的分子水平的研究证实了在三个灰蝶亚家族中蝴蝶的眼组织具有四个视觉色素,而不是三个,认识到这点差异是非常重要的。

为什么蝴蝶的蓝色视蛋白要发生基因加倍进化? *L. rubidus* 中吸收蓝-绿光的视觉色素(P500)的分子进化也许是为了加强光谱中蓝-绿部分的色觉。吸收蓝-绿光的视觉色素是由一个蓝色视蛋白基因(*BRh1*)编码并且相较于它的旁系同源基因(*BRh2*)编码的视觉色素来说,波长还向红外光偏移了 63nm。并且还由长波光视蛋白基因 *LWRh* 编码的吸收绿光的视觉色素(P568)组成(Sison-Mangus et al. 2006)。我们尝试在伊眼灰蝶中验证这个假说(见下文)。另一方面,蓝色视蛋白在白粉蝶(*P. rapae crucivora*)中加倍,使得昆虫得到一个蓝色视蛋白起源的紫光受体得以更好地区别光谱中的短波,这个假说仍然需要在行为学上进行验证。

4. 蛱蝶眼组织中视蛋白的空间表达模式

为了理解这些蓝色视蛋白基因加倍的结果使得灰蝶的眼睛有多么的不同，首先需要对更为简单的蛱蝶眼组织进行描述。蝴蝶的复眼被划分成三个区域，背边缘区（DRA），背部和主视网膜的腹面。蝴蝶的背边缘区，像其他许多昆虫一样，被特化来检测偏振光（Labhart 和 Meyer 1999；Stalleicken，Labhart 和 Mouritsen 2006）。这个区域典型的由眼组织背边的几行小眼组成，这个区域眼组织的微绒毛与眼组织其他区域的微绒毛在结构和方向上有所不同。微绒毛膜或 DRA 上小眼的感杆小体相互之间是以直角排列的。与此不同的是，主视网膜背面和腹面的感杆小体上的微绒毛排列更为随机。与主视网膜相同，每个 DRA 小眼具有九个感光细胞（R1－9）。然而，在蛱蝶眼组织主视网膜和 DRA 之间，光受体特异的亚型视蛋白表达模式存在极大的差异（见下文）。

蛱蝶主视网膜上小眼的视蛋白表达模式与工蜂的类似，在 R1 和 R2 细胞中表达对短波长光（紫外光和蓝光）敏感的视蛋白，而对长波光敏感的视蛋白在六个受体细胞（R3－R8）中表达（Spaethe 和 Briscoe 2005； Wakakuwa，Stavenga 和 Arikawa 2007）。因为对紫外光和蓝光敏感的视蛋白的表达被局限在 R1 和 R2 细胞，依据那些在 R1 和 R2 细胞中视蛋白的表达，三种小眼亚型得以鉴定（图 7.3C），分别是：UV-UV（小眼类型 I），UV-B（类型 II），B-B （类型 III）。烟草天蛾和蛱蝶科的蝴蝶，帝王斑蝶，小红蛱蝶和基红毒蝶无一不证实了这些视蛋白表达的模式（White et al. 2003；Briscoe et al. 2003；Sauman et al. 2005； Zaccardi et al. 2006）。在 DRA 中视蛋白表达的模式是不同于主视网膜上的。在帝王斑蝶中，利用根据东方虎凤蝶中对紫外光敏感的视蛋白 C 端区域的一个短肽产生的抗体，我们发现只有一个对紫外光敏感的视蛋白存在于背侧边缘的 R1－R8 感光细胞中，然而在主视网膜上，对紫外光敏感的视蛋白只在 R1 或者 R2 中表达（Sauman et al. 2005）。来源于 DRA 感光细胞中对紫外偏振光敏感的输入对迁徙的帝王斑蛾的定向是非常重要的（Reppert et al. 2004； Sauman et al. 2005）。

5. 灰蝶 *Lycaena rubidus* 视网膜的性别二态性

在迄今研究的所有蛱蝶中，眼组织中视蛋白 mRNA 表达的模式在性别之间是没有差异的。与此相反的是，我们发现在灰蝶 *L. rubidus* 眼组织的主视网膜上，视蛋白 mRNA 表达的模式是性别二态性的（DRA 还没有被研究），这种模式证实了一个以前关于相同物种中眼组织视觉色素是性别二态性分布的外延微光谱分光光度的报道（Bernard 和 Remington 1991）。在从 *L. rubidus* 克隆了几个视蛋白 cDNAs 后（*UVRh*，*BRh1*，*BRh2* 和 *LWRh*），我们发现所有四个视蛋白 mRNA 在眼组织中的表达不仅具有性别特异的方式，而且也是按背腹方式差异分布的。雄性视网膜背侧只表达 *UVRh* 和 *BRh1*。新小眼类型在雄性视网膜背侧处于支配地位，因为在 R3－R8 细胞中表达的是 *UVRh*，而不是通常在蛱蝶眼组织（迄今所有被调查的蝴蝶眼组织）中见到的 *LWRh* 视蛋白。再者，*UVRh-UVRh* （R1 和 R2 细胞）小眼类型在眼组织背侧的这个部分处于支配地位，与此同时还存在少量 *UVRh-BRh1* 和 *BRh1-BRh1* 类型的小眼。因此雄性的背侧眼很可能在红光范围

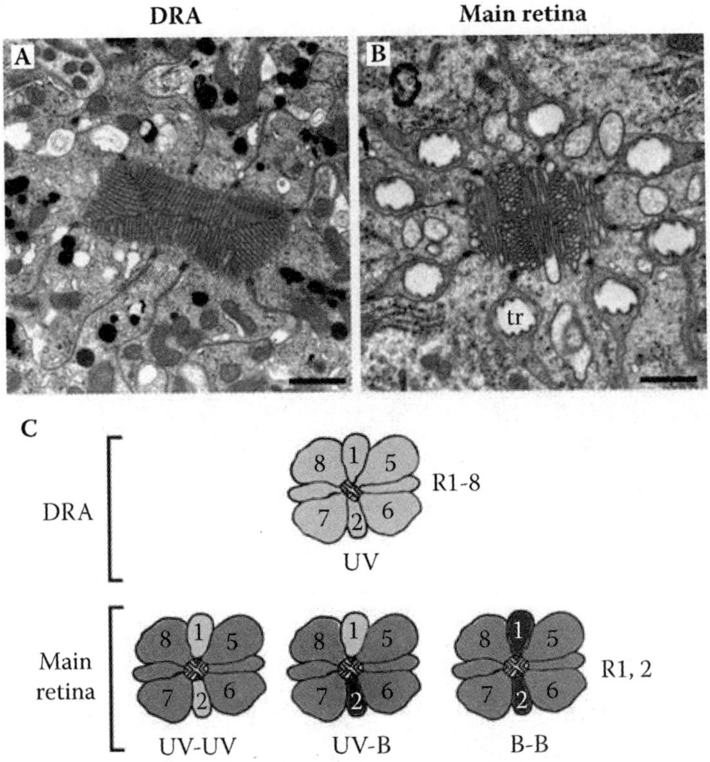

图 7.3　背边缘区的感杆束微绒毛，蝴蝶的主视网膜和典型的蛱蝶眼睛中视蛋白表达模式。(A)灰蝶 *Lycaena rubidus* 背边缘区的小眼是方形的，并且形成了微绒毛便于检测偏振光。(B)*L. rubidus* 的视网膜小眼是圆形的并且也具有微绒毛，但是其不是用来检测偏振光的。修改自 Sison-Mangus et al. (2006)。(C)帝王斑蝶背边缘区的视蛋白表达包含 R1—R8 感光细胞的所有 UV 视蛋白。与此形成对比的是，帝王斑蝶主视网膜上的视蛋白表达包含了所有 R3—R8 感光细胞的 LW 视蛋白，而在 R1 和 R2 感光细胞中要么是 UV-UV 型，UV-B 型或者为 B-B 类型的小眼。

内是色盲，并且只能看到短波光。雌性的背侧视网膜则是另外一回事，*UVRh*，*BRh1* 和 *LWRh* 的 m*RNAs* 都在这一区域表达(图 7.4A-H)。最引人注目的是，这两个视蛋白基因 *BRh*1 和 *LWRh* 在 R3—R8 感光细胞中共表达，使得 *L. rubidus* 在眼组织区域是性别二态性的(图 7.5E)。据我们所知，L. rubidus 是第一个在同一感光细胞中具有两种视觉色素的昆虫，一种吸收短波，一种吸收长波，这两种色素在相同的感光细胞中共表达(Sison-Mangus et al. 2006)。假设这两种视觉色素都参与光信号传导，那么它们在一个单感光细胞中的共表达暗示着这些受体将具有从蓝色到黄色的光谱范围内广泛的灵敏度。但是需要这些共表达视蛋白的细胞内记录来证实这个假说。连同 R1 和 R2 细胞中的 UV 受体，这类小眼类型在背侧眼中处于支配地位并且暗示雌性 *L. rubidus* 能够识别从紫外光到长波范围的光，与此形成鲜明对比的是，雄性的背侧眼是特化来识别从紫外光到蓝光范围内的光谱。有趣的是，雌性背侧眼本身很可能不是特异的、适应性的进化状态，而是反映了在转变成独特的雄性背侧眼进化道路上的一个过渡时期(见 Sison-Mangus et al. 2006 和下面的讨论)。

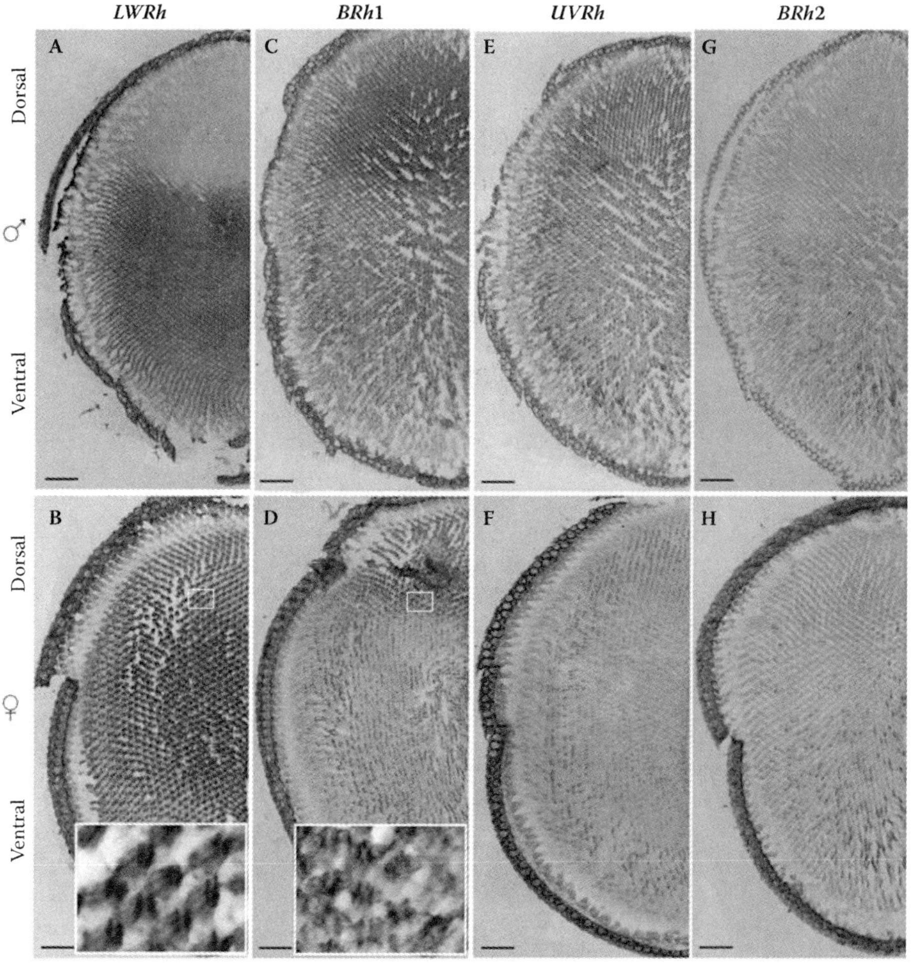

图 7.4 *LWRh* 视蛋白 mRNA 表达的性别二态性，雌性背侧眼中 *LWRh* 和 *BRh1* 视蛋白的共表达，以及 *Lycaena rubidus* 成虫视网膜中 *UVRh* 和 *BRh2* 视蛋白 mRNA 表达存在背腹差异。(A) *LWRh* 只在雄性的腹侧表达。(B)作为对比，*LWRh* 一律沿着雌性的视网膜表达。插图：*LWRh* mRNA 在 R3－R8 雌性背侧眼感光细胞中表达的放大图。(C) 雄性的 *BRh1* 在背侧眼中大量表达而在腹侧表达量较小。(D)类似地，雌性 *BRh1* 的表达量在背侧比在腹侧大。尽管如此，*BRh1* 和 *LWRh* 在雌性的背侧区域是共表达的。插图：*BRh1* mRNA 在 R3－R8 感光细胞中表达的放大图。在雄性(E)和雌性(F)中，*UVRh* 在背侧的表达量比腹侧的要高。在雄性(G)和雌性(H)中，*BRh2* mRNA 不在背侧表达，只在视网膜的腹侧表达。比例尺＝100μm。

另一方面，两性的腹侧眼具有更为典型的视蛋白表达模式，*LWRh* 在 R3－R8 细胞中表达，而对短波敏感的视蛋白 mRNAs(*UVRh*，*BRh1*，*BRh2*)以一种没有重叠的方式在 R1－R2 细胞中表达。尽管如此，由于在腹侧视网膜中，加倍的蓝色视蛋白 *BRh2* 的存在，相较于在蛱蝶眼组织中发现的三种典型小眼类型(UV-UV，UV-B 和 B-B)，腹侧眼具有六种小眼类型(UV-UV，UV-B1，UV-B2，B1-B1，B1-B2，B2-B2)。再者，在表达 *BRh2* 的小眼中，常常会发现一种粉红色过滤色素。想必这会导致另外一种受体的产生，但是需要行为学的实验来断定它是否实际参与了色觉。

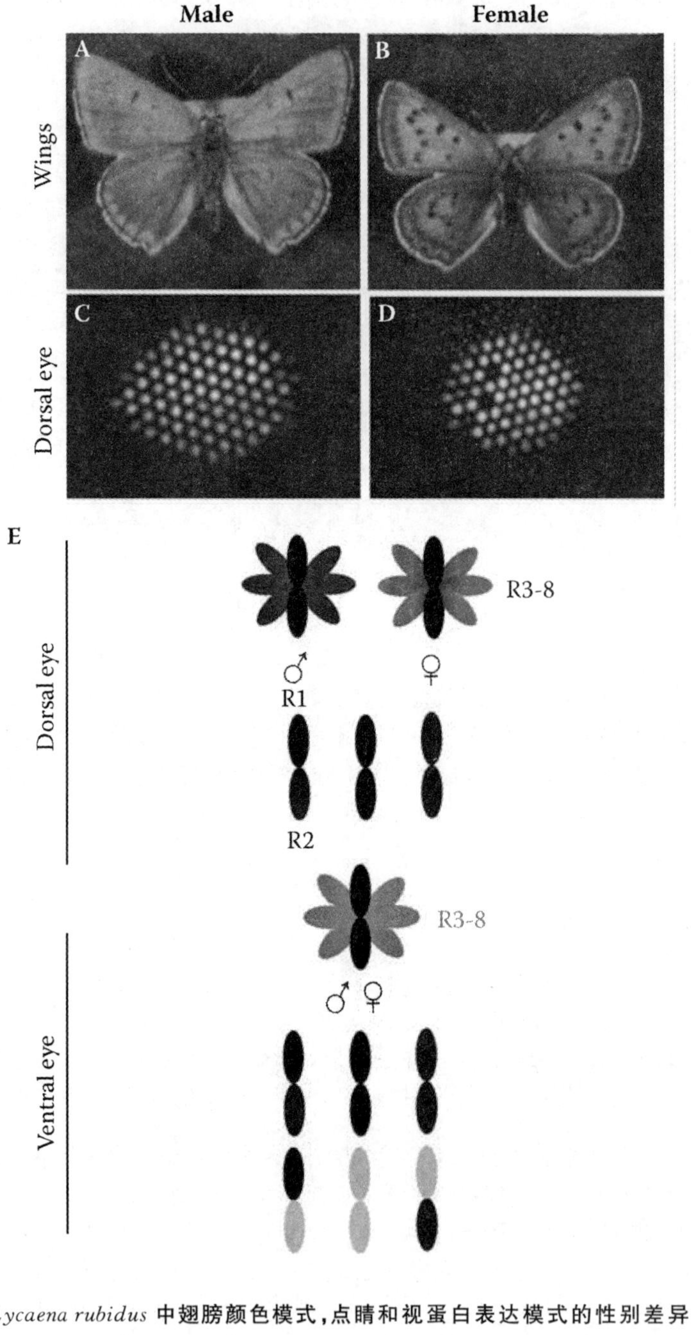

图 7.5　灰蝶 *Lycaena rubidus* 中翅膀颜色模式，点睛和视蛋白表达模式的性别差异（彩色版本图片见附录图 7.5）。（A）在雄性低前翅和后翅边缘反射 UV 的鳞毛（闪光紫色）。（B）在雌性翅膀上不反射 UV 的鳞毛。雄性（C）和雌性（D）背侧眼上的点睛展现出很强的着色性别二态性。（E）概括了 *L. rubidus* 眼中视蛋白表达模式的示意图。深蓝色代表 *BRh1* 视蛋白 mRNA 的表达。橙色代表 *LWRh* 视蛋白 mRNA 的表达。深蓝色和橙色代表 *BRh1* 和 *LWRh* 视蛋白 mRNAs 的共表达。黑色代表 *UVRh* 视蛋白 mRNA 的表达。淡蓝色代表 *BRh2* 视蛋白 mRNA 的表达。修改于 Sison-Mangus 等（2006）。

许多蝴蝶翅膀的颜色模式起着隐形或拟态的功能。但是在类似灰蝶的蝴蝶中，着色在两性中是有差异的。翅膀颜色的差异暗示着性别选择。例如，雄性灰蝶 *L. rubidus* 的翅膀是明亮的，深红铜色，而雌性的翅膀显得是暗褐色的（图 7.5 A，B）。雄性翅膀背侧反射光谱的测量指明它们反射紫外光和红色光谱，而雌性仅仅只反射红色（Bernard 和 Remington 1991）。在新西兰灰蝶 *L. salustris* 物种中也观察到了类似的翅膀反射光谱（Meyer-Rochow 1991）。三个被调查的亚科包括蓝灰蝶亚科（Polyommatinae），线灰蝶亚科（Theclinae）和灰蝶亚科（Lycaeninae）中的雄性灰蝶也反射紫外光（Vertesy et al. 2006）。紫外光是蝴蝶性行为信号中的一个重要部分（Siberglied 1979），并且已经有研究证明紫外光在蓝色灰蝶的择偶过程中起着重要的作用（Burghardt et al. 2000；Knüttel 和 Fiedler 2001）。除此之外，众所周知的是雄性灰蝶会显示其领地（Clark 和 Dickson 1971；McCubbin 1971；Atsatt 1981），这是一种暗示强烈雄性竞争的行为。*BRh1* 视蛋白 mRNA 在 R3－R8 感光细胞中特异性表达，而 *UVRh* 主要在 R1 和 R2 细胞中表达，这种模式暗示雄性灰蝶的背侧眼是特化来辨别闪烁移动物体的，例如其他飞行中的雄性。到目前为止，没有直接的行为证据证实雄性灰蝶背侧眼的功用。辨识雄性背侧眼优于雌性背侧眼的情形对于提供一种与雄性竞争的联系是至关重要的。

粉蝶属白粉蝶（*P. rapae crucivora*）的眼组织也是性别二态性的（Arikawa et al. 2005）。尽管如此，它眼组织的性别二态性不是通过视蛋白表达的变化实现的，而是通过过滤色素的策略来实现的。雄性和雌性白粉蝶（*P. rapae crucivora*）不能通过全光谱颜色进行辨别，但是当用一个紫外光过滤器进行照相时，它们翅膀的反射光谱在两性之间是有差异的。雌性的翅膀强烈地反射紫外光，而雄性的翅膀几乎不反射紫外光；再者，紫外光的反射（在雌性翅膀上更明显）是导致雄性性行为的主要线索（Obara 1970）。鉴于翅膀模式上的这些差异，白粉蝶（*P. rapae crucivora*）具有三种对短波敏感的视蛋白，一种 UV 和一对蓝色视蛋白（PrV 和 PrB）是非常有意思的；但是这一对蓝色视蛋白其中的一个，PrV 对紫色是敏感的。这些短波视蛋白的 mRNAs 独立地在 R1 和 R2 细胞中表达，然而只在三种平常的小眼类型被发现，而 PrV 被局限在 II 型小眼中表达。正是在雄性而不是在雌性的这些小眼中发现了荧光色素，由于这个色素扮演着吸收紫光的光谱过滤器的角色，因此紫色受体的敏感性已经被修改形成一个双峰的蓝光，在紫光范围内有一个小峰和在蓝光范围内有一个高峰。由于昆虫的这种光谱设置，雄性也许能够根据雌性翅膀紫外光的反射而精确地区分同种的雌性，更加确证了在这种粉蝶亚种求偶行为中的观察（Obara 和 Hidaka 1968；Obara 1970）。

这些色素修饰强烈地暗示在白粉蝶（*P. rapae crucivora*）的异性之间和在 *L. rubidus* 的同性之间，性别二态性的眼组织是进化来适应依靠紫外光信号的交流。评估其他性别二态性的蝴蝶的视觉系统是否具有相同的模式将是非常有趣的。例如，豆粉蝶属的蝴蝶已经在性别选择的背景下被广泛地研究其紫外光信号途径（Silberglied 和 Taylor 1978；Silberglied 1979；Rutowski 1985）。已经从菲罗豆粉蝶中克隆出一个对紫光敏感的视蛋白，它与粉蝶属中的 PrV 是同源的（Sison-Mangus et al. 2006），这是这种动物可能具有三种对短波敏感视蛋白很好的体现。在菲罗豆粉蝶中对紫光敏感的视蛋白的存在更加确证了在其他物种 *C. erate*（Eguchi et al. 1982）中发现紫光受体（λ_{max}＝400nm）的报道，这暗示这个受体也是由一个对紫光敏感的视蛋白编码的。另一个值得进一步研究的候选者是性别二态性的蛱蝶科蝴蝶、幻紫斑蛱蝶，这种蝴蝶的雄性在翅膀的背侧具有明亮的、反射紫外光的彩虹色标记(Kemp 和 Macedonia 2006)，并且与雄性相比，雌性具有更明亮和炫彩斑斓的标记，从而对紫外光的反射更加强烈(Kemp 2007)。

6. 蝴蝶中的视觉研究

视觉色素的光谱多样性和它们在蝴蝶眼组织中的排布具有研究色觉的潜力。尽管如此，仅仅复眼中多个受体的出现并不能证明蝴蝶具有色觉。一只蝴蝶要具有色觉必须能够在不考虑强度的情形下辨别具有不同颜色的物体(Goldsmith 1990)。动物具有色觉需要的要求如下：两种具有不同光谱敏感性的视觉色素还需定位在不同的感光细胞上；视叶上中间神经元的存在，这些视叶具有来自于那些受体的对立的信号输入；脑中合适的结构，能够比较不同刺激来源的信号；以及在进行颜色选择实验中做出行为反应(Goldsmith 1990；Kelber，Vorobyev 和 Osorio 2003；Kelber 2006)。因为过滤色素的存在可以改变受体吸收光谱的范围，由一个视觉色素产生的受体和过滤色素能够参与颜色的辨别。

我们对蛱蝶科或灰蝶科蝴蝶的大脑是怎样处理颜色信息的认识是极其不够的(Swihart 1968)，未来在这个领域会产生大量的优秀的研究成果。行为测试已经展示了对蛱蝶科和灰蝶科蝴蝶色觉最清晰的证明。对鳞翅目中真实色觉的实验性测试和证明需要训练一种率先用于实验的昆虫，这种昆虫需要把食物奖励和颜色联系起来。训练或者学习奖励颜色的能力是行为测试中的一个重要部分。这将会给实验人员以研究昆虫感觉能力的机会，因为学习意味着对颜色信息的感知。一旦动物“学会”了奖励颜色，接着让它们在奖励和非奖励色中进行一系列的选择，而颜色刺激的光强度也要进行控制(Kelber 1999；Kelber 和 Pfaff 1999；Zaccardi et al. 2006；Sison-Mangus et al. 2008)或者这些颜色会与变化的灰色色度进行比较(Kinoshita，Shimada 和 Arikawa 1999)。如果昆虫在没受光强度的影响下，选择了奖励颜色，那么它确实具有真正的色觉。

7. 红-绿色觉的选择性策略

利用这种具有严格要求的行为实验已经显示蛱蝶和灰蝶如同其他蝴蝶和蛾一样，具有真正的色觉；在光强度比例在 0.01 到 100 的范围或者与灰色比较的情况下，它们能够分辨出成对不同波长的颜色(Kelber，Balkenius 和 Warrant 2002；表 7.1)。还未彻底地研究清楚蛱蝶科和灰蝶科蝴蝶能够辨别的色彩范围。尽管如此，从已有的实验证据来看，有一点是清楚的，即不同的蝴蝶物种采取不同的策略在绿-红范围具有色觉。例如，蛱蝶科的红纹丽蛱蝶，它的眼组织包含三种典型的视觉色素，吸收紫外光、蓝光和长波光，它能够从橙色光中分辨蓝光(620 vs. 440nm)，但是不能够从橙色光中分辨黄色光(620 vs. 590nm；Zaccardi et al. 2006)。相反的是，蛱蝶科的基红毒蝶，它的眼组织也包含吸收紫外光、蓝光和长波光的视觉色素，尽管它只含有一个吸收长波光的视蛋白，但是能够在红光范围(620 vs. 640nm)识别颜色。正是在这种异质分布的红色过滤色素的帮助下，该物种产生第四种对红光敏感的受体(Zaccardi et al. 2006)。另一方面，灰蝶科的伊眼灰蝶在进食期能够在高达 560nm 的绿光范围内分辨颜色(Sison-Mangus et al. 2008)。尽管如此，它不能够在红光范围(590 vs. 640nm)内分辨颜色，即使它具有对长波光敏感的受体和红色过滤色素所共同产生的反射红光的小眼(图 7.1；Sison-Mangus et al. 2008)。被用于这个工作的光受体很可能是吸收蓝光和长波光的视觉色素，其在同源性上与 *L. rubidus* 的 P500(500nm 时具有最高灵敏度的视觉色素)和 P568(568nm 时具有最高灵敏度的视觉色素)视觉色素是最相似的。然而，对于伊眼灰蝶来说，需要生理学上的数据来更完全地证明蓝光视蛋白基因的加倍确实和长波光视蛋白一起介导了这个行为。

表 7.1　鳞翅目昆虫色觉实验中测试的物种和波长

Table 7.1　Sepciess and Wavelengths Tested in Lepidopteran Color Vision Experiments

Family	Species	Wavelength(nm)/ Color Tested	Color Vision	Sources
Sphingidae	*Macroglossum stellatarum*	380 vs. 360	yes	Kelber and Henique 1999
		380 vs. 420	yes	
		380 vs. 470	yes	
		470 vs. 500	yes	
		500 vs. 420	yes	
		620 vs. 470	yes	
		470 vs. 620	yes	
		595 vs. 620	no	
Sphingidae	*Deilephila elpenor*	blue vs. shades of gray	yes	Kelber et al. 2002
		blue vs. shades of blue	no	
		yellow vs. shades of gray	yes	
		yellow vs. shades of yellow	yes	
Papilionidae	*Papilio aegeus*	430 vs. 590 and 640	yes	Kelber and Pfallf 1999
		640 vs. 430 and 590	yes	
Papilionidae	*Papilio xuthus*	red vs. yellow, green, blue	yes	Kinoshita, Shirnada, and Arikawa 1999
		yellow vs. red, green, blue	yes	
		green vs. red, yellow, blue	yes	
		blue vs. red, yellow, green	no *	
Nymphalidae	*Heliconius erato*	590 vs. 440	yes	Zaccardi et al. 2006
		620 vs. 590	yes	
		620 vs. 640	yes	
Nymphalidae	*Vanessa atatanta*	620 vs. 440	yes	Zaccardi et al. 2006
		620 vs. 590	no	
Lycaenidae	*Polyommatus icarus*	450 vs. 590	yes	Sison-Mangus et al. 2008
		560 vs. 590	yes	
		570 vs. 590	no	

The color or wavelength first listed was used as the training and rewarding color.

* No preference for blue.

伊眼灰蝶和基红毒蝶之间的对比结果暗示在蝴蝶辨别颜色的过程中，过滤色素的影响应该在物种特异性的基础上进行评估。对这些和其他已知具有侧向过滤色素的蝴蝶物种——白粉蛾（*P. rapae crucivora*）（Wakakuwa et al. 2004），帝王斑蝴蝶（Sauman et al. 2005），热带蝴蝶（*Bicyclus anynana*），酢醬灰蝶（*Zizeeria maha*）（Stavenga 2002），大紫蛱蝶（*Sasakia charonda*），and 黄钩蛱蝶（*Polygonium c-aureum*）（Kinoshita，Sato 和 Arikawa 1997）进行进一步的行为学研究，以及生理和分子水平上的研究是必需的，这能够确定过滤色素是否在它们的色觉中起作用。这样的研究应该能够鉴定眼组织中视蛋白的数目和确定侧向的过滤色素是否对单个光受体的光谱敏感性具有影响。对重组视觉色素的最大吸收峰波长的直接测量，通过果蝇或培育细胞中视蛋白转基因表达以及体内测量，这些也将会提供至关重要的实验证据来确定侧边过滤色素对色觉的影响。

8. 结论

漫长的进化塑造了蝴蝶优美的眼睛。灰蝶科和蛱蝶科蝴蝶通过不同的机制（导致相似的性状，例如红-绿色觉）对它们视觉系统进行修饰的事实强烈地暗示了这是适应性进化结果。迄今为止，蓝色视蛋白和长波长视蛋白的基因加倍以及过滤色素的补充是蝴蝶达到其相似生理和行为目标所使用的最常规策略。一些蝴蝶在它们眼组织的特定部位改变了视蛋白的表达模式，一些性别二态性的蝴蝶也发生这样的情况，这种眼组织的改动在性别二态性的蝴蝶中是最显著的，这也暗示着一种发育的开放性。在将来的研究中，弄清楚在此所述的蛱蝶和灰蝶各自的一些独有的特征，这对其能否成为像经典的形态学特征一样，作为这些昆虫家族的界定特征是非常有意思的研究。

9. 致谢

我们对 Steven Reppert 对手稿提出的有帮助的评述表示感谢。这项工作曾受到国家科学基金 IOS－0819936 的部分支持。

参考文献

[1]Arendt，D. 2003. Evolution of eyes and photoreceptor cell types. *Int. J. Dev. Biol.* 47：563－71.

[2]Arikawa，K.，D. G. W. Scholten，M. Kinoshita，and D. G. Stavenga. 1999. Tuning of photoreceptor spectral sensitivities by red and yellow pigments in butterfly *Papilio xuthus*. *Zool. Sci.* 16：17－24.

[3]Arikawa，K.，and D. G. Stavenga. 1997. Random array of colour ilters in the eyes of butterflies. *J. Exp. Biol.* 200：2501－06.

[4]Arikawa，K.，M. Wakakuwa，X. D. Qiu，M. Kurasawa，and D. G. Stavenga. 2005. Sexual dimorphism of short-wavelength photoreceptors in the small white butterfly，*Pieris rapae crucivora*. *J. Neurosci.* 25：5935－42.

[5]Atsatt，P. R. 1981. Lycaenid butterflies and ants—selection for enemy-free space. *Am. Nat.* 118：638－54.

[6]Bennett，R. R.，and P. K. Brown. 1985. Properties of the visual pigments of the moth *Manduca*

sexta and the effects of two detergents,digitonin and chaps. *Vision Res*. 25:1771—81.

[7]Bernard,G. D. ,and W. H. Miller. 1970. What does antenna engineering have to do with insect eyes? *IEEE Student J*. 8:2—8.

[8]Bernard,G. D. ,and C. L. Remington. 1991. Color-vision in *Lycaena* butterflies: Spectral tuning of receptor arrays in relation to behavioral ecology. *Proc. Natl. Acad. Sci. U. S. A*. 88:2783—87.

[9]Briscoe, A. D. 1998. Molecular diversity of visual pigments in the butterfly *Papilio glaucus*. *Naturwissen-schaften* 85:33—35.

[10]Briscoe,A. D. 2008. Reconstructing the ancestral butterfly eye: Focus on the opsins. *J. Exp. Biol*. 211:1805—13.

[11]Briscoe,A. D. ,and G. D. Bernard. 2005. Eyeshine and spectral tuning of long wavelength-sensitive rhodopsins: No evidence for red-sensitive photoreceptors among five Nymphalini butterfly species. *J. Exp. Biol*. 208:687—96.

[12]Briscoe,A. D. ,G. D. Bernard,A. S. Szeto,L. M. Nagy,and R. H. White. 2003. Not all butterfly eyes are created equal: Rhodopsin absorption spectra,molecular identification,and localization of ultraviolet-,blue-,and green-sensitive rhodopsin-encoding mRNAs in the retina of *Vanessa cardui*. *J. Comp. Neurol*. 458:334—49.

[13]Briscoe,A. D. ,and L. Chittka. 2001. The evolution of color vision in insects. *Annu. Rev. Entomol*. 46:471—510.

[14]Burghardt,F. ,H. Knuttel,M. Becker,and K. Fiedler. 2000. Flavonoid wing pigments increase attractiveness of female common blue (*Polyommatus icarus*) butterflies to mate-searching males. *Naturwissenschaften* 87:304—07.

[15]Campbell,D. L. ,A. V. Z. Brower,and N. E. Pierce. 2000. Molecular evolution of the *wingless* gene and its implications for the phylogenetic placement of the butterfly family Riodinidae (Lepidoptera: Papilionoidea). *Mol. Biol. Evol*. 17:684—96.

[16]Clark G. C. ,and C. G. C. Dickson. 1971. *Life histories of South African lycaenid butterflies*. Capetown: Purnell.

[17]Eguchi,E. ,K. Watanabe,T. Hariyama,and K. Yamamoto. 1982. A comparison of electrophysiologically determined spectral responses in 35 species of Lepidoptera. *J. Insect Physiol*. 28:675—82.

[18]Frentiu,F. D. ,G. D. Bernard,M. P. Sison-Mangus,A. V. Z. Brower,and A. D. Briscoe. 2007. Gene duplication is an evolutionary mechanism for expanding spectral diversity in the long-wavelength photopigments of butterflies. *Mol. Biol. Evol*. 24:2016—28.

[19]Goldsmith,T. H. 1990. Optimization constraint and history in the evolution of eyes. *Q. Rev. Biol*. 65:281—322.

[20]Imafuku,M. ,I. Shimizu,H. Imai,and Y. Shichida. 2007. Sexual difference in color sense in a lycaenid butterfly *Narathura japonica*. *Zool. Sci*. 24:611—13.

[21]Kelber,A. 1999. Ovipositing butterflies use a red receptor to see green. *J. Exp. Biol*. 202:2619—30.

[22]Kelber,A. 2006. Invertebrate colour vision. In *Invertebrate vision*,ed. E. J. Warrant and D. E. Nilsson,250—90. Cambridge: Cambridge Univ. Press.

[23]Kelber,A. ,A. Balkenius,and E. J. Warrant. 2002. Scotopic colour vision in nocturnal hawk-

moths. *Nature* 419:922—25.

[24]Kelber, A., and U. Henique. 1999. Trichromatic colour vision in the hummingbird hawkmoth *Macroglossum stellatarum* L. *J. Comp. Physiol. A* 184:535—41.

[25]Kelber, A., and M. Pfaff. 1999. True colour vision in the orchard butterfly *Papilio aegeus*. *Naturwissenschaften* 86:221—24.

[26] Kelber, A., M. Vorobyev, and D. Osorio. 2003. Animal colour vision—behavioural tests and physiological concepts. *Biol. Rev.* 78:81—118.

[27]Kemp, D. J. 2007. Female butterflies prefer males bearing bright iridescent ornamentation. *Proc. Biol. Sci.* 274:1043—47.

[28]Kemp, D. J., and J. M. Macedonia. 2006. Structural ultraviolet ornamentation in the butterfly *Hypolimnas bolina* L. (Nymphalidae): Visual, morphological and ecological properties. *Aust. J. Zool.* 54: 235—44.

[29] Kinoshita, M., M. Sato, and K. Arikawa. 1997. Spectral receptors of nymphalid butterflies. *Naturwissenschaften* 84:199—201.

[30]Kinoshita, M., N. Shimada, and K. Arikawa. 1999. Colour vision of the foraging swallowtail butterfly *Papilio xuthus*. *J. Exp. Biol.* 202:95—102.

[31]Kitamoto, J., K. Sakamoto, K. Ozaki, Y. Mishina, and K. Arikawa. 1998. Two visual pigments in a single photo-receptor cell: Identification and histological localization of three mRNAs encoding visual pigment opsins in the retina of the butterfly *Papilio xuthus*. *J. Exp. Biol.* 201:1255—61.

[32]Knüttel, H., and K. Fiedler. 2001. Host-plant-derived variation in ultraviolet wing patterns inluences mate selection by male butterflies. *J. Exp. Biol.* 204:2447—59.

[33]Labhart, T., and E. P. Meyer. 1999. Detectors for polarized skylight in insects: A survey of ommatidial specializations in the dorsal rim area of the compound eye. *Microsc. Res. Tech.* 47:368—79.

[34]McCubbin, C. 1971. *Australian butterflies*. Melbourne: Thomas Nelson.

[35]Meyer-Rochow, V. B. 1991. Differences in ultraviolet wing patterns in the New Zealand lycaenid butterflies *Lycaena salustius*, *L. rauparaha*, and *L. feredayi* as a likely isolating mechanism. *J. R. Soc. New Zealand* 21:169—77.

[36]Miller, W. H. 1979. Ocular optical iltering. In *Handbook of sensory physiology*, *vol.* VII/6A, ed. H. Autrum, 69—143. Berlin, Heidelberg, New York: Springer.

[37]Nilsson, D. E., M. F. Land, and J. Howard. 1988. Optics of the butterfly eye. *J. Comp. Physiol. A* 162:341—66.

[38]Obara, Y. 1970. Studies on the mating behavior of the white cabbage butterfly, *Pieris rapae crucivora* Boisduval. III. Near ultraviolet reflection as the signal of intraspecific communication. *Z. Vergl. Physiol.* 69:99—116.

[39]Obara, Y., and T. Hidaka. 1968. Recognition of the female by the male on the basis of ultra-violet reflection in the white cabbage butterfly *Pieris rapae crucivora* Boisduval. *Proc. Jpn. Acad.* 44:829—32.

[40]Plachetzki, D. C., D. M. Degnan, and T. H. Oakley. 2007. The origins of novel protein interactions during animal opsin evolution. *PLoS ONE* 2:e1054.

[41]Provencio, I., G. S. Jiang, W. J. De Grip, W. P. Hayes, and M. D. Rollag. 1998. Melanopsin: An opsin in melanophores, brain, and eye. *Proc. Natl. Acad. Sci. U. S. A.* 95:340—45.

[42]Reppert, S. M., H. S. Zhu, and R. H. White. 2004. Polarized light helps monarch butterflies navigate. *Curr. Biol.* 14:155—58.

[43]Rutowski, R. L. 1985. Evidence for mate choice in sulfur butterfly *Colias eurytheme*. Z. Tierpsychol. 70:103—14.

[44]Santillo, S., P. Orlando, L. De Petrocellis, L. Cristino, V. Guglielmotti, and C. Musio. 2006. Evolving visual pigments: Hints from the opsin-based proteins in a phylogenetically old eyeless invertebrate. *Biosystems* 86:3—17.

[45]Sauman, I., A. D. Briscoe, H. S. Zhu, et al. 2005. Connecting the navigational clock to sun compass input in monarch butterfly brain. *Neuron* 46:457—67.

[46]Silberglied, R. E. 1979. Communication in the ultraviolet. *Annu. Rev. Ecol. Syst.* 10:373—98.

[47]Silberglied, R. E., and O. R. Taylor. 1978. Ultraviolet reflection and its behavioral role in courtship of sulfur butterflies *Colias eurytheme* and *Colias philodice* (Lepidoptera, Pieridae). *Behav. Ecol. Sociobiol.* 3:203—43.

[48]Sison-Mangus, M. P., G. D. Bernard, J. Lampel, and A. D. Briscoe. 2006. Beauty in the eye of the beholder: The two blue opsins of lycaenid butterflies and the opsin gene-driven evolution of sexually dimorphic eyes. *J. Exp. Biol.* 209:3079—90.

[49]Sison-Mangus, M. P., A. D. Briscoe, G. Zaccardi, H. Knuttel, and A. Kelber. 2008. The lycaenid butterfly *Polyommatus icarus* uses a duplicated blue opsin to see green. *J. Exp. Biol.* 211:361—69.

[50]Smith, C. W., and T. H. Goldsmith. 1990. Phyletic aspects of the distribution of 3-hydroxyretinal in the Class Insecta. *J. Mol. Evol.* 30:72—84.

[51]Spaethe, J., and A. D. Briscoe. 2005. Molecular characterization and expression of the UV opsin in bumblebees: Three ommatidial subtypes in the retina and a new photoreceptor organ in the lamina. *J. Exp. Biol.* 208:2347—61.

[52]Stalleicken, J., T. Labhart, and H. Mouritsen. 2006. Physiological characterization of the compound eye in monarch butterflies with focus on the dorsal rim area. *J. Comp. Physiol. A* 192:321—31.

[53]Stavenga, D. G. 2002. Colour in the eyes of insects. *J. Comp. Physiol. A Neuroethol. Sens. Neural Behav. Physiol.* 188:337—48.

[54]Stavenga, D. G., and K. Arikawa. 2006. Evolution of color and vision of butterflies. *Arthropod Struct. Dev.* 35:307—18.

[55]Stavenga, D. G., M. Kinoshita, E. C. Yang, and K. Arikawa. 2001. Retinal regionalization and heterogeneity of butterfly eyes. *Naturwissenschaften* 88:477—81.

[56]Swihart, S. L. 1968. Single unit activity in the visual pathway of the butterfly Heliconius erato. *Journal of Insect Physiology* 14:1589—1601.

[57]Takemura, S. Y., D. G. Stavenga, and K. Arikawa. 2007. Absence of eye shine and tapetum in the heterogeneous eye of *Anthocharis* butterflies (Pieridae). *J. Exp. Biol.* 210:3075—81.

[58]Terakita, A. 2005. The opsins. *Genome Biol.* 6:213.

[59]Vanhoutte, K. J. A., D. G. Stavenga. 2005. Visual pigment spectra of the comma butterfly, *Polygonia c-album*, derived from in vivo epi-illumination microspectrophotometry. *J. Comp. Physiol. A* 191:461—73.

[60]Vertesy, Z., Z. Balint, K. Kertesz, J. P. Vigneron, V. Lousse, and L. P. Biro. 2006. Wing scale mi-

crostructures and nanostructures in butterflies—natural photonic crystals. *J. Microsc. Oxford* 224: 108—10.

[61]Wahlberg, N. 2007. That awkward age for butterflies: Insights from the age of the butterfly subfamily Nymphalinae (Lepidoptera: Nymphalidae). *Syst. Biol.* 55:703—14.

[62]Wakakuwa, M., D. G. Stavenga, and K. Arikawa. 2007. Spectral organization of ommatidia in lower-visiting insects. *Photochem. Photobiol.* 83:27—34.

[63]Wakakuwa, M., D. G. Stavenga, M. Kurasawa, and K. Arikawa. 2004. A unique visual pigment expressed in green red and deep-red receptors in the eye of the small white butterfly *Pieris rapae crucivora*. *J. Exp. Biol.* 207:2803—10.

[64]White, R. H., H. H. Xu, T. A. Munch, R. R. Bennett, and E. A. Grable. 2003. The retina of *Manduca sexta*: Rhodopsin expression the mosaic of green- blue- and UV-sensitive photoreceptors and regional specialization. *J. Exp. Biol.* 206:3337—48.

[65]Yagi, N., and N. Koyama. 1963. *The compound eye of Lepidoptera: Approach from organic evolution*. Tokyo: Shinkyo Press.

[66]Zaccardi, G., A. Kelber, M. P. Sison-Mangus, and A. D. Briscoe. 2006. Color discrimination in the red range with only one long-wavelength sensitive opsin. *J. Exp. Biol.* 209:1944—55.

[67]Zhu, H., A. Casselman, and S. M. Reppert. 2008. Chasing migration genes: A brain expressed sequence tag resource for summer and migratory monarch butterflies (*Danaus plexippus*). *PLoS ONE* 3: e1345.

第八章 鳞翅目昆虫的生物钟——从分子到行为

Christine Merlin and Steven M. Reppert

1. 前言

在从蓝细菌到人类的所有物种中，生物钟是内生的计时机制，在分子、细胞、生理、行为等多个水平上控制着有机体的节律。遗传上确定的生物钟通过明暗交替与一天 24h 相同步。昆虫生物学的许多方面都受到生物钟驱使的节律行为的影响，包括卵孵化的时间、成虫的羽化、生殖行为、社交，以及更引人注目的节律的行为，例如在迁徙种群中时间补偿的以太阳为指南针的定向行为(Saunders 2002)。

在分子水平上，昆虫的生物钟由一组定义为生物钟基因形成的自我维持的转录反馈回路所组成。这个生物钟工作机制在黑腹果蝇(*Drosophila melanogaster*)上研究得最为广泛，几十年以来，果蝇也一直被认为是研究昆虫生物钟的模型(Stanewsky 2003; Rosato，Tauber 和 Kyriacou 2006)。尽管如此，近来在帝王斑蝶(*Danaus plexippus*)中发现了一种新奇且古老的生物钟机制，这不仅是对果蝇模型的一种挑战，而且总体而言，也为昆虫生物钟机制的进化提供了一种新的转机(Zhu et al. 2008)。

对鳞翅目昆虫生物钟的研究历史悠久，可追溯到 20 世纪 60 年代 Carroll Williams 对蚕蛾关于蛹滞育的光周期控制中脑调控方面的精细研究，以及 James Truman 和 Lynn Riddiford 20 世纪 70 年代对蚕蛾脑调控成虫羽化时间方面的重要研究。通过脑移植实验，Williams 首次揭示生物钟的感光机制是存在于脑中(Williams 1963; Williams 和 Adkisson 1964)。利用相似的移植方法，Truman 和 Riddiford (1970)确定了物种间成虫羽化的时间是由供体昆虫的脑所控制的，即脑移植不仅重启成虫羽化每日的节律，而且决定了受体蚕蛾的节律状态。这是证明昆虫的脑中确实存在着生物钟的第一个直接证据。至此，这些经典的蚕蛾研究为早期理解驱使昆虫行为的生物钟的定位作出了很大的贡献。

在这里我们概述了关于鳞翅目昆虫中存在的各种类型的生理和行为的节律概况，详细介绍了鳞翅目昆虫生物钟的分子机制，以及确定了细胞生物钟的定位，特别强调了对帝王斑蝶生物钟的研究。为了举例说明鳞翅目中昼夜节律周期的决定性的本质，我们详细说明了两个系统，在这两个系统中，生物钟被用来驾驭物种的生态：在蚕蛾信息素交流过程中的应用(在第十章进行综述)对于适时的繁殖行为是至关重要的，在时间补偿的以太阳为指南针定位方面的应用对于帝王斑蝶在迁徙过程中准确的导航是非常重要的。

2. 鳞翅目昆虫昼夜节律的概况

鳞翅目在生命周期不同阶段的事件中均展示了昼夜节律。其中的一些事件是发育过程的一部分，包括卵孵化，幼虫蜕皮，化蛹时的蜕皮，幼虫消化道的消除以及成虫的羽化(Truman 1992)。成虫的节律行为包括移动、飞行、信息素交流、交配以及产卵。这些成虫的节律行为往往发生在物种特异的时间段内(Saunders 2002 做过综述)。在所报道的这些日常节律中，在一定的条件下，许多具有维持在接近 24h 周期的能力，即它们是真正的昼夜节律的。已经有很多人对上面简要提到的生物钟参与到鳞翅目光周期现象进行了综述(Vas Nunes 和 Saunders 1999; Saunders 2002)，本章不再对其进行进一步的讨论，而我们将重点介绍在鳞翅目各种生理和行为事件中明显昼夜节律的时间控制(表 8.1)。

2.1 发育过程中的节律

卵孵化在鳞翅目生命周期中是一个极其重要的事件。这个程序的正确起始是至关重要的，因为它释放被约束在卵中的幼虫并且使其完成在外界的生活周期(Sauman 和 Reppert 1998)。决定卵孵化时间的行为程序，在至少几个蛾子种群中是由昼夜节律控制的。

依据对卵进行 12h 光照/12h 黑暗的周期处理而出现以破晓后孵化率达到峰值这一特征的孵化节律和对卵进行不间断的光照或黑暗处理会导致节律失调的孵化模式，Minis 和 Pittendrigh (1968)在棉红铃虫(*Pectinophora gossypiella*)中首次表明了鳞翅目卵孵化时间受昼夜节律调控。这个种群的节律显示了昼夜节律的主要特点，那就是，它在中胚层受光的驱使，并且一旦受到驱使，孵化的节律在一定的条件下保持稳定(Minis 和 Pittendrigh 1968)。与之相似，中国柞蚕(Riddiford 和 Johnson 1971; Sauman 和 Reppert 1996,1998; Sauman et al. 1996)，家蚕(Shimizu 和 Matsui 1983)，西南玉米螟(Takeda 1983)卵孵化的昼夜节律已见报道，这些说明功能性的生物钟在早期的发育中就已经开始起作用了。据我们所知，这种卵孵化的昼夜节律在蝴蝶中还未见报道。

表 8.1　鳞翅目中代表性生理和行为的昼夜节律

Table 8.1　Representative Physiological and Behavioral Circadian Rhythms in Lepidoptera

昼夜节律	物种	参考文献
		发育过程
卵孵育	柞蚕	Sauman et al. 1996; Sauman and Reppert 1998
	家蚕	Shimizu and Matsui 1983
	西南玉米螟	Takeda 1983
	棉红铃虫	Minis and Pittendrigh 1968
清肠	蓖麻蚕	Mizoguchi and Ishizaki 1982
成虫羽化	柞蚕	Truman and Riddiford 1970
	家蚕	Truman 1972
	西南玉米螟	Takeda 1983
	帝王斑蝶	Froy et al. 2003
	惜古比天蚕	Truman and Riddiford 1970
	秋幕蛾	Morris and Takeda 1994
		繁殖
信息素的产生与发散	黄地老虎	Rosén 2002
信息素的接收与吸引行为	黄地老虎	Rosén, Han, and Löfstedt 2003
	斜纹夜蛾	Merlin et al. 2007; Silvegren, Löfstedt, and Rosén 2005
交配行为	斜纹夜蛾	Silvegren, Löfstedt, and Rosén 2005
精子滞留	斜纹夜蛾	Bebas, Cymborowski, and Giebultowicz 2002; Bebas et al. 2002
精子释放	苹果蠹蛾	Giebultowicz and Brooks 1998
	舞毒娥	Wergin et al. 1997
	斜纹夜蛾	Syrova, Sauman, and Giebultowicz 2003
产卵	黄地老虎	Byers 1987

续表

昼夜节律	物种	参考文献
	苹果蠹蛾	Riedl and Loher 1980
	玉米螟	Skopik and Takeda 1980
	棉红铃虫	Minis 1965
		运动
飞行	柞蚕	Truman 1974
	秋幕蛾	
	蓖麻蚕	
		迁徙
时间补偿的以太阳为指南针的定向	帝王斑蝶	Froy et al. 2003; Mouritsen and Frost 2002; Perez, Taylor, and Jander 1997
	黄沫粉蝶	Oliveira, Srygley, and Dudley 1998
	大黄蝶	

在中国柞蚕(*A. pernyi*)中进行的移植实验已经说明控制卵孵育行为的生物钟存在于脑中,一种被命名为hatchin的激素因子也许介导了这种昼夜节律的调控(Sauman 和 Reppert 1998)。对在实验上易于操控的昆虫而言,阐明具有这样一种计时活性扩散性物质的功能,有助于在其他动物中鉴定得到类似的物质。

与卵孵化不同,鳞翅目昆虫幼虫期和化蛹时的蜕皮不是直接由生物钟控制的,而是与明显的内分泌事件(激素释放)的昼夜节律相联系的,这在蓖麻蚕(Fujishita 和 Ishizaki 1981,1982)和家蚕(Sakurai 1983)上都有例证。尽管如此,已经证实蓖麻蚕幼虫消化道的退化由生物钟所介导,这个过程发生在五龄幼虫进食的末期以达到消解消化道为化蛹做准备(Mizoguchi 和 Ishizaki 1982)。

正如以前提到的,成虫的羽化是鳞翅目生命过程中又一个重要事件,其发生时间也由生物钟所控制,在这个过程中完全成熟的成虫从蛹壳中脱出。事实上,在许多昆虫物种中,成虫羽化的群体节律已经被广泛用作考察昼夜节律功能的一个可靠标志。在鳞翅目中,包括中国柞蚕(*A. pernyi*),家蚕(*B. mori*),天蚕蛾,惜古比天蚕(*Hyalophora cecropia*),秋幕蛾(*Hyphantria cunea*),西南玉米螟(*D. grandiosella*)以及帝王斑蝶(*D. plexippus*)(Truman 1972; Truman 和 Riddiford 1970; Takeda 1983; Morris 和 Takeda 1994; Froy et al. 2003)的成虫在一天中的羽化的时间是具有物种特异性的。例如,家蚕、帝王斑蝶和惜古比天蚕在光周期的早期出现;中国柞蚕(*A. pernyi*)在一天中的晚些时候出现;西南玉米螟(*D. grandiosella*)和秋幕蛾(*Hyphantria cunea*)在黄昏的时候出现。对于所有这些物种,羽化节律都是以接近24h的周期长度来运行的。

2.2 繁殖行为和生理学上的节律

在蛾子中繁殖行为已经被广泛地研究,这种研究展现了信息素介导的一成不变的行为在一天中发生的时间具有物种特异性。雌蛾在一天中特定的时间产生并放出性激素,这与雄性对性激素敏感性的节律是同步的,这种现象通常发生在夜里(Cardé 和 Minks

1997)。这些行为的昼夜节律本质已经在黄地老虎(*Agrotis segetum*)(Rosén 2002;Rosén, Han 和 Löfstedt 2003)和棉树叶虫中的斜纹夜蛾(Spodoptera littoralis)(Silveggren,Löfstedt 和 Rosén 2005)中进行了研究。其中雌性产生和放出性激素和雄性接收性激素和被吸引的行为都是由生物钟控制的,从而导致在交配行为上协调的节律(近一步的细节见下文)。

蛾子繁殖过程中较不明显的昼夜节律包括睾丸中精子停留的生理节律、精子酸化、以及精子从睾丸中释放的过程。这些结果来自于对苹果蠹蛾(*Cydia pomonella*),舞毒蛾(*Lymantria dispar*)和斜纹夜蛾(*S. littoralis*)的研究(Wergin et al. 1997; Giebultowicz 和 Brooks 1998; Bebas,Cymborowski 和 Giebultowicz 2002; Bebas et al. 2002; Syrova,Sauman 和 Giebultowicz 2003)。不间断的光处理所引起的节律崩溃使得雄性不育充分例证了蛾子精子释放的节律性是非常重要的(Riemann 和 Ruud 1974; Giebultowicz, Ridway 和 Imberski 1990)。而且在黄地老虎(*A. segetum*),苹果蠹蛾(*Cydia pomonella*),以及欧洲玉米蛀虫中的玉米螟(*Ostrinia nubilalis*)和棉红铃虫(*P. gossypiella*)中发现了产卵的节律(Minis 1965; Riedl 和 Loher 1980; Skopik 和 Takeda 1980; Byers 1987)。

2.3 运动和迁徙中的节律

鳞翅目中成虫运动最普通的模型是飞行。在一些蛾子中,单个动物的飞行能力也是处于昼夜节律控制之下的,这些物种包括中国柞蚕(*A. pernyi*),惜古比天蚕(*H. cecropia*)和蓖麻蚕(*Samia cynthia ricini*),它们在夜晚的早期有个简短的爆发式飞行,而在后半夜具有更强的飞行能力(Truman 1974)。因为不间断的光照或黑暗对运动行为有很强的掩蔽性影响,这增加了实施这种实验的难度,仅能对一小部分物种的飞行行为的节律进行考察。尽管如此,许多其他的鳞翅目昆虫的节律性飞行有可能也是处于昼夜节律控制之下。

如果能够监测单个蛾子或蝴蝶的运动节律,那将是追踪生物钟作用活性的理想方法。利用运动测定法将减小单个节律间去同步化的问题,这个问题在监测群体节律的时候是固有的,例如在卵孵化和成虫羽化过程中,目前来说这是监控鳞翅目昆虫中节律作用唯一可信赖的方法。

一个吸引人的例子是生物钟参与了一个极其重要的行为,这个行为是帝王斑蝶利用时间补偿的以太阳为指南针定向为其在秋天的迁徙中正确导航(Perez,Taylor 和 Jander 1997; Mouritsen 和 Frost 2002; Froy et al. 2003)。生物钟提供的这种时间补偿性使得蝴蝶能够根据天光参数(由太阳指南针感觉)随着太阳在白天的天空中移动不停地校正飞行方向以维持一个固定的向南的飞行(详见下文)。新热带粉蝶黄沫粉蝶(*Aphrissa statira*)和大黄蝶(*Phoebis argante*)也是利用时间补偿的以太阳为指南针的方式为其在迁徙过程中定向的(Oliveira,Srygley 和 Dudley 1998)。

3. 鳞翅目昆虫生物钟的分子机制

在果蝇和哺乳动物中,细胞内的生物钟机制涉及转录水平上的反馈回路,用于驱动生物钟关键的组成成分在 mRNA 和蛋白质水平持续的节律(Reppert 和 Weaver 2002; Stanewsky 2003)。转录水平上的负反馈回路对于生物钟的功能是非常重要的,并且在果蝇中转录因子 clock(CLK)和 cycle(CYC)参与了这个过程,这两个转录因子促进 *period* (*per*)和 *timeless* (*tim*)基因的表达。所产生的 PER 和 TIM 蛋白质形成异源二聚体,转位到细胞核内,在细胞核内 PER 抑制 CLK:CYC 介导的其本身的转录。TIM 负责调控 PER 蛋白质的稳定性及核运输,而且 TIM 蛋白是产生生物钟光响应所必需的。果蝇的隐花色素(CRY)是一个参与这种光响应的蓝光受体(Emery et al. 1998,2000; Stanewsky et al. 1998)。与果蝇不同,其他非果蝇属昆虫物种在生物钟机制方面的研究较少。尽管在许多鳞翅目昆虫中生物钟基因和蛋白质已经被用作生物钟标记,但是关于生物钟机制的精细的研究一直被局限在中国柞蚕(*A. pernyi*)和帝王斑蝶(*D. plexippus*)这两个物种中。

3.1 柞蚕生物钟的分子机制

在鳞翅目柞蚕中克隆了果蝇外的第一个 *per* 基因同系物(Reppert et al. 1994)。这一研究促进了对鳞翅目生物钟机制的探索。利用过表达的生物钟蛋白(CLK,CYC,PER 和 TIM)和相关的 *per* 基因启动子元件(Chang et al. 2003),在果蝇 Schneider 2 (S2) 细胞中利用来自于柞蚕的组成成分在体外构建了生物钟反馈回路。利用荧光素报告基因测定法,CLK:CYC 异源二聚体通过 *per* 启动子元件中的一个 E-box 增强子元件激活转录。PER 蛋白通过抑制 CLK:CYC 介导的转录来实现对自己转录的抑制,而 TIM 蛋白适当地增加了 PER 的抑制活性。至此,利用柞蚕生物钟组成成分在体外构建的转录反馈回路表现出果蝇生物钟的主要特征。这暗示着最初在果蝇中报道的生物钟机制很可能在整个昆虫界是广泛存在的。尽管如此,近来在帝王斑蝶生物钟分子机制上的研究却对这个假说提出了挑战。

3.2 帝王斑蝶以隐花色素为中心的生物钟:鳞翅目昆虫生物钟的原型

帝王斑蝶(*D. plexippus*)的分子生物钟机制研究(图 8.1)为理解帝王斑蝶壮观的迁徙发挥了重要作用。类似于果蝇 CRY,被命名为 CRY1 的蛋白主要发挥昼夜节律光感受器的功能 (Zhu et al. 2005,2008; Song et al. 2007),有趣的是,除了 CRY1 之外,帝王斑蝶也表达编码类似于脊椎动物中被定义为 CRY2 的光不敏感蛋白的第二个 *cry* 基因 (Zhu et al. 2005,2008)。CRY2 蛋白在帝王斑蝶的核心生物钟反馈回路中主要起着转录抑制子的作用(图 8.1)。帝王斑蝶的 CRY2 除了在细胞培养中起 CLK:CYC 所介导的强转录抑制子作用,它还在合适的时间通过转运进入帝王斑蝶脑中假定的生物钟细胞的细胞核上抑制基因转录 (Zhu et al. 2008)。帝王斑蝶的 PER 蛋白不抑制 CLK:CYC 介导的转录,但是它能增强 CRY2 的稳定性,协助转运 CRY2 到细胞核中(Zhu et al. 2008;图 8.1)。

在蝴蝶中发现两个功能不同的 *cry* 基因,利用数据库搜寻,在其他鳞翅目昆虫以及迄今未被研究的其他所有非果蝇昆虫中鉴定到了 *cry2* 基因的存在(Zhu et al. 2005; Rubin et al. 2006; Yuan et al. 2007),其中包括中国柞蚕、家蚕和斜纹夜蛾 (Zhu et al. 2005; Merlin et al. 2007; Yuan et al. 2007)。所有的昆虫 CRY2 蛋白(包括蜜蜂和甲虫中的 CRY2 蛋白,见下文)在细胞培养中是 CLK:CYC 所介导的转录的强烈抑制子(Yuan et al. 2007)。

昆虫中 CRY2 的发现为这群节肢动物生物钟的进化提供了新的见解。果蝇只表达

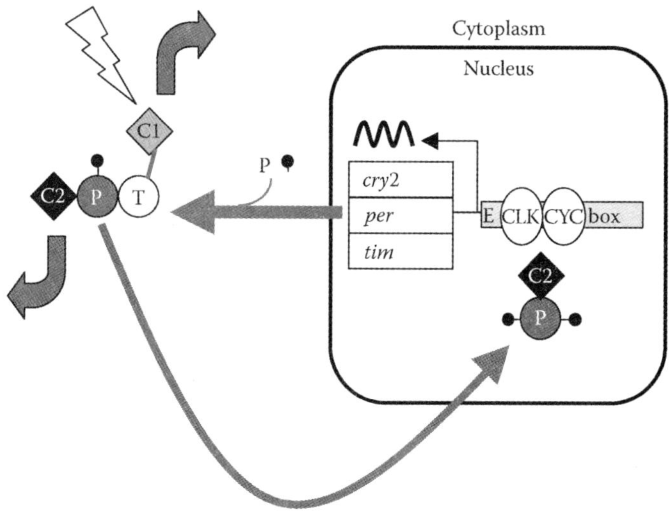

图 8.1 帝王蝴蝶的分子生物钟机制。生物钟机制的主要部分是一个自我调控的转录反馈环，CLK 和 CYC 的异源二聚体通过 E-box 增强子元件启动 *per*，*tim* 和 *cry2* 的转录；CACGTG E-box 元件在蝴蝶 *per*，*tim* 和 *cry2* 基因 1.5kb 中的 5′侧翼区域被发现。TIM (T)，PER (P)和 CRY2 (C2 在细胞质中形成复合物并且 CRY2 穿梭进入细胞核抑制 CLK：CYC 介导的转录。PER 逐步被磷酸化并且可能有助于 CRY2 易位进入细胞核。CRY1 (C1)是一个昼夜节律的光受体，在曝光(闪电符号标注)的情况下引起 TIM 的降解，使得光能够影响中心生物钟机制。深灰色箭头代表 CRY1 和 CRY2 的功能可能将生物钟与包括太阳指南针在内的各种输出联系起来。

CRY1 蛋白，而其他昆虫(蚊子、蝴蝶和蛾子)表达 CYR1 和 CRY2 蛋白。令人吃惊的是，蜜蜂(*Apis mellifera*)和甲虫赤拟谷盗只表达 CRY2 蛋白(Zhu et al. 2005; Rubin et al. 2006)。这暗示着核心的昼夜节律起搏点已经在昆虫谱系中发生进化而产生三类昆虫生物钟(Yuan et al. 2007)：CRY1 和 CRY2 存在于古老的生物钟(在鳞翅目和蚊子中比较明显)里并且两个蛋白在这个生物钟中有着不同的功能；衍生的生物钟(果蝇类型)中，CRY2 已经遗失并且只有 CRY1 起作用；还有一种衍生生物钟(发现于甲虫和蜜蜂中)，CRY1 已经遗失并且只有 CRY2 存在起作用(图 8.2)。有意思的是，通过 CRY2 降解或对其转录活性抑制的阻遏的方法，发现甲虫和蜜蜂中的 CRY2 蛋白在细胞培养中对光不敏感，两者又缺乏 CRY1 蛋白，暗示对它们的生物钟来说，这些物种具有新型光输入途径(Yuan et al. 2007)。在更广泛的背景下，帝王斑蝶的古老生物钟也许是新型生物钟机制的范例，这种新型生物钟由表达 *cry*1 和 *cry*2 基因的非果蝇昆虫所共享。

4. 中枢神经系统的细胞生物钟

在理解生物钟是怎样控制行为输出之前，不仅需要弄清楚生物钟分子机制本身，而且也需要搞清楚细胞生物钟存在于脑的什么部位。在鳞翅目昆虫中，中国柞蚕、帝王斑蝶在细胞生物钟的定位方面被研究得非常广泛，家蚕和烟草天蛾在这方面的研究相对少些。

在中国柞蚕中，*per* 基因的 RNA 和蛋白质是共定位的并且它们在各个脑半球的背

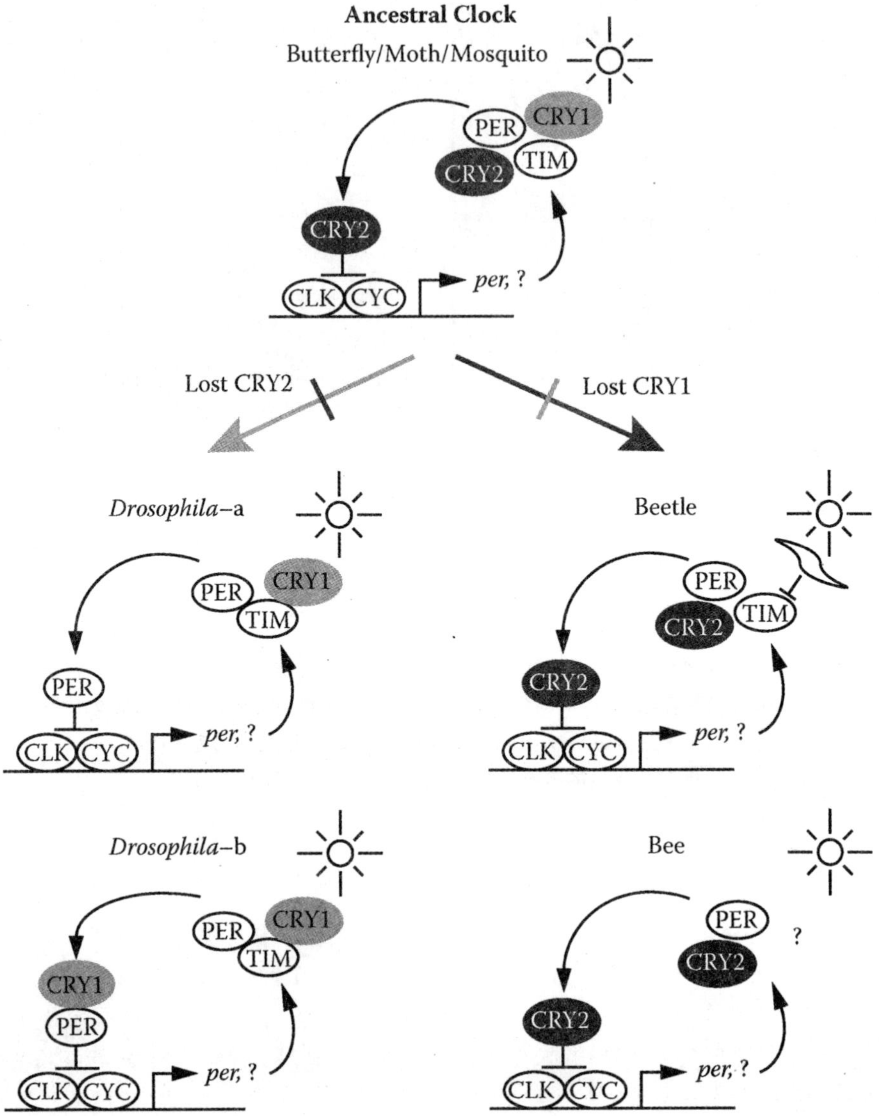

图 8.2 昆虫生物钟模型。昆虫中存在着两种不同功能的 CRYs，从而产生了三种主要的生物钟模型：古老的生物钟(存在于蚊子、帝王蝴蝶和蚕蛾中)，CRY1 和 CRY2 都存在于这种生物钟中，并且各自起着不同的作用；衍生的生物钟(存在于果蝇中)，在这种生物钟中，CRY2 已经遗失，CRY1 在中心脑的生物钟里作为昼夜节律的光受体(panel a; Emery et al. 1998)，而它在外周生物钟里既作为光受体又是中心生物钟的组成部分(panel b; Ivanchenko, Stanewsky 和 Giebultowicz 2001; Krishnan et al. 2001; Levine et al. 2002)；在另一种衍生生物钟里，CRY1 已经遗失，而 CRY2 在生物钟起作用，甲虫和蜜蜂具有这种生物钟。甲虫的 CRY2 在生物钟里作为转录抑制子并且 TIM 的降解可能介导了光的输入。蜜蜂缺乏 TIM (Rubin et al. 2006)，CRY2 也是作为转录抑制子，需要提出新的光输入途径来推动生物钟的发展。修改自 Yuan 等 (2007)。

外侧区域上的两对细胞里表达；这些细胞也表达 TIM 蛋白(Sauman 和 Reppert 1996)。根据早期的移植和切除实验研究的结果(Truman 1972,1974)，在处于正确位置的外侧细胞里安置了生物钟。对这些细胞中的PER和TIM进行染色，得到比较费解的结果，发现经过一天的时间，它们仍在细胞质中，没有证据表明转运到了细胞核（见下文）。最初在外侧细胞中鉴定了 *per* 基因的反义转录物，随后发现它是来自雌特异W染色体的一个基因座（Gotter，Levine 和 Reppert 1999）。因为雄蚕蛾（ZZ）也能正常表现分子和行为的节律性，中国柞蚕中的 *per* 基因反义转录物对于生物钟的功能不是必需的（Gotter，Levine 和 Reppert 1999）。

采用了一种策略对帝王斑蝶进行研究，该策略依赖于主要生物钟蛋白的共表达，同时这些蛋白能被帝王斑蝶特异的抗体（PER，TIM，CRY1 和 CRY2）所追踪。外侧细胞中（脑半球各两个）的4个细胞被鉴定为可能的生物钟位点（图 8.3A；Sauman et al. 2005；Zhu et al. 2008）。TIM 和 CRY2 蛋白在每个脑半球上的四个外侧细胞上表达，并且它们中的两个共表达 PER 和 CRY1 蛋白；在每个外侧上的这两个细胞包含了核心生物钟机制中必需的蛋白（图 8.3A）。

对柞蚕和帝王斑蝶脑定位的研究表明鳞翅目昆虫脑外侧上的生物钟细胞的定位是保守的。而且，几种生物钟蛋白包括 PER，类果蝇的 CRY 和 CYC 蛋白在家蚕每个外侧区域上的4个 $1a_1$ 类型的大型神经分泌细胞中表达（Závodská，Sauman 和 Sehnal 2003；Sehadova et al. 2004）。另外，烟草天蛾外侧上的 PER 阳性细胞也已经被定义为 $1a_1$ 类型细胞，这些 PER 阳性细胞与蚕蛾和蝴蝶外侧上的 PER 阳性细胞是同源的（Wise et al. 2002）。

有意思的是，帝王斑蝶 CRY2 蛋白是迄今为止所有研究的鳞翅目昆虫脑中唯一的生物钟蛋白，它节律性地在外侧细胞核中积累，在适当时间对转录进行抑制，从而终结体内的转录反馈回路（图 8.3B；Zhu et al. 2008）。CRY2 出现在体内细胞核中时，正是最大程度抑制 *per* 基因 RNA 波动的时间，强化了其在转录抑制中所起作用的关联性（图 8.3C）。

PER，TIM，CRY1 和 CRY2 蛋白也在帝王斑蝶脑间部巨大神经分泌细胞里共表达，但是 PER 水平上的昼夜节律控制在那儿不是很明显（Sauman et al. 2005；Zhu et al. 2008）。这些脑间部细胞的功能是不清楚的，但是它们可能是引起迁徙行为或生理过程中昼夜节律网络通路的一部分。

除了有生物钟蛋白的细胞被定位在脑中以外，已经在柞蚕中发现眼组织中的感光细胞拥有生物钟，在这些感光细胞中不仅 *per* 基因的 mRNA 和蛋白质在波动变化，而且 PER 蛋白也进出细胞核（Sauman 和 Reppert 1996）。在柞蚕的感光细胞中，*per* 基因 mRNA 的表达峰值出现在早夜，几小时后 PER 蛋白如同果蝇的同系物一样，转运到核里（Siwicki et al. 1988；Zerr et al. 1990）。这也许能够解释在 S2 细胞上观察到柞蚕 PER 蛋白对由 CLK：CYC 介导的转录活性的抑制（Chang et al. 2003；见上）。与此相反的是，在所调查的帝王斑蝶眼组织中没有生物钟蛋白的表达（Sauman et al. 2005；Zhu et al. 2008）。

考虑到柞蚕和帝王斑蝶中生物钟分子机制的相似性，以及柞蚕、帝王斑蝶和家蚕生物钟在细胞定位上的相似性，提出帝王斑蝶生物钟构成了鳞翅目昆虫生物钟范例的这种假设是合理的。但是 CRY2 蛋白的脑定位以及在外侧细胞的穿核假说需对如同柞蚕和家蚕这样的其他鳞翅目昆虫进行研究来支持。

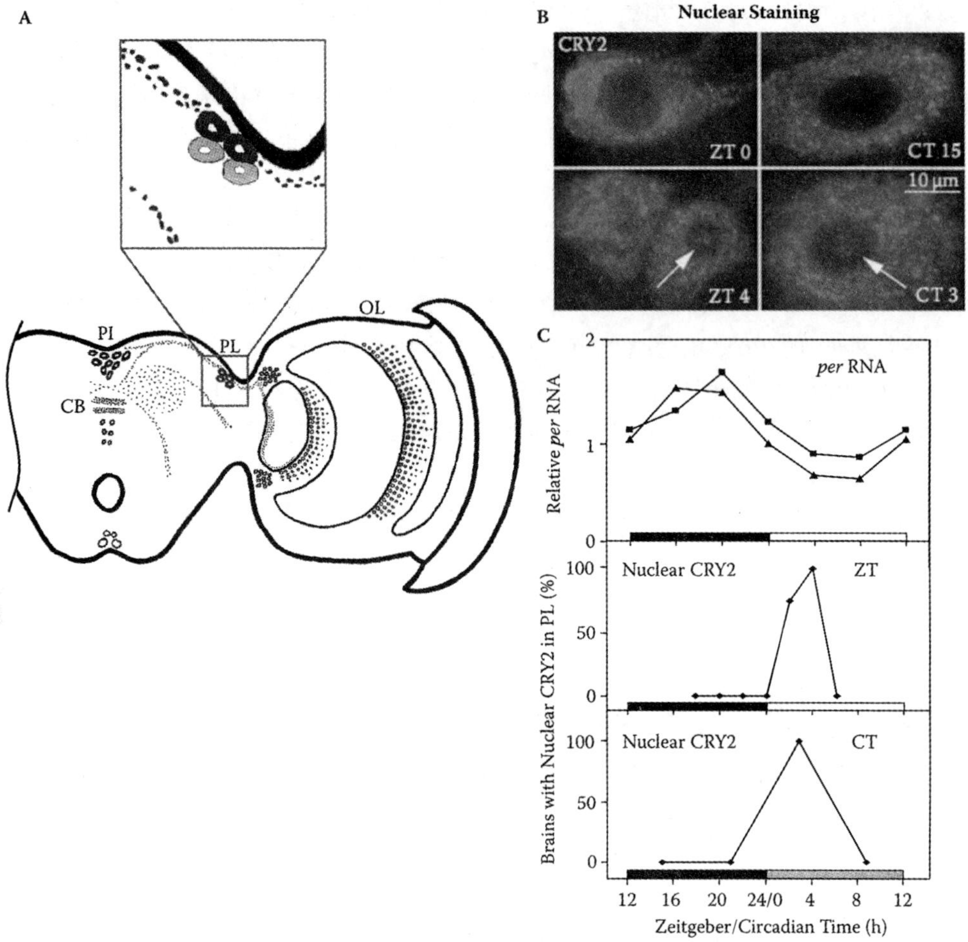

图 8.3　帝王蝴蝶脑中生物钟细胞的细胞定位(彩色版本图片见附录图 8.3)。(A) 通过帝王蝴蝶特异的抗体,前脑切片的局部示意图展示了 CRY2-阳性的细胞和神经投影的图像。(OL)光叶;(PL)外侧部;(PI)脑间部;(CB)中心体。上面放大的 PL 区域显示了对 4 个生物钟蛋白显阳性的细胞;两个红色细胞共表达 PER,TIM,CRY1 和 CRY2,而两个粉红色的细胞共表达 TIM 和 CRY2。修改自 Zhu 等(2008)。(B)PL 细胞中对 CRY2 的细胞核染色。左上,同步时间[ZT]0;左下,同步时间 4;右上,昼夜节律时间[CT]15;右下,昼夜节律时间 3。在 ZT0 或 CT15 时细胞核中的 CRY2 染色没有被发现,但是在 ZT4 和 CT3 时发现在 PL 中的 CRY2 被染色(箭头所指)。来自 Zhu 等 (2008)。(C)脑中 *per*RNA 的表达水平与 PL 中 CRY2 细胞核染色的时间表达模式的比较。在一天中以 4h 为间隔,记录了两组缺乏光受体的脑中 *per* RNA 的表达水平(上部)。在 ZT7(中部)和 CT(下部)时对细胞核中的 CRY2 进行了半定量的免疫染色,图以被检测的脑的百分数绘制(每个时间点使用了 4～5 个脑)。来自 Zhu 等 (2008)。

5. 外周组织的细胞生物钟

尽管在昆虫脑中的生物钟控制许多种行为节律的这个观点是明确的,但是外周组织中的生物钟能够以组织依赖性的方式调控许多生理过程,认识到这一点是非常重要的(有关综述见 Giebultowicz 1999,2000,2001),这个观念最先来自于对果蝇的研究。利用

受 *per* 基因驱使的荧光素酶活性，Plautz 等人（1997）巧妙地证明了 *per* 基因在果蝇体内的触角、喙、翅膀和腿这样的外周器官中表达。当在体外培养的时候，这些器官内的生物荧光能节律性地维持几天，这揭示了外周生物钟自主性的昼夜节律。

通过鉴别器官和细胞内生物钟基因和蛋白的节律性表达以及外围器官中生理输出的节律已经证明了鳞翅目昆虫中外周生物钟的存在。然而，它们的自主程度范围从完全依赖脑生物钟跨度到完全的组织自主。在柞蚕胚胎和一龄幼虫的中肠上皮细胞观察到 *per* 基因的 mRNA 和蛋白质的表达以及 PER 的细胞核转位是节律性的（Sauman et al. 1996；Sauman 和 Reppert 1998）。尽管如此，头部结扎会阻断 PER 蛋白进入中肠的上皮细胞核，这证明了中肠里的 PER 波动是依赖于脑中的某种物质。与此相反，雄蛾的生殖行为受到自主的外周生物钟所调控，控制精子从睾丸里释放进入输精管的上部（UVD）（Wergin et al. 1997；Giebultowicz，Bell 和 Imberski 1988；Giebultowicz et al. 1989；Giebultowicz 和 Brooks 1998；Syrova，Sauman 和 Giebultowicz 2003）并且控制精子在 UVD 中的滞留（Bebas，Cymborowski 和 Giebultowicz 2002；Bebas et al. 2002）。最后，近来有许多发表的文章证明了信息素接收中的生理节律性以及外周触角生物钟的存在，这种生物钟可能负责这些器官中的节律（Flecke et al. 2006；Iwai et al. 2006；Merlin et al. 2006，2007；Schuckel，Siwicki 和 Stengl 2007）。

6. 时间就是一切：节律调控的生理学意义

为了在生态环境下阐明生物钟对蛾子和蝴蝶行为的影响，我们进一步详细地考虑了两个系统：蛾子中的信息素通讯系统和在帝王斑蝶迁徙过程中的以太阳为指南针定向的时间补偿系统。

6.1 交配行为：蛾子信息素通讯系统的节律控制

在大多数昆虫中，特别是像蛾子这种夜出物种，它们的交配行为是由性激素所介导的（鳞翅目昆虫的性交流将在第十章讨论）。这种行为发生在一天的特定时间，依赖于性激素交流的时间。雌性求偶，产生和释放信息素吸引同种雄性以及雄性被性激素所吸引的这些行为在一天中同一时间出现，而这种时间是物种特异的（Raina 和 Menn 1987）。在自然界中，这种时间的物种特异性与使用相似性激素的共栖物种的生殖隔离是极其相关的。

夜出的鳞翅目昆虫的交配行为的时间控制已有详细的记录；交配发生在夜里特定的几个小时内。雌性信息素的产生和放出以及雄性对信息素应激性的每日节律已经在许多物种中有所研究，但生物钟对其调节方面的研究相对较少。尽管如此，最近对黄地老虎和斜纹夜蛾的研究结果（Rosén 2002；Rosén，Han 和 Löfstedt 2003；Silvegren，Löfstedt 和 Rosén 2005；Merlin et al. 2007）让我们明白了蛾子交配行为的节律控制是怎样发生的。黄地老虎中雌性信息素的产生和雄性对信息素的行为反应都是内源性调控的，并且两者都是在夜里的同一时间段内达到峰值（Rosén 2002；Rosén，Han 和 Löfstedt 2003）。两性间的这种同步对成功交配的影响已经在斜纹夜蛾中得到证明（Silvegren，Löfstedt 和 Rosén 2005）。

雄性通过触角上的特化受体探知性激素，触角是雄性对信息素做出行为反应所必需的外周接收器；除去触角会导致吸引力的丧失(Rosén，Han 和 Löfstedt 2003)。黄地老虎在触角对信息素的生理反应过程中没有发现昼夜节律(Rosén，Han 和 Löfstedt 2003)，与此相对的是，斜纹夜蛾在触角上表现出应激性节律，这与已有的烟草天蛾触角上对信息素敏感的感觉器拥有灵敏的日常变化的证据相一致(Flecke et al. 2006)。而且，对包括斜纹夜蛾在内的蛾子触角上生物钟基因和蛋白时空表达模式的研究揭示了生物钟是存在于嗅觉器官内对嗅觉敏感的感受器基底上的(Merlin et al. 2006，2007；Schuckel，Siwicki 和 Stengl 2007)。

信息素接收的昼夜节律是否由触角中的生物钟控制，是否直接参与了雄性对信息素节律性行为的应激性？根据果蝇提供的数据，表明了触角生物钟对嗅觉节律性是必要和充分的(Tanoue et al. 2004)，这种假设具有现实意义，然而只有对离体嗅觉器官的研究才会建立蛾子触角产生的生理节律和这些生物钟的联系。不过更重要的挑战是在体内利用遗传学手段从根本上破坏蛾子触角生物钟，这将确定它们是否确实对信息素吸引的节律性行为负责。

6.2 导航能力：在帝王斑蝶迁徙过程中时间补偿的太阳指南针的定向

美国东北的帝王斑蝶每年秋天要完成一次非凡的迁徙，到墨西哥中部越冬(Urquhart 1987；Brower 1995)；它们基于生物学的导航方式逐渐被人们所知。

生物钟正是通过提供时间补偿的以太阳为指南针的定向在迁徙期间对导航起极其重要的作用(Perez，Taylor 和 Jander 1997；Mouritsen 和 Frost 2002；Froy et al. 2003)。生物钟使得帝王斑蝶能够补偿太阳在白天横跨天空的移动，以至于蝴蝶能不断朝南/西南调整它们的飞行方向 (图 8.4A，upper panel)。利用在持久飞行中蝴蝶上的飞行模拟器来研究蝴蝶的飞行轨道，这种飞行模拟器提供了有说服力的证据表明帝王蝴蝶利用时间补偿的以太阳为指南针 (Mouritsen 和 Frost 2002)。时钟位移实验已经表明了生物钟在调控飞行方向中时间补偿性组成元件的重要性，这个实验中，每天的光暗周期要么被提前，要么被推后，这在飞行方向上引起了可预测的改变(Perez，Taylor 和 Jander 1997；Mouritsen 和 Frost 2002；Froy et al. 2003)。而且，当在实验上用连续的光照去破坏生物钟，帝王斑蝶失去了根据太阳位置调整它们飞行方向的能力，取而代之的是直接向太阳飞去(图 8.4A，lower panel；Froy et al. 2003)。

生物钟怎样与太阳指南针沟通以及方向信息怎样与动力系统整合，弄清楚这两个问题的答案将为帝王斑蝶导航能力提供了全面的认识。连接生物钟和太阳指南针的神经途径信息正逐渐浮现。的确能被 CRY1-染色的神经途径把外侧细胞中的生物钟和通过复眼背侧边缘区域的感光细胞进入脑的偏振光输入连接起来，这对于太阳指南针导航是非常重要的(Sauman et al. 2005；Reppert，Zhu 和 White 2004)。

太阳指南针很可能位于中心复合物上以整合天光信息(包括偏振光；Vitzthum，Muller 和 Homberg 2002；Heinze 和 Homberg 2007)，脑的中线结构是由中枢体构成的，中枢体被划分成上部和下部，以及原脑桥。有趣的是，在帝王斑蝶中，在中枢体中发现了一个密集树枝状的 CRY2 染色区域，也是处于昼夜节律控制之下，在半夜的时候染色很深，而在一天的中间时间染色很少或没有染色(Zhu et al. 2008)。重要的是，其他生物钟蛋白(PER,TIM和CRY1)不在这个位置表达。另外,在CRY2染色的神经途径中观察到的结果表明,在中心复合物中的CRY-2阳性树枝状分支很可能是起源于外侧细胞或间部细胞或者同时起源于这两类细胞(图8.3A)。CRY2要么标记了生物钟指南针神经传导途径,要么调控了中央复杂结构的有节律的神经活动。

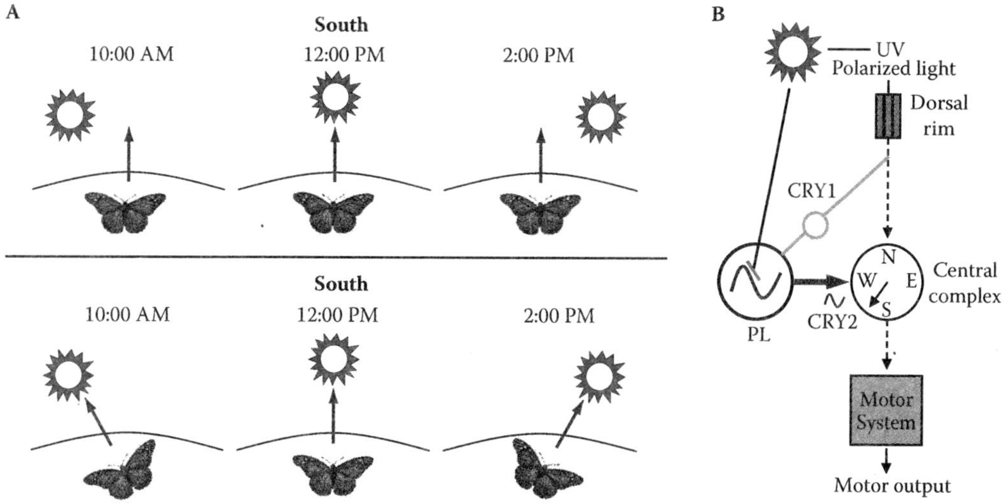

图8.4 时间补偿的以太阳为指南针的定向。(A)没有生物钟导致迷路。(较上面的图)在秋天的迁徙时,东北美帝王蝴蝶利用时间补偿性的以太阳为指南针向南飞。生物钟使得蝴蝶能够补偿太阳移动引起的误差。在白天的时候,即使太阳运行的轨迹与它们迁徙的方向发生了偏移,但是它们仍能够保持一个恒定的向南飞的方向。(较下面的图)帝王蝴蝶在没有功能性生物钟的情况下,一直向着太阳飞。损坏的生物钟会破坏向南飞的迁徙,并且蝴蝶不能够成功地到达它们的越冬栖息地。(B)帝王蝴蝶脑中可能的指南针途径模型。外侧部的生物钟是光通过CRY1启动的,CRY1把生物钟与眼的背侧边缘区域对紫外光和偏振光敏感的光受体上的神经轴突联系起来(Sauman et al. 2005)。阳性的CRY2纤维途径可能将生物钟与太阳指南针(在中心的复合物里)联系起来,这最终控制了运动方向。修改自Zhu等(2008)。

7. 结论

正如在这章中所强调的,鳞翅目昆虫已经为昼夜节律生物学作出了较大贡献并且扩充了我们对昆虫生物钟机制的了解。除了果蝇以外,最核心的发现是在所有迄今被研究的昆虫中鉴定了生物钟的重要组成成分CRY2。果蝇的生物钟是以PER为主要的转录抑制子,与此不同的是大多数其他昆虫的转录活性的抑制可能依赖于CRY2蛋白。而且,迄今为止CRY2是唯一在鳞翅目昆虫体内生物钟细胞核中积累的蛋白。尽管如此,需进一步测试CRY2是不是在体内产生适当行为的必需蛋白。利用家蚕能实现这个目标,因为家蚕是目前能获得的具有确定节律行为(卵孵育和成虫羽化的每日时间点)、在遗传上易入手的唯一鳞翅目昆虫(Sandrelli et al. 2007)。

如前所述CRY2除了在生物钟机制上的功能,它可能还直接参与了帝王斑蝶的时间补偿的以太阳为指南针的定向机制。利用基因沉默对特定脑区域(例如中心复杂结构)或生物钟细胞内 CRY2 表达的选择性操作也许有助于定义其在帝王斑蝶时间补偿-太阳指南针的定向中所起的作用。

开发包括帝王斑蝶在内的非模式鳞翅目物种中基因/遗传工具对于揭开行为的分子基础是至关重要的。家蚕的基因组已经被测序并且有部分注释(Mita et al. 2004; Xia et al. 2004)。在几个其他的鳞翅目昆虫中,基因组方法也在继续向前发展,在 ButterflyBase 中可得到越来越多的基因组信息,这个网站提供所有公开的鳞翅目昆虫基因组数据库(Papanicolaou et al. 2008)。同样的,帝王斑蝶脑 EST 数据现在可供使用(Zhu, Casselman 和 Reppert 2008)。在不久的将来可能获取任何鳞翅目昆虫的基因组数据,这将增加我们对生物钟影响这些轻盈昆虫昼夜节律的理解。

8. 致谢

感谢 Haisun Zhu 对图表的制作,Amy Casselman 和 Emmanuelle Jacquin-Joly 对文稿提出的意见,以及 Reppert 实验室的成员们对手稿提出的有用建议。

参考文献

[1]Bebas, P., B. Cymborowski, and J. M. Giebultowicz. 2002. Circadian rhythm of acidification in insect vas deferens regulated by rhythmic expression of vacuolar H(+)-ATPase. *J. Exp. Biol.* 205:37—44.

[2]Bebas, P., E. Maksimiuk, B. Gvakharia, B. Cymborowski, and J. M. Giebultowicz. 2002. Circadian rhythm of glycoprotein secretion in the vas deferens of the moth, *Spodoptera littoralis*. *BMC Physiol.* 2:15.

[3]Brower, L. P. 1995. Understanding and misunderstanding the migration of the monarch butterfly (Nymphalidae) in North America: 1857—1995. *J. Lep. Soc.* 49:304—85.

[4]Byers, J. A. 1987. Novel fraction collector for studying the oviposition rhythm in the turnip moth. *Chronobiol. Int.* 4 (2):189—94.

[5]Cardé, R. T., and A. K. Minks. 1997. *Insect pheromone research: New directions*. New York: Chapman & Hall.

[6]Chang, D. C., H. G. McWatters, J. A. Williams, A. L. Gotter, J. D. Levine, and S. M. Reppert. 2003. Constructing a feedback loop with circadian clock molecules from the silkmoth, *Antheraea pernyi*. *J. Biol. Chem.* 278:38149—58.

[7]Emery, P., W. V. So, M. Kaneko, J. C. Hall, and M. Rosbash. 1998. CRY, a *Drosophila* clock and light-regulated cryptochrome, is a major contributor to circadian rhythm resetting and photosensitivity. *Cell* 95:669—79.

[8]Emery, P., R. Stanewsky, C. Helfrich-Forster, M. Emery-Le, J. C. Hall, and M. Rosbash. 2000. *Drosophila* CRY is a deep brain circadian photoreceptor. *Neuron* 26:493—504.

[9]Flecke, C., J. Dolzer, S. Krannich, and M. Stengl. 2006. Perfusion with cGMP analogue adapts the action potential response of pheromone-sensitive sensilla trichoidea of the hawkmoth *Manduca sexta* in a day-time-dependent manner. *J. Exp. Biol.* 209:3898—912.

[10]Froy, O., A. L. Gotter, A. L. Casselman, and S. M. Reppert. 2003. Illuminating the circadian clock in monarch butterfly migration. *Science* 300:1303—05.

[11]Fujishita,M. ,and H. Ishizaki. 1981. Circadian clock and prothoracicotropic hormone secretion in relation to the larval-larval ecdysis rhythm of the saturniid,*Samia cynthia ricini*. *J. Insect Physiol*. 28:961—67.

[12]Fujishita,M. ,and H. Ishizaki. 1982. Temporal organization of endocrine events in relation to the circadian clock during larval-pupal development in *Samia cynthia ricini*. *J. Insect Physiol*. 28:77—84.

[13]Giebultowicz,J. M. 1999. Insect circadian clocks: Is it all in their heads? *J. Insect Physiol*. 45: 791—800.

[14] Giebultowicz, J. M. 2000. Molecular mechanism and cellular distribution of insect circadian clocks. *Annu. Rev. Entomol*. 45:769—93.

[15]Giebultowicz,J. M. 2001. Peripheral clocks and their role in circadian timing: Insights from insects. *Philos. Trans. R. Soc. Lond. B Biol. Sci*. 356:1791—99.

[16]Giebultowicz,J. M. ,R. A. Bell,and R. B. Imberski. 1988. Circadian rhythm of sperm movement in the male reproductive tract of the gypsy moth,*Lymantria dispar*. *J. Insect Physiol*. 34:527—32.

[17]Giebultowicz,J. M. ,and N. L. Brooks. 1998. The circadian rhythm of sperm release in the codling moth,*Cydia pomonella*. *Entomol. Exp. Appl*. 88:229—34.

[18]Giebultowicz,J. M. , R. L. Ridway, and R. B. Imberski. 1990. Physiological basis for sterilizing effects of constant light in *Lymantria dispar*. *Physiol. Entomol*. 15:149—56.

[19]Giebultowicz,J. M. ,J. G. Riemann,A. K. Raina,and R. L. Ridgway. 1989. Circadian system controlling release of sperm in the insect testes. *Science* 245:1098—1100.

[20]Gotter,A. L. ,J. D. Levine,and S. M. Reppert. 1999. Sex-linked period genes in the silkmoth,*Antheraea pernyi*:Implications for circadian clock regulation and the evolution of sex chromosomes. *Neuron* 24:953—65.

[21]Heinze,S. ,and U. Homberg. 2007. Maplike representation of celestial E-vector orientations in the brain of an insect. *Science* 315:995—97.

[22]Ivanchenko,M. ,R. Stanewsky, and J. M. Giebultowicz. 2001. Circadian photoreception in *Drosophila*:Functions of *cryptochrome* in peripheral and central clocks. *J. Biol. Rhythms* 16:205—15.

[23]Iwai,S. ,Y. Fukui,Y. Fujiwara,and M. Takeda. 2006. Structure and expressions of two circadian clock genes,*period* and *timeless* in the commercial silkmoth, *Bombyx mori*. *J. Insect Physiol*. 52:625—37.

[24]Krishnan,B. ,J. D. Levine,M. K. Lynch,et al. 2001. A new role for *cryptochrome* in a *Drosophila circadian oscillator*. *Nature* 411:313—17.

[25]Levine,J. D. ,P. Funes,H. B. Dowse,and J. C. Hall. 2002. Advanced analysis of a *cryptochrome* mutation's effects on the robustness and phase of molecular cycles in isolated peripheral tissues of *Drosophila*. *BMC Neurosci*. 3:5.

[26]Merlin,C. , M. C. Francois, I. Queguiner, M. Maibeche-Coisne, and E. Jacquin-Joly. 2006. Evidence for a putative antennal clock in *Mamestra brassicae*: Molecular cloning and characterization of two clock genes—*period* and *cryptochrome*—in antennae. *Insect Mol. Biol*. 15:137—45.

[27]Merlin,C. ,P. Lucas,D. Rochat,M. C. Francois,M. Maibeche-Coisne,and E. Jacquin-Joly. 2007. An antennal circadian clock and circadian rhythms in peripheral pheromone reception in the moth *Spodoptera littoralis*. *J. Biol. Rhythms* 22:502—14.

[28]Minis,D. H. 1965. Parallel pecularities in the entrainment of a circadian rhythm and photoperiodic induction in the pink bollworm (*Pectinophora gossypiella*). In *Circadian clocks*,ed. J. Aschoff,333—43. Amsterdam: North-Holland.

[29]Minis,D. H. ,and C. S. Pittendrigh. 1968. Circadian oscillation controlling hatching: Its ontogeny during embryogenesis of a moth. *Science* 159:534—36.

[30]Mita, K., M. Kasahara, S. Sasaki, et al. 2004. The genome sequence of silkworm, *Bombyx mori*. *DNA Res*. 11:27—35.

[31]Mizoguchi, A., and H. Ishizaki. 1982. Prothoracic glands of the saturniid moth *Samia cynthia ricini* possess a circadian clock controlling gut purge timing. *Proc. Natl. Acad. Sci. U. S. A.* 79:2726—30.

[32]Morris, M. C., and S. Takeda. 1994. The adult eclosion rhythm in *Hyphantria cunea* (Lepidoptera: Arctiidae): Endogenous and exogenous light effects. *Biol. Rhythm Res*. 25:464—76.

[33]Mouritsen, H., and B. J. Frost. 2002. Virtual migration in tethered flying monarch butterflies reveals their orientation mechanisms. *Proc. Natl. Acad. Sci. U. S. A.* 99:10162—66.

[34]Oliveira, E. G., R. B. Srygley, and R. Dudley. 1998. Do neotropical migrant butterflies navigate using a solar compass? *J. Exp. Biol.* 201:3317—31.

[35]Papanicolaou, A., S. Gebauer-Jung, M. L. Blaxter, W. O. McMillan, and C. D. Jiggins. 2008. Butterfly Base: A platform for lepidopteran genomics. *Nucleic Acids Res*. 36:D582—87.

[36]Perez, S. M., O. R. Taylor, and R. Jander. 1997. A sun compass in monarch butterflies. *Nature* 387:29.

[37]Plautz, J. D., M. Kaneko, J. C. Hall, and S. A. Kay. 1997. Independent photoreceptive circadian clocks throughout *Drosophila*. *Science* 278:1632—35.

[38]Raina, A. K., and J. J. Menn. 1987. Endocrine regulation of pheromone production in Lepidoptera. In *Pheromone biochemistry*, ed. G. D. Prestwich and G. J. Blomquist, 159 — 74. Orlando: Academic Press.

[39]Reppert, S. M., T. Tsai, A. L. Roca, and I. Sauman. 1994. Cloning of a structural and functional homolog of the circadian clock gene *period* from the giant silkmoth *Antheraea pernyi*. *Neuron* 13:1167—76.

[40]Reppert, S. M., and D. R. Weaver. 2002. Coordination of circadian timing in mammals. *Nature* 418:935—41.

[41]Reppert, S. M., H. Zhu, and R. H. White. 2004. Polarized light helps monarch butterflies navigate. *Curr. Biol.* 14:155—58.

[42]Riddiford, L. M., and L. K. Johnson. 1971. Synchronization of hatching of *Antheraea pernyi* eggs. In *Proc. XIIIth Int. Congr. Entomol. Moscow* 1:431—32.

[43]Riedl, H., and W. Loher. 1980. Circadian control of oviposition in the codling moth, *Laspeyresia pomonella*, Lepidoptera: Olethreutidae. *Entomol. Exp. Appl.* 27:38—49.

[44]Riemann, J. G., and R. L. Ruud. 1974. Mediterranean flour moth: Effects of continuous light on the reproductive capacity. *Ann. Entomol. Soc. Am.* 67:857—60.

[45]Rosato, E., E. Tauber, and C. P. Kyriacou. 2006. Molecular genetics of the fruit-fly circadian clock. *Eur. J. Hum. Genet.* 14:729—38.

[46]Rosén, W. 2002. Endogenous control of circadian rhythms of pheromone production in the turnip moth, *Agrotis segetum*. *Arch. Insect Biochem. Physiol.* 50:21—30.

[47]Rosén, W. Q., G. B. Han, and C. Löfstedt. 2003. The circadian rhythm of the sex-pheromone-mediated behavioral response in the turnip moth, *Agrotis segetum*, is not controlled at the peripheral level. *J. Biol. Rhythms* 18:402—08.

[48]Rubin, E. B., Y. Shemesh, M. Cohen, S. Elgavish, H. M. Robertson, and G. Bloch. 2006. Molecular and phylogenetic analyses reveal mammalian-like clockwork in the honey bee (*Apis mellifera*) and shed new light on the molecular evolution of the circadian clock. *Genome Res*. 16:1352—65.

[49]Sakurai, S. 1983. Temporal organization of endocrine events underlying larval-larval ecdysis in the silkmoth *Bombyx mori*. *J. Insect Physiol.* 29:919—32.

[50]Sandrelli,F. ,E. Tauber,M. Pegoraro,et al. 2007. A molecular basis for natural selection at the timeless locus in *Drosophila melanogaster*. *Science* 316:1898—900.

[51]Sauman,I. ,A. D. Briscoe,H. Zhu,et al. 2005. Connecting the navigational clock to sun compass input in monarch butterfly brain. *Neuron*46:457—67.

[52]Sauman,I. ,and S. M. Reppert. 1996. Circadian clock neurons in the silkmoth *Antheraea pernyi*: Novel mechanisms of Period protein regulation. *Neuron* 17:889—900.

[53]Sauman,I. ,and S. M. Reppert. 1998. Brain control of embryonic circadian rhythms in the silkmoth *Antheraea pernyi*. *Neuron* 20:741—48.

[54]Sauman,I. ,T. Tsai,A. L. Roca,and S. M. Reppert. 1996. Period protein is necessary for circadian control of egg hatching behavior in the silkmoth *Antheraea pernyi*. *Neuron* 17:901—09.

[55]Saunders,D. S. 2002. *Insect clocks*. 3rd ed. Amsterdam and Boston: Elsevier.

[56]Schuckel,J. ,K. K. Siwicki,and M. Stengl. 2007. Putative circadian pacemaker cells in the antenna of the hawkmoth *Manduca sexta*. *Cell Tissue Res*. 330:271—78.

[57]Sehadova,H. ,E. P. Markova,F. Sehnal,and M. Takeda. 2004. Distribution of circadian clock-related proteins in the cephalic nervous system of the silkworm, *Bombyx mori*. *J. Biol. Rhythms* 19:466—82.

[58]Shimizu,I. ,and K. Matsui. 1983. Photoreceptions in the eclosion of silkworm, *Bombyx mori*. *Photochem. Photobiol.* 37:409—13.

[59]Silvegren,G. ,C. Löfstedt,and W. Q. Rosén. 2005. Circadian mating activity and effect of pheromone pre-exposure on pheromone response rhythms in the moth *Spodoptera littoralis*. *J. Insect Physiol.* 51:277—86.

[60]Siwicki,K. K. ,C. Eastman,G. Petersen,M. Rosbash,and J. C. Hall. 1988. Antibodies to the *period* gene product of *Drosophila* reveal diverse tissue distribution and rhythmic changes in the visual system. Neuron 1:141—50.

[61]Skopik,S. D. ,and M. Takeda. 1980. Circadian control of oviposition activity in *Ostrinia nubilalis*. *Am. J. Physiol.* 239:R259—64.

[62]Song,S. H. ,N. Ozturk,T. R. Denaro,et al. 2007. Formation and function of flavin anion radical in *cryptochrome* 1 blue-light photoreceptor of monarch butterfly. *J. Biol. Chem.* 282:17608—12.

[63]Stanewsky,R. 2003. Genetic analysis of the circadian system in *Drosophila melanogaster* and mammals. *J. Neurobiol.* 54:111—47.

[64]Stanewsky,R. ,M. Kaneko,P. Emery,et al. 1998. The *cryb* mutation identifies cryptochrome as a circadian photoreceptor in *Drosophila*. *Cell* 95:681—92.

[65]Syrova,Z. ,I. Sauman,and J. M. Giebultowicz. 2003. Effects of light and temperature on the circadian system controlling sperm release in moth *Spodoptera littoralis*. *Chronobiol. Int.* 20:809—21.

[66]Takeda,M. 1983. Ontogeny of the circadian system governing ecdysial rhythms in a holometabolous insect,*Diatraea grandiosella* (Pyralidae). *Physiol. Entomol.* 8:321—31.

[67]Tanoue ,S. ,P. Krishnan,B. Krishnan,S. E. Dryer,and P. E. Hardin. 2004. Circadian clocks in antennal neurons are necessary and sufficient for olfaction rhythms in *Drosophila*. *Curr. Biol.* 14:638—49.

[68]Truman,J. W. 1972. Physiology of insect rhythms II. The silkworm brain as the location of the biological clock controlling eclosion. *J. Comp. Physiol.* 81:99—114.

[69]Truman,J. W. 1974. Physiology of insect rhythms IV. Role of the brain in the regulation of flight rhythm of the giant silkmoths. *J. Comp. Physiol.* 95:281—96.

[70]Truman,J. W. 1992. The eclosion hormone system of insects. *Prog. Brain Res*. 92:361—74.

[71]Truman,J. W. ,and L. M. Riddiford. 1970. Neuroendocrine control of ecdysis in silkmoths. *Sci-*

ence 167:1624—26.

[72]Urquhart, F. A. 1987. *The monarch butterfly: International traveler*. Chicago: Nelson-Hall.

[73]Vas Nunes, M., and D. S. Saunders. 1999. Photoperiodic time measurement in insects: A review of clock models. *J. Biol. Rhythms* 14:84—104.

[74]Vitzthum, H., M. Muller, and U. Homberg. 2002. Neurons of the central complex of the locust *Schistocerca gregaria* are sensitive to polarized light. *J. Neurosci.* 22:1114—25.

[75]Wergin, W. P., E. F. Erbe, F. Weyda, and J. M. Giebultowicz. 1997. Circadian rhythm of sperm release in the gypsy moth, *Lymantria dispar*: Ultrastructural study of transepithelial penetration of sperm bundles. *J. Insect Physiol.* 43:1133—47.

[76]Williams, C. M. 1963. Control of pupal diapause by the direct action of light on the insect brain. *Science* 140:386.

[77]Williams, C. M., and P. L. Adkisson. 1964. Physiology of insect diapause. XIV. An endocrine mechanism for the photoperiodic control of pupal diapause in the oak silkworm, *Antheraea pernyi*. *Biol. Bull. Mar. Biol. Lab.* 127:511—25.

[78]Wise, S., N. T. Davis, E. Tyndale, et al. 2002. Neuroanatomical studies of period gene expression in the hawkmoth, *Manduca sexta*. *J. Comp. Neurol.* 447:366—80.

[79]Xia, Q., Z. Zhou, C. Lu, et al. 2004. A draft sequence for the genome of the domesticated silkworm (*Bombyx mori*). *Science* 306:1937—40.

[80]Yuan, Q., D. Metterville, A. D. Briscoe, and S. M. Reppert. 2007. Insect cryptochromes: Gene duplication and loss define diverse ways to construct insect circadian clocks. *Mol. Biol. Evol.* 24:948—55.

[81]Závodská, R., I. Sauman, and F. Sehnal. 2003. Distribution of PER protein, pigment-dispersing hormone, prothoracicotropic hormone, and eclosion hormone in the cephalic nervous system of insects. *J. Biol. Rhythms* 18:106—22.

[82]Zerr, D. M., J. C. Hall, M. Rosbash, and K. K. Siwicki. 1990. Circadian fluctuations of period protein immuno-reactivity in the CNS and the visual system of *Drosophila*. *J. Neurosci.* 10:2749—62.

[83]Zhu, H., A. Casselman, and S. M. Reppert. 2008. Chasing migration genes: A brain expressed sequence tag resource for summer and migratory monarch butterflies (*Danaus plexippus*). *PLoS ONE* 3: e1345.

[84]Zhu, H., I. Sauman, Q. Yuan, et al. 2008. Cryptochromes define a novel circadian clock mechanism in monarch butterflies that may underlie sun compass navigation. *PLoS Biol.* 6:e4.

[85]Zhu, H., Q. Yuan, A. D. Briscoe, O. Froy, A. Casselman, and S. M. Reppert. 2005. The two CRYs of the butterfly. *Curr. Biol.* 15:R953—54.

第九章 鳞翅目昆虫的化学受体

Kevin W. Wanner and Hugh M. Robertson

1. 前言

生物的感知能力以及对周围环境中化学物质的反应能力是基本的进化发展过程。昆虫具有复杂的行为，包括觅食、宿主的追寻及选择、取食、产卵、筑巢、保洁、交配以及对同类的选择行为都是由化学感知所介导的。过去10年的研究揭示了昆虫检测和识别化学刺激的复杂的分子遗传和同样复杂的神经系统。2000年，黑腹果蝇全基因组测序(Adams et al. 2000)的完成对于了解化学感知的分子机理提供了很大的帮助。通过对果蝇全基因组的生物信息学分析第一次鉴定得到了昆虫化学受体基因(Clyne et al. 1999；Gao和Chess 1999；Vosshall et al. 1999)。其他物种全基因组测序的相继完成将有利于将这项研究延伸到其他非模式生物或新兴模式生物中，比如家蚕(Goldsmith，Shimada和Abe 2005；见第二章家蚕基因组)。2004年(Mita et al. 2004；Xia et al. 2004)，家蚕基因组测序草图的公布为鳞翅目昆虫化学受体的序列分析提供了一个新的角度(Wanner et al. 2007；Wanner和Robertson 2008)。虽然这个领域的研究发展迅速，但是目前仍旧处于研究的初级阶段。一旦昆虫基因组测序完成，首要的工作就是对潜在的化学受体进行注释。鳞翅目昆虫中几乎都是植食性并且在世界范围内的食物及纤维作物害虫中占有很大的比例。因此，对鳞翅目昆虫化学受体基因的研究不仅有助于了解昆虫及其宿主植物在进化上的相互关系(见第十一章宿主范围进化)，也将有利于在害虫防治方面寻找一些新的靶基因。在本章中，我们将主要结合果蝇的现有研究来讲述鳞翅目化学受体超家族的系统发生关系。

化学刺激物首先被外周的特化的感觉神经元感知，这种感觉神经元将这些信息传入神经，最终由中枢神经系统加以处理。Buck和Axel于1991年在一项具有里程碑意义的研究中发现一个G蛋白偶联受体的大家族在大鼠的嗅觉上皮中特异地表达。G蛋白偶联受体是受体超家族中的一类，它能通过被特有的配体活化而激发二级信使信号途径。基因组研究表明G蛋白偶联受体超家族在多个独立的物种世系中多次为化学受体的进化研究提供了最原始的研究资源，这些物种世系涵盖了哺乳动物、线虫、棘皮动物(Buck和Axel 1991；Robertson和Thomas 2006；Sea Urchin Genome Sequencing Consortium 2006；Thomas和Robertson 2008)。考虑到昆虫中可能存在相同的情况，Clyne等于1999年利用一种算法预测了果蝇基因组中编码含有7个跨膜结构域的蛋白质的基因，7个跨膜结构域是G蛋白偶联受体的共同特征。在Celera Genomics公司对基因组进行全测序之前，基于率先完成15%的果蝇基因组序列信息，鉴定出了一个可能的嗅觉受体家族(Rockville，Maryland)。Vosshall等于1999年运用了相似的方法发现了同一个家族，而Clyne等于2000年鉴定了一个额外的、具有远亲关系的候选味觉受体家族。化学受体通常用于描述一个包含有嗅觉受体家族(Or)及味觉受体家族(Gr)的超家族。到目前为止，大约有60个属于不同家族的化学受体基因在果蝇中(Robertson，Warr和Carlson 2003)得到鉴定，并且基于配体的特异性、嗅觉神经元及味觉神经元表达的特异性对其中的一些基因做了功能鉴定(参见Rutzler和Zwiebel 2005及其他人最近的综述；Thorne，Bray和Amrein 2005；Hallem，Dahanukar和Carlson 2006)。然而，最近的研究发现嗅觉受体具有与G蛋白偶联受体相反的拓扑结构，其氨基末端在胞外，而羧基末端在胞内，这就对将它们定义为G蛋白偶联受体产生了疑问(Benton et al. 2006; Wistrand, Käll 和 Sonnhammer 2006; Lundin et al. 2007)。而且，更令人惊讶的是目前对嗅觉受体及味觉受体下游的信号途径知之甚少，并且关于这个信号途径中的果蝇突变体还未见报道。因此，昆虫的化学感受信号转导途径尚待进一步研究。最近的两个研究为嗅觉受体的功能研究提供了一个非常有意思的观点，它们证实昆虫的嗅觉受体功能存在一种新的机制，即它们会以异源配体门控的离子通道的形式行使功能，这就为其迅速激活的动力学提供了解释(Sato et al. 2008; Wicher et al. 2008)。更重要的是，Wicher 等于 2008 年提出了气味直接活化异源型嗅觉受体阳离子通道之外，还存在一个相对较慢的、间接的经由G蛋白相互作用，以 cAMP 活化嗅觉受体通道的过程。

尽管在外周神经感觉系统中存在许多不同的基因家族，但是分子遗传学实验表明在模式生物果蝇中，化学受体在化学刺激的感知及识别过程中起了最重要的作用。分布在感觉神经元周围的感觉神经节内含有一些转运蛋白(气味结合蛋白，OBPs)及生物转化酶(细胞色素 P450s，谷胱甘肽 S 转移酶及脂酶(例如，Leal 2005; Rutzler 和 Zwiebel 2005))。在果蝇及蛾类的研究证据表明一些气味结合蛋白会与这些受体相互影响，而不仅是起一个简单的、被动的转运配体功能(Xu et al. 2005; Grosse-Wilde, Svatos 和 Krieger 2006; Smith 2007)。与哺乳动物相比，昆虫的嗅觉受体较少，但其拥有更多的气味结合蛋白(OBPs)，因此，昆虫的 Ors 及 OBPs 可能通过一种相互组合的机制来增加这种配体编译能力。在果蝇 Ors 发现之前就在蛾类触角中鉴定得到了感觉神经元膜蛋白1(SNMP1) (Rogers et al. 1997)，但是对其功能仍知之甚少。最近证实了感觉神经元膜蛋白1在果蝇中的同源基因能在 Or 介导的信号转导中发挥作用(Benton, Vannice 和 Vosshall 2007)。显然，外周神经信号转导会受许多不同基因家族以及它们之间的相互作用关系的影响。然而，可以确信的是，果蝇的研究发现化学受体(Crs)行使了最重要的作用(Hallem, Ho 和 Carlson 2004; Hallem, Dahanukar 和 Carlson 2006)，并且新近的研究发现它们可以形成离子通道进而发挥功能，进一步强化了上述观点(Sato et al. 2008; Wicher et al. 2008)。

在大多数情况下，感觉信号的传入和所导致行为的变化是一个复杂的、级联的反应过程。但是也存在一些感受刺激后直接引起行为变化的情况，比如性外激素引发交配行为，甜、苦味觉介导的取食和避食行为。在大多数(不是全部)情况下，果蝇 Ors 在非嗅觉神经元中也有表达，并且与在神经元中的 Ors 一样具有电生理学反应 (Hallem, Ho 和 Carlson 2004; Hallem, Dahanukar 和 Carlson 2006)。2006 年，Syed 等证明鳞翅目昆虫的性外激素受体在果蝇中也具有相应的功能(Syed et al. 2006)。更进一步的研究发现，性外激素受体在具有远亲关系的家蚕(BmOr1) (Nakagawa et al. 2005)及果蝇(DmOr67d) 中(Ha 和 Smith 2006)具有相同的作用，相应的转基因果蝇(导入家蚕的 BmOr1)会被家蚕的性激素(蚕醇)所吸引，而不对其本身的性激素(vaceyl acetate)起反应(Kurtovic, Widmer 和 Dickson 2007)。另外，Ebbs 等也运用果蝇转基因品系研究了味觉神经元对于甜、苦化合物而直接引起取食刺激或躲避行为(参见 Scott 2005; Ebbs 和

Amrein 2007)。因此,化学受体序列在进化上的变化将影响昆虫的行为及其生态学关系,对这些基因家族的系统发生及进化分析将会是一个非常吸引人的工作。

2. 鳞翅目昆虫化学受体的系统发生

2.1 嗅觉及味觉,嗅觉受体及味觉受体

嗅觉受体(Or)和味觉受体(Gr)家族可依据系统发生上亲缘关系的不同、基因的表达模式以及它们相应识别受体功能的差异来区分。昆虫 Or 基因在特化的嗅觉神经元亚型细胞中表达,这些神经元通过其轴突与触角神经叶相连;而 Gr 基因通常在味觉神经元中表达,其轴突与味觉处理中心(中枢神经系统)相连,比如食下神经节。Ors 趋向于检测一些挥发性的疏水性气味,而 Grs 趋向于检测一些底物可溶性的化学物质。即使如此,嗅觉及味觉之间的区别还不是非常的清楚。昆虫的触角原先定义为一个嗅觉器官,但是它仍旧存在一些味觉感受器,如味觉神经元存在于触角并与食下神经节相连处(Jørgensen et al. 2006)。一部分 Gr 基因,包括果蝇的二氧化碳受体及其在蚊子中的同源基因都相应地在果蝇触角及蚊子触须中的嗅觉神经元中表达(Jones et al. 2007; Kwon et al. 2007; Lu et al. 2007; Robertson 和 Kent 2008)。但是到目前为止,尚没有 Or 基因在味觉神经元中表达的报道。

昆虫的 Ors 通常被认为是在古老的节肢动物祖先过渡到陆生环境过程中由 Gr 家族进化而来的(Robertson, Warr 和 Carlson 2003)。基因的重复和丢失,并伴随着基因序列的趋异进化导致在一些具有亲缘关系的昆虫中 Cr 只具有很弱的系统发生关系。比如,随机地从不同昆虫中选一些 Cr 基因进行序列比对,可以发现它们只有很少的共有氨基酸基序。事实上,除了极少数的例子外,不同昆虫来源的 Crs 几乎是不含有共有氨基酸残基的。为了阐明这点,我们选取了 AgOr1/AgGr、DmOr35a/DmGr8a、AmOr11/AmGr4、BmOr1/BmGr18 和 TcOr6/TcGr140 分别代表双翅目、膜翅目、鳞翅目及鞘翅目昆虫(冈比亚按蚊,Ag; 蜜蜂,Am; 家蚕,Bm; 黑腹果蝇,Dm;赤拟谷盗,Tc)(图 9.1 和图 9.2),进行了序列比对。与 Grs 相比,很多 Ors 的特有配体已经得到鉴定。因此,选择了四个功能得到鉴定的 Ors。AgOr1 会对人类汗液的一种成分产生反应(Hallem et al. 2004);AmOr11 会对雌蜂王性外激素的一个主要成分产生反应(Wanner et al. 2007); BmOr1 会对家蚕的性外激素产生反应(Sakurai et al. 2004;Nakagawa et al. 2005;Syed et al. 2006)以及 DmOr35a 对一种六碳醇有响应(Hallem, Ho 和 Carlson 2004; Wanner et al. 2007)。通过序列比对可以发现,Ors 间只具有 7 个共有的氨基酸残基,而 Grs 只有 6 个(图 9.1B 和图 9.2B)。相应的 Ors 氨基酸序列一致性在 13%~23%之间,而 Grs 的序列一致性在 7%~15%之间(图 9.3B)。

虽然 Or 及 Gr 各自的序列间的分化较大,但它们的序列中仍旧存在一些相类似的结构特征。疏水图表通常被用于界定蛋白的疏水结构域及预测一些潜在的跨膜区域。昆虫的 Ors 和 Grs 都拥有 7 个可能的跨膜结构域(见图 9.1A 及图 9.2A),并且这些区域疏水氨基酸残基出现的频率很高 (见图 9.1B 及图 9.2B)。除氨基酸序列的差异外,Or 和 Gr 基因也具有高度可变的内含子数目、剪切位点及不同的拼接方式(见图 9.1C 及图

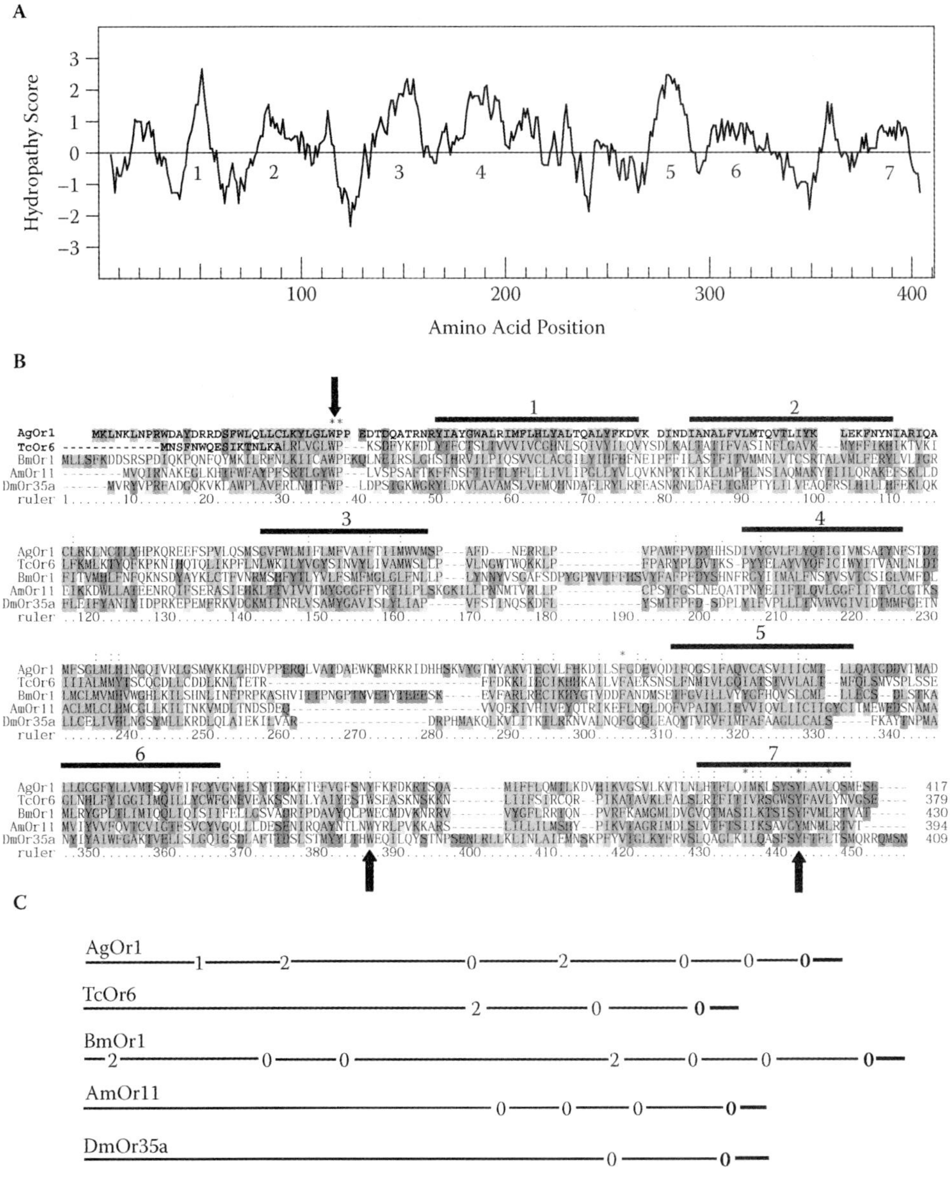

图 9.1 昆虫 Ors 的分子特征。(A)DmOr35a 的疏水图表(Kyte-Doolittle 方法)显示含有 7 个潜在的跨膜结构域。(B)来源于五个不同目昆虫的 Or 序列的 ClustalX 多重比对(AgOr1,冈比亚按蚊,双翅目;TcOr6,赤拟谷盗,鞘翅目;BmOr1,家蚕,鳞翅目;AmOr11,蜜蜂,膜翅目;DmOr35a,黑腹果蝇,双翅目)。带数字的标注线显示的是 7 个预测的跨膜结构域。箭头所示的是保守氨基酸残基。(C)5 个 Ors 基因结构差异示意图。实线代表外显子(刻度代表相应长度);数字代表内含子/外显子的界限及其不同的剪切模式(模式 0,1,2;见文章中剪切模式的解释)。黑色加亮的短实线显示的是最后一个外显子典型的 0 模式剪切。

9.2C)。昆虫 Or 和 Gr 基因内含子数目可以从 1 个到 9 个或 10 个不等,部分甚至全部丢失(与哺乳动物嗅觉受体不同,其在编码区域不具有内含子)。尽管存在这些差异,大多数昆虫 Crs 拥有一个短的、以 0 模式拼接的外显子(见图 9.1C 及图 9.2C),这个特征在 Ors 和 Grs 中是共有的(依据内含子/外显子界限所处编码子位置的不同分为 3 种拼接方式,模式 0:界限位于两个编码子之间;模式 1:界限位于一个编码子的第 1 及第 2 碱基之间的位置;模式 2:界限位于一个编码子的第 2 及第 3 碱基之间的位置)。这种普遍存在于昆虫化学受体的最后一个外显子的拼接方式暗示其行使一项重要的目前还未知的功能。

昆虫 Ors 和 Grs 的 C 末端区域在这两个家族之间及之内也存在一些有迹可寻的保守氨基酸残基(图 9.1B 及图 9.2B)。Ors 和 Grs 最后一个外显子的第 8 个氨基酸残基是一个保守的酪氨酸残基(Y)。在酪氨酸残基之前通常是丝氨酸残基(S)或苏氨酸残基(T)。在酪氨酸残基之后通常是一个短的疏水氨基酸序列,这个特征在 Grs 中尤其明显。Grs(而不是 Ors)拥有位于这段短疏水氨基酸序列之后的保守的谷氨酰胺残基(Q)及苯丙氨酸残基(F),这就构成了 Grs 中的一段 TYhhhhhQF 保守序列,其中 h 代表任何疏水氨基酸(图 9.2B)。Or 家族拥有两个保守的氨基酸区域,分别是位于近 N 末端的 WP 及一个 W 位于 6 和 7 跨膜区域之间的第 3 个胞外环上(图 9.1B)。引人注目的是,这些保守氨基酸残基的具体功能尚不清楚。虽然 Ors 和 Grs 序列间存在很大的差异(Ors 及 Grs 间只有 6%～12%的一致性,图 9.3B),但是仍旧可以利用序列上一些保守的特征,通过系统发生分析去区别 Ors 及 Grs。正如图 9.3A 所示,10 个来源于不同目昆虫的 Crs 序列都能归到相应的 Or 及 Gr 家族中。

2.2 味觉受体家族

在果蝇中,Gr 基因主要在味觉器官的味觉神经元中表达(Clyne,Warr 和 Carlson 2000;Wang et al. 2004;Thorne,Bray 和 Amrein 2005;Slone,Daniels 和 Amrein 2007)。然而,最近的研究发现与 Or 基因相比,一些 Grs 具有更广泛的表达及参与更多的生理学过程(Thorne 和 Amrein 2008)。部分 Grs 可以通过相互协同作用来行使其他功能,而它们在化学感受系统之外的其他功能反映了它们在内部生理学检测及化学营养物质监控中行使一个原始的功能。到目前为止,我们仅仅只能对这些功能进行猜测,但有趣的是与脊椎动物味觉系统一样,Gr 基因也被证实在胃肠道中有表达。另一个有趣的现象是昆虫和脊椎动物的基因组一样,都编码少量的保守性高的 Grs,它们被用于检测一些必要的营养物质,比如糖;另外还编码大量扩充的、趋异性高的味觉受体,它们通常被认为用于检测苦味化合物(Go 2006;Shi 和 Zhang 2006)。我们对家蚕 Gr 家族的注释结果表明鳞翅目 Gr 家族也具有相似的趋势。

最近,我们完成了来源于家蚕基因组数据的 65 个 Gr 基因的注释(Wanner 和 Robertson 2008)。在这之前鳞翅目昆虫仅有三个 Grs 被公布,即 HvCr1,HvCr4 和 HvCr5(Krieger et al. 2002),它们由 Bayer 公司在烟芽夜蛾基因组测序中得到鉴定。基于 Crs 的分类依据,这 3 个基因都属于 Gr 家族,并且具有典型的 Gr 特征(见图 9.2)。BmGrs、HvCr1、HvCr4、HvCr5 及来源于其他类昆虫 Grs 的系统发生关系在图 9.4 中进行了阐述。家蚕 Grs1-10 与其他类昆虫的 Grs 共同形成了三个保守的谱系;它们也就是不同昆虫之间的仅有的相对保守的 Gr 谱系(Robertson 和 Wanner 2006;Wanner 和 Robertson

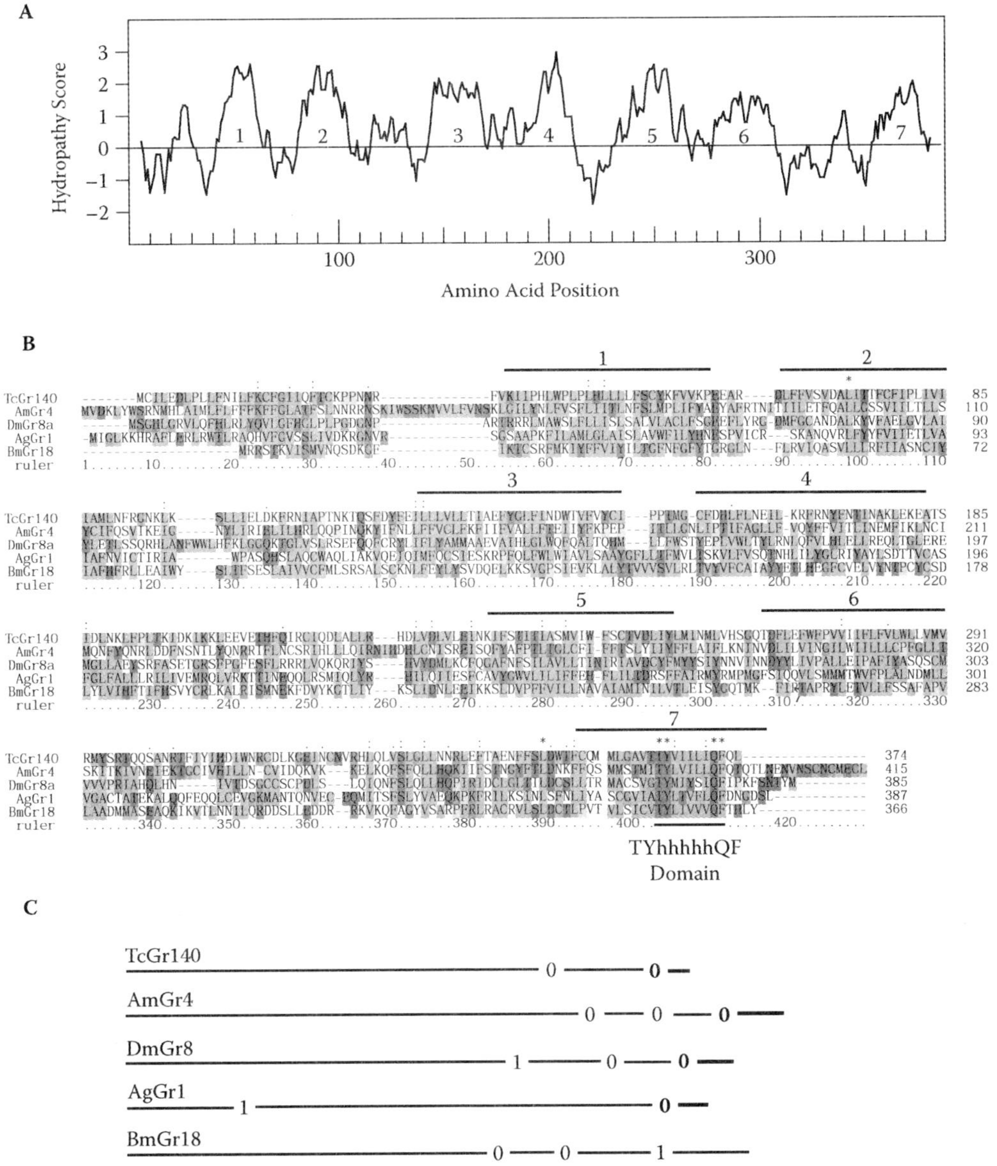

图 9.2 昆虫 Grs 的分子特征。(A) AgGr1 的疏水图表显示含有 7 个潜在的跨膜结构域。(B) 来源于五个不同目昆虫的 Gr 序列的 ClustalX 多重比对(AgGr1,冈比亚按蚊,双翅目; TcGr140,赤拟谷盗,鞘翅目 ; BmGr18,家蚕,鳞翅目; AmGr4,蜜蜂,膜翅目; DmGr35a,黑腹果蝇,双翅目)。带数字的标注线显示的是 7 个预测的跨膜结构域。TYhhhhhQF 代表 C 末端一个保守的结构域。(C) 5 个 Ors 基因结构差异示意图。实线代表外显子(刻度代表相应长度);数字代表内含子/外显子的界限及其不同的剪切模式(模式 0,1,2;见文章中剪切模式的解释)。黑色加亮的短实线显示的是最后一个外显子典型的 0 模式剪切。

2008)。BmGrs1-3 与三分之二果蝇中的二氧化碳受体同源(Jones et al. 2007;Kwon et al. 2007;Lu et al. 2007;Robertson 及 Kent 2008)。根据果蝇糖受体的功能特点,

BmGrs4-8 及 HvCr1 和 HvCr5 都属于昆虫的糖受体亚家族(例如,Slone et al. 2007;Dahanukar et al. 2007;Kent 和 Robertson,2009)。BmGrs9,BmGrs10 及 HvCr4 与 DmGr43a 同源,都属于同一个保守的谱系,这个谱系同时还包括其他一些昆虫,如蜜蜂(AmGr3;Robertson 和 Wanner,2006),蚊子(AgGr25 和 AaGr20;Hill et al. 2002; Kent,Walden 和 Robertson 2008),以及甲虫(TcGr20-28 和 183,Tribolium Genome Sequencing Consortium 2008)。果蝇 DmGr43a 的直系同源基因单独形成一个高度一致的谱系,但它们的功能未知。

A

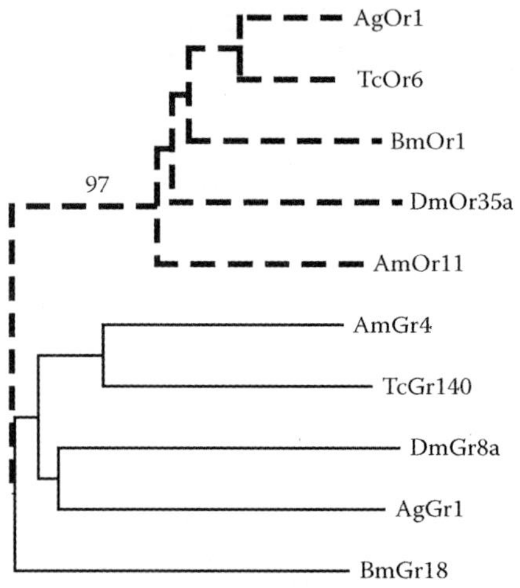

B

	AgOr1	TcOr6	BmOr1	DmOr35a	AmOr11	AmGr4	TcGr140	DmGr8a	AgGr1	BmGr18
AgOr1	**0**	23	14	14	17	10	11	8	8	7
TcOr6	23	**0**	15	13	14	10	11	11	8	9
BmOr1	14	15	**0**	14	15	12	8	7	6	9
DmOr35a	14	13	14	**0**	14	9	10	9	9	9
AmOr11	17	14	15	14	**0**	9	11	11	9	8
AmGr4	10	10	12	9	9	**0**	14	12	13	11
TcGr140	11	11	8	10	11	14	**0**	14	10	11
DmGr8a	8	11	7	9	11	12	14	**0**	15	7
AgGr1	8	8	6	9	9	13	10	15	**0**	11
BmGr18	7	9	9	9	8	11	11	7	11	**0**

图 9.3 Cr 超家族中来源于不同科昆虫的 Ors 及 Grs。(A)5 个来源于不同类昆虫 Ors 及 Grs(见图 9.1 及图 9.2)的系统发生树(AgGr1,冈比亚按蚊,双翅目;TcGr140,赤拟谷盗,鞘翅目;BmGr18,家蚕,鳞翅目;AmGr4,蜜蜂,膜翅目;DmGr35a,黑腹果蝇,双翅目)。尽管其氨基酸序列存在差异性,Or 家族成员仍旧形成一个分支并且与 Gr 家族相分离(bootstrap value equals 97 of 100 replicates)。(B)5 个 Ors 及 5 个 Grs 氨基酸一致性分析(见图 9.1B 和图 9.2B)。

我们在对 65 个 BmGrs 分析的过程中发现了另外一个非常有趣的现象:家蚕中出现了一个单支扩增现象,使得剩下的 55 个 Grs 形成一个与其他昆虫所不同的亚家族(图 9.4)。这 55 个 Grs 氨基酸序列差异很大,它们之间的序列相似性不到 10%。与其他昆虫 Grs 一样,C 端的最末端含有一段保守的 TYhhhhhQF 基序(图 9.2B;Wanner 和 Robertson 2008)。正是由于它们序列的差异性,利用基于距离的系统发生分析并不支持它

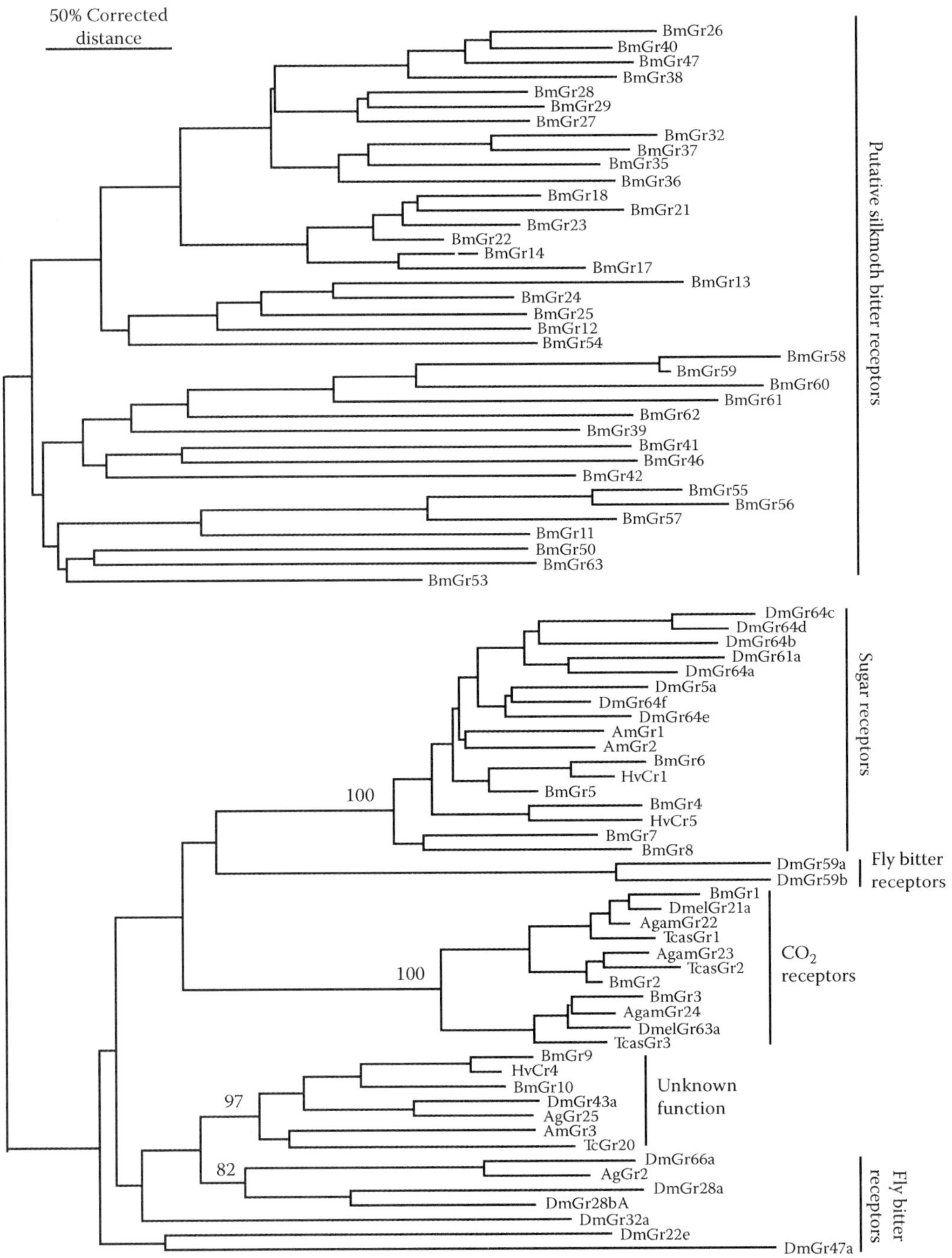

图 9.4　65 个 BmGrs 和 3 个发表的烟芽夜蛾 Crs(HvCr1,4,5)（Krieger et al. 2002)的系统发生树。在蝇(Dm)，蚊子(Ag)，甲虫(Tc)，蜂(Am)中保守的糖和二氧化碳受体同系物及 DmGr43a 谱系也被进行分析以阐明它们之间的保守性。预测的果蝇 bitter 受体(Robertson，Warr 和 Carlson 2003)被证明它们缺乏与蛾类预测 bitter 受体的相关性。二氧化碳受体、糖受体、Dm43a 同系物和候选的 bitter 受体谱系用竖线进行标注。主要分支的数值显示 10000 次独立分析的结果。

们在统计学上的共同起源关系(图 9.4)。但是,它们的共同起源关系可以利用贝叶斯(Bayesian)分析法得到支持(91%bootstrap value; Wanner 和 Robertson 2008)。另外更特别的是,55 个 Grs 中的 54 个含有一致的、独特的基因结构(图 9.2C 对 BmGr18 基因结构做了解释),这就更进一步支持这一大群家蚕特异 Gr 亚家族共同起源关系(第 55 个基因丢失了全部 3 个内含子)。这个亚家族与其他昆虫的 Grs 一样,有一个超过编码序列一半的 N 末端的大外显子,紧随一个大约编码 25 个氨基酸残基的小外显子,第 3 个外显子大约编码 50 个氨基酸,以及最后一个相对较长的 C 末端外显子(图 9.2C)。另外,蛾类中的这一 Gr 谱系缺失了 C 末端典型的模式 0 型内含子/外显子的剪切方式,这就更进一步支持了这一谱系在蛾类中的唯一性(图 9.2C)。与化学受体通常被用于宿主植物的识别、选择及取食、同类的识别及交配、产卵相比,在一些情况下,植食性昆虫对植物的选择也是一个营养物质吞噬刺激和苦味成分检测的平衡过程,当然也存在其他一些特殊的刺激信号(Chapman 2003;参见第十章)。因此,这 55 个 Grs 可能检测一些源于宿主植物或交配过程中的一些特殊信号,但是其主要可能倾向于检测苦味成分和植物的化学信号。

除单支扩增形成预测的苦味受体外,家蚕中的其他 Gr 家族基本都与其他昆虫同源。其总计有 65 个 Grs 与其他昆虫的能很好地匹配,这些昆虫包括蚊子(60～79 Grs; Hill et al. 2002;Kent,Walden,and Robertson 2008)、果蝇(56～60 Grs;Robertson,Warr,and Carlson 2003; Gardiner et al. 2008)及寄生蜂(58 Grs;H. M. Robertson,unpublished results)。但其中有一个区别就是来源于家蚕 Grs 丢失了选择性的剪切过程,而这种过程却在黑腹果蝇、按蚊、伊蚊及赤拟谷盗的 Grs 中存在。在这种情况下,就可以通过选择性剪切方式来不同程度地增加 Gr 蛋白的数目,如在黑腹果蝇等中适度增加了 4～7 个,而按蚊、伊蚊和赤拟谷盗中却多增加了 29～35 个 Gr 蛋白(e. g. Kent, Walden 和 Robertson 2008; Tribolium Genome Sequencing Consoritum 2008)。

2.3 嗅觉受体家族

昆虫的 Or 基因常在一些如触角等嗅觉器官上的嗅觉神经元中表达。到目前为止,在昆虫基因组中发现编码从 60 到 250 个不等的 Or 基因(例如,Robertson,Warr,and Carlson 2003; Tribolium Genome Sequencing Consortium, 2008)。虽然存在少数的例外,在果蝇中研究发现大多数情况下一个嗅觉神经元中只表达 60 个 Or 基因中的一个(其中包括下一节将要提到的一个普遍存在的 Or83b 受体)。更近一步的研究表明,在与触角神经叶中神经纤维球相连的嗅觉神经元轴突中也表达相同的 Or 基因,而这提供了一个神经学机制去检测并放大从周围检测到的气味信号(Vosshall,Wong,and Axel 2000)。一些综述中总结了新近一些关于果蝇嗅觉分子及神经学机理的研究(e. g. ,Rutzler 和 Zwiebel 2005;Hallem,Dahanukar,and Carlson 2006;Vosshall 和 Stocker 2007)。

除特例外,来源于不同目昆虫的 Ors 一般不具有明显的同源性。早期的 Cr 研究发现在不同昆虫中存在高度同源的、单一的 Or 谱系。这个谱系在果蝇中的直系同源基因是 Or83b,并且在果蝇中研究发现 Or83b 是作为一个其他 Ors 的配偶体和伴侣蛋白(e. g. ,Benton et al. 2006)。这一谱系一般称作 DmOr83b 直系同源基因,在鳞翅目、鞘翅目和膜翅目中对应 Or2 基因（Krieger et al. 2003）。昆虫基因组测序后发现仅有一个基因拷贝编码 Or83b 亚家族，并且转基因缺失 Or83b 后会造成果蝇嗅觉功能的广泛丧失（Larsson et al. 2004）。在体外的检测实验同样也证明蜜蜂中的直系同源基因 AmOr2 是 AmOr11 对蜂王性外激素 9-ODA 起反应所必需的（Wanner et al. 2007），以及对赤拟谷盗 TcOr2 基因干涉后会导致其嗅觉受损（Engsontia et al. 2008）。最近，有证据表明 Or83b 形成一个离子通道，离子通道受到具有配基特异性的同伴 Or 调节（Sato et al. 2008；Wicher et al. 2008）。

除了 Or83b 谱系之外，在来源于不同目的昆虫中并不存在其他的直系同源的 Or 谱系。蝇和蚊子代表双翅目的两个不同的亚目，但在这两类昆虫中的 Or 已经发生了较早的系统分离。系统发生分析可以发现它们的 Ors 仅有非常低的同源性，这其中伴随着一部分 Or 谱系在蝇类中扩增（以果蝇为代表），而另一部分谱系在蚊子中扩增（以冈比亚按蚊为代表；Hill et al. 2002）。除同源性很高的果蝇 Or83b 谱系外，缺乏氨基酸序列的相似性的特点阻碍了昆虫中 Ors 的实验发现。比如，66 个鳞翅目昆虫 Ors 仅在全基因组测序完成后才被发现，其中 48 个来源于家蚕基因组（Wanner et al. 2007），18 个来源于烟芽夜蛾基因组序列（Krieger et al. 2002）。然而，通过对 GenBank 进行 BLASTp 搜寻之后，发现了 6 个 Or83b 的直系同源基因，它们分别来自 6 个不同的鳞翅目昆虫中（中国柞蚕、烟青虫、棉铃虫、甘蓝夜蛾、斜纹夜蛾和甜菜夜蛾），它们氨基酸序列的一致性高于 85%。这种氨基酸序列的高度一致性使得可以很容易地从一些未进行全基因组测序的昆虫中发现 Or。

对已经发表的鳞翅目昆虫的 68 个 Ors（图 9.5，Wanner et al. 2007）的分析发现，与双翅目（Hill et al. 2002；Robertson，Warr 和 Carlson 2003）及膜翅目（蜜蜂和丽蝇蛹集金小蜂；K. W. Wannner 和 H. M. Robertson，未发表）昆虫相比，在家蚕（蚕蛾科 Bombycidae）及烟芽夜蛾（夜蛾科 Noctuidae）之间具有非常高的潜在相似性。当比较蝇和蚊子，或黄蜂和蜜蜂，可以发现通常是具有亚目特异性扩增的 Or 谱系占优势。与家蚕 Ors 相比，虽然数目小于一半，但 18 个烟芽夜蛾 Ors 分散存在于一些家蚕的谱系中，其中包括一些潜在直系同源基因对（图 9.5）。我们相信在烟芽夜蛾中更多序列被鉴定及发表之后，这两个物种 Or 同系物的数目将会增加。然而，在将来鳞翅目昆虫基因组的不断测序后，我们期待发现一些由鳞翅目昆虫家族特异扩增而出现的一些谱系，而不是仅仅与果蝇和蚊子进行比较。比如 BmOrs45－48，其中 3 个就很特殊地在雌蛾触角中高量表达，而这就很可能是一个家族特异扩增的结果。事实上，鳞翅目昆虫家族特异 Or 扩增将对理解其气味检测具有重要的生物学意义。

与其他昆虫相比，鳞翅目昆虫中有较高的 Or 同源性的出现，这可能可以简单地反映它们具有更短的进化时间。现有的主要鳞翅目昆虫的进化大约在 1 亿年前（见第一章鳞翅目的系统发生），而果蝇和蚊子、蜜蜂和黄蜂，它们的进化分离大约接近于 2 亿年前（Grimaldi 和 Engel 2005）。然而，事实上几乎所有的鳞翅目昆虫都是以植物为食的，这可能反映出至少与双翅目、鞘翅目及膜翅目昆虫相比，它们具有更加一致的化学感受环境，这显示了鳞翅目昆虫具有更加宽广的生活史。

鳞翅目昆虫的 Or 同源性的增加产生了两个直接的实用结果。第一，利用序列的同

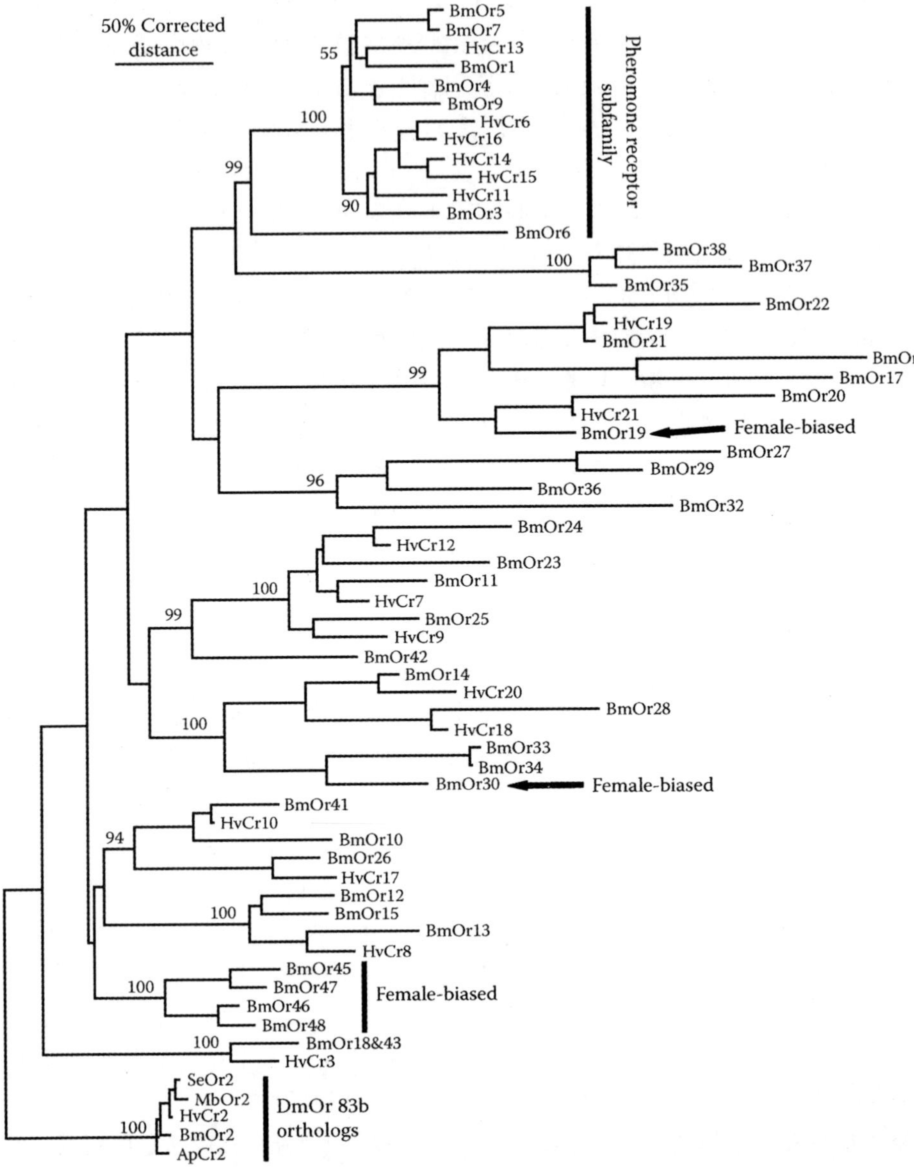

图 9.5　家蚕 Ors (Wanner et al. 2007)及烟芽夜蛾 Ors (Krieger et al. 2002)的系统发生树。鳞翅目的同系物树根所使用的是 DmOr83b 谱系（甜菜夜蛾，Se；甘蓝夜蛾，Mb；烟芽夜蛾，Hv；家蚕，Bm；以及柞蚕，Ap）。性外激素受体超家族，DmOr83b 谱系及具有雌偏向性表达的 Ors 超家族（BmOr45—48）用加亮竖线标注。BmOrs19 及 30（箭头所示）也显示雌偏向性表达。

源性，更容易地发现它们；第二，可能会产生一些保守的功能。基于这种想法，Wanner 等(2007)调查了所知的 48 个家蚕 Ors 在雌雄触角中的表达模式。依据其雄偏向性的表达模式，一些 Ors（包括 BmOrs1，BmOrs3－6）最先被鉴定为性信息素的候选受体（Sakurai et al. 2004; Nakagawa et al. 2005）。BmOrs19 和 30 这两个受体，显示很强的雌偏向性的表达模式，而 BmOrs45－47 显示一种适度的雌偏向性（Wanner et al. 2007）。这些雌偏向性表达模式的受体表明它们是用于检测一些可引起雌蛾生物学行为的气味分子。比如检测用于产卵的宿主植物气味、检测雄性产生的交配信息素。随着不同蛾类中更多的 Or 序列被鉴定，一些具有引起雌特异性行为、功能的保守 Ors 将有可能被验证。

关于雌蛾产生性信息素而吸引处于远距离的雄蛾进行交配的研究在 20 世纪就成为焦点(Schneider 1992)。化学鉴定及合成了几百种蛾的外激素，由此就产生了许多生态环保型的害虫防治方法，这些方法都是基于用这些蛾信息素去诱惑和捕获雄蛾、或者去干扰它们与本地雌蛾的交配能力。第一个蛾性信息素受体的发现及鉴定一直都很受关注。第一个候选的信息素受体由 Krieger 等于 2004 年发现。通过构建触角的 cDNA 文库，并与基因组序列相互比较，他们鉴定了 18 个烟芽夜蛾 Ors，其中的 6 个 Ors(HvCr6，11，13－16)形成一个高度保守的亚家族，并且具有雄蛾触角偏向性的表达模式。利用最近完成的家蚕基因组数据，Sakurai 等于 2004 年同样鉴定了一类具有雄蛾触角偏向性表达模式的 5 个家蚕相关的 Ors(BmOrs1，3－6)。更重要的是，在系统发生分析中，BmOrs1 和 BmOrs3－6 与 HvCr6，11 和 13－16 聚为一个分支(Nakagawa et al. 2005；Krieger et al. 2005)。后续的分析表明，在家蚕中共存在 7 个 BmOrs 属于这个信息素受体家族(图 9.5；Wanner et al. 2007)。事实上，目前已鉴定的所有的信息素受体共同形成了一个保守的亚家族，这就表明在鳞翅目昆虫中，性信息素受体可能具有广泛的保守性，具有重要的应用价值及科研价值(见第十章鳞翅目两性交流的综述)。

图 9.5 的系统发生分析表明在蚕蛾科及夜蛾科的祖先中至少存在两个信息素受体谱系，其中一个谱系在家蚕中扩增(BmOrs1，4，5，7，9 及 HvOr13)，另一个谱系在烟芽夜蛾中扩增(HvCrs6，11，14－16 及 BmOr3)。BmOr1 对信息素中的蚕醇成分产生反应，而 BmOr3 对信息素中的蚕醛成分起反应(Nakagawa et al. 2005)。HvCrs13，14，16 在体外可以对烟芽夜蛾性信息素成分产生反应，而 HvirCr13 特异地对 Z-11-十六碳烯醛产生反应 (Grosse-Wilde et al. 2007)。两个不同蛾科间的性信息素受体的不同扩增与进化过程中 Or 基因典型性的不同扩增及收缩方式相符合，并且这种趋势可能会产生一个相一致的性信息素受体谱系。

3. 总结及展望

蛾类成虫利用它们的化学感受能力去寻求及选择配偶，并且鳞翅目昆虫成虫及幼虫也利用它们的化学感受能力去寻求及选择它们相应的宿主植物。在中枢神经系统处理这些感觉信号过程中，第一个关键步骤就是由外周神经系统去检测这些化学刺激。作为引起一些复杂及重要行为的反应链的第一个关键步骤，因此探寻外周化学感受作用的分子基础一直是受到关注的研究目标。第一个关于哺乳动物 Crs 在大鼠外周神经系统中的发现获得了诺贝尔奖(Buck 和 Axel 1991)。但是在此之后，关于昆虫 Crs 的研究均毫无结果，而直到 9 年后的全基因组测序完成才发现了第一个昆虫的 Crs 家族。随后关于果蝇外周神经中化学信号检测的分子机理的阐明，以及一些新出现的功能基因组分析工具，使得这一研究平台扩展到包括鳞翅目昆虫在内的其他昆虫上。

在其他非模式生物中鉴定和克隆 Cr 基因及进行功能分析存在两个显著的障碍。在鉴定 Cr 基因方面的第一个障碍，已经被一些如 454 焦磷酸测序 (Hudson 于 2008 综述了新的 DNA 测序技术以及在生态学的运用)等重复行好、高通量的 DNA 测序技术所克服。低成本的 DNA 测序进一步加速了全基因组及全转录本的测序工作,所有的这些都会增加如 NCBI 等公共数据库中的 Cr 序列。在非模式生物中 Crs 功能研究方面是第二个障碍,基于细胞水平的受体活性检测技术的发展使得在体外进行受体与配体分析成为可能,这一技术也就克服了在非模式生物中研究 Cr 的不足(Nakagawa et al. 2005; Grosse-Wilde, Svatos 和 Krieger 2006; Grosse-Wilde et al. 2007; Kiely et al. 2007; Wanner et al. 2007)。因此,基于这两个飞跃,就有可能去鉴定及分析鳞翅目昆虫中一些保守的 Cr 谱系与一些重要行为之间的关系。

不断增加的鳞翅目昆虫的 Cr 序列,及一些具有性别及组织特异表达 Cr 基因的鉴定,都将有利于去寻找一些调控交配行为及宿主植物选择行为的 Cr 谱系。如一些表达于雌蛾前睑板及产卵器上的 Crs 就是一个很好的例子。雌性产生的性信息素被认为与鳞翅目物种的形成相关(Cardé 和 Haynes 2004),这也是第一次可以去示踪作为检测性外激素 Or 谱系的起源及进化。许多雄蛾在交配过程中也会产生一些可被雌蛾所检测的性信息素(定义为催欲性外激素)。雄蛾的性信息素腺体具有多样性,这通常被认为在鳞翅目进化过程中发生了多次相互独立的进化(Cardé 和 Haynes 2004)。目前的一些研究关注于鉴定一些在雌蛾触角中表达的、与检测雄蛾产生的性信息素相关的 Ors,以及它们的起源关系(详见第十章)。

另外,鳞翅目中不同科之间这些蛋白受体具有高度的一致性,这就有利于通过实验方法从一些重要的害虫中发现相应的 Ors。比如,性信息素被广泛地运用于害虫的捕集(IPM)、交配的干扰等。目前,就可以用 PCR 简并引物在不同的蛾科中去克隆预测的性信息素受体。体外的检测实验可以用于筛选一些具有信息素促效或抑制作用的化学类似物,这就可以进一步运用于外激素的诱惑实验。其中一个有趣的问题就是在家蚕(蚕蛾科)及烟芽夜蛾(夜蛾科)中发现的 Or 同源性的水平是否能够预示在鳞翅目中存在更宽范围的 Or 同源性。假如存在一些 Or 谱系保留了相应特异性的一类配体(如性信息素受体亚家族),那么,对家蚕全部 Ors 的分析就有可能从一些重要性的经济害虫中预测到具有配体特异性结合的同源的 Ors 基因。从生态学的角度上看,调查世系特异的 Or 扩增和减缩与该类型的迁移和可利用的宿主植物的范围之间的关系是非常有趣的,在果蝇各种类之间已经发现了不同的模式(McBride 2007; McBride, Arguello 和 O'Meara 2007)。

我们对家蚕 Gr 家族的分析得出了一个有趣的结论,就是家蚕中存在一个单支扩增产生 55 个 Grs,并且这些 Grs 大多具有苦味受体的功能。基于其序列间的同源性较低,因此该世系似乎比较古老。随着更多的 Gr 序列从不同的鳞翅目昆虫中鉴定得到,追踪这一世系的系统发生的起源是一项有趣的工作。昆虫 Gr 家族的进化通常认为是对栖息环境变化的一种反应。在蜜蜂基因组中,只发现 10 个 Grs,显然归因于大多数潜在的苦味受体的丢失或基因扩增的缺乏。然而在面粉虫的 Gr 家族中却存在显著的 Gr 谱系扩增,总计存在惊人的 215 个 Gr 基因可能编码了 245 个 Gr 蛋白(Tribolium Genome Sequencing Consortium 2008)。鳞翅目中一个单一 Gr 谱系的扩增可能是在其祖先转变为食草性昆虫的过程中产生的。尽管目前只是猜测,然而鳞翅目昆虫基因组学研究的不断深入一定能够为 Cr 超家族进化模式的研究提供更多线索。

参考文献

[1]Adams, M. D. , S. E. Celniker, R. A. Holt, et al. 2000. The genome sequence of *Drosophila melanogaster*. *Science* 287:2185−95.

[2]Benton, R. , S. Sachse, S. W. Michnick, and L. B. Vosshall. 2006. Atypical membrane topology and heteromeric function of *Drosophila* odorant receptors *in vivo*. *PLoS Biol*. 4:e20.

[3]Benton, R. , K. S. Vannice, and L. B. Vosshall. 2007. An essential role for a CD36-related receptor in pheromone detection in *Drosophila*. *Nature* 450:289−93.

[4]Buck, L. , and R. Axel. 1991. A novel multigene family may encode odorant receptors: A molecular basis for odor recognition. *Cell* 65:175−87.

[5]Cardé, R. T. , and K. F. Haynes. 2004. Structure of the pheromone communication channel in moths. *In Advances in insect chemical ecology*, ed. R. T. Cardé and J. G. Millar, 283−332. Cambridge: Cambridge Univ. Press.

[6]Chapman, R. F. 2003. Contact chemoreception in feeding by phytophagous insects. *Annu. Rev. Entomol.* 48:455−84.

[7]Clyne, P. J. , C. G. Warr, and J. R. Carlson. 2000. Candidate taste receptors in *Drosophila*. *Science* 287:1830−34.

[8]Clyne, P. J. , C. G. Warr, M. R. Freeman, D. Lessing, J. Kim, and J. R. Carlson. 1999. A novel family of divergent seven-transmembrane proteins: Candidate odorant receptors in *Drosophila*. *Neuron* 22:327−38.

[9]Dahanukar, A. , Y. T. Lei, J. Y. Kwon, and J. R. Carlson. 2007. Two Gr genes underlie sugar reception in *Drosophila*. *Neuron* 56:503−16.

[10]Ebbs, M. L. , and H. Amrein. 2007. Taste and pheromone perception in the fruit fly *Drosophila melanogaster*. *Pflugers Arch*. 454:735−47.

[11]Engsontia, P. , A. P. Sanderson, M. Cobb, K. K. Walden, H. M. Robertson, and S. Brown. 2008. The red flour beetle's large nose: An expanded odorant receptor gene family in *Tribolium castaneum*. *Insect Biochem. Mol. Biol.* 38:387−97.

[12]Gao, Q. , and A. Chess. 1999. Identification of candidate *Drosophila* olfactory receptors from genomic DNA sequence. *Genomics* 60:31−39.

[13]Gardiner, A. , D. Barker, R. K. Butlin, W. C. Jordan, and M. G. Ritchie. 2008. *Drosophila* chemoreceptor gene evolution: Selection, specialization and genome size. *Mol. Ecol.* 17:1648−57.

[14]Go, Y. 2006. Lineage-specific expansions and contractions of the bitter taste receptor gene repertoire in verte-brates. *Mol. Biol. Evol.* 23:964−72.

[15]Goldsmith, M. R. , T. Shimada, and H. Abe. 2005. The genetics and genomics of the silkworm, *Bombyx mori*. *Annu. Rev. Entomol.* 50:71−100.

[16]Grimaldi, D. , and M. S. Engel. 2005. *Evolution of the insects*. New York: Cambridge Univ. Press.

[17]Grosse-Wilde, E. , T. Gohl, E. Bouché, H. Breer, and J. Krieger. 2007. Candidate pheromone receptors provide the basis for the response of distinct antennal neurons to pheromonal compounds. *Eur. J. Neurosci.* 25:2364−73.

[18]Grosse-Wilde, E. , A. Svatos, and J. Krieger. 2006. A pheromone-binding protein mediates the bombykol-induced activation of a pheromone receptor *in vitro*. *Chem. Senses* 31:547−55.

[19]Ha, T. S. , and D. P. Smith. 2006. A pheromone receptor mediates 11-cis-vaccenyl acetate-induced responses in *Drosophila*. *J. Neurosci.* 26:8727—33.

[20]Hallem, E. A. , A. Dahanukar, and J. R. Carlson. 2006. Insect odor and taste receptors. *Annu. Rev. Entomol.* 51:113—35.

[21]Hallem, E. A. , A. N. Fox, L. J. Zwiebel, and J. R. Carlson. 2004. Olfaction: Mosquito receptor for human-sweat odorant. *Nature* 427:212—13.

[22]Hallem, E. A. , M. G. Ho, and J. R. Carlson. 2004. The molecular basis of odor coding in the *Drosophila* antenna. *Cell* 117:965—79.

[23]Hill, C. A. , A. N. Fox, R. J. Pitts, et al. 2002. G-protein-coupled receptors in *Anopheles gambiae*. *Science* 298:176—78.

[24]Hudson, M. 2008. Sequencing breakthroughs for genomic ecology and evolutionary biology. *Mol. Ecol. Resources* 8:3—17.

[25]Jones, W. D. , P. Cayirlioglu, I. G. Kadow, and L. B. Vosshall. 2007. Two chemosensory receptors together mediate carbon dioxide detection in *Drosophila*. *Nature* 445:86—90.

[26]Jϕrgensen, K. , P. Kvello, T. J. Almaas, and H. Mustaparta. 2006. Two closely located areas in the suboesophageal ganglion and the tritocerebrum receive projections of gustatory receptor neurons located on the antennae and the proboscis in the moth *Heliothis virescens*. *J. Comp. Neurol.* 496:121—34.

[27]Kent, L. B. , and H. M. Robertson. 2009. Evolution of the sugar receptors in insects. *BMC Evol. Biol.* 9:41.

[28]Kent, L. B. , K. K. Walden, and H. M. Robertson. 2008. The Gr family of candidate gustatory and olfactory receptors in the yellow-fever mosquito *Aedes aegypti*. *Chem. Senses* 33:79—93.

[29]Kiely, A. , A. Authier, A. V. Kralicek, C. G. Warr, and R. D. Newcomb. 2007. Functional analysis of a *Drosophila melanogaster* olfactory receptor expressed in Sf9 cells. *J. Neurosci. Methods* 159:189—94.

[30]Krieger, J. , E. Grosse-Wilde, T. Gohl, et al. 2004. Genes encoding candidate pheromone receptors in a moth (*Heliothis virescens*). *Proc. Natl. Acad. Sci. U. S. A.* 101:11845—50.

[31]Krieger, J. , E. Grosse-Wilde, T. Gohl, and H. Breer. 2005. Candidate pheromone receptors of the silkmoth *Bombyx mori*. *Eur. J. Neurosci.* 21:2167—76.

[32]Krieger, J. , O. Klink, C. Mohl, K. Raming, and H. Breer. 2003. A candidate olfactory receptor subtype highly conserved across different insect orders. *J. Comp. Physiol. A Neuroethol. Sens. Neural Behav. Physiol.* 189:519—26.

[33]Krieger, J. , K. Raming, Y. M. Dewer, et al. 2002 A divergent gene family encoding candidate olfactory receptors of the moth *Heliothis virescens*. *Eur. J. Neurosci.* 16:619—28.

[34]Kurtovic, A. , A. Widmer, and B. J. Dickson. 2007. A single class of olfactory neurons mediates behavioural responses to a *Drosophila* sex pheromone. *Nature* 446:542—46.

[35]Kwon, J. Y. , A. Dahanukar, L. A. Weiss, and J. R. Carlson. 2007. The molecular basis of CO_2 reception in *Drosophila*. *Proc. Natl. Acad. Sci. U. S. A.* 104:3574—78.

[36]Larsson, M. C. , A. I. Domingos, W. D. Jones, et al. 2004. Or83b encodes a broadly expressed odorant receptor essential for *Drosophila* olfaction. *Neuron* 43:703—14.

[37]Leal, W. S. 2005. Pheromone reception. *Top. Curr. Chem.* 240:1—36.

[38]Lu, T. , Y. T. Qiu, G. Wang, et al. 2007. Odor coding in the maxillary palp of the malaria vector mosquito *Anopheles gambiae*. *Curr. Biol.* 17:1533—44.

[39]Lundin, C. , L. Käll, S. A. Kreher, et al. 2007. Membrane topology of the *Drosophila* OR83b o-

dorant receptor. *FEBS Lett*. 581:5601—04.

[40]McBride,C. S. 2007. Rapid evolution of smell and taste receptor genes during host specialization in *Drosophila sechellia*. *Proc. Natl. Acad. Sci. U. S. A.* 104:4996—5001.

[41]McBride,C. S. ,J. R. Arguello,and B. C. O'Meara. 2007. Five *Drosophila* genomes reveal non-neutral evolution and the signature of host specialization in the chemoreceptor superfamily. *Genetics* 177:1395—1416.

[42]Mita,K. ,M. Kasahara,S. Sasaki,et al. 2004. The genome sequence of silkworm, *Bombyx mori*. *DNA Res*. 11:27—35.

[43]Nakagawa,T. ,T. Sakurai,T. Nishioka,and K. Touhara. 2005. Insect sex-pheromone signals mediated by specific combinations of olfactory receptors. *Science* 307:1638—42.

[44]Robertson,H. M. ,and L. B. Kent. 2009. Evolution of the gene lineage encoding the carbon dioxide receptor in insects and other arthropods. *J. Insect Science*. In press.

[45]Robertson, H. M. ,and J. H. Thomas. 2006. The putative chemoreceptor families of *C. elegans* (January 06,2006). In *WormBook*, ed. The *C. elegans* Research Community,doi/10. 1895/wormbook. 1. 66. 1,http://www. wormbook. org.

[46]Robertson,H. M. ,and K. W. Wanner. 2006. The chemoreceptor superfamily in the honeybee *Apis mellifera*: Expansion of the odorant,but not gustatory,receptor family. *Genome Res*. 16:1395—1403.

[47]Robertson,H. M. ,C. G. Warr,and J. R. Carlson. 2003. Molecular evolution of the insect chemoreceptor superfamily in *Drosophila melanogaster*. *Proc. Natl. Acad. Sci. U. S. A.* 100: 14537—42.

[48]Rogers,M. E. ,M. Sun,M. R. Lerner,and R. G. Vogt. 1997. Snmp-1,a novel membrane protein of olfactory neurons of the silk moth *Antheraea polyphemus* with homology to the CD36 family of membrane proteins. *J. Biol. Chem.* 272:14792—99.

[49]Rutzler,M. ,and L. J. Zwiebel. 2005. Molecular biology of insect olfaction: Recent progress and conceptual models. *J. Comp. Physiol. A. Neuroethol. Sens. Neural. Behav. Physiol.* 191:777—90.

[50]Sakurai, T. , T. Nakagawa, H. Mitsuno, et al. 2004. Identification and functional characterization of a sex pheromone receptor in the silkmoth *Bombyx mori*. *Proc. Natl. Acad. Sci. U. S. A.* 101:16653—58.

[51]Sato,K. ,M. Pellegrino,T. Nakagawa,et al. 2008. Insect olfactory receptors are heteromeric ligand-gated ion channels. *Nature* 452:1002—06.

[52]Schneider, D. 1992. 100 years of pheromone research, an essay on Lepidoptera. *Naturwissenschaften* 79:241—50.

[53]Scott,K. 2005. Taste recognition: Food for thought. *Neuron* 48:455—64.

[54]Sea Urchin Genome Sequencing Consortium. 2006. The genome of the sea urchin *Strongylocentrotus purpuratus*. *Science* 314:941—52.

[55]Shi,P. ,and J. Zhang. 2006. Contrasting modes of evolution between vertebrate sweet/umami receptor genes and bitter receptor genes. *Mol. Biol. Evol.* 23:292—300.

[56]Slone,J. ,J. Daniels,and H. Amrein. 2007. Sugar receptors in *Drosophila*. *Curr. Biol.* 17:1809—16.

[57]Smith,D. P. 2007. Odor and pheromone detection in *Drosophila melanogaster*. *Pflugers Arch.* 454:749—58.

[58]Syed,Z. ,Y. Ishida,K. Taylor,D. A. Kimbrell,and W. S. Leal. 2006. Pheromone reception in fruit flies expressing a moth's odorant receptor. *Proc. Natl. Acad. Sci. U. S. A.* 103:16538—43.

[59]Thomas,J. H. ,and H. M. Robertson. 2008. The *Caenorhabditis* nematode chemoreceptor gene families. *BMC Biol*. 6:42.

[60]Thorne, N., and H. Amrein. 2008. Atypical expression of *Drosophila* gustatory receptor genes in sensory and central neurons. *J. Comp. Neurol*. 506:548—68.

[61]Thorne, N., S. Bray, and H. Amrein. 2005. Function and expression of the *Drosophila* Gr genes in the perception of sweet, bitter and pheromone compounds. *Chem. Senses* 30:i270—72.

[62]Tribolium Genome Sequencing Consortium. 2008. The genome of the developmental model beetle and pest *Tribolium castaneum*. *Nature* 452:949—55.

[63]Vosshall, L. B., H. Amrein, P. S. Morozov, A. Rzhetsky, and R. Axel. 1999. A spatial map of olfactory receptor expression in the *Drosophila antenna*. *Cell* 96:725—36.

[64]Vosshall, L. B., and R. F. Stocker. 2007. Molecular architecture of smell and taste in *Drosophila*. *Annu. Rev. Neurosci*. 30:505—33.

[65]Vosshall, L. B., A. M. Wong, and R. Axel. 2000. An olfactory sensory map in the fly brain. *Cell* 102:147—59.

[66]Wang, Z., A. Singhvi, P. Kong, and K. Scott. 2004. Taste representations in the *Drosophila* brain. *Cell* 117:981—91.

[67]Wanner, K. W., A. S. Nichols, K. K. Walden, et al. 2007. A honey bee odorant receptor for the queen substance 9-oxo-2-decenoic acid. *Proc. Natl. Acad. Sci. U. S. A.* 104:14383—88.

[68]Wanner, K. W., and H. M. Robertson. 2008. The gustatory receptor family in the silkworm moth *Bombyx mori* is characterized by a large expansion of a single lineage of putative bitter receptors. *Insect Mol. Biol*. 17:621—29.

[69]Wicher, D., R. Schäfer, R. Bauernfeind, et al. 2008. *Drosophila* odorant receptors are both ligand-gated and cyclic-nucleotide-activated cation channels. *Nature* 452:1007—11.

[70]Wistrand, M., L. Käll, and E. L. Sonnhammer. 2006. A general model of G protein-coupled receptor sequences and its application to detect remote homologs. *Protein Sci*. 15:509—21.

[71]Wu, S. V., N. Rozengurt, M. Yang, et al. 2002. Expression of bitter taste receptors of the T2R family in the gastrointestinal tract and enteroendocrine STC-1 cells. *Proc. Natl. Acad. Sci. U. S. A.* 99:2392—97.

[72]Xia, Q., Z. Zhou, C. Lu, et al. 2004. A draft sequence for the genome of the domesticated silkworm (*Bombyx mori*). *Science* 306:1937—40.

[73]Xu, P., R. Atkinson, D. N. Jones, and D. P. Smith. 2005. *Drosophila* OBP LUSH is required for activity of pheromone-sensitive neurons. *Neuron* 45:193—200.

第十章 鳞翅目昆虫两性交流：交配行为的遗传学、生化及分子生物学

Fred Gould, Astrid T. Groot, Gissella M. Vásquez, and Coby Schal

1. 前言

在夜间飞行的蛾类中，高度特化、具备远距离的信息素交流是实现成功交配及物种生殖隔离的必备条件。雌蛾以精确的比例释放出两种或两种以上的挥发性物质，来吸引同种的雄蛾进行交配(例如，Cardé 和 Haynes 2004)。尽管已知上千种的蛾类，它们都拥有特殊的信息素复合物(例如，Cork 和 Lobos 2003；Witzgall et al. 2004；El-Sayed 2008)，但是仍然不清楚在进化过程中如何产生如此多样的两性交流信号。

在少数情况下，雌蛾会发生突变并影响到所产生的信息素复合物，这样就会降低它的交配成功率，除非相应的雄蛾无法分辨正常与改变后的信息素复合物(Butlin 和 Trickett 1997)。但是，在大多数已报道的研究中，正常雄蛾是能够分辨雌蛾产生的这种非正常的信息素复合物(例如，Zhu et al. 1997)的。同样的，如果雄蛾发生了对信息素反应的突变也会影响其寻找正常雌蛾的效率。这种低效的证据主要来自于对信息素有不同反应蛾子的基因型研究(例如，Linn et al. 1997)。虽然这是一个低概率事件，但针对新生雄雌蛾的交配性状进行选择，将会得到抑制蛾子交配过程中的进化多样性的稳定选择(Butlin 和 Trickett 1997；Phelan 1997)。

关于解释在进化上形成新的交配交流信号及其反应的假说都是基于：信号及其反应的变更是由同一多效基因所控制的(Hoy，Hahn 和 Paul 1977)，或者是雄蛾对变更的信息素的喜好并不亚于正常的信息素。新近的研究结果都不支持这两种猜测(Butlin 和 Ritchie 1989；Butlin 和 Trickett 1997)。然而，蛾两性交流所用化学复合物具有令人惊讶的多样性，这就说明新的信号/反应系统的进化发生是没有受到阻碍的。

这就在生物进化学家之间产生了争论。一部分学者认为在稳定性选择的情况下，这种多样性的变化是不可能的(例如，Coyne，Barton 和 Turelli 1997)；而另一些学者认为尽管面临稳定性选择作用，随机事件(比如，基因漂移)有可能导致这种多样性的产生(Wade 和 Goodnight 1998)。但双方都认同这种多样性可能受一些因素影响：(1)参与最初趋异变化基因的数目；(2)各基因在最适相关表型关系中所受影响的强度大小；(3)等位基因相互作用影响最适相关表型(Coyne 和 Orr 1998；Wade 和 Goodnight 1998；Dieckmann 和 Doebeli 1999；Kondrashov 和 Kondrashov 1999；Whitlock 和 Phillips 2000)。

本章的主要目的就是综合理解涉及蛾交配信息及反应的数量遗传学、生化及分子生物学，这将有利于我们对这一系统趋异化的理解，同时也作为一个模型去研究它在稳定选择情况下产生的新性状。单独的分子及生化研究可以告诉我们有多少、是什么样的酶参与了信息素复合物的生物合成途径。同时也会告诉我们雄蛾是用什么特定的分子去感知这种复合物。数量遗传学研究可以告诉我们在物种内及物种间有多少基因影响这种信号及反应的多样性。然而，只有遗传学及分子生物学研究相互结合才有可能让我们知道是什么样的基因变化(例如：开放阅读框中的单核苷酸多态性，顺式或反式调控的变化)导致蛾交配系统的多样性。这些数据同时又为开放阅读框序列或调控序列中的改变是否在进化的生态适应及多样性中具有更为关键的作用的持续争论提供了信息(Carroll 2005；Hoekstra 和 Coyne 2007)。

在本章中，我们将回顾关于雌蛾信息素产生及雄蛾应答所涉及的遗传学、生物化学及分子生物学中已知和未知的知识。在此过程中，尽管我们的知识中存在很大不足，然而在蛾两性交流这方面已经积累了大量的知识，把这些信息综合起来，我们可以在理解进化过程方面迈出很大一步。在本章的最后部分，我们也指出了一些将来研究的潜在领域。

2. 数量遗传学研究

2.1 雌信息素

目前，关于雌信息素复合物的数量遗传学研究已在一些综述中被详细地描述(例如，Löfstedt 1990，1993；Linn 和 Roelofs 1995；Butlin 1995；Phelan 1997；Roelofs 和 Rooney 2003；Cardé 和 Haynes 2004)，所以我们只选择性地进行讲述。

早期关于两性交流系统的遗传研究集中关注是否存在同一个基因通过多效作用(有时也称为遗传偶联作用)既控制信号产生，也同样控制异性信号的检测。尽管早期实验研究发现一些证据支持在听觉两性交流中遗传偶联的可能作用(Hoy，Hahn 和 Paul 1977)，随后的研究表明这种偶联作用在听觉及化学两性交流中都是很少见的(参见 Butlin 和 Ritchie 1989)。一项研究发现在野外捕获的昆虫的后代中，在雌雄信号/反应特征间存在一种遗传相关性，但是这种相关性在实验室条件下随机交配过程中丧失(Gray 和 Cade 1999)。表明这种相关性是由配子生殖平衡失调，而不是由于基因多效性(或强效连锁基因)引起的。正是由于缺乏遗传偶联的证据，最近的研究多关注于理解多样性信号产生及其信号响应的遗传结构，使得昆虫雌雄两性信号产生协同进化关系(参见 Butlin 和 Trickett 1997；Phelan 1997)。

在两种玉米螟中，两性交流遗传结构差异研究几乎代表了所有昆虫的相关研究。两种玉米螟雌蛾产生的两种醋酸信息素成分(*E*11-14：OAc 和 *Z*11-14：OAc)的比例差别很大(E 品系中，E 型醋酸盐：Z 型醋酸盐比例为 97：3，Z 品系中比例则为 1：99)，并且相应的雄蛾也更喜欢同一种的雌蛾。尽管还不能分析出一些紧密连锁的多基因，但种族之间这种信息素复合物的不同似乎是受常染色体单基因控制(Klun 1975)的。其他一些修饰基因对这种比例产生的影响较小 (Löfstedt et al. 1989)。控制信息素复合物的主效基因并不与控制雄蛾信息素反应行为的遗传区域相连锁(例如，Linn et al. 1999)。草螟蛾(*Ostrinia scapulalis*)在日本有 Z 种及 E 种，最近关于它的研究表明两种间的不同也可发生类似于单基因控制的分离 (Takanashi et al. 2005)。这种草螟蛾系统存在一个很有趣的现象，它是少数雄蛾不能区别来源于不同的两种雌蛾的信息素复合物的昆虫之一(Takanashi et al. 2005)。

在其他鳞翅目昆虫中由杂交产生 F_2 代及回交后代的研究中，也证明单基因控制特定的信息素成分比例。在大多数情况下，如玉米螟，不同种间的差别仅仅是信息素成分同分异构体的不同，并且这些异构体都来源于同一个前体。值得注意的一个例外是来源于实验室的粉纹夜蛾突变系(Haynes 和 Hunt 1990；Jurenka et al. 1994；Zhu et al.

1997)，其单基因控制比例变化与信息素成分无关。在这种情况下，一个截链酶的遗传变化被认为可能是引起成分比例变化的原因，在另一情况下，正常的雄蛾很少受突变雌蛾信息素复合物的吸引(Zhu et al. 1997)。此时，一种信息素成分的增加必然会相应地减少其他的一种成分。

尽管文献中一直强调这种单基因控制的例子，但在不同种属及物种间也存在一些多基因参与的两性交流作用(Cardé 和 Haynes 2004)。多数的研究支持多基因控制，包括简单的分离研究不能为单基因控制提供证据，或者发现单基因假说并不足以解释所观察到的遗传多样性(例如，Teal 和 Tumlinson 1997)。遗憾的是，这些定性的发现说服力不强。

为了更好地理解多基因控制的性状，我们实验室利用了数量性状定位分析(QTL; Remington et al. 1999)这一遗传学方法去分析在烟芽夜蛾(*Heliothis virescens*)和烟青虫(*H. subflexa*)中与决定信息素复合物相关位点的数目(Sheck et al. 2006)。这两个种在信息素成分的比例及有/无上至少存在十个成分的不同，因此它们有一个很好的研究系统。在我们早期的研究中，将这两个种进行杂交并且与烟青虫(*H. subflexa*)回交。对回交后代雌蛾的信息素腺体进行分析，并且每种蛾的 DNA 被用作扩增片段长度多态性(AFLP)分析(Sheck et al. 2006)，从而构建 31 条染色体的遗传图谱(Gahan，Gould 和 Heckel 2001)。不存在重组现象的 F_1 代雌蛾用于回交(Heckel 1993)，所有来源于烟芽夜蛾(*Heliothis virescens*)的染色体都是一个完整的、单独的连锁群(LG)。这种杂交设计可以使我们找到来源于非回归亲本烟芽夜蛾(*Heliothis virescens*)特定的染色体与回交后代雌蛾个体信息素腺体中复合物比例之间的关联。这些杂交的结果见图 10.1。

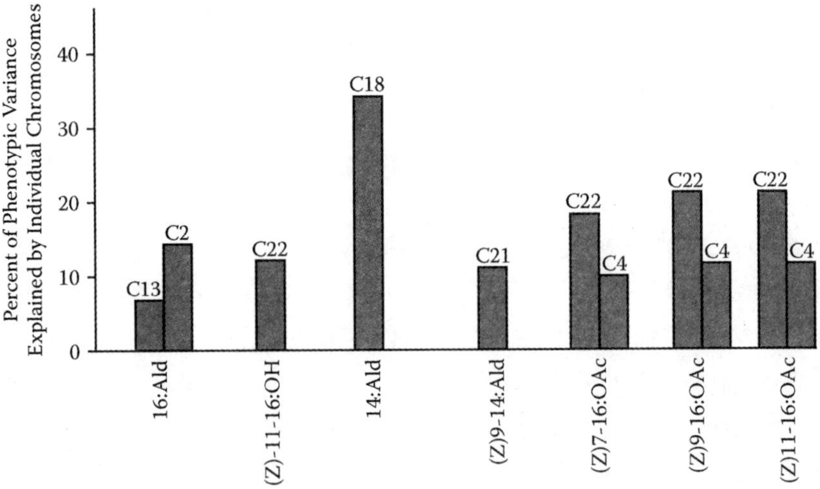

图 10.1 烟芽夜蛾和烟青虫信息素成分的 QTL 分析结果。烟芽夜蛾雌性回交世代中信息素成分比例的多样性，可以解释为存在或丢失烟芽夜蛾的一些特定的染色体(包括 30 条常染色体和 1 条性染色体)。染色体用 C-进行标示。回交世代由 F_1 代(烟芽夜蛾交配烟青虫)雌性与烟青虫雄性交配产生。来源于 Sheck 等(2006)。

这一研究证实，至少有 5 条染色体参与决定这两个种信息素腺体中复合物成分的不同，这也就暗示至少存在 5 个不同的位点影响这两个物种复合物的不同(Sheck et al.

2006)。有两例观察到一个特定的染色体影响一种单一成分的相对含量,而其他染色体却影响 2~4 个成分的相对含量;有四例观察到一种单一的成分受到不止一条染色体的影响。而且,同时含有烟芽夜蛾(*Heliothis virescens*)第 4 及 22 号染色体的雌蛾,含有 3 种仅发现于烟青虫(*H. subflexa*)中的醋酸酯,而且含量最低。这一控制所有的 3 个醋酸化合物的遗传偶联表明相同的代谢过程会影响产物的形成。

为了更好地理解一些表型变化较小而可能会影响进化进程的基因,我们运用烟芽夜蛾和烟青虫的回交系进行数量性状定位分析。尽管增加杂交的代数以及提高回交后代雌蛾的样本数能得到更多可能影响信息素复合物的 QTL,但 QTL 研究仅仅是理解进化压力和导致信息素复合物多样性的遗传途径的第一步工作。最终的目标,都必须从 QTL 分析或其他数量遗传学分析方法转移到分子水平研究,去决定蛾交配系统的多样性所涉及基因和突变的类型。

2.2 雄蛾应答

与所获得的信息素复合物遗传学知识相比,我们关于雄蛾对信息素反应的遗传学了解相对较少。关于这一领域最详尽和吸引人的研究是 Cossé 等人在 1995 年发表的结果,他们发现在分别受到 2 个信息素成分的刺激后,两个不同的信息素种属的欧洲玉米螟(*O. nubilalis*),会在嗅觉受体神经元(ORNs)中传出不同的信号,但是,对信息素做出的不同反应的嗅觉受体神经元被定位在与雄蛾实际行为反应完全不同的基因组上。这一结果强调单方面研究遗传学或受体可能会导致一个错误性的结论。在关于玉米螟物种雄蛾的遗传学以及该种属对信息素反应研究之外,我们仅能找到另外一对蛾类的遗传学实验。在斜纹卷蛾(*Ctenopseustis obliquana*)和 *C. herana* 之间的杂交系中,雄蛾感知信息素复合物几乎是性连锁的(Hansson,Löfstedt 和 Foster 1989)。在不同物种雄蛾应答的遗传学研究之后,将有助于我们更清晰地判断这一模式是否具有共同性。

3. 信息素复合物分子及生化研究

3.1 信息素合成的生化分析

多数蛾类性信息素都是为偶数 C_{10} - C_{18} 直链的不饱和的脂肪酸衍生物,在羧基碳修饰后形成含氧的功能类群(醇、醛或醋酸酯)。自由饱和脂肪酸再生后,在掺合到甘油酯或转为信息素之前,转换成其相应的酰基辅酶 A 硫酯(Foster 2005)。信息素前体酸大多数以三酰基甘油形式存储,少部分以甘油酯或磷脂形式存在(Foster 2005)。在信息素生物合成高峰期,三酰基甘油会水解释放出存储的脂肪酸,由它转化为信息素(Foster 2005)。鳞翅目信息素腺体中最普通的脂肪酸就是硬脂酸(18∶CoA),棕榈酸(16∶CoA)和肉豆蔻酸(14∶CoA; Jurenka 2003)。这些酸经脂肪酸还原酶(Morse 和 Meighen 1987)还原成醇(OH)。醇可以在醇氧化酶的作用下转化为醛,在酰基转移酶的作用下转化为醋酸酯(OAc)。相反的,醛也可以在醛还原酶的作用下转化为醇,而醋酸酯也可以在醋酸脂酶的作用下转变为醇(Tumlinson 和 Teal 1987; Roelofs 和 Wolf 1988)。最通用的鳞翅目性信息素生物合成途径见图 10.2。

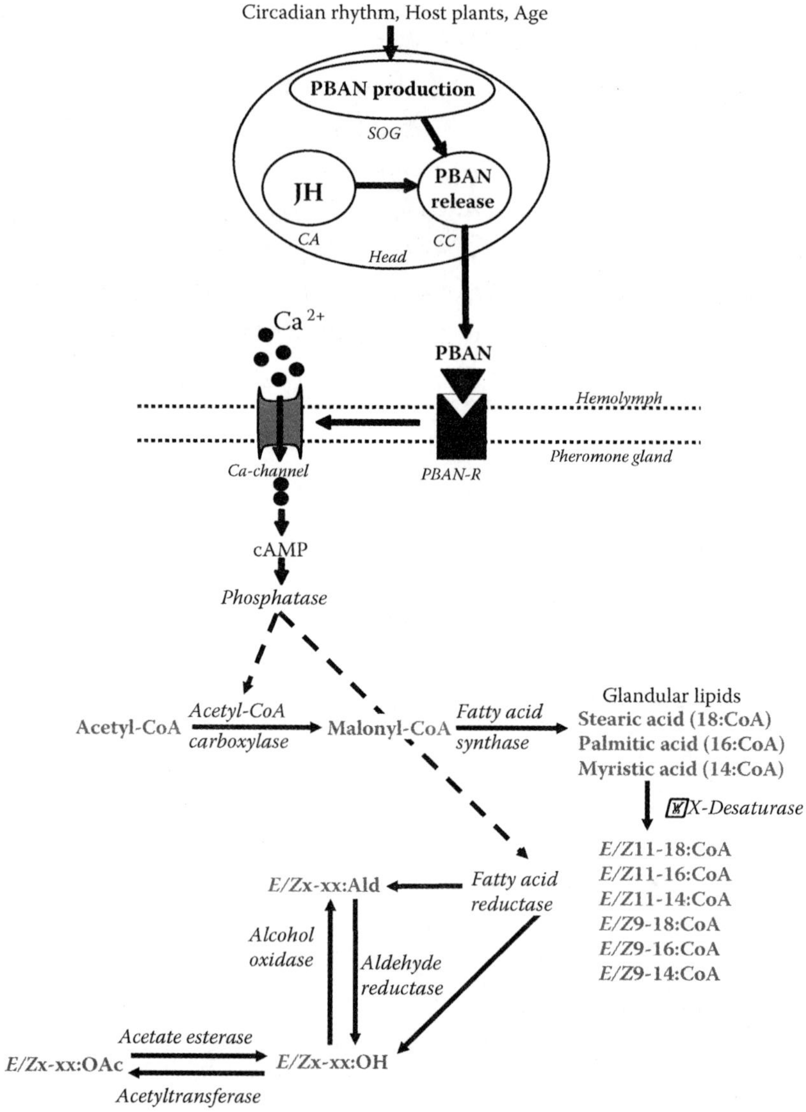

图 10.2 大多数鳞翅目昆虫性信息素生物合成途径涉及酶的示意图。所有可能的酶都列出，尽管家蚕中不存在 cAMP 的产物（Hull et al. 2007）。Δx-去饱和酶可能为 Δ5、Δ9、Δ10、Δ11、Δ12、Δ13 或 Δ14。PBAN：信息素生物合成活化酶；SOG：suboesophagial 神经节；JH：保幼激素；CA：咽侧体；CC：心侧体；OH：醇；Ald：醛；OAc：醋酸酯。来源于 Jurenka（2003），Rafaeli（2005），Ohnishi，Hull 和 Matsumoto（2006）。

3.2 信息素生物合成中所涉及已知和未知的酶及基因

与信息素生物合成途径的大量生化研究相比，对此途径中酶相应的编码基因的分子研究较少。其中去饱和酶是一个主要的例外，多数研究在不同种蛾子中都鉴定了去饱和酶基因（例如，Knipple et al. 1998，2002；Tsfadia et al. 2008）。仅有另外一个酶在后续的研究中得到鉴定，就是家蚕的脂肪酰基还原酶（Moto et al. 2003）。在下文中，我们将介绍参与蛾信息素生物合成的酶及酶学反应。

3.2.1 酰基辅酶 A 羧化酶

酰基辅酶 A(ACCase)可以催化依赖 ATP 的羧化作用,使酰基辅酶 A 转化为丙二酰基辅酶 A,此过程是长链脂肪酸生物合成的限速步骤(Pape,Lopez-Casillas 和 Kim 1988)。在棉铃虫及谷螟雌蛾中,ACCase 受抑制后,性信息素的生物合成也会受到抑制,表明在这些蛾子内,ACCase 是信息素生物合成途径中的一个关键酶(Eliyahu,Applebaum 和 Rafaeli 2003; Tsfadia et al. 2008)。在一些蛾子中,对性信息素成分进行同位素标记掺和实验表明,ACCase 的活化受到信息素生物合成激活神经肽(PBAN)影响(例如,Jurenka,Jacquin 和 Roelofs 1991),但至今未证明该酶是否响应 PBAN 处理而上调表达(Rafaeli 2005)。

3.2.2 脂肪酸合成酶

动物的脂肪酸合成酶(FAS)是已知最大的具有多种功能的蛋白,含有很多的催化结构域(Wakil,Stoops 和 Joshi 1983)。在昆虫中,体壁微粒体中的延伸反应已经在家蝇(Gu et al. 1997),德国土鳖虫(Juárez 2004)以及带原昆虫錐蝽(triatomine bugs)(Juárez 和 Fernández 2007)中得到研究,最初的产物包括长链甲基脂肪酸、醇及碳氢化合物。多功能的 FAS 利用丙二酰基辅酶 A、乙酰基辅酶 A 及 NADPH 合成增加两个碳单位的饱和脂肪酸;甲基丙二酰基辅酶 A 被用于在脂肪族链上插入甲基分支。在鳞翅目系统中,通过醋酸标记研究表明,FAS 的最终产物通常是棕榈酸(16∶0)和硬脂酸(18∶0; e.g.,Jurenka,Jacquin 和 Roelofs 1991)。FASs 已在多个昆虫物种中得到测序(例如,埃及伊蚊,登录号 XM_001658958 和 XM_001654917;黑腹果蝇,登录号 NM_134904)。参与蛾性信息素生物合成的 FAS 酶还未得到确定。

3.2.3 截链酶

改变底物特异性的截链酶会导致信息素复合物的多样性,这已经在松线小卷蛾(*Zeiraphera diniana*)(Baltensweiler 和 Priesner 1988)、红带卷蛾(*Argyrotaenia velutinana*)(Roelofs 和 Jurenka 1996)和黄地老虎(*Agrotis segetum*)(Wu et al. 1998)个体中得到证实。值得注意的是,一个源于粉纹夜蛾突变系的截链酶在体外酶实验中,被发现会引起 *Z*9-14∶OAc 量的增加,而 *Z*9-14∶OAc 在正常粉纹夜蛾雌蛾中是一个微量的信息素复合物成分(Haynes 和 Hunt 1990)。Jurenka 等于 1994 年证实,正常雌蛾信息素腺体中会经过两步截链使得 *Z*11-16∶CoA 变成 *Z*7-12∶CoA,而在突变雌蛾信息素腺体中 *Z*11-16∶CoA 经过一步截链形成 *Z*9-14∶CoA。在昆虫中没有对截链酶进行鉴定或测序,但是它们被认为是脊椎动物过氧化物酶的类似物(Bjostad 和 Roelofs 1983)。

3.2.4 去饱和酶

完整的膜去饱和酶在真核细胞中普遍存在,在调节脂膜物理特性以应对冷处理方面,起着主要的内环境稳定调节作用(Tiku et al. 1996)。在雌蛾信息素生物合成中,去饱和酶会在饱和脂肪酸链中引入一个双键或在单不饱和脂肪酸中引入另一双键。蛾信息素去饱和酶,包括 Δ5、Δ9、Δ10、Δ11、Δ12、Δ13 和 Δ14,具有不同局部和立体的特异性。一些去饱和酶已经被测序及鉴定,并在缺失内源性去饱和酶的酵母细胞中进行表达,用于阐明它们在性信息素生物合成途径中的特殊功能(Knipple et al. 1998; Matoušková, Pichová 和 Svatoš 2007)。

Δ9-Acyl-CoA 去饱和酶普遍存在于动物及真菌组织中(Liu et al. 1999),表明这些去饱和酶是古老的,在不同有机体中起着广泛的作用。这就可以解释为什么 Δ9 去饱和酶序列在不同动物间高度保守的特征(Rodriguez et al. 1992)。在蛾信息素腺体中,已经鉴定了两类 Δ9 去饱和酶:一类底物偏好性为 $C_{16}>C_{18}$;另一类底物偏好性为 $C_{18}>C_{16}$(Rosenfield et al. 2001)。因此,这似乎说明鳞翅目中完整膜去饱和酶基因家族的进化使其不仅具有通常细胞脂代谢的功能,也具有信息素生物合成的功能(Knipple et al. 2002)。

系统发生相关的 Δ11 类去饱和酶具有催化形成 Δ11 脂肪酰基信息素前体的功能,并且它在鳞翅目性信息素腺体中特异表达(Knipple et al. 1998)。这一类群的去饱和酶在不同物种之间存在许多氨基酸位点的高度变异性(Knipple et al. 2002)。其他三种也经常在性信息素腺体中发现的去饱和酶并没有得到功能验证(Knipple et al. 2002)。

一些去饱和酶基因在信息素腺体细胞中存在转录,但不翻译成蛋白(参看 Roelofs 和 Rooney 2003; Xue et al. 2007)。比如,在欧洲玉米螟中,发现了 3 个 Δ14 基因序列和 10 个 Δ11 去饱和酶基因,但是只有其中一个 Δ11 去饱和酶转录本在该物种中具有功能,它会利用 *Z*11= 和 *E*11-14:OAc 信息素成分。亚洲玉米螟(*Ostrinia furnacalis*)利用 *Z*12-和 *E*12-14:OAc 信息素成分,相应的它具有 2 个 Δ14 去饱和酶基因和 5 个 Δ11 基因(Xue et al. 2007)。然而,在亚洲玉米螟的信息素腺体中只发现一个 Δ14 去饱和酶基因的蛋白产物 (Roelofs 和 Rooney 2003)。

3.2.5 脂肪酸还原酶

蛾子中存在两种醛信息素生物合成途径。脂肪酰基辅酶 A 信息素前体在脂肪酸还原酶(FARs)的作用下,被还原为相对应的醇,然后在醇氧化酶的作用下氧化成相对应的醛(例如,Rafaeli 2005)。在另一些情况下,脂肪酰基辅酶 A 在特定的 FAR 作用下可以直接形成醛。在欧洲玉米螟(*O. nubilalis*)的两种种属间,它们信息素复合物的不同似乎源于它们各自脂肪酰基还原酶的不同(Zhu et al. 1996):在 *Z* 品系中的 FAR 对 *Z*11-14:Acyl 有较强的选择性,而在 E 品系中的 FAR 则对 *E*11-14:Acyl 选择性更强。遗憾的是,这些实实在在的酶并没有得到鉴定或分离。

在对家蚕信息素腺体组织匀浆 FAR 活性检测中发现,家蚕 FAR 在将十六酰基辅酶 A 还原为十六醇的过程中,并不释放出醛的中间产物(Ozawa 和 Matsumoto 1996)。随后,Moto 等人于 2003 年在家蚕中鉴定了一个由醇产生的 FAR。该 FAR 序列与植物的 FAR 具有同源性(jojoba),该植物的 FAR 可以将蜡脂肪酸转化为相对应的脂肪醇(Metz et al. 2000)。Ohnishi,Hull 和 Matsumoto 于 2006 年采用 dsRNA 注射蛹的方法去沉默家蚕信息素腺体的 FAR。FAR 的表达沉默会使蚕醇(醇信息素)的产生降到基线水平,这证实了 FAR 在体内信息素产生中具有重要作用(参看 Matsumoto et al. 2007)。除此之外,没有其他的 FARs 在蛾子信息素腺体中得到鉴定。

3.2.6 醛还原酶

在云杉色卷蛾(*Choristoneura fumiferana*)腺体抽提物中能检测到醛还原酶的活性

(Morse 和 Meighen 1986)。要证明这些酶首先生成醛,然后再将其转化为醇是非常困难的,因为醛还原酶也存在催化脂肪醛还原为醇的过程,因此主要产物是醇而不是醛(例如,Fang,Teal 和 Tumlinson 1995)。相反的反应是由醇氧化酶催化的。这两种酶一般都统称为醇脱氢酶。

3.2.7 醇氧化酶

在许多蛾子中,脂肪醇是以信息素中间产物的形式存在于信息素腺体中的,醇氧化酶催化醇形成相应的醛信息素。Fang,Teal 和 Tumlinson 于 1995 年证明,烟草天蛾信息素腺体表皮中的氧化酶可以催化不同碳链长度的醇(C_{14}-C_{17})。2002 年,Hoskovec 等证明烟草天蛾腺体中的氧化酶也可以氧化一些初级的醇,比如芳香族、烯丙基、杂环成分的醇,尽管这些氧化酶更倾向于苄、饱和、烯丙基类型的醇(Luxová 和 Svatoš 2006)。所有这些底物特异性与酵母醇脱氢酶极为相似,但至今为止,还未能成功分离得到此酶。

3.2.8 乙酰基转移酶

这一类酶的功能是在信息素腺体中将脂肪醇转化为醋酸酯;并且该酶已经在云杉色卷蛾(*C. fumiferana*)(Morse 和 Meighen 1987)和红带卷蛾(*A. velutinana*)(Jurenka 和 Roelofs 1989)进行了生化鉴定。在这两个物种中,乙酰基转移酶也仅在信息素腺体中被发现。体外底物偏好性检测实验表明:在红带卷蛾和其他卷叶蛾科中,偏向于 Z 型同分异构体;而在粉纹夜蛾(夜蛾科)或欧洲玉米螟(Pyralidae; Jurenka 和 Roelofs 1989)就存在不一样的情况。值得注意的是,尽管醋酸酯在蛾信息素成分中很普遍,但是目前还未克隆得到乙酰基转移酶基因。

3.2.9 醋酸脂酶

脂类物质的水解作用在信息素合成及降解过程中均存在(Ding 和 Prestwich 1986; Prestwich,Vogt 和 Riddiford 1986)。醋酸脂酶在信息素腺体的活性已经在云杉色卷蛾(*C. fumiferana*)(Morse 和 Meighen 1987),*Hydraecia micacea*,烟芽夜蛾(*Heliothis virescens*)和烟青虫(*H. subflexa*)(Teal 和 Tumlinson 1987)得到检测。在烟青虫中,醋酸酯是信息素复合物的一个成分(例如,Groot et al. 2007);但在烟芽夜蛾腺体中并没有发现醋酸盐,并且在其中存在另一种强烈拮抗作用的复合物(例如,Groot et al. 2006)。1987 年,Teal 和 Tumlinson 认为在烟芽夜蛾腺体中,醋酸酯在合成后就会被快速的转化为醇。

4. 信息素受体反应的分子及生化分析

引导雄蛾飞向雌蛾是通过雌蛾间歇性释放微量雌性信息素完成的(Roelofs 和 Cardé 1977)。雄蛾通过其毛状感觉器来感受这些化学信号(Kaissling 和 Priesner 1970)。这种特化的触角表皮毛发结构通常包括 1 到 3 个特化的,能够精确区别不同的信息素成分的嗅觉受体神经元(例如,Baker et al. 2004)。信息素分子通过表皮小管孔进入包被有信息素结合蛋白(信息素结合蛋白)的毛状感觉器内腔中,这些信息素结合蛋白可以将疏水分子经过感觉器血淋巴,转运到嗅觉受体神经元树突中(见综述 Leal 2005; Rützler 和 Zwiebel 2005; Vogt 2005)。通过与树突膜上负电荷位点相互作用后,信息素结合蛋白释放出信息素,使它能够与定位在嗅觉受体神经元树突表面的信息素受体蛋白(信息素受体蛋白)相互结合(Leal 2005; Rützler 和 Zwiebel 2005)。信息素分子与其受体偶联后会引起局部去极化,扩展至神经元电敏感区域并产生出神经冲动。这种电信号通过嗅觉受体神经元传入大脑,而响应信息素的嗅觉受体神经元在触角神经叶中聚集成宏小球复合体,参与后续的信息处理(Mustaparta 1996)。嗅觉受体神经元可能要通过信息素降解酶(信息素降解酶)将从信息素受体蛋白中释放出来的信息素分子降解后才能复原。这种信号的失活是提高信息素分辨能力所必需的(Vickers 2006)。

目前,不同物种及群体的雄蛾对信息素不同反应的遗传学了解很少,最近在分子生物学上的突破使得对信息素受体蛋白的氨基酸序列和生化特征以及参与信息素感知的信息素加工蛋白(例如,信息素结合蛋白、信息素降解酶和化学感应蛋白)有了更好的理解(Jurenka 2003; Knipple 和 Roelofs 2003; Leal 2005; Rützler 和 Zwiebel 2005; Vogt 2005; Gohl 和 Krieger 2006; Hallem, Dahanukar 和 Carlson 2006; Sato et al. 2008; Wicher et al. 2008)。参与信息素信号处理及传导机制的分子的功能在图 10.3 中被描述。

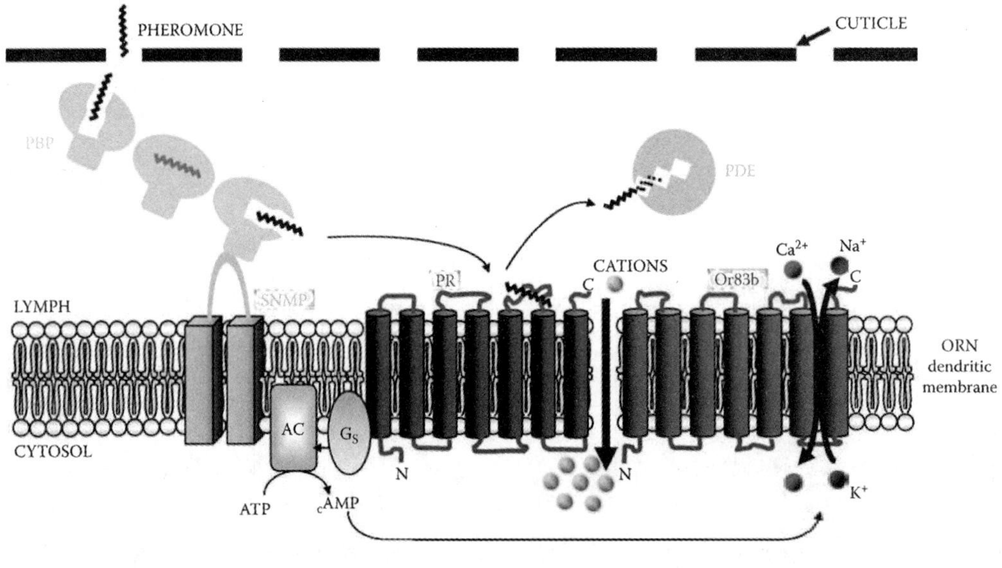

图 10.3 蛾类信息素受体反应的示意图(彩色版本图片见附录图 10.3)。一个信息素分子通过表皮孔进入毛状感觉器的内腔,并结合到信息素结合蛋白(PBP)上,由 PBP 将信息素转运到嗅觉受体神经元(ORN)的树状膜上。一个感觉神经元膜蛋白(SNMP)结合到 PBP-信息素复合体上,或直接与信息素结合,并由其将信息素运送到邻近的信息素受体上(PR)。信息素结合到 PR-Or83b 异源复合物后会导致通过一个离子型通道的快速识别,或经一个代谢 G 蛋白介导的、慢的、延长的、信号放大的检测途径。信息素降解酶(PDEs)会灭活一些未结合的信息素(修改自 Rützler 和 Zwiebel, 2005)。

4.1 信息素受体蛋白

信息素受体蛋白(PRPs)是昆虫趋异性嗅觉受体家族中的成员,含有 7 个跨膜结构域(Mombaerts 1999)。这些信息素受体蛋白与脊椎动物 G 蛋白偶联受体(GPCRs)不存在任何序列的相似性，它们表现出一个氨基酸末端定位在胞外的非典型膜拓扑学结构(Benton et al. 2006)。昆虫的嗅觉受体(ORs)通常有两个亚基形成异源复合物，一个常见的、多样性的嗅觉受体与一个高度保守的、遍在表达的 Or83b 复合受体相偶联(受体的系统发生关系见第九章)。家蚕信息素受体蛋白原位杂交及在非洲蟾蜍卵母细胞中的异源性表达实验表明：鳞翅目信息素受体蛋白会与 Or83b 同源的伴侣蛋白形成一个二聚体(Nakagawa et al. 2005)。然而，在其他一些原位杂交研究(Krieger et al. 2005)中却证实家蚕信息素受体蛋白不存在这种共表达的伴侣蛋白。更进一步的研究表明，在 Flp-In T-REx293/Gα15 细胞系(Große-Wilde，Svatoš 和 Krieger 2006)和果蝇 ab3A 神经元(Syed et al. 2006)中，信息素受体蛋白可受单独的信息素活化，而不需要表达伴侣蛋白(i. e.，Or83b ortholog)。这种相互矛盾的结果可以用不同的标记技术以及使用不同宿主细胞处理信息素受体蛋白来解释。最近外源表达昆虫嗅觉受体异源复合物(包括家蚕信息素受体蛋白－Or83b 同源复合物)的电生理学和可视荧光实验中，表明它们会形成一个非选择性的阳离子通道，直接引导气味或信息素结合到嗅觉受体上(Sato et al. 2008)。除了该离子型信号转导途径信息，一个参与环化核酸活化通道的代谢途径在 Or83b 复合受体中已经被证明(Wicher et al. 2008)。上述新近的研究表明，信息素受体蛋白在辅助蛋白存在下才具有化学信号传导功能。

嗅觉受体神经元通常只表达一个常规的嗅觉受体基因(Vosshall et al. 1999；Mombaerts 2004)，这个基因会决定嗅觉受体神经元的气味反应谱(Hallem，Ho 和 Carlson 2004)，并且信息素受体蛋白通常形成一个受体——一个嗅觉受体神经元的组织形式。然而，不像昆虫通常嗅觉受体那样可以结合一个以上的配体(例如，Hallem，Ho 和 Carlson 2004)，信息素受体蛋白只会精确地结合特异受体(例如，Große-Wilde et al. 2007)。

鳞翅目信息素受体蛋白与昆虫通常的嗅觉受体只具有很少的序列相似性，并且形成一个单一的谱系，这一谱系的蛋白具有很高的序列相似性，在家蚕及烟夜蛾嗅觉受体的系统发生分析中显示其具有保守的功能(Krieger et al. 2005；Nakagawa et al. 2005；Wanner et al. 2007)。这些分析显示信息素受体蛋白形成两个主要的谱系：一个在蚕蛾科中扩展，另一个在夜蛾科中扩展(图 10.4)。这种在谱系内高度的序列一致性表明：这些基因簇可能来源于一个古老信息素受体基因的复制事件。随着更多蛾子信息素受体蛋白的测序，这一模式将会变得更加复杂(具体细节详见第九章)。

J: Krieger 及其同事描述研究了烟芽夜蛾(*Heliothis virescens*)的信息素受体蛋白候选基因。基于同黑腹果蝇嗅觉受体序列的相似性，利用来源于烟夜蛾基因组数据而制备探针，烟夜蛾气味受体(HRs)第一次在触角 cDNA 文库中得到鉴定(Krieger et al. 2002)。后续用编码 HRs 基因和其他昆虫嗅觉受体短序列的探针去扫描触角 cDNA 文库，分别鉴定了 4 个信息素受体蛋白候选基因(HR13，HR14，HR15 和 HR16)，它们都特异地在雄触角下信息素反应感觉器中表达，并且至少有 40% 的氨基酸序列的相似性(Krieger et al. 2004)。HR13 被证明在 A 类型感觉神经元中表达(Gohl 和 Krieger 2006)。此外，免疫组化及外源表达系统中的功能分析清楚地表明：HR13 特异地与烟夜蛾信息素复合物的主要成分 *Z*11-16：Ald 相作用(Gohl 和 Krieger 2006；Große-Wilde et al. 2007)。

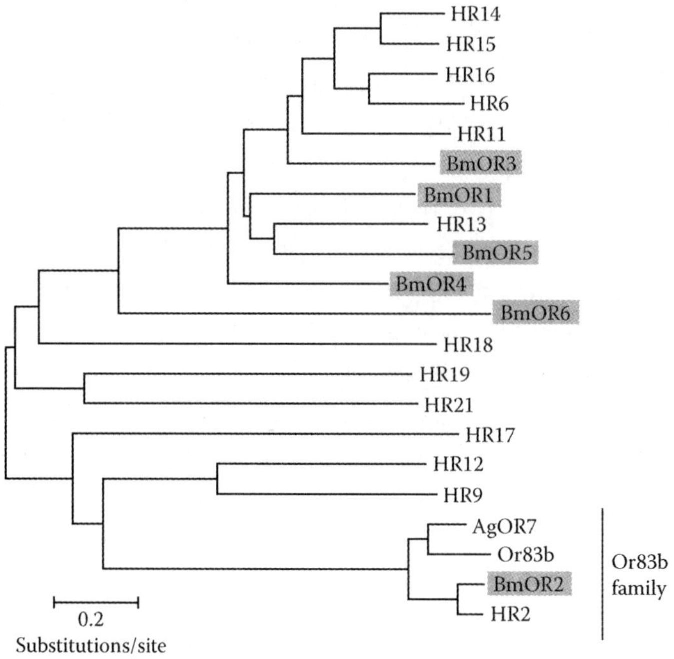

图 10.4 烟芽夜蛾(HR),家蚕(BmOR)和冈比亚按蚊(AgOR)的 ORs 系统发生树。来源于 Nakagawa 等(2005)。

差异筛选雄触角 cDNA 文库分离得到家蚕雄特异的嗅觉受体基因,而家蚕的第一个信息素受体蛋白(家蚕嗅觉受体-1),是基于与其他昆虫嗅觉受体序列相似性而鉴定得到的(Sakurai et al. 2004)。家蚕嗅觉受体-1 在位于毛型感觉器下面的细胞中特异表达,并与烟芽夜蛾的一些受体高度同源。在雌蛾触角以及蟾蜍卵母细胞中的异位表达该受体证明它特异地作用于蚕蛾的性信息素(蚕醇)(Sakurai et al. 2004)。后续在异源表达系统的实验更进一步证实了这一发现(Große-Wilde, Svatoš 和 Krieger 2006; Syed et al. 2006)。原位杂交和异源表达研究同样也证明了家蚕嗅觉受体-2 是蚕醛的受体,蚕醛是蚕醇的氧化形式,它不会引起雄蛾的取向行为(Nakagawa et al. 2005)。

异源表达系统包括蟾蜍卵母细胞(Sakurai et al. 2004),修饰后的 HEK 293 细胞(Große-Wilde, Svatoš 和 Krieger 2006),拥有空的 ab3A 神经元的黑腹果蝇的 *Δhalo* 突变体 (Dobritsa et al. 2003)和嗅觉受体 *Or67d-GAL4* 突变体(Kurtovic, Widmer 和 Dickson 2007),都成功应用于对蛾子候选信息素受体蛋白基因体内的功能分析。另外,这些系统将来也可用于野生型及突变型信息素受体蛋白的比较功能分析来研究信息素受体基因序列的特异变化是否会影响配体的特异性。

4.2 信息素结合蛋白

信息素结合蛋白属于 encapsulin 家族,这些蛋白在水环境中具有助溶疏水化合物的作用(Vogt 2005)。信息素结合蛋白是一个 α 螺旋蛋白:存在一个疏水结构域、一个信号肽以及由 6 个高度保守的半胱氨酸残基形成的 3 对二硫键(例如,Sandler et al. 2000)。

与 OBP 基因家族其他成员不同的是:信息素结合蛋白在长的毛型感觉器中特异或高量表达(例如,Laue 和 Steinbrecht 1997),它由辅助细胞产生,在内腔中具有很高的溶度(Steinbrecht,Ozaki 和 Ziegelberger 1992)。信息素结合蛋白结合、包埋以及运输信息素到嗅觉受体神经元树突膜上的信息素受体蛋白外环上,并且保护其不受信息素降解酶的降解(Krieger 和 Breer 1999; Leal 2005)。与树突膜上负电荷化位点的接触导致另一个 C 末端 α 螺旋的形成,将信息素从信息素结合蛋白中释放出来(Leal 2005 和参考文献)。

信息素结合蛋白,首先在多音天蚕(*Antheraea polyphemus*)中得到鉴定(Vogt 和 Riddiford 1981),后来在许多蛾子中鉴定得到。这些已鉴定的多样信息素结合蛋白亚型显示一定的差异性(32%~39%的氨基酸一致性,例如,Abraham,Löfstedt 和 Picimbon 2005)。系统发生分析表明:多次的基因重复事件导致这些特殊亚型的出现(例如,Robertson et al. 1999; Xiu,Zhou 和 Dong 2008)。鳞翅目信息素结合蛋白可以主要分为 3 群,每一群都包含来源于多个物种的信息素结合蛋白,而夜蛾科信息素结合蛋白形成单独的 3 群(图 10.5),这种现象可能源于两次基因重复事件(Xiu 和 Dong 2007)。

信息素结合蛋白基因的克隆工作先于信息素受体基因的研究;Krieger 等在 1993 年就克隆了一个烟芽夜蛾的信息素结合蛋白。用修饰后的 HEK293 细胞系对烟芽夜蛾信息素结合蛋白(HvirPBP2)进行功能分析表明:HvirPBP2 可以增加 HR13 对 *Z*11-16:Ald 的敏感性和特异性 (Große-Wilde et al. 2007)。然而,异源表达其他两个信息素受体蛋白,HR14 和 HR16,在 HvirPBP1 或 HvirPBP2 的存在下并不会增加。有趣的是,在 HEK293 细胞系异源表达家蚕嗅觉受体 1 显示家蚕信息素结合蛋白的增加对蚕醇的特异性(Große-Wilde,Svatoš 和 Krieger 2006),而家蚕嗅觉受体 1 在蟾蜍卵母细胞和 Δ*halo* 突变体的表达表明:家蚕嗅觉受体信息素结合蛋白并不是对蚕醇反应所必需的(例如,Syed et al. 2006)。后续在烟草天蛾中培养的嗅觉受体神经元研究与先前的研究结果相一致:表明信息素结合蛋白并不是信息素受体蛋白对信息素产生反应所必需的(Stengl et al. 1992)。因此,信息素可单独地(而不是形成信息素结合蛋白-信息素复合体)去活化信息素受体蛋白。然而,信息素结合蛋白在信息素感知动力学及敏感度方面有作用:(1)信息素结合蛋白 PH 依赖性的构象变化与外周信息素在毫秒时间范围内的感知相一致,这对雄蛾的定向导航很必要(Leal et al. 2005);(2)信息素结合蛋白的存在有利于信息素渗透进入感觉器血淋巴,并且会选择性地进行转运(Leal 2005; Syed et al. 2006) ;(3)信息素结合蛋白可以屏蔽一些类群的气味分子,并提高信息素在感觉器血淋巴中的浓度(Pelosi 1996)。通过增加信息素分子的摄入,信息素结合蛋白可以降低信息素反应所需的阈值(van den Berg 和 Zielgelberger 1991)。信息素结合蛋白的配体范围可以从很特异到很宽泛(例如,Rivière et al. 2003),然而,只有那些具有特异性结合的信息素结合蛋白才可能参与分辨信息素成分(Bette,Breer 和 Krieger 2002; Maida,Ziegelberger 和 Kaissling 2003)。信息素结合蛋白为气味系统的敏感性,在一定程度上为其特异性所需。但是,这种特异性的机理仍旧需要进一步验证。

4.3 信息素降解酶

信息素降解酶(PDEs)被认为对一些没有结合到信息素结合蛋白上的信息素,在其

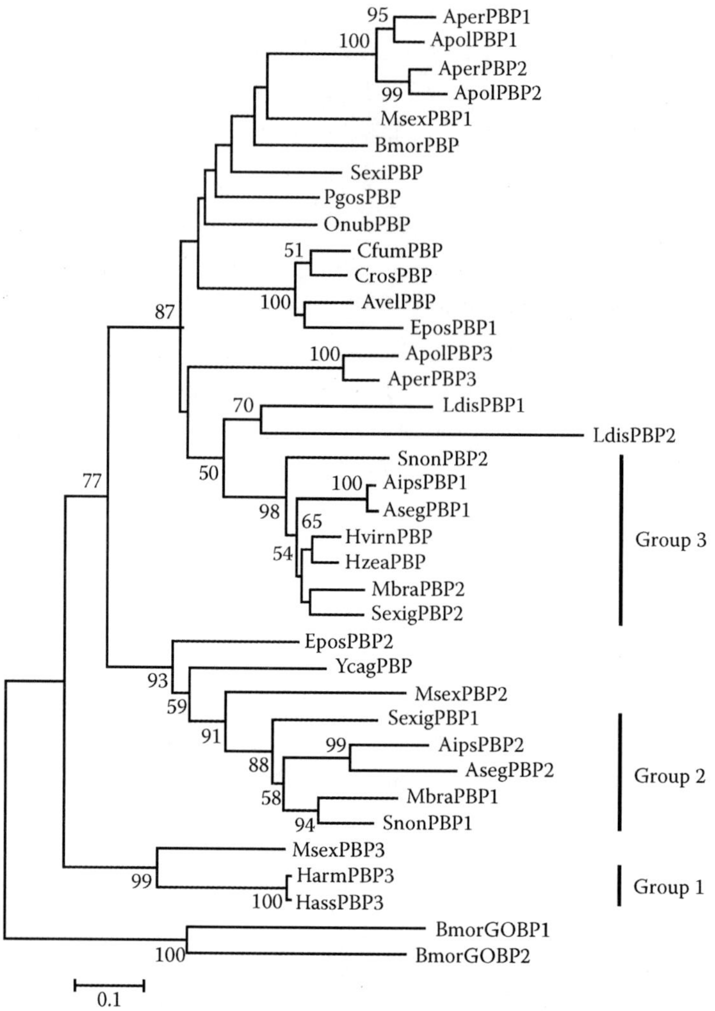

图 10.5　PBP 氨基酸序列的系统发生树。气味结合蛋白(GOBP)序列作为外群。小地老虎（Aips），黄地老虎(Aseg)，柞蚕(Aper)，多音天蚕(Apol)，红带卷蛾（Avel)，家蚕(Bmor)，云杉卷叶蛾(Cfum)，蔷薇斜条卷叶蛾(Cros)，属浅棕苹果蛾(Epos)，棉铃虫(Harm)，烟夜蛾(Hass)，烟芽夜蛾(Hvir)，*H. zea*(Hzea)，舞毒蛾(Ldis)，甘蓝夜蛾(Mbra)，烟草天蛾(Msex)，玉米螟(Onub)，红铃虫(Pgos)，甜菜夜蛾(Sexig)，粉茎螟(Snon)，巢蛾科昆虫(Ycag)。来源于 Xiu 和 Dong(2007)。

到达信息素受体蛋白之前对其具有灭活作用；以及会降解已经激发信息素受体蛋白的那部分信息素分子。信息素降解酶可以对信息素进行化学修饰(Rybczynski，Reagan 和 Lerner 1989)，并且在体外可以在毫秒时间内对信息素分子进行降解(Vogt，Riddiford 和 Prestwich 1985)，尽管在体内这一过程很慢(可能由于信息素受到信息素结合蛋白的保护作用)(Kaissling 2001)。虽然它们在信号灭活中具有重要的作用，但是对信息素降解酶的分子结构了解甚少，并且只有很少的基因得到鉴定(Vogt 2005)。

已经分别在美洲野桑蚕(*A. polyphemus*)（Ishida 和 Leal 2002)和甘蓝夜蛾(*Mamestra brassicae*)（Maïbèche-Coisne et al. 2002)中克隆到雄蛾触角特异的脂酶和细胞色素

P450 酶的 cDNA,然而,还不清楚它们是否具有信息素降解酶功能。一个先前的美洲野桑蚕感觉器脂酶(Vogt,Riddiford 和 Prestwich 1985),*ApolPDE*,已经被分离、克隆,并在杆状病毒载体中得到表达,还用于估计感觉器血淋巴中信息素的浓度(大约低于信息素结合蛋白 20000 倍)和信息素灭活的动力学研究(Ishida 和 Leal 2005)。因此,*ApolPDE* 的序列可以用于鉴定其他蛾子的信息素降解酶。

在家蚕中,醛氧化酶优先在雄蛾触角中表达,具有催化代谢蚕醛的作用(Rybczynski,Vogt 和 Lerner 1990)。最近,在甘蓝夜蛾中(Merlin et al. 2005)发现一个醛氧化基因序列在气味感觉器中特异表达,这也被用来鉴定在家蚕中潜在的醛氧化基因(Pelletier et al. 2007)。一个在家蚕触角中选择性表达的单基因,*BmAOX*2,可能就是编码一个参与蚕醛降解的信息素降解酶;然而还有待后续的功能鉴定。由于系统中缺乏信息素降解酶灭活蚕醇,推测 ab3A 感觉器中异源表达家蚕嗅觉受体 1(*BmorOR*1)具有延展性的神经活性 (Syed et al. 2006)。这一表达系统可以被用于检测 *BmAOX*2 在蚕醛降解中的作用。

4.4 信息素受体蛋白、信息素结合蛋白和信息素降解酶之间的相互作用

信息素结合蛋白、信息素降解酶、信息素受体蛋白,以及中枢神经系统在雄蛾响应信息素时很有可能在物种内行使特异的,而在物种间行使不同的功能。两者的重要性可能与化合物及分类学谱系的不同相关联(Leal 2005; Rützler 和 Zwiebel 2005; Vogt 2005; Hallem,Dahanukar 和 Carlson 2006)。两个"滤片层":信息素结合蛋白和信息素受体蛋白,被认为通过组合作用而参与雄蛾的特异性反应(Leal 2003,2005)。比如,异源表达的家蚕嗅觉受体 1 会对蚕醇和蚕醛产生反应;然而,家蚕信息素结合蛋白可能通过选择性转运蚕醇到其受体使得它的特异性得以增加(Große-Wilde,Svatoš 和 Krieger 2006)。此外,在存在家蚕信息素结合蛋白的 ab3A 空细胞中,可以观察到 *BmOR1* 的反应增高,这可能是通过信息素增溶作用来实现的(Syed et al. 2006)。异源共表达特化的信息素结合蛋白或 OBPs 以及 *BmOR1* 可最终用于家蚕信息素结合蛋白功能的确证。

正如先前的所讲述的一样,嗅觉受体 *Or83b* 同系物通过与信息素受体蛋白相互作用而在信息素分子识别中行使作用。*Or83b* 参与嗅觉受体神经元树突中嗅觉受体的定位以及异源二聚体的形成,而这是受体应答反应和信号转导所必需的(Rützler 和 Zwiebel 2005; Wicher et al. 2008; Sato et al. 2008) 。相类似的,其他一些分子,比如感觉神经元膜蛋白可能与配体相互作用(Vogt 2003; Benton,Vannice 和 Vosshall 2007),或作为一个信息素结合蛋白受体(Rogers et al. 1997; Rogers,Krieger 和 Vogt 2001; Jacquin-Joly 和 Merlin 2004)。

4.5 信号的处理

在雄蛾中,信息素的反应是由外周嗅觉途径的信息素刺激来激活大脑中的行为环路这一过程所主导的。信息素诱导的电生理信号由嗅觉受体神经元的树突传送到轴突中,并由轴突将此信号投射到增大的神经纤维球上(位于触角叶中的宏神经纤维球(MGC)

中)，信号在这里将会被处理并进一步投射到前脑神经元中(Vickers，Poole 和 Linn 2005)。目前所知的是:信息素复合物中的每一种信息素成分都代表一个单独的 MGC 神经纤维球，并且这种组合活化模式要通过许多神经纤维球来对应信息素复合物(Vickers 和 Christensen 2003)。此外，表达相同的信息素受体基因的嗅觉受体神经元被认为会聚集到一个神经纤维球中(例如，Datta et al. 2008) 。在哺乳动物中，嗅觉延髓中嗅觉受体神经元轴突的聚集是依赖受体的，而在昆虫嗅觉受体中却不同，其信息素受体蛋白并不参与引导轴突聚集到其同种神经纤维球中去(Dobritsa et al. 2003)。

在雄性黑腹果蝇中，性信息素 cis-vaccenyl 醋酸(cVA)受体的活化会抑制雄—雄求偶现象，而在雌性中它会促进对雄性的感知力。运用一项结合遗传学和光学神经示踪技术的方法，2008 年，Datta 等发现 cVA 活化一个单一的神经纤维球，在其突触后神经元(PNs)的支配下，在前脑侧角中表现出两性异形的投射模式。一个雄特异的转录因子，Fruitless(Fru^{M})在突触后神经元中的神经纤维球和其他表达 Fru^{M} 的细胞中，可以控制在侧角中雄特异轴突树的形成。行为二态性可能是缘于 3 级神经元从雄性 PNs 中接受到更强的输入或它们的突触在神经纤维球轴突树的雄特异区域受到限制而造成的。在蛾子的前脑中，可能发生相类似的两性异形的神经环路神经。进而，物种间的雄性对同一信息素成分的不同反应可以用其比较解剖学的二态性来解释，这就有可能导致在相近物种中对这些成分形成正向或反向的反应。因此，中枢神经系统拥有处理这些相反反应的能力。

5. 未来的研究方向

与动物其他生态特性相比，我们对信息素的产生及接受的分子及生化基础有相当多的了解，而对于多样的两性交流系统的遗传学基础的了解却很少，但这些基础知识正不断地得到累积。这一领域的每一个研究都有它吸引人的地方，但是将这 3 个领域的知识进行综合将更有助于我们去了解基于信息素的两性交流系统在进化上的多样化这一过程。少数研究者已经开始了这项工作，但是需要更多的人参与进来。下面，我们将概述这项研究的一些可验证的假说和研究方法，另外，也介绍这一领域开创性的研究工作。

5.1 “候选基因”与“基因组网络”假说

据我们所知，基于信息素系统在进化上的多样化必然涉及遗传上的变化，而这种变化会导致信息素产生途径相关酶氨基酸序列的变化，以及一些受体及参与信息素检测的蛋白序列的变化。而这可以称之为“候选基因假说”。而另一极端就是，认为导致多样化的进化上重要的遗传上变化可归因于一套复杂的顺式—反式遗传学因子的变化，而这些因子可以调控候选基因的表达以及对它们的表型进行修饰。而这可以称之为“基因组网络假说”。虽然都可以从不同的性状中找到支持这两种假说的证据，但是，一个重要的争论就是关于:在进化中，这两种机制中的哪一种具有更为重要的作用。然而，目前的数据仍然有限(Wittkopp，Haerum 和 Clark 2004，2008；Carroll 2005；Sambandan et al. 2006；Hoekstra 和 Coyne 2007)。因雄性感受信息素的候选基因已经被详细报道，以及信息素生物合成相关候选基因也指日可待，用蛾子两性交流系统来检测这些假说是可行的，并且这有可能会解决更普遍的争论。

5.2 信息素生成及其反应在进化选择上的分子研究

可用直接和间接的方法验证上面提及的这两种假说。其中的一个方法就是：将来源于不同属或种的蛾（具有信息素复合物的差异性）的所有候选基因构建成一个基因组DNA数据库。这些信息就可以用于检测一些同源基因在不同物种间的变化模式，并且可以去寻找一些关于稳定选择、定向选择、移码、基因重复和缺失的一些证据。这在几年前是一个很繁重的工作，但是近来序列分析技术上的突破使得这种研究更可行了。正如上述的讨论中所提到的那样，Knipple等在2002年用来源于鳞翅目亚家族的去饱和酶cDNA标码序列对这种模式进行评估，做出了重要突破。他们发现并且检测到复杂的6群去饱和酶的不同模式。总之，没有迹象表明存在定向选择，但是在一群去饱和酶中的一些氨基酸的置换可以引起该酶催化位点的变化。为了检验这种可能性，需要更加详尽的结构一功能关系的分析。将鳞翅目和黑腹果蝇的去饱和酶进行比较可以发现：在2.8亿万年前发生的基因重复至少产生了3个类群的去饱和酶(Knipple et al. 2002)。

2002年，Knipple等发现编码序列具有一种普遍保守的模式，这就暗示物种间在产生不同信息素的过程中，去饱和酶基因不同的表达可能比编码序列的变化具有更为重要的作用。D. C. Knipple和他的同事使用的方法也可用于研究信息素生物合成途径中还原酶及其他酶的研究。

关于雄性应答方面，不同物种编码信息素受体蛋白和OBPs的cDNA序列被用作进行简单系统发生相关性分析(图10.4和10.5)。2000年，Willett把研究更深入了一步，在卷叶蛾 *Choristoneura* 中发现了信息素结合蛋白定向选择的证据。令人好奇的是，种间信息素结合蛋白定向变化的幅度与信息素复合物的变化并不具有相关性。2000年，Willett提出如下假说：这种选择压力与信息素复合物无关，而是作用于蛋白序列。

我们认为这小部分研究指出一个未来可用的研究方向。至少在外周感知水平上，我们有许多参与雄蛾信息素感知的很好的候选基因，因此这些原始资料在更详尽的系统发生分析中会被利用。需要注意的是：系统发生分析可以提供很清楚的、有益的信息，但是通常存在不确定性。如果不像Willett那样细致分析的话，就有可能在选择因子上产生一个错误的结论。

W. L. Roelofs及其合作者和同事利用螟属(*Ostrinia*)昆虫结合全系统发生，基因转录以及翻译分析作了开拓性的研究工作。确定了8个螟种的信息素组成。所有物种中(除虎杖螟(*O. latipennis*)外，它只使用Z11-14：OH的主要成分都是不饱和醋酸(参看Roelofs和Rooney 2003；Xue et al. 2007)。如本章所述，一个令人惊奇的结果是：对两个种的详尽分析发现两者都有许多去饱和酶基因转录但不进行翻译。在具有较远关系的家蚕中，也发现并不是所有在信息素腺体细胞中转录的去饱和酶基因都会产生活性蛋白(Moto et al. 2004)。尽管目前已经清楚去饱和酶产物的不同参与 *Ostrinia* 信息素多样化，但许多差异归因于转录后的后加工过程，所以，简单的候选基因分析可能会丢失这些重要因素。

在进化及作用机理水平上，什么样类型的基因组DNA序列变化决定哪些去饱和酶mRNA转录本将翻译为功能蛋白变得尤为重要。在这点上，我们并不能回答在不同物种间参与信息素生物合成的酶存在多少遗传变化或遗传网路的不同。有可能是靠单个、小的、关键的候选基因序列或一个关键的顺式调控元件调控单个酶的活化和浓度来控制多样性；另一方面，非相关的和非连锁的基因也可能调控那些产生有活性酶产物的mRNA。

在雄蛾对信息素反应方面，W. L. Roelofs团队再次率先揭示了候选基因之内或之外的变化可能负责雄蛾对不同信息素复合物的不同反应。他们在玉米螟上早期的研究清楚地显示：当暴露在E11-14：OAc的E和Z异构体中，E品系及Z品系的雄蛾产生不同振幅的神经元反应峰；E品系暴露在*E*11-14：OAc中会有更高振幅的峰，而Z品系暴露在*Z*11-14：OAc中会有更高振幅的峰(Roelofs et al. 1987)。F_1杂种具有一个中间振幅的峰，而F_2代雄性峰形表明这种差异遗传在一个常染色体上(Roelofs et al. 1987)。最简单的假说就是：这种峰形振幅的不同至少代表不同的反应行为。F_2子代的风洞实验分析发现，遗传有编码E品系峰形变化的常染色体的雄性并不比具有编码Z品系峰形变化的同源染色体的雄性具有更高的E品系信息素复合物反应强度。进一步的遗传分析表明编码不同行为反应的基因是性连锁的(Roelofs et al. 1987)。我们仍旧不知道相应基因的序列，只是清楚它们不是由顺式基因来决定峰形变化的。

当W. L. Roelofs及其合作者进行这些早期的遗传工作时，他们可能满足于对一条常染色体和一条性染色体的分析，因为这样可以使得分析更为容易些。今天，AFLP和微卫星标记使得在玉米螟中定位这些控制信息素反应的位点变得很容易(例如，Dopman et al. 2005)，并且玉米螟的全基因组测序有望在随后几年里得以完成，这就会提供更加详细的关于这两个品系序列差异的信息。结合基因组大量的信息，我们可以确定性染色体上造成雄蛾不同反应的等位基因。同样的，我们也能鉴定一些在两个品系中造成Z和E异构体比例变化的酶编码序列及调控序列。一旦全部差异序列得到鉴定，这就可能去分析是哪些核苷酸的不同对这种表型的变异具有更为重要的作用。这些信息有助于我们进一步去理解玉米螟两性交流在遗传进化上的多样性。

5.3 信息素复合物及其反应遗传结构上的不同在物种间及物种内是否一致

进化生物学上的一个普遍问题是导致微进化(种内变化)和宏观进化的基因是否是相同的(种间和高级分类单位间的差异)。2008年，Weber等综述了一些种群被平等地选择作为性状变化的研究。他们得出对于一些性状，基因对选择的响应在种群间是完全不同的，而对于其他一些性状，它们似乎是相似的结论。遗传学响应总是相似的情况下，那么很有可能宏观进化的变化与微观进化利用的基因是相同的。如同蛾子性交流这种性状，种群间差异起源的详细研究将有助于理解蛾两性交流宏观进化的多样性。尽管如此，作为一个忠告，在Gleason和Ritchie (2004)的研究中发现影响两个果蝇种间求爱差异的遗传区域与黑腹果蝇内的遗传区域是不一样的。如果在其他两性交流系统中也存在这种情况，那么在群体水平上的研究将不会对宏观进化过程产生强干扰。

5.4 分子和遗传研究技术的结合及行为方面的研究

鉴定那些DNA序列变化影响蛾子表现型是十分重要的。然而,如果我们要弄清楚在自然选择中特异的序列是怎样传播的,我们必须把序列研究与它们怎样影响雌雄交配适应度的检测相联系。理想的方法就是检测单个基因的改变对野外释放个体适应度的影响。尽管这个方法在一些物种内是可能的,但是在大多数蛾子中是不可行的。为了解决这个问题,Groot等人于2006年,利用野外和笼子的组合研究,估算了引起烟青虫信息素复合物中醋酸盐变化基因的适应度。野外研究用于估计对长距离吸引力的影响,而笼子研究测定了交配和精子传递的可能性。相似的方法也可以被用于其他蛾子中与交配适合度相关联的表型的测定。掌握这些数据后,最终就有可能去检测两性交流性状编码基因在雌雄中的单一突变所将受到的选择压力的强度。

6. 结论

因为蛾子两性交流系统具有精确性、高效性和难以置信的多样性,它吸引了分子生物学家、生物化学家和进化生物学家对它进行研究。尽管在各个学科工作的研究者可以独立地对其进行研究,但带来突破性进展的跨学科工作的时代已经到来,这对于理解蛾子两性交流具有非常重要的作用,并使作为整体去理解进化过程将会变得更加普遍。

参考文献

[1]Abraham, D., C. Löfstedt, and J. F. Picimbon. 2005. Molecular characterization and evolution of pheromone binding protein genes in *Agrotis moths*. *Insect Biochem. Mol. Biol.* 35:1100－11.

[2]Baker, T. C., S. A. Ochieng, A. A. Cossé, S. G. Lee, and J. L. Todd. 2004. A comparison of responses from olfactory receptor neurons of *Heliothis subflexa* and *Helitohis virescens* to components of their sex pheromone. *J. Comp. Physiol. A* 190:155－65.

[3]Baltensweiler, W., and E. Priesner. 1988. Studien zum Pheromon-polymorphismus von *Zeiraphera diniana* Gn. (Lep., Tortricidae). *J. Appl. Entomol.* 106:217－31.

[4]Benton, R., S. Sachse, S. W. Michnick, and L. B. Vosshall. 2006. Atypical membrane topology and heteromeric function of *Drosophila* odorant receptors in vivo. *PLoS Biol.* 4:e20.

[5]Benton, R., K. S. Vannice, and L. B. Vosshall. 2007. An essential role for a CD36-related receptor in pheromone detection in *Drosophila*. *Nature* 450:289－93.

[6]Bette, S., H. Breer, and J. Krieger. 2002. Probing a pheromone binding protein of the silkmoth *Antheraea polyphemus* by endogenous tryptophan fluorescence. *Insect Biochem. Mol. Biol.* 32:241－46.

[7]Bjostad, L. B., and W. L. Roelofs. 1983. Sex pheromone biosynthesis in *Trichoplusia ni*: key steps involve delta11 desaturation and chain shortening. *Science* 220:1387－89.

[8]Butlin, R. 1995. Genetic variation in mating signals and responses. In *Speciation and the recognition concept: Theory and application*, ed. D. M. Lambert and H. G. Spencer, 327－66. Baltimore: Johns Hopkins Univ. Press.

[9]Butlin, R., and M. G. Ritchie. 1989. Genetic coupling in mate recognition systems: What is the ev-

idence? *Biol. J. Linn. Soc.* 37:237—46.

[10]Butlin,R. ,and A. J. Trickett. 1997. Can population genetic simulations help to interpret pheromone evolution? In *Insect pheromone research*:*New directions*,ed. R. T. Cardé and A. K. Minks,548—62. New York: Chapman and Hall.

[11]Cardé, R. T. , and K. F. Haynes. 2004. Structure of the pheromone communication channel in moths. In *Advances in insect chemical ecology*, ed. R. T. Cardé and J. G. Millar, 283—332. Cambridge: Cambridge Univ. Press.

[12]Carroll,S. B. 2005. Evolution at two levels: On genes and form. *PLoS Biol.* 3:1159—66.

[13]Cork, A. , and E. A. Lobos. 2003. Female sex pheromone components of *Helicoverpa gelotopoeon*:First heliothine pheromone without (Z)-11-hexadecenal. *Entomol. Exp. Appl.* 107:201—06.

[14]Cossé,A. A. ,M. Campbell,T. J. Glover,et al. 1995. Pheromone behavioral responses in unusual male European corn borer hybrid progeny not correlated to electrophysiological phenotypes of their pheromone-*specific* antennal neurons. *Experientia* 51:809—16.

[15]Coyne,J. A. , N. H. Barton, and M. Turelli. 1997. Perspective: A critique of Sewall Wright's shifting balance theory of evolution. *Evolution* 51:643—71.

[16]Coyne,J. A. ,and H. A. Orr. 1998. The evolutionary genetics of speciation. *Philos. Trans. R. Soc. Lond. B Biol. Sci.* 353:287—305.

[17]Datta,S. R. ,M. L. Vasconcelos,V. Ruta,et al. 2008. The *Drosophila* pheromone cVA activates a sexually dimorphic neural circuit. *Nature* 452:473—77.

[18]Dieckmann,U. ,and M. Doebeli. 1999. On the origin of species by sympatric speciation. *Nature* 400:354—57.

[19]Ding,Y. , and G. D. Prestwich. 1986. Metabolic transformations of tritium labeled pheromone by tissues of *Heliothis virescens moths*. *J. Chem. Ecol.* 12:411—29.

[20]Dobritsa,A. A. ,W. van der Goes van Naters,C. G. Warr,RA. Steinbrecht,and J. R. Carlson. 2003. Integrating the molecular and cellular basis of odor coding in the *Drosophila antennae*. *Neuron* 37: 827—41. 188

[21]Dopman,E. B. ,L. Perez,S. M. Bogdanowicz,and R. G. Harrison. 2005. Consequences of reproductive barriers for genealogical discordance in the European corn borer. *Proc. Natl. Acad. Sci. U. S. A.* 102:14706—11.

[22]Eliyahu,D. ,S. W. Applebaum,and A. Rafaeli. 2003. Moth sex pheromone biosynthesis is inhibited by the herbicide diclofop. *Pestic. Biochem. Physiol.* 77:75—81.

[23]El-Sayed,A. M. 2008. The Pherobase: Database of insect pheromones and semiochemicals. http://www. pherobase. com (accessed September 12,2008).

[24]Fang,N. , P. E. A. Teal, and J. H. Tumlinson. 1995. PBAN regulation of pheromone biosynthesis in female tobacco hornworm moths,*Manduca sexta* (*L.*). *Arch. Insect Biochem. Physiol.* 29:35—44.

[25]Foster,S. P. 2005. Lipid analysis of the sex pheromone gland of the moth *Heliothis virescens*. *Arch. Insect Biochem. Physiol.* 59:80—90.

[26]Gahan,L. J. ,F. Gould,and D. G. Heckel. 2001. Identification of a gene associated with Bt resistance in *Heliothis virescens*. *Science* 293:857—60.

[27]Gleason,J. M. ,and M. G. Ritchie. 2004. Do quantitative trait loci (QTL) for a courtship song difference between *Drosophila simulans* and *D. sechellia* coincide with candidate genes and *intraspecific*

QTL? *Genetics* 166:1303—11.

[28]Gohl, T., and J. Krieger. 2006. Immunolocalization of a candidate pheromone receptor in the antennae of the male moth, *Heliothis virescens*. *Invert. Neurosci.* 6:13—21.

[29]Gray, D. A., and W. H. Cade. 1999. Quantitative genetics of sexual selection in the field cricket, *Gryllus integer*. *Evolution* 53:848—54.

[30]Groot, A. T., J. Bennett, J. Hamilton, R. G. Santangelo, C. Schal, and F. Gould. 2006. Experimental evidence for interspecific directional selection on moth pheromone communication. *Proc. Nat. Acad. Sci. U. S. A.* 103:5858—63.

[31]Groot, A. T., R. G. Santangelo, E. Ricci, C. Brownie, F. Gould, and C. Schal. 2007. Differential attraction of *Heliothis subflexa* males to synthetic pheromone lures in Eastern US and Western Mexico. *J. Chem. Ecol.* 33:353—68.

[32]Große-Wilde, E., T. Gohl, E. Bouché, H. Breer, and J. Krieger. 2007. Candidate pheromone receptors provide the basis for the response of distinct antennal neurons to pheromonal compounds. *Eur. J. Neurosci.* 25:2364—73.

[33]Große-Wilde, E., A. Svatoš, and J. Krieger. 2006. A pheromone-bindin protein mediated the bombykol-induced activation of a pheromone *in vitro*. *Chem. Senses* 31:547—55.

[34]Gu, P., W. H. Welch, L. Guo, K. M. Schegg, and G. J. Blomquist. 1997. Characterization of a novel microsomal fatty acid synthetase (FAS) compared to a cytosolic FAS in the housefly, *Musca domestica*. *Comp. Biochem. Physiol.* 118B:447—56.

[35]Hallem, E. A., A. Dahanukar, and J. R. Carlson. 2006. Insect odor and taste receptors. *Annu. Rev. Entomol.* 51:113—35.

[36]Hallem, E. A., M. G. Ho, and J. R. Carlson. 2004. The molecular basis of odor coding in the *Drosophila* antennae. *Cell* 117:965—79.

[37]Hansson, B. S., C. Löfstedt, and S. P. Foster. 1989. Z-linked inheritance of male olfactory response to sex pheromone components in two species of tortricid moths, *Ctenopseusis obliquana and Ctenopseusis sp*. *Entomol. Exp. Appl.* 53:137—45.

[38]Haynes, K. F., and R. E. Hunt. 1990. A mutation in the pheromonal communication system of the cabbage looper moth, *Trichoplusia ni*. *J. Chem. Ecol.* 16:1249—57.

[39]Heckel, D. G. 1993. Comparative genetic linkage mapping in insects. *Annu. Rev. Entomol.* 38:381—408.

[40]Hoekstra, H. E., and J. A Coyne. 2007. The locus of evolution: Evo devo and the genetics of adaptation. *Evolution* 61:995—1016.

[41]Hoskovec, M., A. Luxová, A. Svatoš, and W. Boland. 2002. Biosynthesis of sex pheromones in moths: Stereochemistry of fatty alcohol oxidation in *Manduca sexta*. *Tetrahedron* 58:9193—9201.

[42]Hoy, R. R., J. Hahn, and R. C. Paul. 1977. Hybrid cricket auditory behavior: Evidence for genetic coupling in animal communication. *Science* 195:82—84.

[43]Hull, J. J., R. Kajigaya, K. Imai, and S. Matsumoto. 2007. The *Bombyx mori* sex pheromone biosynthetic pathway is not mediated by cAMP. *J. Insect Physiol.* 53:782—93.

[44]Ishida, Y., and W. S. Leal. 2002. Cloning of putative odorant-degrading enzyme and integumental esterase cDNAs from the wild silkmoth, *Antheraea polyphemus*. *Insect Biochem. Mol. Biol.* 32:1775—80.

[45]Ishida, Y., and W. S. Leal. 2005. Rapid inactivation of a moth pheromone. *Proc. Natl. Acad. Sci.*

U. S. A. 102:14075—79.

[46]Jacquin-Joly, E., and C. Merlin. 2004. Insect olfactory receptors: Contributions of molecular biology to chemical ecology. *J. Chem. Ecol.* 30:2359—97.

[47]Juárez, M. P. 2004. Fatty acyl-CoA elongation in *Blatella germanica* integumental microsomes. *Arch. Insect Biochem. Physiol.* 56:170—78.

[48]Juárez, M. P., and G. C. Fernández. 2007. Cuticular hydrocarbons of triatomines. *Comp. Biochem. Physiol. A Mol. Integr. Physiol.* 147:711—30.

[49]Jurenka, R. 2003. Biochemistry of female moth sex pheromones. In *Insect pheromone biochemistry and molecular biology*, ed G. J. Blomquist and R. C. Vogt, 54—80. Amsterdam: Elsevier.

[50]Jurenka, R. A., K. F. Haynes, R. O. Adolf, M. Bengtsson, and W. L. Roelofs. 1994. Sex pheromone component ratio in the cabbage looper moth altered by a mutation affecting the fatty acid chain-shortening reactions in the pheromone biosynthetic pathway. *Insect Biochem. Mol. Biol.* 24:373—81.

[51]Jurenka, R. A., E. Jacquin, and W. L. Roelofs. 1991. Control of the pheromone biosynthetic pathway in *Helicoverpa zea* by the pheromone biosynthesis activating neuropeptide. *Arch. Insect Biochem. Physiol.* 17:81—91.

[52]Jurenka, R. A., and W. L. Roelofs. 1989. Characterization of the acetyltransferase used in pheromone biosynthesis in moths: Specificity for the Z isomer in Tortricidae. *Insect Biochem.* 19:639—44.

[53]Kaissling, K. —E. 2001. Olfactory perireceptor and receptor events in moths: A kinetic model. *Chem. Senses* 26:125—50.

[54]Kaissling, K. —E., and E. Priesner. 1970. Die Riechschwelle des Seidenspinners. *Naturwissenschaften* 57:23—28.

[55]Klun, J. A. 1975. Insect sex pheromones: Intraspecific pheromonal variability of *Ostrinia nubilalis* in North America and Europe. *Environ. Entomol.* 4:891—94.

[56]Knipple, D. C., and W. L. Roelofs. 2003. Molecular biological investigations of pheromone desaturases. In: Insect pheromone biochemistry and molecular biology, ed. G. J. Blomquist and R. C. Vogt, 81—106. London: Elsevier Academic Press.

[57]Knipple, D. C., C. —L. Rosenield, S. J. Miller, et al. 1998. Cloning and functional expression of a cDNA encoding a pheromone gland-specific acyl-CoA Δ^{11}-desaturase of the cabbage looper moth, *Trichoplusia ni*. *Proc. Natl. Acad. Sci. U. S. A.* 95:15287—92.

[58]Knipple, D. C., C. —L. Rosenield, R. Nielsen, K. M. You, and S. E. Jeong. 2002. Evolution of the integral membrane desaturase gene family in moths and flies. *Genetics* 162:1737—52.

[59]Kondrashov, A. S., and F. A. Kondrashov. 1999. Interactions among quantitative traits in the course of sympatric speciation. *Nature* 400:351—54.

[60]Krieger, J., and H. Breer. 1999. Olfactory reception in invertebrates. *Science* 286:720—23.

[61]Krieger, J., H. Gaenssle, K. Raming, and H. Breer. 1993. Odorant binding proteins of *Heliothis virescens*. *Insect Biochem. Mol. Biol.* 23:449—56.

[62]Krieger, J., E. Große-Wilde, T. Gohl, and H. Breer. 2005. Candidate pheromone receptors of the silkmoth *Bombyx mori*. *Eur. J. Neurosci.* 21:2167—76.

[63]Krieger, J., E. Große-Wilde, T. Gohl, Y. M. E. Dewer, K. Raming, and H. Breer. 2004. Genes encoding candidate pheromone receptors in a moth (*Heliothis virescens*). *Proc. Natl. Acad. Sci. U. S. A.* 101:11845—50.

[64]Krieger,J. ,K. Raming,Y. M. E. Dewer,S. Bette,S. Conzelmann,and H. Breer. 2002. A divergent gene family encoding candidate olfactory receptors of the moth *Heliothis virescens*. *Eur. J. Neurosci.* 16: 619—28.

[65]Kurtovic,A. ,A. Widmer,and B. J. Dickson. 2007. A single class of olfactory neurons mediates behavioural responses to a *Drosophila* sex pheromone. *Nature* 446:542—46.

[66]Laue,M. ,and R. A. Steinbrecht. 1997. Topochemistry of moth olfactory sensilla. *Int. J. Insect Morphol. Embryol.* 26:217—28.

[67]Leal,W. S. 2003. Proteins that make sense. In *Insect pheromone biochemistry and molecular biology*,ed. G. J. Blomquist and R. G. Vogt,447—76. London: Elsevier Academic Press.

[68]Leal,W. S. 2005. Pheromone reception. *Topics Curr. Chem.* 240:1—36.

[69]Leal,W. S. ,A. M. Chen,Y. Ishida,et al. 2005. Kinetics and molecular properties of pheromone binding and release. *Proc. Natl. Acad. Sci. U. S. A.* 102:5386—91.

[70]Linn,C. Jr. ,K. Poole,A. Zhang,and W. Roelofs. 1999. Pheromone-blend discrimination by European corn borer moths with inter-race and inter-sex antennal transplants. *J. Comp. Physiol. A* 184:273—78.

[71]Linn,C. E. Jr. ,and W. L. Roelofs. 1995. Pheromone communication in moths and its role in the speciation process. In *Speciation and the recognition concept*:*Theory and application*,ed. D. M. Lambert and H. G. Spencer,263—300. Baltimore: Johns Hopkins Univ. Press.

[72]Linn,C. E. ,M. S. Young,M. Gendle,et al. 1997. Sex pheromone blend discrimination in two races and hybrids of the European corn borer moth,*Ostrinia nubilalis*. *Physiol. Entomol.* 22:212—23.

[73]Liu,W. ,P. W. K. Ma,P. Marsella-Herrick,C. —L. Rosenfield,D. C. Knipple,and W. L. Roelofs. 1999. Cloning and functional expression of a cDNA encoding a metabolic acyl-CoA Δ9-desaturase of the cabbage looper moth,*Trichoplusia ni*. *Insect Biochem. Mol. Biol.* 29:435—43.

[74]Löfstedt,C. 1990. Population variation and genetic control of pheromone communication systems in moths. *Entomol. Exp. Appl.* 54:199—218. 19

[75]Löfstedt,C. 1993. Moth pheromone genetics and evolution. *Philos. Trans. R. Soc. Lond. B Biol. Sci.* 340:161—77.

[76]Löfstedt,C. ,B. S. Hansson,W. L. Roelofs,and B. O. Bengtsson. 1989. No linkage between genes controlling female pheromone production and male pheromone response in the European corn borer,*Ostrinia nubilalis* Hübner (Lepidoptera: Pyralidae). *Genetics* 123:553—56.

[77]Luxová,A. ,and A. Svatoš. 2006. Substrate specificity of membrane-bound alcohol oxidase from the tobacco hornworm moth (*Manduca sexta*) female pheromone glands. *J. Mol. Catal. B Enzymatic* 38: 37—42.

[78]Maïbèche-Coisne,M. ,E. Jacquin-Joly,M. C. François,and P. Nagnan-Le Meillour. 2002. cDNA cloning of biotransformation enzymes belonging to the cytochrome P450 family in the antennae of the noctuid moth *Mamestra brassicae*. *Insect Mol. Biol.* 11:273—81.

[79]Maida,R. ,G. Ziegelberger,and K. E. Kaissling. 2003. Ligand binding to six recombinant pheromone-binding proteins of *Antheraea polyphemus and Antheraea pernyi*. *J. Comp. Physiol. B* 173:565—73.

[80]Matoušková,P. ,I. Pichová,and A. Svatoš. 2007. Functional characterization of a desaturase from the tobacco hornworm moth (*Manduca sexta*) with bifunctional Z11- and 10,12-desaturase activity. *Insect Biochem. Mol. Biol.* 37:601—10.

[81]Matsumoto,S. ,J. J. Hull,A. Ohnishi,K. Moto,and A. Fonagy. 2007. Molecular mechanisms un-

derlying sex pheromone production in the silkmoth, *Bombyx mori*: Characterization of the molecular components involved in bombykol biosynthesis. *J. Insect Physiol.* 53:752—59.

[82]Merlin, C., M. C. François, F. Bozzolan, J. Pelletier, E. Jacquin-Joly, and M. Maïbèche-Coisne. 2005. A new aldehyde oxidase selectively expressed in chemosensory organs of insects. *Biochem. Biophys. Res. Commun.* 332:4—10.

[83]Metz, J. G., M. R. Pollard, L. Anderson, T. R. Hayes, and M. W. Lassner. 2000. Purification of a jojoba embryo fatty acyl-Coenzyme A reductase and expression of its cDNA in high erucic acid rapeseed. *Plant Physiol.* 122:635—44.

[84]Mombaerts, P. 1999. Seven-transmembrane proteins as odorant and chemosensory receptors. *Science* 286:707—11.

[85]Mombaerts, P. 2004. Odorant receptor gene choice in olfactory sensory neurons: The one receptor-one neuron hypothesis revisited. *Curr. Opin. Neurobiol.* 14:31—36.

[86]Morse, D., and E. Meighen. 1986. Pheromone biosynthesis and the role of functional groups in pheromone specificity. *J. Chem. Ecol.* 12:335—51.

[87]Morse, D., and E. Meighen. 1987. Pheromone biosynthesis: Enzymatic studies in Lepidoptera. In *Pheromone biochemistry*, ed. G. D. Prestwich and G. J. Blomquist, 212—15. New York: Academic Press.

[88]Moto, K., M. G. Suzuki, J. J. Hull, et al. 2004. Involvement of a bifunctional fatty-acyl desaturase in the biosynthesis of the silkmoth, *Bombyx mori*, sex pheromone. *Proc. Natl. Acad. Sci. U. S. A.* 101: 8631—36.

[89]Moto, K., T. Yoshiga, M. Yamamoto, et al. 2003. Pheromone gland-specific fatty-acyl reductase of the silkmoth, *Bombyx mori*. *Proc. Natl. Acad. Sci. U. S. A.* 100:9156—61.

[90]Mustaparta, H. 1996. Central mechanisms of pheromone information processing. *Chem. Senses* 21:269—75.

[91]Nakagawa, T., T. Sakurai, T. Nishioka, and K. Touhara. 2005. Insect sex-pheromone signals mediated by specific combinations of olfactory receptors. *Science* 307:1638—42.

[92]Ohnishi, A., J. J. Hull, and S. Matsumoto. 2006. Targeted disruption of genes in the *Bombyx mori* sex pheromone biosynthetic pathway. *Proc. Natl. Acad. Sci. U. S. A.* 103:4398—4403.

[93]Ozawa, R., and S. Matsumoto. 1996. Intracellular signal transduction of PBAN action in the silkworm, *Bombyx mori*: Involvement of acyl-CoA reductase. *Insect Biochem. Mol. Biol.* 26:259—65.

[94]Pape, M. E., F. Lopez-Casillas, and K. —H. Kim. 1988. Physiological regulation of Acetyl-CoA carboxylase gene expression: Effects of diet, diabetes, and lactation on acetyl-CoA carboxylase mRNA. *Arch. Biochem. Biophys.* 267:104—09.

[95]Pelletier, J., F. Bozzolan, M. Solvar, M. —C. François, E. Jacquin-Joly, and M. Maïbèche-Coisne. 2007. Identification of candidate aldehyde oxidases from the silkworm *Bombyx mori* potentially involved in antennal pheromone degradation. *Gene* 404:31—40.

[96]Pelosi, P. 1996. Perireceptor events in olfaction. *J. Neurobiol.* 30:3—19.

[97]Phelan, P. L. 1997. Genetic and phylogenetics in the evolution of sex pheromones. In *Insect pheromone research: New directions*, ed. R. T. Cardé and A. K. Minks, 563—79. New York: Chapman and Hall.

[98]Prestwich, G. D., R. G. Vogt, and L. M. Riddiford. 1986. Binding and hydrolysis of radiolabeled pheromone and several analogs by male-specific antennal proteins of the moth *Antheraea polyphemus*. *J.*

Chem. Ecol. 12:323—33.

[99]Rafaeli, A. 2005. Mechanisms involved in the control of pheromone production in female moths: Recent developments. *Entomol. Exp. Appl.* 115:7—15.

[100]Remington, D. L., R. W. Whetten, B. H. Liu, and D. M. O'Malley. 1999. Construction of genetic map with nearly complete genome coverage in *Pinus taeda*. *Theor. Appl. Genet.* 98:1279—92.

[101]Rivière, S., A. Lartigue, B. Quennedey, et al. 2003. A pheromone-binding protein from the cockroach *Leucophaea maderae*: Cloning, expression and pheromone binding. *Biochem. J.* 371:573—79.

[102]Robertson, H. M., R. Martos, C. R. Sears, E. Z. Todres, K. K. Walden, and J. B. Nardi. 1999. Diversity of odourant binding proteins revealed by an expressed sequence tag project on male *Manduca sexta* moth antennae. *Insect Mol. Biol.* 8:501—18.

[103]Rodriguez, F., D. L. Hallahan, J. A. Pickett, and F. Camps. 1992. Characterization of the delta-11 palmitoyl-CoA-desaturase from *Spodoptera littoralis* (Lepidoptera, Noctuidae). *Insect Biochem. Mol. Biol.* 22:143—48.

[104]Roelofs, W. L., and R. T. Cardé. 1977. Responses of Lepidoptera to synthetic sex-pheromone chemicals and their analogs. *Annu. Rev. Entomol.* 22:377—405.

[105]Roelofs, W. L., T. Glover, X. H. Tang, et al. 1987. Sex-pheromone production and perception in European corn-borer moths is determined by both autosomal and sex-linked genes. *Proc. Natl. Acad. Sci. U. S. A.* 84:7585—89.

[106]Roelofs, W. L., and R. A. Jurenka. 1996. Biosynthetic enzymes regulating ratios of sex pheromone components in female redbanded leafroller moths. *Bioorg. Med. Chem.* 4:461—66.

[107]Roelofs, W. L., and A. P. Rooney. 2003. Molecular genetics and evolution of pheromone biosynthesis in Lepidoptera. *Proc. Natl. Acad. Sci. U. S. A.* 100:9179—84.

[108]Roelofs, W. L., and W. A. Wolf. 1988. Pheromone biosynthesis in Lepidoptera. *J. Chem. Ecol.* 14:2019—31.

[109]Rogers, M. E., J. Krieger, and R. G. Vogt. 2001. Antennal SNMPs (sensory neuron membrane proteins) of Lepidoptera define a unique family of invertebrate CD36—like proteins. *J. Neurobiol.* 49:47—61.

[110]Rogers, M. E., M. Sun, M. R. Lerner, and R. G. Vogt. 1997. Snmp-1, a novel membrane protein of olfactory neurons of the silk moth *Antheraea polyphemus* with homology to the CD36 family of membrane proteins. *J. Biol. Chem.* 272:14792—99.

[111]Rosenfield, C. —L., K. M. You, P. Marsella-Herrick, W. L. Roelofs, and D. C. Knipple. 2001. Structural and functional conservation and divergence among acyl-CoA desaturases of two noctuid species, the corn earworm, *Helicoverpa zea*, and the cabbage looper, *Trichoplusia ni*. *Insect Biochem. Mol. Biol.* 31:949—64.

[112]Rützler, M., and L. J. Zwiebel. 2005. Molecular biology of insect olfaction: Recent progress and conceptual models. *J. Comp. Physiol.* A 191:777—90.

[113]Rybczynski, R., J. Reagan, and M. R. Lerner. 1989. A pheromone-degrading aldehyde oxidase in the antennae of the moth *Manduca sexta*. *J. Neurosci.* 9:1341—53.

[114]Rybczynski, R., R. G. Vogt, and M. R. Lerner. 1990. Antennal-specific pheromone-degrading aldehyde oxidases from the moths *Antheraea polyphemus* and *Bombyx mori*. *J. Biol. Chem.* 265:19712—15.

[115]Sakurai, T., T. Nakagawa, H. Mitsuno, et al. 2004. Identification and functional characterization of a sex pheromone receptor in the silkmoth *Bombyx mori*. *Proc. Natl. Acad. Sci. U. S. A.* 101:16653—58.

[116]Sambandan, D., A. Yamamoto, J. J. Fanara, T. F. C. Mackay, R. R. H. Anholt. 2006. Dynamic genetic interactions determine odor-guided behavior in *Drosophila melanogaster*. *Genetics* 174:1349—63.

[117]Sandler, B. H., L. Nikonova, W. S. Leal, and J. Clardy. 2000. Sexual attraction in the silkworm moth: Structure of the pheromone-binding-protein-bombykol complex. *Chem. Biol.* 7:143—51.

[118]Sato, K., M. Pellegrino, T. Nakagawa, T. Nakagawa, L. B. Vosshall, and K. Touhara. 2008. Insect olfactory receptors are heteromeric ligand-gated ion channels. *Nature* 452:1002—06.

[119]Sheck, A. L., A. T. Groot, C. M. Ward, et al. 2006. Genetics of sex pheromone blend differences between *Heliothis virescens* and *Heliothis subflexa*: A chromosome mapping approach. *J. Evol. Biol.* 19: 600—17.

[120]Steinbrecht, R. A., M. Ozaki, and G. Ziegelberger. 1992. Immunocytochemical localization of pheromone-binding protein in moth antennae. *Cell Tissue Res.* 270:287—302.

[121]Stengl, M., F. Zufall, H. Hatt, and J. G. Hildebrand. 1992. Olfactory receptor neurons from antennae of developing male *Manduca sexta* respond to components of the species-specific sex-pheromone *in vitro*. *J. Neurosci.* 12:2523—31.

[122]Syed, Z., Y. Ishida, K. Taylor, D. A. Kimbrell, and W. S. Leal. 2006. Pheromone reception in fruit flies expressing a moth's odorant receptor. *Proc. Natl. Acad. Sci. U. S. A.* 103:16538—43.

[123]Takanashi, T., Y. P. Huang, K. R. Takahasi, S. Hoshizaki, S. Tatsuki, and Y. Ishikaw. 2005. Genetic analysis and population survey of sex pheromone variation in the adzuki bean borer moth, *Ostrinia scapulalis*. *Biol. J. Linn. Soc.* 84:143—60.

[124]Teal, P. E. A., and J. H. Tumlinson. 1987. The role of alcohols in pheromone biosynthesis by two noctuid moths that use acetate pheromone components. *Arch. Insect Biochem. Physiol.* 4: 261—69. 192

[125]Teal, P. E. A., and J. H. Tumlinson. 1997. Effects of interspecific hybridization between *Heliothis virescens* and *Heliothis subflexa* on the sex pheromone communication system. In *Insect pheromone research: New directions*, ed. R. T. Cardé and A. K. Minks, 535—47. New York: Chapman and Hall.

[126]Tiku, P. E., A. Y. Gracey, A. I. Macartney, R. J. Beyton, and A. R. Crossins. 1996. Cold-induced expression of δ9-desaturase in carp by transcriptional and posttranslational mechanisms. *Science* 271:815—18.

[127]Tsfadia, O., A. Azrielli, L. Falach, A. Zada, W. Roelofs, and A. Rafaeli. 2008. Pheromone biosynthetic pathways: PBAN-regulated rate-limiting steps and differential expression of desaturase genes in moth species. *Insect Biochem. Mol. Biol.* 38:552—67.

[128]Tumlinson, J. H., and P. E. A. Teal. 1987. Relationship of structure and function to biochemistry in insect pheromone systems. In *Pheromone biochemistry*, ed. G. D. Prestwich and G. J. Blomquist, 3—26. New York: Academic Press.

[129]van den Berg, M. J., and G. Zielgelberger. 1991. On the function of the pheromone binding protein in the olfactory hairs of *Antheraea polyphemus*. *J. Insect Physiol.* 37:79—85.

[130]Vickers, N. J. 2006. Inheritance of olfactory preferences I. Pheromone-mediated behavioral responses of *Heliothis subflexa x Heliothis virescens* hybrid male moths. *Brain Behav. Evol.* 68:63—74.

[131]Vickers, N. J., and T. A. Christensen. 2003. Functional divergence of spatially conserved olfactory glomeruli in two related moth species. *Chem. Senses* 28:325—38.

[132]Vickers, N. J., K. Poole, and C. E. Linn. 2005. Plasticity in central olfactory processing and pheromone blend discrimination following interspecies antennal imaginal disc transplantation. *J. Comp.*

Neurol. 491:141—56.

[133]Vogt, R. G. 2003. Biochemical diversity of odor detection: OBPs, ODEs and SNMPs. In *Insect pheromone biochemistry and molecular biology*, ed. G. J. Blomquist and R. G. Vogt, 391—445. London: Elsevier Academic Press.

[134]Vogt, R. G. 2005. Molecular basis of pheromone detection in insects. In *Comprehensive insect physiology, biochemistry, pharmacology and molecular biology*, *Vol*. 3, *Endocrinology*, ed. L. I. Gilbert, K. Iatrou, and S. Gill, 753—804. London: Elsevier.

[135]Vogt, R. G., and L. M. Riddiford. 1981. Pheromone binding and inactivation by moth antennae. *Nature* 293:161—63.

[136]Vogt, R. G., L. M. Riddiford, and G. D. Prestwich. 1985. Kinetic properties of a pheromone degrading enzyme: The sensillar esterase of *Antheraea polyphemus*. *Proc. Natl. Acad. Sci. U. S. A.* 82:8827—31.

[137]Vosshall, L. B., H. Amrein, P. S. Morozov, A. Rzhetsky, and R. Axel. 1999. A spatial map of olfactory receptor expression in the *Drosophila antennae*. *Cell* 96:725—36.

[138]Wade, M. J., and C. J. Goodnight. 1998. Perspective: The theories of Fisher and Wright in the context of metapopulations: When nature does many small experiments. *Evolution* 52:1537—53.

[139]Wakil, S. J., J. K. Stoops, and V. C. Joshi. 1983. Fatty acid synthesis and its regulation. *Annu. Rev. Biochem*. 52:537—79.

[140]Wanner, K. V., A. R. Anderson, S. C. Trowell, D. A. Theilmann, H. M. Robertson, and R. D. Newcomb. 2007. Female-biased expression of odourant receptor genes in the adult antennae of the silkworm, *Bombyx mori*. *Insect Mol. Biol*. 16:107—19.

[141]Weber, K. E., R. J. Greenspan, D. R. Chicoine, K. Fiorentino, M. H. Thomas, and T. L. Knight. 2008. Microarray analysis of replicate populations selected against a wing-shape correlation in *Drosophila melanogaster*. *Genetics* 178:1093—1108.

[142]Whitlock, M. C., and P. C. Phillips. 2000. The exquisite corpse: A shifting view of the shifting balance. *Trends Ecol. Evol*. 15:347—48.

[143]Wicher, D., R. Schäfer, R. Bauernfeind, et al. 2008. *Drosophila* odorant receptors are both ligand-gated and cyclic-nucleotide-activated cation channels. *Nature* 452:1007—11.

[144]Willett, C. S. 2000. Evidence for directional selection acting on pheromone-binding proteins in the genus *Choristoneura*. *Mol. Biol. Evol*. 17:553—62.

[145]Wittkopp, P. J., B. K. Haerum, and A. G. Clark. 2004. Evolutionary changes in *cis* and *trans* regulation. *Nature* 430:85—88.

[146]Wittkopp, P. J., B. K. Haerum, and A. G. Clark. 2008. Independent effects of *cis*- and *trans*-regulatory variation on gene expression in *Drosophila melanogaster*. *Genetics* 178:1831—35.

[147] Witzgall, P., T. Lindblom, M. Bengtsson, and M. Tóth. 2004. The Pherolist. http://www-pherolist. slu. se (accessed September 12, 2008).

[148]Wu, W—Q., J. —W. Zhu, J. Millar, and C. Löfstedt. 1998. A comparative study of sex pheromone biosynthesis in two strains of the turnip moth, *Agrotis segetum*, producing ratios of sex pheromone components. *Insect Biochem. Mol. Biol*. 28:895—900.

[149]Xiu, W. —M., and S. —L. Dong. 2007. Molecular characterization of two pheromone binding proteins and quantitative analysis of their expression in the beet armyworm, *Spodoptera exigua* Hübner. *J. Chem. Ecol*. 33:947—61.

[150]Xiu,W. —M. ,Y. —Z. Zhou,and S. —L. Dong. 2008. Molecular characterization and expression pattern of two pheromone-binding proteins from *Spodoptera litura*(Fabricius). *J. Chem. Ecol.* 34:487—98.

[151]Xue,B. ,A. P. Rooney,M. Kajikawa,N. Okada,and W. L. Roelofs. 2007. Novel sex pheromone desaturases in the genomes of corn borers generated through gene duplication and retroposon fusion. *Proc. Natl. Acad. Sci. U. S. A.* 104:4467—72.

[152]Zhu,J. W. ,B. B. Chastain,B. G. Spohn,and K. F. Haynes. 1997. Assortative mating in two pheromone strains of the cabbage looper moth,*Trichoplusia ni*. *J. Insect Behav.* 10:805—17.

[153]Zhu,J. W. ,C. H. Zhao,F. Lu,M. Bengtsson,and C. Löfstedt. 1996. Reductase specificity and the ratio regulation of E/Z Isomers in pheromone biosynthesis of the European corn borer,*Ostrinia nubilalis* (Lepidoptera: Pyralidae). *Insect Biochem. Molec. Biol.* 26:171—76.

第十一章 鳞翅目昆虫宿主范围的遗传学

Sara J. Oppenheim and Keith R. Hopper

1. 前言

尽管对复杂并且和生态相关的性状是如何进化的这一问题，人们的研究兴趣一直是盎然的，但对任何一种生物体来说，这些性状的遗传基础仍然未能得到很好的阐释。紧密相关的物种在它们对环境的适应过程中分化迥异，基于此，这个问题就格外具有挑战性。蛾子和蝴蝶对植物宿主选择上的差异提供了一些这方面的例子；尽管密切相关的物种倾向于选择类似的宿主，仍然有很多例子表明，同属物种有迥然相异的宿主范围。对遗传变化如何导致宿主选择上变迁的研究工作在不同体系上正在开展。尽管这些变化有可能对资料周详的食叶类昆虫外观没什么作用，但了解宿主范围的遗传结构对了解鳞翅目昆虫的进化是最根本的。增强对宿主范围的遗传学认识，对安全实施生物控制和培育能持续抵抗害虫侵染的植物有非常关键的实用意义。认识鳞翅目昆虫的宿主范围进化，将需要其遗传结构的知识，即哪些基因参与，这些基因是如何相互作用的，以及每一个基因需要多大的变化才能造成宿主范围的变迁。

食叶类昆虫的宿主范围涉及很多特征。为了能利用某种植物，昆虫必须找到该植物，在上面产卵、生长并发育到成虫。因此宿主范围是多因素决定的，在宿主的选择上必须整合每一个阶段中相互竞争的需要。尽管“宿主范围”这个名词在文献中用于不同的语境下，我们在此指的是宿主植物的名册，在宿主植物上草食性昆虫能排卵且其幼虫能有机会完成整个发育过程。宿主范围是动态的，因为一个给定的宿主既依赖于外部因素（例如当地宿主是否可获得、竞争）也依赖于内部因素（例如产卵、年龄、以前的经历）。宿主范围是一个不断变迁的目标，而且决定宿主范围的基因参与多种多样的途径，因此很难对导致宿主范围差异的遗传差异有很明晰的了解。

昆虫幼虫能摄食的植物物种仅限于那些其母代在上面产卵的植物。大多数新生幼虫在它们饥饿前仅能移动数米去寻找食物（尽管如气球状膨胀的新生幼虫能迁移得更远），甚至晚龄幼虫活动范围也只能达到数米至几十米，而相比较而言，鳞翅目昆虫成虫，依靠自身或风力，能飞行上万米远。因此雌性成虫相对于其子孙在选择适合的宿主植物方面有更大的机会。幼虫能否发育成为成虫并为它们的子孙寻找宿主取决于幼虫在它们所找到的植物宿主上的摄食以及表现。因此，宿主范围涉及成虫化学感应和神经内处理的相关基因，以及幼虫化学感应、消化和营养代谢的相关基因，这些基因决定了幼虫觅食、生长和生存。

鳞翅目昆虫染色体数目较多以及具有性别限制的重组，从而使其成为研究宿主范围和复杂性状的遗传基础的模式生物。鳞翅目昆虫中，染色体内重组仅限于雄性，因此来源于母本的染色体得以完整保留（Suomalainen 1969；Marec 1996；见第三章对雌性个体无重组的详细论述）。遗传连锁图因来源于母本的“连锁群”，实际上就是完整的染色体而简化，并且任何假设的重组都能归结为记录错误（在重组发生的体系中，解决真实重组中的记录错误是一个很大的困扰）。尽管鳞翅目昆虫染色体数目能从 $n=5$ 变化到 $n=223$（White 1973；De Prins 和 Saitoh 2003；见第三章对鳞翅目昆虫染色体特点的详细论述），但是大多数鳞翅目昆虫有 28 条到 32 条小染色体，这些染色体具有较为均一的大小（Suomalainen 1969；Robinson 1971）。但数目众多的小染色体使它们在细胞学上难以区分，这也意味着每一个染色体是由基因组中比例较小的一部分组成。如果鳞翅目昆虫染色体大小的分布和总基因数都类似于家蚕（Xia et al. 2004；Yoshido et al. 2005），那么每个染色体将包含全基因组的2%～5%或者300到1000个基因。这就是说，鳞翅目昆虫染色体的分辨率至少和那些拥有染色体内重组的体系具有相等的精准度。更精细尺度上的分辨率由两步法获得，首先用带有雌性信息的标记将基因绘制到染色体上，然后再用带有雄性信息的标记在染色体内进行定位（Heckel et al. 1999）。这个方法能将注意力集中到携带所研究基因的染色体上，成为精细绘图和定位克隆的主要优势。

截至2002年，已发表的关于植物－昆虫相互作用的论文和图书，已超过5000篇（Scriber 2002），这个数目还在逐年增加。现有的对草食性昆虫－植物相互作用的进化生物学知识，包括系统发生、生物化学、行为和进化，在最近做了一个透彻的回顾（Tilmon 2008）。尽管有丰富的研究和上面所论述过的鳞翅目昆虫遗传学的优势，但对于任何一种蛾子或蝴蝶，我们仍然对其详细的宿主范围的遗传结构知之甚少。在这一章中，我们将回顾已知的关于鳞翅目昆虫宿主范围的遗传学知识，讨论宿主范围的生物学基础及其遗传结构，并提出一些具有前景的研究线路。尽管我们多以现在从事的，因而也是我们知道得最清楚地——泛化种，烟芽夜蛾 *Heliothis virescens* 以及其紧密相关的专一种，烟青虫 *Heliothis subflexa* 做全面的论述，我们也在不同的深浅程度上顾及到其他一些体系。贯穿文献的主题和问题将在这个回顾中得到清楚的阐述：成虫排卵倾向和幼虫表现的比较；在宿主物种间的折中表现；宿主范围的进化速度；多数和少数基因的比较；常染色体基因和性染色体基因的比较；种间特异和种内特异基础的差异；宿主范围的扩增或缩小和宿主范围变迁的比较；从泛化种到专一种进化的方向性或者相反；以及宿主变迁在物种形成中的作用。

2. 宿主特异性的遗传学

在这一节中，我们将依不同属总结现有的关于宿主范围遗传学的证据：(1)系统发生模式；(2)群体和品系比较，尤其是在通常的园圃实验中；(3)对人工或自然选择的应答；(4)寄主和物种间的交配；(5)亲缘关系（亲子交配退化，全同胞家系，半同胞家系）；(6)基于分子标记的数量性状的定位；(7)参与化学感受，解毒和对植物化学分子的同化作用的蛋白质序列和表达的差异。另外两类证据——定位克隆和候选基因沉默，也将很快获得。

2.1 实夜蛾（*Heliothis*）

烟芽夜蛾 *Heliothis virescens* 复合群由南美洲和北美洲至少13种密切相关的物种组成，这些物种在宿主特异性和地理分布上均有差异（Mitter，Poole 和 Matthews 1993）。在这些成员中，研究界对两个种系尤为关注：烟芽夜蛾 *H. virescens* 和烟青虫 *H. subflexa*。烟芽夜蛾 *H. virescens* 是最主要的农业害虫，已成为许多研究的素材（在所参考的刊物中有超过1300篇文献报道）。烟青虫 *H. subflexa* 不是害虫，但和烟芽夜蛾 *H. virescens* 亲缘关系很近：在做过对比的基因中，99%的序列都是相似的（Cho et al. 1995；Fang et al. 1997）。它们的地理分布也有大范围的重叠（Mitter，Poole 和 Matthews

1993)，并且两物种的形态上也相似，致使烟青虫 *H. subflexa* 在 1941 年才得以确定成为一个独立的物种（McElvare 1941）。在实验室，烟芽夜蛾 *H. virescens* 和烟青虫 *H. subflexa* 可以交配，产生可育的雌性 F_1 代和不育的雄性 F_1 代（雄性的育性经过数代回交后得到恢复；Karpenko 和 Proshold 1977）。这两个物种被认为是在近期才从一个共同的泛化种祖先分化而来的（Mitter，Poole 和 Matthews 1993；Poole，Mitter 和 Huettel 1993；Fang et al. 1997）。尽管它们之间存在相似性，但是它们在宿主的范围上差异很大。烟芽夜蛾 *H. virescens* 有非常广泛的宿主范围，至少能以 14 个目中的 37 种植物为食，包括 *Nicotiana tabacum*（烟草），*Gossypium hirsutum*（棉花），*Glycine max*（大豆）和其他一些粮食作物（Sheck 和 Gould 1993）；但是烟青虫 *H. subflexa* 仅以 *Physalis*（酸浆）属（如地莓 *P. pruinosa*；Laster，Pair 和 Martin 1982）为食。有意思的是，在野外烟芽夜蛾 *H. virescens* 是不以酸浆属的植物为食的。因为遗传差异最可能集中在参与宿主选择的位点（Sheck 和 Gould 1993）和交配识别上（Groot et al. 2004），因此，烟芽夜蛾 *H. virescens* 和烟青虫 *H. subflexa* 组合是用于研究导致不同宿主范围的遗传差异是如何进化的好模型。

已经有研究检查了烟芽夜蛾 *H. virescens* 和烟青虫 *H. subflexa* 宿主范围的遗传基础。Sheck 和 Gould（1993）通过将烟芽夜蛾 *H. virescens*、烟青虫 *H. subflexa*、它们的杂交 F_1 代和烟青虫 *H. subflexa* 回交代置于棉花、大豆、烟草（*H. virescens* 的宿主）和毛酸浆 *Physalis pubescens*（*H. subflexa* 的宿主）上，分析了它们幼虫的表现，发现两物种在它们自己的宿主植物上成活且体重增加但在非宿主植物上表现很差。杂交 F_1 代在所有的宿主植物上都能成活，但在被测的四种植物上体重增加水平中等。烟青虫 *H. subflexa* 回交代幼虫在棉花、大豆和烟草上的成活率不及在毛酸浆 *P. pubescens* 上的存活率；幼虫在棉花和烟草上体重的增加不及在大豆和毛酸浆 *P. pubescens* 上。对该结果的分析表明来源于烟芽夜蛾 *H. virescens* 的基因对幼虫在棉花和烟草上的存活和体重增加起部分主导作用，但对在大豆上的存活和体重增加起附加作用。来源于烟青虫 *H. subflexa* 的基因对于在毛酸浆 *P. pubescens* 上的存活和体重增加起主导作用，因此回交代可以更好地存活并且和烟青虫 *H. subflexa* 一样能增加体重。但是上位效应或者基因－环境互作也显然参与其间，因为仅仅是附加和主导作用不能解释这些结果。在后续的研究中，对和烟青虫 *H. subflexa* 不断回交产生的世代基于其幼虫在大豆上的表现水平进行筛选，来鉴定对多种植物选用的遗传结构（Sheck 和 Gould 1996）。在经过数代的筛选后，幼虫对棉花、大豆、烟草和毛酸浆 *P. pubescens* 的取向和表现进行了测试。尽管在大豆上的表现得到了改善，但是在棉花、烟草或毛酸浆 *P. pubescens* 的表现无相关变化，这表明在这些植物上的表现有独立的遗传基础。有趣的是，幼虫对大豆的取向，尽管并不是选择的标尺，但是也得到了提高，说明幼虫对大豆的取向和表现具有相同的遗传基础。幼虫在毛酸浆 *P. pubescens* 的表现和烟青虫 *H. subflexa* 在毛酸浆 *P. pubescens* 的表现无异，表明选择大豆的基因掺入到烟青虫 *H. subflexa* 基因背景下并不涉及对选择毛酸浆 *P. pubescens* 能力的折中。

Sheck 和 Gould（1995）还测试了烟芽夜蛾 *H. virescens*，烟青虫 *H. subflexa* 和它们的杂交 F_1 代的排卵行为。在实验中，雌性成虫被置于棉花、大豆、烟草和毛苦蘵 *Physa-*

lis angulata(一种 *H. subflexa* 喜爱的宿主)上。烟芽夜蛾 *H. virescens* 大多数在烟草上产卵,而有些在其他植物上产卵;烟青虫 *H. subflexa* 大多数在毛苦蘵 *Physalis angulata* 产卵,但偶尔也能在非宿主植物上产卵;在以烟芽夜蛾 *H. virescens* 和烟青虫 *H. subflexa* 互为父本或母本的杂交雌性 F_1 代,倾向于在烟草上产卵,表明来源于 *H. virescens* 的基因起主导作用。遗传是来自于常染色体,因为没有任何证据表明性别连锁的基因影响产卵表现、幼虫行为或者幼虫取向(Sheck 和 Gould 1993,1995)。

在最近对种间杂交代的实验中(作者和 F. Gould 未发表的合作研究),我们进一步研究了烟芽夜蛾 *H. virescens* 和烟青虫 *H. subflexa* 宿主范围的遗传基础。在实验室中,我们通过回交将来源于一个物种的基因掺入到另一个物种中,并在棉花和毛苦蘵 *P. angulata* 上对回交后代进行检测。运用扩增片断长度多态性(AFLP)标记和已发表的基因序列多态性,我们构建了覆盖烟芽夜蛾 *H. virescens* 和烟青虫 *H. subflexa* 31 条染色体的连锁图谱(关于染色体数目,参见 Chen 和 Graves 1970; Sheck et al. 2006),并运用了数量性状位点(QTL)分析来构建幼虫行为差异的遗传结构。在棉花上的实验,我们作了五代回交。为获得一到四代,雌性杂种和烟青虫 *H. subflexa* 雄性交配;这些交配产生的源自雌性的分子标记使我们能鉴定导致表型差异的被掺入染色体。为获得回交五代,雄性杂种和烟青虫 *H. subflexa* 雌性回交,使我们得到源自雄性的分子标记用于染色体内图谱构建。对第一代回交幼虫的初步分析,幼虫觅食的逐步退化对比有无来源于烟芽夜蛾 *H. virescens* 的染色体确定了 6 条可共同解释幼虫摄食棉花所表现出的 39%的差异的染色体(未发表的数据)。其中 4 条烟芽夜蛾 *H. virescens* 染色体增加了对棉花的摄食量,携带有这四条染色体的幼虫和烟芽夜蛾 *H. virescens* 的表型无差异。被掺入的两条烟芽夜蛾 *H. virescens* 染色体产生了未预计到的效果:它们不是增加而是降低了回交幼虫的棉花摄食量。可能这些染色体携带有摄食其他宿主植物的基因和摄食棉花的基因存在基因上位作用。因为回交幼虫携带烟青虫 *H. subflexa* 的纯合等位基因或者携带烟芽夜蛾 *H. virescens* 和烟青虫 *H. subflexa* 的杂合等位基因,来自于烟芽夜蛾 *H. virescens* 的掺入基因至少起附加作用,也可能起主导作用。性染色体中的一条也是增加棉花摄食量的染色体之一,尽管它的影响并不比常染色体大。在杂种雌性(W_VZ_S)和烟青虫 *H. subflexa* 雄性(Z_SZ_S)回交代(BC)中,所有的雌性后代都有来自于烟青虫 *H. subflexa* 的 Z 染色体和来自于烟芽夜蛾 *H. virescens* 的 W 染色体,但是所有雄性后代只有来自于烟青虫 *H. subflexa* 的两条性染色体。这说明可能有位于 W_V 的基因能够增加棉花的摄食,但这个解释可能性不大,因为 W 染色体上能表达的基因很少(参见第三章和第四章,有关 W 连锁的候选基因和 W 染色体的分子组成);或者在不同性别上存在整体的差异。尽管我们还没有对 BC_5 幼虫进行 QTL 定位,但是具有和烟芽夜蛾 *H. virescens* 相似表型的 BC_5 幼虫出现频率表明几个基因就能解释很大一部分棉花摄食差异。最近的理论和证据显示寻找几个解释大部分数量性状差异的 QTL 的设想并不奇怪(Orr 2001,2005; Remington,Ungerer 和 Purugganan 2001)。

为研究幼虫在毛苦蘵 *P. angulata* 上行为的遗传学,我们通过和烟芽夜蛾 *H. virescens* 回交将烟青虫 *H. subflexa* 的基因引入烟芽夜蛾 *H. virescens* 遗传背景中。当幼虫以 *P. angulata* 果实为食,烟青虫 *H. subflexa* 同化作用的效率(幼虫所增加的体重/消耗的每克果实)比烟芽夜蛾 *H. virescens* 高 30 倍,尽管烟芽夜蛾 *H. virescens* 幼虫也愿意以毛苦蘵 *P. angulata* 为食。回交幼虫的表型在烟芽夜蛾 *H. virescens* 相似表型和烟青虫 *H. subflexa* 相似表型范围内变化。5 条被引入的染色体影响了回交幼虫在毛苦蘵 *P. angulata* 上的行为,共同解释了 45%的同化作用效率的差异。和棉花实验的结果相似,3 条染色体增加了同化作用效率,而另外两条降低了同化作用效率。这 3 条染色体的存在导致了和烟青虫 *H. subflexa* 相似的表型(未发表的数据)。

很多研究精力都放在弄清楚烟芽夜蛾 *H. virescens* 如何检测和选择宿主植物和配偶这些问题上。21 个编码嗅觉受体蛋白的基因——每个蛋白都来源于不同的嗅觉神经细胞群——已经在烟芽夜蛾 *H. virescens* 中测定了序列(Krieger et al. 2002,2004)。触角神经叶的结构和神经分布模式显示至少有 30 到 60 种嗅觉神经细胞和其对应的嗅觉受体蛋白(Mustaparta 2002; Rostelien et al. 2005)。而且,基于对植物气味的电生理反应,16 种嗅觉神经细胞已经得到了确定,这些细胞对独特的植物气味作出细微的感应(Rostelien et al. 2005)。这些受体基因将提供对两物种间宿主特异性差异的可能解释。

2.2 铃夜蛾(*Helicoverpa*)

和实夜蛾 *Heliothis* 一样,铃夜蛾属 *Helicoverpa*(对该属的详细信息见第十二章)包括具有广泛宿主范围的物种,例如对棉铃虫 *H. armigera*,有记录的宿主涵盖多个目超过 150 种植物(Zalucki et al. 1994),对玉米夜蛾 *H. zea*,有记录的植物宿主涵盖 11 个目至少 34 种植物(Sudbrink 和 Grant 1995);还包括具有狭窄宿主范围的物种,例如棉铃虫 *H. assulta*,有记录的植物宿主仅限于茄科 Solanaceae 中的某几种物种(Fitt 1989)。对这些物种的群体差异、产卵趋向和幼虫行为的遗传性在实验室研究中都得到检测。来自于澳大利亚不同区域的棉铃虫 *H. armigera* 群体对宿主植物(玉米、高粱、土豆、棉花、豇豆、苜蓿)的产卵趋向上并没有差异,但是群体内的雌性却表现出对这些植物的遗传差异(亲子交配退化,Jallow 和 Zalucki 1996)。除了表现出在这些植物上产卵的遗传差异,雌性棉铃虫 *H. armigera* 还表现出学习能力:雌性优先选择在曾经接触过的植物上产卵(Cunningham et al. 1998)。另一个实验室对澳大利亚棉铃虫 *H. armigera* 的研究发现,对比一种不受欢迎的宿主植物陆地棉 *Gossypium hirsutum* (锦葵科)(Gu,Cao 和 Walter 2001),全同胞亲子交配退化表现出棉铃虫 *H. armigera* 在一种本土地理范围中喜爱的宿主植物苦苣菜 *Sonchus oleraceus* (紫菀科) 上产卵的时候,高遗传性(60%)(Gu 和 Walter 1999)。尽管棉铃虫 *H. armigera* 幼虫在苦苣菜 *S. oleraceus* 上,比在陆地棉 *G. hirsutum* 上存活更好,体重增加更多,但幼虫的表现和产卵的取向在遗传上并没有联系(Gu,Cao 和 Walter 2001)。在对澳大利亚棉铃虫 *H. armigera* 全同胞/半同胞的实验中,幼虫增加更多的体重(对初生幼虫达到 73%,对三龄幼虫达到 23%),但是对 *Cicer arietinum*(鹰嘴豆;Cotter 和 Edwards 2006)的抵抗性或者易感性上并不表现出生存差异。遗传力对抗性或易感性不同的品系的幼虫体重的增加作用很强,但是对产卵能力毫无作用,并且不同品系的雌性在产卵上没有差别。

在实验室研究中,来自于北美洲不同区域的棉铃虫 *Helicoverpa zea* 群体(对这些地域,蛾子是本地固有的)对宿主植物的物种和品系上表现出产卵差异(多毛的对比光滑的大豆和棉花），在一个群体中，被检测到的排卵倾向性上是可遗传的（虽然差异很大；Ward et al. 1993）。

在有关 *Helicoverpa* 属宿主范围的遗传学研究最有趣的实验中，泛化性棉铃虫 *H. armigera* 和专一性棉铃虫 *H. assulta* 种间交配（F_1，F_2 和回交代）表明，至少一个主要的常染色体基因参与了幼虫对棉花的摄食而且棉铃虫 *H. armigera* 的等位基因对于 *H. assulta* 的等位基因是部分显性的（Tang et al. 2006）。

2.3 凤蝶（*Papilio*）

广泛定义的 *Papilio* 属包括大约 205 个物种（实际上，这可能代表了 6 个属），它们的祖先摄食 Rutaceae（芸香科）的物种；80%的 *Papilio* 属物种现在仍然摄食这个属的植物（Zakharov，Caterino 和 Sperling 2004）。但是，数个进化枝已经从 Rutaceae 的摄取中分化开来，包括 *glaucus* 复合群（*Papilio*[*Pterourus*] *glaucus*，*P. canadensis* 和相关物种），它们摄取至少 8 个目的植物物种（Bossart 和 Scriber 1995a）；*machaon* 复合群（*Papilio machaon*，*P. zelicaon*，*P. oregonius* 和相关物种），它们摄食 Apiaceae（Umbelliferae，伞状花科）和 Asteraceae（紫菀科，Sperling 和 Harrison 1994）的物种。

在紫菀科和芸香科的超过 60 个物种上，都有报道黄色黑斑凤蝶 *Papilio zelicaon*（Wehling 和 Thompson 1997），但是 *P. oregonius* 仅在紫菀科的单一物种上（龙蒿 *Artemisia dracunculus*）有报道（Thompson 1988）。在对 *P. zelicaon*，*P. oregonius* 和它们相互交配的 F_1 代的实验室研究中，每种雌性对合适的野外宿主都表现强烈的产卵倾向性；但是它们的杂种表现的倾向性和它们的父本相似，这表明 Z 性染色体上存在一个主要的位点，尽管有来自常染色的基因修饰了这种倾向性（Thompson 1988）。每个物种的幼虫在合适的宿主植物上存活率都很高，但是杂交幼虫对父母本的宿主植物的存活能力都处中等水平，说明常染色体基因遗传有附加效应（Thompson，Wehling 和 Podolsky 1990）。在另一方面，杂种蛹的重量，以及在次要程度上，发育时间和它们的母本相似，表明了母本基因的影响，但并不是性连锁的（Thompson，Wehling 和 Podolsky 1990）。

在针对 5 个 *machaon* 复合群物种对 5 个伞状花科和紫菀科物种的产卵倾向性的等级划分研究中发现，成虫（蝴蝶）表现出从窄到广的倾向等级（Thompson 1998）。一对姐妹种（*P. machaon*/*P. oregonius*）在倾向上等级分明，*P. machaon* 在几乎所有的宿主植物上产卵而 *P. oregonius* 只在一种几乎不会被 *P. machaon* 用到的植物上产卵，但另一对姐妹种（*P. polyxenes*/*P. zelicaon*）对植物的倾向上表现相似（Thompson 1998）。*P. machaon* 物种内的群体，以及在次要程度上，*P. polyxenes* 和 *P. zelicaon* 物种内的群体，对植物倾向性的等级划分上有一定差异，而这些差异可能对宿主范围的变迁提供原始依据（Thompson 1998）。例如，一些 *P. machaon* 种内群体中雌性个体，在龙蒿上只产少量的卵，而龙蒿是 *P. oregonius* 唯一的宿主。*P. oregonius* 转变到在这种植物上产卵是比较容易的，因为 *P. oregonius* 的产卵倾向性是性连锁的而且可能只涉及极少的几个位点（Thompson 1988，1998）。但是，这并没有解释这个变化为什么导致 *P. oregonius* 在宿主范围内放弃了其他的宿主植物。尽管对 *P. zelicaon* 群体所在地的植物表现出倾向性，但蝴蝶并非强烈选择当地的植物，而在这些群体内倾向性的遗传差异，与所测定的全同胞家系内的差异一样(Thompson 1993；Wehling 和 Thompson 1997)。对当地植物缺乏强烈兴趣可能是由涉及对某些植物倾向性的共适应基因复合体所导致的，由阻止选择应答的基因在群体中流动所导致，或者由缺乏对当地植物有强烈选择适应的基因所导致(Thompson 1993；Wehling 和 Thompson 1997)。

Bossart(1998,2003)，Bossart 和 Scriber(1995a，b，1999)针对地理上不同分布的 *P. glaucus* 群体产卵倾向性差异和在三种植物物种(金边马褂木 *Liriodendron tulipifera*，北美木兰 *Magnolia virginiana* 和黑野樱 *Prunus serotina*)上幼虫的表现差异开展了一系列实验室工作，这些群体在野外对这三种植物有不同的接触。来自于佛罗里达的雌性——在那里，北美木兰 *M. virginiana* 常见而金边马褂木 *L. tulipifera* 少——相比那些来自于金边马褂木 *L. tulipifera* 常见而北美木兰 *M. virginiana* 少(佐治亚)或根本没有(俄亥俄)的地方，在北美木兰 *M. virginiana* 上产卵更多(Bossart 和 Scriber 1995a)。但是，正如加州的 *P. zelicaon*(Thompson 1993)，佛罗里达的 *P. glaucus* 并没有对当地的植物表现出强烈的产卵倾向性。俄亥俄的群体在金边马褂木 *L. tulipifera* 和北美木兰 *M. virginiana* 间表现出强烈的产卵倾向遗传性(0.81)，有些家族在两者上都能产卵，而有些仅在金边马褂木 *L. tulipifera* 上产卵(Bossart 和 Scriber 1999)。来自于佛罗里达和佐治亚的幼虫，相比来自于俄亥俄的幼虫，在北美木兰 *M. virginiana* 上表现更好(通过发育时间和蛹的重量进行测定)，表明尽管来自三个地域的幼虫在金边马褂木 *L. tulipifera* 上都表现最好，仍然发展了对当地可获得的植物的倾向性(Bossart 和 Scriber 1995a；Bossart 2003)。这些地理分布不同的群体在同种异型酶的频率上并没有差异，表明在群体间的基因流被当地的选择所抵消，从而维持了产卵倾向性和幼虫表现上的差异(Bossart 和 Scriber 1995a)。对全同胞家系幼虫表现上的比较发现，来自佛罗里达的群体，其幼虫在北美木兰 *M. virginiana* 上的表现没有遗传性，但另两个群体，对这种宿主植物的幼虫表现有很显著的遗传性，而且三个群体在黑野樱 *P. serotina* 上的幼虫表现也有很强的遗传性(Bossart 1998)。在三个宿主上的幼虫表现或者是遗传上不相关或者是正相关，表明在宿主植物适合度上没有折中妥协(Bossart 1998)。在俄亥俄杂食性群体中，若母代倾向在金边马褂木 *L. tulipifera* 上产卵，那么其幼虫比母代更倾向在北美木兰 *M. virginiana* 上产卵的幼虫表现更好，而无论它们是在哪种宿主植物上喂养，这揭示了对北美木兰 *M. virginiana* 倾向性和表现力是负相关的(Bossart 2003)。考虑到俄亥俄的群体没有和北美木兰 *M. virginiana* 接触过，这个结果并不令人奇怪，因为缺失了对倾向性一表现力关系的选择。对于佛罗里达单食性群体，产卵的倾向性和幼虫表现的 4 个测量指标中的三个无关，但和第 4 个指标呈负相关(Bossart 2003)，这表明选择同时改变了产卵倾向性和幼虫表现力(Bossart 和 Scriber 1995a)，但是还不足以导致正相关的关系。尽管倾向性一表现力关系呈现出具有遗传基础，但是这并非优化的产卵理论所期待的，可能是因为来自于控制倾向性和表现力的基因共适应，基因多效性或者基因上位作用的限制(Bossart 2003)。

普遍来说，细胞色素 P450 单氧化酶在调节和活性方面的差异被认为是构成鳞翅目昆虫宿主选择差异的因素，特别是 Papilionids(凤蝶科，参见综述 Berenbaum 和 Feeny 2008)。香芹黑凤蝶 *Papilio polyxenes* 特异性针对伞状花科和芸香科中具有高浓度特异性的 furanocumarins(xanthotoxin——花椒毒素和 angelicin——当归根素)的物种,并且具有专门针对这些异源激素且活性很高的 P450,这些 P450 对代谢其他异源激素的效率不高;青灰叶下珠、*P. glaucus* 和四季杨 *P. canadensis* 具有更为广阔的宿主范围,其 P450 可以代谢多种异源激素,虽然效率相对较低但能被强烈地诱导(Li,Schuler 和 Berenbaum 2007)。青灰叶下珠、*P. glaucus* 和四季杨 *P. canadensis* 之间 P450 在调节和活性方面的差异可能造成它们在宿主选择上的不同(Li,Schuler 和 Berenbaum 2007)。

2.4 金蝶(*Euphydryas*)

Euphydryas 属(蛱蝶)在广泛意义上来说,具有 14 个种(Zimmermann,Wahlberg 和 Descimon 2000)。它们的幼虫以产生 5 个环烯醚萜类的植物为食,Neartic *Euphydryas* 特异性地针对玄参科 Scrophulariaceae 和车前科 Plantaginaceae 为目的植物,这些植物产生蝴蝶隐蔽的环烯醚萜类糖苷(Zimmermann,Wahlberg 和 Descimon 2000,以及该文中的参考文献)。这些蝴蝶在宿主植物利用上分散少并且显示了群体间的差异。这些特征,以及在野外可以操控和测量的产卵行为,导致了一系列针对遗传学和宿主选择的进化学的研究工作,特别是针对帝网蛱蝶 *E. editha* 的研究。在文献中都有记录 *E. editha* 不同群体快速改变宿主植物种群(参见综述 Singer et al. 2008)。一个群体中大多数雌性在本地的一种植物 *Collinsia parviflora*(Scrophulariaceae,玄参科)上产卵,变迁到大多数雌性在一种人为引入的植物长叶车前 *Plantago lanceolata*(Plantagenaceae)上产卵。这个改变发生得很快,在 8 个世代中从 5%增长到 53%的雌性表现出光照后产卵的倾向性(Singer,Thomas 和 Parmesan 1993)。在改变的早期(1983～1984),野外和实验室测量的光照后产卵倾向性基于母子回归,表现出 0.90 的遗传力,尽管在存在母本效应的情况下,遗传力可能被高估(Singer,Ng 和 Thomas 1988)。很显然,幼虫在外来的植物上适应良好,因此对幼虫的行为来说遗传上的改变可能根本不需要(Thomas et al. 1987)。的确,幼虫在外来植物上比在 *C. parviflora* 上表现更好,因为外来植物和蝴蝶的物候关系更相配(Singer 1984)。到 1985 年,排卵倾向性和宿主植物的相互作用解释了 32%的幼虫行为上的差异(以体重增加为尺度),幼虫在它们母本选择产卵的植物上表现更好(Singer,Ng 和 Thomas 1988)。在外来植物上更好的幼虫表现,排卵倾向性的高遗传性,加上产卵倾向性和幼虫行为之间弱相关性,解释了这个改变为什么会这么快(Singer et al. 2008)。另一个 *E. editha* 的群体,从大多数雌性倾向于在一种本地的植物 *Pedicularis semibarbata* (Scrophulariaceae)上产卵,呈指数降低到大多数雌性选择到一个不同的本地植物 *Collinsia torreyi*(Scrophulariaceae)上产卵,与该植物的适合度呈指数增长(Singer,Thomas 和 Parmesan 1993)。随之改变的是光照后产卵的倾向性,而且在 12 个世代中改变很快(Singer 和 Thomas 1996)。但是,这些倾向性在不同的区片类型间(岩石地表上的 *P. semibarbata* 对比林木覆盖地域的 *C. torreyi*)有所差别,雌性倾向于在最丰富和最适合它们的区片类型的植物上产卵(Singer 和 Thomas 1996)。有趣的是,当林木覆盖区随后与 *C. torreyi* 不相适合时,产卵频率回复到原点。根据 *E. editha* 种群快速进化的现象,可以得出两个有关宿主范围遗传结构的结论:第一,选择是强烈的;第二,在起始的群体中或者存在一定数量的遗传差异,或者突变支持了这些差异;如果是后者,那么就涉及少数简单相互作用的基因。

2.5 其他系统

在 *Depressaria patinacella*（Oecophoridae，织叶蛾科）全同胞家系的实验室研究中，其幼虫表现出以它们的原始宿主欧独活 *Pastinaca sativa*（Apiaceae）果实为饲料的存活率上的遗传差异，遗传差异同样表现在以一种新植物短毛牛防风 *Heracleum lanatum*（Apiaceae）果实为饲料上（Berenbaum 和 Zangerl 1991），以及代谢不同浓度的蔓荆喃香豆素类上（Berenbaum 和 Zangerl 1992）。但是幼虫在喂食倾向性上没有遗传差异，这说明对植物异源激素的适应是生理上而非行为上的（Berenbaum 和 Zangerl 1991，1992）。

对于水稻和玉米品系的草地贪夜蛾 *Spodoptera frugiperda*（Noctuidae）的实验室研究中，所有品系的幼虫在水稻上表现最好，水稻品系在玉米上表现差而玉米品系在两种宿主上都表现很好（Prowell，McMichael 和 Silvain 2004）。通过品系内全同胞家系的环境互作的基因型分析表明变异能促进宿主相关的差别（Pashley 1988）。

以 Larch（落叶松）和 Pine（松树）为宿主的松线小卷蛾 *Zeiraphera diniana*（Tortricidae）对称交配（Emelianov et al. 2003）。两种群相互杂交的全基因组变异说明对宿主的选择存在于松线小卷蛾 *Z. diniana* 基因组中很小的区域中，提示只有有限数目的基因参与选择其他的宿主（Emelianov，Marec 和 Mallet 2004）。这两个种群在落叶松上产卵相对于在松树上产卵存在差异，但是对于这些植物的气味有同样的电触角反应；可是这些植物在数目上和在能引起反应的刺激浓度上是不同的（Syed，Guerin 和 Baltensweiler 2003）。因此两寄主能区别两植物物种，但它们根据这些信息将采取什么样的决定是不同的。

以雪松和柏树为宿主的 *Mitoura*（Lycaenidae，灰蝶科）属，排卵倾向性和幼虫的行为关系紧密（Forister 2004），但是这一相关性在宿主不同的种群间相互交配的 F_1 代上丢失了（Forister 2005）。杂种幼虫在柏树上的存活率和以柏树为宿主的种群是一样的，但是杂种幼虫在雪松上的存活率比以雪松为宿主的种群低 30%。杂种雌性倾向于在雪松上产卵，相同的宿主导致了杂种幼虫存活率的降低。因此对雪松的产卵倾向性是显性的，杂种和以雪松为宿主的种群的存活率是没有分别的；但是在雪松上幼虫的表现是隐性的，杂种和柏树群体的幼虫表现没有区别。

比较宿主可能最终导致比较物种。例如，玉米螟蛾 *Ostrinia nubilalis* 在艾属植物和玉米上的宿主群体在遗传上可以分离（Martel et al. 2003；Bethenod et al. 2005），对称交配（Malausa et al. 2005），并且产生只吸引来自于同一宿主种群的雄性的不同的信息素（Pelozuelo et al. 2004）。最近这些玉米螟蛾 *O. nubilalis* 被确定是不同的物种（Frolov，Bourguet 和 Ponsard 2007）。不论它们是不同宿主还是同形种，研究宿主选择差异的遗传基础是有用的；的确，拥有不同的宿主范围而又紧密相关的物种间杂交可能是最有助于决定宿主范围遗传结构的手段。

对三个紧密相关的 *Y ponomeuta*（巢蛾科）物种，杂交 F_1 代和回交代的实验室研究中，杂交 F_1 代和回交代在欧卫矛 *Euonymus europaeus*（Celastraceae，卫矛科）——*Y. cagnagellus* 的正常宿主上产卵的特性，对于在 *Y. padellus* 的正常宿主黑刺李 *Prunus spinosa*（Rosaceae)上产卵和在 *Y. malinellus* 的正常宿主苹果 *Malus domestica*(Rosaceae)上产卵是部分显性的(Hora,Roessingh 和 Menken 2005)。相互间的交配结果一样,表明参与的基因是常染色体连锁的而非性染色体连锁的。在这些实验中,*Y. padellus* 和 *Y. malinellus* 在非宿主欧卫矛 *E. europaeus* 上可产一些卵,可能是因为还保存了在作为 *Y. ponomeuta* 原始宿主——卫矛科(Celastraceae)物种上产卵的能力(Menken 1996; Hora,Roessingh 和 Menken 2005)。

苜蓿黄蝶 *Colias eurytheme* 和黄色蝴蝶 *C. philodice*(Pieridae,粉蛾科)显然是截然不同的物种,因协调交配而维持着可检测到的差异。不过,它们能杂交,这可能导致它们对于一些新的,被引入的宿主植物物种的适应性上没有差异(Porter 和 Levin 2007)。只是,遗传上的相关性和对宿主植物适应度成分的可遗传性上的差异表明它们在宿主选择的遗传结构上可能不同(Porter 和 Levin 2007)。

2.6 结论

当前对鳞翅目昆虫宿主范围的遗传架构的了解,大多数都局限在对遗传力的估计、支配关系和基因的定位上。尽管 QTL 定位的研究正在进行中,现在对基因的定位只能判定是处于常染色体上还是性染色体上。排卵倾向性和幼虫表现的遗传力可能很高(例如 60%~90%),也可能是零。支配关系可以从附加到完全主导。幼虫表现倾向于由常染色体基因控制,而排卵倾向性则由性染色体基因控制,但是这个趋势仍是比较弱的。最后,不断积累有关相近物种间和宿主间宿主范围上差异的证据,这些证据显示宿主范围有相对简单的遗传构架,涉及极少几个分离因子(少于 10 个)可能基因的上位作用。

3. 幼虫和成虫性状的整合

因为排卵倾向性和幼虫表现对宿主范围的进化和物种形成有作用,所以投入了很多对这两者之间关系的研究。除了极少见的情况,鳞翅目昆虫成虫饮用花蜜,有时候也食花粉(如果它们完全摄食),但是它们的幼虫吸食植物的组织。成虫依靠视觉、气味和味道选择它们的食物(Ramaswamy 1988; Fitt 1991)并且可以不适于幼虫发育的多种宿主植物的花蜜为食。雌性成虫的排卵部分由视觉外形决定,但主要由完整植物表面的气味和味道所决定。幼虫摄食则大部分由完整植物表面和被浸软的组织的气味和味道所决定,尽管幼虫在从一株植物迁移到另一株植物的时候也可能用到视觉方面的线索。幼虫能否在宿主植物上存活并发育成健康的成虫取决于它们的消化系统,包括对植物化学物质的解毒能力(Berenbaum 和 Zanger(1992; Berenbaum,Cohen 和 Schuler 1992; Hung et al. 1995; Rose et al. 1997; Stevens et al. 2000; Li et al. 2002; Wittstock et al. 2004; Zagrobelny et al. 2004; Berenbaum 和 Feeny 2008)和它们所消化的植物组织之间的关系,以及它们的营养需要(Lee,Behmer 和 Simpson 2006),特别是对必需营养或它们自己不能产生的防御性化学物质的需要(Engler-Chaouat 和 Gilbert 2007)。对宿主的选用包括两组性状之间的平衡:成虫的性状(例如在一个相对大的地域里宿主的位置,对适合的排卵位置的识别和接受,成功地寻找到配偶)和幼虫的性状(例如在合适的宿主上摄食,一旦失去定居地之后在宿主植物上的重新形成群落,宿主植物被耗尽后在新的宿主植物上定居下来，以及对植物防御性成分的适应能力）。在历史上，排卵选择和幼虫表现一直被认为是有联系的，因此雌性成虫倾向于在那些能支持幼虫最好表现的宿主植物上产卵，即排卵倾向性被幼虫的宿主所影响（例如，Darwin 1909）。但是，考虑到这些性状有不同的选择成分所影响，而且可能被不同组的基因所控制，排卵倾向性和幼虫表现完全的整合是不可能的（Scheirs 和 De Bruyn 2002；Quental，Patten 和 Pierce 2007）。

所观测到的排卵倾向性和幼虫表现之间的对应关系可以从极好到很差（参见综述，Thompson 和 Pellmyr 1991）。最近对烟青虫 *H. subflexa* 的研究揭示了就算是最极端的专一种也不总是在那些支持幼虫表现最大化的宿主植物上排卵。在涉及 7 个酸浆属 *Physalis* 物种的一个普通的花圃试验中，野生的烟青虫 *H. subflexa* 雌性产卵倾向和幼虫表现没有关系（Benda 2007）。毛酸浆 *Physalis pubescens* 是成虫产卵倾向性最强的植物，但烟青虫 *H. subflexa* 的幼虫在 *P. angulata* 和 *P. philadelphica* 上摄食最好，但这两种植物并不是成虫倾向于产卵的宿主。在墨西哥自然生成的 17 种 *Physalis* 属物种（*Physalis* 属多样性的中心），烟青虫 *H. subflexa* 幼虫在毛酸浆 *P. pubescens*，苦蘵 *P. angulata* 和 *P. philade phica* 上的密度没有区别，并且大大高于另外 10 种被烟青虫 *H. subflexa* 摄食的 *Physalis* 属物种（Bateman 2006）。在实验室的检验中，烟青虫 *H. subflexa* 幼虫在苦蘵 *P. angulata* 上成活最好（46％的新生幼虫能成活到蛹期）但是在毛酸浆 *P. pubescens* 上（34％）和 *P. philadelphica* 上（30％）上都相对较差。有趣的是，差的宿主选择并不限于成虫：幼虫摄食同样不能可信赖地反映其行为。对 13 个 *Physalis* 属物种的检验中，幼虫因饥饿（没有摄食倾向）而死亡的比例很高，在宿主植物上从 25％到 83％（Bateman 2006）。如果幼虫仅仅不摄食那些它们在其上表现很差的植物，我们就可能得出这样的结论：拒绝在这些次优的宿主植物上摄食是一种适应。实际上，愿意摄食和成活到蛹期并不一致：只有 52％的新生幼虫尝试摄食苦蘵 *P. angulata*，但是 89％的幼虫成活到蛹期；相反的，75％的幼虫试图摄食 *P. philadlephica*，可只有 43％的幼虫活到蛹期。特别在没有其他选择的情况下不清楚为什么幼虫拒绝摄食合适的物种。当供给人工饲料，95％的幼虫摄食，并且它们极不愿意摄食植物材料，这可能反映了幼虫对物种特异性的植物成分的敏感（Bateman 2006）。在任何一种情况下，对于烟青虫 *H. subflexa*，产卵倾向性或幼虫摄食和幼虫表现都不是一致的。

考虑到幼虫和成虫选择上的不同，得出不同的基因控制着产卵倾向性和幼虫表现这一结论并不吃惊（Thompson，Wehling 和 Podolsky 1990；Sheck 和 Gould 1993，1995，1996）。就算对于排卵倾向性和幼虫表现间存在非常强烈联系的例子，这个联系似乎反映的是对这些性状的独立地选择，而非共同拥有一个遗传基础。如果幼虫和成虫对宿主选用的性状被同一染色体上的基因所控制，物理上的连锁可能造成它们协同进化。但是，影响幼虫表现的基因一贯被定位在常染色体上（Hagen 1990；Thompson，Wehling 和 Podolsky 1990；Sheck 和 Gould 1996；Forister 2005），而影响产卵倾向性的基因就相对不一致，有时候定位在性染色体上，有时候定位在常染色体上（Sheck 和 Gould 1995；Forister 2005；Hora，Roessingh 和 Menken 2005）。许多性状和成虫的相关行为是性连锁的（如雄性对性激素的反应：*Ostrinia*［Dopman et al. 2005］；雌性交配的选择：*Colias*［Grula 和 Taylor 1980］，Arctiidae［Iyengar，Reeve 和 Eisner 2002］；雌性产卵倾向性：

Papilio[Thompson 1988; Scriber, Giebink 和 Snider 1991], *Polygonia*[Nygren, Nylin 和 Stefanescu 2006])。尤其对雄(Z)性染色体连锁,表明在 Z 染色体上找到的基因可能对生殖隔离起不相称的作用,因此对物种形成很重要(Sperling 1994; Prowell 1998)。但是,产卵倾向性的性连锁可能依赖于地理尺度上的比较:Janz(1998)发现具有不同宿主特异性的 *Polygonia c-album* 两个群体间的差异是性连锁的,而 Nylin 等(2005)发现在单一群体中雌性的产卵倾向性存在巨大的差异,但没有证据表明这是性连锁的。不管产卵倾向性是常染色体还是性染色体决定的,至今所有的研究都表明产卵倾向性和幼虫表现是由不同染色体上的基因所控制的。

4. 宿主范围的神经生物学

尽管一些控制宿主使用的基因(例如,那些有关幼虫去除植物异源信息素毒性的基因)可能只影响一个生命阶段,但另一些基因不仅作用于幼虫还影响成虫。最值得注意的是,幼虫和成虫能通过气味和味道去评估潜在的宿主,因此涉及嗅觉和味觉的基因最可能既影响产卵也影响幼虫摄食。成虫和幼虫的化学感受系统是统一的,这个被雌性成虫往往在其幼虫喜欢摄食的宿主上产卵这一事实所证明。对于凤蝶 *Papilio*,成虫产卵和幼虫摄食都被相同的化学物质所刺激(或抑制),这表明相同的化学感受基因导致了成虫和幼虫的宿主选择(Ono, Kuwahara 和 Nishida 2004; Nishida 2005)。而且,P450 降解凤蝶 *Papilio* 成虫的气味,还在幼虫体内去除植物异源信息素的毒性(Ono, Ozaki 和 Yoshikawa 2005),因此可能为成虫产卵和幼虫表现提供联系(Berenbaum 和 Feeny 2008)。

化学受体被广泛地分为嗅觉(Or)受体或味觉(Gr)受体亚家族成员。嗅觉过程由分泌到感觉器的淋巴液中的嗅觉分子结合蛋白(OBP)和嵌入到刺激感觉器的嗅觉受体神经元(ORN)细胞膜上的嗅觉受体蛋白(ORP)所介导。本章没有对昆虫嗅觉的神经生物学作一个全面的综述(关于最近的综述,参见 Mustaparta 2002; Chyb 2004; Rützler 和 Zwiebel 2005; Hallem, Dahanukar 和 Carlson 2006;以及第九章对鳞翅目昆虫化学受体基因系统的讨论)。简单来说,气味分子通过触角和颚部触须(主要和次要的嗅觉器官)上嗅觉感官的孔径,由 OBP 运输到 ORN 的细胞膜上,在此处它们结合 ORP,诱导沿 ORN 轴突传播的动作电位。ORN 的树突刺激感官,轴突则伸入到触觉神经叶的小球内。因此,外周的刺激能快速传递到中枢神经系统中等级较高的处理区域。

ORP 存在高度的多样性,很多 ORP 氨基酸序列的相似性不到 20%。多样性阻碍了对昆虫 ORP 的发现,以和哺乳动物已知的 ORP 相似性为基础的检索在昆虫中并不成功。利用生物信息学(Clyne et al. 1999)和遗传学(Vosshall et al. 1999)相结合的方法最终鉴定出了包括 60 个果蝇 Or 基因的 ORP。确定鳞翅目昆虫的嗅觉受体更具有挑战性,因为鳞翅目昆虫 ORP 的序列和其他生物的 Or 基因相似性都很低。Krieger 等(2002)通过对烟芽夜蛾 *H. virescens* 触角 cDNA 文库所编码的具有部分序列和果蝇 ORP 相似的蛋白进行筛选,发现了 9 个候选 ORP,并且通过原位杂交找到那些 ORP 在 ORN 中表达。这些新发现的蛋白和其他已鉴定的 ORP 在氨基酸序列上相似度很低,而且同源性仅限于果蝇 ORP 的某些小的区段上。寻找鳞翅目昆虫成虫化学感受基因的研究工作,属 Wanner 等(2007)在家蚕中的发现最为成功。一旦烟芽夜蛾 *H. virescens* Or 基因被确定，它们使用传统的序列相似性比较，在新发布的家蚕基因组中确定了 41 个候选 ORP，这其中许多都是烟芽夜蛾 *H. virescens* ORP 的同源基因（见第九章详述）。

现在还不清楚 OBP 和 ORP 在决定鳞翅目昆虫宿主范围中所扮演的角色。在果蝇中，一些 ORP 仅感知单一的气味，而一些 ORP 则能感知更广泛气味。每个 ORP 表达在由 3 到 50 个 ORN 组成的亚群上，而 ORN 的反应特性来源于它们所表达的 ORP 的差异（de Bruyne 和 Warr 2006）。许多 ORN 对相同的气味作出反应，因此同一种气味一般能激活多个受体（Hallem，Ho 和 Carlson 2004；Goldman et al. 2005）。这个结果有助于解释缺失一个嗅觉受体的果蝇通常表现出正常的嗅觉诱导行为（Elmore et al. 2003）。果蝇对气味反应的特异性有赖于综合性地区分气味分子，这个现象在神经元整合的第一个中心——触角神经叶——上被观测到（Ng et al. 2002；Wang et al. 2003），可能鳞翅目昆虫也是这样的。行为上的变化能被选择性的嗅觉神经元亚群失活所诱导（Suh et al. 2004），并且最近一个研究显示超表达单一的气味受体减弱了对苯甲醛——一种参与某些鳞翅目昆虫宿主逃避和吸引的物质——的逃避行为（Stortkuhl et al. 2005）。

OBP 最先在多音天蚕 *Antheraea polyphemus* 蛾子的触角中被发现（Vogt 和 Riddiford 1981），继而在烟芽夜蛾 *H. virescens*（Krieger et al. 2002）、甜菜夜蛾 *S. exigua*（Xiu 和 Dong 2007）、欧洲玉米螟 *O. nubialis*（Coates，Hellmich 和 Lewis 2005）、烟草天蛾 *M. sexta*（Vogt et al. 2002）和其他一些物种上被发现。对 OBP 多种生化作用作出了很多推测，包括输送气味分子通过感觉器淋巴到 ORP 以及在受体激活后对气味分子去活化（Park et al. 2000）；OBP 是否参与了鳞翅目昆虫多个过程仍然需要去探索。对于果蝇，最近由 Matsuo 等（2007）所作的研究暗示 OBP 可能参与了宿主选择。他们检测了 *D. sechellia* 果蝇——海巴戟 *Morinda citrifolia* 的专一种——产卵选择的遗传基础。海巴戟 *M. citrifolia* 对相近物种黑腹果蝇（*D. melanogaster*）是有毒的。海巴戟 *Morinda citrifolia* 的果实吸引 *D. sechellia* 果蝇，而黑腹果蝇 *D. melanogaster* 则排斥这种气味。通过定位基因敲除和替换，他们将黑腹果蝇 *D. melanogaster* 的两个 OBP 基因（*Obp57d* 和 *Obp57e*）替换成 *D. sechellia* 果蝇的基因。这些被改造的果蝇，产卵倾向性极类似于 *D. sechellia* 果蝇。但是这种基因改造可能对任意鳞翅目昆虫种类还不可行，但 OBP 和 ORP 是解释宿主植物选择差异的最具吸引力的候选基因。

对宿主植物接纳的讨论中，经常忽视幼虫对植物的选择。尽管幼虫相对于成虫来说，运动性较弱，但它们可能对寻找最佳宿主更具有驱动性。和成虫一样，在幼虫的宿主识别和接纳的神经生理学研究上取得了进展。幼虫的宿主选择似乎是以一小群触角、颚部触须和咽上部的味觉受体为基础的（Hanson 和 Dethier 1973；de Boer 1993，2006；Glendinning，Valcic 和 Timmermann 1998；Schoonhoven 和 van Loon 2002；Schoonhoven 2005）。在成虫也表达的味觉感受器，由四个味觉受体神经元（GRN）所支配，每个神经元只对甜味、盐分或水的刺激作出反应（Dethier 1976）。GRN 的轴突伸入到中央神经系统的食管下神经节中，食管下神经节也是大脑中第一个味道处理的接力中心。味觉感官在多种昆虫，包括蛾和蝴蝶中都有研究（Zăcharuk 1980）。迄今，Gr 基因只在两种鳞翅目昆虫：烟芽夜蛾 *H. virescens*（Krieger 2002）和家蚕（*B. mori*，参见第九章详述）中得以确定。在果蝇中，已经鉴定了 60 个 Gr 基因（Robertson，Warr 和 Carlson 2003），但仅有几个 Gr 确定了受体的特异性(即糖受体,Dahanukar et al. 2007;二氧化碳受体,Jones et al. 2007;Kwon et al. 2007;苦味受体,Moon et al. 2006)。

Gr 基因所编码的蛋白,比如 OBP 和 ORP,在序列上具有极端的多样,氨基酸序列的相似度只有 8%(Scott et al. 2001)。和 Or 基因一样,Gr 基因在鳞翅目昆虫内的序列相似度比鳞翅目昆虫和其他物种间的相似度要高,例如鳞翅目和双翅目。因此不断增大的鳞翅目昆虫特异性资源(例如 ButterflyBase,Papanicolaou et al. 2008;*B. mori* 基因组项目,Mita et al. 2004;Xia et al. 2004)应该使鉴定多样的鳞翅目昆虫基因更为容易。

考虑到宿主选择/接纳是由诱食信号和戒食信号之间的平衡所决定的(Schoonhoven 1987),那么不可能对宿主范围进行简单的解释(即单基因)。在很多例子中,昆虫身处环境、经历和它的基因组相互作用,从而产生了可观测到的宿主范围。菜粉蝶 *Pieris rapae* 和烟草天蛾 *Manduca sexta* 的幼虫在孵化时是杂食性的,在接触到宿主特异性的物质(对于 *P. rapae* 来说是一种葡糖异硫氰酸盐[Renwick 和 Lopez 1999],对烟草天蛾 *M. sexta* 来说是羟基吲哚 D[del Campo et al. 2001])后变成寡食性的。在这次接触后,推测缺少相关物质的植物被排斥(或者,只有具有这种物质的植物具有刺激性)。对于烟草天蛾 *M. sexta*,外周神经系统活动上的变化发生在接触羟基吲哚 D 后(del Campo 和 Miles 2003),但是这些变化影响幼虫行为的机制仍不清楚。

行为水平上的证据支持宿主范围的变化可能是由对排斥或刺激性化学物质敏感度的变化所引起的(Bernays 和 Chapman 1987;Bernays et al. 2000)。例如,苯甲醛对昆虫摄食欧洲卫矛 *Y. cagnagellus* 没有影响,*Y. cagnagellus* 是一种仍保留着和卫矛科(不含苯甲醛)原初祖先联系的物种,但是刺激那些摄食含有苯甲醛的蔷薇科的昆虫,这些昆虫在进化上处于一个派生出的分化枝上(Roessingh,Xu 和 Menken 2007)。有趣的是,这些在外周敏感度上对宿主相关的化学物质的变化,似乎是宿主范围变化的结果而非原因。在对幼虫和成虫的研究中,拥有广泛多样的宿主范围的物种,具有相似的受体神经元敏感度。烟青虫 *H. subflexa* 和烟芽夜蛾 *H. virescens* 的幼虫,种间行为上的差异不能归功于对排斥或刺激性化学物质的感受神经元反应的不同(Bernays 和 Chapman 2000;Bernays et al. 2000)。对于泛化种棉铃虫 *H. armigera* 和烟芽夜蛾 *H. virescens* 和专一种烟青虫 *H. assulta*,每个物种雌性成虫拥有的 4 种 ORN,对 4 种相同的挥发性植物化学物质作出反应,并且虽然每种类型的 ORN 仅对应单一的分子,但对类似的分子会作出某些相似的反应(Stranden et al. 2003)。幼虫和成虫一样,物种特异的宿主接纳似乎取决于对输入信号的中央处理上的差异,因此宿主范围的变化可能有赖于中枢神经系统的变化(Bernays 和 Chapman 1987;Bernays et al. 2000;Chyb 2004)。

尽管对于何种机制导致不同宿主范围的鳞翅目昆虫对相同的外周输入信号产生出不同的行为反应还没有确定,但已经在果蝇上开展了这个方面的研究。Melcher 和 Pankratz(2005)已经鉴定了一个在味觉中间神经元表达的神经肽(由 *hugin* 基因编码),味觉中间神经元将外周受体神经元和腹部神经索以及咽部器官上的运动神经元连接起来。从这些表达 *hugin* 基因的中间神经元输入的信号,整合了味道、内分泌系统、更高一级的脑部以及改变摄食行为的运动信号。和嗅觉基因以及味觉基因一样,确定鳞翅目昆虫神经肽可能需依赖于发展和运用鳞翅目昆虫特有的资源。例如,在寻找刺激神经肽受体基因的信息素生物合成的研究中发现，鳞翅目昆虫和果蝇基因序列的相似度很低，但是鳞翅目昆虫物种间的相似度较高(Zheng et al. 2007)。

5. 宿主范围进化的方向性

大多数鳞翅目昆虫具有相对较窄的宿主范围，它们仅以能够获取的植物中很小的一部分为食。专一种的优势可能反映了宿主相关的适应性妥协(Jaenike 1990)，被天敌所选择(Bernays 和 Graham 1988)，或者受神经约束(Bernays 2001)。神经约束(即限制信息处理速度的昆虫神经系统的局限性；Dukas 1998)可能是宿主范围执行中许多监控环节运行的手段，因为具有广阔宿主范围的昆虫相对低效地正确接纳或排斥植物(和选择那些使它们的适应力降低的植物)或者需要更长的时间去选择宿主(因此增加了它们暴露于天敌的机会)。雌性成虫选择最佳产卵位置的能力依赖于精确地评估宿主的品质，在这一点上，泛化种似乎差于专一种。在对处于不同品质的荨麻上蛱蝶科三个专一种和两个泛化种的研究中发现，雌性专一种倾向于在高品质的荨麻上产卵，而雌性泛化种则不是这样(Janz 和 Nylin 1997)。所有的幼虫在低品质的荨麻上都表现很弱，因此由雌性泛化种作出的不恰当选择降低了它们生殖的成功率。

幼虫表现同样在专一种和泛化种间有差异，这个现象符合神经约束的猜想。Bernays 等(2000)发现专一种烟青虫 *H. subflexa* 幼虫在未咬食或者在仅咬食一口后就排斥有毒食物，但是泛化种烟芽夜蛾 *H. virescens* 幼虫只会在大量摄食后才会拒绝有毒食物。很显然，专一种依赖对食物的快速感受评价，而泛化种依赖负面的消化后效果。烟芽夜蛾 *H. virescens* 幼虫这种"现在吃，以后再决定"的策略极大地增加了它们摄入致死剂量毒素的风险。无效的选择同样能导致摄食机会降低：对灯蛾两种专一种和两种泛化种幼虫的觅食行为的研究发现，泛化种幼虫需要耗费更多的时间去接纳或拒绝一种植物，并且还拒绝了很多适合的宿主植物(Bernays，Chapman 和 Singer 2004)。

尽管 Mayr(1963)认为宿主范围进化是无方向性的也是不可逆的，这样泛化种成为专一种，专一化形成就是进化的终结，但是近来越来越多的研究工作显示这个论点是错误的(Nosil 和 Mooers 2005)。相反的，从泛化种变成专一种或者从专一种变成泛化种是自由发生的，并不受系统进化的限制(Winkler 和 Mitter 2008)。例如，蛱蝶属系统进化上的宿主选用性状的优化暗示在荨麻上的远祖专一化之后，宿主范围的频繁扩增和收缩(Janz，Nyblom 和 Nylin 2001)。并且，许多种类的幼虫能摄食当下宿主范围之外的植物，且强烈趋向于被蛱蝶属其他物种所使用的宿主(Janz，Nyblom 和 Nylin 2001)。这个趋向和大多数鳞翅目昆虫在宿主使用上惊人的保守性是一致的。相近种类的昆虫经常使用相近的宿主植物，这个现象被系统发生重建(例如，Mitter 和 Farrell 1991；Winkler 和 Mitter 2008)所支持，且可能由保留宿主使用的祖先基因所导致。但是，创新性也是宿主使用进化的一个常见主题。对于蛱蝶属，发生了一些"极端的"宿主范围变化，宿主转变成既不是蛱蝶属祖先的宿主范围也不是其相近的 *Nymphalis-Polygonia* 分化枝的祖先宿主范围。类似的现象在凤蝶科的 Troidini 部落上也观测到了，宿主范围既不反映宿主植物的系统发生也不反映植物的次要化学(Silva-Brandao 和 Solferini 2007)。相反的，Troidini 宿主范围是严格机会性的，地域范围的增加和所使用的植物物种数量的增加是紧密相关的。

宿主范围的进化可能既反映系统发生的限制，也反映了自然选择的建设性影响，即对于特定的植物宿主可获得性以及可适应性价值的反应上的变化。有趣的是，转变到新的植物宿主和 *Polygonia* 种类丰富度的增加是相关的，转变到新宿主范围的分化枝比起那些仅使用祖先宿主群体的姐妹分化枝来说，种类更丰富（Weingartner，Wahlberg 和 Nylin 2006）。该模式和宿主范围的扩增是一致的，驱动在鳞翅目昆虫和其他食叶类昆虫上观测到的上升的多样化速度。在最近对 145 个食叶类物种形成事件的分析发现，整整一半的事件伴随着向新植物宿主物种的转变（Winkler 和 Mitter 2008）。

6. 未来展望

现在我们对宿主范围的遗传架构的理解是有限的。尽管紧密相关的物种和宿主间宿主范围的差异似乎具有相对简单的遗传架构，这个结论还有待对实际涉及的基因进行更具体的分析。解毒酶类和化学感受蛋白序列及表达上的差异被推测是构成宿主范围差异的基础，但它们的完整功能仍需要被确认。我们对此问题更进一步的理解将需要更大规模的实验或者更新的手段，也许两者都需要。宿主范围不同而又密切相关的物种间或种族间的杂交提供强烈的表型差异和独特的分子标记，而表型差异和分子标记将极大地有助于确认那些导致宿主范围差异的基因。并且，那些最近才分化的群体和物种间的宿主范围差异更有趣。该方案最具有前景的系统包括金蝶 *Euphydryas*，铃夜蛾 *Helicoverpa*，实夜蛾 *Heliothis* 和凤蝶 *Papilio* 属的物种和群体。

最有希望更详细地解析宿主范围差异的遗传架构的三个主要策略是：(1)更多和更精细的 QTL 遗传定位，包括结合遗传和物理图谱；(2)对宿主范围不同而又密切相关的物种或种族基因序列和表达差异的分析；(3)定位敲除候选基因后的效果。

QTL 遗传定位是确定影响数量性状（详见综述，Lynch 和 Walsh 1998）的基因位置的数量和互作的有力技术。但是，QTL 定位可能更应该称为 QTR（数量性状区域）定位，因为间隔 QTL 的分子标记之间存在的基因数目可能很大。许多研究者强烈呼吁要比 QTL 定位走得更远以确认作为适应性分化基础的特异遗传变化（例如，Remington，Ungerer 和 Purugganan 2001；Orr 2005）。从 QTL 到候选基因可能挑战性很大，特别是当一些不同的 QTL 影响表型的情况下。即使只有一个 QTL 强烈影响表型，对一个新的模式生物测定间隔 QTL 的分子标记间的序列可能也是困难的。

全基因组整合物理/遗传图谱手段的发展（Chang et al. 2001；Yamamoto et al. 2006）可能显著地加快以图谱为基础的（定位）QTL 克隆。并且，在引进例如基因组测序仪 FLX 系统（454-Life Sciences/Roche Applied Science，Indianapolis，Indiana，USA）、Illumina 基因组分析仪（Illumina，San Diego，California，USA）和 SOLiD 系统（Applied Biosystems，Foster City，California，USA）等超高通量技术后，测序近年来变得更为容易和便宜。Vera 等（2008）使用 454 热测序仪从格兰维尔豹纹蝶、庆网蛱蝶 *Melitaea cinxia* (Nymphalidae) 基因组生成了大约 50 万个高质量的读点。对家蚕 *B. mori* 基因组的 BLAST 搜索产生了大约 9000 个结果。如果，根据对家蚕的估计，大多数鳞翅目昆虫有约 18000 个基因，那么庆网蛱蝶 *M. cinxia* 至少一半的基因和家蚕 *B. mori* 有高相似度。

Vera 等(2008)能检测到数目很大的序列多态性,而序列多态性为 QTL 定位和群体遗传分析所需的遗传标记提供了有价值的资源。而且,大规模的转录组学信息使构建物种特异的芯片成为可能。组装来自于这些新技术的短读点仍是个问题,但使用例如配对一末端测序和对相近物种的基因结构的借鉴等一些手段应该使组装更容易(Goldsmith,Shimada 和 Abe 2005)。

利用基因组细菌人工染色体(BACs)作为探针改善荧光原位杂交(FISH)技术,使鳞翅目昆虫细胞遗传学复兴,它将昆虫小且均一的、没有分化的染色体转变成分析基因组结构的有力工具(Yoshido et al. 2005; Yasukochi et al. 2006; Sahara et al. 2007)。遗传定位加上 BAC-FISH 已经揭示了多种鳞翅目昆虫的同线性(Jiggins et al. 2005; Kaplan et al. 2006; Lee 和 Heckel 2007; Sahara et al. 2007)。结合遗传/物理图谱和同线性使利用已经很好定位和测序的家蚕 *B. mori* 基因组在由寻常锚定位点勾勒的染色体片断上寻找候选基因成为可能。

表达分析结合遗传定位可以确定候选基因是否定位到相同的区段上,因为 QTL 和表型差异是相关的。解毒酶和嗅觉受体蛋白基因序列上的差异可能非常大,因此它们应该在表达分析上很明显地被区分开。另一个检测基因功能的办法是用 RNA 干涉(RNAi)使候选基因表达沉默,RNAi 是通过诱导胞内酶,降解和所引入的双链 RNA 同源的内源 mRNA 从而实现功能的(Bettencourt,Terenius 和 Faye 2002)。一些鳞翅目昆虫物种被遗传改造(Tamura et al. 2000; Thomas et al. 2002; Imamura et al. 2003; Marcus 2005),打开了在合适的组织或发育阶段发展内源性表达 RNAi 的潜能。

分子遗传学上更新的和正在开发的技术具有将鳞翅目昆虫遗传学研究突破模式生物的限制,提升到更好反映鳞翅目昆虫生物学广泛多样性层面上的潜力。和早期的昆虫模式生物系统不同,现在对鳞翅目昆虫的研究是基于探索对那些使昆虫适应各种环境性状的进化的理解。随着快速发展的前沿遗传学和基因组学技术,在过去的半个世纪中,多种鳞翅目昆虫的行为、生态和系统发生学知识不断积累,为我们在不久的将来,对鳞翅目昆虫宿主范围的遗传架构上的研究中做出更多激动人心的进展赋予了希望。

参考文献

[1]Bateman, M. L. 2006. Impact of plant suitability, biogeography, and ecological factors on associations between the specialist herbivore *Heliothis subflexa* G. (Lepidoptera: Noctuidae) and the species in its host genus, *Physalis* L. (Solanaceae), in west-central Mexico. PhD diss., North Carolina State Univ.

[2]Benda, N. D. 2007. Host location by adults and larvae of specialist herbivore *Heliothis subflexa* G. (Lepidoptera: Noctuidae). PhD diss., North Carolina State Univ.

[3] Berenbaum, M. R., M. B. Cohen, and M. A. Schuler. 1992. Cytochrome-P450 monooxygenase genes in oligo-phagous Lepidoptera. *ACS Symp. Ser.* 505:114—24.

[4]Berenbaum, M. R., and P. P. Feeny. 2008. Chemical mediation of host-plant specialization-the papillionid paradigm. In *Specialization, speciation and radiation: The evolutionary biology of herbivorous insects*, ed. K. J. Tilmon, pp. 3—19. Berkeley: Univ. California Press.

[5]Berenbaum, M. R., and A. R. Zangerl. 1991. Acquisition of a native hostplant by an introduced oligophagous herbivore. *Oikos* 62:153—59.

[6]Berenbaum, M. R., and A. R. Zangerl. 1992. Genetics of physiological and behavioral resistance to host furanocoumarins in the parsnip webworm. *Evolution* 46:1373—84.

[7]Bernays, E. A. 2001. Neural limitations in phytophagous insects: Implications for diet breadth and evolution of host affiliation. *Annu. Rev. Entomol.* 46:703—27.

[8]Bernays, E. A., and R. F. Chapman. 1987. The evolution of deterrent responses in plant-feeding insects. In *Perspectives in chemoreception and behavior*, ed. R. E. Chapman, E. A. Bernays, and J. G. Stoffolano, 159—74. New York: Springer.

[9]Bernays, E. A., and R. F. Chapman. 2000. A neurophysiological study of sensitivity to a feeding deterrent in two sister species of *Heliothis* with different diet breadths. *J. Insect Physiol.* 46:905—12.

[10]Bernays, E. A., R. F. Chapman, and M. S. Singer. 2004. Changes in taste receptor cell sensitivity in a polyphagous caterpillar reflect carbohydrate but not protein imbalance. *J. Comp. Physiol. A Neuroethol. Sens. Neural Behav. Physiol.* 190:39—48.

[11]Bernays, E., and M. Graham. 1988. On the evolution of host specificity in phytophagous arthropods. *Ecology* 69:886—92.

[12]Bernays, E. A., S. Oppenheim, R. F. Chapman, H. Kwon, and F. Gould. 2000. Taste sensitivity of insect herbivores to deterrents is greater in specialists than in generalists: A behavioral test of the hypothesis with two closely related caterpillars. *J. Chem. Ecol.* 26:547—63.

[13]Bethenod, M. T., Y. Thomas, F. Rousset, et al. 2005. Genetic isolation between two sympatric host plant races of the European corn borer, *Ostrinia nubilalis* Hubner. II: Assortative mating and host-plant preferences for oviposition. *Heredity* 94:264—70.

[14]Bettencourt, R., O. Terenius, and I. Faye. 2002. Hemolin gene silencing by ds-RNA injected into *Cecropia* pupae is lethal to next generation embryos. *Insect Mol. Biol.* 11:267—71.

[15]Bossart, J. L. 1998. Genetic architecture of host use in a widely distributed, polyphagous butterfly (Lepidoptera: Papilionidae): Adaptive inferences based on comparison of spatio-temporal populations. *Biol. J. Linn. Soc. Lond.* 65:279—300.

[16]Bossart, J. L. 2003. Covariance of preference and performance on normal and novel hosts in a locally monophagous and locally polyphagous butterfly population. *Oecologia* 135:477—86.

[17]Bossart, J. L., and J. M. Scriber. 1995a. Maintenance of ecologically significant genetic variation in the tiger swallowtail butterfly through differential selection and gene flow. *Evolution* 49:1163—71.

[18]Bossart, J. L., and J. M. Scriber. 1995b. Genetic variation in oviposition preference in tiger swallowtail butterflies: Interspecific, interpopulation and interindividual comparisons. In *Swallowtail butterflies: Their ecology and evolutionary biology*, ed. J. M. Scriber, Y. Tsubaki, and R. C. Lederhouse, 183—93. Gainesville: Scientific Publishers.

[19]Bossart, J. L., and J. M. Scriber. 1999. Preference variation in the polyphagous tiger swallowtail butterfly (Lepidoptera: Papilionidae). *Environ. Entomol.* 28:628—37.

[20]Chang, Y.-L., Q. Tao, C. Scheuring, K. Ding, K. Meksem, and H.-B. Zhang. 2001. An integrated map of *Arabidopsis thaliana* for functional analysis of its genome sequence. *Genetics* 159:1231—42.

[21]Chen, G. T., and J. B. Graves 1970. Spermatogenesis of the tobacco budworm. *Ann. Entomol. Soc. Am.* 63:1095—104.

[22]Cho, S. W., A. Mitchell, J. C. Regier, et al. 1995. A highly conserved nuclear gene for low-level phylogenetics: Elongation factor-1 alpha recovers morphology-based tree for heliothine moths. *Mol. Biol. Evol.* 12:650—56.

[23]Chyb, S. 2004. *Drosophila* gustatory receptors: From gene identification to functional expression. *J. Insect Physiol*, 50:469—77.

[24]Clyne, P. J. , C. G. Warr, M. R. Freeman, D. Lessing, J. H. Kim, and J. R. Carlson. 1999. A novel family of divergent seven-transmembrane proteins: Candidate odorant receptors in *Drosophila. Neuron* 22:327—38.

[25]Coates, B. S. , R. L. Hellmich, and L. C. Lewis. 2005. Two differentially expressed ommochrome binding protein-like genes (*obpl* and *obp2*) in larval fat body of the European corn borer, *Ostrinia nubilalis. J. Insect Sci.* 5:19.

[26]Cotter, S. C. , and O. R. Edwards. 2006. Quantitative genetics of preference and performance on chickpeas in the noctuid moth, *Helicoverpa armigera. Heredity* 96:396—402.

[27]Cunningham, J. P. , M. F. A. Jallow, D. J. Wright, and M. P. Zalucki. 1998. Learning in host selection in *Helicoverpa armigera* (Hubner) (Lepidoptera: Noctuidae). *Animal Behav.* 55:227—34.

[28]Dahanukar A. , Y. T. Lei, J. Y. Kwon, and J. R. Carlson. 2007. Two Gr genes underlie sugar reception in *Drosophila. Neuron* 56:503—16.

[29]Darwin, C. 1909. Essay of 1844. In *The foundations of the origin of species*, ed. F. Darwin, 127. Cambridge: Cambridge Univ. Press.

[30]de Boer, G. 1993. Plasticity in food preference and diet-induced differential weighting of chemosensory information in larval *Manduca sexta. J. Insect Physiol.* 39:17—24.

[31]de Boer, G. 2006. The role of the antennae and maxillary palps in mediating food preference by larvae of the tobacco hornworm, *Manduca sexta. Entomol. Exp. Appl.* 119:29—38.

[32]de Bruyne, M. , and C. G. Warr. 2006. Molecular and cellular organization of insect chemosensory neurons. *Bioessays* 28:23—34.

[33]De Prins, J. , and K. Saitoh. 2003. Karyology and sex determination. In *Lepidoptera, moths and butterflies: Morphology, physiology, and development*, ed. N. P. Kristensen, 449—68. Berlin: Walter de Gruyter.

[34]del Campo, M. L. , and C. I. Miles. 2003. Chemosensory tuning to a host recognition cue in the facultative specialist larvae of the moth *Manduca sexta. J. Exp. Biol.* 206:3979—90.

[35]del Campo, M. L. , C. I. Miles, F. C. Schroeder, C. Mueller, R. Booker, and J. A. Renwick. 2001. Host recognition by the tobacco hornworm is mediated by a host plant compound. *Nature* 411:186—89.

[36]Dethier, V. G. 1976. *The hungry fly: A physiological study of the behavior associated with feeding*. Cambridge: Harvard Univ. Press.

[37]Dopman, E. B. , L. Perez, S. M. Bogdanowicz, and R. G. Harrison. 2005. Consequences of reproductive barriers for genealogical discordance in the European corn borer. *Proc. Natl. Acad. Sci. U. S. A.* 102:14706—11.

[38]Dukas, R. 1998. Constraints on information processing and their effects on behavior. In *Cognitive ecology*, ed. R. Dukas, 89—127. Chicago: Chicago Univ. Press.

[39]Elmore, T. , R. Ignell, J. R. Carlson, and D. P. Smith. 2003. Targeted mutation of a *Drosophila* odor receptor defines receptor requirement in a novel class of sensillum. *J. Neurosci.* 23:9906—12.

[40]Emelianov, I. , F. Marec, and J. Mallet. 2004. Genomic evidence for divergence with gene low in host races of the larch budmoth. *Proc. Biol. Sci.* 271:97—105.

[41]Emelianov, I. , F. Simpson, P. Narang, and J. Mallet. 2003. Host choice promotes reproductive isolation between host races of the larch budmoth *Zeiraphera diniana. J. Evol. Biol.* 16:208—18.

[42]Engler-Chaouat, H. S., and L. E. Gilbert. 2007. De novo synthesis vs. sequestration: Negatively correlated metabolic traits and the evolution of host plant specialization in cyanogenic butterflies. *J. Chem. Ecol.* 33:25—42.

[43]Fang, Q. Q., S. Cho, J. C. Regier, et al. 1997. A new nuclear gene for insect phylogenetics: DOPA carboxylase is informative of relationships within Heliothinae (Lepidoptera: Noctuidae). *Syst. Biol.* 46:269—83.

[44]Fitt, G. P. 1989. The ecology of *Heliothis* species in relation to agroecosystems. *Annu. Rev. Entomol.* 34:17—53.

[45]Fitt, G. P. 1991. Host selection in Heliothinae. In *Reproductive behaviour of insects: Individuals and populations*, ed. W. J. Bailey and J. Ridsdill-Smith, 172—200. London: Chapman & Hall.

[46]Forister, M. L. 2004. Oviposition preference and larval performance within a diverging lineage of lycaenid butterflies. *Ecol. Entomol.* 29:264—72.

[47]Forister, M. L. 2005. Independent inheritance of preference and performance in hybrids between host races of *Mitoura* butterflies (Lepidoptera: Lycaenidae). *Evolution* 59:1149—55.

[48]Frolov, A. N., D. Bourguet, and S. Ponsard. 2007. Reconsidering the taxomony of several *Ostrinia* species in the light of reproductive isolation: A tale for Ernst Mayr. *Biol. J. Linn. Soc. Lond.* 91:49—72.

[49]Glendinning, J. I., S. Valcic, and B. N. Timmermann. 1998. Maxillary palps can mediate taste rejection of plant allelochemicals by caterpillars. *J. Comp. Physiol. A Neuroethol. Sens. Neural Behav. Physiol.* 183:35—43.

[50]Goldman, A. L., W. V. van Naters, D. Lessing, C. G. Warr, and J. R. Carlson. 2005. Coexpression of two functional odor receptors in one neuron. *Neuron* 45:661—66.

[51]Goldsmith, M. R., T. Shimada, and H. Abe. 2005. The genetics and genomics of the silkworm, *Bombyx mori*. *Annu. Rev. Entomol.* 50:71—100.

[52]Groot, A. T., C. Ward, J. Wang, et al. 2004. Introgressing pheromone QTL between species: Towards an evolutionary understanding of differentiation in sexual communication. *J. Chem. Ecol.* 30:2495—2514.

[53]Grula, J. W., and O. R. Taylor. 1980. The effect of X-chromosome inheritance on mate-selection behavior in the sulfur butterflies, *Colias eurytheme* and *Colias philodice*. *Evolution* 34:688—95.

[54]Gu, H., A. Cao, and G. H. Walter. 2001. Host selection and utilisation of *Sonchus oleraceus* (Asteraceae) by *Helicoverpa armigera* (Lepidoptera: Noctuidae): A genetic analysis. *Ann. Appl. Biol.* 138:293—99.

[55]Gu, H., and G. H. Walter. 1999. Is the common sowthistle (*Sonchus oleraceus*) a primary host plant of the cotton bollworm, *Helicoverpa armigera* (Lep., Noctuidae)? Oviposition and larval preference. *J. Appl. Entomol.* 123:99—105.

[56]Hagen, R. H. 1990. Population structure and host use in hybridizing subspecies of *Papilio glaucus* (Lepidoptera, Papilionidae). *Evolution* 44:1914—30.

[57]Hallem, E. A., A. Dahanukar, and J. R. Carlson. 2006. Insect odor and taste receptors. *Annu. Rev. Entomol.* 51:113—35.

[58]Hallem, E. A., M. G. Ho, and J. R. Carlson. 2004. The molecular basis of odor coding in the *Drosophila* antenna. *Cell* 117:965—79.

[59]Hanson, F. E., and V. G. Dethier. 1973. Role of gustation and olfaction in food plant discrimina-

tion in tobacco hornworm, *Manduca sexta*. *J. Insect Physiol.* 19:1019—34.

[60]Heckel, D. G., L. J. Gahan, Y. -B. Liu, and B. E. Tabashnik. 1999. Genetic mapping of resistance to *Bacillus thuringiensis* toxins in diamondback moth using biphasic linkage analysis. *Proc. Natl. Acad. Sci. U. S. A.* 96:8373—77.

[61]Hora, K. H., P. Roessingh, and S. B. J. Menken. 2005. Inheritance and plasticity of adult host acceptance in Yponomeuta species: Implications for host shifts in specialist herbivores. *Entomol. Exp. Appl.* 115:271—81.

[62]Hung, C. F., T. L. Harrison, M. R. Berenbaum, and M. A. Schuler. 1995. CYP6B3: A second furanocoumarin-inducible cytochrome P450 expressed in *Papilio polyxenes*. *Insect Mol. Biol.* 4:149—60.

[63]Imamura, M., J. Nakai, S. Inoue, G. X. Quan, T. Kanda, and T. Tamura. 2003. Targeted gene expression using the GAL4/UAS system in the silkworm *Bombyx mori*. *Genetics* 165:1329—40.

[64]Iyengar, V. K., H. K. Reeve, and T. Eisner. 2002. Paternal inheritance of a female moth's mating preference. *Nature* 419:830—32.

[65]Jaenike, J. 1990. Host specialization in phytophagous insects. *Annu. Rev. Ecol. Syst.* 21:243—73.

[66]Jallow, M. F. A., and M. P. Zalucki. 1996. Within- and between-population variation in host-plant preference and specificity in Australian *Helicoverpa armigera* (Hubner) (Lepidoptera: Noctuidae). *Aust. J. Zool.* 44:503—19.

[67]Janz, N. 1998. Sex-linked inheritance of host-plant specialization in a polyphagous butterfly. *Proc. R. Soc. Lond. B. Biol. Sci.* 265:1675—78.

[68]Janz, N., K. Nyblom, and S. Nylin. 2001. Evolutionary dynamics of host-plant specialization: A case study of the tribe Nymphalini. *Evolution* 55:783—96.

[69]Janz, N., and S. Nylin. 1997. The role of female search behaviour in determining host plant range in plant feeding insects: A test of the information processing hypothesis. *Proc. R. Soc. Lond. B. Biol. Sci.* 264:701—07.

[70]Jiggins, C. D., J. Mavarez, M. Beltrán, W. O. McMillan, J. S. Johnston, and E. Bermingham. 2005. A genetic linkage map of the mimetic butterfly *Heliconius melpomene*. *Genetics* 171:557—70.

[71]Jones, W. D., P. Cayirlioglu, I. G. Kadow, and L. B. Vosshall. 2007. Two chemosensory receptors together mediate carbon dioxide detection in Drosophila. *Nature* 445:86—90.

[72]Kaplan, D. D., N. S. Flanagan, A. Tobler, R. Papa, and R. D. Reed. 2006. Localization of Mullerian mimicry genes on a dense linkage map of *Heliconius erato*. *Genetics* 173:735—57.

[73]Karpenko, C. P., and F. I. Proshold. 1977. Fertility and mating performance of interspecific crosses between *Heliothis virescens* and *H. subflexa* (Lepidoptera: Noctuidae) backcrossed for 3 generations to *H. subflexa*. *Ann. Entomol. Soc. Am.* 70:737—40.

[74]Krieger, J., E. Grosse-Wilde, T. Gohl, Y. M. E. Dewer, K. Raming, and H. Breer. 2004. Genes encoding candidate pheromone receptors in a moth (*Heliothis virescens*). *Proc. Natl. Acad. Sci. U. S. A.* 101:11845—50.

[75]Krieger, J., K. Raming, Y. M. E. Dewer, S. Bette, S. Conzelmann, and H. Breer. 2002. A divergent gene family encoding candidate olfactory receptors of the moth *Heliothis virescens*. *Eur. J. Neurosci.* 16:619—28.

[76]Kwon, J. Y., A. Dahanukar, L. A. Weiss, and J. R. Carlson. 2007. The molecular basis of CO_2 reception in *Drosophila*. *Proc. Natl. Acad. Sci. U. S. A.* 104:3574—78.

[77]Laster, M. L., S. D. Pair, and D. F. Martin. 1982. Acceptance and development of *Heliothis sub-*

flexa and *Heliothis virescens* (Lepidoptera, Noctuidae), and their hybrid and backcross progeny on several plant species. *Environ. Entomol.* 11:979—80.

[78]Lee, K. P., S. T. Behmer, and S. J. Simpson. 2006. Nutrient regulation in relation to diet breadth: A comparison of *Heliothis* sister species and a hybrid. *J. Exp. Biol.* 209:2076—84.

[79]Lee, S. F., and D. G. Heckel. 2007. Chromosomal conservation in Lepidoptera: Synteny versus collinearity. In Iatrou, K., and P. Couble 2007. 7th International Workshop on the Molecular Biology and Genetics of the Lepidoptera. August 20—26, 2006. Orthodox Academy of Crete, Kolympari, Crete, Greece. *J. Insect Sci.* 7:29.

[80]Li, W., R. A. Petersen, M. A. Schuler, and M. R. Berenbaum. 2002. CYP6B cytochrome P450 monooxygenases from *Papilio canadensis* and *Papilio glaucus*: Potential contributions of sequence divergence to host plant associations. *Insect Mol. Biol.* 11:543—51.

[81]Li, W. M., M. A. Schuler, and M. R. Berenbaum. 2003. Diversification of furanocoumarin-metabolizing cytochrome P450 monooxygenases in two papilionids: Specificity and substrate encounter rate. *Proc. Natl. Acad. Sci. U. S. A.* 100:14593—98.

[82]Li, X. C., M. A. Schuler, and M. R. Berenbaum. 2007. Molecular mechanisms of metabolic resistance to synthetic and natural xenobiotics. *Annu. Rev. Entomol.* 52:231—53.

[83]Lynch, M., and B. Walsh. 1998. *Genetics and analysis of quantitative traits*. Sunderland: Sinauer.

[84]Malausa, T., M. T. Bethenod, A. Bontemps, D. Bourguet, J. M. Cornuet, and S. Ponsard. 2005. Assortative mating in sympatric host races of the European corn borer. *Science* 308:258—60.

[85]Marcus, J. M. 2005. Jumping genes and AFLP maps: Transforming lepidopteran color pattern genetics. *Evol. Dev.* 7:108—14.

[86]Marec, F. 1996. Synaptonemal complexes in insects. *Int. J. Insect Morphol. Embryol.* 25:205—33.

[87]Martel, C., A. Rejasse, F. Rousset, M. T. Bethenod, and D. Bourguet. 2003. Host-plant-associated genetic differentiation in Northern French populations of the European corn borer. *Heredity* 90:141—49.

[88]Matsuo, T., S. Sugaya, J. Yasukawa, T. Aigaki, and Y. Fuyama. 2007. Odorant-binding proteins OBP57d and OBP57e affect taste perception and host-plant preference in *Drosophila sechellia*. *PLoS Biol.* 5:985—96.

[89] Mayr, E. 1963. *Animal species and evolution*. Cambridge: Belknap Press of Harvard Univ. Press.

[90]McElvare, R. R. 1941. Validity of the species *Heliothis subflexa*. *Bull. Brooklyn Entomol. Soc.* 36:29—30.

[91]Melcher, C., and M. J. Pankratz. 2005. Candidate gustatory interneurons modulating feeding behavior in the *Drosophila* brain. *PLoS Biol.* 3:1618—29.

[92]Menken, S. B. J. 1996. Pattern and process in the evolution of insect-plant associations: *Yponomeuta* as an example. *Entomol. Exp. Appl.* 80:297—305.

[93]Mita, K., M. Kasahara, S. Sasaki, et al. 2004. The genome sequence of silkworm, *Bombyx mori*. *DNA Res.* 11:27—35.

[94]Mitter, C., and B. D. Farrell. 1991. Macroevolutionary aspects of insect-plant relationships. In *Insect/plant interactions*, *vol.* 3, ed. E. A. Bernays, 35—78. Boca Raton: CRC Press.

[95]Mitter, C., R. W. Poole, and M. Matthews. 1993. Biosystematics of the Heliothinae (Lepidoptera: Noctuidae). *Annu. Rev. Entomol.* 38:207—25.

[96]Moon, S. J., M. Kottgen, Y. C. Jiao, H. Xu, and C. Montell. 2006. A taste receptor required for the caffeine response *in vivo*. *Curr. Biol.* 16:1812—17.

[97]Mustaparta, H. 2002. Encoding of plant odour information in insects: Peripheral and central mechanisms. *Entomol. Exp. Appl.* 104:1—13.

[98]Ng, M., R. D. Roorda, S. Q. Lima, B. V. Zemelman, P. Morcillo, and G. Miesenbock. 2002. Transmission of olfactory information between three populations of neurons in the antennal lobe of the fly. *Neuron* 36:463—74.

[99]Nishida, R. 2005. Chemosensory basis of host recognition in butterflies—Multi-component system of oviposition stimulants and deterrents. *Chem. Senses* 30, Suppl. 1:i293—94.

[100]Nosil, P., and A. O. Mooers. 2005. Testing hypotheses about ecological specialization using phylogenetic trees. *Evolution* 59:2256—63.

[101]Nygren, G. H., S. Nylin, and C. Stefanescu. 2006. Genetics of host plant use and life history in the comma butterfly across Europe: Varying modes of inheritance as a potential reproductive barrier. *J. Evol. Biol.* 19:1882—93.

[102]Nylin, S., G. H. Nygren, J. J. Windig, N. Janz, and A. Bergstrom. 2005. Genetics of host-plant preference in the comma butterfly *Polygonia c-album* (Nymphalidae), and evolutionary implications. *Biol. J. Linn. Soc. Lond.* 84:755—65.

[103]Ono, H., Y. Kuwahara, and R. Nishida. 2004. Hydroxybenzoic acid derivatives in a nonhost rutaceous plant, *Orixa japonica*, deter both oviposition and larval feeding in a Rutaceae-feeding swallowtail butterfly, *Papilio xuthus* L. *J. Chem. Ecol.* 30:287—301.

[104]Ono, H., K. Ozaki, and H. Yoshikawa. 2005. Identification of cytochrome P450 and glutathione-S-transferase genes preferentially expressed in chernosensory organs of the swallowtail butterfly, *Papilio xuthus* L. *Insect Biochem. Mol. Biol.* 35:837—46.

[105]Orr, H. A. 2001. The genetics of species differences. *Trends Ecol. Evol.* 16:343—50.

[106]Orr, H. A. 2005. Theories of adaptation: What they do and don't say. *Genetica* 123:3—13.

[107]Papanicolaou, A., S. Gebauer-Jung, M. L. Blaxter, W. O. McMillan, and C. D. Jiggins. 2008. Butterfly Base: A platform for lepidopteran genomics. *Nucleic Acids Res.* 36:D582—87.

[108]Park, S. K., S. R. Shanbhag, Q. Wang, G. Hasan, R. A. Steinbrecht, and C. W. Pikielny. 2000. Expression patterns of two putative odorant-binding proteins in the olfactory organs of *Drosophila melanogaster* have different implications for their functions. *Cell Tissue Res.* 300:181—92.

[109]Pashley, D. P. 1988. Quantitative genetics, development and physiological adaptation in host strains of fall armyworm. *Evolution* 42:93—102.

[110]Pelozuelo, L., C. Malosse, G. Genestier, H. Guenego, and B. Frerot. 2004. Host-plant specialization in pheromone strains of the European corn borer *Ostrinia nubilalis* in France. *J. Chem. Ecol.* 30:335—52.

[111]Poole, R. W., C. Mitter, and M. D. Huettel. 1993. A revision and cladistic analysis of the *Heliothis virescens* species group (Lepidoptera: Noctuidae) with a preliminary morphometric analysis of *H. virescens*. *In Mississippi Agriculture and Forestry Experiment Station Technical Bulletin* 185:1—51.

[112]Porter, A. H., and E. J. Levin. 2007. Parallel evolution in sympatric, hybridizing species: Performance of Colias butterflies on their introduced host plants. *Entomol. Exp. Appl.* 124:77—99.

[113]Prowell, D. P. 1998. Sex linkage and speciation in Lepidoptera. In *Endless forms: Species and speciation*, ed. S. Berlocher and D. Howard, 309—19. New York: Oxford Univ. Press.

[114]Prowell, D. P., M. McMichael, and J. F. Silvain. 2004. Multilocus genetic analysis of host use,

introgression, and speciation in host strains of fall armyworm (Lepidoptera: Noctuidae). *Ann. Entomol. Soc. Am.* 97:1034—44.

[115]Quental, T. B., M. M. Patten, and N. E. Pierce. 2007. Host plant specialization driven by sexual selection. *Amer. Nat.* 169:830—36.

[116]Ramaswamy, S. B. 1988. Host finding by moths: Sensory modalities and behaviours. *J. Insect Physiol.* 34:235—49.

[117]Remington, D. L., M. C. Ungerer, and M. D. Purugganan. 2001. Map-based cloning of quantitative trait loci: Progress and prospects. *Genet. Res.* 78:213—18.

[118]Renwick, J. A. A., and K. Lopez. 1999. Experience-based food consumption by larvae of *Pieris rapae*: Addiction to glucosinolates? *Entomol. Exp. Appl.* 91:51—58.

[119]Robertson, H. M., C. G. Warr, and J. R. Carlson. 2003. Molecular evolution of the insect chemoreceptor gene superfamily in *Drosophila melanogaster*. *Proc. Natl. Acad. Sci. U. S. A.* 100:14537—42.

[120]Robinson, R. 1971. *Lepidoptera genetics*. New York: Pergamon Press.

[121]Roessingh, P., S. Xu, and S. B. J. Menken. 2007. Olfactory receptors on the maxillary palps of small ermine moth larvae: Evolutionary history of benzaldehyde sensitivity. *J. Comp. Physiol. A Neuroethol. Sens. Neural Behav. Physiol.* 193:635—47.

[122]Rose, R. L., D. Goh, D. M. Thompson, et al. 1997. Cytochrome P450 (CYP) 9A1 in *Heliothis virescens*: The first member of a new CYP family. *Insect Biochem. Mol. Biol.* 27:605—15.

[123]Rostelien, T., M. Stranden, A. K. Borg-Karlson, and H. Mustaparta. 2005. Olfactory receptor neurons in two heliothine moth species responding selectively to aliphatic green leaf volatiles, aromatic compounds, monoterpenes and sesquiterpenes of plant origin. *Chem. Senses* 30:443—61.

[124]Rützler, M., and L. J. Zwiebel. 2005. Molecular biology of insect olfaction: Recent progress and conceptual models. *J. Comp. Physiol. A Neuroethol. Sens. Neural Behav. Physiol.* 191:777—90.

[125]Sahara, K., A. Yoshido, F. Marec, et al. 2007. Conserved synteny of genes between chromosome 15 of *Bombyx mori* and a chromosome of *Manduca sexta* shown by five-color BAC-FISH. *Genome* 50:1061—65.

[126]Scheirs, J., and L. De Bruyn. 2002. Temporal variability of top-down forces and their role in host choice evolution of phytophagous arthropods. *Oikos* 97:139—44.

[127]Schoonhoven, L. M. 1987. What makes a caterpillar eat? The sensory codes underlying feeding behaviour. In *Advances in chemoreception and behavior*, ed. R. Chapman, E. Bernays, and J. Stoffolano, 69—97. New York: Springer.

[128]Schoonhoven, L. M. 2005. Insect-plant relationships: The whole is more than the sum of its parts. *Entomol. Exp. Appl.* 115:5—6.

[129]Schoonhoven, L. M., and J. J. A. van Loon. 2002. An inventory of taste in caterpillars: Each species its own key. *Acta Zool. Acad. Sci. Hung.* 48:215—63.

[130]Scott, K., R. Brady, A. Cravchik, et al. 2001. A chemosensory gene family encoding candidate gustatory and olfactory receptors in *Drosophila*. *Cell* 104:661—73.

[131]Scriber, J. M. 2002. Evolution of insect-plant relationships: Chemical constraints, coadaptation, and concordance of insect/plant traits. *Entomol. Exp. Appl.* 104:217—35.

[132]Scriber, J. M., B. L. Giebink, and D. Snider. 1991. Reciprocal latitudinal clines in oviposition behavior of *Papilio glaucus* and *P. canadensis* across the Great Lakes hybrid zone: Possible sex-linkage of oviposition preferences. *Oecologia* 87:360—68.

[133]Sheck, A. L., and F. Gould. 1993. The genetic basis of host range in *Heliothis virescens*: Larval

survival and growth. *Entomol. Exp. Appl.* 69:157—72.

[134]Sheck, A. L. , and F. Gould. 1995. Genetic analysis of differences in oviposition preferences of *Heliothis virescens* and *Heliothis subflexa* (Lepidoptera, Noctuidae). *Environ. Entomol.* 24:341—47.

[135]Sheck, A. L. , and F. Gould. 1996. The genetic basis of differences in growth and behavior of specialist and generalist herbivore species: Selection on hybrids of *Heliothis virescens* and *Heliothis subflexa* (Lepidoptera). *Evolution* 50:831—41.

[136]Sheck, A. L. , A. T. Groot, C. M. Ward, et al. 2006. Genetics of sex pheromone blend differences between *Heliothis virescens* and *Heliothis subflexa*: A chromosome mapping approach. *J. Evol. Biol.* 19: 600—17.

[137]Silva-Brandao, K. L. , and V. N. Solferini. 2007. Use of host plants by Troidini butterflies (Papilionidae, Papilioninae): Constraints on host shift. *Biol. J. Linn. Soc. Lond.* 90:247—61.

[138]Singer, M. C. 1984. Butterfly-host plant relationships. In *The biology of butterflies, Symposium of the Royal Entomological Society XIII, London*, ed. R. L. Vane-Wright and P. R. Ackery, 81—88. London: Royal Entomological Society.

[139]Singer, M. C. , and C. D. Thomas. 1996. Evolutionary responses of a butterfly metapopulation to human- and climate-caused environmental variation. *Amer. Nat.* 148:S9—S39.

[140]Singer, M. C. , D. Ng, and C. D. Thomas. 1988. Heritability of oviposition preference and its relationship to offspring performance within a single insect population. *Evolution* 42:977—85.

[141]Singer, M. C. , C. D. Thomas, and C. Parmesan. 1993. Rapid human-induced evolution of insect host associations. *Nature* 366:681—83.

[142]Singer, M. C. , B. Wee, S. Hawkins, and M. Butcher. 2008. Rapid natural and antropogenic diet evolution: Three examples from checkerspot butterflies. In *Specialization, speciation and radiation: The evolutionary biology of herbivorous insects*, ed. K. J. Tilmon, pp. 311—24. Berkeley: Univ. California Press.

[143]Sperling, F. A. H. 1994. Sex-linked genes and species differences in Lepidoptera. *Canad. Entomol.* 126:807—18.

[144]Sperling, F. A. H. , and R. G. Harrison. 1994. Mitochondrial DNA variation within and between species of the *Papilio machaon* group of swallowtail butterflies. *Evolution* 48:408—22.

[145]Stevens, J. L. , M. J. Snyder, J. F. Koener, and R. Feyereisen. 2000. Inducible P450s of the CYP9 family from larval *Manduca sexta* midgut. *Insect Biochem. Mol. Biol.* 30:559—68.

[146]Stortkuhl, K. F. , R. Kettler, S. Fischer, and B. T. Hovemann. 2005. An increased receptive field of olfactory receptor Or43a in the antennal lobe of Drosophila reduces benzaldehyde-driven avoidance behavior. *Chem. Senses* 30:81—87.

[147]Stranden, M. , T. Rostelien, I. Liblikas, T. J. Almaas, A. K. Borg-Karlson, and H. Mustaparta. 2003. Receptor neurones in three heliothine moths responding to floral and inducible plant volatiles. *Chemoecology* 13:143—54.

[148]Sudbrink, D. L. , and J. F. Grant. 1995. Wild host plants of *Helicoverpa zea* and *Heliothis virescens* (Lepidoptera, Noctuidae) in eastern Tennessee. *Environ. Entomol.* 24:1080—85.

[149]Suh, G. S. B. , A. M. Wong, A. C. Hergarden, et al. 2004. A single population of olfactory sensory neurons mediates an innate avoidance behaviour in *Drosophila. Nature* 431:854—59.

[150]Suomalainen, E. 1969. Chromosome evolution in the Lepidoptera. *Chromosomes Today* 2:132—38.

[151]Syed, Z. , P. M. Guerin, and W. Baltensweiler. 2003. Antennal responses of the two host races of

the larch bud moth, *Zeiraphera diniana*, to larch and cembran pine volatiles. *J. Chem. Ecol.* 29: 1691—1708.

[152]Tamura,T. ,C. Thibert,C. Royer,et al. 2000. Germline transformation of the silkworm *Bombyx mori* L. using a *piggyBac* transposon-derived vector. *Nature Biotechnol.* 18:81—84.

[153]Tang,Q. B. ,J. W. Jiang,Y. H. Yan,J. J. A. van Loon,and C. Z. Wang. 2006. Genetic analysis of larval host-plant preference in two sibling species of *Helicoverpa*. *Entomol. Exp. Appl.* 118:221—28.

[154]Thomas, J. L. , M. Da Rocha, A. Besse, B. Mauchamp, and G. Chavancy. 2002. 3xP3-EGFP marker facilitates screening for transgenic silkworm *Bombyx mori* L. from the embryonic stage onwards. *Insect Biochem. Mol. Biol.* 32:247—53.

[155]Thomas,C. D. ,D. Ng,M. C. Singer,J. L. B. Mallet,C. Parmesan,and H. L. Billington. 1987. Incorporation of a European weed into the diet of a North American herbivore. *Evolution* 41:892—901.

[156]Thompson,J. N. 1988. Evolutionary genetics of oviposition preference in swallowtail butterflies. *Evolution* 42:1223—34.

[157]Thompson,J. N. 1993. Preference hierarchies and the origin of geographic specialization in host use in swallowtail butterflies. *Evolution* 47:1585—94.

[158]Thompson,J. N. 1998. The evolution of diet breadth: Monophagy and polyphagy in swallowtail butterflies. *J. Evol. Biol.* 11:563—78.

[159]Thompson,J. N. ,and O. Pellmyr. 1991. Evolution of oviposition behavior and host preference in Lepidoptera. *Annu. Rev. Entomol.* 36:65—89.

[160]Thompson,J. N. ,W. Wehling,and R. Podolsky. 1990. Evolutionary genetics of host use in swallowtail butterflies. *Nature* 344:148—50.

[161]Tilmon,K. J. 2008. *Specialization,speciation,and radiation: The evolutionary biology of herbivorous insects*. Berkeley: Univ. California Press.

[162]Vera,J. C. ,C. W. Wheat,H. W. Fescemyer,et al. 2008. Rapid transcriptome characterization for a nonmodel organism using 454 pyrosequencing. *Mol. Ecol.* 17:1636—47.

[163]Vogt,R. G. ,and L. M. Riddiford. 1981. Pheromone binding and inactivation by moth antennae. *Nature* 293:161—63.

[164]Vogt,R. G. , M. E. Rogers, M. D. Franco, and M. Sun. 2002. A comparative study of odorant binding protein genes: differential expression of the PBP1—GOBP2 gene cluster in *Manduca sexta* (Lepidoptera) and the organization of OBP genes in *Drosophila melanogaster* (Diptera). *J. Exp. Biol.* 205:719—44.

[165]Vosshall,L. B. ,H. Amrein,P. S. Morozov,A. Rzhetsky,and R. Axel. 1999. A spatial map of olfactory receptor expression in the *Drosophila* antenna. *Cell* 96:725—36.

[166]Wang,J. W. ,A. M. Wong,J. Flores,L. B. Vosshall,and R. Axel. 2003. Two-photon calcium imaging reveals an odor-evoked map of activity in the fly brain. *Cell* 112:271—82.

[167]Wanner,K. W. ,A. R. Anderson,S. C. Trowell,D. A. Theilmann,H. M. Robertson,and R. D. Newcomb. 2007. Female-biased expression of odourant receptor genes in the adult antennae of the silkworm, *Bombyx mori*. *Insect Mol. Biol.* 16:107—19.

[168]Ward,K. E. ,J. L. Hayes,R. C. Navasero,and D. D. Hardee. 1993. Genetic variability in oviposition preference among and within populations of the cotton bollworm (Lepidoptera,Noctuidae). *Ann. Entomol. Soc. Am.* 86:103—10.

[169]Wehling,W. F. ,and J. N. Thompson. 1997. Evolutionary conservatism of oviposition preference

in a widespread polyphagous insect herbivore, *Papilio zelicaon*. *Oecologia* 111:209—15.

[170]Weingartner, E., N. Wahlberg, and S. Nylin. 2006. Dynamics of host plant use and species diversity in *Polygonia* butterflies (Nymphalidae). *J. Evol. Biol.* 19:483—91.

[171]White, M. J. D. 1973. *Animal cytology and evolution*. Cambridge: Cambridge Univ. Press.

[172]Winkler, I. S., and C. Mitter. 2008. Phylogenetic dimension of insect-plant interactions. In *Specialization, speciation and radiation: The evolutionary biology of herbivorous insects*, ed. K. J. Tilmon, 240—63. Berkeley: Univ. California Press.

[173]Wittstock, U., N. Agerbirk, E. J. Stauber, et al. 2004. Successful herbivore attack due to metabolic diversion of a plant chemical defense. *Proc. Natl. Acad. Sci. U. S. A.* 101:4859—64.

[174]Xia, Q., Z. Zhou, C. Lu, et al. 2004. A draft sequence for the genome of the domesticated silkworm (*Bombyx mori*). *Science* 306:1937—40.

[175]Xiu, W. M., and S. L. Dong. 2007. Molecular characterization of two pheromone binding proteins and quantitative analysis of their expression in the beet armyworm, *Spodoptera exigua* Hübner. *J. Chem. Ecol.* 33:947—61.

[176]Yamamoto, K., J. Narukawa, K. Kadono-Okuda, et al. 2006. Construction of a single nucleotide polymorphism linkage map for the silkworm, *Bombyx mori*, based on bacterial artificial chromosome end sequences. *Genetics* 173:151—61.

[177]Yasukochi, Y., L. A. Ashakumary, K. Baba, A. Yoshido, and K. Sahara. 2006. A second-generation integrated map of the silkworm reveals synteny and conserved gene order between lepidopteran insects. *Genetics* 173:1319—28.

[178]Yoshido, A., H. Bando, Y. Yasukochi, and K. Sahara. 2005. The *Bombyx mori* karyotype and the assignment of linkage groups. *Genetics* 170:675—85.

[179]Zacharuk, R. Y. 1980. Ultrastructure and function of insect chemosensilla. *Annu. Rev. Entomol.* 25:27—47.

[180]Zagrobelny, M., S. Bak, A. V. Rasmussen, B. Jorgensen, C. M. Naumann, and B. L. Moller. 2004. Cyanogenic glucosides and plant-insect interactions. *Phytochemistry* 65:293—306.

[181]Zakharov, E. V., M. S. Caterino, and F. A. H. Sperling. 2004. Molecular phylogeny, historical biogeography, and divergence time estimates for swallowtail butterflies of the genus *Papilio*(Lepidoptera: Papilionidae). *Syst. Biol.* 53:193—215.

[182]Zalucki, M. P., D. A. H. Murray, P. C. Gregg, G. P. Fitt, P. H. Twine, and C. Jones. 1994. Ecology of *Helicoverpa armigera* (hubner) and *Heliothis punctigera* (wallengren) in the inland of Australia: Larval sampling and host-plant relationships during winter and spring. *Aust. J. Zool.* 42:329—46.

[183]Zheng, L., C. Lytle, C. N. Njauw, M. Altstein, and M. Martins-Green. 2007. Cloning and characterization of the pheromone biosynthesis activating neuropeptide receptor gene in *Spodoptera littoralis* larvae. *Gene* 393:20—30.

[184]Zimmermann, M., N. Wahlberg, and H. Descimon. 2000. Phylogeny of *Euphydryas* checkerspot butterflies (Lepidoptera: Nymphalidae) based on mitochondrial DNA sequence data. *Ann. Entomol. Soc. Am.* 93:347—55.

第十二章 主要农作物害虫夜蛾属的遗传学和分子生物学

Karl Gordon, Wee Tek Tay, Derek Collinge, Adam Williams, and Philip Batterham

1. 前言

夜蛾属 *Helicoverpa*(Hardwick 1965)是鳞翅目昆虫中研究最为广泛的种群之一。这个属包括了世界上几种最主要的农作物害虫。但是这个属的大多数物种具有有限的地理分布和狭窄的宿主植物范围,因此它们为研究泛化种食叶类昆虫的进化提供有价值的模式系统,而泛化种食叶类昆虫的进化展现出作为害虫最为核心的特性:多产、散布力强和杂食性。该章将针对如何应用现代分子生物学工具去研究这个属的昆虫从而解决基本问题进行了综述。

夜蛾属 *Helicoverpa* 需要在它们所属的亚科——Heliothinae 的范畴下进行研究。实夜蛾亚科 Heliothinae 被认为是在大约 2000 万年前的新生代下半段分化出来的(Cho et al. 1995;第一章中有对夜蛾属昆虫的杂食性关系的讨论)。在这个亚科 365 个物种中横跨超过十二个属。*Heliothis* 包括 *Helicoverpa*(Matthews 1999;Cho et al. 2008)和它的姐妹属 *Australothis*,以仅在澳大利亚、印度尼西亚和新西兰找到的 4 个物种为代表(Matthews 1999)。在这个组内,以地域命名的属,*Heliothis* Oschenheimer,因其合并分散到 *Helicoverpa*/*Australothis* 分化枝中(Cho et al. 2008)而在如今代表了最主要的变异,但是,一个仅在新时代发现的包含了两个 *Heliothis* 物种的家系,即最主要的农作物害虫烟芽夜蛾 *Heliothis virescens* 和它的姐妹种烟青虫 *H. subflexa*,在系统发生上和 *Helicoverpa*/*Australothis* 分化枝密切相关,Cho et al. (2008)取名为重大害虫家系(图 12.1)。重大害虫家系的起源不明确,但可能发生在澳大利亚,因为基础 *Helicoverpa* 种 *H. punctigera* 和其他很多处于 *Heliocheilus* 分化枝上的早期分化的近亲是在澳大利亚发现的(Matthews 1999)。烟芽夜蛾 *H. virescens* 家系从澳大利亚消失,而 *Australothis*/*Helicoverpa* 的祖先保留了下来。这些主要的农作物害虫是在澳大利亚开始分化的,这个推论令人吃惊,可能是因为它们主要栖息地的干燥状态,而干燥状态表现为多变的降雨量和在广袤面积上多样的且无法预测的植被覆盖。

以上对实夜蛾亚科全面的系统发生学研究为这个亚科宿主范围特异性进化提供了一个坚实的基础。在这个亚科的物种中,包括最早分化的家系在内的大多数物种(ca. 200),都是宿主特异型昆虫,反映了所推断的祖先夜蛾昆虫宿主范围的简约性(Cho et al. 2008)。*Heliothis* 组,包括重大害虫家系,还包括大多数杂食性昆虫(*Heliothines* 的 30%)。这可能代表该亚科中杂食性昆虫中不多的几个所熟知的种属中的一个(Cho et al. 2008)。在 *Heliothis* 组中,在一些特殊情况下发生了宿主特异性的逆转(Mitter,Poole 和 Matthews 1993;Cho 1997;Matthews 1999)。这些逆转发生在 *Helicoverpa* 属和包括 *Heliothis* 属在内的一些家系上(图 12.1;Cho et al. 2008)。

如今铃夜蛾 *Helicoverpa* 属被公认有二十个已经被鉴定的成员(Matthews 1999)。在确立铃夜蛾 *Helicoverpa* 为一个新的属时,Hardwick(1965)结合了来源广泛的在形态、饲养和杂交研究工作中得到的结果,并且将这些蛾子按照推测的物种组群进化的顺序来排列。他对铃夜蛾 *Helicoverpa* 属的论断后来被 Matthews(1999)使用自己所开发的复杂的形态学方法而做的细致研究所支持。这些物种，如表 12.1 所示，包括世界上对农作物最具有摧毁力的害虫，*H. armigera*（Hübner，棉铃虫）和 *H. zea*（Boddie，玉米夜蛾），这两者分别是旧世代和新世代中主要的害虫，由 Hardwick（1965）第一次划分成两个物种。这些昆虫是真正的主要害虫，可能是棉铃虫 *H. armigera* 的奠基者事件导致了它们的全球分布，同时也导致了玉米夜蛾 *H. zea* 在新世代的建立（Mallet et al. 1993; Behere et al. 2007）。其他一些铃夜蛾 *Helicoverpa* 昆虫也是农作物害虫，但它们或者宿主范围有限或者在地理分布上有限（Cho et al. 2008）。绝大多数铃夜蛾 *Helicoverpa* 昆虫是寡食性的，因此并不认为是农业害虫。事实上，有些物种非常稀少，它们甚至被列入

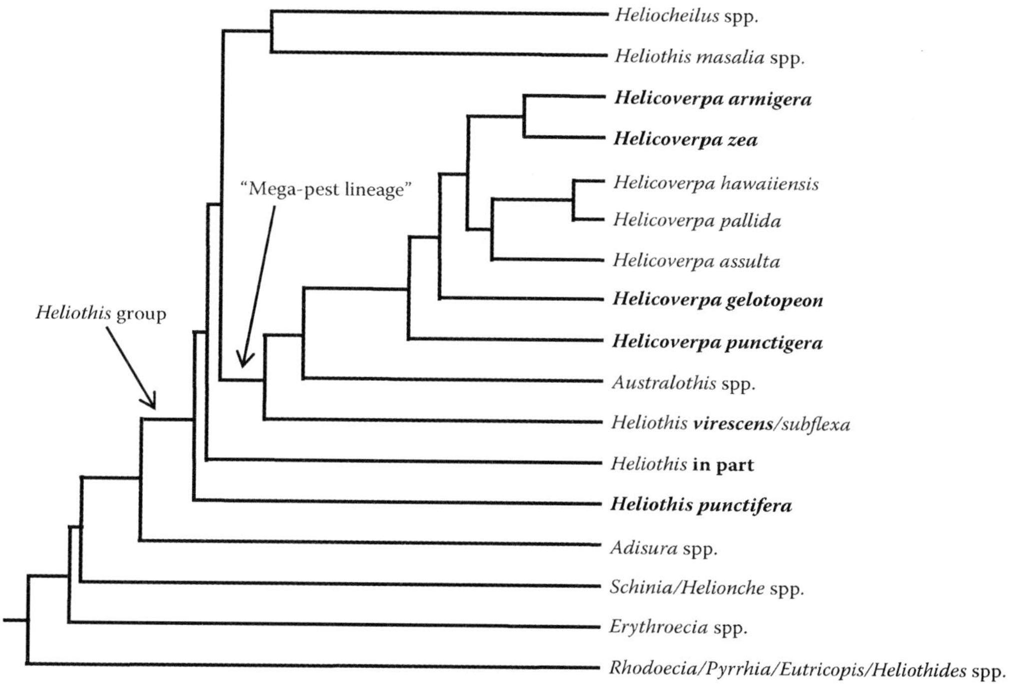

图 12.1 夜蛾亚科的系统分化。Cho 等人在 2008 年根据三个基因（EF-1α，DDC 和 COI）序列的极大似然分析所得的结构图。从完整的设置中仅选择了有代表性的世系（参见 Cho et al. 2008）。*Heliothis* 群落包括 *Heliothis* 种（并系群），澳大利亚姐妹属 *Australothis*，铃夜蛾 *Helicoverpa* 和 *Heliocheilus*。黑体表示杂食性种，有部分包含在其他的种中。详细描述参见 Cho et al. 2008。

濒临灭绝的名单中，还有可能已经灭绝了。不幸的是，这个属的系统发生学还没有被彻底地弄清楚，只对八种昆虫开展了严格的分子系统发生学的分析（Cho et al. 2008; 表 12.1）。这个属最基础的物种是澳洲棉铃虫 *H. punctigera*（Wallengren），是仅在澳大利亚广泛分布的杂食性物种，偶然成为棉花的主要害虫。分子系统发生学（Cho et al. 2008）支持了早期对澳洲棉铃虫 *H. punctigera* 的形态学和以同种异型酶为基础的评价（Matthews 1999）。并且任何一个其他的物种，如果也作过分子系统发生学分析，不太可能被证明比澳洲棉铃虫 *H. punctigera* 更基础。该属随后的分化以迁移和在世界各处建立起隔离的

物种为特征，被推测是由偶然的奠基者事件所导致的。在澳大利亚之外找到的最古老的分枝物种是 *H. geloteon*，在南美建立，仍然是禾本科植物特异性的摄食者。它孕育了一些新的物种，而这些物种都在南美分布(表 12.1)。铃夜蛾 *Helicoverpa* 家系后来分化产生了棉铃虫 *H. armigera* 和烟青虫 *H. assulta*(Guenée)，而这两者可能最先都是在澳大利亚分化的(Matthews 1999)。在以形态学为基础的特征上，烟青虫 *H. assulta* 与 *H. hardwickii* 和 *H. prepodes* 更接近，而后两者只在澳大利亚被找到；对它们之间关系的进一步确认需要分子水平上的定性。然而烟青虫 *H. assulta* 在东方获得了非常广泛的分布，比如在亚洲、非洲和澳大利亚的本土化(Cho et al. 1995)，它仍然是针对茄科植物的寡食性昆虫，因此并不被认为是主要的害虫。最可能的是，由烟青虫 *H. assulta* 产生了另一个截然不同的夏威夷物种(*H. pacifica*)，而 *H. pacifica* 因为后来引入的烟芽夜蛾 *H. virescens*，现在濒临灭绝或者已经消失了踪迹。

表 12.1　铃夜蛾 *Helicoverpa* 属的种群

Table 12.1　Species of the Genus *Helicoverpa*

	Distribution										
Species	Australia	Chr. I s.	NZ	Asia	Africa	Hawaii	South America	North America	Phagy	Hosts	Pest
H. armigera armigera	+		+	+	+				P		**
H. armigera conferta			+						P		**
H. zea							+	+	P		**
H. confusa						+					
H. helenae											
H. toddi					+						
H. tibetensis				+							
H. prepodes	+								O		
H. pacifica						+					
H. fletcheri					+				P		
H. assulta	+			+	+				O	Solan	*
H. hardwicki	+								O		
H. pauliana		+									
H. minuta						+					
H. hawaiiensis						+			O	Malv	
H. pallida						+			O	Cheno	
H. gelotopoeon							+		P		**
H. titicacae							+		P		
H. bracteae							+				
H. atacamae							+		O	Poac	
H. punctigera	+								P		**

根据 Mitter，Poole 和 Matthews 1993；Cho 1997；Matthews 1999；Cho 等人 2008 的文献；以及同 M. Matthews 的私人通讯的数据而编。*H. conferta* 和 *H. confusa* 可能已灭绝。*H. helenae* 仅存在于 *St. Helena*，可能是 *H. armigera* 的亚种。*H. pauliana* 在圣诞节岛上发现。寡食性种被标注为 O，泛化型摄食者被标注为 P。以下为宿主植物的简写 Solan：Solanaceae；Malv，Malvaceae；Cheno，Chenopodiaceae；Poac，Poaceae.

Heliothis 群是研究害虫关键特征进化的绝佳模式系统。研究工作主要是通过把在不同 heliothine 属找到的专一型和泛化型摄食者和所能获得的与其最为接近的物种进行配对。而这些研究的目的是为解决排卵、宿主倾向性、摄食和适应的遗传学问题（第十一章对宿主范围遗传学做了探讨）。其中一个焦点放在两个物种棉铃虫 *H. armigera* 和烟青虫 *H. assulta* 上。高度杂食性的棉铃虫 *H. armigera* 对新宿主的适应，部分源自新孵化幼虫所具有的、能在新宿主上摄食和发育的高度遗传力。年长幼虫的摄食和排卵倾向性不具有那么高的遗传力，因此可能在对宿主植物的适应进化上只扮演一个小的角色（Cotter 和 Edwards 2006）。雌性烟青虫 *H. assulta* 对适合幼虫摄食的茄科植物宿主上表现出明确的产卵倾向性（Cai，Konno 和 Matsuda 2002）。有趣的是，有一种宿主植物（烟草），它可以作为这两个物种中的任何一个的摄食对象。尽管由于烟青虫 *H. assulta* 的摄食导致了更高水平的过氧化酶和尼古丁，以及更低水平的叶多酚氧化酶，但是它表现出了对同一个反应系统（茉莉酸）的诱导（Zong 和 Wang 2007）。这两个物种可以在有限的程度上交配：雄性烟青虫 *H. assulta* 和雌性棉铃虫 *H. armigera* 交配产生可育的雄性；剩下的后代是不育的并且形态上异常（Zhao et al. 2005）。尽管反过来交配在最初的报道中产生了可育后代，但并没有被确认（S. C. Cotter 和 O. R. Edwards，unpubl.）。这两个物种经常在野外同样的宿主植物上出现（Cai，Konno 和 Matsuda 2002；Zong 和 Wang 2007），但是它们各自的信息素混合物决定了它们的生殖隔离（Ming，Yan 和 Wang 2007）。

雄性杂交后代和父母品系回交产生的摄食特征使确定控制宿主选择的基因成为可能。这些研究已经开始但仍然有限，因为缺少对这些物种详细的分子和遗传学知识（Wang et al. 2004；Tang et al. 2006；S. C. Cotter 和 O. R. Edwards，unpubl.）。杂食性烟芽夜蛾 *H. virescens* 和专一种烟青虫 *H. subflexa* 都能产生杂种后代，并被用于对宿主植物的适应研究中（如第十一章所述）。有趣的是，专一种烟青虫 *H. subflexa* 表现出对宿主植物（*Physalis*，浆酸属植物）的行为适应，而 *Physalis* 也是烟芽夜蛾 *H. virescens* 的宿主。烟青虫 *Heliothis subflexa* 幼虫"藏"在 *Physalis* 果实膨胀的花萼内摄食，因此躲避了寄生蜂的袭击（Oppenheim 和 Gould 2002）。相反的，烟芽夜蛾 *H. virescens* 在摄食的时候是暴露在外的，因此在 *Physalis* 上烟青虫 *H. subflexa* 的成活率比烟芽夜蛾 *H. virescens* 要高得多。对广阔范围的植物适应能力被推测可能排除了宿主特异的适应。

广泛分布的棉铃虫 *H. armigera* 越来越多地成为大多数在 Heliothines 上开展的遗传和分子生物学研究的主体。其贪婪的杂食性幼虫造成农作物产量显著的降低（Fitt 1989）并产生了对杀虫剂的抗性（Daly 1993；Heckel et al. 1998；Yang et al. 2004）。对澳大利亚（Daly 和 Gregg 1985）、西非（Nibouche et al. 1998）和东地中海区域（Zhou et al. 2000）的群体遗传变异性研究发现，它们是相对无组织的，并带有一定程度的杂合性。这些研究为基因跨越长远的距离流动提供了证据，并和其他这方面的证据——如成蛾是有效的飞行者，尤其在一些有利气候条件下能长距离迁徙（Farrow 1984；Pedgley 1985；Gregg et al. 1993；Hmimina et al. 1993）——是相符合的。因此棉铃虫 *H. armigera* 的随机交配群体覆盖了很广泛的区域。而且，雌性棉铃虫 *H. armigera* 是多产的，虽然一次最多能产生 3000 个卵（Mitter，Poole 和 Matthews 1993），但是雌蛾和雄蛾能多次交配。在合适的气候条件下,一年内棉铃虫 *H. armigera* 能产生 6～7 代,在极端温度下蛹能滞育。确实,Nibouche 等(1998)发现在南欧(法国和葡萄牙)的群体表现出更有组织性,可能是因为在这些天气环境下滞育较高的相关效应。这些特征使昆虫具有很高的适应能力。

2. 铃夜蛾(*Helicoverpa armigera*)遗传和基因组资源的开发

2.1 遗传作图和基因组学

近年来的研究工作逐渐包括了广泛的分子生物学手段,我们对铃夜蛾 *Helicoverpa* 物种的了解也逐渐成形。其中最主要的一步是通过遗传连锁图谱来确定拟除虫菊酯抗性基因的运用(Heckel et al. 1998)。另一个相似的手段随后被用于将 *Bacillus thuringiensis*(Bt)抗性定位于 *Heliothis* 的钙黏着蛋白(cadherin)基因上(Gahan,Gould 和 Heckel 2001)。这个工作也显示出鳞翅目昆虫基因组上广泛的微观线性(参见 Iatrou 和 Couble 2007 中摘要, d'Alencon et al. ,Lee 和 Heckel),并使得在毫无基因组信息的物种中寻找基因变得容易。

在撰写这篇文章的时候,可用的铃夜蛾 *Helicoverpa* 属物种的基因组信息很少。我们所知道的仅限于有限的基因组 DNA 序列,且绝大多数都是被认为是 Bt 抗性的基因,还有一些 EST 序列(expressed sequence tags,表达序列标签)。但是,很多实验室积累了很多未发表的材料,并成为提议中的国际铃夜蛾 *Helicoverpa* 基因能组计划的资源。这些材料包括细菌人工染色体(BAC)文库和 EST 数据库(参见 Iatrou 和 Couble 2007 中摘要, d'Alencon et al. ,Lee 和 Heckel)。BAC 序列尤其得到关注,因为它们包含了很大一部分编码关键解毒酶的基因,例如细胞色素 P450s(Cyp450s),谷胱甘肽-S-转移酶(GSTs; R. Feyereisen,P. Fournier et al. ,pers. comm),并很大地拓展了来自 EST 研究所获得的数据。

对 Cyp450s 能力(参见综述 Feyereisen 2006)的研究开始揭示这些基因,尤其是在控制铃夜蛾 *Helicoverpa* 属物种宿主选择中所扮演的角色。来自于玉米夜蛾 *H. zea* 的某些 Cyp450s 表现出结构上的变化,使昆虫能消化各种各样的植物化感物质(Li,Berenbaum 和 Schuler 2002; Rupasinghe et al. 2007)。现在仍不清楚在何种程度上杂食性昆虫基因组中基因的多样性是如何附和消化能力的多样性。对现在已有的 BAC 序列进行比较,为棉铃虫 *H. armigera* 以及另一个夜蛾属害虫草地贪夜蛾 *Spodoptera frugiperda* 的解毒基因做出了更完整的描绘。从对基因组序列和 cDNA 文库的研究中发现:夜蛾属昆虫的基因重复和分化是很普遍的,*Spodoptera* 和 *Helicoverpa* 都表现出不同的基因复制。相反的,寡食性昆虫家蚕 *Bombyx mori* 的基因组只有少量编码这些参与宿主植物变异化学物质解毒的关键酶基因(Mita et al. 2004; Xia et al. 2004)。

对编码 Bt δ-内毒素受体的基因的确立,发展出对很大一部分 *Heliothine* 基因组的分析,包括编码钙黏着蛋白(cadherin)和氨肽酶(aminopeptidase)的基因。在抗 Bt Cryl Ac 的棉铃虫 *H. armigera* 中发现了一定数目的钙黏着蛋白基因的缺失突变(Xu,Yu 和 Wu

2005)。一定数目的结合 Bt Cryl Ac 的氨肽酶基因也得到确定，并作为棉铃虫 *H. armigera* 中对这种蛋白家族完整鉴定的一部分(Angelucci *et al.* 2008)。比较易感性棉铃虫 *H. armigera* 和抗性棉铃虫 *H. armigera* 氨肽酶 3 的序列发现一小部分突变可能和抗性有关(Wang et al. 2005)。氨肽酶属于位于棉铃虫 *H. armigera* 围食膜上的蛋白(Campbell et al. 2008)；其他一些主要的蛋白家族组成了防御性的膜，比如昆虫肠道黏蛋白和几丁质去乙酰酶。

2.2 群体生物学和遗传学的新标记

棉铃虫 *H. armigera* 是害虫控制最主要也是最独特的挑战。控制该物种侵袭作物的方法主要依赖于杀虫剂(例如有机磷酸酯，氨基甲酸酯，spinosyn，人工合成的拟除虫菊酯)和转基因作物，如 Bt 棉花。在亚洲(如印度，Armes et al. 1992; Kranthi et al. 2001; 中国，Yang et al. 2004)为控制棉铃虫 *H. armigera* 而大规模使用杀虫剂已经导致了抗杀虫剂群体的产生；但是，在澳大利亚因整合涉及 Bt 棉花等多种防虫策略的协调努力已经推迟了抗 Bt 群体的产生。

杂食性和幼虫能快速产生杀虫剂抗性的能力使得我们迫切需要了解棉铃虫 *H. armigera* 群体遗传结构。虽然棉铃虫 *H. armigera* 作为一种显著的害虫，并且我们使用杀虫剂付出了极其沉重的社会经济代价(即杀虫剂相关的健康问题，环境污染和对有益昆虫及拟寄生物生态的负面影响；Fitt 1989)，但是很多受其影响的国家的研究人员对棉铃虫 *H. armigera* 群体遗传结构的了解仍然很少。其部分原因是由于缺乏例如微卫星等有效的分子遗传标记，所以我们需要确定更好的标记以及在该领域进行更多的研究。

开发作为遗传标记的微卫星 DNA(Tautz 1989)已经快速取代了群体和进化遗传学研究中的异型酶标记。和大多数物种的研究所不同的是，开发鳞翅目昆虫简单序列重复(SSR)标记的工作仅仅获得了少数成功(如，Butcher，Wright 和 Cook 2001; Meglécz et al. 2004; Perera et al. 2007)。尽管 SSR DNA 的频率最初被认为在鳞翅目昆虫中是很低的(Nève 和 Meglécz 2000)，但在家蚕 *Bombyx mori* 中已经报道了大量的 SSR DNA 标记(Reddy，Abraham 和 Nagaraju 1999; Prasad et al. 2005a，b)，并且被用于制作包含超过 500 个标记的遗传连锁图谱(Miao et al. 2005; 见第二章)。迄今，我们在棉铃虫 *H. armigera* 中已开发了 20 个 SSR 标记(Tan et al. 2001; Ji et al. 2003; Scott et al. 2004; Ji，Wu 和 Zhang 2005)，并且其中 13 个标记同样被 Perera 等(2007)在相近的物种(Mitter，Poole 和 Matthews 1993; Behere et al. 2007)新世代害虫玉米夜蛾 *H. zea* 中所报道。对使用异型酶，微卫星 DNA，随机扩增的多态性 DNA(RAPD)标记以及如在棉铃虫 *H. armigera* 群体遗传学研究中所使用的钠离子通道基因(Daly 和 Gregg 1985; McKechnie et al. 1993; Stokes，McKechnie 和 Forrester 1997; Nibouche et al. 1998; Zhou et al. 2000; Behere et al. 2007)所获得的结果进行比较发现，SSR DNA 标记在对澳大利亚群体研究中产生了矛盾的结果(Scott et al. 2003，2005a，b，2006; Endersby et al. 2007)。通过对微卫星 DNA 数据的仔细分析，Endersby 等(2007)发现来自于澳大利亚东海岸的棉铃虫 *H. armigera* 群体并不展现出显著的群体亚结构模式，这个发现和以前运用其他的

DNA 标记系统对澳大利亚棉铃虫 *H. armigera* 群体遗传学研究的结果(如,Daly snd Gregg 1985; McKechnie et al. 1993; Stokes,McKechnie 和 Forrester 1997)相似。相反的,Scott 等(2003,2005a,b,2006)基于 SSR DNA 标记一直都检测到澳大利亚产棉区棉铃虫 *H. armigera* 群体亚结构的模式,因此我们怀疑 SSR DNA 存在一些问题(如缺少杂合性;Scott et al. 2005b)。

为了解导致 Scott 等(2003,2005a,b,2006)和 Endersby 等(2007)矛盾结果的原因,就有必要先考虑现在绝大多数鳞翅目昆虫 SSR DNA 标记中所遇到的问题。尽管鳞翅目昆虫 SSR DNA 的低频率可以通过微卫星富集分离方法(如,Keyghobadi,Roland 和 Strobeck 1999; Scott et al. 2004)来增加,一个更重要的阻止广泛使用鳞翅目昆虫 SSR DNA 标记的因素是双边和单边微卫星 DNA 家族的出现(Ji 和 Zhang 2004; Meglécz et al. 2004,2007; van't Hof et al. 2007),即在 5'和/或 3'端间隔序列上具有显著的相似性的位点。因此对这些间隔区域设定的引物可能扩增了基因组中其他位点,产生了多种 DNA 条带模式(Meglécz et al. 2004)。其他一些影响鳞翅目昆虫微卫星位点的典型因素,包括过度纯合性、高频率无效等位基因和等位剔除,以及缺少 Hardy-Weinberg 平衡可能都是常见的原因;对它们的澄清需要对微卫星 DNA 家族的产生做更细致的研究。例如,RTE(反式转座因子)簇中的反式转座因子的非长末端重复(LTR)3'端最近由 Tay 和其同事发现(W. T. Tay,pers. comm.)是导致包括棉铃 *H. armigera*(如,HarSSR3,Ji et al. 2003; HaD47,Scott et al. 2004; HarSSR7,Ji,Wu 和 Zhang 2005)和玉米夜蛾 *H. zea*(HzMS1-6,Perera et al. 2007)在内的至少 22 个鳞翅目昆虫物种中超过 190 个单边微卫星 DNA 位点的原因。这些发现肯定了以前对单边微卫星 DNA 家族和转移因子之间具有联系的怀疑(Ji 和 Zhang 2004; Meglécz et al. 2004,2007; van't Hof et al. 2007),与由 van't Hof 等(2007)提出的在 DNA 重组中不平等交换机制相异。导致产生新 RTE-相关的 SSR 的机制可能涉及在宿主基因组中低复杂性的区域上,反转座因子完整或部分拷贝的不准确和非位点特异性的插入(W. T. Tay,pers. comm.)。微卫星 DNA 和非-LTR RTE 转座因子之间的联系可以解释为什么诸如棉铃虫 *H. armigera* HaD47 标记在群体遗传学分析中显示出高频率的无效等位基因(如,Endersby et al. 2007)或在杂合性观测水平上的显著缺失(如,*H. armigera* HarSSR7 marker,Ji,Wu 和 Zhang 2005; *H. zea* HzMS1-6 marker,Perera et al. 2007)。

一些棉铃虫 *H. armigera* SSR 标记同样存在于双边微卫星 DNA 家族中(如,Ji 和 Zhang 2004 图 1),即间隔 SSR 序列的 5'和 3'端在序列间具有高度的同源性和在单独的序列内不具有同源性。对已发表的棉铃虫 *H. armigera* HarSSR1(Ji et al. 2003)和 HarSSR8(Ji,Wu 和 Zhang 2005)DNA 标记更进一步的检测发现这些标记存在于毫无质疑的家族中(W. T. Tay,unpubl.),因此潜在地暗示了对它们的使用同样存在和那些已经确定和非-LTR RTE1 反转座因子相关的标记(W. T. Tay,pers. comm.)相似的困难。导致鳞翅目昆虫双边微卫星 DNA 家族产生的因素还没有被确定,但可能和新的转座因子相关。因此需要谨慎地对待在群体遗传学研究中运用已发表的鳞翅目昆虫微卫星 DNA 标记(如,Endersby et al. 2007)。

在绝大多数鳞翅目昆虫物种中微卫星 DNA 家族的存在以及相关的标记问题可能限制了它们在进化和群体遗传学研究中的有效性。因此需要另一种鳞翅目昆虫核 DNA 标记。Tay 等(2008)基于棉铃虫 *H. armigera* 多巴脱羧酶(*Dopa decarboxylase Ddc*)和核糖体蛋白(*ribosomal protein rp*)基因的外显子长度多态性(Palumbi 和 Baker 1994；Palumbi 1996)开发了以外显子为引物的内含子交换(EPIC)的 DNA 标记。EPIC DNA 标记中,引物退火区位于保守的编码序列内,因此极大地降低了因诸如 INDEL(插入和/或缺失)等随机突变和非编码区 SNP(单核苷酸多态性)导致的等位基因剔除和无效等位基因的发生可能。对 EPIC 标记的仔细设计以避免重复基因、处于选择压力下的基因和已知的假基因可以大大地减小例如典型的和鳞翅目昆虫 SSR DNA 标记相关的多条带模式等不良效果。对这些棉铃虫 *H. armigera* EPIC DNA 标记的家族性研究没有检测到等位基因剔除或无效等位基因;而且,EPIC 标记在澳大利亚、中国和印度的群体中都展现出一致性的 PCR 扩增。棉铃虫 *H. armigera* EPIC DNA 标记在杂合性观测水平上和 SSR DNA 标记所期待的水平相似。以可以作为锚定位点的保守基因为基础,棉铃虫 *H. armigera* EPIC DNA 标记将成为研究这种重要的农作物害虫,以及相关物种如烟青虫 *H. assulta*,玉米夜蛾 *H. zea* 和澳洲棉铃虫 *H. punctigera* 进化和群体遗传结构的重要工具,而且同时为鳞翅目昆虫群体和进化遗传学研究提供了一条向前发展的道路。

3. 改造实夜蛾(*Heliothine*):挑战和展望

改造棉铃虫 *H. armigera* 品系可以带来许多益处。转基因过量表达可以用来确定抗杀虫剂候选基因,而基于转座子的突变能快速从头确定抗性基因。这些知识有助于更合理地设计杀虫剂以及预测和监控自然群体抗性等位基因(Vreysen et al. 2007)。不育昆虫技术(SIT):每年在加州使用 SIT 控制粉螟蛉,这项技术通过雄性不育品系的转基因开发将得到极大的帮助。转基因标记,例如绿色和红色荧光蛋白(EGFP 和 DsRed)同样有助于监控为优化 SIT 策略而释放的昆虫(Alphey 2002；Vreysen et al. 2007)。

绝大多数的昆虫的改造都是通过微注射在卵里引入 II 类“剪切—黏接”转座子,这些转座子以发育中的胚胎(G_0；Handler 和 James 2000)祖先生殖系“极细胞”为靶点。任何一个 G_0 生殖细胞的改造将在下一代(G_1)中产生基因组已被改造的昆虫。

起源于鳞翅目昆虫的转座子 *piggyBac*,因其成功地改造了包括鳞翅目昆虫在内的 4 个昆虫目,已成为近年来研究 *Heliothine* 改造工作的焦点(Handler 和 James 2000)。转座子 *piggyBac* 由一个被倒置末端重复(ITRs)间隔的无内含子转座酶基因所组成,ITR 被转座酶所识别从而允许精确的基因组切除和整合(Fraser et al. 1996)。转座子元件先被克隆到细菌质粒中得以扩增,然后其转座酶基因被目标 DNA 所替代形成“整合”质粒。和另一个包含转座酶基因但去除了 ITR 的“辅助”质粒共同注射,这个“辅助”质粒提供转座酶的活性,但是它自己却无法整合到基因组中(Handler 和 James 2000)。

不少鳞翅目昆虫已经被 *piggyBac* 所改造,包括:棉铃虫,*Pectinophora gossypiella* (Peloquin et al. 2000);热带蝴蝶 *Bicyclus anynana*(Marcus,Ramos 和 Monteiro 2004);苹果小卷蛾 *Cydia pomonella*(Neven 2007)。家蚕 *B. mori* 转基因在超过 15 个研究中得

到了应用(如,Tamura et al. 2000),而转化频率(TF:被注射的且可育的并产生改造的后代的存活者的比例)现在一般为 5%~10%。另一些鳞翅目昆虫 TF 为 3.5%~5%,这些数据表明所有的鳞翅目昆虫物种都能被正常的改造,因此该载体应该也适用于棉铃虫 *H. armigera*。但是,用 *piggyBac* 和其他一些载体改造棉铃虫 *H. armigera*,玉米夜蛾 *H. zea* 和烟芽夜蛾 *H. virescens* 被证实是失败的(A. M. Handler, T. A. Miller pers. comm.; A. Williams, D. Collinge, unpubl.)。PBLE(类 *piggyBac* 元件)最近在这些物种中都得到了确定(Wang et al. 2006; Zimowska 和 Handler 2006; Sun et al. 2008)。许多 PBLE 是全长的,因此可能是具有活性的,阻止了转座子元件的整合或强制再活化。但是,多个 PBLE 存在于 *P. gossypiella*(T. A. Miller, pers. comm.)和家蚕 *B. mori*(Xu et al. 2006)中,而这两个物种都能被改造。尽管家蚕 *B. mori* 的 *piggyBac* 改造率在三个品系间变动最明显,从 1.5%到 14%,现在还不知道这是否和 PBLE 的拷贝数目相关(Zhong et al. 2007)。因此,在一些鳞翅目昆虫中多个同源元件并不阻碍稳定的整合,但是在 Heliothines 中,它们可能起阻碍作用,因此转基因的成功率或有效性可能是品系特异的。

3.1 启动子活性

除一个好的载体之外,强启动子是启动表达转座酶和选择标记所需要的。前者对载体整合是重要的,而后者对随后筛选转化的细胞是重要的。有些时候这些启动子扮演这两个方面的作用(如,*Actin A3*, Tamura et al. 2000)。*Hsp70*"热激"启动子注射到卵里后在没有热激的情况下起作用(Tamura et al. 2000),尽管当其整合到 *B. mori* 基因组后,它仍然是正常的热诱导启动子(Uhlirova et al. 2002)。

Actin A3 和 *Hsp70 piggyBac* 辅助质粒连接荧光报告蛋白在棉铃虫 *H. armigera* 中都作了检测,没有成功地分离到 G_1 转化子(D. Collinge, A. Williams, unpubl.)。和上述原因相同,转基因失败可能是由 PBLE 抑制所导致的,尽管报告蛋白的信号可能和下面将讨论的情况一样是问题所在。两个主要的启动子,*Actin A3* 和 *3xP3*,被用来在鳞翅目昆虫转基因中启动最常用的选择标记,增强的 GFP(EGFP)和 DsRed。当注射到 15~30min 龄长的棉铃虫 *H. armigera* 胚胎中,EGFP 信号可在 4~6h 后被检测到。两种启动子在 16h 内 80%的被注射卵中都启动了 EGFP 的表达,在 48h 达到高峰,能持续长至第 5 天,直到第一龄期早期(A. Williams, unpubl.)。

对棉铃虫 *H. armigera* 卵,大多数 EGFP 都在首次排泄中被分泌了,在此之后没有胚胎组织的荧光,尽管很弱的 EGFP 转录信号能用 RT-PCR 在第一龄晚期被检测到(A. Williams, unpubl.)。这个和家蚕 *B. mori* 中的情况一致:卵里大量的瞬时表达仅限于胚胎外的噬卵黄细胞,这些细胞参与了卵黄蛋白的代谢随后被隔离到中肠中,在卵孵化后被排出(Coulon-Bublex et al. 1993)。虽然质粒上的 *Actin A3* 启动子在家蚕 *B. mori* 卵里强烈表达,因此其作为辅助质粒启动子很有用,但一旦整合到基因组中,在卵里的表达量很低,但在幼虫肠道里强烈地表达(Tamura et al. 2000)。*Actin A3* 启动肠道中 EGFP 的表达,因此通过家蚕(如,Tamura et al. 2000)和粉螟蛉 *P. gossypiella*(如,Peloquin et al. 2000)的幼虫表皮就能很容易地确定转化子。

3xP3 是眼部和神经系统启动子(Berghammer, Klingler 和 Wimmer 1999),被用于家蚕(Thomas et al. 2002)和 *Bicyclus*(Marcus, Ramos 和 Monteiro 2004)中启动 EGFP 的表达。作为一个真核生物保守的 *Pax6*(无眼)转录因子的人工随机结合位点,3xP3 在家蚕幼虫侧眼中强烈表达(Thomas et al. 2000);但是,在瞬时转染后,孵化时表达大量 EGFP 的棉铃虫 *H. armigera* 侧眼并不发荧光(A. Williams, unpubl.),而且用 *3xP3-EGFP* 来筛选生殖系细胞,并没有筛选到侧眼或成虫眼部发荧光的转化子。其原因是否是由于眼部色素阻碍了信号或者这个启动子在 *Heliothine* 中并不是一个真正的眼部启动子还不能确定。前者看起来更可能些,因为 *3xP3* 甚至在原始的涡虫(扁平虫)中表达(Gonzalez-Estevez et al. 2003),表明它具有高保守性。在下面会讨论到褪色突变对 *3xP3* 筛选的辅助。

酵母双杂交 GAL4-UAS 系统使模块化的组织特异性基因成为可能(见综述 Handler 和 James 2000)。家蚕 *B. mori* 的 GAL4 由 *Actin A*3 或 *3xP3* 启动(Imamura et al. 2003)来驱动 EGFP 或保幼激素酯酶(Tan et al. 2005)的表达。*H. armigera* 卵中,Imamura 载体(Imamura et al. 2003)在 GAL4 存在情况下能特异性地启动 EGFP 的表达,这表明它们将对异位驱动,例如抗杀虫剂候选基因等,是很有作用的(A. Williams, unpubl.)。

一个携带黑腹果蝇 *D. melanogaster* 多聚泛素(*PUb*)启动子——用以驱动带有一个核定位信号的 EGFP(Handler 和 Harrell 1999)表达——的 *piggyBac* 载体,被注射到棉铃虫 *H. armigera* 胚胎中,产生了定位于卵壳下胚盘细胞细胞核中的荧光(D. Collinge, unpubl.)。两天后荧光强度达到顶峰,但在孵化前却观测不到。有趣的是,存活的 G_0 代交配后,两个公认的 G_1 代转基因幼虫通过第一龄期的荧光得到确定,荧光随时间而消退。不幸的是,这些借助 *Actin A*3 辅助质粒得到的转化子,是不能生殖的。另一个潜在的候选启动子是鳞翅目昆虫病毒 *hr5-iel* 启动子,该启动子对于某些鳞翅目昆虫物种而言是很好的启动子(Mohammed 和 Coates 2004),它能在棉铃虫 *H. armigera* 卵中启动 EGFP 的表达,尽管其能力一开始显得比较弱(A. Williams, unpubl.)。

3.2 荧光筛选

一系列选择性荧光标记,因其在早先的如 white^{+} 等着色基因互补系统中被提高了敏感度,现在已经在昆虫改造系统中被普遍使用,而且不止一个报告基因可以组合到一个昆虫上。并且,一个预先存在的着色突变并不是绝对需要的,使得荧光标记能应用于广泛的物种改造(Berghammer, Klingler 和 Wimmer 1999; Handler 和 Harrell 1999; Horn, Jaunich 和 Wimmer 2000; Horn et al. 2003)。

在迄今为止的所有鳞翅目昆虫改造中,只有 EGFP 和 DsRed 得到使用,一些研究室已经注意到 EGFP 和胚胎/幼虫自发荧光的波谱重叠。棉铃虫 *H. armigera* 一龄幼虫通常在孵化后消耗掉它们的卵壳,而这个具有反射性的物质很容易被误认为是 EGFP 荧光,从而导致 G_1 假阳性(A. Williams 和 D. Collinge, pers. obs.)。相反的,DsRed. T3 变体和德克萨斯红滤片一同使用,自发荧光很低,因此对于棉铃虫 *H. armigera* 的改造而言，其是这两种选择标记中较好的一种。

尽管很容易在转化的野生型幼虫着色的表皮观测到由 *Actin A*3 启动的 EGFP 荧光，我们通常在 G_1 代成虫眼部开展 3*xP*3 筛选，但在眼部，荧光被野生型眼色素严重干扰（如，Marcus，Ramos 和 Monteiro 2004）。最佳观测 3*xP*3 的方法是在浅色眼突变的背景下进行（Berghammer，Klingler 和 Wimmer 1999；Handler 和 Harrell 1999；Horn Jaunich 和 Wimmer 2000；Thomas et al. 2002）。这部分解释了为什么不能通过使用 3*xP*3-*EGFP*"看到"潜在的棉铃虫 *H. armigera* 和其他 heliothine G_1 转化子。

从澳大利亚棉铃虫 *H. armigera* Toowoomba 品系中分离到黄眼突变，缺少眼色素，从而成虫眼从绿色变成黄色，幼虫侧眼从棕色变成白色（A. Williams，pers. obs.）。通过和另一个 Toowoomba 深色眼突变杂交获得黄眼纯合子从而可以使成虫眼色素降为零（A. Williams，pers. obs.）。褪色的侧眼在胚胎晚期荧光侧眼筛选中尤其有用，在用 *3xP3* 启动子进行家蚕 *B. mori* 转基因研究（如，Thomas et al. 2002）中使用，可以节省广泛的 G_1 代饲养和成虫筛选。

3.3 DNA 输送方法

现在最常用的将 DNA 输送到昆虫中的方法是卵的微注射（Handler 和 James 2000；Handler 2002）。和黑腹果蝇 *D. melanogaster* 所不同的是，在鳞翅目昆虫微注射中实施去绒毛膜是很罕见的，去除绒毛膜导致卵的脆性增大。已有多个研究小组运用机械和微操控技术来优化胚胎的存活率和针头的寿命（Peloquin et al. 1997；Tamura et al. 2000）。对于棉铃虫 *H. armigera* 而言，与果蝇微注射相比，一根锋利的硼酸硅针头可注射多达 300 个卵，存活率大于 50%（A. Williams，unpubl.）。如果在棉铃虫 *H. armigera* 胚胎产出 15 到 30min 后进行注射，将改善载体的扩散，这与其他一些鳞翅目昆虫研究中的报道（Coulon-Bublex et al. 1993；Bossin et al. 2007）相一致。早期注射应该能使载体扩散到包括最关键的极细胞在内的所有卵区域。

另一些将 DNA 输送到鳞翅目昆虫胚胎的方法包括基因枪、电穿孔和病毒介导等。后两种方法已经成功地产生出转化子。电穿孔在用 *hobo* 转化玉米夜蛾 *H. zea* 中得到使用（DeVault et al. 1996），也是唯一一个改造 *Heliothine* 生殖系的报道。转化频率为 1.7%，报告基因 *lacZ* 在 5 个世代中用 Southern 印迹得到检测。值得注意的还有，用 PCR 方法检测存活的 G_0 个体蛹期 *lacZ* 基因，每次都预测到将产生生殖系转化子的个体。尽管这个技术上的成功在 *Heliothine* 昆虫上没有得到重复，电穿孔在理论上仍是一个远比微注射更有效的 DNA 输送方法，近来已经用于将 DNA 输送到家蚕胚胎中（Guo et al. 2004）。

家蚕 *B. mori* 最早的生殖系改造通过杆状病毒载体获得（Mori et al. 1995；Yamao et al. 1999）。Yamao 等（1999）提到，不到 300 个五龄幼虫注射了病毒，该病毒系统地扩散到生殖细胞并继而产生转基因个体，TF 为 2.7%。另一个让这篇文章引人关注的原因是，它获得了同源重组，即和载体上的丝蛋白轻链第 7 外显子融合的 EGFP，在一小部分转化子中，与内源性的丝蛋白轻链基因发生了同源重组，这在昆虫中是首次的发现。和

DeVault 等(1996)一样，Yamao 等也采用了基于 PCR 技术的筛选策略，其结果得到 Southern 和 Western 印迹结果的支持。尽管家蚕 *B. mori* 能幸免于重组杆状病毒(特别是 Baculogold AcNPV)的感染，但是这个方法通常被认为对其他大多数鳞翅目昆虫是致死的。

基因枪(例如用 DNA 包被的微粒“轰炸”)已经被用于将 DNA 输送到家蚕 *B. mori* 胚胎、组织(Thomas et al. 2001)和被解剖的丝腺中(Takahashi et al. 2003)。在棉铃虫 *H. armigera* 胚胎上尝试了用 *piggyBac* DNA 包被的金微粒基因枪(D. Collinge, unpubl.)。外卵壳成为一个无法穿越的屏障，需要部分去除才能允许金微粒的通过。结果胚胎的组织完整性被破坏了，导致了高死亡率。利用基因枪的任何尝试都没有获得转基因棉铃虫 *H. armigera*。

为提高整合频率、提供常规和有效的位点特异性以及一致性的基因组打靶，更复杂的转化方法在黑腹果蝇 *D. melanogaster* 上使用，例如 phiC31 整合酶，但仍然依赖于最基础的 II 型转座系统将这些强大的重组系统整合到基因组中(Bischof et al. 2007)。不幸的是，基础的转化系统是方法改善的前提。

3.4 对改造棉铃虫 *H. armigera* 的未来展望

我们逐渐在更广范的鳞翅目昆虫转化研究中理解了 *Heliothine* 转化上许多“神秘”的失败的原因。开发一个功能性的棉铃虫 *H. armigera* 转基因系统所面对的挑战包括(1)在转化前筛选和排除那些携带有例如 PBLE 等内源性转座酶的个体；(2)通过分离本身具有的启动子使转座酶和报告基因表达最大化；(3)通过褪色突变改善表现型观测和筛选。

Southern 印迹或 PCR 筛选无 PBLE 黄眼和深色眼突变品系有可能确定 *piggyBac* *3xP3* 的一个潜在品系，以及 DsRed 将成为所选择的报告基因。本身固有的、强有力的卵特异性启动子包括卵成色素基因 *white3*(A. Williams, unpubl.)，生殖系特异基因 *vasa*，或本身表达的核糖体基因，如 *RpS3A*。对 G_0 或 G_1 采取原始的 PCR 筛选再用 Southern 印迹确证转化子(DeVault et al. 1996; Yamao et al. 1999)也可以绕开任何观测报告基因的问题，并在一定程度上使筛选过程自动化。

因为 *hobo* 仍然是现在为止唯一一个转化 *Heliothine* 的载体，也许有利于重新利用 *hobo*(DeVault et al. 1996)。尽管 *Hermes* 筛选在 *3xP3-EGFP* *H. armigera* 是失败的(A. Williams, unpubl.)，但是在热带蝴蝶 *B. anynana* 中则是成功的(Marcus, Ramos 和 Monteiro 2004)，因此在用一个本身固有的启动子“重建”后还是值得尝试的。Minos 在家蚕 *B. mori* 上的成功(Uchino et al. 2007)强调了它的潜力，而 *Mariner* 也转化了众多门类的生物，因此也可能是有用的(Hartle et al. 1997; Wang, Swevers 和 Iatrou 2000)。考虑到 PBLE 潜在的负面效果和相关的转座子拷贝/遗迹，转座试验(Handler 和 James 2000)将有助于测试新的载体和品系，同样，转座酶 RNA 的共注射也将绕开启动子的限制。

尽管有挫折，棉铃虫 *H. armigera* 转基因的未来似乎是光明的。许多技术基础已到位：微注射效率和存活率都很好；EGFP、DsRed 和各种各样启动子是有活性的；GAL4-

UAS系统是有用的。利用有力的卵特异性自身启动子、无转座子和褪色品系可协助转基因筛选和增强转化率。克服这些困难应该使heliothine转化能和家蚕 *B. mori* 匹敌，并提供更多抗性和产生SIT群落的工具。

4. RNA干涉和在昆虫中的研究

所谓RNAi，是用dsRNA触发序列特异的细胞mRNA的降解，由此使基因表达沉默。最初是在植物中被发现的，也称为转录后基因沉默(PTGS)，在一些动物，尤其是在线虫 *C. elegans*(Guo和Kemphues 1995；Fire et al. 1998)上，也被观测到了。在这些系统中，细胞内dicer类酶切割dsRNA至短片段；然后这些短RNA通过结合包含有RNA酶Argonaute(见综述Meister和Tuschl 2004)的由RNA诱导的沉默复合体触发对细胞mRNA的切割。在植物和线虫中，诱导更进一步沉默的dsRNA扩增由依赖于细胞RNA的RNA聚合酶(RdRps)实施，RdRps能够识别短RNA并将它们作为mRNA的引物，导致基因沉默的系统性扩散(见概述Baulcombe 2007；Gordon和Waterhouse 2007)。RNAi机制在病毒防御上扮演核心作用(Ding和Voinnet 2007)。最开始这被认为是RNAi的最主要作用，但现在RNAi机制在细胞生物学其他方面，包括转座子沉默、配对敏感的沉默、端粒功能、染色质绝缘活性、核仁稳定性和异染色质形成等，都显示出作用；它还参与了由miRNA调控的基因表达(见综述Nilsen 2007；Filipowicz，Bhattacharyya和Sonenberg 2008；Kavi et al. 2008)。利用RNAi作全基因组筛选在例如线虫 *C. elegans*(Kim et al. 2005)和黑腹果蝇 *D. melanogaster*(Mathey-Prevot和Perrimon 2006)等模式生物上被广泛应用。对于非模式生物的研究者，RNAi被证实为一种通过基因表达的特异性敲除来研究基因功能的独特机会。它不需要对靶基因的全面了解：通常，200～500bp的基因片段就是产生能触发RNAi反应的dsRNA所需要的。

在包括鳞翅目昆虫在内的昆虫中使用RNAi，通常涉及dsRNA的注射。系统性的RNAi，即基因沉默扩散到整个昆虫中，在鞘翅目昆虫，如赤拟谷盗 *Tribolium*(Ober和Jokusch 2006；Tomoyasu et al. 2008)上观察到；父母辈的RNAi(即影响下一代)已经在这些昆虫中通过使三个发育基因沉默实现(Bucher，Scholten和Klingler 2002；Schröder 2003)。在 *Nasonia* 胡蜂上也获得了相似的观测(Lynch和Desplan 2006)。对蚱蜢 *Schistocerca americana* 的研究，针对一龄幼虫眼睛颜色基因vermilion的RNAi启动了长达两个龄期的对眼部色素形成的抑制(Dong和Friedrich 2005)。尽管在双翅目昆虫中对系统性RNAi的尝试还没有成功(Roignant et al. 2003)，组织特异的和时期特异的RNAi在黑腹果蝇 *Drosophila* 上通过转基因来实现是常规的(Kennerdell和Carthew 2000)，使细致地研究RNAi通路成为可能(Kavi et al. 2008)。比较基因组学研究确定了大多数昆虫具有几乎所有RNAi所需的基因，包括(双翅目除外)候选dsRNA转运蛋白(SID-1)；它们缺乏扩增所需的RdRp(蜜蜂基因组测序联会 2006；Gordon和Waterhouse 2007；Tomoyasu et al. 2008)。

RNAi已经在多种鳞翅目昆虫中被证明，包括家蚕 *B. mori*(Quan，Kanda和Tamura 2002；Uhlirova et al. 2003；Huang et al. 2007)，*Plodia interpunctella*(Fabrick，Kanost

和 Baker 2004)，地老虎 *Spodoptera litura*(Rajagopal et al. 2002)，甜菜夜蛾 *Spodoptera exigua*(Herrero et al. 2005)，惜比古天蚕 *Hyalophora cecropia*(Betterncourt，Terenius 和 Faye 2002)和人工培养的烟草天蛾 *Manduca sexta* 神经元细胞(Vermehren，Oazi 和 Trimmer 2001)。对于家蚕 *B. mori*，卵注射后 *white* 基因沉默产生了一个清楚的表型(白卵，半透明的幼虫表皮；Quan，Kanda 和 Tamura 2002)。另一个家蚕 *B. mori* 中改变表型的 RNAi 实验涉及 Sindbis 病毒载体的使用，该病毒载体输送引起广泛一复杂转录因子(Br-C; Uhlirova et al. 2003)沉默的 dsRNA；幼虫表现出一系列清楚的形态缺陷，表明了 Br-C 在变态发育中的保守作用。为解答表型效应是否可以在成虫中观测到，在蚕蛹中注射引起 *bursicon* 基因沉默的 dsRNA，表现出对翅膀扩展的影响(Huang et al. 2007)。大多数 RNAi 研究都集中在为功能分析而做基因敲除，而不是为防治的目的而作基因敲除。但是，它们激励了为其他鳞翅目昆虫开发基于 RNAi 的防治方法。

最近对棉铃虫 *H. armigera* 胚胎的研究工作展现了功能性的 RNAi 机制和获得至少一个瞬时 RNAi 效果的可能性。在转基因系缺乏的条件下，通过向前胚盘期胚胎注射 EGFP 报告载体开发了瞬时沉默实验。为达到基因沉默的目的，不同形式的 dsRNA 和报告基因载体共注射到胚胎中，然后对比得到的 EGFP 表达水平。这个方法表现了 EGFP 瞬时表达可以被共注射的 dsRNA 所沉默(D. Collinge，unpubl.)。对于这个工作，A. Williams 作了延伸(unpubl.)，他发现 *white* 基因能被注射到前胚盘期胚胎中的 dsRNA 瞬时沉默，导致了和在家蚕 *B. mori*(Quan，Kanda 和 Tamura 2002)上所观测到的结果相似的幼虫表型。

另外，与褪色表型相关的有力的 mRNA 敲除在胚胎中注射针对棉铃虫 *H. armigera white* 基因的 2 个 22bp 双重短干扰 RNA(siRNA)中的一个后能观测到，而且表型能长至第一龄期结束。用 siRNA，EGFP 和拟除虫菊酯目标基因 *para* 获得约 90%的相似敲除。siRNA 和两个载体的混合和配对同时在同一个卵里为多达 4 个基因提供了清楚的表型(A. Williams，pers. obs.)。因此，这成为一个具有前景的用以有效地分析例如烟碱酸乙酰胆碱受体的亚基成分(新烟碱杀虫剂的靶基因，在昆虫中尚不明)等复杂冗余的综合系统功能的方法。这些研究能有助于改善杀虫剂设计。

为主要害虫，如棉铃虫 *H. armigera*，开发一种有效的、口服的 RNAi 系统将不仅为反式遗传学提供一个强大的工具，还能作为一个具有潜力的物种特异性防治害虫的方法。转基因编码的 RNAi 长期以来被认为对植物是可行的(Waterhouse，Graham 和 Wang 1998)，而口服 dsRNA 能引起线虫的 RNAi(Timmons 和 Fire 1998)。尽管这些发现引发了追问植物能否被表达针对害虫基因的 dsRNA 的基因来改造，从而保护自身不受植食类昆虫侵害，但缺乏对这个方法的成功报道似乎说明在植物细胞中表达发夹 RNA 不能提供足够的完整 dsRNA 以在昆虫服食后引发潜在的 RNAi。的确，直到最近，唯一一个在鳞翅目昆虫中口服引发系统性 RNAi 的报道仅表现在浅棕色苹果蛾 *Epiphyas postvittana* 食用大量的 dsRNA 后，基因得到了部分沉默(Turner et al. 2006)。

最近两个文献引人注目地报道了通过转基因编码的发夹 RNA 赋予农作物抵御害虫的潜力。第一个报道——棉铃虫 *H. armigera* 可能对可服用的转基因编码 dsRNA 是易感的——来自于 Mao 等(2007),他们最初在棉铃虫 *H. armigera* 中鉴定了一个细胞色素 P450 单加氧酶(Cyp6AE12),Cyp6AE12 由产生自棉花(*Gossypium hirsutum*)的主要防御性变异化学物质棉子酚所诱导。然后它们产生表达针对这个基因序列的发夹 dsRNA 的转基因拟南芥(*Arabidopsis thaliana*)。这个 dsRNA 能使摄食该植物的幼虫中该基因沉默,因此幼虫对添加了棉子酚的植物是易感的。Mao 等(2007)继而用拟南芥 *A. thaliana* 的突变系来解答在摄食幼虫中引发 RNAi 的 dsRNA 分子是长的发夹 RNA 而不是 siRNA。这些 siRNA 通常是由植物的 dicer 样(dcl)酶产生的:携带有 *Dcl2*、*Dcl3* 和 *Dcl4* 突变的 *Arabidopsis* 类植物产生更多完整的发夹 RNA 且导致了更显著的 P450 基因沉默。在这些植物中仍然可以检测到 siRNA,可能是由于另一个没有突变的 dicer 基因导致的——*Arabidopsis* 有 4 个这样的基因(Margis et al. 2006)。这些观测表现了通过植物产生的长 dsRNA 所造成的昆虫基因沉默的有效性可能极大地取决于在转基因植物中生成发夹 dsRNA 速度比它们被植物 dicer 酶加工的速度更快(Gordon 和 Waterhouse 2007),尽管这些植物本身所产生的 siRNA 的重要性也不能忽略。这些对 *H. armigera* 基因沉默的观测得到了在同一时间由 Baum 等(2007)发表的一篇文献的确认,Baum 发现玉米能有效地防御鞘翅目昆虫害虫,如西部玉米食根虫(*Diabrotica virgifera*)、南部玉米食根虫(*Diabrotica undecimpunctata*)和科罗拉多土豆甲虫(*Leptinotarsa decemlineata*)。

尽管仍处于初期,这个工作却提出了振奋人心的害虫防治新方法,特别是在具有 *Helicoverpa* 类昆虫更全面的遗传信息甚至更多的完整基因组数据的情况下。但是这个方案面临许多重要的问题,不仅是其机制还有在这个领域中采用这种防治策略所面对的挑战(Gordon 和 Waterhouse 2007)。特别是 Heliothinae 类害虫,最显著的问题是植物介导的可服食 RNAi 的供应将被害虫群落中的序列多态性所逃避。对鞘翅目昆虫害虫,实验室内的成功很可能转化成田间的有效害虫防治,因为实验室内的研究包含了测量对农作物本身根部的破坏。对 *Helicoverpa* 来说,问题是如何有效地保护例如棉花等农作物。

5. 结论和展望

对 *Helicoverpa* 属昆虫的研究,逐渐应用分子生物学、遗传学和基因组学的工具,将越来越靠近探索它们在昆虫生物学基础研究和宿主进化研究中全部潜力的时刻。尽管对一些基因和基因家族已经做过了较好的研究,在不久的将来,*H. armigera* 基因组分析的完成将具有最大的潜力来增强我们对这个重要的害虫属的了解以及转变我们对其研究的能力。研究者将有能力研究群体基因组学和定位抗性基因以及为例如杀虫剂抗性、滞育倾向和宿主选择等 QTL 性状。通过对包括基因组可知的宿主植物的使用,研究者能够研究对宿主植物适应(例如通过对植物变异化学物质的解毒)的进化。比较紧密相关的物种和那些可能种间交配的物种(例如 *H. assulta*)将使细致地分析泛化种和专一种进食者(宿主特异性在第十一章中讨论到)成为可能。迁徙模式将更精确地得以研究,使严格评测群体抑制选择和对抗性管理的庇护价值成为可能。

参考文献

[1]Alphey, L. 2002. Re-engineering the sterile insect technique. *Insect Biochem. Mol. Biol.* 32:1243—47.

[2]Angelucci, C., G. A. Barrett-Wilt, D. F. Hunt, et al. 2008. Diversity of aminopeptidases, derived from four lepidopteran gene duplications, and polycalins expressed in the midgut of *Helicoverpa armigera*: Identification of proteins binding the δ-endotoxin, Cry1Ac of *Bacillus thuringiensis*. *Insect Biochem. Mol. Biol.* 38:685—96.

[3]Armes, N. J., D. R. Jadhav, G. S. Bond, and A. B. S. King. 1992. Insecticide resistance in *Helicoverpa armigera* in South India. *Pestic. Sci.* 34:355—64.

[4]Baulcombe, D. C. 2007. Amplified silencing. *Science* 315:199—200.

[5]Baum, J. A., T. Bogaert, W. Clinton, et al. 2007. Control of coleopteran insect pests through RNA interference. *Nat. Biotechnol.* 25:1322—26.

[6]Behere, G. T., W. T. Tay, D. A. Russell, et al. 2007. Mitochondrial DNA analysis of field populations of *Helicoverpa armigera* (Lepidoptera: Noctuidae) and of its relationship to *H. zea*. *BMC Evol. Biol.* 7:17.

[7]Berghammer, A. J., M. Klingler, and E. A. Wimmer. 1999. A universal marker for transgenic insects. *Nature* 402:370—71.

[8]Bettencourt, R., O. Terenius, and I. Faye. 2002. Hemolin gene silencing by ds-RNA injected into *Cecropia* pupae is lethal to next generation embryos. *Insect Mol. Biol.* 11:267—71.

[9]Bischof, J., R. K. Maeda, M. Hediger, F. Karch, and K. Basler. 2007. An optimized transgenesis system for *Drosophila* using germ-line-specific phiC31 integrases. *Proc. Natl. Acad. Sci. U. S. A.* 104:3312—17.

[10]Bossin, H., R. B. Furlong, J. L. Gillett, M. Bergoin, and P. D. Shirk. 2007. Somatic transformation efficiencies and expression patterns using the JcDNV and *piggyBac* transposon gene vectors in insects. *Insect Mol. Biol.* 16:37—47.

[11]Bucher, G., J. Scholten, and M. Klingler. 2002. Parental RNAi in *Tribolium* (Coleoptera). *Curr. Biol.* 12: R85—86.

[12]Butcher, R. D. J., D. J. Wright, and J. M. Cook. 2001. Development and assessment of microsatellites and AFLPs for *Plutella xylostella*. In *The management of diamondback moth and other crucifer pests*, Proc. 4th Int. Workshop, ed. N. M. Endersby and P. M. Ridland, 87—93. Melbourne: The Regional Institute.

[13]Cai, C. Y., Y. Konno, and K. Matsuda. 2002. Studies on the ovipositional preferences and feeding preferences in *Helicoverpa assulta* and *Helicoverpa armigera*. *Tohoku J. Agric. Res.* 53:11—24.

[14]Campbell, P. M., A. T. Cao, E. R. Hines, P. D. East, and K. H. J. Gordon. 2008. Proteomic analysis of the peritrophic matrix from the gut of the caterpillar, *Helicoverpa armigera*. *Insect Bioch. Mol.* Biol. 38:950—8.

[15]Cho, S. 1997. Molecular phylogenetics of the Heliothinae (Lepidoptera: Noctuidae) based on the nuclear genes for elongation factor—1α and dopa decarboxylase. PhD diss., Univ. Maryland.

[16]Cho, S., A. Mitchell, C. Mitter, J. Regier, M. Matthews, and R. Robertson. 2008. Molecular phylogenetics of heliothine moths (Lepidoptera: Noctuidae: Heliothinae), with comments on the evolution of host range and pest status. *Syst. Entomol.* 33:581—94.

[17]Cho, S., A. Mitchell, J. C. Regier, et al. 1995. A highly conserved nuclear gene for low-level phylogenetics: Elongation factor-1α recovers morphology-based tree for heliothine moths. *Mol. Biol. Evol.* 12:

650—56.

[18]Cotter, S. C. , and O. R. Edwards. 2006. Quantitative genetics of preference and performance on chickpeas in the noctuid moth, *Helicoverpa armigera*. *Heredity* 96:396—402.

[19]Coulon-Bublex, M. , N. Mounier, P. Couble, and J. C. Prudhomme. 1993. Cytoplasmic *Actin A3* gene promoter injected as supercoiled plasmid is transiently active in *Bombyx mori* embryonic vitellophages. *Roux's Arch. Dev. Biol.* 203:123—27.

[20]Daly, J. C. 1993. Ecology and genetics of insecticide resistance in *Helicoverpa armigera*: Interactions between selection and gene flow. *Genetica* 90:217—26.

[21]Daly, J. C. , and P. Gregg. 1985. Genetic variation in *Heliothis* in Australia: Species identification and gene flow in the two species *H. armigera* (Hübner) and *H. punctigera*(Wallengren) (Lepidoptera: Noctuidae). *Bull. Entomol. Res.* 75:169—84.

[22]DeVault, J. D. , K. J. Hughes, R. A. Leopold, O. A. Johnson, and S. K. Narang. 1996. Gene transfer into corn earworm (*Helicoverpa zea*) embryos. *Genome Res.* 6:571—79.

[23]Ding, S. W. , and O. Voinnet. 2007. Antiviral immunity directed by small RNAs. *Cell* 130:413—26.

[24]Dong, Y. , and M. Friedrich. 2005. Nymphal RNAi: Systemic RNAi mediated gene knockdown in juvenile grasshopper. *BMC Biotechnol.* 5:25.

[25]Endersby, N. M. , A. A. Hoffmann, S. W. Mckechnie, and A. R. Weeks. 2007. Is there genetic structure in populations of *Helicoverpa armigera* from Australia? *Entomol. Exp. Appl.* 122:253—63.

[26]Fabrick, J. A. , M. R. Kanost, and J. E. Baker. 2004. RNAi-induced silencing of embryonic tryptophan oxygenase in the pyralid moth, *Plodia interpunctella*. *J. Insect Sci.* 4:15.

[27]Farrow, R. A. 1984. Detection of transoceanic migration of insects to a remote island in the Coral Sea, Willis Island. *Aust. J. Ecol.* 9:253—72.

[28]Feyereisen, R. 2006. Evolution of insect P450. *Biochem. Soc. Trans.* 34:1252—55.

[29]Filipowicz, W. , S. N. Bhattacharyya, and N. Sonenberg. 2008. Mechanisms of post-transcriptional regulation by microRNAs: Are the answers in sight? *Nat. Rev. Genet.* 9:102—14.

[30]Fire, A. , S. Xu, M. K. Montgomery, S. A. Kostas, S. E. Driver, and C. C. Mello. 1998. Potent and specific genetic interference by double-stranded RNA in *Caenorhabditis elegans*. *Nature* 391:806—11.

[31]Fitt, G. P. 1989. The ecology of *Heliothis* species in relation to agroecosystems. *Annu. Rev. Entomol.* 34:17—52.

[32]Fraser, M. J. , T. Ciszczon, T. Elick, and C. Bauser. 1996. Precise excision of TTAA-specific lepidopteran transposons *piggyBac* (IFP2) and tagalong (TFP3) from the baculovirus genome in cell lines from two species of Lepidoptera. *Insect Mol. Biol.* 5:141—51.

[33]Gahan, L. J. , F. Gould, and D. G. Heckel. 2001 Identification of a gene associated with Bt resistance in *Heliothis virescens*. *Science* 293:857—60.

[34]Gonzalez-Estevez, C. , T. Momose, W. J. Gehring, and E. Salo. 2003. Transgenic planarian lines obtained by electroporation using transposon-derived vectors and an eye-specific GFP marker. *Proc. Natl. Acad. Sci. U. S. A.* 100:14046—51.

[35]Gordon, K. H. J. , and P. M. Waterhouse. 2007. RNAi for insect-proof plants. *Nat. Biotechnol.* 25:1231—32.

[36]Gregg, P. C. , G. P. Fitt, M. Coombs, and G. S. Henderson. 1993. Migrating moths (Lepidoptera)

collected in tower-mounted light traps in northern New South Wales, Australia: Species composition and seasonal abundance. *Bull. Entomol. Res.* 83:53—78.

[37]Guo, X. Y. , L. Dong, S. P. Wang, T. Q. Guo, J. Y. Wang, and C. D. Lu. 2004. Introduction of foreign genes into silkworm eggs by electroporation and its application in transgenic vector test. *Acta Biochim. Biophys. Sin.* (*Shanghai*) 36:323—30.

[38]Guo, S. , and K. J. Kemphues. 1995. *par-1*, a gene required for establishing polarity in *C. elegans* embryos, encodes a putative Ser/Thr kinase that is asymmetrically distributed. *Cell* 81:611—20.

[39]Handler, A. M. 2002. Use of the *piggyBac* transposon for germ-line transformation of insects. *Insect Biochem. Mol. Biol.* 32:1211—20.

[40]Handler, A. M. , and R. A. Harrell II. 1999. Germline transformation of *Drosophila melanogaster* with the *piggyBac* transposon vector. *Insect Mol. Biol.* 8:449—57.

[41]Handler, A. M. , and A. A. James, Eds. 2000. *Insect transgenesis: Methods and applications*. Boca Raton: CRC Press.

[42]Hardwick, D. F. 1965. The corn earworm complex. *Mem. Entomol. Soc. Canada* 40:1—248.

[43]Hartl, D. L. , E. R. Lozovskaya, D. I. Nurminsky, and A. R. Lohe. 1997. What restricts the activity of mariner-like transposable elements? *Trends Genet.* 13:197—201.

[44]Heckel, D. G. , L. J. Gahan, J. C. Daly, and S. Trowell. 1998. A genomic approach to understanding *Heliothis* and *Helicoverpa* resistance to chemical and biological insecticides. *Phil. Trans. R. Soc. Lond. B* 353:1713—22.

[45]Herrero, S. , T. Gechev, P. L. Bakker, W. J. Moar, and R. A. de Maagd. 2005. *Bacillus thuringiensis* Cry1Ca-resistant *Spodoptera exigua* lacks expression of one of four Aminopeptidase N genes. *BMC Genomics* 6:96.

[46]Hmimina, M. , S. Poitout, and R. Buès. 1993. Variabilité des potentialités diapausantes intra et interpopulations chez *Heliothis armigera* Hb (Lep: Noctuidae). *J. Appl. Entomol.* 116:273—83.

[47]Honeybee Genome Sequencing Consortium. 2006. Insights into social insects from the genome sequence of the honeybee *Apis mellifera*. *Nature* 443:931—49.

[48] Horn, C. , B. Jaunich, and E. A. Wimmer. 2000. Highly sensitive, fluorescent transformation marker for *Drosophila* transgenesis. *Dev. Genes Evol.* 210:623—29.

[49]Horn, C. , N. Offen, S. Nystedt, U. Hacker, and E. A. Wimmer. 2003. *piggyBac*-based insertional mutagenesis and enhancer detection as a tool for functional insect genomics. *Genetics* 163:647—61.

[50]Huang, J. , Y. Zhang, M. L. S. Wang, et al. 2007. RNA interference-mediated silencing of the bursicon gene induces defects in wing expansion of silkworm. *FEBS Lett.* 581:697—701.

[51]Iatrou, K. , and P. Couble. 2007. 7th International Workshop on the Molecular Biology and Genetics of the Lepidoptera. August 20—26, 2006. Orthodox Academy of Crete, Kolympari, Crete, Greece. 52 pp. *J. Insect Sci.* 7:29.

[52]Imamura, M. , J. Nakai, S. Inoue, G. X. Quan, T. Kanda, and T. Tamura. 2003. Targeted gene expression using the GAL4/UAS system in the silkworm *Bombyx mori*. *Genetics* 165:1329—40.

[53]Ji, Y. —J. , and D. —X. Zhang. 2004. Characteristics of microsatellite DNA in lepidopteran genomes and implications for their isolation. *Acta Zool. Sin.* 50:608.

[54]Ji, Y. —J. , Y. —C. Wu, and D. —X. Zhang. 2005. Novel polymorphic microsatellite markers developed in the cotton bollworm *Helicoverpa armigera* (Lepidoptera: Noctuidae). *Insect Sci.* 12:331—34.

[55]Ji, Y. —J., D. —X. Zhang, G. M. Hewitt, L. Kang, and D. —M. Li. 2003. Polymorphic microsatellite loci for the cotton bollworm *Helicoverpa armigera* (Lepidoptera: Noctuidae) and some remarks on their isolation. *Mol. Ecol. Notes* 3:102—04.

[56]Kavi, H. H., H. Fernandez, W. Xie, and J. A. Birchler. 2008. Genetics and biochemistry of RNAi in *Drosophila. Curr. Top. Microbiol. Immunol.* 320:37—75.

[57]Kennerdell, J. R., and R. W. Carthew. 2000. Heritable gene silencing in *Drosophila* using double-stranded RNA. *Nat. Biotechnol.* 18:896—98.

[58]Keyghobadi, N., J. Roland, and C. Strobeck. 1999. Influence of landscape on the population genetic structure of the alpine butterfly *Parnassius smintheus* (Papilionidae). *Mol. Ecol.* 8:1481—95.

[59]Kim, J. K., H. W. Gabel, R. S. Kamath, et al. 2005. Functional genomic analysis of RNA interference in *C. elegans. Science* 308:1164—67.

[60]Kranthi, K. R., D. Jadhav, R. Wanjari, S. Kranthi, and D. Russell. 2001. Pyrethroid resistance and mechanisms of resistance in field strains of *Helicoverpa armigera* (Lepidoptera: Noctuidae). *J. Econ. Entomol.* 94:253—63.

[61]Li, X., M. R. Berenbaum, and M. A. Schuler. 2002. Plant allelochemicals differentially regulate *Helicoverpa zea* cytochrome p450 genes. *Insect Mol. Biol.* 11:343—51.

[62]Lynch, J. A., and C. Desplan. 2006. A method for parental RNA interference in the wasp *Nasonia vitripennis. Nat. Protoc.* 1:486—94.

[63]Mallet, J., A. Korman, D. Heckel, and P. King. 1993. Biochemical genetics of *Heliothis* and *Helicoverpa* (Lepidoptera: Noctuidae) and evidence for a founder event in *Helicoverpa zea*. *Ann. Entomol. Soc. Am.* 86:189—97.

[64]Mao, Y. —B., W. —J. Cai, J. —W. Wang, et al. 2007. Silencing a cotton bollworm P450 monooxygenase gene by plant-mediated RNAi impairs larval tolerance of gossypol. *Nat. Biotechnol.* 25:1307—13.

[65]Marcus, J. M., D. M. Ramos, and A. Monteiro. 2004. Germline transformation of the butterfly *Bicyclus anynana*. *Proc. Biol. Sci.* 27:S263—65.

[66]Margis, R., A. F. Fusaro, N. A. Smith, et al. 2006. The evolution and diversification of Dicers in plants. *FEBS Lett.* 580: 2442—50.

[67]Mathey-Prevot, B., and N. Perrimon. 2006. *Drosophila* genome-wide RNAi screens: Are they delivering the promise? *Cold Spring Harb. Symp. Quant. Biol.* 71:141—48.

[68]Matthews, M. 1999. *Heliothine moths of Australia: a guide to pest bollworms and related noctuid groups*. Melbourne: CSIRO Publishing.

[69]McKechnie, S. W., M. E. Spackman, N. E. Naughton, I. V. Kovacs, M. Ghosn, and A. A. Hoffman. 1993. Assessing budworm population structure in Australia using the A-T rich region of mitochondrial DNA. In *Proc. Beltwide Cotton Conf.*, ed. D. J. Herber and D. J. Richter, 838—840. New Orleans, LA. January 10—14, 1993. Natl. Cotton Council of Am., Memphis, TN.

[70]Meglécz, E., S. J. Anderson, D. Bourguet, et al. 2007. Microsatellite flanking region similarities among different loci within insect species. *Insect Mol. Biol.* 16:175—85.

[71]Meglécz, E., F. Petenian, E. Danchin, A. C. D'Acier, J. —Y. Rasplus, and E. Faure. 2004. High similarity between flanking regions of different microsatellites detected within each of two species of Lepidoptera: *Parnassius apollo and Euphydryas aurinia*. *Mol. Ecol.* 13:1693—700.

[72]Meister, G., and T. Tuschl. 2004. Mechanisms of gene silencing by double-stranded RNA. *Na-*

ture 431:343—49.

[73]Miao,X. —X. ,S. —J. Xu,M. — H. Li,et al. 2005. Simple sequence repeat-based consensus linkage map of *Bombyx mori*. *Proc. Natl. Acad. Sci. U. S. A.* 102:16303—08.

[74]Ming,Q. L. ,Y. H. Yan,and C. Z. Wang. 2007. Mechanisms of premating isolation between *Helicoverpa armigera* (Hübner) and *Helicoverpa assulta* (Guenee) (Lepidoptera: Noctuidae). *J. Insect Physiol.* 53:170—78.

[75]Mita,K. ,M. Kasahara,S. Sasaki,et al. 2004. The genome sequence of silkworm, *Bombyx mori*. *DNA Res.* 11:27—35.

[76]Mitter,C. ,R. W. Poole,and M. Matthews. 1993. Biosystematics of the Heliothinae (Lepidoptera: Noctuidae). *Annu. Rev. Entomol.* 38:207—25.

[77]Mohammed,A. ,and C. J. Coates. 2004. Promoter and *piggyBac* activities within embryos of the potato tuber moth,*Phthorimaea operculella* ,Zeller (Lepidoptera: Gelechiidae). *Gene* 342:293—301.

[78]Mori,H. ,M. Yamao, H. Nakazawa,et al. 1995. Transovarian transmission of a foreign gene in the silkworm, *Bombyx mori* ,by *Autographa californica* nuclear polyhedrosis virus. *Biotechnology*(*NY*) 13:1005—07.

[79]Nève,G. ,and E. Meglécz. 2000. Microsatellite frequencies in different taxa. *Trends Ecol. Evol.* 15:376—77.

[80] Neven, L. 2007. Development of transgenic codling moth for ABC and SIT. *Entomol. Res.* 37:A54.

[81]Nibouche,S. ,R. Bues,J. —F. Toubon,and S. Poitout. 1998. Allozyme polymorphism in the cotton bollworm *Helicoverpa armigera* (Lepidoptera: Noctuidae): Comparison of African and European populations. *Heredity* 80:438—45.

[82]Nilsen,T. W. 2007. Mechanisms of microRNA-mediated gene regulation in animal cells. *Trends Genet.* 23:243—49.

[83]Ober,K. A. ,and E. L. Jockusch. 2006. The roles of wingless and decapentaplegic in axis and appendage development in the red flour beetle,*Tribolium castaneum*. *Dev. Biol.* 294:391—405.

[84]Oppenheim,S. J. ,and F. Gould. 2002. Behavioral adaptations increase the value of enemy-free space for *Heliothis subflexa* ,a specialist herbivore. *Evolution* 56:679—89.

[85]Palumbi,S. R. 1996. Nucleic acids II: The polymerase chain reaction. In *Molecular systematics*, ed. D. M. Hillis,C. Moritz,and B. K. Mable,205—47. Sunderland,MA: Sinauer Associates.

[86]Palumbi,S. R. ,and C. S. Baker. 1994. Contrasting population structure from nuclear intron sequences and mtDNA of humpback whales. *Mol. Biol. Evol.* 11:426—35.

[87]Pedgley,D. E. 1985. Windborne migration of *Heliothis armigera* (Hübner) (Lepidoptera: Noctuidae) to the British Isles. *Entomol. Gaz.* 36:15—20.

[88]Peloquin,J. J. ,S. T. Thibault,L. P. Schouest,Jr. ,and T. A. Miller. 1997. Electromechanical microinjection of pink bollworm *Pectinophora gossypiella* embryos increases survival. *Biotechniques* 22:496—99.

[89]Peloquin,J. J. , S. T. Thibault, R. Staten, and T. A. Miller. 2000. Germ-line transformation of pink bollworm (Lepidoptera: Gelechiidae) mediated by the *piggyBac* transposable element. *Insect Mol. Biol.* 9:323—33.

[90]Perera,O. P. ,C. A. Blanco,B. E. Scheffler,and C. A. Abel. 2007. Characteristics of 13 polymor-

phic microsatellite markers in the corn earworm, *Helicoverpa zea* (Lepidoptera: Noctuidae). *Mol. Ecol. Notes* 7:1132—34.

[91]Prasad, M. D., M. Muthulakshmi, K. P. Arunkumar, et al. 2005a. SilkSatDb: A microsatellite database of the silkworm, *Bombyx mori*. *Nucleic Acids Res*. 33:D403—06.

[92]Prasad, M. D., M. Muthulakshmi, M. Madhu, S. Archak, K. Mita, and J. Nagaraju. 2005b. Survey and analysis of microsatellites in the silkworm, *Bombyx mori*: frequency, distribution, mutations, marker potential and their conservation in heterologous species. *Genetics* 169:197—214.

[93]Quan, G. X., T. Kanda, and T. Tamura. 2002. Induction of the white egg 3 mutant phenotype by injection of the double-stranded RNA of the silkworm white gene. *Insect Mol. Biol.* 11:217—22.

[94]Rajagopal, R., S. Sivakumar, N. Agrawal, P. Malhotra, and R. K. Bhatnagar. 2002. Silencing of midgut amino-peptidase N of *Spodoptera litura* by double-stranded RNA establishes its role as *Bacillus thuringiensis* toxin receptor. *J. Biol. Chem.* 277:46849—51.

[95]Reddy, K. D., E. G. Abraham, and J. Nagaraju. 1999. Microsatellites in the silkworm, *Bombyx mori*: Abundance, polymorphism, and strain characterization. *Genome* 42:1057—65.

[96]Roignant, J. Y., C. Carré, B. Mugat, D. Szymczak, J. A. Lepesant, and C. Antoniewski. 2003. Absence of transitive and systemic pathways allows cell-specific and isoform-specific RNAi in *Drosophila*. *RNA* 9:299—308.

[97]Rupasinghe, S. G., Z. Wen, T. —L. Chiu, and M. A. Schuler. 2007. *Helicoverpa zea* CYP6B8 and CYP321A1: Different molecular solutions to the problem of metabolizing plant toxins and insecticides. *Protein Eng. Des. Sel.* 20:615—24.

[98]Schröder, R. 2003. The genes orthodenticle and hunchback substitute for bicoid in the beetle *Tribolium*. *Nature* 422:621—25.

[99]Scott, K. D., C. L. Lange, L. J. Scott, and L. J. Gahan. 2004. Isolation and characterization of microsatellite loci from *Helicoverpa armigera* Hübner (Lepidoptera: Noctuidae). *Mol. Ecol. Notes* 4:204—05.

[100]Scott, K. D., N. Lawrence, C. L. Lange, et al. 2005a. Assessing moth migration and population structuring in *Helicoverpa armigera* (Lepidoptera: Noctuidae) at the regional scale: Example from the Darling Downs, Australia. *J. Econ. Entomol.* 98:2210—19.

[101]Scott, L. J., N. Lawrence, C. L. Lange, et al. 2006. Population dynamics and gene glow of *Helicoverpa armigera* (Lepidoptera: Noctuidae) on cotton and grain crops in the Murrumbidgee Valley, Australia. *J. Econ. Entomol.* 99:155—63.

[102]Scott, K. D., K. S. Wilkinson, N. Lawrence, et al. 2005b. Gene-flow between populations of cotton bollworm *Helicoverpa armigera* (Lepidoptera: Noctuidae) is highly variable between years. *Bull. Entomol. Res.* 95:381—92.

[103]Scott, K. D., K. S. Wilkinson, M. A. Merritt, et al. 2003. Genetic shifts in *Helicoverpa armigera* Hübner (Lepidoptera: Noctuidae) over a year in the Dawson/Callide Valleys. *Aust. J. Agric. Res.* 54:739—44.

[104]Stokes, N. H., S. W. McKechnie, and N. W. Forrester. 1997. Multiple allelic variation in sodium channel gene from populations of Australian *Helicoverpa armigera* (Hübner) (Lepidoptera: Noctuidae) detected via temperature gel electrophoresis. *Aust. J. Entomol.* 36:191—96.

[105]Sun, Z. C., M. Wu, T. A. Miller, and Z. J. Han. 2008. *piggyBac*-like elements in cotton boll-

worm, *Helicoverpa armigera* (Hubner). *Insect Mol. Biol.* 17:9—18.

[106]Takahashi, M., K. Kikuchi, S. Tomita, et al. 2003. Transient in vivo reporter gene assay for ecdysteroid action in the *Bombyx mori* silk gland. *Comp. Biochem. Physiol. B, Biochem. Mol. Biol.* 135:431—37.

[107]Tamura, T., C. Thibert, C. Royer, et al. 2000. Germline transformation of the silkworm *Bombyx mori* L. using a *piggyBac* transposon-derived vector. *Nat. Biotechnol.* 18:81—84.

[108]Tan, S., X. Chen, A. Zhang, and D. —M. Li. 2001. Isolation and characterization of DNA microsatellite from cotton bollworm (*Helicoverpa armigera*, Hubner). *Mol. Ecol. Notes* 1:243—44.

[109]Tan, A., H. Tanaka, T. Tamura, and T. Shiotsuki. 2005. Precocious metamorphosis in transgenic silkworms overexpressing juvenile hormone esterase. *Proc. Natl. Acad. Sci. U. S. A.* 102:11751—56.

[110]Tang, Q. B., J. W. Jiang, Y. H. Yan, J. J. A. Van Loon, and C. Z. Wang. 2006. Genetic analysis of larval host-plant preference in two sibling species of *Helicoverpa*. *Entomol. Exp. Appl.* 118:221—28.

[111]Tautz, D. 1989. Hypervariability of simple sequences as a general source for polymorphic DNA markers. *Nucleic Acids Res.* 17:6463—71.

[112]Tay, W. T., G. T. Behere, D. G. Heckel, S. F. Lee, and P. Batterham. 2008. Exon-primed intron-crossing (EPIC) PCR markers of *Helicoverpa armigera* (Lepidoptera: Noctuidae). *Bull. Entomol. Res.* 98:509—18.

[113]Thomas, J. L., J. Bardou, S. L'Hoste, B. Mauchamp, and G. Chavancy. 2001. A helium burst biolistic device adapted to penetrate fragile insect tissues. *J. Insect Sci.* 1:9.

[114]Thomas, J. L., M. Da Rocha, A. Besse, B. Mauchamp, and G. Chavancy. 2002. 3xP3-EGFP marker facilitates screening for transgenic silkworm *Bombyx mori* L. from the embryonic stage onwards. *Insect Biochem. Mol. Biol.* 32:247—53.

[115]Thomas, D. D., C. A. Donnelly, R. J. Wood, and L. S. Alphey. 2000. Insect population control using a dominant, repressible, lethal genetic system. *Science* 287:2474—6.

[116]Timmons, L., and A. Fire. 1998. Specific interference by ingested dsRNA. *Nature* 395:854.

[117]Tomoyasu, Y., S. C. Miller, S. Tomita, M. Schoppmeier, D. Grossmann, and G. Bucher. 2008. Exploring systemic RNA interference in insects: A genome-wide survey for RNAi genes in *Tribolium*. *Genome Biol.* 9:R10.

[118]Turner, C. T., M. W. Davy, R. M. MacDiarmid, K. M. Plummer, N. P. Birch, and R. D. Newcomb. 2006. RNA interference in the light brown apple moth, *Epiphyas postvittana* (Walker) induced by double-stranded RNA feeding. *Insect Mol. Biol.* 15:383—91.

[119]Uchino, K., M. Imamura, K. Shimizu, T. Kanda, and T. Tamura. 2007. Germ line transformation of the silkworm, *Bombyx mori*, using the transposable element *Minos*. *Mol. Genet. Genomics* 277:213—20.

[120]Uhlirova, M., M. Asahina, L. M. Riddiford, and M. Jindra. 2002. Heat-inducible transgenic expression in the silkmoth *Bombyx mori*. *Dev. Genes Evol.* 212:145—51.

[121]Uhlirova, M., B. D. Foy, B. J. Beaty, K. E. Olson, L. M. Riddiford, and M. Jindra. 2003. Use of Sindbis virus-mediated RNA interference to demonstrate a conserved role of Broad-Complex in insect metamorphosis. *Proc. Natl. Acad. Sci. U. S. A.* 100:15607—12.

[122]van't Hof, A. E., P. M. Brakefield, I. J. Saccheri, and B. J. Zwaan. 2007. Evolutionary dynamics of multi-locus microsatellite arrangements in the genome of the butterfly *Bicyclus anynana*, with implications for other Lepidoptera. *Heredity* 95:320—28.

[123]Vermehren, A. , S. Qazi, and B. A. Trimmer. 2001. The nicotinic α subunit MARA1 is necessary for cholinergic evoked calcium transients in *Manduca* neurons. *Neurosci. Lett.* 313:113—16.

[124]Vreysen, M. J. B. , A. S. Robinson, and J. Hendrichs, Eds. 2007. *Area-wide control of insect pests: From research to field implementation*. Dordrecht: Springer.

[125]Wang, C. Z. , J. F. Dong, D. L. Tang, J. H. Zhang, W. Li, and J. Qin. 2004. Host selection of *Helicoverpa armigera* and *H. assulta* and its inheritance. *Prog. Nat. Sci.* 14:880—84.

[126]Wang, G. R. , G. M. Liang, K. M. Wu, and Y. Y. Guo. 2005. Gene cloning and sequencing of aminopeptidase N3, a putative receptor for *Bacillus thuringiensis* insecticidal Cry1Ac toxin in *Helicoverpa armigera* (Lepidoptera: Noctuidae). *Eur. J. Entomol.* 102:13—19.

[127]Wang, J. , X. Ren, T. A. Miller, and Y. Park. 2006. *piggyBac*-like elements in the tobacco budworm, *Heliothis virescens* (Fabricius). *Insect Mol. Biol.* 15:435—43.

[128]Wang, W. , L. Swevers, and K. Iatrou. 2000. *Mariner* (*Mos*1) transposase and genomic integration of foreign gene sequences in *Bombyx mori* cells. *Insect Mol. Biol.* 9:145—55.

[129]Waterhouse, P. M. , M. W. Graham, and M. B. Wang. 1998. Virus resistance and gene silencing in plants is induced by double-stranded RNA. *Proc. Natl. Acad. Sci. U. S. A.* 95:13959—63.

[130]Xia, Q. , Z. Zhou, C. Lu, et al. 2004. A draft sequence for the genome of the domesticated silkworm (*Bombyx mori*). *Science* 306:1937—40.

[131]Xu, H. F. , Q. Y. Xia, C. Liu, et al. 2006. Identification and characterization of *piggyBac*-like elements in the genome of domesticated silkworm, *Bombyx mori*. *Mol. Genet. Genomics* 276:31—40.

[132]Xu, X. , L. Yu, and Y. Wu. 2005. Disruption of a cadherin gene associated with resistance to Cry1Ac δ-endotoxin of *Bacillus thuringiensis* in *Helicoverpa armigera*. *Appl. Environ. Microbiol.* 71:948—54.

[133]Yamao, M. , N. Katayama, H. Nakazawa, et al. 1999. Gene targeting in the silkworm by use of a baculovirus. *Genes Dev.* 13:511—16.

[134]Yang, Y. , Y. Wu, S. Chen, et al. 2004. The involvement of microsomal oxidases in pyrethroid resistance in *Helicoverpa armigera* from Asia. *Insect Biochem. Mol. Biol.* 34:763—73.

[135]Zhao X. C. , J. F. Dong, Q. B. Tang, et al. 2005. Hybridization between *Helicoverpa armigera* and *Helicoverpa assulta* (Lepidoptera: Noctuidae): Development and morphological characterization of F1 hybrids. *Bull. Entomol. Res.* 95:409—16.

[136]Zhong, B. , J. Li, J. Chen, J. Ye, and S. Yu. 2007. Comparison of transformation efficiency of piggyBac transposon among three different silkworm *Bombyx mori* Strains. *Acta Biochim. Biophys. Sin. (Shanghai)* 39:117—22.

[137]Zhou, X. , O. Faktor, S. W. Applebaum, and M. Coll. 2000. Population structure of the pestiferous moth *Helicoverpa armigera* in the eastern Mediterranean using RAPD analysis. *Heredity* 85:251—56.

[138]Zimowska, G. J. , and A. M. Handler. 2006. Highly conserved *piggyBac* elements in noctuid species of Lepidoptera. *Insect Biochem. Mol. Biol.* 36:421—28.

[139]Zong, N. , and C. Z. Wang. 2007. Larval feeding induced defensive responses in tobacco: Comparison of two sibling species of *Helicoverpa* with different diet breadths. *Planta* 226:215—24.

第十三章 鳞翅目昆虫对杀虫剂抗性的分子遗传学研究

David G. Heckel

1. 前言

1.1 抗性是一种进化现象

现代农业中，在控制昆虫害虫对粮食作物和棉花损害方面，化学和生物杀虫剂是必不可少的。这种对于杀虫剂的依赖性导致的众多的问题之一是广泛出现和频频发生的杀虫剂抗性。抗性被定义为一个群体由于长期、过度地暴露于一种毒素而导致的基于遗传的对该毒素敏感性下降的一种现象。这种敏感性下降的机制是该群体对于由这种毒素导致的不同死亡率在进化上的一种响应，这种机制经过若干代的时间会导致群体内控制抗性机制的等位基因频率的增加。杀虫剂抗性是一个全球性的问题，超过 450 种害虫对至少一种杀虫剂表现出抗性（Mota-Sanchez，Bills 和 Whalon 2002）。

一个引人注目的例子是小菜蛾（*Plutella xylostella*），它是一种十字花科的鳞翅目昆虫害虫。它很早就进化出了对广泛使用的杀虫剂 DDT 的抗性（Ankersmit 1953），而且继续表现出对多种杀虫剂的抗性，抗性通常在这些杀虫剂上市两到三年就产生了（Talekar 和 Shelton 1993）。除了现在常用的靶向神经系统的有机磷杀虫剂，氨基甲酸酯，环二烯和拟除虫菊酯外，小菜蛾对较少使用的昆虫生长调节剂例如几丁质合成抑制剂也产生了抗性（Morishita 1998）。在 20 世纪 90 年代早期，基于苏云金芽孢杆菌的晶体/孢子制备剂（Bt）的喷雾成功地控制了小菜蛾的危害，然而几年之后，小菜蛾（*P. xylostella*）同样对其表现出了抗性（Tabashnik et al. 1990）。

尽管分子遗传学的方法仅在近些年才应用到杀虫剂抗性研究中，但是整个发展过程还是令人鼓舞的。对不同物种中与毒素敏感性降低相关的基因结构、功能和导致抗性突变知识的积累使得抗性研究领域开始了理论上的统一，这是以前基于表现型的研究方法所不能及的。这是因为其中的许多基因是相同的，或者说这些基因是同源的。这意味着，暴露于相同的选择压力下的不同的物种提供了独立的对这一选择压力响应的进化上的重复，最终导致抗性的产生。这一进化观点产生一个实际的好处是，基于这一观点的害虫防止策略可能对很多物种有效。

1.2 抗性产生机制概述

以下机制中的一种或几种导致经典化学杀虫剂抗性的产生：躲避行为、表皮渗透性的降低、解毒代谢的增强或者靶标敏感性的降低等等。后两种是实践中最常遇到的。解毒能力的增强经常是因为特定的基因家族的一个或几个成员的突变或者上调表达造成的，这些基因家族包括 P450 酶家族、羧酸酯酶家族以及谷胱甘肽转移酶家族。这种抗性往往是广谱的，因为以上这些酶可以代谢解毒不同种类的化合物。这也是交叉抗性的一个基础，对一种杀虫剂的敏感性的降低与该群体接触的杀虫剂是不一样的。靶向位点突变导致的对杀虫剂敏感性的降低常常是对一类具有共同靶点的化合物是特异的。然而，当前开发新的指向共同靶点中的亚位点的趋势很可能导致对于不同类别化合物的交叉抗性的产生。

到目前，抗性发展到可以检测的水平，两种或者更多种不同的针对相同杀虫剂的抗性机制可能共同存在于一个群体中，例如，增强的羧酸酯酶的活性水平与不敏感的乙酰胆碱酯酶活性。这些机制之间可能会产生双重的、多重的或者其他方式的相互作用(Raymond,Heckel 和 Scott 1989),极大地增加了生化和生理分析的复杂性。为了解释观察到的抗性水平,针对每种抗性机制的遗传学和分子基础的研究是必须的,但是这一目标在实践中很少能实现。此外,一个群体可能由于同时或连续面对不同的杀虫剂,而对其中许多杀虫剂产生多重抗性。

1.3 鳞翅目昆虫在抗性研究中的地位

大部分杀虫剂对广泛的昆虫种类是有毒的,因为这些杀虫剂位于神经系统的靶点在进化上是高度保守的,同时对所有昆虫的生存是非常关键的。因此,存在于不同目的昆虫中的甚至是不同的节肢动物之间的抗性机制都具有相似性,而且从一族昆虫里得到的数据结果对于启发和加速其他族昆虫中的相关研究是非常有用的。因为历史的原因,分子遗传学的研究工具在双翅目中比在鳞翅目昆虫中具有更长的研究历史。大部分当前鳞翅目昆虫中所知的抗性相关基因最初在果蝇中克隆得到,在双翅目昆虫中得到了广泛研究。

另外,一些重要的抗性产生机制是从属于其他目的昆虫的研究中观察到的,但是到目前为止,在鳞翅目昆虫中还没有在分子水平上得到鉴定。

因而,一个关于鳞翅目中化学杀虫剂抗性机制的分子遗传学方面的综述必须提供在其他目的昆虫研究中收集所得的关于理解该种机制的最基本的背景信息。而且,为了更加完整地理解这些抗性机制,如果仅仅把例子限制在鳞翅目昆虫内,将无法看到全貌。以上这些因素可能解释了为什么虽然以往有许多关于杀虫剂抗性分子方面的研究综述,却很少有集中探讨鳞翅目昆虫的综述。本综述除了试图阐述鳞翅目昆虫如何加深我们对杀虫剂抗性机制的理解,还突出强调了利用鳞翅目昆虫可以作出特殊贡献的领域。与此相反,关于对苏云金芽孢杆菌的杀虫毒素的抗性的研究方面,鳞翅目昆虫已经发挥了突出的作用,而且关于这方面研究已有很多综述文章(Ferré 和 Van Rie 2002; Griffitts 和 Aroian 2005; Bravo,Gill 和 Soberon 2007; Heckel et al. 2007; Pigott 和 Ellar 2007),在此不再赘述。

2. 对化学杀虫剂的靶向位点抗性

与除草剂和杀真菌剂相比,杀虫剂的开发集中在少数的分子靶标上,其中大部分在神经系统发挥作用。这些靶点的优点是它们可以非常快速地使昆虫发生行为反应或者死亡,因为针对这些靶点的毒素可以在非常低的浓度下表现出高的亲和力。它们在神经系统内的正常功能常常具有严格的结构要求,即使在面对进化多样性的压力下仍然是保守的,所以这些靶点在系统发育多样性的昆虫中仍然表现出对毒素相同的亲和力。这些靶点的大部分结构突变体因此很少在已经施用过杀虫剂的群体中存在。然而,若一些个体偶然间被施用的其中一种杀虫剂所选择,选择下来的靶点形式可能或多或少地对同类的化学杀虫剂表现出抗性。在本综述中,利用四种主要的化学杀虫剂的靶点包括一个酶和三个离子通道,阐述了结构保守性和突变性之间的对立关系。

2.1 乙酰胆碱酯酶

乙酰胆碱酯酶(AChE)定位于昆虫中枢神经系统的胆碱性突触里,在神经递质乙酰胆碱扩散过突触间隙完成激活乙酰胆碱受体之后，酯酶将启动水解乙酰胆碱。水解的结果使得相同的神经递质分子群体不能继续激活突触后神经元。有机磷（OP）和氨基甲酸酯型杀虫剂可以抑制乙酰胆碱酯酶阻止它的活性水解，最终导致昆虫的死亡。然而，有机磷（OP）和氨基甲酸酯也会抑制脊椎动物 AChE 在中枢神经系统和神经肌肉接头的功能，所以它们对人类也是有毒性的。

有机磷（OP）和氨基甲酸酯的抑制作用可以通过 AChE 中的氨基酸替换来解除，这种类型的靶点杀虫剂抗性在节肢动物中已经发生多次演变。因为靶点是一种酶，所以这种抗性机制可以通过直接的体外生化实验来验证。利用一种检测底物例如乙酰胆碱来测定部分纯化的酶在抑制剂存在或不存在时的活性。在可以水解和结合杀虫剂的干扰蛋白质不存在的情况下，这种方法为测定靶点不敏感程度提供了一种直接和定量的方法，这种不敏感性不能被其他杀虫剂的靶点干扰。这种方法对具有多重抗性机制的昆虫品系尤为有用。

2.2 乙酰胆碱酯酶的抗性及其活性受抑制的表现型

在多种昆虫中对不敏感的 AChE 进行调查后，Russell 等人（2004）分辨出了两种主要的靶位点抗性模式。在第一种模式中，对氨基甲酸酯的不敏感性强于有机磷；在第二种模式中，对两者的不敏感性相差不大。在许多案例中没有充足的杀虫剂敏感性的数据可以通过表 13.1 来区分它们究竟属于哪一类，两种类型在鳞翅目昆虫中都有发现。下面将对几个有充足比较数据的案例进行描述。

表 13.1　鳞翅目中乙酰胆碱酯酶的不敏感性的模式

Table 13.1　Acetylcholinesterase Insensitivity Patterns in Lepidoptera

物种	不敏感性模式[a]	地理位置	机理	文献
草地夜蛾	Ⅱ	美国佛罗里达州	两个异构酶	Yu 1992；Yu, Nguyen, and Abo-Elghar 1993；Yu 2006
甜菜夜蛾	Ⅰ	荷兰；美国加州	—	Vanlaecke, Smagghe, and Degheele 1995；Byrne and Toscano 2001
斜纹夜蛾	Ⅱ	中国	—	Huang and Han 2007
烟夜蛾	Ⅰ	美国南部	乙酰胆碱酯酶不敏感，常染色体，2 号染色体	Brown and Bryson 1992；Heckel, Bryson, and Brown 1998
棉铃虫	Ⅰ，Ⅱ	澳大利亚，印度	2 或 3 各不同的 R 等位基因	Gunning, Moores, and Devonshire 1996, 1998；Srinivas et al. 2004
梨小食心虫	Ⅰ	加拿大安大略省	性别连锁	Kanga et al. 1997
小菜蛾	?	韩国	乙酰胆碱酯酶-1 氨基酸替换，A201S，G227A[b]	Baek et al. 2005；Lee et al. 2007
苹果蠹蛾	Ⅰ	西班牙	乙酰胆碱酯酶-1 氨基酸替换，F290V[b]	Cassanelli et al. 2006

a. 模式Ⅰ，乙酰胆碱酯酶对氨基甲酸酯类的不敏感性强于有机磷类；模式Ⅱ，乙酰胆碱酯酶对氨基甲酸酯类和有机磷类的敏感性是相当的。

b. 氨基酸替换的位置的编号是相对于鱼雷夜蛾的成熟的乙酰胆碱酯酶的氨基酸的位置来说的。

在 1990 年，在美国佛罗里达的玉米上收集到的草地贪夜蛾（*Spodoptera frugiperda*）(J. E. Smith)对多种杀虫剂都有抗性，其中包括有机磷类的甲基对硫磷（517 倍）和氨基甲酸酯类的西维因（507 倍）。除了一系列解毒相关酶的活性的增强，敌敌畏对 AChE 的抑制活性比非抗性的低 4 到 8 倍（Yu 1992）。一个在 10 年后收集到的品系也显示出相似的高抗性，另外通过成虫头部匀浆组分的测量，发现对甲基对氧磷和西维因的敏感性分别下降了 9 倍和 85 倍（Yu，Nguyen 和 Abo-Elghar 2003）。这些差异是通过凝胶过滤和亲和层析纯化得到的 AchE 来确定的，双分子速率常数的比较发现抗性酶对甲基对氧磷和西维因的不敏感性分别是 218 倍和 345 倍（Yu 2006）。通过非变性的聚丙烯酰胺凝胶电泳，发现有两种异构酶存在，主要形式为 66.1kD，而在 63.7kD 位置有另一条带。易感的和抗性品系具有相同的电泳条带模式，并且所有的含有这两种形式的纯化产物的抑制曲线都是线性的，暗示了这两种异构酶具有相同的生化性质。这个例子应该归为第二种模式。

一个来自荷兰温室的甜菜夜蛾（*Spodoptera exigua*）(Hükner)，除了具有解毒抗性机制之外，灭多威和敌敌畏对其 AChE 的抑制活性分别下降了 1.9 倍和 2.7 倍（Vanlaecke，Smagghe 和 Degheele 1995）。一个来自美国加利福尼亚的棉花品系对氨基甲酸酯灭多威的抗性增加了 68 倍却并没有对毒死蜱产生交叉抗性；另外，它的 AChE 对灭多威的抑制敏感性下降了 30 倍而对毒死蜱氧磷的敏感性只下降了 6 倍（Byrne 和 Toscano 2001）。这些都符合第一种模式的特征。

来自中国大豆和白菜普通地老虎（*Spodoptera litura*）对氨基甲酸酯类和有机磷类杀虫剂的抗性分别增强了 5.7 倍和 26 倍，同时对拟除虫菊酯也表现出高抗性。细胞色素 P450 和酯酶或许可以解释对这些不同种类杀虫剂的交叉抗性的产生，而 AChE 对有机硫代辛硫磷和氨基甲酸酯灭多威的敏感性也下降了 2～3 倍（Pattern II; Huang 和 Han 2007）。

在 20 世纪 60 到 70 年代，美国南部棉花田的烟芽夜蛾（*Heliothis virescens*）对甲基对硫磷表现出抗性。一个来自美国南卡罗来纳抗性品系的成虫和幼虫的 AChE 对甲基对硫磷的活性形式——甲基对氧磷的敏感性下降了 20 到 25 倍（Brown 和 Bryson 1992）。残杀威的双分子速率常数比一些有机磷类要低 10 到 100 倍，因此，Russell 等人(2004)把它归为第一种模式；然而，它对其他氨基甲酸酯类的抑制常数却与有机磷类相当。一些抗耐药性的化合物，像有机磷类的杀虫剂久效磷和百治磷能更多地抑制甲基对氧磷耐药性酶的活性。通过将烟青虫头部匀浆液分成单一组后用甲基对氧磷或久效磷进行测试发现，可以从田间采集的样本中得到三种抗性表型。杂交试验表明这些表型对应了三种不同的基因型：*RR*，*RS* 和 *SS*；这个位点是一个常染色体显性遗传位点，名为 *AceIn*，其中 R 编码一种甲基对氧磷耐受和久效磷敏感的 AChE 形式（Brown 和 Bryson 1992）。这种遗传鉴定阐明了遗传连锁与甲基对氧磷抗性表型的关系（Gilbert，Bryson，Brown 1996），同时也将 *AceIn* 基因定位到第 2 连锁群上（Heckel，Bryson 和 Brown 1998）。因为烟芽夜蛾（*H. virescens*）常染色体上的同工酶标记基因 IDH-2 在许多卷蛾中都是性别连锁的，Heckel，Bryson 和 Brown (1998)因此推测非敏感的 AChE 基因在卷蛾科中可能是性别连锁的。此外，辨别现场采集的个体的基因型，产生了一个有 R 等位基因频率检测的独特监视程序(Brown et al. 1996)。

通过对一个澳大利亚棉铃虫群体氨基甲酸酯抗性的研究,鉴定出了 AChE 的一种不敏感的形式,它对灭多威和硫双威耐受但是对有机磷类敏感(Gunning,Moores 和 Devonshire 1996),这是第一种模式杀虫剂的又一个案例。然而,随后的研究发现一个对有机磷类的丙溴磷和甲基对硫磷耐受而对氨基甲酸酯类敏感的品系,它的 AChE 对甲基对氧磷抑制的敏感性下降了 100 倍却对毒死蜱氧磷毫无抗性(Gunning,Moores 和 Devonshire 1998)。第二种类型不符合以上两种模式的任何一种,这暗示了在澳大利亚的该物种的不同群体中存在两种不同类型的非敏感 AChE。Srinivas 等人(2004)报道了一种来自印度棉铃虫群体的 AChE,它对久效磷和甲基对氧磷的敏感性下降了 3 到 4 倍,使人想到第二种模式以及可能存在第三种模式。

来自加拿大 Ontario 的一个梨小食心虫(*Grapholita molesta*)的种群对氨基甲酸酯类和有机磷类杀虫剂同时表现出抗性,其中对有机磷类的 gluthoxon 的敏感性下降了 3 到 6 倍,对氨基甲酸酯类的敏感性下降了 3 到 44 倍,属于第一种模式的杀虫剂(Kanga et al. 1997)。通过活性实验证明,在易感品系和抗性品系杂交的 F_1 代梨小食心虫对克百威的抗性参差不齐,另外,回交 F_2 代得到的雌雄后代之间的抗性也有所不同。在西维因和残杀威对 AChE 抑制性下降方面表现出相似的模式,暗示了该物种 AChE 对杀虫剂不敏感可能是由单一性别连锁的基因座控制的。再次在至少一种卷蛾中印证了以前的推测(Heckel,Bryson 和 Brown 1998);是否其他的卷蛾科的害虫例如苹果蠹蛾(*Cydia pomonella*)具有性别连锁不敏感的 AChE 还未见报道。

其他的鳞翅目昆虫的研究提供了一些证据,证明被改变的 AChE 与抗性相关。一个来自中国的二化螟(*Chilo suppressalis*)品系对三唑磷的抗性增强了 700 倍,与一个易感品系相比,它的 AChE 的酶动力学的 V_{max} 下降了 32%,K_m 值下降了 65%,对三唑磷的敏感性下降了 2.5 倍(Qu et al. 2003)。两个利用稻丰散筛选出来的小菜蛾的品系的 AChE 对乙酰甲胺磷和稻丰散的敏感性下降了 10 倍左右(Noppun,Miyata 和 Saito 1987)。一个来自西班牙的经过保棉磷选择出来的苹果蠹蛾(*C. pomonella*)的 AChE 对保棉磷和西维因的敏感性分别下降了 1.7 倍和 14 倍(Cassanelli et al. 2006)。

已有的证据表明 AChE 不敏感导致的杀虫剂抗性已经在鳞翅目昆虫中被广泛发现。种类间的相互比较因为在不同物种中研究杀虫剂的多样性而变得复杂,到目前为止,还无法将杀虫剂抗性的两种模式与其他的生物学特征联系起来。对机制的解释要依赖对 AChE 中涉及的特定的氨基酸替代以及它们怎样与酶的结构相联系的理解;并且对一些鳞翅目昆虫种类来说目前才刚刚获得这些信息。

2.3 乙酰胆碱酯酶的结构、序列以及抗性基因型

通过在脊椎动物中的广泛研究,对乙酰胆碱酯酶的结构和功能有了充分的了解,其中包括从富含乙酰胆碱酯酶的太平洋电鳐(*Torpedo californica*)的放电器官中分离得到蛋白质并确定了它的 3-D 结构(Sussman et al. 1991)。果蝇的 AChE 的基因序列是首先利用染色体步移技术确定的基因之一(Hall 和 Spierer 1986)。该酶的结构已经解析(Harel et al. 2000),这也为在昆虫中建立靶点突变模型提供了基础。与大部分酯酶一

样，乙酰胆碱酯酶属于 α/β 水解酶折叠结构超家族(Ollis et al. 1992)。该结构的核心是一系列 11 个几乎平行的 β-片层形成轻微扭曲的构象，因此第一个和最后一个 β-片层约成 90°角。另外在环中有 14 个 α-螺旋来连接这些 β-折叠。活性位点深埋在酶的内部，通过催化小沟与外界连接，催化小沟是一个 20Å 长几埃 Å 宽的通道，在通道内内衬的芳香族氨基酸残基使底物能有效地进入催化小沟。

活性位点包括四个亚位点，其中的三个容纳底物的三个不同区域。P1 位点或者叫做阴离子位点容纳带正电的胆碱基团，P2 位点或者叫做酰基结合口袋容纳甲基。此外，氧离子孔面对羰基氧，稳定在反应过程中流向酶的负电荷。第四个区域就是所谓的催化三联体，包括 238 位的丝氨酸、480 位的组氨酸和 367 位的天冬氨酸。虽然在一级结构上这三个位点距离比较远，但是蛋白质的折叠将它们拉近到一起形成特定的排列方式，这种特定的折叠方式对酶的催化活性至关重要。β-折叠和 α-螺旋之间狭小一块被称为亲和肘部，丝氨酸从这里移动到活性位点。丝氨酸的核亲和性通过与被天冬氨酸导向的组氨酸相互作用而被增强。与丝氨酸的氧形成短暂过渡的共价键产生了反应循环中的酰基一酶中间体，如果与这个氧形成稳定持久的共价键就会导致反应过程的抑制现象，这就是杀虫剂的作用原理(Oakeshott et al. 2005a)。

乙酰胆碱的水解需要两步反应。第一步，丝氨酸的氧原子亲核攻击底物的羰基碳，形成一个取代醇羟基的共价键。这些与碳原子形成的价键采取了一种四面体的构象，氧离子孔使这种四面体结构保持稳定。在第二步反应中，水分子中的氧原子对同一个碳原子发起亲核攻击，将酰基一丝氨酸键水解，同时释放自由酸，使酶恢复到原始的反应状态。

杀虫剂通过阻止第一步反应之后的反应进程来抑制乙酰胆碱酯酶的活性。丝氨酸的氧原子亲核攻击有机磷杀虫剂中的对氧磷形式的磷酸盐或者氨基甲酸酯类杀虫剂中的羰基碳，形成一种非常稳定的共价键，以致第二步的水解过程无法完成。结果，通过与"自杀抑制剂"共价连接导致乙酰胆碱酯酶不可逆地被束缚在一种没有活性的状态。因此，靶位点 AChE 杀虫剂抗性最终可解释为相关的特异氨基酸替换产生无活性的磷一酰基一酶或者氨基甲酰基一酶。目前还没有关于对 AChE 靶向位点抗性的产生是因为酶一抑制剂复合物的水解无法完成的报道，然而，在双翅目昆虫铜绿蝇(*Lucilia cuprina*, Newcomb et al. 1997)和家蝇(*Musca domestica*, Claudianos, Russell 和 Oakeshott 1999)中报道了一个与此相似的因为特定的羧酸酯酶一抑制剂复合物无法水解造成的抗性类型。

果蝇和家蝇的突变模式研究表明，在靠近催化沟附近的六个较大的氨基酸中的任何一个发生替换都会阻止抑制剂进入活性位点(Oakeshott et al. 2005a)。然而，在将果蝇中的研究结果推广到蚊子中的时候却遇到了意想不到的困难，因为在蚊子中基于序列相似性克隆得到的 AChE 基因与果蝇中显示了不同的氨基酸替换模式，甚至在蚊子中与 AChE 不敏感造成的抗性表型相对应的基因在染色体上的位置与果蝇中都是不同的(Malcolm et al. 1998)。通过研究发现蚊子和大部分昆虫中有两个相似但是不同的基因，现在命名为 *Ace*-1 和 *Ace*-2(Weill et al. 2002)，每个基因分别编码一个具有 AChE 活性的不同蛋白质，最终解决了这些问题。在多数昆虫中，研究发现 *Ace*-1 基因的产物在胆碱突触起作用,如果发生突变将直接影响 AChE 的不敏感性;而 *Ace*-2 基因的功能还不清楚。在高等双翅目如果蝇和家蝇中(Huchard et al. 2006),*Ace*-1 基因已经被丢失,但是 *Ace*-2 基因的产物替代 *Ace*-1 基因在胆碱突触中起作用。所以,在其他昆虫(Fournier 2005; Oakeshott et al. 2005b)中的相对于 *Ace*-1 基因的突变(而不是 *Ace*-2 基因)与高等双翅目中的 *Ace*-2 基因具有同样的作用原理:在该分子许多部分的突变能够减缓特异性的酶一抑制剂复合物的形成,但是这种效应具有杀虫剂特异性而且常常能加快其他抑制剂的这种效应。

第一个在鳞翅目昆虫中克隆得到的 AChE 基因是 *Ace*-2 基因的同源基因。Ni 等人(2003)在一个小菜蛾易感品系中克隆了一个 *Ace*-2 基因的全长。Ren,Han 和 Wang(2002)比较了一个来自中国的对久效磷耐受品系的棉铃虫品系的 5 个个体的和来自 6 个易感品系个体的全长 *Ace*-2 基因序列,发现除了 585 位的丙氨酸变为苏氨酸之外,品系内和品系间的广泛的多态性与抗性之间没有明确的关联;然而,一个苏氨酸存在于其他物种的易感品系,这个位置在鳞翅目昆虫中并不是高度保守的。

最近,已经克隆了来自几个鳞翅目昆虫种类的 *Ace*-1 和 *Ace*-2 基因并进行了比较研究,其中包括烟青虫(*Helicoverpa assulta*)(Lee et al. 2006);长蠹蛾(*C. pomonella*)(Cassanelli et al. 2006);小菜蛾(*P. xylostella*)(Lee et al. 2007);家蚕(*Bombyx mori*)(Seino et al. 2007; Shang et al. 2007)等。在后续的研究中,用细胞表达来自杀虫剂易感家蚕品系的这两种蛋白质,发现都具有 AChE 的活性,同时 *Ace*-2 蛋白质比 *Ace*-1 蛋白对依色林和对氧磷更加敏感。异源表达一个来自韩国小菜蛾丙硫磷抗性品种的 *Ace*-1 蛋白比来自一个易感品系的 *Ace*-1 蛋白对对氧磷更加不敏感,该研究是在鳞翅目昆虫中第一次进行这样的比较(Lee et al. 2006)。

到目前为止,*Ace*-1 基因的突变可能导致对氨基甲酸酯和有机磷类的不敏感已经在两个鳞翅目昆虫种类中得到鉴定。Baek 等人(2005)比较了一个来自韩国的丙硫磷抗性小菜蛾品系与一个来自日本易感品系的部分 *Ace*-1 基因序列,发现三个氨基酸位点的替换。其中的两个位点也在 Lee 等人(2007)关于来自同一个抗性品系与另一个不同的韩国小菜蛾易感品系中的全长 *Ace*-1 基因的比较研究中得以发现。第一个位点出现在 201 位的丙氨酸(Tc:根据太平洋电鳐(*Torpedo californica*)的活性形式进行序列编号),该丙氨酸是氧离子孔的三个氨基酸之一,在氨基酸序列一级结构中,该位点紧跟着催化丝氨酸位点,结构基序为 GESAG。Lee 等人 (2007)在一个抗性品系中发现该位置是 S,所以结构基序变为 GESSG;一个基于来自抗性个体 DNA 库的序列色谱分析研究表明在这个位点存在多态性,有些等位基因编码 A,而另一些则编码 S(Baek et al. 2005)。在一个棉蚜虫(*Aphis gossypii*)抗性品系的 *Ace*-1 基因中发现这个位置上也是丝氨酸(Andrews et al. 2004; Li 和 Han 2004; Toda et al. 2004);在所有其他已知的来自节肢动物的 *Ace*-1 基因序列中,该位置上都是丙氨酸。第二个被以上两个研究小组同时发现的替换位点出现在高度保守的 227 位甘氨酸上(Tc;根据太平洋电鳐(*Torpedo californica*)的活性形式序列进行编号),在抗性品系中该位置的甘氨酸被丙氨酸替换。相同的替换位点还出现在黑腹果蝇(Mutero et al. 1994)和米家蝇(*M. domestica*,Kozaki et al. 2001; Walsh et al. 2001)的抗性品系中的 *Ace*-2 基因中,异源表达来自几个双翅目昆虫的重组

AchE,造成它对许多杀虫剂不敏感(参见 Fournier 2005)。Lee 等人(2007)报道的第三个突变位置在 131 位(Tc;在小菜蛾的 *Ace*-1 序列的 D229G 位置),根据模型认为该位置的突变与抗性无关;另外,该位置的突变在 Baek 等人(2005)的研究中未见报道。Baek 等人(2005)报道的第三个突变的位点位于催化三联体的组氨酸之后;一个丙氨酸出现在易感品系中,而抗性品系中在该位点是丙氨酸或者甘氨酸。Lee 等人(2007)报道在两个研究的品系中,在该位点都是甘氨酸;一个关于昆虫该基因序列的调查统计发现,大部分鳞翅目昆虫在该位点是丙氨酸,而甘氨酸则出现在大部分非鳞翅目昆虫中。因而,在对丙硫磷耐受的小菜蛾品系中的这些突变中的两个与其他物种的抗性品系中的突变是同源的。

Cassanelli 等人(2006)研究了来自西班牙的苹果蠹蛾(*C. pomonella*)的 Raz 品系,该品系对谷硫磷-甲基的耐受性提高了 6.7 倍,同时对西维因的耐受性增强了 130 倍,这种抗性表型属于第一种模式,也可以通过其 AChE 对谷硫磷-甲基对氧磷和西维因的敏感性分别降低了 1.7 倍和 14 倍得到反映。Raz 品系与易感品系相比,*Ace*-2 基因并没发生氨基酸的变化而 *Ace*-1 基因中出现了一个氨基酸替换。研究者开发了一种基于 PCR 的实验来检测存在于个体 DNA 中的突变用于进行群体遗传学研究。位点的替换是 F290V(Tc),是一个位于 P2 位点(酰基结合口袋)的残基,在苹果蠹蛾序列中的位点编号是 F399V。在果蝇(Mutero et al. 1994)和家蝇(Walsh et al. 2001)中,同一个位点的缬氨酸替换为酪氨酸,与第二种模式的抗性相关,这种突变增加了空间位阻,限制了大的杀虫剂无法到达活性位点,而仍然允许乙酰胆碱到达活性位点。Villatte 等人(2000)通过体外定点突变实验来替换 290 位的酪氨酸(F290),检测了果蝇基因在 9 种氨基酸替代形式下,19 种化合物对它的抑制作用。缬氨酸替换酪氨酸即 F290V(如果是果蝇该酶的前体,则是在 368 位发生替换,即 F368V)的个体对谷硫磷-甲基对氧磷和西维因的敏感性分别下降了 1.9 倍和 3.7 倍,对于大多数其他的氨基甲酸酯类的敏感性下降得更多,这种模式与第一种模式的抗性表型非常相似。

当前在小菜蛾和苹果蠹的研究进展对于将来破解鳞翅目昆虫 AChE 的靶向位点特异性抗性的分子基础提供了很好的范例。在一系列已经进行生物化学研究的抗性品系中,检测 *Ace*-1 基因序列的突变可能产生更多的突变,对这些突变的研究将会使我们更加深入地理解结构和功能之间的关系。这可以推动对特定杀虫剂敏感性增加的利用,当耐受性由一种进化为另一种时,这种敏感性必须跟着进化,同时促使发现和开发抗耐受的有机磷类和氨基甲酸酯类杀虫剂。虽然果蝇对 *Ace*-2 基因的研究提供了基础,但是该世系中 *Ace*-1 基因的缺失导致研究者们将精力集中在其他昆虫"错误"的酶上面,可能已经短暂地延迟了对 *Ace*-1 基因的研究。到目前为止,还没有一个昆虫的 *Ace*-1 蛋白的 3-D 结构得到解析,也没有用定点突变方法系统性地去研究鳞翅目昆虫抗性品系 *Ace*-1 基因中发现的由于特定氨基酸的替换较相同种类的易感品系抑制性的效果,毕竟,这种突变是在杀虫剂的选择下立即产生的。

3. 电压门控钠通道

药理学和毒理学的证据显示电压门控的钠通道是 DDT 和拟除虫菊酯类杀虫剂的主

要靶标(Narahashi 1996)。这一靶标的脆弱性已经被神经毒素的多种进化表现所证实，神经毒素参与了化学物质在体内的快速沉积和天敌捕食(见 Soderlund 2005 年的综述)。人们在农业生产中采用新方法去利用它，如把它作为拟除虫菊酯类杀虫剂茚虫威的靶标(Wing et al. 2005)以及用杆状病毒重组表达蝎毒毒素作为生物农药(Zlotkin et al. 1995)。

尽管钠通道突变可能在抵抗 DDT 方面发挥了很好的作用，但是，毕竟 DDT 使用的全盛时期出现在利用分子遗传学去研究钠通道之前。在 20 世纪 70 年代，环境和安全问题与抗性问题共同促使人们停止了对 DDT 的使用；然而，在非洲为了控制传播疟疾的蚊子的数量上升，DDT 的使用又开始呈上升趋势。大部分当前关于钠通道在杀虫剂抗性中的研究是由合成的拟除虫菊酯广泛耐受性所驱动的。这类杀虫剂在天然除虫菊酯杀虫剂的结构基础上发展而来，它们的出现是化学防控昆虫、农作物保护以及对由昆虫传播的动物和人类疾病控制上的一个主要的进步(Elliott 1996)。拟除虫菊酯类杀虫剂在全球的积极使用带来的后果是种类繁多的节肢动物产生了抗药性。

拟除虫菊酯抗性的产生常常由是多种因素造成的，包括解毒机制和靶向位点抗性。梳理这些机制需要间接的方法。一种方法是关注杀虫剂中毒后在行为学和神经生理学水平产生的症状，可能是由于靶向位点的原因而发生变化，而不是代谢抗性水平的变化。另一种方法是利用杀虫剂激动剂，杀虫剂激动剂是一类可以抑制一大类解毒相关酶的化学物质，利用激动剂抑制解毒代谢产生的抗性以及使用任何残留的抗性去评价靶向位点抗性。一幅以家蝇抗性下调和抗性超级下调的品系为典型代表的关于 DDT 和拟除虫菊酯靶向位点抗性的画卷浮出水面。出现在其他物种中的相似的抗性综合征被称作“抗性下调类”。把它作为靶向位点耐受性的最好评判来自神经生理学研究，包括早期对鳞翅目昆虫害虫的研究。

3.1 钠通道抗性与神经生理学表型

关于内在神经不敏感性作为鳞翅目昆虫中拟除虫菊酯抗性的早期证据来自 Gammon(1980)关于一个抗氯菊酯而不抗氯氰菊酯的斜纹夜蛾(*Spodoptera littoralis*)的研究。经阈值浓度的氯菊酯处理一段时间后，分离出的幼虫神经索在一次单一的电刺激之后显示出增强的射频爆发。来自抗性品系的幼虫神经比易感品系需要更长的处理时间，但是这种效应在氯氰菊酯处理后没有出现。

于 1983 年在澳大利亚东部发现了耐受氰戊菊酯的棉铃虫(Gunning et al. 1984)，而且调查了多种抗性机制(Gunning et al. 1991)。将从腹部肌肉局部解剖得来的外周神经暴露在 10 μM 氰戊菊酯的检测浓度中，然后测定自发的重复射频的发作时间来进行神经不敏感性的定量分析。采集于 1983 年的一个抗性品系的幼虫产生响应的时间比一个易感品系长了 5.1 倍；三分之一的抗性品系对这个浓度的氰戊菊酯没有任何响应(Gunning et al. 1991)。F_1 代杂交幼虫的响应时间平均增加了 3.7 倍。然而，来自一个采集于 1987 年的抗性品系幼虫的响应时间平均增加了 2.8 倍，另外，在抗性样本中的不敏感个体的频率在 1989 年下降至不足 20%。神经不敏感性机制效力和频率明显下降的同时代谢抗性方面的机制反而在增强，这种代谢机制可以被细胞色素 P450 抑制剂胡椒基丁醚所抑制(PBO; Gunning et al. 1991)。与世界上其他大部分棉花种植区不同,因早期检测到对拟除虫菊酯的抗性,澳大利亚在这个时期通过了一个抗性管理策略,高度限制拟除虫菊酯在棉田的使用(Forrester et al. 1993)。多年的持续使用使得拟除虫菊酯抗性稳定增加,而限制的选择压力或许可以部分地解释神经敏感性机制的先增强然后下降,当杀虫剂不存在时,可能存在适应度代价。

使用相似的神经生理学方法研究发现,神经不敏感性一直是棉铃虫对拟除虫菊酯抗性机制的重要组成部分,这种现象在多个地区都有发现,其中包括泰国(Ahmad, Gladwell 和 McCaffery 1989),印度的安德拉邦(McCaffery et al. 1997),中国的江苏省(Tan 和 McCaffery 1999)和河北省(Ru et al. 1998)。这些品系中,大部分还兼有代谢抗性机制,因为经过 PBO 处理幼虫的抗性水平有一定程度的下降。

在美国,另一种棉花害虫棉铃虫夜蛾(*Helicoverpa virescens*)也产生了对拟除虫菊酯多种抗性之一的神经不敏感性(Nicholson 和 Miller 1985)。Church 和 Knowles (1993)研究了来自成虫头部的神经膜的粗提物,测定了放射性标记的箭毒蛙毒素的结合活性,箭毒蛙毒素是一种钠通道选择性的神经毒素,它的功能是增加拟除虫菊酯的浓度。尽管在抗性品系和易感品系之间存在一些差异,但是他们还无法对作用机理做出解释。通过将幼虫的周围神经组织提取物置于浓度逐渐升高的氯菊酯中,通过幼虫(McCaffery, Holloway 和 Gladwell 1995)或成虫(Holloway 和 McCaffery 1996)的多单元活性的胞外记录的方法设计累积剂量响应实验来测定不敏感性程度在群体的分布频率。这是与 PBO 处理后残留的抗性程度相关联的,并且在棉花生长季节有所增加。Ottea 和 Holloway (1998)利用烯丙菊酯在相同的物种中得到了相似的结果。Lee 等人(1999)培养了来自成虫胸部和腹部神经节的神经元细胞,利用全细胞膜片钳技术对来自易感品系和 Pyr-R 抗性品系的钠通道的特性进行了比较。Pyr-R 抗性品系的钠通道具有不同的门控特性,在电压依赖中具有 13mV 的正向转移电压,当活化阈值电压提高了 11mV 后激活反应降低。另外,在延长尾电流时,它们对氯菊酯的敏感性下降了 20 倍。Pyr-R 是一个纯合品系,其钠通道的 410 位的缬氨酸被替换为甲硫氨酸(见下文)。Park 等人 (2000) 将相似的单细胞膜片钳测量结果与成虫下调生物活性实验的结果和对来自一系列杂交后代的幼虫神经肌肉连接的胞外记录结果联系起来,神经肌肉连接的胞外记录结果显示它们是因为钠通道中的一个突变导致的遗传分离,这个突变就是 1014 位的亮氨酸替换为组氨酸(见下文)。虽然以上三种测量方法的结果可以通过基于该突变的三种基因型联系起来,但是在每种神经不敏感性的基因型中仍然存在其他的突变形式,这暗示在钠通道中其他部位的突变或者在其他基因中的突变同时也会影响这种性状。

研究者利用来自小菜蛾(*P. xylostella*)幼虫腹部内部纵向肌肉细胞的胞内 EPSP(兴奋性突触后电位)进行了测定,自发的 mini-EPSP 或者 EPSP 会被阶段性神经电刺激所诱发(Schuler et al. 1998)。当暴露在溴氰菊酯中时,mini-EPSP 的活性会增强,而 EPSP 的振幅会下降,另外基于这两种响应的混合测量方法估计该品系的 EC50 值比来自台湾的 FEN 抗性品系大 390000 倍。

3.2 钠通道的结构、序列及其抗性个体的基因型

随着在电鳗(Noda et al. 1984)中的一个电压敏感的钠通道的基因克隆以及在其他脊椎动物中广泛开展的钠通道研究，在果蝇中已经发现了两个同源基因。第一个被克隆的是 *dsc1* 基因(Salkoff et al. 1987)，它与拟除虫菊酯的抗性并不相关，而是在处理嗅觉信息方面有重要作用(Kulkarni et al. 2002)，也有研究者认为它是个钙通道基因(Zhou et al. 2004)。第二个钠通道基因是在 *para*^{ts} 位点发现的，定义为温度敏感的麻痹表现型(Loughney，Kreber 和 Ganetzky 1989)。*Vssc1* 基因是家蝇(*M. domestica*)中 *para* 基因的直系同源基因，是在保存 kdr 和 super-kdr 突变品系时发现的，随后在至少 26 种节肢动物中发现 *para* 同源基因存在于更多的突变中(见综述 Davies et al. 2007)。

钠通道是由一个很大的蛋白质组成的，其中含有大约 2100 个氨基酸残基，包括 4 个内部串联的同源结构域(I-IV)，每一个结构域都含有 6 个跨膜螺旋(S1－S6)以及膜内和膜外环状结构。来自电鳗的钠通道蛋白的低分辨率 3-D 构象已经被解析出来(Sato et al. 2001)。这四个结构域通过各自的指向中间的 S5 和 S6 跨膜螺旋连接在一起组装成四聚体，离子通道孔内嵌在四聚体内部。在 S5 和 S6 之间的短 P-环深入到细胞膜的胞外侧面用于选择钠离子。钠离子是在通道处于开启状态时，通过钠通道孔进入细胞的。从静息状态到开启状态是由于细胞膜去极化造成的，而去极化是由钠通道结构中的 S1－S4 的结构域感知的。依赖电压激活的“m-gate”面向胞外侧开启，在几毫秒内允许钠离子流进入细胞；紧接着的是钠通道孔的胞内侧端偶联的延迟的“h-gate” 来关闭抑制激活。这时的钠通道处于抑制状态。随着细胞膜的复极性化，m-gate 关闭，h-gate 重新开启，接着恢复到静息构象，准备进入下一次细胞膜去极化激活状态(Davies et al. 2007)。

放射配位和功能研究已经在钠通道上至少发现了 10 个不同的功能位点，这些位点可以结合神经毒素或者杀虫剂，对钠通道激活和抑制的不同方面产生影响(见综述 Soderlund 2005)。结合拟除虫菊酯使得开启构象更加稳定，从而阻止了钠通道的抑制活动，所以在复极性化时产生了一个比较大的尾电流。在苄基醇位点上拥有一个 α-氰基的 II 型拟除虫菊酯比缺少该基团的 I 型拟除虫菊酯对钠通道具有更强的影响。关于钠通道对 DDT 和拟除虫菊酯的结合位点已经在突变体抗性中进行了大量的研究。

通过将果蝇的 *para* 基因的全长转录本与基因组序列进行比较，发现该基因具有复杂的选择性拼接模式，7 个可选的外显子出现在胞内的环状结构中，两对外显子出现在转膜结构区域(见综述 Dong 2007)。这些序列在多数昆虫中是保守的，可能在功能上具有重要的意义。除此之外，还有证据显示 RNA 编辑也涉及其中，在转录后修饰中，mRNA 中的 A 可能转化为 I，或者 U 转化为 C，在转膜区域和胞内侧区域至少发现了 8 个氨基酸替换位点。选择性拼接和 RNA 编辑这两种机制使得功能蛋白和基因水平的突变检测之间的关系更加复杂。

然而，已经研究了许多来自 DNA 突变导致氨基酸替换而产生的抗性，这项工作始于对负责家蝇下调抗性表型的 *Vssc1* 基因突变的定位和测序研究(在接下来的叙述中，替换位点的编号是根据其在果蝇 *para* 蛋白中的氨基酸的序号)。kdr 突变对所有拟除虫菊酯类的杀虫剂产生了低的但是广谱的抗性，这种抗性产生的内在机制是由于在 IIS6 结构域

中 1029 位的亮氨酸被苯丙氨酸替换(L1029F)(L1014F in the *Musca Vssc1* sequence; Ingles et al. 1996; Miyazaki et al. 1996; Williamson et al. 1996)。在节肢动物中,这是最普通的抗性位点,其中亮氨酸被一个苯丙氨酸、组氨酸或者丝氨酸所替换,这样的替换出现在 11 个物种中,包括蟑螂、蚜虫、蚊子、甲虫、蓟马、跳蚤和飞蛾。家蝇的 *super-kdr* 抗性品系除了 L1029F 之外,还在 933 位上甲硫氨酸被替换为苏氨酸,该位置处于 IIS4 和 S5 的连接处(M918T in *Vssc1*; Williamson et al. 1996),这样额外的替换导致该品系主要对 II 型拟除虫菊酯表现出较高的抗性。在角蝇、蚜虫、斑潜蝇和粉虱也检测到同样位置的突变。另外还有 8 个突变位点出现在 IIS4-S6 的结构域内,其他的 18 个突变位点出现在一个或者多个对拟除虫菊酯有抗性的节肢动物种类中(见综述 Davies et al. 2007)。到目前为止,已有 5 个突变位点在鳞翅目昆虫中发现,其中三个是独立存在的。因为研究中很少给出分别来自抗性和易感品系基因的全长序列,将导致突变样本可能存在偏差。而且,许多研究仅仅限制于检测其他物种内已经存在的突变。

在棉铃虫夜蛾(*Helicoverpa virescens*)中,利用一个内含子内的多态性,首先将拟除虫菊酯的抗性定位到 *para*-同源的钠通道基因座 *hscp* 上(Taylor et al. 1993)。对这个物种的进一步的测序发现 1029 位的亮氨酸被替换为组氨酸,这与家蝇的 *kdr* 抗性品系的 IIS6 结构域的突变出现在相同的位点(Park 和 Taylor 1997)。这个多态性被用作确定生物活性实验和神经不敏感性实验个体的基因型,进一步加强了与抗性之间的联系。在一个抗性个体中,保留了易感的 1029 位的亮氨酸,而一个 421 位的缬氨酸被甲硫氨酸替换,这个位点位于 IS6 结构域之内,IS6 结构域与 *kdr* 突变体的 IIS6 结构域是类似的(Park,Taylor 和 Feyereisen 1997)。到目前为止,这个发生在 421 位的氨基酸替换还没有在任何其他节肢动物中有报道。Park,Taylor 和 Feyereisen (1999)随后对一个含有大部分 *hscp* 基因序列进行了测序,发现一些与早先提到的果蝇的 *para* 基因的选择性拼接模式相类似的地方,同时也有一些差异。Head,McCaffery 和 Callaghan (1998)专门寻找发生在胞膜内侧的位于结构域 III 和 IV 之间的环状结构区域内的突变,他们发现了两个位点:1561 位的天冬氨酸被缬氨酸替换和 1565 位的谷氨酸突变为甘氨酸。这两个突变都出现在一个对拟除虫菊酯显示出抗性并表现出神经不敏感的棉铃虫夜蛾(*Helicoverpa virescens*)品系中,而且也出现在一个来自中国江苏省的神经不敏感的棉铃虫(*Helicoverpa armigera*)品系中,但是在其他种类的易感品系中并没有出现(Head,McCaffery 和 Callaghan 1998)。迄今为止,虽然在其他有拟除虫菊酯抗性的节肢动物种类中还没有发现突变,但是已经在蟑螂中发现了出现在结构域连接区域 DI-DII 的突变和 *kdr* 突变。然而,遗憾的是在该研究中并没有检测 1029 位的亮氨酸和 421 位的缬氨酸的状态。

对溴氰菊酯显示出明显神经不敏感性的小菜蛾的 FEN 品系用于筛选抗性相关的突变,从 cDNA 扩增出一个 350bp 的片段,这个片段覆盖了前面提到的存在于家蝇 *kdr* 突变发生的区域(Schuler et al. 1998)。在该区域中发现了 L1029F 突变,另外还发现了一个新的突变,944 位的苏氨酸被异亮氨酸所替换,12 个已经测序的纯合幼虫在神经生理学实验中都表现出神经不敏感性。后一个突变与 *super-kdr* 突变位点 M933T 非常接近,当与 *kdr* 突变同时存在时可以极大地提高昆虫的神经不敏感性。在 944 位的突变随后也在西花蓟马、猫蚤、烟粉虱和头虱中发现,但是在每一种昆虫中 944 位突变后的氨基酸都是不一样的。

因为钠通道蛋白的基因非常大而且内含子数目多，以及 *kdr* 和 *super-kdr* 突变在抗性物种中出现的频率很高、距离很近，因此许多其他的研究在比较抗性极易感品系时仅仅检查了上述的那个区域。一个在对溴氰菊酯耐受的果蠹蛾（*C. pomonella*）的研究发现 L1029F 位点的突变确实存在，然而在 M933 和 T944 位点并没有发生改变（Brun-Barale et al. 2005）。

3.3 抗性相关的钠通道基因型的功能检验

为了探明氨基酸替换对钠通道蛋白功能的影响，在爪蟾卵母细胞中将钠通道蛋白与 TipE 蛋白一起表达。以微注射进入爪蟾卵母细胞的 mRNA 为模板，翻译的蛋白质被整合进入这些细胞的外膜，在拟除虫菊酯和其他配基存在和不存在的情况下检测钠通道的电生理记录。因为不容易获得该基因的全长 cDNA 序列，因此同源的突变位点被引入到果蝇、家蝇或者蟑螂的 cDNA 中。以上三个物种的 cDNA 在爪蟾系统中可以很好的使用，这也加速了对引入突变的比较。另一方面，是否被引入的突变在野生型背景物种中也具有同样的效应还不清楚；例如，与以上三个物种相比，棉铃虫夜蛾的基因显示出 15%～20%的序列差异。

Lee 和 Soderlund (2001)将在棉铃虫夜蛾中发现的 V421M 的替换形式引入到家蝇的 *Vssc*1 的 cDNA 中，并且将这种引入形式与家蝇的野生型个体和 L1029F 突变的 *kdr* 个体进行了比较。结果显示，与 L1029F 突变的 *kdr* 个体类似，经 V421M 修饰的家蝇的钠通道的中间点电位比野生型的上调了 9mV。同时，与野生型相比，在延长尾电流的情况下，经 V421M 修饰的家蝇的对顺菊酯敏感性下降了 20 倍。这些与野生型的比较结果与 Lee 等人 (1999)在来自棉铃虫夜蛾的 V421M 的纯合体神经细胞培养的研究中得到的结果类似。与 L1029F 突变相比，爪蟾表达系统中含有 V421M 突变的 *Vssc*1 基因表现出的对顺菊酯的不敏感的程度和模式确实与 L1029F 突变个体非常类似，这暗示 V421M 突变在体内能够产生像经典的 *kdr* 突变一样的抗性水平。

Zhao，Park 和 Adams (2000)将 V421M 突变或者来自棉铃虫夜蛾的 L1029H 突变引入到果蝇的野生型 *para* 通道中，然后在爪蟾卵母细胞中进行表达。在突变体中观察到一些差异。在 V421M 突变体的中间点电位上调了 6.6mV，而在 L1029H 突变体中是它的两倍。稳定状态的抑制点电位在 V421M 和 L1029H 突变体中分别是 4.5 mV 和 2.1 mV，也有所上调。在 10 μM 氯菊酯存在的情况下，两种突变体的失活速率与野生型相比都加快了 45 倍，失活速率提高的好处是氯菊酯对钠通道的修饰效应缩短了钠离子流入细胞完成去极性化的时间。另外，通过尾电流与最大电流电导的比值看，在 10 μM 氯菊酯环境中，L1029H 突变体具有更强的保护效应。作者推测，在氯菊酯不存在的情况下这些差异将会转化为携带这些突变的个体的降低的适合度消耗，从这个角度看，V421M 突变的适合度消耗更大。虽然钠通道的基因型没有被直接研究，有证据显示蛾子对拟除虫菊酯耐受的神经不敏感，其繁殖力比较低，而且雌性个体的性激素产量也比较低，同时被雄性性激素吸引的比率同样比较低（Campanhola et al. 1991）。从爪蟾表达系统的结果看，作者预测 L1029H 突变体比 V421M 突变体具有竞争优势，因为 L1029H 突变体赋予了昆虫在杀虫剂存在的情况下更好的保护能力，而且在杀虫剂不存在的情况下适合度代价较低。

V421M 突变体与其他昆虫中的另外的突变是否存在组合的效应也已经做过检测。在蟑螂（*Blattella germanica*）中，*kdr* 突变 L1029F 与另外两个出现在胞内 DI-DII 结构域的连接环内（在蟑螂蛋白质序列中的编号是 E434K 和 C764R）。L1029F 突变单独存在时使得在爪蟾系统中对溴氰菊酯的敏感性下降了 5 倍，而另外两个突变没有任何效应。当 L1029F 突变与其他两个突变中任何一个同时存在时，对溴氰菊酯的敏感性下降了 100 倍，当三种突变同时存在时，敏感性下降了 500 倍之多（Tan et al. 2002）。为了检验是否出现在与 *kdr* 品系中的 IIS6 类似的 IS6 位置的 V421M 突变是否具有同样的增强效应，蟑螂的序列中引入 V421M 突变，结果显示 V421M 突变与其他两种突变组合时，对溴氰菊酯敏感性下降了 10 倍。该研究显示，V421M 突变单独与其他任何一种存在时没有增强效应，除非其他两种突变同时与 V421M 突变组合才会使敏感性下降 100 倍（Liu et al. 2002）。如果这些发现可以对棉铃虫野生型钠通道的背景进行推论，这些研究结果可能暗示了相对于棉铃虫这种增强效应的突变连续的累积更可能发生在蟑螂中。在这种背景下，检测在棉铃虫中，发生在 DIII-DIV 连接区域的突变 D1549V 和 E1553G 是否对 V421M 和 L1029H 突变具有增强效应将会非常有意思。

3.4 钠通道抗性基因型的群体遗传学

在研究钠通道的序列突变方面，基于 DNA 的对个体的基因分型技术已经用于检测选择和实时追踪等位基因频率的动态变化。在 *kdr* 突变被鉴定之前，Taylor，Shen 和 Kreitman（1995）利用通过 *hscp* 基因中的非编码序列中的突变定义的单倍体的群体频率去验证了杀虫剂对来自美国棉花种植区的棉铃虫夜蛾的选择性。单倍型是基于在变性梯度凝胶电泳中迁移率的差异而进行鉴定的，这种方法可以检测到约 500bp 的 PCR 产物中的大部分单核苷酸替换。在 1990 年，利用信息素陷阱从美国四州一共收集到约 660 个雄性成虫用于检测 *kdr* 突变。具有最高抗性的样本具有最低的单倍型多样性，这与少数共同单倍型的选择性去除相一致。与同工酶和另一个中性遗传标记 *Hejs* 相反，多数共同 *hscp* 单倍型的频率显示出固定的地理差异。1995 年收集的样本在 *hscp* 基因单倍型中仍显示出重要的地理区域异质性，而在 *Hejs* 基因则不然（Taylor，Park 和 Shen 1996）。在 1990 和 1995 年的样本中，一个单倍型在拟除虫菊酯选择之后强烈增加，暗示了在这段时间内的选择压力下，这个单倍型和实际的位点之间的平衡失调。发现导致抗性的突变之后，对地域样本的回溯分析显示，随着时间的推移 L1029H 和 V421M 突变的频率发生了改变。在 1990 年，两者在一个路易斯安那的群体中的频率都是 20%；到 1997 年，L1029H 的频率增加到了 78%，没有检测到 V421M 的突变（Zhao，Park 和 Adams 2000）。这种“突变的进化演变”模式与基于在爪蟾系统所预测的特征是相一致的。

相似的方法被用于其他物种的研究，其中包括利用温度梯度凝胶电泳检测在澳大利亚棉铃虫的单倍型突变体（Stokes，McKechnie 和 Forrester 1997）。虽然发现了广泛的地域突变体，但是没有发现任何单倍体与氰戊菊酯耐受性之间的联系。这与 8 年前报道的发生在澳大利亚的神经不敏感机制的频率下降是相一致的（Gunning et al. 1991）。另一方面，对苹果蠹蛾的群体调查的确显示，通过 PCR 对选择性的等位基因(PASA)特异性扩增确定的 *kdr* 突变 L1014F 的频率与一个法国果园里经过拟除虫菊酯处理的频率之间存在关联，但是没有对应，这种模式的中性微卫星标记(Franck et al. 2007)。

在小菜蛾中出现的两个突变为选择动态的比较提供了新的机会。L1029F *kdr* 突变存在于一个日本的抗性品系中(Tsukahara et al. 2003)，而另外三个突变 V421M，D1561V 和 E1565G 并没有在该品系中被发现。以前 Schuler 等人(1998)在这个品系中发现的 T944I 替换也在这个研究中得以发现。这个区域被选择性拼接，因此两个可选择的外显子 A1 和 A2 中只有一个将会出现在成熟的 mRNA 中，另外 T944I 的替换出现在抗性品系 A2 的外显子中。因而，选择性拼接可能会影响抗性水平(Sonoda et al. 2006)。

依靠针对 L1029F 和 T944I 替换而设计的 PASA 实验，Kwon，Clark 和 Lee (2004)通过比较现在与 1974 年到 1995 年间收集到的种类的两个突变等位基因的频率，从一个韩国品系中得到了一些启示。令人惊奇的是，L1029F 突变在 1974 年就已经存在了，甚至在利用拟除虫菊酯控制 DBM 之前就已经存在了，可能是在更早之前面对 DDT 的选择而产生的。T944I 突变是在 1995 年在拟除虫菊酯广泛使用之后才产生的。现在，这两个突变都以很高的基因频率在韩国的品系群体中出现。PASA 技术可能对于评价地理品系群体中的抗性突变的基因频率有作用，这样可以预测喷洒拟除虫菊酯除虫的有效性。

与鳞翅目昆虫相比，在其他目的昆虫中点突变呈现了更广泛的多样性，这暗示了我们可能低估了这种存在于鳞翅目昆虫中的真实的点突变多样性。即使是相同的突变所表现出的效应也是不可预测的，在许多重要方面的种属特异性足够影响它们的进化动态。因此对抗性突变的抗性谱研究仍在继续，同时也得到一些关于作用模式的基础信息，用于提高杀虫剂的有效性来对抗昆虫的抗性。通过这种研究，可以提高对 *kdr* 以及其他已经在至少一个案例中有报道的突变的适合度代价的认识，最终达到降低和根本消除钠通道靶向位点抗性的目的。

4. 烟碱型乙酰胆碱受体

乙酰胆碱受体(AChRs)分为烟碱型乙酰胆碱受体和毒蕈碱型乙酰胆碱受体。毒蕈碱型乙酰胆碱受体属于异源三聚体组成的 7 次跨膜结构域的 G 蛋白偶联受体超家族。烟碱型乙酰胆碱受体(nAChRs)属于半胱氨酸环配体门控离子通道超家族(LGIC)，它是尼古丁、新烟碱杀虫剂(吡虫啉，烯啶虫胺)、放线菌发酵产生的多杀菌素的作用靶标。烟碱型乙酰胆碱受体能引起快速的突触传递，这种传递是在胆碱能突触处被神经递质乙酰胆碱(ACh)所激活。

具有功能的受体由 5 个同源亚基组成，每个亚基的 N 末端胞外结构域都有一个 13 个氨基酸的半胱氨酸环和四个跨膜结构域(TM1-TM4)，在第二个跨膜结构域上具有离子通道(Corringer，Le Novere 和 Changeux 2000)。在 TM3 和 TM4 的胞内额外的环起着受体调节和定位的功能。当五聚体组装时，每个亚基都有两面与邻近的亚基相接触。A—F 六个环形成 ACh 的结合位点。A－C 在主面上，D－F 在补充面上(Arias 2000)。如果 C 环包括结合 Ach 所需的两个邻近 Cys，这个亚基就叫 α 亚基(Kao 和 Karlin

1986)。神经递质 ACh 的结合发生在一个 α 亚基的 A 和 C 环和邻近亚基 D 和 F 环的接触面上。在五聚体中,受体要起作用至少需要两个 α 亚基。

果蝇基因组编码 7 个 α 亚基基因(Dα1－Dα7)和 3 个 β 亚基基因(Dβ1－Dβ3; Sattelle et al. 2005)。尽管 Dβ2 缺少定义为 α 亚基的两个邻近 Cys,但是看起来仍然是蜜蜂和蚊子 α 亚基的同源物。Dα4 和 Dα6 的选择性剪接,以及 Dα6 (Grauso et al. 2002)、Dα5、Dβ1 和 Dβ2 mRNA 的编辑(Hoopengardner et al. 2003),极大地增加了异构体潜在的数量。在爪蟾卵母细胞异源表达的方法已经用于研究果蝇和其他昆虫的 α 亚基的功能;发现正确组装需要共表达一个脊椎动物的 β 亚基。

与钠离子通道和 AchE 不同,昆虫中只发现了很少的 nAChRs 靶标抗性位点。在褐飞虱的一个抗吡虫啉抗性系中显示出膜对氚标记的吡虫啉结合的降低。对 5 个 nAChRs 亚基的测序显示在抗性昆虫 α1、α3 亚基上高度保守的 B 环的 151 位发生了 Y 到 S 氨基酸的突变(根据太平洋电鳐(*Torpedo californica*)nAChRs 成熟肽形式进行序列编号)(Liu et al. 2005)。等位基因特异 PCR 检测到在 α1 突变揭示吡虫啉抗性和 Y151S 等位基因频率的正相关性。当在果蝇 S2 细胞中共表达老鼠的 β2 亚基时,突变系产生与氚标记的吡虫啉更低的结合(Liu et al. 2005)。对爪蟾卵母细胞的功能研究表明新烟碱杀虫剂降低了抑制最大电流的敏感性,使其他新烟碱杀虫剂促效剂剂量应答曲线向右迁移(Liu et al. 2006)。因此,当氨基酸的替换发生在 Ach 结合高度保守和重要的区域的两个不同 α 亚基时,表现出更高的抗性。

为确定果蝇中可能的烟碱杀虫剂靶标亚基,Perry 等(2008)通过甲磺酸乙酯细胞突变诱导来选择,对烯啶虫胺有抗性的杂交群在包括 3 个 nAchR 亚基基因的 96A 细胞学区域有缺失。缺失 Dα1 的 TM4 的突变产生抗性。Dβ2 的一个突变也产生这种抗性,可能是其中一个环 A 和其他远离 Ach 结合位点的突变,干扰了亚基的组装和通道激活。通过在已知包含 nAChR 基因的染色体区域筛选半合子缺陷品系,发现一个 Dα6 基因的断裂产生了对刺糖菌素 1180 倍的抗性(Perry, McKenzie 和 Batterham 2007)。因此,点突变和亚基基因的断裂都能产生对靶位的抗性,这种抗性产生的原因可能是在一类相对冗余的亚基中的一些敏感性较低的位点上发生替换。

直到最近才开始对鳞翅目昆虫的 nAChR 亚基进行研究。通过家蚕基因组序列,Shao, Dong 和 Zhang 等(2007)鉴定出 9 个 α-type 和 3 个 β-type 亚基。Bmα1－Bmα4, Bmα6, Bmα7 和 Bmβ1 与果蝇具有同源性的亚基而命名。Bmα8 可能是 Dβ2(α8 来自蜜蜂和蚊子)的同源物。其他亚基的关系还不太清楚。Bmα6 和其在果蝇的同源物一样,进行 RNA 编辑;而 Bmα4, Bmα6 和 Bmα8 有选择性剪接。

Eastham 等(1998)从烟草天蛾中克隆一个 α-subunit(Bmα3 的同源物),烟草天蛾以烟草为食,能忍受尼古丁。α-银环蛇毒素(α-btx)的氨基酸残基和尼古丁结合位点(包括 Y151)与其他 α-btx 敏感的 nAChRs 一样是保守的。幼虫和成虫脑膜被用来检测对放射性标记的 α-btx 的结合,并用来竞争性地置换胆碱配基吡虫啉、尼古丁和 Ach。检测结果与尼古丁敏感的昆虫相似。氚标记的地棘蛙素用来标记 α-btx 不敏感的结合位点,没有显示特异的结合。作者因此认为这个物种的尼古丁耐受可能不是由于尼古丁受体结合力降低引起的。

包括烟草天蛾的天蛾科有尼古丁耐受和尼古丁敏感的种类。Wink 和 Theile (2002) 通过注射尼古丁到血腔中定量检测忍耐力，其中烟草天蛾耐受力最强。基于 16S rRNA 序列的系统发生树被用来定位尼古丁耐受力的进化关系。所有的天蛾科(Sphinginae)和长喙天蛾亚科(Macroglossinae)测试均有耐受力，而所有的 Smerenthinae 姊妹群都是敏感性的。Msα1(与 Bmα1 同源)和 Msα2(与 Bmα3 同源)亚基的部分序列从烟草天蛾中克隆得到。包围环 C 的 31 个氨基酸的区域从 14 个其他的尼古丁敏感的天蛾亚科中检测到。在这些物种中没有发现编码的不同，排除了这两个亚基在这个区域序列的改变引起尼古丁耐受的可能性。

Adamczewski，Ollers 和 Schulte(2005)从烟芽夜蛾中克隆出两个亚基，α7-1(Bmα7 同源物)和 α7-2 (Bmα6 同源物)。通过将烟芽夜蛾的 N-末端的 AChR 和老鼠 5-羟色胺受体 C-末端一起在 HEK293 细胞中融合表达，暴露在 500uMde 尼古丁中用 Fura-2 染色，测量钙的吸收来进行功能研究。烟芽夜蛾的另外四个亚基的序列已提交到 NCBI，但并未见相关的报道。Grauso 等 (2002) 鉴定了两个烟芽夜蛾的 α7 亚基会发生 RNA 编辑。果蝇的 Dα6 亚基有 7 个点可能发生 A－I 的编辑，其中 5 个邻近环 E 上的 Ach 结合位点。它们中的 4 个在烟芽夜蛾 α7-2 同源序列中存在 RNA 编辑，但没有证据显示 α7-1 会发生 RNA 编辑。

疏有证据表明在鳞翅目昆虫中有对尼古丁靶区特异的抗性位点，但发现含有多杀菌素的抗性位点。Moulton，Pepper 和 Dennehy(2000)报道了从泰国的一个甜菜夜蛾品系中发现对多杀菌素有 85 倍的抗性，但未调查其机制。一个烟芽夜蛾的敏感品系在实验室被用来测试对多杀菌素的抗性潜力，局部用药产生 1068 倍的抗性，添食产生 314 倍抗性，血腔注射产生 163 倍抗性(Young，Bailey 和 Roe 2003)。杂交显示常染色体的部分隐性基因对抗性起作用(Wyss et al. 2003)。抗性与代谢和排泄的增加，渗透的减少无关(Young et al. 2000)，但抗性品系在低浓度的多杀菌素下，表现出神经敏感性的减少，暗示了靶区位点可能产生响应(Young et al. 2001)。

田间小菜蛾群体对多杀菌素表现出明显的抗性。马来西亚的 CH1 品系对多杀菌素有 20000 倍以上的抗性，杂交显示抗性可能是由常染色体上单位点基因控制，在高浓度下完全隐性，在低浓度下表现为共显性(Sayyed，Omar 和 Wright 2004)。在夏威夷连续两年使用多杀菌素，在 2000 年发现田间控制失败，在几个种群中发现了不同的抗性水平(Mau 和 Gusukuma-Minuto 2004)。Pearl-Sel 品系在实验室选育一代后抗性达到 13100 倍。抗性通过常染色体遗传，不受增效剂 S，S，S-三丁基 DEF 和 PBO 影响(Zhao et al. 2002)。影响抗性的一个主要基因定位在 AFLP 图谱的 LG 5，在 Bmα6 同源物 *Plutella* 基因 4.2cM 以内(Baxter 2005)。RNA 编辑和先前在果蝇和烟芽夜蛾中发现的一样，发生在 *Plutella* 基因四个同样的位点上。这些结果暗示了靶位点抗性的存在，但还未鉴定得到相应的突变。

我们已经发现对新杀虫剂的靶位点抗性正在 AChR 中出现，特别是鳞翅目昆虫。nAChR 亚基对一种杀虫剂的不同内部敏感性，与 AChE 或钠离子通道不同，可能极大增加获得抗性突变的范围。敏感亚基的单氨基酸替换能够降低它与杀虫剂的结合。另外，敏感亚基进一步的破坏性突变能干扰它的合成和组装成受体，提供其他敏感性较低功能相似的亚基来替代它并借此获得抗性，证明也能发生其他功能相似的低敏感亚基。这些亚基替换的适合度代价决定了它们在自然种群中的重要性。

5. GABA 门控的氯离子通道

昆虫的 GABA 门控的氯通道（又称 $GABA_A$ 受体）是长期使用的环二烯类杀虫剂（如狄氏剂和硫丹）和六氯环己烷类杀虫剂（如林丹）的靶点，另外它还是最近开发的吡唑类杀虫剂氟虫腈的靶标分子。它们介导了快速抑制性的突触传递，使得氯离子流响应神经递质与 GABA（γ-氨基丁酸）的结合。GABA 门控的氯离子通道是五聚体，每个亚基包括四个跨膜结构域，其中的第二个结构域负责孔道和一个含有 GABA 结合位点的巨大 N-末端氨基端胞外结构域的连接。许多亚基类型是根据脊椎动物中的研究得知的；在昆虫中的研究较少，其亚基的研究仍在进行当中。与 nAChR 家族一样，它们属于半胱氨酸-环状 LGIC 超家族。

20 年前，几乎 60%的关于害虫对杀虫剂抗性的文献报道的案例都与狄氏剂及其类似物相关。尽管如此，抗性的分子基础仍不清楚，直到从黑腹果蝇的一个地理品系中克隆得到了 *Rdl* 基因（ffrench-Constant et al. 1993）。*Rdl* 基因编码了 $GABA_A$ 受体的一个亚基，在第二个跨膜结构域中的一个单氨基酸的突变 A302S（实际上在 301 位）导致了对狄氏剂的抗性（ffrench-Constant et al. 1993）。这种机制通过在野生型的 cDNA 中将该位置上的丙氨酸定点突变为丝氨酸，通过在爪蟾卵母细胞中表达系统的研究表明，突变后的蛋白质对狄氏剂和 $GABA_A$ 的激动剂反应的敏感性确实降低了。接下来的分析表明，*Rdl* 基因在果蝇基因组中只有一个拷贝，但是这个基因由于选择性拼接最终可以产生四种不同的蛋白质形式（ffrench-Constant 和 Rocheleau 1993）。

由于这些探索性的研究，A302S（或者 A302G）的替换形式已经在多种昆虫中被找到，其中包括对环二烯具有抗性的咖啡浆果钻心虫（*Hypothenemus hampei*），粉虱（*Bemesia tabaci*），蚜虫（*Nasonovia ribisnigri*）和桃蚜虫（*Myzus persicae*），以及赤拟谷盗（*Tribolium castaneum*）（见综述 Buckingham 和 Sattelle 2005）。明显地，鳞翅目昆虫不在其列。

在一个对环二烯和氟虫腈易感品系中的棉铃虫夜蛾中克隆得到了 *Rdl* 基因的一个同源基因，它们之间具有 HVRDL 同源区域（Wolff 和 Wingate 1998）。除了在跨膜结构域 3 和 4 之间的胞内连接环区域内的相似性较低外，这个基因与果蝇的 *Rdl* 基因具有大概 95%的相似性。令人好奇的是，棉铃虫的基因在 285 位编码的是一个丝氨酸，与狄氏剂抗性的果蝇中的 A302S 突变位点相对应。通过点突变技术，将该位点突变为丙氨酸，将产生的含有 HVRDL-A285 的突变类型和 HVRDL-S285 的野生类型分别注射爪蟾的卵母细胞中用于同源五聚体的电生理研究。同时对果蝇的 Rdl-A302 和 Rdl-S302 也进行了研究。每种蛋白质的丙氨酸形式针对狄氏剂和印防己毒内酯对氯离子流的抑制作用比对应的丝氨酸形式敏感。HVRDL-Ala285 蛋白质对氟虫腈的敏感性比 HVRDL-Ser285 蛋白质大约高 15 倍；然而果蝇蛋白质的两种形式之间并没有什么区别，与棉铃虫的两种形式的蛋白质相比，在相似的 GABA 浓度条件下，果蝇中的两种形式的蛋白质对氟虫腈都表现出中等程度敏感性。因此，A302S的替换对狄氏剂抗性来说是普遍的，对氟虫腈的抗性来说并不是关键的，它们对氟虫腈的抗性取决于该亚基其他部分的序列和结构。

在棉铃虫氟虫腈易感品系的285位出现的丝氨酸替换比果蝇模式中显示了更高的复杂性。的确，来自相同棉铃虫夜蛾易感品系(Halling 和 Yuhas 2001)的三种蛋白质的每一种形式(GenBank accessions：a1：AF006190；a2：AF006189；a3：AF006192)在这个位点上都具有不同的氨基酸，分别是：谷氨酰胺、丝氨酸和丙氨酸。在棉铃虫的硫丹抗性和易感品系中都有发现在该位点上是谷氨酰胺的替换形式(ffrench-Constant et al. 1996)。另有证据显示，在家蚕基因组中，这三种形式的蛋白质分别是由三个不同基因编码的，来自家蚕*Dazao*品系(没有暴露在环二烯和氟虫腈中)的不完全拼接的基因组序列信息显示这三种形式的蛋白质在285位分别为谷氨酰胺(GenBank AADK01004540)，丝氨酸(AADK01022486)和丙氨酸(AADK01003208)。然而，在其他物种中，通过全基因组检索发现在基因组中只有一个拷贝的昆虫有：黑腹果蝇(*D. melanogaster*)和11个已经测序的果蝇种类，包括库蚊(*Culex quinquefasciatus*)，埃及伊蚊(*Aedes aegypti*)，膜翅目的蜜蜂(*Apis mellifera*)和蛹集金小蜂(*Nasonia vitripennis*)，以及鞘翅目的赤拟谷盗(*T. castaneum*)。所有这些物种在285位上都是丙氨酸。而蚜虫(*Acyrthosiphon pisum*)含有两个拷贝，一个(ABLF01022071)在该位点上是丝氨酸，另一个(ABLF01036703)在该位点上是丙氨酸。Anthony 等人 (1998)提出的证据显示，在桃蚜虫(*M. persicae*)中的两个基因中，一个在该位点上是丝氨酸，另一个存在多态性或者是丙氨酸或甘氨酸，而后一种形式与硫丹的抗性相关。

小菜蛾中显示出相似的关系。用氟虫腈筛选出来的小菜蛾SZ－F品系对氟虫腈的抗性增加了300倍，而对环二烯和硫丹的抗性仅分别增加了3.5倍和6.5倍(Li et al. 2006)。利用PCR从基因组上扩增出了一段与果蝇*Rdl*基因同源的含有54个氨基酸的区域，在该区域中包含了A302。在两个易感品系中该位置都是丙氨酸，而在抗性品系中该位点上表现出丙氨酸或者丝氨酸的多态性。通过对来自同一个SZ－F品系幼虫的PCR产物测序显示该位置编码丝氨酸的等位基因的频率是30%，在经过可以杀死25%的SZ－F的氟虫腈的浓度处理后，存活的群体中的在该位置编码丝氨酸的等位基因的频率增加到57%。被调查的54个氨基酸的区域除了一些多态性的位点其他三个家蚕的基因中是相似的，因此该基因的家蚕直系同源基因还是未知的。在这项研究中，未探讨这个基因中其他位置的突变。

当氟虫腈被投放市场的时候，在广泛的抗性存在的情况下它的持续使用性受到人们的广泛关注，这种抗性与环二烯抗性的作用靶点是一样的，而且其抗性早在环二烯抗性产生的几十年前就存在了。然而，早期的研究表明这两种杀虫剂的作用位点是不同的，而且对狄氏剂耐受的品系同时对氟虫腈的抗性很低(见 Bloomquist 2001年的综述)。其他存在于环二烯抗性品系中的突变位点对氟虫腈的抗性来说也是很重要的。在一个实验室选择出来的花椒果蝇(*Drososophila simulans*)的品系中，当另一个突变位点T351M与A302G共同存在时确实强烈地增强了该品系对氟虫腈的抗性，而且在爪蟾卵母细胞表达系统中，尽管未达到活性实验结果中反映出的不敏感程度，具有双突变位点的通道蛋白质对氟虫腈的敏感性的确比单突变蛋白质要低(Le Goff et al. 2005)。

在鳞翅目昆虫中有 3 个 *Rdl* 旁系同源基因,与其他只有一个 *Rdl* 基因的昆虫相比,该基因家族在功能上的冗余性或许对氟虫腈的抗性进化模型方面具有重要的意义。如果在 302 位上含有一个丙氨酸的旁系同源个体是最敏感的,那么任何可以降低它表达的突变形式都会降低该昆虫的敏感性,而通道蛋白的其他旁系同源个体会对其进行补偿。例如花椒果蝇中的 T351M 突变出现在对狄氏剂和氟虫腈更强耐受性的已经具有一个丝氨酸突变的鳞翅目昆虫亚基中,与那些只有一个 *Rdl* 基因的昆虫相比,已经存在的功能冗余性将会加速氟虫腈抗性的进化。另一方面,突变位点是谷氨酰胺的通道蛋白还没有在爪蟾卵母细胞系统中得到验证。如果这种含有谷氨酰胺的形式比含有丙氨酸的蛋白质形式更加敏感,昆虫把两种突变隐藏起来以避开杀虫剂,功能的冗余性从而延迟抗性的发展。在任何情况下,鳞翅目昆虫对杀虫剂的选择性进化上的响应可能比果蝇对狄氏剂抗性中推论出来的机制更加复杂。

6. 对化学杀虫剂的代谢抗性

在鳞翅目昆虫中,分子遗传学在代谢抗性方面取得的进步远远落后于靶向位点抗性的研究。后者的优势是在昆虫中高度保守的靶向位点数目很少,在不同种类之间是高度保守的,其研究结果可以参照相同的基本分子结构。另一方面,解毒酶具有数量巨大而且涉及多样性的基因家族。它们的巨大数量和进化上快速的演变,除非是在非常相近的种类之间进行比较,几乎不可能将它们与预测值进行深入比较。到目前为止,研究还未涉及鳞翅目昆虫。我们简要综述了三个在杀虫剂抗性方面比较重要的解毒酶家族,包括谷胱甘肽转移酶、羧酸酯酶和细胞色素 P450 酶。

6.1 谷胱甘肽转移酶

谷胱甘肽转移酶(GSTs)在解毒反应的第二阶段催化谷胱甘肽与亲电化合物形成共轭连接,保护细胞免受亲电化合物的攻击以及加速它们的排泄。谷胱甘肽是一个三肽,γ-谷氨酰基-L-半胱氨酸-甘氨酸,简写为 GSH 是为了强调巯基(—SH)在反应中的重要性。它在细胞中无处不在,维持细胞的氧化还原稳态并通过共轭结合亲电物质进而将它们清除。大多数 GSTs 是二聚体的细胞质蛋白。通过序列相似性对它们进行分类,在昆虫中已经划分出了六类 GSTs,包括:Delta,Epsilon,Omega,Sigma,Theta 和 Zeta。在昆虫中也发现了一些微粒 GSTs(MAPEG 酶);它们在溶解状态下具有不同的结构。GSTs 通过 O-烷基化或者 O-去乙酰化对有机磷杀虫剂进行解毒;另外 DDT 的脱氯化酶也是一种 GST。GSTs 也可以清除拟除虫菊酯,但是研究证明 GSTs 在去除拟除虫菊酯诱导产生的脂质过氧化产物中发挥更为重要的作用(Vontas,Small 和 Hemingway 2001)。

尽管一系列生物化学研究的证据表明 GSTs 在抗性方面的重要意义(Ranson 和 Hemingway 2005),但是在鳞翅目昆虫中只有一个 GST 基因被克隆。Chiang 和 Sun (1993)从一个对甲基对硫磷和几丁质合成抑制剂伏虫隆耐受的小菜蛾的台湾品系中利用亲和层析技术在谷胱甘肽-琼脂糖凝胶柱上分离了三个 GST 酶。因为共轭反应的产物是 S-(4-硝基苯基)-谷胱甘肽，与其他两个 GST 酶相比，GST-3 对模式底物 CDNB 的催化活性并不高，而是对 DCNB 和有机磷类杀虫剂对硫磷、对氧磷和甲基对硫磷的活性比较高(Chiang 和 Sun 1993)。第四种异构酶，GST-4 是随后在同一个品系中分离得到的，它具有对 DCNB 和有机磷类杀虫剂更高的特异性(Ku et al. 1994)。通过 GST-3 制备的多克隆抗体可以与 GST-4 发生交叉反应，而且在台湾抗性品系中具有很高的蛋白含量，这与来自法国的通过甲基对硫磷选择的抗性品系相类似。

随后克隆了编码 GST-3 的基因，并在大肠杆菌中表达，表达产生的 GST-3 与纯化得到的酶具有相似的共轭特性(Huang et al. 1998)。抗性品系中的 mRNA 的表达水平比易感品系中要高很多，然而通过 Southern 杂交证明，这种 mRNA 的增加不是由于基因拷贝数的增加造成的。Sonoda 和 Tsumuki (2005)研究了日本的一个对氯菊酯和几丁质合成抑制剂定虫隆耐受的品系中的同一个基因，该基因同样高水平表达 mRNA，而没有增加基因拷贝数。在抗性和易感品系中，两者的蛋白质序列没有任何差异，而且在基因组中该基因的侧翼区域也几乎是相同的。这个基因在起始密码子的上游四个核苷酸处有一内含子(Sonoda，Ashfaq 和 Tsumuki 2006)。Enayati，Ranson 和 Hemingway (2005)将该基因归为 Epsilon 家族。这四种异构酶存在于所有研究的品系中，但是在抗性品系中 GST-3 的含量较高。在没有连锁图谱的情况下，是否存在顺式或者反式介导的上调机制还不得而知。

6.2 羧酸酯酶

羧酸酯酶是 α/β-水解酶折叠超家族的水解酶，AChE 也是这个家族的成员。它们通过两步反应切割羧酸酯键，与 AChE 类似，活性位点丝氨酸中的氧原子对底物的羰基碳原子发起亲和攻击，形成一种酰基-酶的链接取代了醇羟基。随后水分子发起类似的亲和攻击，取代丝氨酸，释放产物酸以及重新产生自由的酶。这些酶和它们的没有催化活性的类似物形成了一个巨大的超家族，还没有标准对它们进行命名；然而，在一个昆虫酯酶的详尽综述中，Oakeshott 等人 (2005a)基于序列相似性已经定义了 14 个分支。

在双翅目和半翅目中，许多参与到有机磷杀虫剂抗性的酯酶已经被克隆了，而且研究多次证明基因加倍导致高水平的基因表达，通过去除而不是水解杀虫剂来保护昆虫(见 Oakeshott 在 2005 年的综述)。在活性位点的丝氨酸磷酸化的酶，被困在反应链的第一步，使得杀虫剂的分子远离靶标 AChE，这一过程要消耗一个蛋白质分子。在分支 E 中的酯酶是在半翅目中克隆得到的，其中包括褐飞虱(*N. lugens*)和桃蚜虫(*M. persicae*)，通过基因扩增产生的酯酶在量上可以达到蚜虫可溶蛋白质总量的 1%。分支 C 中的酯酶同样在库蚊和按蚊中也发生了基因的加倍。一个在分支 B 中的酶，αE7，在米家蝇(*M. domestica*)和湖羊蝇(*L. cuprina*)中没有发生基因扩增，而是发生了几个点突变从而增加了其有机磷的水解活性。有机磷的水解进行得很慢，但是这种速率足够用来产生足够的抗性水平。许多生物化学研究为鳞翅目昆虫对有机磷和拟除虫菊酯的抗性方面研究提供了证据，其中包括许多过表达的酶。然而，在鳞翅目昆虫中仍然缺乏对这类基因的克隆研究。

6.3 细胞色素 P450

细胞色素 P450 是一类大量的、广泛存在的亚铁血红素-硫醇盐蛋白，它们能够共同催化大量的反应，其中许多对外源物质诸如杀虫剂的解毒起重要作用。这个蛋白具有一个亚铁血红素基团，在其中心有一个铁原子，在氧化状态下蛋白的变化对于其所参与的反应循环非常重要，在反应循环中氧分子的一个氧原子被转移到底物当中，另一个氧原子被还原成水。这个循环的精确度可以发生很大的改变，根据酶和底物的不同，可以发生很多不同的反应。膜相关的 P450 需要从 NADPH 或 NADH 转移来的电子，这种转移是通过膜相关的氧化还原配体实现的，包括 NADPH-细胞色素 P450 还原酶。这些特点不论是从感兴趣的有机体中提取膜碎片还是进行异源表达，都对体外研究 P450 的酶活性提出了特别的挑战。P450 的生物化学和分子生物学研究是一个庞大的课题，与昆虫细胞色素 P450 相关的资料已经得到综述(Scott 和 Wen 2001；Feyereisen 2005；Li，Schuler 和 Berenbaum 2007)。

昆虫当中的 P450 起到很多作用，包括激素和信息素的合成，脂肪酸代谢和外源物质(如寄生植物的毒素和杀虫剂等)的解毒作用。许多解毒类型是由外源物质诱导的；即在正常调节的情况下转录速率维持在较低的水平上，但当有外源或内源性的物质出现的时候转录水平会增加。尽管这可能是对环境中有害物质进行解毒的功能相适应，但是诱导物并不一定是必须要除去的毒素。以芳香烃受体(AHR)在多环芳香烃诱导 CYP1A1 上调过程中的作用为例，哺乳动物中的诱导过程的研究最为详尽。存在于细胞质中的蛋白复合体包含 AHR 和分子伴侣。当结合上合适的诱导物时，AHR 向细胞核移动，与相关的芳香烃受体核转运蛋白(ARNT)形成异源二聚体，在 CYP1A1 和其他基因的启动子附近特异的增强子序列处和 DNA 结合，通过和转录因子及其他共激活因子的相互作用来促进转录。在昆虫当中存在这些蛋白的类似物，但是并不清楚它们在调节昆虫 P450 中所起的作用。

研究 P450 功能的生物学方法包括利用模式底物来监测特定类型反应的过程，如 O-去甲基作用，P450 酶抑制剂如 PBO，诱导剂如苯巴比妥。许多研究已经用这些方法比较了一些抗性和敏感性的品系，但是这些研究都缺乏对单一种酶的特异性，再加上昆虫基因组中 P450 基因的多样性(例如，埃及伊蚊中有 160 个，Strode et al. 2008；家蚕中有 87 个，Kozaki et al. 2008)，这使得推断一个特异性的 P450 基因非常困难。对棉铃虫(Forrester et al. 1993；Kranthi et al. 2001；Yang et al. 2004)、草地夜蛾(Yu 1992)、海灰翅夜蛾(Huang 和 Han 2007)、小菜蛾 (Yu 和 Nguyen 1992)、烟芽夜蛾(Rose et al. 1995)和果蠹蛾(Bouvier et al. 2002)等鳞翅目昆虫的研究表明 P450 是代谢上杀虫剂抗药性的一个成分。

对 P450 的这些证据的后续研究得出了一个普遍的范例。通过在敏感的昆虫当中表达一个或多个 P450 或许能够解除杀虫剂的毒性，但却不够快速而抵抗致死效应。诱导机制同样可能反应太慢或根本没有反应。因此，通过杀虫剂处理，能筛选出来任意一种能够提高 P450mRNA 水平和蛋白水平的突变种。这些突变可能包括抑制 P450 转录的调节蛋白的功能缺失，破坏这些抑制蛋白与 P450 基因启动子的结合位点，在基因的附近创造新的增强子位点，或者增加基因的拷贝数。其中的许多变化可以破坏诱导系统所需要的反馈调控；因此，诱导能力的缺失常常伴随着很高的组成型表达。其中的许多突变发生在基因上而不是 P450 蛋白上，这就增加了鉴定它们的难度。有证据表明 P450 蛋白本身的氨基酸替换能够增强抵抗杀虫剂的活性（果蝇属（*Drosophila*）的 Cyp6a2，Amichot et al. 2004），但是迄今为止的研究表明，在大多数情况下，活性增加之前就存在最低活性酶的表达更为重要。

昆虫当中第一个被克隆的 P450 基因是家蝇的 CYP6A1，在这个系统中关键的发现作为范例已被综述（见综述 Feyereisen 2005）。在敏感型品系中 CYP6A1 表达水平很低并且可以被苯巴比妥所诱导，在敌匹硫磷抗性的避虫酮品系和其他一些以 P450 为基础的抗性品系中 CYP6A1 表达水平很高并且是组成型表达。通过免疫方法检测到了蛋白水平的增加，由此可以推断出 mRNA 水平的增加。当在大肠杆菌中表达 CYP6A1 和它的氧化还原配体时，CYP6A1 能够使敌匹硫磷以很高的周转速率代谢。CYP6A1 基因位于第 5 号染色体上；然而，增加其表达水平的因子位于第 2 号染色体上。不同的因子表明在抗性品系和敏感性品系中是反式作用方式，并不是 CYP6A1 基因本身的一部分。另一种情况中，黑腹果蝇 CYP6G1 的顺式作用过表达是由于基因本身的突变；在编码序列上游由转座子介导的插入引起了 P450 的高量表达（Daborn et al. 2002），这种高量表达能够促进 DDT 和吡虫啉的代谢（Joussen et al. 2008）。

P450 催化的反应也能够激活杀虫剂前体促使其变成杀虫剂。例如，昆虫的 P450 的氧化脱硫作用能够将非活性的磷硫酰形式的 OP 杀虫剂中的 P＝S 转变成抑制 AChE 形式的 P＝O，使其转变为杀虫剂。这种现象或许能为负面的交叉抗性提供根据，即一种品系对某一种杀虫剂有抗性是因为增加了对其的解毒作用，同时对另一种杀虫剂变得更加敏感，这可能是由于相同的酶系统增加了对另一种杀虫剂的激活作用。在西非的棉铃虫种群中，发现了除虫菊素类和三唑磷之间这种负面的交叉抗性，对纯化的重组的果蝇的 AChE 蛋白的抑制性进行分析，发现用三唑磷处理的除虫菊素类抗性的品系中具有更高水平的三唑磷（Martin et al. 2003）。烟芽夜蛾的一个品系对除虫菊素类有抗性部分上是由于 P450 的活性在遗传上对吡咯杀虫剂虫螨腈的高敏感性相连锁，它必须在解耦联的氧化磷酸化之前就进行生物激活（Pimprale et al. 1997）。

鳞翅目昆虫中第一个被克隆的和杀虫剂相关的 P450 属于 CYP6B 亚家族，在研究除虫菊素类抗性中从澳大利亚棉铃虫中克隆得到 CYP6B2（Wang 和 Hobbs 1995），CYP6B6 和 CYP6B7（Ranasinghe 和 Hobbs 1998）。从田间收集的除虫菊素类抗性的幼虫中的 CYP6B7 的 mRNA 水平比敏感型品系有所增加（Ranasinghe，Campbell 和 Hobbs 1998），并且在培养的脂肪体器官中，当暴露于苯巴比妥和除虫菊素类时其 mRNA 水平可被诱导。这些作者假设在澳大利亚田间种群中 CYP6B7 的过表达是引起抗性的主要原因。支持这个观点的事实是，研究表明在相关的物种 *H. zea* 中，一个相关的基因 CYP6B8 可被苯巴比妥所诱导（Li，Berenbaum 和 Schuler 2002），并且能够引起除虫菊素类氯氰菊酯的代谢（Li et al. 2004）。

Grubor 和 Heckel（2007）在棉铃虫 AN02 品系中通过连锁图谱检验了这个假设，结果表明由于一个半显性的抗性基因 *RFen1*，幼虫和成虫都表现出 50 倍的 PBO 可抑制的氰戊菊酯抗性。对基因组 BAC 克隆的测序表明这三个基因排列在一个群上，排列顺序是 CYP6B7-CYP6B6-CYP6B2。这个群位于第 14 连锁群，然而 *RFenl* 基因位于第 13 连锁群（Heckel et al. 1998）；因此，由 *RFenl* 控制的抗性并不是由于这三个 CYP6B 基因编码序列或顺式调控序列的改变引起的。为了检验 *RFenl* 是否能够编码一个反式作用因子来调控这三个 CYP6B 基因的组成型表达，它们的 mRNA 水平用实时定量 RT-PCR 进行测量，测量的个体没有用除虫菊素类处理，但是 mRNA 水平和 *RFenl* 基因型之间并没有发现对应关系。因此，在 AN02 品系中，CYP6B 基因表达水平并没有因为抗性而改变。

为了在 AN02 品系中筛选出哪些基因的表达和除虫菊素类抗性有共变，Wee 等(2008)利用 cDNA-AFLP 技术作为一种差异显示方法来寻找携带有 *Rfen*1 抗性等位基因的个体中上调或下调的 mRNA。抗性个体中的一个基因表现出明显的上调，这是一个新的 P450，CYP337B1，它也位于第 13 连锁群上，与 *RFenl* 之间的距离不超过 1cM。CYP337B1 对除虫菊素类的代谢能力仍需检验。

Yang 等（2006）从中国山东省的棉铃虫 YGF 品系中克隆了两个 P450 基因，CYP9A12 和 CYP9A14，这个品系通过 14 代的筛选具有了对氰戊菊酯 1690 倍的抗性。实时定量 RT-PCR 表明，与非选择的亲本相比，YGF 品系的 CYP9A12 在中肠有 19 倍的增加，在脂肪体有 433 倍的增加，CYP9A14 在中肠有 4 倍的增加，在脂肪体有 59 倍的增加。然而，在这些研究中相关的抗性基因并未被定位。

另外的鳞翅目昆虫的 P450 在抗性品系中表现出更高的表达水平，包括从烟芽夜蛾克隆的 CYP9 家族的第一个成员（Rose et al. 1997）。和相邻的敏感性品系相比，北卡罗来纳的硫双威抗性品系的 CYP9A1 在 RNA 印迹中表现出 29 倍的表达。Bautista，Tanaka 和 Miyata (2007)从日本品系的小菜蛾中克隆了四个 P450 基因，这个品系对扑灭司林有 200 倍的抗性，能够被 PBO 部分抑制。在抗性品系中，CYP6BG1 和 CYP6BG2 的 mRNA 水平分别高出了 5 倍和 4 倍。这些和另外两个，CYP6AE13 和 CYPBF1v4，在敏感性品系中能被低水平的扑灭司林诱导。此外的一个从中国小菜蛾品系中克隆的 P450 是 CYP9G2，这个品系对扑灭司林有 30 倍的抗性（Shen et al. 2004）。

我们需要从已经存在的鳞翅目昆虫具有代谢多样性的酶中，建立一个选择性驱动上调 P450 的范例。其他的 P450 的克隆增加了检验这种观点的机会。鳞翅目昆虫的一个重要特点是它具有很长的昆虫-植物相互作用的共进化历史。P450 具有对很多化学物质的解毒功能，而这些化学物质正是植物为了避免被食而合成的，并且其中许多 P450 的诱导模式和代谢能力已经被研究。这里仅仅举一个例子，CYP6B 酶在寡食性的北美黑凤蝶（*Papilio polyxenes*）中能被喃香豆素类所诱导，并且引起喃香豆素类的代谢，而喃香豆素类正是由其宿主植物所产生；而其他的 CYP6B 酶，在杂食性的灰绿凤蝶（*Papilio glaucus*）中具有宽广范围的底物，多尾凤蝶是一个中间类型，取食含有呋喃香豆素的宿主植物，这种蝶类拥有从寡食性到杂食性诱导活性的 CYP6B 酶（Mao，Schuler 和 Berenbaum 2007）。

以上的例子与其他的昆虫与宿毛植物之间的古老的化学相互作用可以理解为昆虫面临现代杀虫剂的挑战之前的预适应（见综述 Despres，David 和 Gallet 2007）。是否如 Krieger，Feeny 和 Wilkinson (1971)假设的那样杂食性昆虫宿主-植物间获得了一系列具有宽广代谢范围的P450，是否意味着与寡食性昆虫相比，它们会更快地获得杀虫剂抗性？为了回答这个问题，需要将经典的定向抗性研究和化学生态学研究进行更好的整合。

7. 结论与展望

对大多数化学杀虫剂来说，在作用靶点和解毒酶方面具有足够的相似性，来自其他目的昆虫的研究结果可以广泛地应用到鳞翅目昆虫中。然而，这种情况下，往往忽视了许多种属差异性和鳞翅目昆虫特异性方面的信息。在杀虫剂靶点中导致抗性的突变似乎在不同目的昆虫中反复出现，然而仍然表现出了不同的效应，说明这些突变出现的蛋白质的突变位点外的其余部分一定程度上可能决定了抗性的差异。在感兴趣的物种中，基于序列的定点突变的研究变得更有可行性，而且应该经常运用到研究中。对于新出现的抗性品系的研究，应该着眼于整个靶标分子，而不是仅仅将目光集中在其他群体中已经发现的突变位点上。

系统研究鳞翅目昆虫抗性机制的资源需求陡增，很大程度上是因为家蚕基因组信息的大量涌现(International Silkworm Genome Consortium 2008)。家蚕和其他在接下来的几年中会出现的鳞翅目昆虫(包括棉铃虫)的基因组序列将帮助我们理解靶标和解毒酶的多样性，它们大部分出现在基因家族中，而且这些基因家族都比果蝇的要大。另外，在填补我们已知的靶标及其旁系同源基因方面，比较基因组学的研究将会发现其他一些基因家族，它们具有足够相似的广谱控制鳞翅目昆虫的效应，而且又可以与其他脊椎动物和其他目的昆虫相区别，从而尽可能地减少脱靶效应。

对昆虫与植物共进化关系的全面了解对理解植物怎样减少对植食性动物的吸引力和取食，以及昆虫怎样共进化以克服植物对它们的防御有作用(见第十一章，关于宿主范围的遗传学)。将来对抗性发展应该着眼于"过去选择的阴影"，因为它们塑造了我们今天所看到的关系。关于这点的一个很好的例子是，植物中蛋白质抑制剂的高度多样性(Ryan 1990)，与此相对应的是在昆虫中产生的大量蛋白酶(Jongsma et al. 1995)。

在一些发达国家，因为它们稳定的或者呈逐步下降趋势的人口，关于化学杀虫剂抗性的研究力量已经有一些转移。对食物需求正逐步增加的发展中国家应该开始担起领导的责任，并提供与之相对应的正迅速完备的科学设施。发展中国家对过去产生的抗性问题首当其冲，必须对这些问题进行充分了解，从而在相同的靶点上开发新的亚位点。这些过去的相同靶位点包括如前面所述的几种杀虫剂(氟虫腈、茚虫威和新烟碱类)的位点。在采纳新开发的化学杀虫剂的时候，不应该忽视由于过去对杀虫剂的滥用造成的抗性问题、环境问题以及经济上的不利影响。虽然不能够对其绝对地定量，来自DDT的抗性选择的阴影无疑加速了拟除虫菊酯的选择响应。更多的关注应该集中在发现那些改进的化合物，这些化合物能够对先前进化的抗性机制具有抗耐药性。

分子遗传学应该在延长已有杀虫剂的有效性以及为安全有效的杀虫剂管理策略提供信息方面做出贡献，对杀虫剂管理的目的是为农作物提供持续的、安全的保护。如果有一个可管理的抗性突变谱，可以检测到这些突变在害虫群体中频率很低的基因，而不是在对其控制失败之后才能对高频率的基因进行检测，将对于昆虫杀虫剂的抗性管理非常有用。通过抗性植物的管理对抵制害虫抗性也是非常有用的，例如通过转基因方法将 *Bt* 导入棉花来抵抗昆虫，这种策略在美国和澳大利亚仍在使用。大部分发展中国家没有种植这样的植物，因而在这种情况下，无法为抗性的发展提供可预料的控制，除非在等位基因频率方面的增加能够得到及时的检测和纠正。

就像在苏云金芽孢杆菌毒素的案例中所看到的一样，出现了许多针对抗性研究的新机会，一系列的毒素或生物活性蛋白被转入植物，仅举几例，其中包括新的细菌毒素，来自蜘蛛和蝎毒的蛋白质、蛋白酶和其他的消化酶类抑制剂，以及与神经肽偶联的凝集素。通过 RNA 干涉技术控制害虫的基因表达目前还处于探索的阶段，但是在某些目的昆虫却可能容易实施，例如鞘翅目(Baum et al. 2007)。鳞翅目昆虫中的 RNA 干涉实验的成功率不高，其中的原因还不清楚，但是 RNA 干涉在这方面还是有很大的潜力。在棉铃虫中，利用其食用的植物产生的双链 RNA 对其一个 P450 基因起到了很好的抑制作用(Mao et al. 2007)。抗性问题经常在新技术的应用及其早期的成功中迷失了方向。但是，进化的进程却从未停止。关于这点一个清醒的提示是，一种杀虫剂颗粒体病毒在田间施用后，昆虫对它产生了抗性，该种病毒在过去被用来控制苹果蠹蛾(Asser-Kaiser et al. 2007)，这种广泛使用的控制策略却成为了产生抗性的条件。在"正确的"情况下，抗性不是不可能发生的，而是不可避免的。

参考文献

[1]Adamczewski, M., N. Oellers, and T. Schulte. 2005. Nucleic acids encoding insect acetylcholine receptor subunits. US Patent No. 6933131.

[2]Ahmad, M., R. T. Gladwell, and A. R. McCaffery. 1989. Decreased nerve sensitivity is a mechanism of resistance in a pyrethroid resistant strain of *Heliothis armigera* from Thailand. *Pestic. Biochem. Physiol.* 35:165—71.

[3]Amichot, M., S. Tares, A. Brun-Barale, et al. 2004. Point mutations associated with insecticide resistance in the *Drosophila* cytochrome P450 Cyp6a2 enable DDT metabolism. *Eur. J. Biochem.* 271:1250—57.

[4]Andrews, M. C., A. Callaghan, L. M. Field, et al. 2004. Identification of mutations conferring insecticide-insensitive AChE in the cotton-melon aphid, *Aphis gossypii* Glover. *Insect Mol. Biol.* 13:555—61.

[5]Ankersmit, G. W. 1953. DDT-resistance in *Plutella maculipennis* (Curt.) (Lep.) in Java. *Bull. Entomol. Res.* 44:421—23.

[6]Anthony, N., T. Unruh, D. Ganser, et al. 1998. Duplication of the *Rdl* GABA receptor subunit gene in an insecticide-resistant aphid, *Myzus persicae*. *Mol. Gen. Genet.* 260:165—75.

[7]Arias, H. R. 2000. Localization of agonist and competitive agonist binding sites on nicotinic acetylcholine receptors. *Neurochem. Int.* 36:595—645.

[8]Asser-Kaiser, S., E. Fritsch, K. Undorf-Spahn, et al. 2007. Rapid emergence of baculovirus resistance in codling moth due to dominant, sex-linked inheritance. *Science* 317:1916—18.

[9]Baek, J. H., J. I. Kim, D. W. Lee, et al. 2005. Identification and characterization of *Ace*1-type acetylcholinesterase likely associated with organophosphate resistance in *Plutella xylostella*. *Pestic. Biochem. Physiol.* 81:164—75.

[10]Baum, J. A. , T. Bogaert, W. Clinton, et al. 2007. Control of coleopteran insect pests through RNA interference. *Nat. Biotechnol.* 25:1322—26.

[11]Bautista, M. A. M. , T. Tanaka, and T. Miyata. 2007. Identification of permethrin-inducible cytochrome P450s from the diamondback moth, *Plutella xylostella* (L.) and the possibility of involvement in permethrin resistance. *Pestic. Biochem. Physiol.* 87:85—93.

[12]Baxter, S. W. 2005. Molecular and genetic analysis of Bt and spinosad resistance in diamondback moth, *Plutella xylostella*. PhD diss. Univ. Melbourne.

[13]Bloomquist, J. 2001. GABA and glutamate receptors as biochemical sites for insecticide action. In *Biochemical sites of insecticide action and resistance*, ed. I. Ishaaya, 17—41. Berlin: Springer-Verlag.

[14]Bouvier, J. C. , T. Boivin, D. Beslay, et al. 2002. Age-dependent response to insecticides and enzymatic variation in susceptible and resistant codling moth larvae. *Arch. Insect Biochem. Physiol.* 51:55—66.

[15]Bravo, A. , S. S. Gill, and M. Soberon. 2007. Mode of action of *Bacillus thuringiensis* Cry and Cyt toxins and their potential for insect control. *Toxicon* 49:423—35.

[16]Brown, T. M. , and P. K. Bryson. 1992. Selective inhibitors of methyl parathion-resistant acetylcholinesterase from *Heliothis virescens*. *Pestic. Biochem. Physiol.* 44:155—64.

[17]Brown, T. M. , P. K. Bryson, F. Arnette, et al. 1996. Surveillance of resistant acetylcholinesterase in *Heliothis virescens*. In *Molecular genetics and evolution of pesticide resistance*, ed. T. M. Brown, 149—57. Washington, D. C. : American Chemical Society.

[18]Brun-Barale, A. , J. C. Bouvier, D. Pauron, et al. 2005. Involvement of a sodium channel mutation in pyrethroid resistance in *Cydia pomonella* L, and development of a diagnostic test. *Pest Manag. Sci.* 61: 549—54.

[19]Buckingham, S. D. , and D. B. Sattelle 2005. GABA receptors of insects. In *Comprehensive molecular insect science*, Vol. 5, ed. L. Gilbert, K. Iatrou, and S. Gill, 107—42. Oxford: Elsevier.

[20]Byrne, F. J. , and N. C. Toscano. 2001. An insensitive acetylcholinesterase confers resistance to methomyl in the beet armyworm *Spodoptera exigua* (Lepidoptera: Noctuidae). *J. Econ. Entomol.* 94:524—28.

[21]Campanhola, C. , B. F. McCutchen, E. H. Baehrecke, et al. 1991. Biological constraints associated with resistance to pyrethroids in the tobacco budworm (Lepidoptera, Noctuidae). *J. Econ. Entomol.* 84: 1404—11.

[22]Cassanelli, S. , M. Reyes, M. Rault, et al. 2006. Acetylcholinesterase mutation in an insecticide-resistant population of the codling moth *Cydia pomonella* (L.). *Insect Biochem. Mol. Biol.* 36:642—53.

[23]Chiang, F. M. , and C. N. Sun. 1993. Glutathione transferase isozymes of diamondback moth larvae and their role in the degradation of some organophosphorus insecticides. *Pestic. Biochem. Physiol.* 45: 7—14.

[24]Church, C. J. , and C. O. Knowles. 1993. Relationship between pyrethroid enhanced batrachotoxinin A 20-alpha-benzoate binding and pyrethroid toxicity to susceptible and resistant tobacco budworm moths *Heliothis virescens*. *Comp. Biochem. Physiol. C Pharmacol. Toxicol. Endocrinol.* 104:279—87.

[25]Claudianos, C. , R. J. Russell, and J. G. Oakeshott. 1999. The same amino acid substitution in orthologous esterases confers organophosphate resistance on the house fly and a blowfly. *Insect Biochem. Mol. Biol.* 29:675—86.

[26]Corringer, P. J., N. Le Novere, and J. P. Changeux. 2000. Nicotinic receptors at the amino acid level. *Annu. Rev. Pharmacol. Toxicol.* 40:431—58.

[27]Daborn, P. J., J. L. Yen, M. R. Bogwitz, et al. 2002. A single P450 allele associated with insecticide resistance in *Drosophila*. *Science* 297:2253—56.

[28]Davies, T. G. E., L. M. Field, P. N. R. Usherwood, et al. 2007. DDT, pyrethrins, pyrethroids and insect sodium channels. *IUBMB Life* 59:151—62.

[29]Despres, L., J. P. David, and C. Gallet. 2007. The evolutionary ecology of insect resistance to plant chemicals. *Trends Ecol. Evol.* 22:298—307.

[30]Dong, K. 2007. Insect sodium channels and insecticide resistance. *Invertebr. Neurosci.* 7:17—30.

[31]Eastham, H. M., R. J. Lind, J. L. Eastlake, et al. 1998. Characterization of a nicotinic acetylcholine receptor from the insect *Manduca sexta*. *Eur. J. Neurosci.* 10:879—89.

[32]Elliott, M. 1996. Synthetic insecticides related to natural pyrethrins. In *Crop protection agents from nature: Natural products and analogues*, ed. L. Copping, 254—300. Cambridge: Royal Society of Chemistry.

[33]Enayati, A. A., H. Ranson, and J. Hemingway. 2005. Insect glutathione transferases and insecticide resistance. *Insect Mol. Biol.* 14:3—8.

[34]Ferré, J., and J. Van Rie. 2002. Biochemistry and genetics of insect resistance to *Bacillus thuringiensis*. *Annu. Rev. Entomol.* 47:501—33.

[35]Feyereisen, R. 2005. Insect cytochrome P450. In *Comprehensive molecular insect science*, Vol. 4, ed. L. Gilbert, K. Iatrou, and S. Gill, 1—77. Oxford: Elsevier.

[36]ffrench-Constant, R. H., N. Anthony, D. Andreev, et al. 1996. Single versus multiple origins of insecticide resistance: Inferences from the cyclodiene resistance gene *Rdl*. In *Molecular genetics and evolution of pesticide resistance*. *ACS Symposium Series* 645, ed. T. M. Brown, 106—11. Washington, D. C.: American Chemical Society.

[37]ffrench-Constant, R. H., D. P. Mortlock, C. D. Shaffer, et al. 1991. Molecular cloning and transformation of cyclodiene resistance in *Drosophila*: An invertebrate γ-aminobutyric acid subtype A receptor locus. *Proc. Natl. Acad. Sci. U. S. A.* 88:7209—13.

[38]ffrench-Constant, R. H., and T. A. Rocheleau. 1993. Drosophila γ-aminobutyric acid receptor gene *Rdl* shows extensive alternative splicing. *J. Neurochem.* 60:2323—26.

[39]ffrench-Constant, R. H., T. A. Rocheleau, J. C. Steichen, et al. 1993. A point mutation in a *Drosophila* GABA receptor confers insecticide resistance. *Nature* 363:449—51.

[40]Forrester, N. W., M. Cahill, L. J. Bird, et al. 1993. Management of pyrethroid and endosulfan resistance in *Helicoverpa armigera* (Lepidoptera, Noctuidae) in Australia. *Bull. Entomol. Res.*, Suppl. 1:R1—132.

[41]Fournier, D. 2005. Mutations of acetylcholinesterase which confer insecticide resistance in insect populations. *Chem. Biol. Interact.* 157:257—61.

[42]Franck, P., M. Reyes, J. Olivares, et al. 2007. Genetic architecture in codling moth populations: Comparison between microsatellite and insecticide resistance markers. *Mol. Ecol.* 16:3554—64.

[43]Gammon, D. W. 1980. Pyrethroid resistance in a strain of *Spodoptera littoralis* is correlated with decreased sensitivity of the CNS *in vitro*. *Pestic. Biochem. Physiol.* 13:53—62.

[44]Gilbert, R. D., P. K. Bryson, and T. M. Brown. 1996. Linkage of acetylcholinesterase insensitivity

to methyl parathion resistance in *Heliothis virescens*. *Biochem. Genet*. 34:297—312.

[45]Grauso, M., R. A. Reenan, E. Culetto, et al. 2002. Novel putative nicotinic acetylcholine receptor subunit genes, Dα5, Dα6 and Dα7 in *Drosophila melanogaster* identify a new and highly conserved target of adenosine deaminase acting on RNA-mediated A-to-I pre-mRNA editing. *Genetics* 160:1519—33.

[46]Griffitts, J. S., and R. V. Aroian. 2005. Many roads to resistance: How invertebrates adapt to Bt toxins. *Bioessays* 27:614—24.

[47]Grubor, V. D., and D. G. Heckel. 2007. Evaluation of the role of CYP6B cytochrome P450s in pyrethroid resistant Australian *Helicoverpa armigera*. *Insect Mol. Biol*. 16:15—23.

[48]Gunning, R. V., C. S. Easton, M. E. Balfe, et al. 1991. Pyrethroid resistance mechanisms in Australian *Helicoverpa armigera*. *Pestic. Sci*. 33:473—90.

[49]Gunning, R. V., C. S. Easton, L. R. Greenup, et al. 1984. Pyrethroid resistance in *Heliothis armigera* (Hübner) (Lepidoptera, Noctuidae) in Australia. *J. Econ. Entomol*. 77:1283—87.

[50]Gunning, R. V., G. D. Moores, and A. L. Devonshire. 1996. Insensitive acetylcholinesterase and resistance to thiodicarb in Australian *Helicoverpa armigera* Hübner (Lepidoptera: Noctuidae). *Pestic. Biochem. Physiol*. 55:21—28.

[51]Gunning, R. V., G. D. Moores, and A. L. Devonshire. 1998. Insensitive acetylcholinesterase and resistance to organophosphates in Australian *Helicoverpa armigera*. *Pestic. Biochem. Physiol*. 62:147—51.

[52]Hall, L. M. C., and P. Spierer. 1986. The *Ace* locus of *Drosophila melanogaster*: Structural gene for acetylcholinesterase with an unusual 5' leader. *EMBO J*. 5:2949—54.

[53] Halling, B., and D. Yuhas. 2001. Lepidopteran GABA-gated chloride channels. US Patent No. 6329516.

[54]Harel, M., G. Kryger, T. L. Rosenberry, et al. 2000. Three-dimensional structures of *Drosophila melanogaster* acetylcholinesterase and of its complexes with two potent inhibitors. *Protein Sci*. 9:1063—72.

[55]Head, D. J., A. R. McCaffery, and A. Callaghan. 1998. Novel mutations in the *para*-homologous sodium channel gene associated with phenotypic expression of nerve insensitivity resistance to pyrethroids in Heliothine Lepidoptera. *Insect Mol. Biol*. 7:191—96.

[56]Heckel, D. G., P. K. Bryson, and T. M. Brown. 1998. Linkage analysis of insecticide-resistant acetylcholinesterase in *Heliothis virescens*. *J. Hered*. 89:71—78.

[57]Heckel, D. G., L. J. Gahan, S. W. Baxter, et al. 2007. The diversity of Bt resistance genes in species of Lepidoptera. *J. Invertebr. Pathol*. 95:192—97.

[58]Heckel, D. G., L. J. Gahan, J. C. Daly, et al. 1998. A genomic approach to understanding *Heliothis* and *Helicoverpa* resistance to chemical and biological insecticides. *Philos. Trans. R. Soc. Lond. B Biol. Sci*. 353:1713—22.

[59]Holloway, J. W., and A. R. McCaffery. 1996. Nerve insensitivity to cis-cypermethrin is expressed in adult *Heliothis virescens*. *Pestic. Sci*. 47:205—11.

[60]Hoopengardner, B., T. Bhalla, C. Staber, et al. 2003. Nervous system targets of RNA editing identified by comparative genomics. *Science* 301:832—36.

[61]Huang, H. S., N. T. Hu, Y. E. Yao, et al. 1998. Molecular cloning and heterologous expression of a glutathione S-transferase involved in insecticide resistance from the diamondback moth, *Plutella xylostella*. *Insect Biochem. Mol. Biol*. 28:651—658.

[62]Huang, S. J., and Z. J. Han. 2007. Mechanisms for multiple resistances in field populations of common cutworm, *Spodoptera litura* (Fabricius) in China. *Pestic. Biochem. Physiol.* 87:14—22.

[63]Huchard, E., M. Martinez, H. Alout, et al. 2006. Acetylcholinesterase genes within the Diptera: Takeover and loss in true flies. *Proc. Biol. Sci.* 273:2595—2604.

[64]Ingles, P. J., P. M. Adams, D. C. Knipple, et al. 1996. Characterization of voltage-sensitive sodium channel gene coding sequences from insecticide-susceptible and knockdown-resistant house fly strains. *Insect Biochem. Mol. Biol.* 26:319—26.

[65]International Silkworm Genome Consortium. 2008. The genome of a lepidopteran model insect, the silkworm *Bombyx mori*. *Insect Biochem. Mol. Biol.* 38:1036—45.

[66]Jongsma, M. A., P. L. Bakker, J. Peters, et al. 1995. Adaptation of *Spodoptera exigua* larvae to plant proteinaseinhibitors by induction of gut proteinase activity insensitive to inhibition. *Proc. Nat. Acad. Sci. U. S. A.* 92:8041—45.

[67]Joussen, N., D. G. Heckel, M. Haas, et al. 2008. Metabolism of imidacloprid and DDT by P450 GYP6G1 expressed in cell cultures of *Nicotiana tabacum* suggests detoxification of these insecticides in Cyp6g1-overexpressing strains of *Drosophila melanogaster*, leading to resistance. *Pest Manag. Sci.* 64:65—73.

[68]Kanga, L. H. B., D. J. Pree, J. L. van Lier, et al. 1997. Mechanisms of resistance to organophosphorus and carbamate insecticides in oriental fruit moth populations (*Grapholita molesta* Busck). *Pestic. Biochem. Physiol.* 59:11—23.

[69]Kao, P. N., and A. Karlin. 1986. Acetylcholine receptor binding site contains a disulfide cross-link between adjacent half-cystinyl residues. *J. Biol. Chem.* 261:8085—88.

[70]Kozaki, T., H. Sezutsu, R. Feyereisen, et al. 2008. The *Bombyx mori* P450s. In *Ninth international symposium on cytochrome P450 biodiversity and biotechnology*, ed. R. Feyereisen, 43. Nice, France: Institut National de la Recherche Agronomique.

[71]Kozaki, T., T. Shono, T. Tomita, et al. 2001. Fenitroxon insensitive acetylcholinesterases of the housefly, *Musca domestica* associated with point mutations. *Insect Biochem. Mol. Biol.* 31:991—97.

[72]Kranthi, K. R., D. Jadhav, R. Wanjari, et al. 2001. Pyrethroid resistance and mechanisms of resistance in field strains of *Helicoverpa armigera* (Lepidoptera: Noctuidae). *J. Econ. Entomol.* 94:253—63.

[73]Krieger, R. I., P. P. Feeny, and C. F. Wilkinson. 1971. Detoxication enzymes in the guts of caterpillars: An evolutionary answer to plant defenses? *Science* 172:579—81.

[74]Ku, C. C., F. M. Chiang, C. Y. Hsin, et al. 1994. Glutathione transferase isozymes involved in insecticide resistance of diamondback moth larvae. *Pestic. Biochem. Physiol.* 50:191—97.

[75]Kulkarni, N. H., A. H. Yamamoto, K. O. Robinson, et al. 2002. The DSC1 channel, encoded by the smi60E locus, contributes to odor-guided behavior in *Drosophila melanogaster*. *Genetics* 161:1507—16.

[76]Kwon, D. H., J. M. Clark, and S. H. Lee. 2004. Estimation of knockdown resistance in diamondback moth using real-time PASA. *Pestic. Biochem. Physiol.* 78:39—48.

[77]Le Goff, G., A. Hamon, J. B. Bergé, and M. Amichot. 2005. Resistance to fipronil in *Drosophila simulans*: Influence of two point mutations in the RDL GABA receptor subunit. *J. Neurochem.* 92:1295—1305.

[78]Lee, D. W., J. Y. Choi, W. T. Kim, et al. 2007. Mutations of acetylcholinesterase1 contribute to prothiofos-resistance in *Plutella xylostella* (L.). *Biochem. Biophys. Res. Commun.* 353:591—97.

[79]Lee, D. W. , S. S. Kim, S. W. Shin, et al. 2006. Molecular characterization of two acetylcholinesterase genes from the oriental tobacco budworm, *Helicoverpa assulta* (Guenee). *Biochim. Biophys. Acta* 1760:125—33.

[80]Lee, D. , Y. Park, T. M. Brown, et al. 1999. Altered properties of neuronal sodium channels associated with genetic resistance to pyrethroids. *Mol. Pharmacol.* 55:584—93.

[81]Lee, S. H. , and D. M. Soderlund. 2001. The V410M mutation associated with pyrethroid resistance in *Heliothis virescens* reduces the pyrethroid sensitivity of housefly sodium channels expressed in *Xenopus oocytes*. *Insect Biochem. Mol. Biol.* 31:19—29.

[82] Li, X. C. , J. Baudry, M. R. Berenbaum, et al. 2004. Structural and functional divergence of insect CYP6B proteins: From specialist to generalist cytochrome P450. *Proc. Natl. Acad. Sci. U. S. A.* 101: 2939—2944.

[83]Li, X. C. , M. R. Berenbaum, and M. A. Schuler. 2002. Plant allelochemicals differentially regulate Helicoverpa zea cytochrome P450 genes. *Insect Mol. Biol.* 11:343—51.

[84]Li, F. , and Z. J. Han. 2004. Mutations in acetylcholinesterase associated with insecticide resistance in the cotton aphid, *Aphis gossypii* Glover. *Insect Biochem. Mol. Biol.* 34:397—405.

[85]Li, X. C. , M. A. Schuler, and M. R. Berenbaum. 2007. Molecular mechanisms of metabolic resistance to synthetic and natural xenobiotics. *Annu. Rev. Entomol.* 52:231—53.

[86]Li, A. G. , Y. H. Yang, S. W. Wu, et al. 2006. Investigation of resistance mechanisms to fipronil in diamondback moth (Lepidoptera: Plutellidae). *J. Econ. Entomol.* 99:914—19.

[87]Liu, Z. Q. , J. G. Tan, S. M. Valles, et al. 2002. Synergistic interaction between two cockroach sodium channel mutations and a tobacco budworm sodium channel mutation in reducing channel sensitivity to a pyrethroid insecticide. *Insect Biochem. Mol. Biol.* 32:397—404.

[88]Liu, Z. W. , M. S. Williamson, S. J. Lansdell, et al. 2005. A nicotinic acetylcholine receptor mutation conferring target-site resistance to imidacloprid in *Nilaparvata lugens* (brown planthopper). *Proc. Natl. Acad. Sci. U. S. A.* 102:8420—25.

[89]Liu, Z. W. , M. S. Williamson, S. J. Lansdell, et al. 2006. A nicotinic acetylcholine receptor mutation (Y151S) causes reduced agonist potency to a range of neonicotinoid insecticides. *J. Neurochem.* 99: 1273—81.

[90]Loughney, K. , R. Kreber, and B. Ganetzky. 1989. Molecular analysis of the *para* locus, a sodium-channel gene in *Drosophila*. *Cell* 58:1143—54.

[91]Malcolm, C. A. , D. Bourguet, A. Ascolillo, et al. 1998. A sex-linked *Ace* gene, not linked to insensitive acetylcholinesterase-mediated insecticide resistance in *Culex pipiens*. *Insect Mol. Biol.* 7:107—20.

[92]Mao, Y. B. , W. J. Cai, J. W. Wang, et al. 2007. Silencing a cotton bollworm P450 monooxygenase gene by plantmediated RNAi impairs larval tolerance of gossypol. *Nat. Biotechnol.* 25:1307—13.

[93]Mao, W. , M. A. Schuler, and M. R. Berenbaum. 2007. Cytochrome P450s in *Papilio multicaudatus* and the transition from oligophagy to polyphagy in the Papilionidae. *Insect Mol. Biol.* 16:481—90.

[94]Martin, T. , O. G. Ochou, M. Vaissayre, et al. 2003. Oxidases responsible for resistance to pyrethroids sensitize *Helicoverpa armigera* (Hübner) to triazophos in West Africa. *Insect Biochem. Mol. Biol.* 33:883—87.

[95]Mau, R. F. L. , and L. Gusukuma-Minuto. 2004. Diamondback moth, *Plutella xylostella* (L.),

resistance management in Hawaii. In *The management of diamondback moth and other crucifer pests: Proceedings of the 4th international workshop*, ed. N. Endersby and P. Ridland, 307－11. Melbourne: Victorian Department of Primary Industries.

[96]McCaffery, A. R., D. J. Head, T. Jianguo, et al. 1997. Nerve insensitivity resistance to pyrethroids in heliothine Lepidoptera. *Pestic. Sci.* 51:315－20.

[97]McCaffery, A. R., J. W. Holloway, and R. T. Gladwell. 1995. Nerve insensitivity resistance to cypermethrin in larvae of the tobacco budworm *Heliothis virescens* from USA cotton field populations. *Pestic. Sci.* 44:237－47.

[98]Miyazaki, M., K. Ohyama, D. Y. Dunlap, et al. 1996. Cloning and sequencing of the *para*-type sodium channel gene from susceptible and *kdr*-resistant German cockroaches (*Blattella germanica*) and house fly (*Musca domestica*). *Mol. Gen. Genet.* 252:61－68.

[99]Morishita, M. 1998. Changes in susceptibility to various pesticides of diamondback moth (*Plutella xylostella* L.) in Gobo, Wakayama Prefecture. *Jap. J. Appl. Entomol. Zool.* 42:209－13.

[100]Mota-Sanchez, D., P. S. Bills, and M. E. Whalon. 2002. Arthropod resistance to pesticides: Status and overview. In *Pesticides in agriculture and the environment*, ed. W. B. Wheeler, 241－72. New York: Marcel Dekker Inc.

[101]Moulton, J. K., D. A. Pepper, and T. J. Dennehy. 2000. Beet armyworm (*Spodoptera exigua*) resistance to spinosad. *Pest Manag. Sci.* 56:842－48.

[102]Mutero, A., M. Pralavorio, J. M. Bride, et al. 1994. Resistance-associated point mutations in insecticide-insensitive acetylcholinesterase. *Proc. Nat. Acad. Sci. U. S. A.* 91:5922－26.

[103]Narahashi, T. 1996. Neuronal ion channels as the target sites of insecticides. *Pharmacol. Toxicol.* 79:1－14.

[104]Newcomb, R. D., P. M. Campbell, D. L. Ollis, et al. 1997. A single amino acid substitution converts a carboxylesterase to an organophosphorus hydrolase and confers insecticide resistance on a blowfly. *Proc. Nat. Acad. Sci. U. S. A.* 94:7464－68.

[105]Ni, X. Y., T. Tomita, S. Kasai, et al. 2003. cDNA and deduced protein sequence of acetylcholinesterase from the diamondback moth, *Plutella xylostella* (L.) (Lepidoptera: Plutellidae). *Appl. Entomol. Zool.* 38:49－56.

[106]Nicholson, R. A., and T. A. Miller. 1985. Multifactorial resistance to transpermethrin in field-collected strains of the tobacco budworm *Heliothis virescens* F. *Pestic. Sci.* 16:561－70.

[107]Noda, M., S. Shimizu, T. Tanabe, et al. 1984. Primary structure of *Electrophorus electricus* sodium channel deduced from cDNA sequence. *Nature* 312:121－27.

[108]Noppun, V., T. Miyata, and T. Saito. 1987. Insensitivity of acetylcholinesterase in phenthoate resistant diamondback moth, *Plutella xylostella*. *Appl. Entomol. Zool.* 22:116－18.

[109]Oakeshott, J., C. Claudianos, P. Campbell, et al. 2005a. Biochemical genetics and genomics of insect esterases. In *Comprehensive molecular insect science*, Vol. 5, ed. L. Gilbert, K. Iatrou, and S. Gill, 309－81. Oxford: Elsevier.

[110]Oakeshott, J. G., A. L. Devonshire, C. Claudianos, et al. 2005b. Comparing the organophosphorus and carbamate insecticide resistance mutations in cholin- and carboxyl-esterases. *Chem. Biol. Interact.* 157:269－75.

[111]Ollis, D. L. , E. Cheah, M. Cygler, et al. 1992. The alpha/beta hydrolase fold. *Protein Eng*. 5: 197—211.

[112]Ottea, J. A. , and J. W. Holloway. 1998. Target-site resistance to pyrethroids in *Heliothis virescens*(F.) and *Helicoverpa zea* (Boddie). *Pestic. Biochem. Physiol.* 61:155—67.

[113]Park, Y. S. , D. W. Lee, M. F. J. Taylor, et al. 2000. A mutation Leu1029 to His in *Heliothis virescens* F. : *hscp* sodium channel gene associated with a nerve-insensitivity mechanism of resistance to pyrethroid insecticides. *Pestic. Biochem. Physiol.* 66:1—8.

[114]Park, Y. , and M. F. J. Taylor. 1997. A novel mutation L1029H in sodium channel gene hscp associated with pyrethroid resistance for *Heliothis virescens* (Lepidoptera: Noctuidae). *Insect Biochem. Mol. Biol.* 27:9—13.

[115]Park, Y. , M. F. J. Taylor, and R. Feyereisen. 1997. A valine421 to methionine mutation in IS6 of the hscp voltagegated sodium channel associated with pyrethroid resistance in *Heliothis virescens* F. *Biochem. Biophys. Res. Commun.* 239:688—91.

[116]Park, Y. , M. F. J. Taylor, and R. Feyereisen. 1999. Voltage-gated sodium channel genes *hscp* and hDSC1 of *Heliothis virescens* F. : Genomic organization. *Insect Mol. Biol.* 8:161—70.

[117]Perry, T. , D. G. Heckel, J. A. McKenzie, et al. 2008. Mutations in Dα1 or Dβ2 nicotinic acetylcholine receptor subunits can confer resistance to neonicotinoids in *Drosophila melanogaster*. *Insect Biochem. Mol. Biol.* 38:520—28.

[118]Perry, T. , J. A. McKenzie, and P. Batterham. 2007. A Dα6 knockout strain of *Drosophila melanogaster* confers a high level of resistance to spinosad. *Insect Biochem. Mol. Biol.* 37:184—88.

[119]Pigott, C. R. , and D. J. Ellar. 2007. Role of receptors in *Bacillus thuringiensis* crystal toxin activity. *Microbiol. Mol. Biol. Rev.* 71:255—81.

[120]Pimprale, S. S. , C. L. Besco, P. K. Bryson, et al. 1997. Increased susceptibility of pyrethroid-resistant tobacco budworm (Lepidoptera: Noctuidae) to chlorfenapyr. *J. Econ. Entomol.* 90:49—54.

[121]Qu, M. J. , Z. J. Han, X. J. Xu, et al. 2003. Triazophos resistance mechanisms in the rice stem borer (*Chilo suppressalis* Walker). *Pestic. Biochem. Physiol.* 77:99—105.

[122]Ranasinghe, C. , B. Campbell, and A. A. Hobbs. 1998. Over-expression of cytochrome P450 CYP6B7 mRNA and pyrethroid resistance in Australian populations of *Helicoverpa armigera* (Hübner). *Pestic. Sci.* 54:195—202.

[123]Ranasinghe, C. , and A. A. Hobbs. 1998. Isolation and characterization of two cytochrome P450 cDNA clones for CYP6B6 and CYP6B7 from *Helicoverpa armigera* (Hübner): Possible involvement of CYP6B7 in pyrethroid resistance. *Insect Biochem. Mol. Biol.* 28:571—80.

[124]Ranasinghe, C. , and A. A. Hobbs. 1999. Induction of cytochrome P450 CYP6B7 and cytochrome b5 mRNAs from *Helicoverpa armigera* (Hübner) by pyrethroid insecticides in organ culture. *Insect Mol. Biol.* 8:443—47.

[125]Ranson, H. , and J. Hemingway. 2005. Glutathione transferases. In *Comprehensive molecular insect science*, Vol. 5, ed. L. Gilbert, K. Iatrou, and S. Gill, 383—402. Oxford: Elsevier.

[126]Raymond, M. , D. G. Heckel, and J. G. Scott. 1989. Interactions between pesticide resistance genes: Model and experiment. *Genetics* 123:543—51.

[127]Ren, X. X. , Z. J. Han, and Y. C. Wang. 2002. Mechanisms of monocrotophos resistance in cotton

bollworm, *Helicoverpa armigera* (Hübner). *Arch. Insect Biochem. Physiol.* 51:103—10.

[128]Rose, R. L., L. Barbhaiya, R. M. Roe, et al. 1995. Cytochrome P450-associated insecticide resistance and the development of biochemical diagnostic assays in *Heliothis virescens*. *Pestic. Biochem. Physiol.* 51:178—91.

[129]Rose, R. L., D. Goh, D. M. Thompson, et al. 1997. Cytochrome P450 CYP9A1 in *Heliothis virescens*: The first member of a new CYP family. *Insect Biochem. Mol. Biol.* 27:605—15.

[130]Ru, L. J., C. Wei, J. Z. Zhao, et al. 1998. Differences in resistance to fenvalerate and cyhalothrin and inheritance of knockdown resistance to fenvalerate in *Helicoverpa armigera*. *Pestic. Biochem. Physiol.* 61:79—85.

[131]Russell, R. J., C. Claudianos, P. M. Campbell, et al. 2004. Two major classes of target site insensitivity mutations confer resistance to organophosphate and carbamate insecticides. *Pestic. Biochem. Physiol.* 79:84—93.

[132]Ryan, C. A. 1990. Protease inhibitors in plants: Genes for improving defenses against insects and pathogens. *Ann. Rev. Phytopathol.* 28:425—49.

[133]Salkoff, L., A. Butler, A. Wei, et al. 1987. Genomic organization and deduced amino acid sequence of a putative sodium channel gene in *Drosophila*. *Science* 237:744—49.

[134]Sato, C., Y. Ueno, K. Asai, et al. 2001. The voltage-sensitive sodium channel is a bell-shaped molecule with several cavities. *Nature* 409:1047—51.

[135]Sattelle, D. B., A. K. Jones, B. M. Sattelle, et al. 2005. Edit, cut and paste in the nicotinic acetylcholine receptor gene family of *Drosophila melanogaster*. *Bioessays* 27:366—76.

[136]Sayyed, A. H., D. Omar, and D. J. Wright. 2004. Genetics of spinosad resistance in a multi-resistant fieldselected population of *Plutella xylostella*. *Pest Manage. Sci.* 60:827—32.

[137]Schuler, T. H., D. Martinez-Torres, A. J. Thompson, et al. 1998. Toxicological, electrophysiological, and molecular characterisation of knockdown resistance to pyrethroid insecticides in the diamondback moth, *Plutella xylostella* (L.). *Pestic. Biochem. Physiol.* 59:169—82.

[138]Scott, J. G., and Z. M. Wen. 2001. Cytochromes P450 of insects: The tip of the iceberg. *Pest Manag. Sci.* 57:958—67.

[139]Seino, A., T. Kazuma, A. J. Tan, et al. 2007. Analysis of two acetylcholinesterase genes in *Bombyx mori*. *Pestic. Biochem. Physiol.* 88:92—101.

[140]Shang, J. Y., Y. M. Shao, G. J. Lang, et al. 2007. Expression of two types of acetylcholinesterase gene from the silkworm, *Bombyx mori*, in insect cells. *Insect Sci.* 14:443—49.

[141]Shao, Y. M., K. Dong, and C. X. Zhang. 2007. The nicotinic acetylcholine receptor gene family of the silkworm, *Bombyx mori*. *BMC Genomics* 8:324—34.

[142]Shen, B. C., D. X. Zhao, C. L. Qiao, et al. 2004. Cloning of CYP9G2 from the diamondback moth, *Plutella xylostella* (Lepidoptera: Yponomeutidae). *DNA Sequence* 15:228—33.

[143]Soderlund, D. 2005. Sodium channels. In *Comprehensive molecular insect science*, Vol 5. ed. L. Gilbert, K. Iatrou, and S. Gill, 1—24. Oxford: Elsevier.

[144]Sonoda, S., M. Ashfaq, and H. Tsumuki. 2006. Genomic organization and developmental expression of glutathione S-transferase genes of the diamondback moth, *Plutella xylostella*. *J. Insect Sci.* 6:1—9.

[145]Sonoda, S., C. Igaki, M. Ashfaq, et al. 2006. Pyrethroid-resistant diamondback moth expresses

alternatively spliced sodium channel transcripts with and without T929I mutation. *Insect Biochem. Mol. Biol.* 36:904—10.

[146]Sonoda,S. ,and H. Tsumuki. 2005. Studies on glutathione S-transferase gene involved in chlorfluazuron resistance of the diamondback moth,*Plutella xylostella* L. (Lepidoptera: Yponomeutidae). *Pestic. Biochem. Physiol.* 82:94—101.

[147]Srinivas,R. ,S. S. Udikeri,S. K. Jayalakshmi,et al. 2004. Identification of factors responsible for insecticide resistance in *Helicoverpa armigera*. *Comp. Biochem. Physiol. C Toxicol. Pharmacol.* 137:261—69.

[148]Stokes,N. H. ,S. W. McKechnie,and N. W. Forrester. 1997. Multiple allelic variation in a sodium channel gene from populations of Australian *Helicoverpa armigera* (Hübner) (Lepidoptera: Noctuidae) detected via temperature gradient gel electrophoresis. *Aust. J. Entomol.* 36:191—96.

[149]Strode,C. ,C. S. Wondji,J. P. David,et al. 2008. Genomic analysis of detoxification genes in the mosquito *Aedes aegypti*. *Insect Biochem. Mol. Biol.* 38:113—23.

[150]Sussman,J. L. ,M. Harel,F. Frolow,et al. 1991. Atomic structure of acetylcholinesterase from *Torpedo californica*: A prototypic acetylcholine-binding protein. *Science* 253:872—79.

[151]Tabashnik,B. E. ,N. L. Cushing,N. Finson,and M. W. Johnson. 1990. Field development of resistance to *Bacillus thuringiensis* in diamondback moth (Lepidoptera,Plutellidae). *J. Econ. Entomol.* 83: 1671—76.

[152]Talekar,N. S. ,and A. M. Shelton. 1993. Biology,ecology,and management of the diamondback moth. *Annu. Rev. Entomol.* 38:275—301.

[153]Tan,J. ,Z. Liu,T. D. Tsai,et al. 2002. Novel sodium channel gene mutations in *Blattella germanica* reduce the sensitivity of expressed channels to deltamethrin. *Insect Biochem. Mol. Biol.* 32:445—54.

[154]Tan,J. G. ,and A. R. McCaffery. 1999. Expression and inheritance of nerve insensitivity resistance in larvae of *Helicoverpa armigera* (Lepidoptera: Noctuidae) from China. *Pestic. Sci.* 55:617—25.

[155]Taylor,M. F. J. ,D. G. Heckel,T. M. Brown,et al. 1993. Linkage of pyrethroid insecticide resistance to a sodium channel locus in the tobacco budworm. *Insect Biochem. Mol. Biol.* 23:763—75.

[156]Taylor,M. F. J. ,P. Park,and Y. Shen. 1996. Molecular population genetics of sodium channel and juvenile hormone esterase markers in relation to pyrethroid resistance in *Heliothis virescens* (Lepidoptera: Noctuidae). *Ann. Entomol. Soc. Am.* 89:728—38.

[157]Taylor,M. F. J. ,Y. Shen,and M. E. Kreitman. 1995. A population genetic test of selection at the molecular level. *Science* 270:1497—99.

[158]Toda,S. ,S. Komazaki,T. Tomita,et al. 2004. Two amino acid substitutions in acetylcholinesterase associated with pirimicarb and organophosphorous insecticide resistance in the cotton aphid,*Aphis gossypii* Glover (Homoptera: Aphididae). *Insect Mol. Biol.* 13:549—53.

[159]Tsukahara,Y. ,S. Sonoda,Y. Fujiwara,et al. 2003. Molecular analysis of the *para*-sodium channel gene in the pyrethroid-resistant diamondback moth,*Plutella xylostella* (Lepidoptera: Yponomeutidae). *Appl. Entomol. Zool.* 38:23—29.

[160]Vanlaecke,K. ,G. Smagghe,and D. Degheele. 1995. Detoxifying enzymes in greenhouse and laboratory strains of beet armyworm (Lepidoptera,Noctuidae). *J. Econ. Entomol.* 88:777—81.

[161]Villatte,F. ,P. Ziliani,V. Marcel,et al. 2000. A high number of mutations in insect acetylcholinesterase may provide insecticide resistance. *Pestic. Biochem. Physiol.* 67:95—102.

[162]Vontas, J. G. , G. J. Small, and J. Hemingway. 2001. Glutathione S-transferases as antioxidant defence agents confer pyrethroid resistance in *Nilaparvata lugens*. *Biochem. J*. 357:65—72.

[163]Walsh, S. B. , T. A. Dolden, G. D. Moores, et al. 2001. Identification and characterization of mutations in housefly (*Musca domestica*) acetylcholinesterase involved in insecticide resistance. *Biochem. J*. 359:175—81.

[164]Wang, X. P. , and A. A. Hobbs. 1995. Isolation and sequence analysis of a cDNA clone for a pyrethroid inducible cytochrome P450 from *Helicoverpa armigera*. *Insect Biochem. Mol. Biol*. 25:1001—09.

[165]Wee, C. W. , S. F. Lee, C. Robin, et al. 2008. Identification of candidate genes for fenvalerate resistance in *Helicoverpa armigera* using cDNA-AFLP. *Insect Mol. Biol*. 17:351—60.

[166]Weill, M. , P. Fort, A. Berthomieu, et al. 2002. A novel acetylcholinesterase gene in mosquitoes codes for the insecticide target and is non-homologous to the *Ace* gene in *Drosophila*. *Proc. Biol. Sci*. 269:2007—16.

[167]Williamson, M. S. , D. MartinezTorres, C. A. Hick, et al. 1996. Identification of mutations in the housefly *para-type* sodium channel gene associated with knockdown resistance (*kdr*) to pyrethroid insecticides. *Mol. Gen. Genet*. 252:51—60.

[168]Wing, K. , J. Andaloro, S. McCann, et al. 2005. Indoxacarb and the sodium channel blocker insecticides: Chemistry, physiology, and biology in insects. In *Comprehensive molecular insect science*, Vol. 5. , ed. L. Gilbert, K. Iatrou, and S. Gill, 31—53. Oxford: Elsevier.

[169]Wink, M. , and V. Theile. 2002. Alkaloid tolerance in *Manduca sexta* and phylogenetically related sphingids (Lepidoptera: Sphingidae). *Chemoecology* 12:29—46.

[170]Wolff, M. A. , and V. P. M. Wingate. 1998. Characterization and comparative pharmacological studies of a functional γ-aminobutyric acid (GABA) receptor cloned from the tobacco budworm, *Heliothis virescens* (Noctuidae: Lepidoptera). *Invertebr. Neurosci*. 3:305—15.

[171]Wyss, C. F. , H. P. Young, J. Shukla, et al. 2003. Biology and genetics of a laboratory strain of the tobacco budworm, *Heliothis virescens* (Lepidoptera: Noctuidae), highly resistant to spinosad. *Crop Protect*. 22:307—14.

[172]Yang, Y. H. , S. Chen, S. W. Wu, et al. 2006. Constitutive overexpression of multiple cytochrome P450 genes associated with pyrethroid resistance in *Helicoverpa armigera*. *J. Econ. Entomol*. 99:1784—89.

[173]Yang, Y. , Y. Wu, S. Chen, et al. 2004. The involvement of microsomal oxidases in pyrethroid resistance in *Helicoverpa armigera* from Asia. *Insect Biochem. Mol. Biol*. 34:763—73.

[174]Young, H. P. , W. D. Bailey, and R. M. Roe. 2003. Spinosad selection of a laboratory strain of the tobacco budworm, *Heliothis virescens* (Lepidoptera: Noctuidae), and characterization of resistance. *Crop Protect*. 22:265—73.

[175]Young, H. P. , W. D. Bailey, R. M. Roe, et al. 2001. Mechanism of resistance and cross-resistance in a laboratory, spinosad-selected strain of the tobacco budworm and resistance in laboratory-selected cotton bollworms. In *Proceedings of the 2001 Beltwide Cotton Conference*, 1167—71. Memphis: National Cotton Council of America.

[176]Young, H. P. , W. D. Bailey, C. F. Wyss, et al. 2000. Studies on the mechanisms of tobacco budworm resistance to spinosad (Tracer). In *Proceedings of the 2000 Beltwide Cotton Conference*, 1197—1201. Memphis: National Cotton Council of America.

[177]Yu,S. J. 1992. Detection and biochemical characterization of insecticide resistance in fall armyworm (Lepidoptera,Noctuidae). *J. Econ. Entomol.* 85:675—82.

[178]Yu,S. J. 2006. Insensitivity of acetylcholinesterase in a field strain of the fall armyworm, *Spodoptera frugiperda* (J. E. Smith). *Pestic. Biochem. Physiol.* 84:135—42.

[179]Yu,S. J. ,and S. N. Nguyen. 1992. Detection and biochemical characterization of insecticide resistance in diamondback moth. *Pestic. Biochem. Physiol.* 44:74—81

[180]Yu,S. J. ,S. N. Nguyen,and G. E. Abo-Elghar. 2003. Biochemical characteristics of insecticide resistance in the fall armyworm, *Spodoptera frugiperda* (J. E. Smith). *Pestic. Biochem. Physiol.* 77:1—11.

[181]Zhao,J. Z. ,Y. X. Li, H. L. Collins,et al. 2002. Monitoring and characterization of diamondback moth (Lepidoptera: Plutellidae) resistance to spinosad. *J. Econ. Entomol.* 95:430—36.

[182]Zhao,Y. ,Y. Park,and M. E. Adams. 2000. Functional and evolutionary consequences of pyrethroid resistance mutations in S6 transmembrane segments of a voltage-gated sodium channel. *Biochem. Biophys. Res. Commun.* 278:516—21.

[183]Zhou,W. ,I. B. Chung,Z. Q. Liu,et al. 2004. A voltage-gated calcium-selective channel encoded by a sodium channel-like gene. *Neuron* 42:101—12.

[184]Zlotkin,E. ,H. Moskowitz,R. Herrmann,et al. 1995. Insect sodium channel as the target for insect selective neurotoxins from scorpion venom. In *Molecular action of insecticides on ion channels*, ed. J. M. Clark,56—85. Washington,D. C. : American Chemical Society.

第十四章　烟草天蛾的先天免疫应答

Michael R. Kanost and James B. Nardi

1. 前言

昆虫免疫系统需要面对从寄生虫,拟寄生虫到病原体的一系列外源物的攻击。据估计拟寄生昆虫占大约100万已知昆虫种类的15%。病原体和寄生虫,包括大量的病毒、细菌、微孢子、真菌、线虫、原虫(比如孢子虫、簇虫、球虫)。这些生物体的种类每年都在增加。寄生虫和病原体之所以能够成功繁殖,很大程度上是因为它们能够完全避免遭遇昆虫免疫系统细胞(Kathirithamby,Ross 和 Johnston 2003;Manfredini et al. 2007),或者能够产生可以防止宿主血细胞防御启动的物质。这些防御物质可以是:(a)抑制免疫应答的病毒或分子;(b)模拟宿主抗原使其识别为自身抗原的分子;(c)能够掩盖侵入物抗原的分子,从而使拟寄生虫和病原体不被宿主免疫细胞识别为外源或非内源物质(例如,Salt 1070; Rizki 和 Riziki 1984; Davies 和 Vinson 1986; Schmidt et al. 1990; Strand 和 Noda 1991; Strand 和 Wong 1991; Hoek et al. 1996; Lavine 和 Backage 1996;Asgari et al. 1998; Galibert et al. 2003; Beck 和 Strand 2005; Wang 和 St. Leger 2006)。

要明白拟寄生虫和病原体怎样防止先天性免疫系统中细胞的体液免疫或细胞免疫被外源物激活,需要知道哪些外源或内源信号被血浆或细胞表面受体识别,这些信号怎样激活先天免疫应答。对黑腹果蝇的免疫应答遗传学和应答途径方面的研究取得了广泛进展(Ferrandon et al. 2007; Lemaitre 和 Hoffmann 2007);蚊子是传播人类疾病的载体,其基因组信息也为免疫应答实验奠定了坚实的基础(Michel 和 Kafatos 2005; Waterhouse et al. 2007)。由于各种原因,鳞翅目昆虫被证明是研究昆虫免疫的重要实验物种。毛虫大量的血淋巴和血细胞数为血浆蛋白的生化研究及使用现代生物技术调查血细胞功能提供了充足的实验材料。对蚕蛾,特别是家蚕(Ashida 和 Brey 1997; Ponnurel 和 Yamakawa 2002)和刻克罗普斯蚕蛾(Boman et al. 1991; Su et al. 1998)、蜡螟、夜蛾(包括大豆夜蛾)(Lavine 和 Strand 2002)的研究,引起在细胞和体液免疫方面重要的发现,随后证实这些发现在其他昆虫,甚至脊椎动物里具有广泛的保守性。本章我们主要关注烟草天蛾免疫系统的一些最近进展(Kanost, Jiang 和 Yu 2004; Jiang 2008)。烟草天蛾在血淋巴蛋白和血细胞吸附分子研究中是最好的模式系统之一(见表 4.1)。

2. 模式识别受体引起昆虫和哺乳动物的先天免疫应答

昆虫先天性系统免疫应答包括体液免疫和细胞免疫。体液免疫包括黑化、抗菌肽以及其他行使免疫功能的血浆蛋白的诱导合成(Gillespie, Kanost 和 Trenczek 1997; Nappi 和 Christensen 2005; Ferrandon et al. 2007)。细胞免疫是指包囊作用、吞噬、瘤块化(Lavine 和 Strand 2002; Jiravanichpaisal, Lee 和 Söderhäll 2006)。体液和细胞免疫相互影响并被同样或相似的现象触发。触发的现象是昆虫识别正常自身物质的策略。Medzhitov 和 Janeway(2002)强调,区别自身和外源物需要识别:(a)微生物特征;(b)自身特征;(c)修饰的自身。

免疫系统利用模式识别受体(PRRs)识别外源保守的分子模式或有高度保守基序的病原特征分子模式(PAMPs)来区分外源物。病原特征分子模式包括微生物的结构,比如在真核非微生物细胞表面没有，但在微生物表面存在的脂多糖、脂膜酸、肽聚糖、β-1,3-葡聚糖(Janeway 和 Medzhitov 2002; Sansonetti 2006)。然而，在没有病原特征分子模式存在时，也能激活静止的、非黏性的昆虫血细胞。免疫系统区分自身或外源物不只是识别病原特征分子模式或微生物，也识别缺失的及变化的自身物质(Medzhitov 和 Janeway 2002)。先天免疫应答识别缺失的及变化的自身物质包括识别受伤、发育中产生的衰老细胞(变化的自身)、缺失病原特征分子模式的外源表面(比如无活性和拟寄生表面)。此外，这三种不同的先天免疫系统的识别策略都能用一种免疫模型概括。该模型假说认为免疫系统事实上除了已研究透彻的对病原发出的外源信号作出反应外；还对内源危险信号、应急和损伤产生的物质或死亡细胞也作出反应(Aderem 和 Ulevitch 2000; Matzinger 2002)。

模式识别受体一旦结合到特异病原特征分子后，将激活调控系统免疫应答的信号途径。由于最初对先天免疫的研究集中于体液应答，模式识别受体识别病原菌引起免疫应答的这部分信息快速累积(Kanost，Jiang 和 Yu 2004; Michel 和 Kafatos 2005; Nappi 和 Christensen 2005; Wang et al. 2005; Ferrandon et al. 2007)。然而，对模式识别受体如何和血细胞相互作用激活免疫应答的研究很少。此外，在缺失任何微生物污染物和无菌的情况，引起细胞免疫应答的原因尚不清楚(通过识别缺失的自身和改变后的自身)。

烟草天蛾已经鉴定出 6 个家族的模式识别受体。这些血浆蛋白结合到微生物表面引起免疫应答，包括前酚氧化物酶的激活、抗菌肽的诱导、血细胞的反应如吞噬、瘤化(Yu et al. 2002)。免疫凝集素是有 2 个糖结合结构域(CRD)的 34kD 的 C-类凝集素(Yu 和 Kanost 2008)。这种串联的 CRD 结构域在其他鳞翅目昆虫包括家蚕(Koizumi et al. 1999; Watanabe et al. 2006)，美国白蛾(Shin et al. 1998)，棉铃虫(登录号：ABF83203)，*Lonomia oblique*(登录号：AAV91450)中都有发现，但还没在双翅目昆虫果蝇和按蚊，膜翅目的蜜蜂中发现。在赤拟谷盗的 16 个 C-类凝集素基因中，只有一个编码双 CRD 结构(Zou et al. 2007)。可能 2-CRD 在鳞翅目昆虫免疫系统中起着相当独特的作用。免疫凝集素结合细菌或真菌表面的脂多糖，参与细胞和体液免疫应答，包括酚氧化物酶的激活、吞噬、细胞包囊作用(Yu 和 Kanost 2008)。

在烟草天蛾中鉴定出两个 β-1,3-葡聚糖识别蛋白(GRP) (Ma 和 Kanost 2000; Jiang et al. 2004)。这种首先在家蚕中发现的 52kDa 蛋白(Ochiai 和 Ashida 1988, 2000)，在很多节肢动物中都存在。烟草天蛾中的 β-1,3-葡聚糖识别蛋白能紧密结合真菌表面的 β-1,3-葡聚糖和革兰氏阳性菌表面的脂膜酸。β-1,3-葡聚糖识别蛋白由两个不同的结构域组成，它们都能结合 β-1,3-葡聚糖(Fabrick，Baker 和 Kanost 2004)。它们氨基末端的结构域只与无脊椎动物其他具有葡聚糖识别蛋白类似结构域相似，并以一个连接序列与羧基端相连。羧基端结构域与葡聚糖酶有序列相似性，但缺少水解活性。β-1,3-葡聚糖识别蛋白和多糖结合，通过与血淋巴的丝氨酸蛋白酶 HP14 的相互作用(Wang 和 Jiang 2006)，刺激前酚氧化物酶的激活，这部分将在下面做进一步的描述。烟草天蛾葡聚糖识别蛋白 1(GRP-1)在脂肪体内组成性表达，葡聚糖识别蛋白 2(GRP-2)在未感染的幼虫中无表达，但在感染的情况下，在脂肪体中上调表达。葡聚糖识别蛋白 2 也受发育调控，处于未感染的情况下，在预蛹的脂肪体和中肠中都有表达(Jiang et al. 2004)。

表 14.1 烟草天蛾体液和血球蛋白在免疫中的功能

Table 14.1 Proteins from *M. sexta* Plasma or Hemocytes with Functions in Immune Responses[a]

Protein	Function	Microbe-Induced Expression[b]	References
	Pattern Recognition Proteins		
Immulectins 1–4	LPS binding, proPO activation encapsulation, phagocytosis, nodule formation	+	Yu, Gan, and Kanost 1999; Yu, Prakash, and Kanost 1999; Yu and Kanost 2000, 2004; Yu et al. 2002, 2003, 2006; Eleftherianos et al. 2006; Ling and Yu 2006
β-1,3-glucan recognition protein-1	β-1,3-glucan, LTA binding proPO activation	–	Ma and Kanost 2000
β-1,3-glucan recognition protein-2	β-1,3-glucan, LTA binding proPO activation	+	Jiang et al. 2004; Wang and Jiang 2006
peptidoglycan recognition protein-1	binds to gram-negative bacteria	+	Zhu et al. 2003a; E. Ragan and M. R. Kanost, unpublished
hemolin	LPS, LTA binding/ phagocytosis	+	Ladendorff and Kanost 1991; Yu and Kanost 1999, 2002; Zhao and Kanost 1996; Eleftherianos et al. 2007
leureptin	LPS binding	+	Zhu et al. 2003a; Y. Zhu and M.R. Kanost, unpublished
Induced protein-1 (hdd-11/ noduler)	bind to microbes/nodule formation	+	Zhu et al. 2003a
	Antimicrobial Peptides/Proteins		
lysozyme	bacteriolytic	+	Mulnix and Dunn 1994
cecropins	antibacterial	+	Dickinson, Russell, and Dunn 1988; Zhu et al. 2003a
attacins	antibacterial	+	Kanost et al. 1990; Zhu et al. 2003a
lebocins	antibacterial	+	Zhu et al. 2003a
moricin	antibacterial	+	Zhu et al. 2003a
gloverin	antibacterial	+	Zhu et al. 2003a
	Proteinases		
proPO activating proteinase-1	PPO activation	+	Jiang, Wang, and Kanost 1998; Gupta, Wang, and Jiang 2005a,b; Zou, Wang, and Jiang 2005
proPO activating proteinase-2	PPO activation	+	Jiang et al. 2003a; Wang, Zou, and Jiang 2006
proPO activating proteinase-3	PPO activation	+	Jiang et al. 2003b; Wang and Jiang 2004a; Zou and Jiang 2005a
hemolymph proteinase-6	activate proHP8 and proPAP1	–	Jiang et al. 2005; C.J. An, J. Ishibashi, H.B. Jiang, and M.R. Kanost, unpublished
hemolymph proteinase-8	activate spätzle	–	Jiang et al. 2005; C.J. An and M.R. Kanost, unpublished
hemolymph proteinase-14	activate proHP21	+	Ji et al. 2004; Wang and Jiang 2006, 2007
hemolymph proteinase-21	activate proPAP2 and proPAP3	+	Gorman et al. 2007; Wang and Jiang 2007
scolexin	unknown	+	Kyriakides, McKillip, and Spence 1995; Finnerty and Granados 1997; Finnerty, Karplus, and Granados 1999;
serine proteinase homologues-1 and 2	PPO activation	+	Yu et al. 2003; Wang and Jiang 2004a; Gupta, Wang, and Jiang 2005b; Lu and Jiang 2008

续表

Protein	Function	Microbe-Induced Expression[b]	References
		Proteinase Inhibitors	
serpin-1 (12 splicing isoforms)	regulation of PPO activation (serpin-1J), others unknown	–	Kanost, Prasad, and Wells 1989; Jiang, Wang, and Kanost 1994; Jiang, Mulnix, and Kanost 1995; Kanost et al. 1995; Jiang et al. 1996; Jiang and Kanost 1997; Li et al. 1999; Ye et al. 2001
serpin-2	unknown	+	Gan et al. 2001
serpin-3	regulate PAPs	+	Zhu et al. 2003b
serpin-4	regulate HP1, HP6, and HP21	+	Tong, Jiang, and Kanost 2005; Tong and Kanost 2005
serpin-5	regulate HP1 and HP6	+	Tong, Jiang, and Kanost 2005; Tong and Kanost 2005
serpin-6	regulate PAP-3, HP8	+	Wang and Jiang 2004b; Zou and Jiang 2005b
serpin-7		+	C Suwanchaichinda, R Ochieng, and M.R. Kanost, unpublished results
		Other Enzymes	
proPO-1, proPO-2	oxidation of catechols, melanization	–	Hall et al. 1995; Jiang et al. 1997; Zhao et al. 2007; Lu and Jiang 2007
tyrosine hydroxylase	synthesis of DOPA	+	Gorman, An, and Kanost 2007
DOPA decarboxylase	synthesis of dopamine	+	Hiruma, Carter, and Riddiford 1995; Zhu et al. 2003a
Carboxylesterases	unknown	+	Zhu et al. 2003a
		Hemocyte-Modulating Proteins	
hemolin	inhibit hemocyte aggregation opsonin	+	Ladendorff and Kanost 1991; Zhao and Kanost 1996; Yu and Kanost 2002; Eleftherianos et al. 2007
hemocyte aggregation-inhibiting protein	inhibit hemocyte aggregation	–	Kanost et al. 1994; Scholz 2002
plasmatocyte spreading peptide	stimulate plasmatocyte adherence and spreading	–	Wang, Jiang, and Kanost 1999; Yu, Prakash, and Kanost 1999
lacunin	predicted function in adhesion	+	Nardi, Gao, and Kanost 2001; Nardi et al. 2005
hemocytin	predicted function in adhesion	?	Scholz 2002; J.B. Nardi, unpublished results
		Hemocyte Membrane Proteins	
Integrins	hemocyte adhesion, encapsulation	–	Levin et al. 2005; Zhuang et al. 2007b, 2008
tetraspanin D76	modulate integrin function	–	Zhuang et al. 2007b
neuroglian	hemocyte adhesion, encapsulation	?	Nardi et al. 2006; Zhuang et al. 2007a
Toll	microbe-induced gene expression	+	Ao, Ling, and Yu 2008

[a] Abbreviations: DOPA, dihydroxyphenylalanine; HP, hemolymph proteinase; LPS, lipopolysaccharide; LTA, lipoteichoic acid; PAP, prophenoloxidase-activating proteinase; proPO, prophenoloxidase

[b] Increased expression after injection of bacteria determined by immunoblot, northern blot, or RT-PCR analysis.

肽聚糖识别蛋白也是首先在家蚕中发现的能结合肽聚糖并激活前酚氧化物酶的蛋白（Yoshida，Kinoshita 和 Ashida 1996；Ochiai 和 Ashida 1999）。从粉纹夜蛾（*Trichoplusia ni*）

中鉴定得到最早报道的肽聚糖识别蛋白序列，它与噬菌体的溶菌酶和哺乳动物的同源蛋白具有相似性（Kang et al. 1998）。在烟草天蛾中鉴定出两个相似的肽聚糖识别蛋白cDNA，可能是同一基因的等位基因（Zhu et al. 2003a）。我们最近发现这个肽聚糖识别蛋白与革兰氏阴性菌的表面选择性结合（E. Ragan 和 M. R. Kanost，未发表数据）。这种19kDa 的鳞翅目昆虫血浆肽聚糖识别蛋白的序列与其他鳞翅目的这类蛋白相似，而且缺少溶菌酶活性。哺乳动物的肽聚糖识别蛋白和其他昆虫的肽聚糖识别蛋白由于能水解肽聚糖而具有直接的抗菌活性。鳞翅目昆虫肽聚糖识别蛋白的发现，将促进脊椎和非脊椎动物先天免疫中重要蛋白家族的鉴定（Royet 和 Dziarski 2007）。

Hemolin 是烟草天蛾和其他鳞翅目种类中由微生物诱导含量最丰富的 45kDa 血浆蛋白（Faye 和 Kanost 1998）。Hemolin 由四个免疫球蛋白（Ig）结构域组成。它首先从天蚕蛾和烟草天蛾的 cDNA 序列克隆得到，是第一个在昆虫中鉴定得到的属于免疫球蛋白（Ig）家族的蛋白（Sun et al. 1990；Ladendorff 和 Kanost 1991）。尽管编码 Hemolin 的cDNA 已经从不止 10 个鳞翅目物种中克隆得到，但到目前为止，还没有在其他非鳞翅目和昆虫基因组序列中发现。Hemolin 似乎是一个鳞翅目特异蛋白，可能是由编码含有 6 个 Ig 结构域的细胞黏性蛋白 Neruoglian 的基因经历进化上的部分复制事件衍生而来的。具有四个氨基端 Ig 结构域的 Neruoglian 在序列上和 Hemolin 显示出惊人的相似性（Faye 和 Kanost 1998）。最近发现的 Hemolin 的免疫功能会在下面有关章节进行描述。Hemolin 结合细菌表面的脂多糖和脂膜酸，也结合到血细胞表面，提供了一个促进细菌吞噬的调理素的功能（Zhao 和 Kanost 1996；Yu 和 Kanost 2002）。Hemolin 已经显示出其在形成细胞瘤从而捕获和杀死细菌的作用（Eleftherianos et al. 2007）。Hemolin 和血细胞表面结合的免疫相关功能也将在下面进行讨论。

在烟草天蛾中纯化和克隆了一个和 Toll 受体相关的蛋白（Zhu 2001；Zhu et al. 2003a）。它由 12 个亮氨酸重复组成，与 Toll 的胞外配体结合区域相似，命名为 Leureptin。Leureptin 结合于革兰氏阴性菌表面，连接血细胞表面，暗示其在血细胞对细菌感染应答反应中的一个功能。细菌感染后，Leureptin 在脂肪体的 mRNA 水平上调，而血浆中的蛋白含量下降，暗示其在免疫应答中可能被消耗以完成其功能。尚未在其他昆虫中发现 Leureptin。

一个编码具有 reeler 结构域（PFAM domain 02014）的 16kDa 的蛋白，在烟草天蛾幼虫脂肪体注射细菌后，该基因表达量明显上调（Zhu et al. 2003a）。这个命名为“诱导蛋白1”的蛋白，与美国白蛾 *H. cunea* 的免疫诱导蛋白 Hdd1（Shin et al. 1998）很相似。尽管 reeler 结构域出现在哺乳动物蛋白中，但它们的功能还不清楚。最近在天蚕蛾中鉴定了一个命名为 nodular 的同源蛋白，它能结合细菌，形成细胞瘤来应对感染（Gandhe，John 和 Nagaraju 2007）。因此，看起来，它在鳞翅目脂肪体中是主要的免疫转录组成成分，具有 reeler 结构域的血浆蛋白，可能是促使小瘤形成的模式识别蛋白。在其他昆虫基因组中存在的类似蛋白还没有在实验水平上进行功能验证。

3. 抗菌肽和抗菌蛋白

采用抑制性消减杂交技术检测烟草天蛾的脂肪体在细菌感染后基因表达情况，检测出包括有杀菌活性肽相关序列在内的许多上调基因(Zhu et al. 2003a)。采用这种方法鉴定的抗菌分子包括先前已经研究过的在感染后诱导表达的溶菌酶(Kanost, Dai 和 Dunn 1988; Mulnix 和 Dunn 1994, 1995)。在诱导上调的序列中也鉴定出抗菌肽几个家族的新成员。烟草天蛾最少产生了 5 个 cecropins(Dickinson, Russell 和 Dunn 1988)，4 个 attacins，2 个 lebocins，1 个 gloverin 和 1 个 moricin(Zhu et al. 2003a)。这些在血淋巴中能裂解细菌的肽和溶菌酶复合物能提供持续两至三天有力的抗菌保护，这种保护能力取决于注射细菌的剂量。

4. 前酚氧化物酶的激活

由微生物或昆虫多细胞寄生虫感染而普遍存在的保守反应是在侵入的微生物表面包裹黑色素。激活的酚氧化酶把血液中的多巴胺氧化，产生的醌进一步反应生成不溶的黑色素(Nappi 和 Christensen 2005)。酚氧化酶在血淋巴中以前体形式(前酚氧化酶)存在，通过在特异位点水解而激活(Ashida 和 Brey 1997)。昆虫免疫应答的酚氧化酶激活方面的研究最近已经进行了综述总结(Kanost 和 Gorman 2008)。

在烟草天蛾中鉴定出两个前酚氧化物酶基因(proPO)，其蛋白质在血浆中以二聚体形式存在(Hall et al. 1995; Jiang et al. 1997)。烟草天蛾的 proPO 在类绛色细胞中组成性合成(Jiang et al. 1997; Gorman, An 和 Kanost 2007)，由于缺少信号肽，所以通过裂解细胞释放蛋白。最近通过对另一种鳞翅目昆虫甜菜夜蛾(*Spodoptera exigua*)的研究表明，这种类绛色细胞破裂是由类花生酸推动的(Shrestha 和 Kim 2008)。

在烟草天蛾中鉴定出 3 个酚氧化酶激活蛋白酶(PAPs)(Jiang, Wang 和 Kanost 1998; Jiang et al. 2003a, b)。它们以酶原形式(proPAP 1－3)存在，不同的蛋白酶在特异位点切割得到有活性的酶。在感染和损伤的应答中，蛋白酶级联路径末端被激活，这与哺乳动物的补体系统或血液凝集级联类似(图 14.1)。酚氧化酶激活蛋白酶前体(proPAP)包括一个氨基末端丝氨酸结构域和一个(proPAP-1)或两个(proPAP-2, proPAP-3)氨基末端 clip 结构域。Clip 结构域有 35～55 个氨基酸包括 3 对保守的二硫键，可能介导蛋白酶级联通路中蛋白相互作用。蛋白酶可能包括一到多个氨基末端 clip 结构域，由一个 20～100 残基的连接序列连接催化蛋白酶结构域。具有氨基末端 clip 结构域的蛋白酶在免疫应答和发育中起作用(Jiang 和 Kanost 2000)，氨基末端 clip 结构域是节肢动物特有的结构。烟草天蛾 PAP-2 clip 结构域的 3D 结构(Huang et al. 2007)解析对它们的功能研究提供了重要的信息。

另一类包括氨基末端 clip 结构域的蛋白是丝氨酸蛋白酶类似物(SPH)，因为它们与 S1(胰凝乳蛋白酶)家族的丝氨酸蛋白酶类似，但因催化位点的丝氨酸突变成其他氨基酸，通常是甘氨酸，从而导致水解活性缺失。PAPs 在没有 SPH 存在的情况下可以完全水解小肽底物，但要有效地激活 proPO，需要与 SPH1 和 SPH2 相互作用（Yu et al. 2003）。SPH 是 proPO 激活的辅助因子，活性形式的 SPH 被血淋巴的丝氨酸蛋白酶激活（Yu et al. 2003；Lu 和 Jiang 2008）。烟草天蛾的 SPHs 刺激 proPO 激活，结合免疫凝集素-2，proPO 和 PAP（Yu et al. 2003），形成大的多聚体（～800kDa）（Wang 和 Jiang 2004a）。凝集素和 proPO 的激活复合物可能定位在侵入细菌的表面合成黑色素。烟草天蛾活化的 SPH 结合 proPO 去激活 PAPs，但机制尚不清楚（Wang 和 Jiang 2004a；Gupta，Wang 和 Jiang 2005b）。

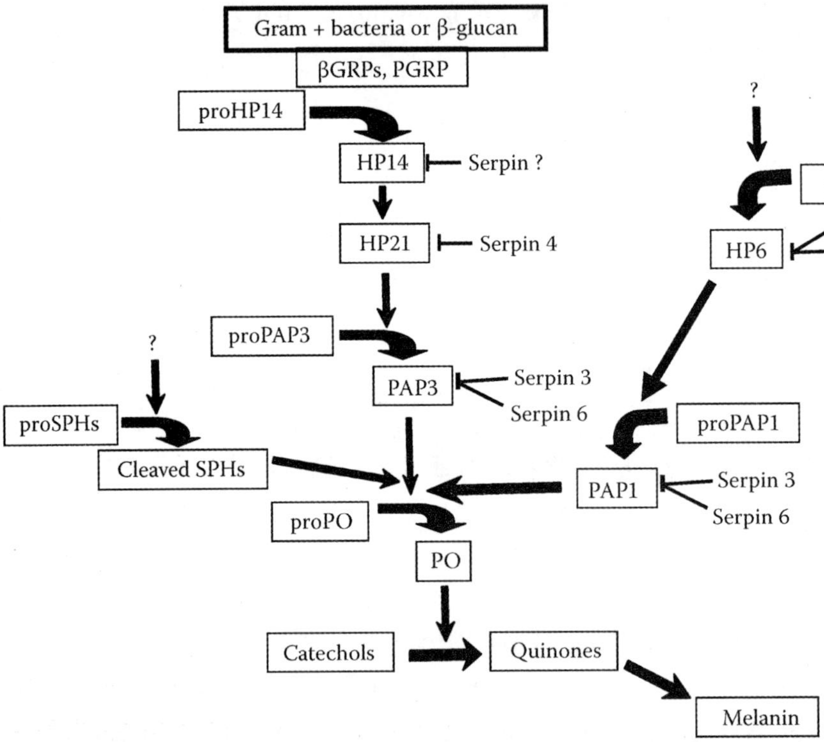

图 14.1 烟草天蛾酚氧化物酶激活路径模型。在血浆中，包括 β-葡聚糖识别蛋白、肽聚糖识别蛋白在内的模式识别蛋白，结合到微生物表面的多糖上，促使血淋巴中蛋白酶级联反应的激活。起始激活蛋白，血浆蛋白 14（HP14），在目前还不清楚的机制下自激活，然后激活血浆蛋白 21（HP21），HP21 再切割激活前酚氧化物酶激活蛋白酶 3（PAP3）。前酚氧化物酶激活蛋白酶 2（PAP2）也可以在相同的机制下被激活。PAP3 与水解激活的丝氨酸蛋白酶同源物（SPH）相互作用，切割激活前酚氧化物酶（proPO）。前酚氧化物酶激活蛋白酶 1（PAP1）是通过级联反应的一个分支，被血液蛋白酶 6（HP6）所激活。血浆丝氨酸蛋白酶抑制剂同过抑制蛋白酶来调控这些信号通路，从而将生理上的黑化反应限制在一定时间的局部范围内。

除了 PAPs，烟草天蛾至少还有 11 个在脂肪体和血细胞中表达的具有 clip 结构域的蛋白酶（Jiang et al. 2005）。我们已经开始明白其中一些的功能。ProPAP-2 和 ProPAP-3 被具有一个 clip 结构域的 HP21 激活（Gorman，An 和 Kanost 2007；Wang 和 Jiang

2007)。HP21 被起始这个级联反应的具有多复杂结构的 HP14 激活。在对革兰氏阳性菌或 β-1,3-葡聚糖的反应中,HP14 引起 proPO 的激活(Ji et al. 2004)。HP14 包括一个羧基末端蛋白酶结构,5 个低密度脂蛋白受体 class-A 重复,一个 sushi 结构域和一个独特的富含半胱氨酸的区域。它在 β-1,3-葡聚糖和葡聚糖受体存在下自动激活(Wang 和 Jiang 2006)。果蝇和按蚊 HP14 的同源序列(ortholog),也被认为是起始级联反应的蛋白酶。

那些产生双酚底物并被酚氧化物酶氧化的酶,是酚氧化物酶和黑化途径另外的成分。包括酪氨酸羟化酶、DOPA 脱羧酶、黑色素多聚体前体及抗菌氧化产物。在用细菌注射烟草天蛾后,酪氨酸羟化酶在脂肪体和血细胞中上调表达(Gorman,An 和 Kanost 2007),从而增加 DOPA 的产生。DOPA 脱羧酶在感染后的脂肪体中高诱导表达(Zhu et al. 2003a),它能产生被 PO 有效氧化的底物-多巴胺(Zhao et al. 2007)。

5. 蛋白酶抑制剂

昆虫血淋巴中含有不同基因家族的高浓度丝氨酸蛋白酶抑制剂,包括已知的 serpins 类的 45kDa 蛋白酶抑制剂(Kanost 1999,2007; Silverman et al. 2001)。烟草天蛾中发现 7 个 serpin 基因(Jiang et al. 1996; Gan et al. 2001; Zhu et al. 2003b; Tong 和 Kanost 2005; Tong,Jiang 和 Kanost 2005; Zou 和 Jiang 2005b),部分与靶标蛋白酶相互作用的 serpin 活性位点是暴露在 serpin 序列羧基末端附近的环。一些昆虫的 serpin 基因具有独特的结构,通过编码活性环的一个外显子相互地选择性剪接,产生具有不同选择性的几个抑制剂。这种现象首先在烟草天蛾的 serpin-1 中观察到,serpin-1 基因的第九外显子有 12 个拷贝。第九外显子的每一个版本编码一个不同的活性环,抑制一系列不同的蛋白酶(Jiang,Wang 和 Kanost 1994; Jiang et al. 1996; Jiang 和 Kanost 1997)。两个 serpin-1 剪接体的结构已经进行了 X-ray 晶体解析(Li et al. 1999; Ye et al. 2001)。在其他几种昆虫中也发现与烟草天蛾 serpin-1 类似的在同样位点具有选择性的外显子(Kanost 和 Clarke 2005)。

3 个烟草天蛾的血淋巴丝氨酸蛋白酶抑制剂 serpin (serpin-1J,serpin-3,serpin-6)通过抑制 PAPs 来调节其活性(Jiang et al. 2003b; Zhu et al. 2003b; Wang 和 Jiang 2004b)。重组的 serpin-4 和 serpin-5 能阻断 proPO 的活性,但不会抑制 PAPs,暗示它们是抑制 PAPs 上游的蛋白酶(Tong 和 Kanost 2005)。serpin-4 抑制 HP1、HP6 和 HP21。serpin-5 抑制 HP1 和 HP6(Tong,Jiang 和 Kanost 2005)。由于 serpin-4 和 serpin-5 都能抑制血浆中 proPO 的活性,HP1、HP6 和 HP21 便是 proPO 激活途径的候选成分。这与 HP21 能够激活 proPAP-2 和 proPAP-3 的观察结果是相一致的(Gorman,An 和 Kanost 2007; Wang 和 Jiang 2007)。

6. 血细胞介导的免疫应答:从非黏附到黏附态的转换

血细胞参与吞噬、瘤化、包囊作用等免疫应答。这些需要细胞结合到病原菌表面或细胞黏附在一起。瘤化聚集血细胞,捕捉大量的细菌。包囊作用是血细胞片层黏附到外源物上形成的紧密结构,从而包围杀死寄生虫。瘤化和包囊作用通常会产生黑化,暗示细胞免疫和体液免疫的协调反应(Stanley,Miller 和 Howard 1998; Yu 和 Kanost 2004; Rodríguez-Pérez et al. 2005)。两种类型的血细胞、颗粒细胞和浆血细胞参与了烟草天蛾的细胞免疫。它们可以通过特异的凝集素和抗体标记(Nardi et al. 2003,2005,2006; Willott et al. 1994)以及不同的细胞倍性水平(图 14.2)轻易地进行区分。在细胞介导的免疫应答中,游离的非黏性血细胞转变为黏性的伸展细胞(图 14.3)。正如 Matzinger (2002)假设的外源信号(PAMP)或损伤、应激细胞释放的信号,将激活昆虫免疫系统的细胞。类花生酸就是这类危险信号,能推动血细胞的聚集和瘤的形成(Stanley 2006)。另外,颗粒细胞释放的蛋白或血浆蛋白的产物,能引起浆血细胞的细胞免疫:包括吞噬、包囊作用、瘤化(Ratcliffe 和 Gagen 1977; Schmit 和 Ratcliffe 1977; Gillespie,Kanost 和 Trenczek 1997; Loret 和 Strand 1998; Lavine 和 Strand 2002; Dean et al. 2004)。在烟草天蛾和其他鳞翅目昆虫中,一个类似细胞因子肽的血细胞延伸肽也能刺激浆血细胞的黏附和延伸(Clark,Pech 和 Strand 1997; Wang,Jiang 和 Kanost 1999)。伸展肽以前体形式存在于血浆中,当细胞受到损伤后,被特异蛋白水解而激活,结合到浆血细胞表面受体上发挥作用(Clark et al. 2004)。

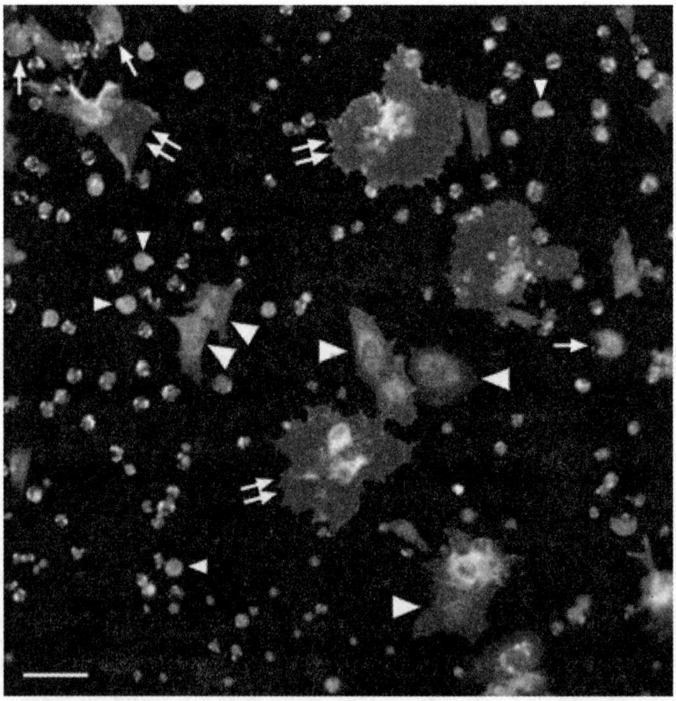

图 14.2 烟草天蛾血细胞在 Grace's 培养基中培养 60min(彩色版本图片见附录图 14.2)。一些重要的特征如下: DAPI(蓝色)将浆血细胞大的多倍体核与小的颗粒细胞的二倍体核区别开来。PNA(花生凝集素)- 若丹明(橙红色)标记血凝素(hemocytin)。注意小的未伸展的浆血细胞覆盖有若丹明和 beta1-整合素(小箭头),而伸展到底物上的浆血细胞则没被若丹明覆盖。一般标记颗粒细胞用若丹明。Alexa Fluor 488(绿色)标记大的(双箭头)和小的(大箭头)伸展浆血细胞的 beta1-整合素。Alexa Fluor 647(红紫色)标记大的伸展细胞的神经胶质蛋白和大的球状、未伸展的浆血细胞(小箭头)。参照条表示 20μm。

6.1 血细胞表面黏附蛋白及与颗粒细胞释放蛋白的相互作用

黏附细胞的几个超家族控制昆虫和哺乳动物免疫细胞的细胞应答。细胞黏附分子控制哺乳动物细胞的相互作用：包括炎症反应时白细胞和内皮细胞相互作用，T 细胞和抗原呈递细胞（巨噬细胞，树状细胞）相互作用，激活 T 淋巴细胞，起始获得性免疫（Aderem 和 Ulevitch 2000；Janeway et al. 2001）。以下介绍在鳞翅目昆虫中发现的 3 个家族的细胞黏性分子。

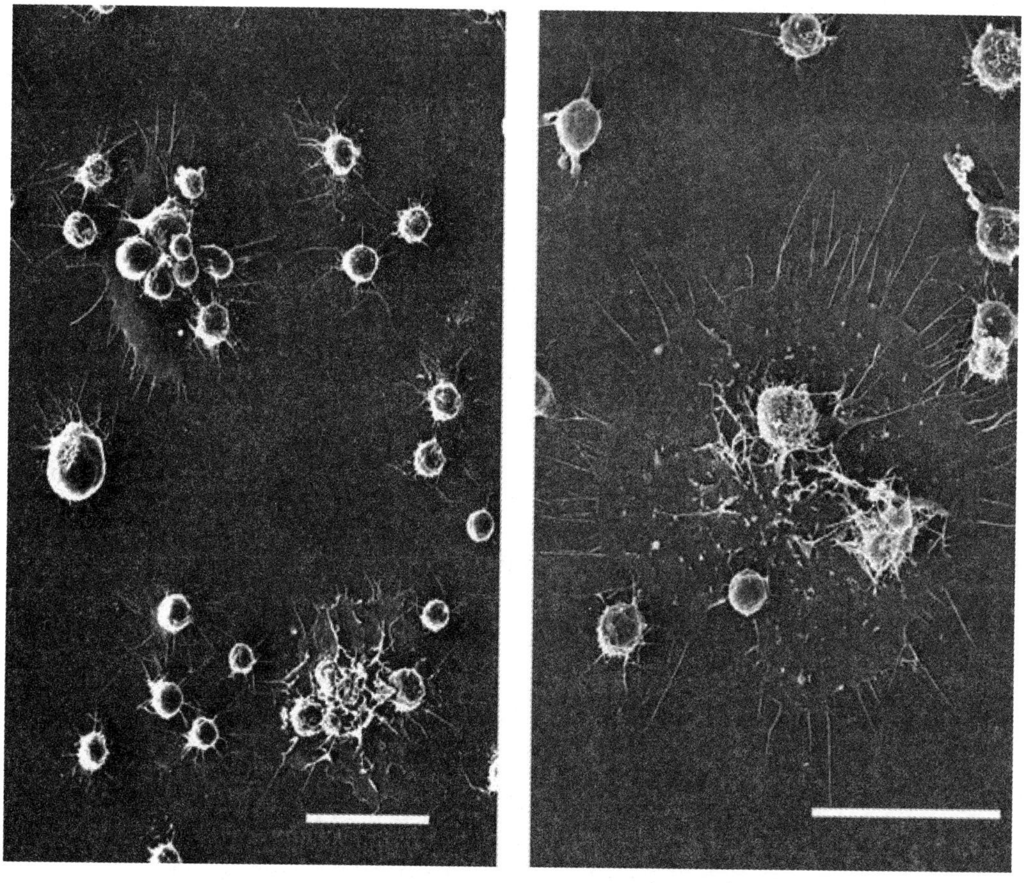

图 14.3　扫描电镜展示烟草天蛾血细胞的黏附性和伸展性。大的浆血细胞在外源物表面伸展，并形成一个黏附中心，使其他浆血细胞和颗粒细胞吸附并生出微丝扩展开来。

6.1.1 整合素

昆虫血细胞整合素的功能研究刚起步（Lavine 和 Strand 2003；Irving et al. 2005；Levin et al. 2005；Moita et al. 2006）。大豆夜蛾（*P. includens*）血细胞上已鉴定的多整合素亚基包括：3 个 α 亚基（αPi1－3）和 1 个 β 亚基（βPi1）（Lavine 和 Strand 2003）。在包囊作用中，大豆夜蛾的两个亚基（αPi2 和 βPi1）表达增加，暗示一到多个整合素在调控血细胞黏附中起着重要作用。2 个标记烟草天蛾血浆蛋白，阻断延伸和包囊作用的单抗

(MS13,MS34)(Willott et al. 1994; Wiegand et al. 2000),能够识别β整合素(β1)的配体结合位点(Levin et al. 2005)。通过抗体MS13结合整合素β1,来纯化血细胞膜上相应的整合素异源二聚体,从而产生α亚基(α1)。这个整合素(α1β1)只在血细胞中表达,称为血细胞特异整合素(HS)(Zhuang et al. 2007b)。通过上面提到的特异β1抗体和β1-整合素的小分子RNA(siRNA)能阻断包囊作用,证明了HS整合素对细胞免疫应答的重要性(Levin et al. 2005)。对α整合素保守结构域用简并引物进行PCR扩增,得到多于两个α亚基的序列(Zhuang et al. 2008)。

不同血细胞表达不同的整合素可能反映了血细胞的功能分化。三种不同血细胞上β1-整合素时空表达的显著差异可能与烟草天蛾中血细胞存在的三种不同的α亚基有关。不同血细胞的定向吸附性与整合素的时空表达模式有关。(1)小浆血蛋白同时吸附于其他细胞及底物上。这些非吸附和吸附血细胞都被β1-整合素抗体标记。(2)颗粒细胞吸附底物和浆血细胞,不吸附其他颗粒细胞;这些血细胞能被β1-整合素抗体标记为吸附血细胞,而不是游离的非吸附血细胞。(3)大的神经胶质蛋白-阳性浆血蛋白在延伸和吸附其他浆血蛋白和颗粒细胞前首先吸附底物。只有在开始延伸到底物时,浆血蛋白才被β1-整合素抗体标记。因此免疫标记的整合素(用来标记)在细胞激活过程中是伴随着自身构象变化的(Hynes 1992)。

6.1.2 Tetraspanins

在脊椎动物和昆虫中,血吸虫表膜蛋白Tetraspanin在发育和免疫应答中介导细胞相互作用(Hemler, Mannion 和 Berditchevski 1996; Kopczynski, Davis 和 Goodman 1996; Maecker, Todd 和 Levy 1997; Berditchevski 2001; Boucheix 和 Rubinstein 2001; Zhuang et al. 2007b)。基于EST数据,在烟草天蛾Tetraspanin超家族中发现四个成员(Todres, Nardi 和 Robertson 2000)。在果蝇基因组,存在这个超家族37个完整膜蛋白(Todres, Nardi 和 Robertson 2000)。烟草天蛾有4个已研究的Tetraspanin,对血细胞表面表达的一个蛋白(D76)进行功能分析,表明这个蛋白的胞外环(LEL)结合到整合素HS上(Zhuang et al. 2007b)。用单抗和dsRNA阻断Tetraspanin和HS整合素的功能,及对两个蛋白的结合实验,暗示整合素和Tetraspanin在血细胞表面呈反式相互作用。这种鳞翅目血细胞的反式相互作用与已知的哺乳动物Tetraspanin和整合素α亚基的顺式相互作用相反(Zhuang et al. 2007b)。

6.1.3 神经胶质蛋白

在烟草天蛾中,神经细胞吸附蛋白即神经胶质蛋白有6个免疫球蛋白结构域。该蛋白不仅在神经系统中表达,(Chen et al. 1997)在那些为血细胞聚集的生成细胞的血细胞亚型中也有表达(Nardi et al. 2006)。这些血细胞表面的神经胶质蛋白以同嗜性的方式与相对的血细胞表面神经胶质蛋白结合,也可以异嗜性的方式与HS整合素结合。与其他免疫球蛋白(Ig)超家族的成员一样,神经胶质蛋白也可以作为整合素的配基。比如,哺乳动物T细胞的整合素,至少可以结合Ig超家族中的三个蛋白:ICAM-1,ICAM-2,ICAM-3(Janeway et al. 2001)。用单抗或双链RNA(dsRNA)阻碍血细胞表面神经胶质蛋白的同嗜性、异嗜性相互作用，阻断了烟草天蛾血细胞的包囊作用(Zhuang et al. 2007a)，这就暗示神经胶质蛋白参与了血细胞包囊化的细胞吸附事件。

我们推测抑血细胞聚集素(hemolin)可能作为神经胶质蛋白和 HS 整合素的配基起作用。这可以解释 hemolin 妨碍血细胞吸附性的能力(Ladendorff 和 Kanost 1991)，是通过竞争性的与神经胶质蛋白在邻近细胞的同嗜性结合或干扰神经胶质蛋白和 HS 整合素的结合，从而达到阻止血细胞聚集的作用。这种相互作用也可解释 hemolin 促进吞噬的调理素作用，在结合细菌的同时，通过与 HS 整合素或神经胶质蛋白相互作用结合到血细胞表面。

6.2 颗粒细胞释放蛋白促进浆血细胞的吸附

颗粒细胞的颗粒释放蛋白看起来促进了瘤的形成和包囊作用(Ratcliffe 和 Gagen 1977; Schmit 和 Ratcliffe 1977; Pech 和 Strand 1996; Loret 和 Strand 1998)。我们正在尝试从烟草天蛾颗粒细胞中鉴定这些蛋白，并研究在血细胞吸附中的作用。现在已经找到两个候选蛋白。Lacunin 是颗粒细胞产生的大的胞外基质蛋白(Nardi，Gao 和 Kanost 2001)。幼虫与细菌的 PAMP 孵育能诱导其表达(M. R. Kanost 未发表)。它从颗粒细胞释放并结合到浆血细胞表面(Nardi et al. 2005)，定位于底层和黏附浆血细胞表面的中间层(Nardi et al. 2005)。在聚集体中，以 β1-整合素抗体标记均值细胞表面，沿着细胞轮廓，可以看出这个蛋白斑片状分布于黏附的浆血蛋白表面。因此，颗粒细胞释放 Lacunin 可能促进浆血细胞的吸附。

最早在家蚕中发现的 340kDa 的高分子蛋白 hemocytin(Kotani et al. 1995)是颗粒细胞产生，同样在烟草天蛾中也存在该蛋白(Scholz 2002; J. B. Nardi，未发表数据)。Hemocytin 结合血细胞，在蛾和蝇中参与了血淋巴凝结。此蛋白被损伤强烈诱导，在血淋巴凝结中处于优势地位(Karlsson et al. 2004; Scherfer et al. 2004)，同时可作为免疫激活的信号。在果蝇中，用 RNA 干扰技术(RNAi)抑制 hemolectin(果蝇 hemocytin 的相似物)的表达，能阻止凝血的发生(Goto et al. 2003)。

在损伤响应中，血细胞释放血凝素(hemocytin)和 lacunin，并结合到浆血细胞表面(Li et al. 2002; Theopold et al. 2004; J. B. Nardi，未发表; 图 14.4)。这两个颗粒细胞释放蛋白结构域的信息为血细胞表面可能存在的受体提供了预测信息。它们都有至少一个或是多达三个的以下结构域：凝血酶致酶蛋白 1，免疫球蛋白，von willebrand 因子和 Arg-Gly-Asp 基序。每一个结构域都具有潜在的作为配基与整合素结合的功能(Humphries，Byron 和 Humphries 2006)。

颗粒细胞和/或脂肪体细胞释放的某些蛋白被推测在血细胞表面作为多价配基相互交联，并形成受体多聚类物。多价配体蛋白结合表面受体，比如整合素、神经胶质蛋白(Nardi et al. 2005)，会诱导其在膜上聚集。受体多聚类物能促进表面受体的结合能力，比如整合素、神经胶质蛋白，将细胞从非吸附向吸附细胞转化(Pech 和 Strand 1996; Xavier et al. 1998; Corinti et al. 1999; van Kooyk 和 Figdor 2000; Fahmy et al. 2001;

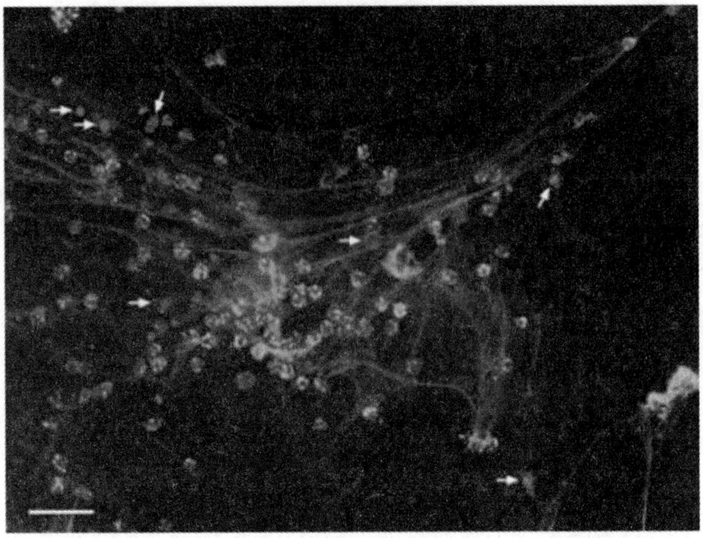

图 14.4 在凝结的烟草天蛾血细胞中标记黏附分子(彩色版本图片见附录图 14.4)。在固定和免疫标记以前,将最后一龄幼虫血细胞在无菌的盖玻片上放置 5min。(花生凝集素 FITC 标记血凝素)绿色标记颗粒细胞及其周围的纤维和基质。小的浆血细胞用 TRITC-MS13(抗 β1-整合素的抗体)标记为红色。细胞核用 DAPI 标记为蓝色。大的细胞核(箭头)是浆血细胞的。聚集处(图片中心的两个红色块)的两个明显的点包括颗粒细胞(绿色颗粒,二倍体核),小浆血细胞(红色的膜,多倍体)和大的、神经胶质蛋白阳性的浆血细胞(淡绿色标记,非红色,多倍体)。参照条表示 20μm。

Hynes 2003; Lavine 和 Strand 2003; Nardi et al. 2005)。受体多聚类物也可能激活多信号通路,影响多种细胞蛋白活性:包括各种蛋白激酶、细胞骨架构建、离子通道、生长受体(Giancotti 和 Ruoslahti 1999)。

脊椎动物中存在的游离黏着受体、非黏附血小板和白细胞通常在它们与配基结合前处于失活状态。激活过程伴随着整合素分子的构象变化(Hynes 1992)。类似的现象可能在烟草天蛾的非黏附血细胞中发生。尽管 β1-整合素参与了包囊作用(Wiegand et al. 2000; Levin et al. 2005),但只有血细胞暴露于非自我表面才能像一个黏附受体。多价 PRR 配基可能在血细胞表面聚集整合素,将具有未激活的整合素受体的非黏附细胞向激活受体的黏附细胞转化。

在血细胞的黏附性转化中,颗粒细胞释放蛋白并结合到所有血细胞表面。在培养基中,几分钟内就能检测到浆血细胞表面与蛋白的结合。在颗粒细胞表面观察到结合的蛋白以环形斑排列(Nardi et al. 2005)。这些斑可能代表哺乳动物免疫球蛋白突触进化的前身(Davis 2002)。

7. 结论和展望

烟草天蛾和其他昆虫的免疫应答是一个保护昆虫不受外界感染的复杂而多反应的免疫系统。血细胞的活动很快,能在感染后 1h 内明显地减少病原物的数量。酚氧化物酶激活由于使用组成性表达产生血浆蛋白，反应十分迅速。再通过蛋白酶级联途径放大起始信号，几分钟内便在病原菌或寄生虫表面产生活性醌和黑色素。几小时内就可诱导合成抗菌肽、抗菌蛋白及其他免疫应答蛋白，清除那些逃脱血细胞防御的微生物，并且在以后几天内保护机体免受感染。我们正在开展研究来了解烟草天蛾中参与血细胞黏附病原菌、胞外基质和其他血细胞的分子，以及其他很多在免疫应答中起作用的血浆蛋白。然而，这些研究尤其是对调控免疫应答的了解，还远远不够。

进一步的研究包括更全面地鉴定在免疫中起作用的基因，这需要烟草天蛾基因组测序和多方面的实验工作。尽管我们已经鉴定了由细菌和真菌激发的模式识别受体，但是对于如何启动对真核寄生虫和病毒感染的免疫应答，仍然知之甚少。在发育，特别是胚胎发育和变态过程中免疫系统的变化，应该是烟草天蛾作为一个模式系统重点研究的地方，而且现在已经有了一个小规模的但很有趣的开始(Russell 和 Dunn 1996; Nardi et al. 2003; Nardi 2004; Gorman，Kankanala 和 Kanost2004; Beetz et al. 2008)。此外，在今后的研究中我们需要明确免疫应答如何被调控(如细胞因子的作用)，并探究多个免疫途径相互关联和交叉的程度。

8. 致谢

著者实验室的研究工作曾受到 National Institutes of Health 基金 GM41247 和 HL64657 的资助。伊利诺大学贝克曼实验室图像技术小组的 Charles Mark Bee 帮助完成了血球的图像处理。

参考文献

[1]Aderem，A. and R. J. Ulevitch. 2000. Toll-like receptors in the induction of the innate immuine response. *Nature* 406:782－787.

[2]Ao，J. Q.，E. Ling，and X. Q. Yu. 2008. A Toll receptor from *Manduca sexta* is in response to *Escherichia coli* infection. *Mol. Immunol.* 45:543－552.

[3]Asgari，S.，U. Theopold，C. Wellby，and O. Schmidt. 1998. A protein with protective properties against the cellular defense reactions in insects. *Proc. Natl. Acad. Sci. U. S. A.* 95:3690－3695.

[4]Ashida，M. and P. Brey. 1997. Recent advances in research on the insect prophenoloxidase cascade. In *Molecular mechanisms of immune responses in insects*, ed. P. Brey and D. Hultmark，135－171. New York: Chapman and Hall.

[5]Beck，M.，and M. R. Strand. 2005. Glc1. 8 from *Microplitis demolitor* bracovirus induces a loss of adhesion and phagocytosis in insect High Five and S2 cells. *J. Virol.* 79:1861－1870.

[6]Beetz，S.，T. K. Holthusen，J. Koolman，and T. Trenczek. 2008. Correlation of hemocyte counts with different developmental parameters during the last larval instar of the tobacco hornworm，*Manduca sexta*. *Arch. Insect Biochem. Physiol.* 67:63－75.

[7]Berditchevski，F. 2001. Complexes of tetraspanins with integrins: More than meets the eye. *J. Cell*

Sci. 114:4143—4151.

[8]Boman, H. G. , I. Faye, G. H. Gudmundsson, J. Y. Lee, and D. A. Lidholm. 1991. Cell-Free immunity in Cecropia—A model system for antibacterial proteins. *Eur. J. Biochem.* 201:23—31.

[9]Boucheix, C. , and E. Rubinstein. 2001. Tetraspanins. *Cell. Mol. Life Sci.* 58:1189—1205.

[10]Chen, C. L. , D. J. Lampe, H. M. Robertson, and J. B. Nardi. 1997. Neuroglian is expressed on cells destined to form the prothoracic glands of *Manduca* embryos as they segregate from surrounding cells and rearrange during morphogenesis. *Dev. Biol.* 181:1—13.

[11]Clark K. D. , S. F Garczynski, A. Arora, J. W. Crim, and M. R. Strand. 2004. Specific residues in plasmatocyte-spreading peptide are required for receptor binding and functional antagonism of insect immune cells. *J. Biol. Chem.* 279:33246—33252.

[12]Clark, K. D. , L. L. Pech, and M. R. Strand. 1997. Isolation and identification of a plasmatocyte-spreading peptide from the hemolymph of the lepidopteran insect *Pseudoplusia includens*. *J. Biol. Chem.* 272:23440—23447.

[13]Corinti, S. , E. Fanales-Belasio, C. Albanesi, A. Cavani, P. Angelisova, and G. Girolomoni. 1999. Cross-linking of membrane CD43 mediates dendritic cell maturation. *J. Immunol.* 162:6331—6336.

[14]Davies, D. H. , and S. B Vinson. 1986. Passive evasion by eggs of braconid parasitoid Cardiochiles nigriceps of encapsulation in vitro by haemocytes of host *Heliothis virescens*. Possible role for fibrous layer in immunity. *J. Insect Physiol.* 32:1003—1010.

[15]Davis, D. M. 2002. Assembly of the immunological synapse for T cells and NK cells. *Trends Immunol.* 23:356—363.

[16]Dean, P. , U. Potter, E. H. Richards, J. P Edwards, A. K. Charnley, and S. E. Reynolds. 2004. Hyperphagocytic haemocytes in *Manduca sexta*. *J. Insect Physiol.* 50:1027—1036.

[17]Dickinson, L. , V. W. Russell, and P. E. Dunn. 1988. A family of bacteria-regulated cecropin D-like peptides from *Manduca sexta*. *J. Biol. Chem.* 263:19424—19429.

[18]Eleftherianos, I. , F. Gokcen, G. Felfoldi, et al. 2007. The immunoglobulin family protein Hemolin mediates cellular immune responses to bacteria in the insect *Manduca sexta*. *Cellular Microbiol.* 9:1137—1147.

[19]Eleftherianos, I. , P. J. Millichap, R. H. ffrench-Constant, and S. E. Reynolds. 2006. RNAi suppression of recognition protein mediated immune responses in the tobacco hornworm *Manduca sexta* causes increased susceptibility to the insect pathogen Photorhabdus. *Dev. Comp. Immunol.* 30:1099—1107.

[20]Fabrick, J. A. , J. E. Baker, and M. R. Kanost. 2004. Innate immunity in a pyralid moth: Functional evaluation of domains from a β-1,3-glucan recognition protein. *J. Biol. Chem.* 279:26605—26611.

[21]Fahmy, T. M. , J. G. Bieler, M. Edidin, and J. P. Schneck. 2001. Increased TCR Avidity after T cell activation: A mechanism for sensing low-density antigen. *Immunity* 14:135—143.

[22]Faye, I. , and M. R. Kanost. 1998. Function and regulation of hemolin. In *Molecular mechanisms of immune responses in insects*, ed. P. T. Brey and D. D. Hultmark, 173—188. New York: Chapman and Hall.

[23]Ferrandon, D. , J. —L. Imler, C. Hetru, and J. A. Hoffmann. 2007. The *Drosophila* systemic immune response: Sensing and signaling during bacterial and fungal infections. *Nature Rev. Immunol.* 7:862—874.

[24]Finnerty, C. M., and R. R. Granados. 1997. The plasma protein scolexin from *Manduca sexta* is induced by Baculovirus infection and other immune challenges. *Insect Biochem. Mol. Biol.* 27:1—7.

[25]Finnerty, C. M., P. A. Karplus, and R. R. Granados. 1999. The insect immune protein scolexin is a novel serine proteinase homolog. *Protein Sci.* 8:242—248.

[26]Galibert, L., J. Rocher, M. Ravallec, M. Duonon-Cerutti, B. A. Webb, and A. N. Volkoff. 2003. Two *Hyposoter didymator* ichnovirus genes expressed in the lepidopteran host encode secreted or membrane-associated serine-and threonine-rich proteins in segments that may be nested. *J. Insect Physiol.* 49:441—451.

[27]Gan, H., Y. Wang, H. B. Jiang, K. Mita, and M. R. Kanost. 2001. A bacteria-induced, intracellular serpin in granular hemocytes of *Manduca sexta*. *Insect Biochem. Mol. Biol.* 31:887—898.

[28]Gandhe, A. S., S. H. John, and J. Nagaraju. 2007. Noduler, a novel immune up-regulated protein mediates nodulation response in insects. *J. Immunol.* 179:6943—6951.

[29]Giancotti, F. G., and E. Ruoslahti. 1999. Integrin signaling. *Science* 285:1028—1032.

[30]Gillespie, J. P., M. R Kanost, and T. Trenczek. 1997. Biological mediators of insect immunity. *Annu. Rev. Entomol.* 42:611—643.

[31]Gorman, M. J., C. J. An, and M. R. Kanost. 2007. Characterization of tyrosine hydroxylase from *Manduca sexta*. *Insect Biochem. Mol. Biol.* 37:1327—1337.

[32]Gorman, M. J., P. Kankanala, and M. R. Kanost. 2004. Bacterial challenge stimulates innate immune responses in extra-embryonic tissues of tobacco hornworm eggs. *Insect Mol. Biol.* 13:19—24.

[33]Gorman, M. J., Y. Wang, H. B. Jiang, and M. R. Kanost. 2007. *Manduca sexta* hemolymph proteinase 21 activates prophenoloxidase-activating proteinase 3 in an insect innate immune response proteinase cascade. *J. Biol. Chem.* 282:11742—11749.

[34]Goto, A., T. Kumagai, C. Kumagai, et al. 2003. *Drosophila* hemolectin gene is expressed in embryonic and larval hemocytes and its knockdown causes bleeding defects. *Dev. Biol.* 264:582—591.

[35]Gupta, S., Y. Wang, and H. B. Jiang. 2005a. Purification and characterization of *Manduca sexta* prophenoloxidase-activating proteinase-1, an enzyme involved in insect immune responses. *Protein Express. Purif.* 39:261—268.

[36]Gupta, S., Y. Wang, and H. B. Jiang. 2005b. *Manduca sexta* prophenoloxidase (proPO) activation requires proPO-activating proteinase (PAP) and serine proteinase homologs (SPHs) simultaneously. *Insect Biochem. Mol. Biol.* 35:241—248.

[37]Hall, M., T. Scott, M. Sugumaran, K. Söderhäll, and J. H. Law. 1995. Proenzyme of *Manduca sexta* phenol oxidase: Puriication, activation, substrate specfiicity of the active enzyme, and molecular cloning. *Proc. Natl. Acad. Sci. U. S. A.* 92:7764—7768.

[38]Hemler, M., B. A. Mannion, and F. Berditchevski. 1996. Association of TM4SF proteins with integrins: Relevance to cancer. *Biochim. Biophys. Acta* 1287:67—71.

[39]Hiruma, K., M. S. Carter, and L. M. Riddiford. 1995. Characterization of the dopa decarboxylase gene of *Manduca sexta* and its suppression by 20-hydroxyecdysone. *Dev. Biol.* 169:195—209.

[40]Hoek, R. M., A. B. Smit, H. Frings, J. M. Vink, M. deJong-Brink, and W. P. M. Geraerts. 1996. A new Ig-superfamily member, molluscan defence molecule (MDM) from *Lymnaea stagnalis*, is down-regu-

lated during parasitosis. *Eur. J. Immunol.* 26:939—944.

[41]Huang,R. D. ,Z. Q. Lu,H. E. Dai,D. V. Velde,O. Prakash,and H. B. Jiang. 2007. The solution structure of clip domains from *Manduca sexta* prophenoloxidase activating proteinase-2. *Biochemistry* 46: 11431—11439.

[42]Humphries,J. D. , A. Byron, M. J. Humphries. 2006. Integrin ligands at a glance. *J. Cell Sci.* 119: 3901—3903.

[43]Hynes,R. O. 1992. Integrins: Versatility,modulation,and signaling in cell adhesion. *Cell* 69:11—25.

[43]Hynes,R. O. 2003. Changing partners. *Science* 300:755—756.

[45]Irving,P. ,J. M. Ubeda,D. Doucet,et al. 2005. New insights into *Drosophila* larval haemocyte functions through genome-wide analysis. *Cell. Microbiol.* 7:335—350.

[46]Janeway,C. A. ,and R. Medzhitov. 2002. Innate immune recognition. *Annu. Rev. Immunol.* 20: 197—216.

[47]Janeway,C. A. ,P. Travers,M. Walport,and M. J. Shlomchik. 2001. *Immunobiology: The immune system in health and disease*. New York: Garland Publishing.

[48]Ji,C. Y. ,Y. Wang,X. P. Guo,S. Hartson,and H. B. Jiang. 2004. A pattern recognition serine proteinase triggers the prophenoloxidase activation cascade in the tobacco hornworm,*Manduca sexta*. *J. Biol. Chem.* 279:34101—34106.

[49]Jiang, H. 2008. The biochemical basis of antimicrobial responses in Manduca sexta. *Insect Science*. 15:53—66.

[50]Jiang H. ,and M. R. Kanost. 1997. Characterization and functional analysis of twelve naturally occurring reactive site variants of serpin-1 from *Manduca sexta*. *J. Biol. Chem.* 272:1082—1087.

[51]Jiang,H. ,and M. R. Kanost. 2000. The clip-domain family of serine proteinases in arthropods. *Insect Biochem. Molec. Biol.* 30:95—105.

[52]Jiang,H. B. ,C. C. Ma,Z. Q. Lu,and M. R. Kanost. 2004. β-1,3-Glucan recognition protein-2 (β GRP-2) from *Manduca sexta*: An acute-phase protein that binds β-1,3-glucan and lipoteichoic acid to aggregate fungi and bacteria and stimulate prophenoloxidase activation. *Insect Biochem. Mol. Biol.* 34:89—100.

[53]Jiang H. B. ,A. B. Mulnix,and M. R. Kanost. 1995. Expression and characterization of recombinant *Manduca sexta* serpin-1B and site-directed mutants that change its inhibitory selectivity. *Insect Biochem. Molec. Biol.* 25:1093—1100.

[54]Jiang,H. B. ,Y. Wang,Y. L. Gu,et al. 2005. Molecular identification of a bevy of serine proteinases in *Manduca sexta* hemolymph. *Insect Biochem. Mol. Biol.* 35:931—943.

[55]Jiang H. ,Y. Wang,Y. Huang,et al. 1996. Organization of serpin gene-1 from *Manduca sexta*: Evolution of a family of alternate exons encoding the reactive site loop. *J. Biol. Chem.* 271:28017—28023.

[56]Jiang H,Y. Wang,and M. R. Kanost. 1994. Mutually exclusive exon use and reactive center diversity in insect serpins. *J. Biol. Chem.* 269:55—58.

[57]Jiang,H. ,Y. Wang,and M. R. Kanost. 1998. Pro-phenol oxidase activating proteinase from an nsect,*Manduca sexta*: A bacteria-inducible protein similar to *Drosophila* easter. *Proc. Natl. Acad. Sci. U. S. A.* 95:12220—12225.

[58]Jiang, H. , Y. Wang, C. Ma, and M. R. Kanost. 1997. Subunit composition of pro-phenol oxidase from *Manduca sexta*: Molecular cloning of subunit proPO-p1. *Insect Biochem. Molec. Biol.* 27:835—850.

[59]Jiang, H. B. , Y. Wang, X. Q. Yu, and M. R. Kanost. 2003a. Prophenoloxidase-activating proteinase-2 from hemolymph of *Manduca sexta*. *J. Biol. Chem.* 278:3552—3561.

[60]Jiang, H. B. , Y. Wang, X. Q. Yu, Y. F. Zhu, and M. Kanost. 2003b. Prophenoloxidase-activating proteinase-3 (PAP-3) from *Manduca sexta* hemolymph: A clip-domain serine proteinase regulated by serpin-1J and serine proteinase homologs. *Insect Biochem. Mol. Biol.* 33:1049—1060.

[61]Jiravanichpaisal, P. , B. L. Lee, and K. *Söderhäll*. 2006. Cell-mediated immunity in arthropods: Hematopoiesis, coagulation, melanization and opsonization. *Immunobiol.* 211:213—236.

[62]Kang, D. W. , G. Liu, A. Lundstrom, E. Gelius, and H. Steiner. 1998. A peptidoglycan recognition protein in innate immunity conserved from insects to humans. *Proc. Nat. Acad. Sci. U. S. A.* 95:10078—10082.

[63]Kanost, M. R. 1999. Serine proteinase inhibitors in arthropod immunity. *Dev. Comp. Immunol.* 23:291—301.

[64]Kanost, M. R. 2007. Serpins in a Lepidopteran insect, *Manduca sexta*. In The *serpinopathies: Molecular and cellular aspects of serpins and their disorders*, ed. G. A. Silverman and D. A. Lomas, 229—242. Singapore: World Scientific Publishing Co.

[65]Kanost, M. R. , and T. Clarke. 2005. Proteases. In *Comprehensive molecular insect science*, Vol. 4, ed. L. I. Gilbert, K. Iatrou, and S. Gill, 247—266. San Diego: Elsevier.

[66]Kanost, M. R. , W. Dai, and P. E. Dunn. 1988. Peptidoglycan fragments elicit antibacterial protein synthesis in larvae of *Manduca sexta*. *Arch. Insect Biochem. Physiol.* 8:147—164.

[67]Kanost, M. R. , and M. G. Gorman. 2008. Phenoloxidases in insect immunity. In *Insect immunology*, ed. N. Beckage, 69—96. San Diego: Academic Press/Elsevier.

[68]Kanost, M. R. , H. Jiang, and X. —Q. Yu. 2004. Innate immune responses of a lepidopteran insect, *Manduca sexta*. *Immunol. Rev.* 198:97—105.

[69]Kanost, M. R. , J. K. Kawooya, J. H. Law, R. O. Ryan, M. C. Van Heusden, and R. Ziegler. 1990. Insect Haemolymph proteins. *Adv. Insect Physiol.* 22:299—396.

[70]Kanost, M. R. , S. V. Prasad, Y. Huang, and E. Willott. 1995. Regulation of serpin gene-1 in *Manduca sexta*. *Insect Biochem. Mol. Biol.* 25:285—291.

[71]Kanost M. R. , S. V. Prasad, and M. A. Wells. 1989. Primary structure of a member of the serpin superfamily of proteinase inhibitors from an insect, *Manduca sexta*. *J. Biol. Chem.* 264:965—972.

[72]Kanost, M. R. , M. K. Zepp, N. E. Ladendorff, and L. A. Andersson. 1994. Isolation and characterization of a hemocyte aggregation inhibitor from hemolymph of *Manduca sexta* larvae. *Arch. Insect Biochem. Physiol.* 27:123—136.

[73]Karlsson, C. , A. M. Korayem, C. Scherfer, O. Loseva, M. S. Dushay, and U. Theopold. 2004. Proteomic analysis of the Drosophila larval hemolymph clot. *J. Biol. Chem.* 279:52033—52041.

[74]Kathirithamby, J. , L. D. Ross, and J. S. Johnston. 2003. Masquerading as self? Endoparasitic Strepsiptera (Insecta) enclose themselves in host-derived epidermal bag. *Proc. Natl. Acad. Sci. U. S. A.* 100:7655—7659.

[75]Koizumi, N. , M. Imamura, T. Kadotani, K. Yaoi, H. Iwahana, and R. Sato. 1999. The lipopolysaccharide binding protein participating in hemocyte nodule formation in the silkworm *Bombyx mori* is a novel member of the C-type lectin superfamily with two different tandem carbohydrate-recognition domains. *FEBS Letters* 443:139—143.

[76]Kopczynski, C. C. , G. W. Davis, and C. S. Goodman. 1996. A neural tetraspanin, encoded by late bloomer, that facilitates synapse formation. *Science* 271:1867—1870.

[77]Kotani, E. , M. Yamakawa, S. Iwamoto, et al. 1995. Cloning and expression of the gene of hemocytin, an insect humoral lectin which is homologous with the mammalian von Willebrand factor. *Biochim. Biophys. Acta* 1260:245—258.

[78]Kyriakides, T. R. , J. L. McKillip, and K. D. Spence. 1995. Biochemical characterization, developmental expression, and induction of the immune protein scolexin from *Manduca sexta*. *Arch. Insect Biochem. Physiol.* 29:269—80.

[79]Ladendorff, N. E. , and M. R. Kanost. 1991. Bacteria-induced protein P4 (hemolin) from *Manduca sexta*: A member of the immunoglobulin superfamily which can inhibit hemocyte aggregation. *Arch. Insect Biochem. Physiol.* 18:285—300.

[80]Lavine, M. D. , and N. E. Beckage 1996. Temporal pattern ofparasitism-induced immunosuppression in *Manduca sexta* larvae parasitized by *Cotesia congregata*. *J. Insect Physiol.* 42:41—51.

[81]Lavine, M. D. , and M. R. Strand. 2002. Insect hemocytes and their role in immunity. *Insect Biochem. Mol. Biol.* 32:1295—1309.

[82]Lavine, M. D. , and M. R. Strand. 2003. Haemocytes from *Pseudoplusia includens* express multiple α and β integrin subunits. *Insect Molec. Biol.* 12:441—452.

[83]Lemaitre, B. , and J. Hoffmann. 2007. The host defense of *Drosophila melanogaster*. *Annu. Rev. Immunol.* 25:697—743.

[84]Levin, D. , L. N. Breuer, S. Zhuang, S. A. Anderson, J. B. Nardi, and M. R. Kanost. 2005. A hemocyte-speciic integrin required for hemocytic encapsulation in the tobacco hornworm, *Manduca sexta*. *Insect Biochem. Mol. Biol.* 35:369—380.

[85]Li, D. , C. Scherfer, A. M. Korayem, Z. Zhao, O. Schmidt, and U. Theopold. 2002. Insect hemolymph clotting: Evidence for interaction between the coagulation system and the prophenoloxidase activating cascade. *Insect Biochem. Mol. Biol.* 32:919—928.

[86]Li J. , Z. Wang, B. Canagarajah, H. Jiang, M. R. Kanost, and E. J. Goldsmith. 1999. The structure of active serpin K from *Manduca sexta* and a model for serpin-protease complex formation. *Structure* 7: 103—109.

[87]Ling, E. , and X. —Q. Yu. 2006. Cellular encapsulation and melanization are enhanced by immulectins, pattern recognition receptors from the tobacco hornworm *Manduca sexta*. *Dev. Comp. Immunol.* 30:289—299.

[88]Loret, S. M. , and M. R. Strand. 1998. Follow-up of protein release from *Pseudoplusia includens* hemocytes: A first step toward identification of factors mediating encapsulation in insects. *Eur. J. Cell Biol.* 76:146—155.

[89]Lu, Z. , and H. Jiang. 2007. Regulation of phenoloxidase activity by high- and low-molecular-weight inhibitors from the larval hemolymph of *Manduca sexta*. *Insect Biochem. Mol. Biol.* 37:478—485.

[90]Lu, Z. , and H. Jiang. 2008. Expression of *Manduca sexta* serine proteinase homolog precursors in insect cells and their proteolytic activation. *Insect Biochem. Mol. Biol.* 38:89—98.

[91]Ma, C. , and M. R. Kanost. 2000. A beta 1,3-glucan recognition protein from an insect, *Manduca sexta*, agglutinates microorganisms and activates the phenoloxidase cascade. *J. Biol. Chem.* 275:7505—7514.

[92]Maecker, H. T. , S. C. Todd, and S. Levy. 1997. The tetraspanin superfamily: Molecular facilitators. *FASEB J*. 11:428—442.

[93]Manfredini, F. , F. Giusti, L. Beani, and R. Dallai. 2007. Developmental strategy of the endoparasite *Xenos vesparum* (Strepsiptera, Insecta): Host invasion and elusion of its defense reactions. *J. Morph.* 268:588—601.

[94]Matzinger, P. 2002. The danger model: A renewed sense of self. *Science* 296:301—305.

[95]Medzhitov, R. , and C. A. Janeway. 2002. Decoding the patterns of self and nonself by the innate immune system. *Science* 296:298—300.

[96]Michel, K. , and F. C. Kafatos. 2005. Mosquito immunity against *Plasmodium*. *Insect Biochem. Mol. Biol.* 35:677—89.

[97]Moita, L. F. , G. Vriend, V. Mahairaki, C. Louis, and F. C. Kafatos. 2006. Integrins of *Anopheles gambiae* and a putative role of a new integrin, BINT2, in phagocytosis of E. coli. *Insect Biochem. Mol. Biol.* 36:282—290.

[98]Mulnix, A. B. , and P. E. Dunn. 1994. Structure and induction of a lysozyme gene from the tobacco hornworm, *Manduca sexta*. *Insect Biochem. Mol. Biol.* 24:271—281.

[99]Mulnix, A. B. , and P. E. Dunn. 1995. Molecular biology of the immune response. In *Molecular model systems in the lepidoptera*, ed. M. R. Goldsmith and A. S. Wilkins, 369—395. New York: Cambridge University Press.

[100]Nappi, A. J. , and B. M. Christensen. 2005. Melanogenesis and associated cytotoxic reactions: Applications to insect innate immunity. *Insect Biochem. Mol. Biol.* 35:443—59.

[101]Nardi, J. B. 2004. Embryonic origins of the two main classes of hemocytes—granular cells and plasmatocytes—in *Manduca sexta*. *Dev. Genes Evol.* 214:19—28.

[102]Nardi, J. B. , C. Gao, and M. R. Kanost. 2001. The extracellular matrix protein lacunin is dynamically expressed by a subset of the hemocytes involved in basal lamina morphogenesis. *J. Insect Physiol.* 47:997—1006.

[103]Nardi, J. B. , B. Pilas, C. M. Bee, K. Garsha, S. Zhuang, and M. R. Kanost. 2006. Neuroglian-positive plasmatocytes and the initiation of hemocyte attachment to foreign surfaces. *Dev. Comp. Immunol.* 30:447—462.

[104]Nardi, J. B. , E. Ujhelyi, B. Pilas, K. Garsha, and M. R. Kanost. 2003. Hematopoietic organs of *Manduca sexta* and hemocyte lineages. *Dev. Genes Evol.* 213:477—491.

[105]Nardi, J. B. , S. Zhuang, B. Pilas, C. M. Bee, and M. R. Kanost. 2005. Clustering of adhesion receptors following exposure of insect blood cells to foreign surfaces. *J. Insect Physiol.* 51:555—564.

[106]Ochiai,M. ,and M. Ashida. 1988. Puriication of a β-1,3-glucan recognition protein in the prophenoloxidase activating system from hemolymph of the silkworm, *Bombyx mori*. *J. Biol. Chem.* 263:12056—12062.

[107]Ochiai,M. ,and M. Ashida. 1999. A pattern recognition protein for peptidoglycan—Cloning the cDNA and the gene of the silkworm, *Bombyx mori*. *J. Biol. Chem.* 274:11854—11858.

[108]Ochiai,M. ,and M. Ashida. 2000. A pattern recognition protein for β-1,3-glucan. *J. Biol. Chem.* 275:4995—5002.

[109]Pech,L. L. ,and M. R. Strand. 1996. Granular cells are required for encapsulation of foreign targets by insect hemocytes. *J. Cell Sci.* 109:2053—2060.

[110]Ponnuvel,K. M. , and M. Yamakawa. 2002. Immune responses against bacterial infection in *Bombyx mori* and regulation of host gene expression. *Curr. Sci.* 83:447—454.

[111]Ratcliffe,N. A. ,and S. J. Gagen. 1977. Studies on the in vivo cellular reactions of insects: An ultrastructural analysis of nodule formation in *Galleria mellonella*. *Tissue Cell* 9:73—85.

[112]Rizki,R. M. ,and T. M. Rizki. 1984. The cellular defense system of *Drosophila melanogaster*. In *Insect ultra-structure*,Vol. 2,ed. R. C. King and H. Akai,579—604. New York: Plenum.

[113]Rodríguez-Pérez,M. A. ,R. F. Dumpit,J. M. Lenz,E. N. Powell,S. Y. Tam,and N. E. Beckage. 2005. Host refractoriness of the tobacco hornworm,*Manduca sexta*,to the braconid endoparasitoid *Cotesia lavipes*. *Arch. Insect Biochem. Physiol.* 60:159—71.

[114]Royet,J. ,and R. Dziarski. 2007. Peptidoglycan recognition proteins: Pleiotropic sensors and effectors of antimicrobial defences. *Nat. Rev. Microbiol.* 5:264—277.

[115]Russell,V. ,and P. E. Dunn 1996. Antibacterial proteins in the midgut of *Manduca sexta* during metamorphosis. *J. Insect Physiol.* 42:65—71.

[116]Salt,G. ,1970. *The cellular defence reactions of insects*. Cambridge,UK: Cambridge Univ. Press.

[117]Sansonetti,P. J. 2006. The innate signaling of dangers and the dangers of innate signaling. *Nature Immunol.* 7:1237—1242.

[118]Scherfer,C. ,C. Karlsson,O. Loseva,et al. 2004. Isolation and characterization of hemolymph clotting factors in *Drosophila melanogaster* by a pullout method. *Curr. Biol.* 14:625—629.

[119]Schmidt,O. ,K. Andersson,A. Will,and I. Schuchmann-Feddersen. 1990. Viruslike particle proteins from a hymenopteran endoparasitoid are related to a protein component of the immune system in the lepidopteran host. *Arch. Insect Biochem. Physiol.* 13:107—115.

[120]Schmit,A. R. ,and N. A. Ratcliffe. 1977. The encapsulation of foreign tissue implants in *Galleria mellonella* larvae. *J. Insect Physiol.* 23:175—184.

[121]Scholz,F. 2002. Aktivierung von Hämozyten des *Tabakschwärmers Manduca sexta* nach bakteriellen Infektionen. PhD dissertation,Justus-Liebig Universität,Giessen.

[122]Shin,S. W. ,S. S. Park,D. S. Park,et al. 1998. Isolation and characterization of immune-related genes from the fall webworm,*Hyphantria cunea*,using PCR-based differential display and subtractive cloning. *Insect Biochem. Mol. Biol.* 28:827—837.

[123]Shrestha,S. ,and Y. Kim. 2008. Eicosanoids mediate prophenoloxidase release from oenocytoids

in the beet armyworm *Spodoptera exigua*. *Insect Biochem. Mol. Biol.* 38:99—112.

[124]Silverman, G. A., P. I. Bird, R. W. Carrell, et al. 2001. The serpins are an expanding superfamily of structurally similar but functionally diverse proteins: Evolution, mechanism of inhibition, novel functions, and a revised nomenclature. *J Biol. Chem.* 276:33293—33296.

[125]Stanley, D. 2006. Prostaglandins and other eicosanoids in insects: *Biological signiicance. Annu. Rev. Entomol.* 51:25—44.

[126]Stanley, D. W., J. S. Miller, and R. W. Howard. 1998. The influence of bacterial species and intensity of infections on nodule formation in insects. *J. Insect Physiol.* 44:157—164.

[127]Strand, M. R., and T. Noda. 1991. Alterations in the haemocytes of Pseudoplusia includens after parasitism by *Microplitis demolitor*. *J. Insect Physiol.* 37:839—850.

[128]Strand, M. R., and E. A. Wong. 1991. The growth and role of *Microplitis demolitor* teratocytes in parasitism of *Pseudoplusia includens*. *J. Insect Physiol.* 37:503—515.

[129]Su, X. —D., L. N. Gastinel, D. E. Vaughn, I. Faye, P. Poon, and P. J. Bjorkman. 1998. Crystal structure of hemolin: A horseshoe shape with implications for homophilic adhesion. *Science* 281:991—995.

[130]Sun, S. C., I. Lindström, H. G. Boman, I. Faye, and O. Schmidt. 1990. Hemolin: An insect-immune protein belonging to the immunoglobulin superfamily. *Science* 250:1729—1732.

[131]Theopold, U., O. Schmidt, K. Söderhäll, and M. S. Dushay. 2004. Coagulation in arthropods: Defence, wound closure and healing. *Trends Immunol.* 25:289—294.

[132]Todres, E., J. B. Nardi, and H. M. Robertson. 2000. The tetraspanin superfamily in insects. *Insect Mol. Biol.* 9:581—590.

[133]Tong, Y. R., H. B. Jiang, and M. R. Kanost. 2005. Identification of plasma proteases inhibited by *Manduca sexta* serpin-4 and serpin-5 and their association with components of the prophenol oxidase activation pathway. *J. Biol. Chem.* 280:14932—14942.

[134]Tong, Y. R., and M. R. Kanost. 2005. *Manduca sexta* serpin-4 and serpin-5 inhibit the prophenol oxidase activation pathway. *J. Biol. Chem.* 280:14923—14931.

[135]Van Kooyk, Y., and C. G. Figdor. 2000. Avidity regulation of integrins: The driving force in leukocyte adhesion. *Curr. Opinion Cell Biol.* 12:542—547.

[136]Wang, X., J. F. Fuchs, L. C. Infanger, et al. 2005. Mosquito innate immunity: Involvement of S 1,3-glucan recognition protein in melanotic encapsulation immune responses in *Armigeres subalbatus*. *Mol. Biochem. Parasitol.* 139:65—73.

[137]Wang, Y., and H. B. Jiang. 2004a. Prophenoloxidase (proPO) activation in *Manduca sexta*: An analysis of molecular interactions among proPO, proPO-activating proteinase-3, and a cofactor. *Insect Biochem. Mol. Biol.* 34:731—742.

[138]Wang, Y., and H. B. Jiang. 2004b. Puriication and characterization of *Manduca sexta* serpin-6: A serine proteinase inhibitor that selectively inhibits prophenoloxidase-activating proteinase-3. *Insect Biochem. Mol. Biol.* 34:387—395.

[139]Wang, Y., and H. Jiang. 2006. Interaction of beta-1,3-glucan with its recognition protein activates hemolymph proteinase 14, an initiation enzyme of the prophenoloxidase activation system in *Mandu-*

ca sexta. *J*. *Biol*. *Chem*. 281:9271—9278.

[140]Wang, Y., and H. Jiang. 2007. Reconstitution of a branch of the *Manduca sexta* prophenoloxidase activation cascade in vitro: Snake-like hemolymph proteinase 21 (HP21) cleaved by HP14 activates prophenol oxidase- activating proteinase-2 precursor. *Insect Biochem*. *Mol*. *Biol*. 37:1015—1025.

[141]Wang, Y., H. Jiang, and M. R. Kanost. 1999. Biological activity of *Manduca sexta* paralytic and plasmatocyte spreading peptide and primary structure of its hemolymph precursor. *Insect Biochem*. *Molec*. *Biol*. 29:1075—1086.

[142]Wang, C., and R. J. St. Leger. 2006. A collagenous protective coat enables *Metarhizium anisopliae* to evade insect immune responses. *Proc*. *Natl*. *Acad*. *Sci*. *U*. *S*. *A*. 103:6647—6652.

[143]Wang, Y., Z. Zou, and H. B. Jiang. 2006. An expansion of the dual clip-domain serine proteinase family in *Manduca sexta*: Gene organization, expression, and evolution of prophenoloxidase-activating proteinase-2, hemolymph proteinase 12, and other related proteinases. *Genomics* 87:399—409.

[144]Watanabe, A., S. Miyazawa, M. Kitami, H. Tabunoki, K. Ueda, and R. Sato. 2006. Characterization of a novel C-type lectin, *Bombyx mori* multibinding protein, from the B. mori hemolymph: Mechanism of wide range microorganism recognition and role in immunity. *J*. *Immunol*. 177:4594—4604.

[145]Waterhouse, R. M., Z. Y. Xi, E. Kriventseva, et al. 2007. Evolutionary dynamics of immune-related genes and pathways in disease vector mosquitoes. *Science* 316:1738—1743.

[146]Wiegand, C., D. Levin, J. P. Gillespie, E. Willott, M. R. Kanost, and T. Trenczek. 2000. Monoclonal antibody MS13 identifies a plasmatocyte membrane protein and inhibits encapsulation and spreading reactions of *Manduca sexta* hemocytes. *Arch*. *Insect Biochem*. *Physiol*. 45:95—108.

[147]Willott, E., T. Trenczek, L. W. Thrower, and M. R. Kanost. 1994. Immunochemical identification of insect hemocyte populations: Monoclonal antibodies distinguish four major hemocyte types in *Manduca sexta*. Eur. J. Cell Biol. 65:417—423.

[148]Xavier, R., T. Brennan, Q. Li, C. McCormack, and B. Seed. 1998. Membrane compartmentation is required for efficient T cell activation. Immunity 8:723—732. Ye, S., A. L. Cech, R. Belmares, et al. 2001. The structure of a Michaelis serpin-protease complex. *Nature Struct*. *Biol*. 8:979—983.

[149]Yoshida, H., K. Kinoshita, and M. Ashida. 1996. Purification of a peptidoglycan recognition protein from hemolymph of the silkworm, *Bombyx mori*. *J*. *Biol*. *Chem*. 271:13854—13860.

[150]Yu, X. Q., H. Gan, and M. R. Kanost. 1999. Immulectin, an inducible C-type lectin from an insect, *Manduca sexta*, stimulates activation of plasma prophenol oxidase. *Insect Biochem*. *Mol*. *Biol*. 29:585—597.

[151]Yu, X. Q., H. B. Jiang, Y. Wang, and M. R. Kanost. 2003. Nonproteolytic serine proteinase homologs are involved in prophenoloxidase activation in the tobacco hornworm, *Manduca sexta*. *Insect Biochem*. *Mol*. *Biol*. 33:197—208.

[152]Yu, X. Q., and M. R. Kanost. 1999. Developmental expression of *Manduca sexta* hemolin. *Arch*. *Insect Biochem*. *Physiol*. 42:198—212.

[153]Yu, X. —Q., and M. R. Kanost. 2000. Immulectin-2, alipopolysaccharide-speciic lectin from an insect, *Manduca sexta*, is induced in response to Gram-negative bacteria. *J*. *Biol*. *Chem*. 275: 37373—37381.

[154]Yu,X. —Q. ,and M. R. Kanost. 2002. Binding of hemolin to bacterial lipopolysaccharide and lipoteichoic acid—An immunoglobulin superfamily member from insects as a pattern-recognition receptor. *Eur. J. Biochem.* 269:1827—1834.

[155]Yu,X. —Q. ,and M. R. Kanost. 2003. *Manduca sexta* lipopolysaccharide-speciic immulectin-2 protects larvae from bacterial infection. *Dev. Comp. Immunol.* 27:189—196.

[156]Yu,X. —Q. ,and M. R. Kanost. 2004. Immulectin-2,a pattern recognition receptor that stimulates hemocyte encapsulation and melanization in the tobacco hornworm,*Manduca sexta*. *Dev. Comp. Immunol.* 9:891—900.

[157]Yu,X. —Q. ,and M. R. Kanost. 2008. Activation of lepidopteran insect innate immune responses by C-type immulectins. In *Animal lectins: A functional view*,ed. G. R. Vasta and H. A. Ahmed. Boca Raton: CRC Press. pp. 383—396.

[158]Yu,X. —Q. ,E. Ling,M. E. Tracy,and Y. Zhu. 2006. Immulectin-4 from the tobacco hornworm *Manduca sexta* binds to lipopolysaccharide and lipoteichoic acid. *Insect Mol. Biol.* 15:119—128.

[159]Yu,X. —Q. ,O. Prakash,and M. R. Kanost. 1999. Structure of a paralytic peptide from an insect,*Manduca sexta*. *J. Peptide Res.* 54:256—261.

[160]Yu,X. —Q. ,Y. Zhu,C. Ma,J. A. Fabrick,and M. R. Kanost. 2002. Pattern recognition proteins in *Manduca sexta* plasma. *Insect Biochem. Molec. Biol.* 32:1287—1293.

[161]Zhao,L. ,and M. R. Kanost. 1996. In search of a function for hemolin,a hemolymph protein from the immunoglobulin superfamily. *J. Insect Physiol.* 42:73—79.

[162]Zhao,P. C. ,J. J. Li,Y. Wang,and H. B. Jiang. 2007. Broad-spectrum antimicrobial activity of the reactive compounds generated in vitro by *Manduca sexta* phenoloxidase. *Insect Biochem. Mol. Biol.* 37:952—959.

[163]Zhu,Y. 2001. Identification of immune-related genes from the tobacco hornworm,*Manduca sexta*,and characterization of two immune-inducible proteins,serpin-3 and leureptin. Ph. D. dissertation,Kansas State University,Manhattan,KS.

[164]Zhu,Y. ,T. J. Johnson,A. A. Myers,and M. R. Kanost. 2003a. Identification by subtractive suppression hybridization of bacteria-induced genes expressed in *Manduca sexta* fat body. *Insect Biochem. Mol. Biol.* 33:541—559.

[165]Zhu,Y. F. ,Y. Wang,M. J. Gorman,H. B. Jiang,and M. R. Kanost. 2003b. *Manduca sexta* serpin—3 regulates prophenoloxidase activation in response to infection by inhibiting prophenoloxidase—activating proteinases. *J. Biol. Chem.* 278:46556—46564.

[166]Zhuang,S. ,L. Kelo,J. B. Nardi,and M. R. Kanost. 2007a. Neuroglian on hemocyte surfaces is involved in homophilic and heterophilic interactions of the innate immune system of *Manduca sexta*. *Dev. Comp. Immunol.* 31:1159—1167.

[167]Zhuang,S. ,L. Kelo,J. B. Nardi,and M. R. Kanost. 2007b. An integrin-tetraspanin interaction required for cellular innate immune responses of an insect,*Manduca sexta*. *J. Biol. Chem.* 282:22563—22572.

[168]Zhuang,S. ,L. Kelo,J. B. Nardi,and M. R. Kanost. 2008. Multiple α subunits of integrin are involved

in cell mediated responses of the *Manduca sexta* immune system. *Dev. Comp. Immunol.* 32:365—79.

[169]Zou,Z. ,J. D. Evans,Z. Lu,et al. 2007. Comparative genomic analysis of the *Tribolium* immune system. *Genome Biol.* 8:R177.

[170]Zou,Z. and H. Jiang. 2005a. Gene structure and expression proile of *Manduca sexta* prophenoloxidase activating proteinase-3 (PAP—3),an immune protein containing two clip domains. *Insect Mol. Biol.* 14:433—442.

[171]Zou,Z. ,and H. B. Jiang. 2005b. *Manduca sexta* serpin-6 regulates immune serine proteinases PAP-3 and HP8—cDNA cloning,protein expression,inhibition kinetics,and function elucidation. *J. Biol. Chem.* 280:14341—14348.

[172]Zou,Z. ,Y. Wang,and H. B. Jiang. 2005. *Manduca sexta* prophenoloxidase activating proteinase-1(PAP-1) gene: Organization, expression, and regulation by immune and hormonal signals. *Insect Biochem. Mol. Biol.* 35:627—636.

第十五章 鳞翅目昆虫作为人类病原体的微型宿主和抗生素肽的来源

Andreas Vilcinskas

1. 引言

在人类历史长河中,传染性疾病一直折磨着我们,并且构成了我们人类主要的威胁。由于对新的治疗策略的渴望,人们先前用小鼠、大鼠和兔子等哺乳动物作为模型,研究并详细阐明了人类病原体和它们的致病因子与宿主模型防御系统之间的相互作用,这些哺乳动物为复制人类感染提供了一个有力的实验系统。但以哺乳动物作为宿主模型,涉及许多问题:比如伦理问题、高成本问题、逻辑问题以及哺乳动物获得性免疫和先天性免疫系统之间复杂的联系。所以,我们迫切需要找到一种可以替代的宿主模型(Mylonakis,Casadevall 和 Ausubel 2007)。

同无脊椎动物一样,脊椎动物具有先天性免疫。然而,在脊椎动物中还保留了获得性免疫和抗体。如果我们的目的是为了研究病原体和宿主先天性免疫系统之间的相互作用,那么我们可以用缺乏获得性免疫和抗体的无脊椎动物作为模型。通过研究在遗传上易处理及繁殖周期短的无脊椎动物模型,比如果蝇(Lemaitre 和 Hoffmann 2007)和线虫(Sifiri et al. 2003;Mylonakis 和 Aballay 2005),我们对先天性免疫系统有了一定的了解。用线虫和果蝇作为模型的优势在于,它们的基因组已经完全测序以及我们可以方便地得到其基因芯片文库、RNA 干涉文库和突变品系。这些有利于在分子水平上分析宿主与病原体之间的相互作用。另一方面,由于鳞翅目幼虫个体较大,所以能精确地向其体内注射抗生素和病原体,另外用蛋白质组学等方法来研究病理生理学时,采集其组织和血淋巴样品也容易操作。许多鳞翅目昆虫,比如烟草天蛾(Silva et al. 2002;Kanost,Jiang 和 Yu 2004),家蚕(Cheng et al. 2006)和大蜡蛾(Vilcinskas 和 Götz 1999) 已作为经典模型用于研究昆虫病原微生物和它们的宿主之间的相互作用。在本书第十四章中,对鳞翅目昆虫先天性免疫作了全面的综述,本章着重强调使用鳞翅目昆虫作为人类病原体的微型宿主模型和作为抗生素肽的储存宿主,研究其在医学和植物保护方面的治疗潜能。

2. 大蜡蛾 *Galleria mellonella* 作为人类病原体的微型宿主

最近几年,大蜡蛾的幼虫作为引起人类某些疾病病原体的微型宿主备受青睐,大蜡蛾也作为一种实验系统来研究药物疗效的有效性(Kavanagh 和 Reeves 2004;Scully 和 Bidochka 2006) 。大蜡蛾作为实验模型比家蚕、烟草天蛾等其他鳞翅目昆虫更具优势。第一,大蜡蛾幼虫体积小,饲养成本低,用这种廉价的鳞翅目昆虫个体使高通量的感染测定变得很方便(Mylonakis,Casadevall 和 Ausubel 2007)。第二,大蜡蛾可以很方便地在市场上买到,因为它们常被用作鱼饵和宠物(爬行动物)的食物。第三,最近文献上记载的大蜡蛾作为宿主模型的主要优势是它能够适应人体的生理温度(37℃)。模拟哺乳动物的生理条件是必需的,因为人类病原体能够适应人体的生理温度,而它们需要在这种生理条件下合成和释放致病因子(Fuchs 和 Mylonakis 2006)。因此,为了研究在人体温度病原体感染和宿主有效反应之间的关系,我们必须选择像大蜡蛾一样的耐热模型系统(Mylonakis,Casadevall 和 Ausubel 2007)。第四,细菌和真菌在大蜡蛾和小鼠中的致病性成正相关(Jander et al. 2000;Brennan et al. 2002)。这些优势使得更多的研究者赞成用大蜡蛾作为引起人类某些疾病的主要细菌和真菌的微型宿主模型。这些致病菌包括:蜡样芽孢杆菌 (Fedhila et al. 2006),粪球肠菌 (Park et al. 2007),土拉热杆菌 (Aperis et al. 2007),绿脓假单胞菌 (Miyata et al. 2003),金黄色葡萄球菌 (Garcia-Lara,Needham 和 Foster 2005),白色念珠菌 (Bergin et al. 2006),新型隐球菌 (Mylonakis et al. 2005)。本质上,这些研究说明了大蜡蛾能够快速地筛选病原体突变体库,并且能够探究微生物的致病性和宿主反应之间在进化上的保守性。

2.1 大蜡蛾作为致病细菌的宿主模型

许多昆虫病原微生物细菌与人类病原体相关,它们能够感染包括昆虫和人在内的宿主。比如,作为控制害虫的苏云金芽孢杆菌和以炭疽为感染因子的炭疽杆菌关系密切。据报道,两种病菌使用一套相似的致病因子,在这些致病因子中,金属蛋白酶起了重要的作用(Silva et al. 2002;Chung et al. 2006)。一种随机的并经常随食物传播能引起传染性胃炎的蜡样芽孢杆菌与苏云金芽孢杆菌和炭疽杆菌关系紧密。它们都有荚膜并能产生毒素等共同点。大蜡蛾对蜡样芽孢杆菌同样敏感,并被用作鉴定蜡样芽孢杆菌的致病因子(Fedhila et al. 2006)。最近的另一个研究用大蜡蛾对来自人类致病细菌粪肠球菌产生的蛋白水解酶致病因子的功能进行了分析。肠球菌是通常的医院获得性病原体,能引起人类各种疾病。在它的致病因子中,有一种金属蛋白酶,是粪肠球菌在大蜡蛾血淋巴和人体血清中释放的一种致病因子,它能摧毁宿主防御分子(Park et al. 2007)。在分类学上非常复杂且极重要的一个细菌属伯克霍尔德菌,在生态学和病理学方面包括 9 个种,它们的聚集体叫做洋葱伯克霍尔德菌复合体(Bcc)。最近,用大蜡蛾建立了洋葱伯克霍尔德菌复合体的感染模型。与野生型相比,遗传突变的洋葱伯克霍尔德菌复合体对大蜡蛾幼虫致病力并发性减少,大蜡蛾幼虫存活率增加,因此遗传突变的洋葱伯克霍尔德菌复合体品系能被检测出来(Seed 和 Dennis 2008)。

据报道,尽管感染昆虫和人体的病原体非常相似,但也应该考虑它们之间的很多不同。比如,大蜡蛾被用于研究由绿脓假单胞菌通过Ⅲ型蛋白分泌系统所释放的毒素的作用。绿脓假单胞菌是能引起植物、线虫、昆虫、小鼠和人类疾病的通用病原体(Miyata et al. 2003)。而土壤昆虫病原微生物细菌——假单胞菌虱缺乏Ⅲ型蛋白分泌系统(Vodovar et al. 2006)。但是Ⅲ型蛋白分泌系统存在于其他昆虫病原体中,比如,发光杆菌。发光杆菌分泌的毒素和鼠疫耶氏菌分泌的毒素非常相似(Joyce,Watson 和 Clarke 2006)。然而,绿脓假单胞菌Ⅲ型蛋白分泌系统在大蜡蛾中的作用和在哺乳动物组织培养系统中的作用高度关联,这使得用大蜡蛾作为致病细菌的感染模型非常有效(Miyata et al. 2003)。Mylonakis 和他的合作者用大蜡蛾的幼虫作为有效的宿主系统来研究被美国疾病控制与防御中心认为是 A 类生物恐怖主义的由土拉热杆菌引起的兔热病的抗生素的有效性(Aperis et al. 2007)。这些研究说明,在用哺乳动物模型测试抗生素以前,大蜡蛾作为高通量、全动物模型系统具有很大的潜力。

2.2 大蜡蛾作为致病真菌的宿主模型

大蜡蛾不但可以检测细菌病原体,还可以作为一种微型宿主模型研究真菌的发病机理和致病力(Chamilos et al. 2007)。比如,大蜡蛾被用作建立人类致病真菌烟曲霉分泌的曲霉菌素在致病力方面的作用(Reeves et al. 2004)。烟曲霉对大蜡蛾幼虫的致病力取决于其分生孢子的萌发阶段。不萌发的分生孢子是没有毒性的,将被幼虫的免疫活性细胞吞噬;相反,萌发的分生孢子不能被吞噬,反而能杀死宿主(Renwick et al. 2006)。大蜡蛾幼虫也被当作研究隐球菌宿主和病原体相互作用的一种有力工具(London,Orozco 和 Mylonakis 2006)。新型隐球菌能引起器官移植的接受者和艾滋病感染者等有免疫缺陷的个体发病和死亡。CAP59,GRA1,RAS1 和 PKA1 等与它毒性相关的基因与感染哺乳动物有关,这些基因在与大蜡蛾的免疫系统相互作用中也起到了一定的作用。和在哺乳动物中发现的一样,*MFα1* 基因是在感染的增值阶段被诱导的(London,Orozco 和 Mylonakis 2006)。

鳞翅目微型宿主也被用作研究白色念珠菌的致病性。这些致病真菌的致病力在小鼠和大蜡蛾中一致(Brennan et al. 2002)。此外,大蜡蛾幼虫提前暴露在真菌中,可以避免白色念珠菌造成的致命感染,因为大蜡蛾表达了更多的抗菌肽(Bergin et al. 2006)。一致认为,大蜡蛾提前暴露在真菌和细菌细胞中,将会提高暴露在致死剂量真菌的存活率,或者提高暴露在巴西安白僵菌和金龟子绿僵菌等昆虫病原微生物真菌的存活率。提前暴露在细菌和真菌中激发由先天性免疫应答所产生的保护作用是因为在血淋巴中诱导和分泌了抗菌肽以及微生物蛋白水解酶抑制因子(Vilcinskas 和 Matha 1997a;Vilcinskas 和 Wedde 1997)。总之,大蜡蛾是一种特别有希望成为致病真菌微型模型的鳞翅目昆虫。

3. 对大蜡蛾先天性免疫应答的定性和定量分析

3.1 对大蜡蛾细胞防御机理的分析

在过去的 10 年,已经分离、克隆和描述了大蜡蛾中许多与免疫相关的效应因子。在强调它们在医学和植物保护方面的治疗潜能之前,也发表了定性和定量检测大蜡蛾细胞和体液免疫应答的方法。

大蜡蛾的细胞免疫应答和其他鳞翅目昆虫一样,包括吞噬作用和对侵入血腔中微生物的包囊作用(见第十四章)。昆虫的这种防御机制是由于在血淋巴中循环着重要的浆细胞和颗粒细胞。两种有免疫活性的血细胞能区别微生物和宿主表面,并且能够黏附病源相关分子模式,使得血细胞能将其从血淋巴中清除。用针扎其腹足,从体积相对较大的大蜡蛾中采集血淋巴样本也比较方便。通过用通常计数哺乳动物血液样本的血细胞计数器可以计算出从血淋巴样本中分离出来的血细胞密度。在人类致病真菌和昆虫病原微生物感染期间,大蜡蛾血淋巴样本中血细胞数目的波动对定量细胞防御作用提供了一个有价值的参数(Vilcinskas,Matha 和 Götz 1997a;Bergin,Brennan 和 Kavanagh 2003)。

如果其血淋巴中被注射和自然感染很多细菌和真菌细胞，以至于免疫活性血细胞不能将其吞噬，那么，这些细菌和真菌细胞将会被浆细胞和颗粒细胞包裹成多分子层。这个多分子层的包囊作用是一个复杂的过程，在这个过程中它分离侵入宿主组织的病原体和寄生虫，它是在凝聚物的黑化过程中完成的。在大蜡蛾中，通过解剖幼虫和计数其组织的黑色颗粒可以很容易地定量微生物黑化陷阱的形成。用荧光标记细胞（比如，FITC）和荧光显微术（Rohloff，Wiesner 和 Götz 1994）可以检测在大蜡蛾中对微生物的吞噬和包囊作用。除了在体内可以用这些方法来定量细胞的防御反应，在体外，大蜡蛾也被用作研究昆虫和人类病原体与免疫活性血细胞的相互作用。

一个简单的方法可以分离在血淋巴中具有吞噬微生物作用的浆细胞，并且能在原代培养基中培养它们（Wiesner 和 Götz 1993）。浆细胞吞噬微生物的能力取决于它们黏附和散布在外来物表面的能力。许多结合测试可以定量体外培养的浆细胞的数量和活性，以及黏附、散布和分离外来物表面的能力。这些测试可以探究昆虫病原微生物真菌和它们的致病因子对细胞防御的影响（Vilcinskas，Matha 和 Götz 1997a，b；Griesch 和 Vilcinskas 1998）。在感染早期，体内和体外的结合测试说明通过其表皮直接感染昆虫的病原微生物真菌能产生所谓的被浆细胞吞噬了细胞壁的菌丝体。浆细胞没有破坏被吞噬的真菌细胞，相反，浆细胞被当作真菌细胞侵入宿主组织的载体（Vilcinskas，Matha 和 Götz 1997b）。在感染后期，当细胞外的菌丝体侵入血淋巴时，免疫活性血细胞的吞噬、黏附、散布能力将会受到损害。在感染的宿主中，昆虫病原微生物真菌抑制先天性免疫系统是因为它们的蛋白水解酶和免疫抑制毒素的结合活性所致（Vilcinskas 和 Götz 1999）。和人类相比，病原体真菌损害大蜡蛾和人类的防御系统的方式非常相似（Reeves et al. 2004）。

研究人类病原体和它们的致病因子在大蜡蛾中细胞免疫反应的另一个原因是，大蜡蛾的血细胞呈现出与在人体内第一道防线起关键作用的中性粒细胞相似的功能。比如，它们在吞噬作用和微生物杀伤作用中具有相似的动力学。吞噬细胞杀死被吞入的微生物是因为呼吸作用的突然爆发，呼吸作用的突然爆发需要像超氧化物和过氧化氢等自由基的酶联产物，这些自由基能依次减少被吞噬的微粒。大量的人类中性粒细胞 NADPH 氧化酶复合体蛋白的同源蛋白在大蜡蛾的血细胞中被发现，这些中性粒细胞能产生超氧化物（Bergin et al. 2005）。此外，在呼吸作用中，人类中性粒细胞超氧化物形成氧化酶的活化需要细胞质蛋白——$p47^{phox}$和 $p67^{phox}$，这两种蛋白易位到原生质膜才能活化超氧化物形成氧化酶，活化此酶需要刺激细胞和活化此酶系统的氧化—还原中心。然而，在大蜡蛾的血细胞中，人们也发现了 $p47^{phox}$和 $p67^{phox}$的同源蛋白，以及发现这两种蛋白易位到原生质膜同样需要刺激细胞（Renwick et al. 2007）。

另外，大蜡蛾最近被用作阐明受伤的细胞和组织释放的细胞内的核酸如何发挥免疫相关的功能。宿主衍生的细胞外的核酸增强先天性免疫，引起血凝固，增加感染的大蜡蛾的存活率。有趣的是，人们发现一种类绛色血细胞能够破坏与之相连的外来物的表面，接着外来物释放核酸，然后核酸形成纤维结构网住细菌，这和我们知道的人类中性粒细胞相似,人类中性粒细胞形成纤维结构网包括 DNA 及其与之相连的蛋白(Altincicek et al. 2008)。人类中性粒细胞和大蜡蛾血细胞的共性说明大蜡蛾是一种有效的微型宿主模型。

3.2 用蛋白质组学和转录组学的方法对大蜡蛾先天性免疫进行分析

鳞翅目昆虫有多种多样的免疫相关作用分子,其中,最主要的是抗菌肽和蛋白水解酶抑制因子(详见第十四章)。大量的微生物,如,革兰氏阴性菌,大肠杆菌和革兰氏阳性菌,藤黄微球菌和真菌,酿酒酵母被用作抑菌圈分析检测,此法能够检测和定量鳞翅目昆虫血淋巴和组织样本中所有的抑制活性(Vilcinskas 和 Matha 1997b)。另外,运用阿佐科劳(Azocoll)比色法,使用生色底物可以测量蛋白酶的活性,并且能定性和定量检测血淋巴中广泛分布的微生物蛋白水解酶(Wedde et al. 1998; Fröbius et al. 2000)。这种对抗菌肽和蛋白水解酶抑制因子活性快速而简单的检测方法,能更加容易地发现和分离相应的效用因子。纯化和鉴定抗菌肽需要大量的血淋巴和组织样品,这些样品可以很容易地从鳞翅目昆虫的幼虫和蛹中获得。用蛋白质组学的方法可以分析大蜡蛾幼虫血淋巴蛋白形式在先天性免疫活化试验或者在感染人类病原体期间的变化。通过双向电泳技术比较其未处理的和经免疫活化的幼虫血淋巴样本,人们发现,注射过微生物激发子的幼虫释放了至少三百种肽和抗生素。通过质谱法和埃德曼氨基酸测序法对新发现的和丰度较高的点进行分析,结果发现很多没有鉴定的和免疫相关的肽和蛋白质,其中溶菌酶、gallerimycin、葛佬素已经被鉴定(Altincicek et al. 2007)。蛋白质组学的方法除了用于分析大蜡蛾血淋巴样本,还被用于研究人类致病真菌对其感染者先天性免疫的影响(Bergin et al. 2006)。蛋白质组学方法能作为一种补充技术,探究致病微生物对昆虫血淋巴蛋白形式的影响。

3.3 用转录组学的方法对大蜡蛾先天性免疫进行分析

除了蛋白质组学方法,鳞翅目幼虫的先天性免疫反应也可以用转录组学的方法来分析,虽然芯片分析没有被广泛使用。抑制消减杂交技术被独立地用作对烟草天蛾脂肪体(Zhu et al. 2003)和大蜡蛾血淋巴(Seitz et al. 2003)免疫活化后上调基因的筛选,此技术以 PCR 为基础,选择性地扩增不同表达水平的 cDNAs,抑制像管家基因这类通常的 cDNAs。虽然抑制消减杂交技术不能完全对鳞翅目免疫相关转录组进行分析,但是,运用这种方法在大蜡蛾中发现了新的抗菌肽——gallerimycin (Schuhmann et al. 2003)。有趣的是,在用细菌脂多糖活化大蜡蛾免疫相关基因时,发现了一个新的抗生素肽,Gall-6-tox,它有六个稳定保守串联重复的半胱氨酸 α-β 基序(CS-αβ),蝎子毒素和无脊椎动物防御素具有这种结构基序(Seitz et al. 2003)。在另外一些鳞翅目昆虫,如在家蚕和甜菜夜蛾中,Gall-6-tox 同源蛋白的 CS-αβ 基序的串联重复数量不同。它们是新的非典型的由防御衍生出来的免疫相关蛋白家族,这个蛋白家族存在于鳞翅目昆虫中,并被叫作 x-tox (Girard et al. 2008)。总之,这些研究的成功说明抑制消减杂交技术对发现新基因是一种有效的工具,当然,如果基因序列已知,那么芯片分析是一种更好的用于检测其表达的技术。

大蜡蛾免疫相关基因表达的定量分析为研究微生物突变体库对宿主防御系统的影响提供了十分重要的信息。为扩增可诱导的免疫相关基因而设计的引物和定量PCR最近被用于探测抗菌肽的表达，这些抗菌肽的产生是大蜡蛾先天性免疫系统对人类致病真菌、白色念珠菌等微生物激发因子作出的应激反应（Bergin et al. 2006；Altincicek et al. 2007）。相似的方法可以用来研究人类的其他病原体与大蜡蛾先天性免疫系统之间的相互作用。然而，就像引言提及的一样，比起果蝇和新杆状线虫，用大蜡蛾作为模型宿主有许多缺点，比如，它缺乏基因组和芯片分析数据（Mylonakis，Casadevall 和 Ausubel 2007）。为了跨越这个障碍，我们已经完成了整个大蜡蛾转录组的测序，并在人类致病菌所感染的这种鳞翅目微型宿主模型中，用基因芯片技术来研究免疫相关的表达模式。利用刻克罗普斯蚕蛾，已经发表了关于运用RNAi诱发基因沉默的文章。最后，也是很重要的一点，我们希望用新的方向进一步扩展鳞翅目昆虫免疫相关基因的功能分析（Bettencourt，Terenius 和 Faye 2002；Terenius et al. 2007）。

4. 大蜡蛾作为抗生素肽的储存宿主

4.1 来自大蜡蛾的抗菌肽

鳞翅类先天性免疫系统可以作为昆虫先天性免疫系统的经典模型，一些主要与免疫相关的分子、肽或者蛋白质已从血淋巴中首先被检测和分离；然而，果蝇的抗菌肽主要是在基因水平上发现的（Bulet 和 Stocklin 2005）。最先报道的昆虫抗菌肽是大蜡蛾的溶菌酶，40年前它已经得到鉴定（Mohrig 和 Messner 1968），其结构和C—（鸡）型溶菌酶很相似（Jolles et al. 1979），通过水解肽聚糖N-乙酰葡萄糖胺和N-乙酰胞壁酸的β1-4糖苷键，从而抑制革兰氏阳性菌（Powning 和 Davidson 1976），而对革兰氏阴性菌只有轻度的抑制作用（Yu et al. 2002）。大蜡蛾溶菌酶在体外具有抗真菌活性（Vilcinskas 和 Matha 1997a），这和人类抗致病真菌：白色念珠菌（Kamaya 1970）、粗球孢子菌（Collins 和 Papagianis 1974）的溶菌酶非常相似。昆虫病原微生物真菌能成功地感染和杀死大蜡蛾幼虫，在感染期间，它能分泌蛋白水解酶消化和抑制溶菌酶的合成（Vilcinskas 和 Matha 1997a，b）。相应地，人类致病菌、粪肠球菌在大蜡蛾血淋巴和人类血清中都能释放消化宿主防御分子的蛋白水解酶（Park et al. 2007）。

第一个线性和两亲的α-螺旋结构的抗菌肽叫作天蚕抗菌肽，是从惜古比天蚕的血淋巴中发现和分离的（Steiner et al. 1981）。之后，我们在其他昆虫中找到了麻蝇毒素，它是用褐尾麻蝇来命名的，并且是天蚕抗菌肽的同源蛋白（Okada 和 Natori 1983）。大蜡蛾中的天蚕抗菌肽同源蛋白以前体形式合成，其前体有一个22个氨基酸残基的信号肽，一个四个残基的前肽和一个包含39个氨基酸残基，分子量为4.3kDa道尔顿的成熟肽。来自其他昆虫的天蚕抗菌肽同源蛋白同样具有抑制革兰氏阴性菌和阳性菌的活性（Kim et al. 2004）。

在鳞翅目昆虫家蚕中发现了另一两亲的α-螺旋结构抗菌肽家族——moricins 家族(Hara 和 Yamakawa 1995),moricins 家族能够抑制对二甲氧基苯青霉素有抗性的金黄色葡萄球菌(Hara,Asaoka 和 Yamakava 1996)。由于在大蜡蛾中发现因基因复制而产生的八个 moricins 家族的同源蛋白,moricins 家族在体外也能抑制革兰氏阴性菌、革兰氏阳性菌、酵母和丝状真菌(Brown et al. 2008)。在大蜡蛾中,也发现了另外两种富含半胱氨酸的抗菌肽,它们是动物防卫素样抗真菌家族(Lee et al. 2004)和 gallerimycin 家族(Schuhmann et al. 2003),能专一地抑制丝状真菌。至少,gallerimycin 家族有助于调节大蜡蛾幼虫对由人类致病真菌——白色念珠菌所引起的致命感染的抗性(Bergin et al. 2006);此外,gallerimycin 家族能使转基因作物对由真菌引起的疾病具有抗性(Langen et al. 2006)。除了动物防卫素样抗真菌家族,在大蜡蛾血淋巴中也纯化和鉴定了另外五种抗菌肽家族,其中两种是富含脯氨酸的抗菌肽家族和一种具有革兰氏阳性菌抗性的阴离子抗菌肽家族(Cytrynska et al. 2006)。

总而言之,与其他鳞翅目昆虫一样,大蜡蛾运用先天性免疫系统产生广谱抗菌肽,具有抗细菌和真菌的活性。并且它们之间在不断地协同进化着。这些抗菌肽为新型抗生素的筛选提供了新的来源。

4.2 鳞翅目昆虫抗菌肽的治疗潜能

抗生素的销售量在药物销售量中排名第三(Breithaupt 1999)。对抗生素的滥用,使病原体对以前能抑制它们的药物的耐药性不断增强。作为全球最为紧迫之一的公共健康问题已证实了这一点,这强烈需要新的治疗手段,由此产生了所谓的第二代抗生素。属于昆虫防卫素的阴离子抗菌肽,是很有希望的一种抗生素,因为它们能有效地抵抗一些产生了耐药性的细菌,还不曾发现细菌对这些抗菌肽的抗药性。动物防卫素通过肽一脂相互作用破坏细菌细胞膜,细胞膜结构对这些防卫素不敏感突变的概率是很低的,因为影响整个细胞膜结构突变的发生率远低于酶或者把抗性基因转移至细胞内的变异(Saido-Sakanaka et al. 2005)。但是,把防卫素作为抗菌药物受到了很多限制。比如,生产成本高,一些防卫素和现在临床所使用的抗生素相比,抗菌活性很低。然而,昆虫抗菌肽能被修饰为人们所期待的,并能减少或者消除人们所不期望的特性。比如,包括九种氨基酸的根据昆虫防卫素活性位点而设计生产的合成肽在小鼠体内显示为对有致命感染和对二甲氧基苯青霉素有抗药性的金黄色葡萄球菌有抑制作用,并且不会引起其他的毒副作用(Saido-Sakanaka et al. 2005)。

自从昆虫被视为新药物的潜在来源以来,昆虫抗菌肽作为临床使用的治疗药物的商业开发就开始了。在临床前期研究中,鳞翅目昆虫烟芽夜蛾的海利霉素首先被测试,因为在免疫抑制的病人中,海利霉素对来自医院的威胁生命的真菌有抑制潜能(Zasloff 2002)。1999 年于法国的 Strasbourgh 成立第一个私人公司 EntoMed S. A.,主要研发昆虫作为抗菌肽的储存宿主。虽然 EntoMed S. A 已不再从事商业活动,但它形成了由昆虫来源的能治疗人类疾病药物的发现与发展途径。

5. 来自大蜡蛾的致病因子抑制剂

昆虫和人类病原体能产生致病因子，这些致病因子能破坏宿主组织，消化宿主防御分子，激活宿主调控蛋白，获取宿主营养素(Maeda 1996；Gillespie et al. 2000)。大多数致病因子是细胞外蛋白水解酶，在这些蛋白水解酶中，嗜热菌蛋白酶家族的金属蛋白酶属于M4家族，它在致病力方面扮演着主要角色。据报道，由人类致病菌所产生的主要致病因子，比如，金霉素、芽孢菌黏素、pseudolysin，以及弧菌溶血素，能导致被感人群血管渗透性增强，出血性水肿，组织坏死。这些致病因子已作为第二代抗生素的开发靶标(Travis 和 Potempa 2000)。从经免疫活化的大蜡蛾血淋巴中发现并提纯了这些致病因子的抑制肽(Wedde et al. 1998)。这种昆虫金属蛋白水解酶抑制剂(IMPI)的氨基酸与任何已知肽没有相似性。分离得到的IMPI分子量是8300Da，包含五个内部二硫键的多肽，这些二硫键对它的热稳定性起重要作用。重组体IMPI能抑制嗜热菌蛋白酶家族的金属蛋白酶，由人类病原体所产生的中温酶蛋白酶和弧菌溶血素是重要的微生物金属蛋白酶，它们已被当作靶标蛋白(Clermont et al. 2004；Wedde et al. 2007)。在最近的文献中记载，越来越多的M4家族金属蛋白水解酶被认为是医学上人类致病细菌和真菌重要的致病因子，IMPI已被当作第二代抗生素的模板，引起了越来越多的注意。因此，已经研究了它的三维结构，生产出其合成类似物(A. Vilcinskas，unpubl. data)。

6. 大蜡蛾作为转基因来源为植物提供抗病能力

植物病原体是可怕的，是逐年增加的世界范围内作物损失的原因(每年300亿～500亿)，因此这与不正确使用抗生素而造成的人类病原体产生耐药性相似，严重威胁人类的营养。在农业上最热门的问题之一是植物致病菌迅速对曾经能控制害虫的杀虫剂产生耐药性。农业生产成本的显著增长导致公众担忧农用化学药品对人类健康和环境造成的有害影响(Osusky et al. 2000)。因此，迫切需要在现代化植物保护方式中寻找新的途径，作为可持续农业生产的一种优势策略，现在人们改变作物的自身防御机制，使它们能抗病(Kogel 和 Langen 2005)，这是可以实现的。例如，可以通过转移一个编码抗菌肽的基因而不改变植物基因组特有价值的特征。来源于植物的抗菌基因的表达仅对抗病菌提供了一定的抗性，因为它们在协同进化中已经获得了对植物抗菌肽的耐药性，所以，作为创造抗病作物的有力工具，已经出现源自昆虫的抗菌肽的转基因表达(Jaynes et al. 1987；Vilcinskas 和 Gross 2005；Yevtushenko et al. 2005；Coca et al. 2006)。

毫不奇怪，第一个来源于昆虫的基因是编码攻击素和天蚕抗菌肽的基因，这两个基因是为抗病作物而设计的。攻击素结合细菌脂多糖，然后抑制细菌外膜蛋白的合成(Carlsson et al. 1998)。在攻击素转基因苹果(Ko et al. 2000)和梨中(Reynoird et al. 1999)，具有抗火烧病的抗性。除了抗细菌活性，天蚕抗菌肽在体外也具有抗真菌的活性(Ekengren 和 Hultmark 1999)，因此，它被用作转基因给作物提供抗真菌活性。由于翻译后产物的降解，第一次尝试失败了，天蚕抗菌肽在作物中只能维持很短时间，这大大阻碍了它的利用(Mills et al. 1994)。为了弥补这一缺陷，人们合成了稳定性更高的修饰了的天蚕抗菌肽(Owens 和 Heutte 1997)和天蚕抗菌肽——二甲双胍杂合体(Huang et al. 1997;Yevtushenko et al. 2005),这些稳定性更高的抗菌肽基因被转入作物内,对植物致病细菌和真菌有更好的抑制作用(Coca et al. 2006)。

与其他昆虫中已被鉴定的抗菌肽相比,发现能够专一抑制真菌的抗菌肽数目是极其有限的。来自果蝇的 drosomycin(Fehlbaum et al. 1994)和棉铃虫的海利霉素(Lamberty et al. 1999)与植物防卫素具有相似的氨基酸序列,并且和敏感真菌的结合位点相同(Thevissen et al. 2004)。在与植物宿主协同进化过程中,致病真菌似乎获得了对植物防卫素的抗药性,由于 drosomycin 和海利霉素与植物防卫素相似(Banzet et al. 2002),在转基因烟草中,它们的表达只稍稍地提高了对真菌病原体的抗性。相比而言,来源于大蜡蛾的一种新的抗真菌肽,gallerimycin(Schuhmann et al. 2003),最近被用作一种传递真菌抗性的有力工具,将抗性从昆虫转移到植物。Gallerimycin 作为转基因特别合适,因为,(1)它属于已知的数量有限的能专一抑制菌丝体的抗菌肽。(2)从昆虫转移到烟草中,它比其他抗真菌肽具有更高的抗真菌活性。(3)它具有信号序列,这个信号序列能被植物识别并指导其在细胞内合成,并且不容易被内源性蛋白酶分解(Langen et al. 2006)。此外,最近的一项种系发生的分析研究证明,与昆虫防卫素 drosomycin(Altincicek 和 Vilcinskas 2007)相比,gallerimycin 与真菌的关系更为密切。

在植物中,为了防止伴随昆虫的抗真菌肽组成性转基因表达产生的影响,比如,影响生长和减产,设计了一种植物转化载体,这个载体含有 gallerimycin 的编码序列,该序列的启动子是根瘤土壤杆菌甘露碱合酶 *mas* p2'启动子。因为这个启动子在烟草有伤口和真菌感染时开启,它提供了人们所期望的昆虫转基因表达,并排除了植物中的致病菌对抗真菌肽产生抗药性的可能(Langen et al. 2006)。为了进一步探究从昆虫转移到植物的抗菌肽的抗病性,现在我们研究所已经详细阐述了该转基因技术,该技术能使几个鳞翅目昆虫抗菌肽经病原体结合和诱导的表达模式在作物中实现。gallerimycin 基因和其他抗菌肽与昆虫来源的蛋白水解酶抑制剂有不同的作用方式,它们在转基因植物中的表达能有效防止植物致病菌抗性选择的产生。昆虫抗菌肽的转基因表达已开始发展,并为植物病原体的进化适应性保持领先提供了工具(Vilcinskas 和 Gross 2005)。

参考文献

[1]Altincicek, B. , M. Linder, D. Linder, K. Preissner, and A. Vilcinskas. 2007. Microbial metalloproteinases mediate sensing of invading pathogens and activate innate immune responses in the lepidopteran model host *Galleria mellonella*. *Infect. Immun.* 75:175－83.

[2]Altincicek, B. , S. Stötzel, M. Wygrecka, K. Preissner, and A. Vilcinskas. 2008. Host-derived extracellular nucleic acids enhance innate immune responses, induce coagulation, and prolong survival upon infection in insects. *J. Immunol.* 181:2705－12.

[3]Altincicek, B. , and A. Vilcinskas. 2007. Identification of immune-related genes from an apterygote insect, the firebrat *Thermobia domestica*. *Insect Biochem. Mol. Biol.* 37:726－31.

[4]Aperis, G. , B. Fuchs, C. Anderson, J. Warner, S. Calderwood, and E. Mylonakis. 2007. *Galleria mellonella* as a model host to study infection by the *Francisella tularensis* live vaccine strain. *Microbes In-*

fect. 9:729—34.

[5]Banzet, N., M. Latorse, P. Bulet, E. Francois, C. Derpierre, and M. Dubal. 2002. Expression of insect cystein-rich antifungal peptides in transgenic tobacco enhances resistance to a fungal disease. *Plant Sci*. 162:995—1006.

[6]Bergin, D., M. Brennan, and K. Kavanagh. 2003. Fluctuations in haemocyte density and microbial load may be used as indicators of fungal pathogenicity in larvae of *Galleria mellonella*. *Microbes Infect*. 5:1389—95.

[7]Bergin, D., L. Murphy, J. Keenan, M. Clynes, and K. Kavanagh. 2006. Pre-exposure to yeast protects larvae of *Galleria mellonella* from a subsequent lethal infection by *Candida albicans* and is mediated by the increased expression of antimicrobial peptides. *Microbes Infect*. 8:2105—12.

[8]Bergin, D., E. Reeves, J. Renwick, F. Wientjes, and K. Kavanagh. 2005. Superoxide production in *Galleria mellonella* hemocytes: Identification of proteins homologous to the NADPH oxidase comples of human neutrophils. *Infect. Immun*. 73:4161—70.

[9]Bettencourt, R., O. Terenius, and I. Faye. 2002. *Hemolin* silencing by ds-RNA injected into Cecropia pupae is lethal to next generation embryos. *Insect Mol. Biol*. 11:267—71.

[10]Breithaupt, M. 1999. The new antibiotics. *Nat. Biotechnol*. 17:1165—69.

[11]Brennan, M., D. Y. Thomas, M. Whiteway, and K. Kavanagh. 2002. Correlation between virulence of *Candida albicans* mutants in mice and in *Galleria mellonella* larvae. *FEMS Immunol. Med. Microbiol*. 34:153—57.

[12]Brown, S., A. Howard, A. Kasprzak, K. Gordon, and P. East. 2008. The discovery and analysis of a diverged family of novel antifungal moricin-like peptides in the wax moth *Galleria mellonella*. *Insect Biochem. Mol. Biol*. 38:201—12.

[13]Bulet, P., and R. Stocklin. 2005. Insect antimicrobial peptides: Structures, properties and gene regulation. *Protein Pept. Lett*. 12:3—11.

[14]Carlsson, A., T. Nystrom, H. de Cock, and H. Bennich. 1998. Attacin—an insect immune protein—binds LPS and triggers the specific inhibition of bacterial outer membrane protein synthesis. *Microbiology* 144:2179—88.

[15]Chamilos, G., M. Lionakis, R. Lewis, and D. Kontojiannis. 2007. Role of mini-host models in the study of medically important fungi. *Lancet Infect. Dis*. 7:42—55.

[16]Cheng, T., P. Zhao, C. Liu, et al. 2006. Structures, regulatory regions, and inductive expression patterns of antimicrobial peptide genes in the silk worm *Bombyx mori*. *Genomics* 87:356—65.

[17]Chung, M. —C., T. Popova, B. Millis, et al. 2006. Secreted neutral metalloproteases of *Bacillus anthracis* as candidate pathogenic factors. *J. Biol. Chem*. 281:31408—18.

[18]Clermont, A., M. Wedde, V. Seitz, L. Podsiadlowski, M. Hummel, and A. Vilcinskas. 2004. Cloning and expression of an inhibitor against microbial metalloproteinases from insects (IMPI) contributing to innate immunity. *Biochem. J*. 382:315—22.

[19]Coca, M., G. Penas, J. Gomez, et al. 2006. Enhanced resistance to the rice blast fungus *Magnaporthe grisea* conferred by expression of a cecropin A gene in transgenic rice. *Planta* 233:392—406.

[20]Collins, M. S., and D. Papagianis. 1974. Inhibition by lysozyome on the growth of the sperule phase of *Coccidoides immitis in vitro*. *Infect. Immun*. 10:616—23.

[21]Cytrynska, M., P. Mak, A. Zdybicka-Barabas, P. Suder, and T. Jakubowicz. 2006. Purification and

characterization of eight peptides from *Galleria mellonella* immune hemolymph. *Peptides* 28:533—46.

[22]Ekengren, S., and D. Hultmark. 1999. Drosophila cecropin as an antifungal agent. *Insect Biochem. Mol. Biol.* 29:965—72.

[23]Fedhila, S., N. Daou, D. Lereclus, and C. Nielsen-LeRoux. 2006. Identification of *Bacillus cereus* internalin and other candidate virulence genes specifically induced during oral infection in insects. *Mol. Microbiol.* 62:339—55.

[24]Fehlbaum, P., P. Bulet, L. Michaut, et al. 1994. Insect Immunity. Septic injury of *Drosophila* induces the synthesis of a potent antifungal peptide with sequence homology to plant antifungal peptides. *J. Biol. Chem.* 269:33159—63.

[25]Fröbius, A., M. Kanost, P. Götz, and A. Vilcinskas. 2000. Isolation and characterization of novel inducible serine protease inhibitors from larval hemolymph of the greater wax moth, *Galleria mellonella*. *Eur. J. Biochem.* 267:2046—53.

[26]Fuchs, B., and E. Mylonakis. 2006. Using non-mammalian hosts to study fungal virulence and host defense. *Curr. Opin. Microbiol.* 9:346—51.

[27]Garcia-Lara, J., A. Needham, and S. Foster. 2005. Invertebrates as animal models for *Staphylococcus aureus* pathogenesis: A window into host-pathogen interaction. *FEMS Immunol. Med. Microbiol.* 43:311—23.

[28]Gillespie, J., A. Bailey, B. Cobb, and A. Vilcinskas. 2000. Fungal elicitors of insect immune responses. *Arch. Insect Biochem. Physiol.* 44:49—68.

[29]Girard, P. — A., Y. Boublik, C. Wheat, et al. 2008. X—tox: An atypical defensin derived family of immune-related proteins specific to Lepidoptera. *Dev. Comp. Immunol.* 32:575—84.

[30]Griesch, J., and A. Vilcinskas. 1998. Proteases released by entomopathogenic fungi impair phagocytic activity, attachment and spreading of plasmatocytes isolated from hemolymph of the greater wax moth *Galleria mellonella*. *Biocontrol Sci. Technol.* 8:517—31.

[31]Hara, S., A. Asaoka, and M. Yamakava. 1996. Effect of moricin, a novel antibacterial peptide of *Bombyx mori* (Lepidoptera, Bombycidae) on the growth of methicillin-resistant *Staphylococcus aureus*. *Appl. Entomol. Zool.* 31:465—66.

[32]Hara, S., and M. A. Yamakawa. 1995. Moricin, a novel type of antibacterial peptide isolated from the silkworm, *Bombyx mori*. *J. Biol. Chem.* 270:29923—27.

[33] Huang, Y., R. Nordeen, M. Di, L. Owens, and J. McBeath. 1997. Expression of engineered cecropin gene cassette in transgenic tobacco plants confers resistance to *Pseudomonas syringae pv. tabaci*. *Phytopathology* 87:494—99.

[34]Jander, G., L. Rahme, F. Ausubel, and E. Drenkard. 2000. Positive correlation between virulence of *Pseudomonas aeroginosa* mutants in mice and insects. *J. Bacteriol.* 182:3843—45.

[35]Jaynes, J., K. Xanthopoulos, L. Destefano-Beltran, and J. Dodds. 1987. Increasing bacterial disease resistance in plants utilizing antibacterial genes from insects. *BioEssays* 6:263—70.

[36]Jolles, J., F. Schoentgen, G. Croizier, L. Croizier, and P. Jolles. 1979. Insect lysozymes from three species of Lepidoptera: Their structural relatedness to the C (chicken) type lysozyme. *J. Mol. Evol.* 14:267—71.

[37]Joyce, S., R. Watson, and D. Clarke. 2006. The regulation of pathogenicity and mutualism in *Photorhabdus*. *Curr. Opin. Microbiol.* 9:127—32.

[38]Kanost,M. ,H. Jiang,and X. —Q. Yu. 2004. Innate immune response of a lepidopteran insect, *Manduca sexta*. *Immunol. Rev.* 198:97—105.

[39]Kamaya,T. 1970. Lytic action of lysozyme on *Candida albicans*. *Mycopathol. Mycologia Appl.* 37:320—30.

[40]Kavanagh,K. ,and E. P. Reeves. 2004. Exploiting the potential of insects for the in vivo pathogenicity testing of microbial pathogens. *FEMS Microbiol*. 28:101—12.

[41]Kim,C. H. ,J. H. Lee,I. Kim,et al. 2004. Purification and cDNA cloning of a cecropin-like peptide from the greater wax moth *Galleria mellonella*. *Mol. Cells* 17:262—66.

[42]Ko,K. ,J. L. Norelli,J. P. Reynoird,W. W. Boresjza,S. K. Brown,and H. S. Aldwinckle. 2000. Effect of untranslated leader sequence of AMV RNA 4 and signal peptide of pathogenesis related protein 1b on attacin gene expression and resistance to fire blight in transgenic apple. *Biotechnol. Lett.* 22:373—81.

[43]Kogel,K. H. ,and G. Langen. 2005. Induced disease resistance and gene expression in cereals. *Cell. Microbiol.* 7:1555—64.

[44]Lamberty,M. ,S. Ades,J. S. Uttenweiler,et al. 1999. Insect immunity. Isolation from the lepidopteran *Heliothis virescens* of a novel insect defensin with potent antifungal activity. *J. Biol. Chem.* 274:9320—26.

[45]Langen,G. ,J. Imani,B. Altincicek,G. Kieseritzky,K. — H. Kogel,and A. Vilcinskas. 2006. Transgenic expression of gallerimycin,a novel antifungal insect defensin from the greater wax moth *Galleria mellonella* ,confers resistance against pathogenic fungi in tobacco. *Biol. Chem.* 387:549—57.

[46]Lee,Y. S. ,E. K. Yun,W. S. Jang,et al. 2004. Purification,cDNA cloning and expression of an insect defensin from the great wax moth,*Galleria mellonella*. *Insect Mol. Biol.* 13:65—72.

[47]Lemaitre,B. ,and J. Hoffmann. 2007. The host defence of *Drosophila melanogaster*. *Annu. Rev. Immunol.* 25:697—743.

[48]London,R. ,B. Orozco,and E. Mylonakis. 2006. The pursuit of cryptococcal pathogenesis: Heterologous hosts and the study of cryptococcal host-pathogen interactions. *FEMS Yeast Res*. 6:567—73.

[49]Maeda,H. 1996. Role of microbial proteases in pathogenesis. *Microbiol. Immunol.* 40:685—99.

[50]Mills,D. ,F. A. Hammerschlag,R. O. Nordeen,and L. D. Owens. 1994. Evidence for the breakdown of cecropin B by proteinases in the intercellular fluid of peach leaves. *Plant Sci*. 104:17—22.

[51]Miyata,S. ,M. Casey,D. Frank,F. Ausubel,and E. Drenkard. 2003. Use of the *Galleria mellonella* caterpillar as a model host to study the role of type III secretion system in Pseudomonas aeroginosa pathogenesis. *Infect. Immun.* 71:2404—13.

[52]Mohrig,W. ,and B. Messner. 1968. Immunreaktionen bei Insekten. I. Lysozyme als grundlegender antibakterieller Faktor im humoralen Abwehrgeschehen. *Biol. Zentralbl.* 87:439—47.

[53]Mylonakis,E. ,and A. Aballay. 2005. Worms and flies as genetically tractable animal models to study host-pathogen interactions. *Infect. Immun.* 73:3833—41.

[54]Mylonakis,E. ,A. Casadevall,and F. Ausubel. 2007. Exploiting amoeboid and non-vertebrate animal model systems to study the virulence of human pathogen fungi. *PLoS Pathogens* 3:0859—0865.

[55]Mylonakis,E. ,R. Moreno,J. El Khoury,et al. 2005. *Galleria mellonella* as a model system to study *Cryptococcus neoformans* pathogenesis. *Infect. Immun.* 73:3842—50.

[56]Okada,M. ,and S. Natori. 1983. Purification and characterization of an antibacterial protein from

haemolymph of *Sarcophaga peregrina* (flesh-ly) larvae. *Biochem. J*. 211:727—34.

[57]Osusky, M. , G. Zhou, L. Osuska, R. E. Hancock, W. W. Kay, and S. Misra. 2000. Transgenic plants expressing cationic peptide chimeras exhibit broad-spectrum resistance to phytopathogens. *Nat. Biotechnol*. 18:1162—66.

[58]Owens, L. D. , and T. M. Heutte. 1997. A single amino acid substitution in the antimicrobial defense protein cecropin B is associated with diminished degradation by leaf intercellular fluid. *Mol. Plant Microbe Interact*. 10:525—28.

[59]Park, S. , K. M. Kim, J. H. Lee, S. J. Seo, and I. H. Lee. 2007. Extracellular gelatinase of *Enterococcus faecalis* destroys a defense system in insect hemolymph and human serum. *Infect. Immun*. 75:1861—69.

[60]Powning, R. F. , and W. J. Davidson. 1976. Studies on the insect bacteriolytic enzymes-II. Some physical and enzymatic properties of lysozme from haemolymph of *Galleria mellonella*. *Comp. Biochem. Physiol*. 55:221—28.

[61]Reeves, E. P. , C. G. Messina, S. Doyle, and K. Kavanagh. 2004. Correlation between gliotoxin production and virulence of *Aspergillus fumigatus* in *Galleria mellonella*. *Mycopathologia* 158:73—79.

[62]Renwick, J. , P. Daly, E. Reeves, and K. Kavanagh. 2006. Susceptibility of larvae of *Galleria mellonella* to infection by *Aspergillus fumigatus* is dependent upon stage of conidial germination. *Mycopathologia* 161:377—84.

[63]Renwick, J. , E. Reeves, F. Wientjes, and K. Kavanagh. 2007. Translocation of proteins homologous to human neutrophil p47phox and p67phox to cell membrane in activated hemocytes of *Galleria mellonella*. *Dev. Comp. Immunol*. 31:347—59.

[64]Reynoird, J. , F. Mourgues, J. Norelli, H. S. Aldwinckle, M. Brisset, and E. Chevreau. 1999. First evidence for improved resistance to fire blight in transgenic pear expressing the *attacin E* gene from *Hyalophora cecropia*. *Plant Sci*. 149:23—31.

[65]Rohloff, L. , A. Wiesner, and P. Götz. 1994. A fluorescence assay demonstrating stimulation of phagocytosis by haemolymph molecules of *Galleria mellonella*. *J. Insect Physiol*. 40:1045—49.

[66]Saido-Sakanaka, H. , J. Ishibashi, E. Momotani, and M. Yamakawa. 2005. Protective effects of synthetic anti-bacterial oligopeptides based on the insect defensins on Methicillin-resistant *Staphylococcus aureus* in mice. *Dev. Comp. Immunol*. 29:469—77.

[67]Schuhmann, B. , V. Seitz, A. Vilcinskas, and L. Podsiadlowski. 2003. Cloning and expression of gallerimycin, an antifungal peptide expressed in immune response by the greater wax moth, *Galleria mellonella*. *Arch. Insect Biochem. Physiol*. 53:125—33.

[68]Scully, L. , and M. Bidochka. 2006. Developing insects as models for current and emerging human pathogens. *FEMS Microbiol. Lett*. 263:1—9.

[69]Seed, K. , and J. Dennis. 2008. Development of *Galleria mellonella* as an alternative infection model for the *Burkholderia cepacia* complex. *Infect. Immun*. 76:1267—75.

[70]Seitz, V. , A. Clermont, M. Wedde, et al. 2003. Identification of immunorelevant genes from greater wax moth (*Galleria mellonella*) by a subtractive hybridization approach. *Dev. Comp. Immunol*. 27:207—15.

[71]Sifiri, C. D. , J. Begun, F. M. Ausubel, and S. Calderwood. 2003. *Caenorhabditis elegans* as a model host for *Staphylococcus aureus* pathogenesis. *Infect. Immun*. 71:2208—17.

[72]Silva, C. P. , N. Waterfield, P. Daborn, et al. 2002. Bacterial infection of a model insect: *Photorhabdus luminescens* and *Manduca sexta*. *Cell. Microbiol.* 4:329—39.

[73]Steiner, H. , D. Hultmark, A. Engstom, H. Bennich, and H. G. Boman. 1981. Sequence and specficity of two antibacterial proteins involved in insect immunity. *Nature* 292:246—48.

[74]Terenius, O. , R. Bettencourt, S. Y. Lee, W. Li, K. Söderhäll, and I. Faye. 2007. RNA interference of hemolin causes depletion of phenoloxidase activity in *Hyalophora cecropia*. *Devel. Comp. Immunol.* 31: 571—75.

[75]Thevissen, K. , D. C. Warnecke, I. E. Francois, et al. 2004. Defensins from insects and plants interact with fungal glucosylceramides. *J. Biol. Chem.* 279:3900—05.

[76]Travis, J. , and J. Potempa. 2000. Bacterial proteinases as targets for development of second generation antibiotics. *Biochim. Biophys. Acta* 1477:35—50.

[77]Vilcinskas, A. , and P. Götz. 1999. Parasitic fungi and their interactions with the insect immune system. *Adv. Parasitol.* 43:267—13.

[78]Vilcinskas, A. , and J. Gross. 2005. Drugs from bugs: The use of insects as a valuable source of transgenes with potential in modern plant protection strategies. *J. Pest Sci.* 78:187—91.

[79]Vilcinskas, A. , and V. Matha. 1997a. Effect of the entomopathogenic fungus *Beauveria bassiana* on humoral immune response of *Galleria mellonella* larvae (Lepidoptera: Pyralidae). *Eur. J. Entomol.* 94: 461—72.

[80]Vilcinskas, A. , and V. Matha. 1997b. Antimycotic activity of lysozyme and its contribution to antifungal humoral defence reactions in *Galleria mellonella*. *Animal Biol.* 6:13—23.

[81]Vilcinskas, A. , V. Matha, and P. Götz. 1997a. Effects of the entomopathogenic fungus *Metarhizium anisopliae* and its secondary metabolites on morphology and cytoskeleton of plasmatocytes isolated from *Galleria mellonella*. *J. Insect Physiol.* 43:1149—59.

[82]Vilcinskas, A. , V. Matha, and P. Götz. 1997b. Inhibition of phagocytic activity of plasmatocytes isolated from *Galleria mellonella* by entomogenous fungi and their secondary metabolites. *J. Insect Physiol.* 43:475—83.

[83]Vilcinskas, A. , and M. Wedde. 1997. Inhibition of *Beauveria bassiana* proteases and fungal development by inducible protease inhibitors in the haemolymph of *Galleria mellonella* larvae. *Biocontrol Sci. Technol.* 7:591—601.

[84]Vodovar, N. , D. Vallenet, S. Cruveiller, et al. 2006. Complete genome sequence of the entomopathogenic and metabolically versatile soil bacterium *Pseudomonas entomophila*. *Nat. Biotechnol.* 24: 673—79.

[85]Wedde, M. , C. Weise, P. Kopacek, P. Franke, and A. Vilcinskas. 1998. Purification and characterization of an inducible metalloprotease inhibitor from the hemolymph of greater wax moth larvae, *Galleria mellonella*. *Eur. J. Biochem.* 255:534—43.

[86]Wedde, M. , C. Weise, C. Nuck, B. Altincicek, and A. Vilcinskas. 2007. The insect metalloproteinase inhibitor gene of the lepidopteran *Galleria mellonella* encodes two distinct inhibitors. *Biol. Chem.* 388: 119—27.

[87]Wiesner, A. , and P. Götz. 1993. Silica beads induce cellular and humoral immune responses in *Galleria mellonella* larvae and in isolated plasmatocytes, obtained by a newly adapted nylon wool separation. *J. Insect Physiol.* 39:865—76.

[88]Yevtushenko, D. P. , R. Romero, B. S. Forward, R. E. Hancock, W. W. Kay, and S. Misra. 2005. Pathogen-induced expression of a cecropin A-melittin antimicrobial peptide gene confers antifungal resistance in transgenic tobacco. *J. Exp. Bot.* 56:1685—95.

[89]Yu, K. H. , K. N. Kim, J. H. Lee, et al. 2002. Comparative study on characteristics of lysozymes from the hemoymph of three lepidopteran larvae, *Galleria mellonella* , *Bombyx mori* , *Agrius convolvuli*. *Dev. Comp. Immunol.* 26:707—13.

[90]Zasloff, M. 2002. Antimicrobial peptides of multicellular organisms. *Nature* 415:389—95.

[91]Zhu, Y. , T. Johnson, A. Meyers, and M. Kanost. 2003. Identification by subtractive suppression hybridization of bacteria-induced genes expressed in *Manduca sexta* fat body. *Insect Biochem. Mol. Biol.* 33:541—59.

第十六章 体腔内毒素与鳞翅目害虫防治

Nina Richtman Schmidt and Bryony C. Bonning

1. 前言

对经济作物来说，鳞翅目昆虫是最主要的害虫，它们严重影响农业生产。据估计，每年鳞翅目害虫造成全球 25%的作物损失（Oerke 1994）。治理鳞翅目害虫最常用的方法是采用传统的化学杀虫剂或种植表达苏云金（芽孢）杆菌（Bt）衍生毒素的转基因农作物，这些毒素能作用于昆虫中肠上皮细胞（Huang et al. 2002；Pray et al. 2002；Toenniessen，O'Toole 和 DeVries 2003；Lawrence 2005）。虽然 Bt 转基因农作物被越来越多的种植户所接纳，但昆虫对 Bt 毒素产生的抗性、Bt 毒素对非靶标物种以及对环境的影响受到人们持续的关注（Bates et al. 2005；Heckel et al. 2007；Rosi-Marshall et al. 2007；Sisterson et al. 2007）。通过转基因导入作用于鳞翅目昆虫的其他毒素是一种能延缓昆虫产生抗性的好方法（这种方法也称为叠加法或金字塔法）（Cao et al. 2002），或通过转基因方法对基因改造（GM：genetically modified）作物进行维护。推广使用其他毒素的农作物品种引起的市场竞争也能减少种植者在种子上的花费。

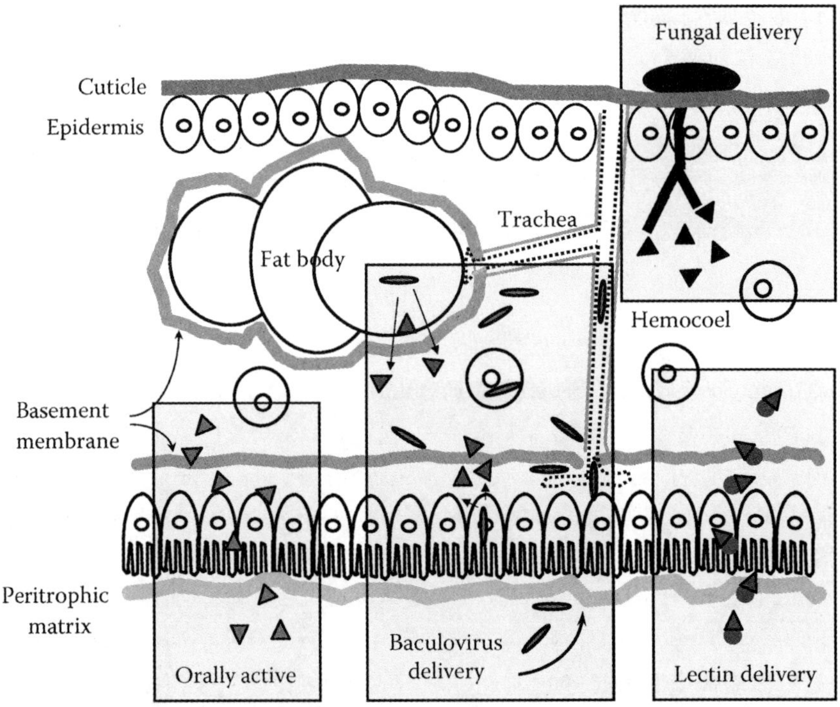

图 16.1 体腔内毒素的传递与鳞翅目害虫防治（彩色版本图片见附录图 16.1）。在一些案例中（例如采自蜘蛛毒液的 atracotoxin Hvla（Khan et al. 2006）），体腔内毒素（以三角形表示）具有口服活性，可以直接通过转基因植物传递。由于体腔内毒素本身不具有口服活性，人们可以用凝集素，诸如雪花莲凝集素 *Galanthus nivalis agglutin*（GNA，以环状表示）把毒素传递到体腔中（Fitches et al. 2004；Trung，Fitches 和 Gatehouse 2006），或者把昆虫病原体当作病毒传递的媒介，然后分泌到被感染昆虫的体腔中（Harrison 和 Bonning 2000b）。昆虫杆状病毒感染昆虫的中肠上皮细胞和其他一些组织（比如脂肪体），然后表达病毒的毒素分泌到体腔中杀死昆虫宿主（Kamita et al. 2005）。绿僵菌（*Metarhizium anisopliae*）等昆虫致病性真菌（如图右上所示）也可用来传递体腔内毒素到体腔（Wang 和 Leger 2007）。

人们已经利用转基因技术对肠活性的效应物，如 Bt 毒素和蛋白酶抑制剂进行了研究和利用，那些仅作用于体腔而不作用于肠的毒素，为鳞翅目害虫的治理提供了新的资源。在过去的 20 年中，人们一直对昆虫杆状病毒快速杀虫机理进行遗传优化研究，已经成功地筛选出一些通过病毒传递到鳞翅目昆虫幼虫体腔内的效应因子(Kamita et al. 2005)。研究人员也已经证实可用昆虫致病性真菌传递系统来传递体腔内毒素(Wang 和 Leger 2007)。根据其作用靶点，体腔内毒素通过口服不会对鳞翅目昆虫产生作用。但也有例外，源自蜘蛛毒液的 atracotoxin，经口服食下后仍具有稳定活性(King 2007)。凝集素能从肠进入到体腔，对那些在体腔中起作用的毒素而言，人们可以利用凝集素把毒素运输到体腔内的作用靶区(Fitches et al. 2002，2004；Trung，Fitches 和 Gatehouse 2006)。以下通过重组昆虫杆状病毒杀虫剂，昆虫致病性真菌和转基因植物把体腔内毒素传递到鳞翅目害虫体内的几种方式进行综述(图 16.1)。

2. 昆虫杆状病毒介导的体腔内毒素的传递

杆状病毒是专门针对昆虫的病毒，主要感染鳞翅目昆虫(Adams 和 McClintock 1991)。杆状病毒的两个属，核型多角体病毒(NPV)和粒状体病毒(GV)在病毒包涵体外形上不同，前者为多角体，后者是单个的颗粒。被视为安全和选择性的杀虫剂，杆状病毒已用于多种农业和森林害虫的治理(Federici 1999；Moscardi 1999)。杆状病毒杀虫剂比传统的化学杀虫剂起效慢，作用的宿主范围窄(在特定的种植系统内，不一定能包括所有的害虫种类)，使它在害虫防治上的进一步使用受到限制。在过去的 20 年里，杆状病毒杀虫剂已在遗传上得到了优化以加快杀虫速度(Kamita et al. 2005；Inceoglu，Kamita 和 Hammock 2006)。根据病毒一宿主的不同组合，病毒使宿主致死的时间不等，一般为几天到几周。现在主要采用的方法是把编码毒素或其他生理效应物的基因与杆状病毒基因组整合在一起。杆状病毒在宿主体内复制(繁殖)时，毒素随之产生，这些毒素会影响被感染昆虫的正常生长甚至致死(图 16.2)。因此，杆状病毒是以毒素传递系统起作用的。通过这种方法，人们已经鉴定了大约 30 种体腔内毒素的杀虫活性(图 16.3；Kamita et al. 2005)，在能承受的经济损失之下，一些重组的杆状病毒杀虫剂在害虫种群数量控制上与传统的化学杀虫剂不相上下。

2.1 激素和酶

对昆虫早期的研究主要是通过利用杆状病毒在昆虫发育过程中不恰当的时期或者过度表达昆虫的激素或酶，来破坏昆虫的生理水平。这些例子中包括杆状病毒表达利尿激素破坏宿主的水平衡(Maeda 1989)，表达蜕皮激素引起过早的蜕皮、羽化(Eldridge et al. 1991)，表达促前胸腺激素(PTTH)产生过多的蜕皮素(O'Reilly et al. 1995)，以及表达保幼激素酯酶过早引发的保幼激素滴度的降低和变态的提前(Hammock et al. 1990；Bonning et al. 1997)。这些转基因系统对缩减宿主昆虫的存活期效果相对较差，有些是因为存在有效的调控系统，使保幼激素或酶维持在适当的水平，其余的则是由于转基因表达本身有影响。比如，杆状病毒表达促前胸腺激素(PTTH)就抑制了杆状病毒的致病性(O'Reilly et al. 1995)。

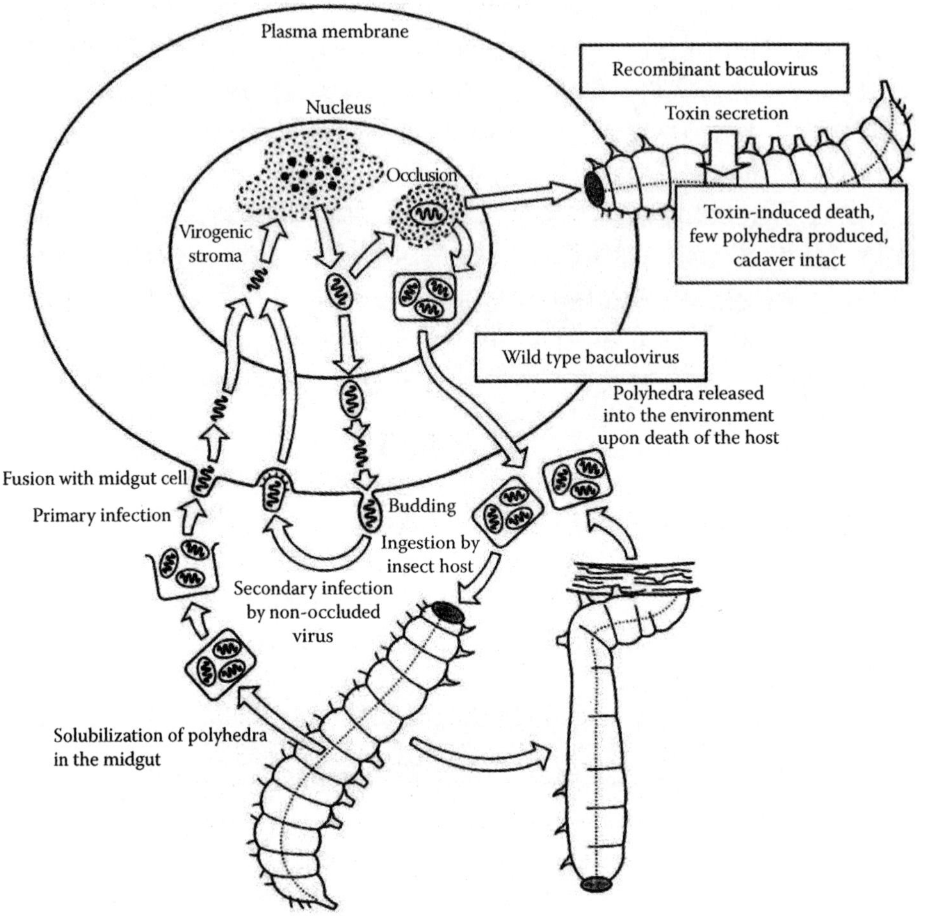

图 16.2　核型多角体病毒的生活周期。鳞翅目幼虫摄入的多角体在中肠的碱性环境中分解，释放出有感染性的包埋型病毒(ODV)。ODV 感染中肠细胞然后释放出杆状核衣壳(NC)，杆状核衣壳进入到细胞核。在细胞核内的病毒发生基质中，病毒开始复制。最初，感染细胞产生出芽病毒(BV)。出芽病毒(BV)在宿主体内传播然后又感染其他细胞和组织。随后，新合成的保留在细胞核内的核壳体被包埋进一个多角体蛋白基质，最后生成多角体。就野生型病毒的感染而言，宿主昆虫被感染后，在死之前爬上一个显眼的位置用前足悬挂起来。随后外界因素使虫尸易碎的表皮破裂，释放出多达 10^{10} 个多角体。与之相反，如果采用重组昆虫杆状病毒表达体腔内毒素，毒素会在昆虫幼虫成熟前杀死它，而不是让它因自身感染病毒而亡。感染表达麻痹神经毒素的幼虫会从植物上掉下。与感染野生型病毒形成鲜明对照的是这些虫尸很完整而且其中存在的多角体很少。

食下神经节分泌的信息素合成激活肽(PBAN)在一些鳞翅目昆虫中有赖于信息素的合成和释放(信息素的生物合成参考第十章)。与感染野生型病毒的幼虫相比，感染表达玉米夜蛾(*Helicoverpa zea*)性信息素合成激活肽(PBAN)在内的多种神经肽的重组杆状病毒的幼虫的存活期缩减至 26%(Ma et al. 1998)。但是具体是由哪个神经肽起的作用还不清楚。

与上文提到的杆状病毒表达激素与酶来调控昆虫的发育相比，一种最快的重组杆状

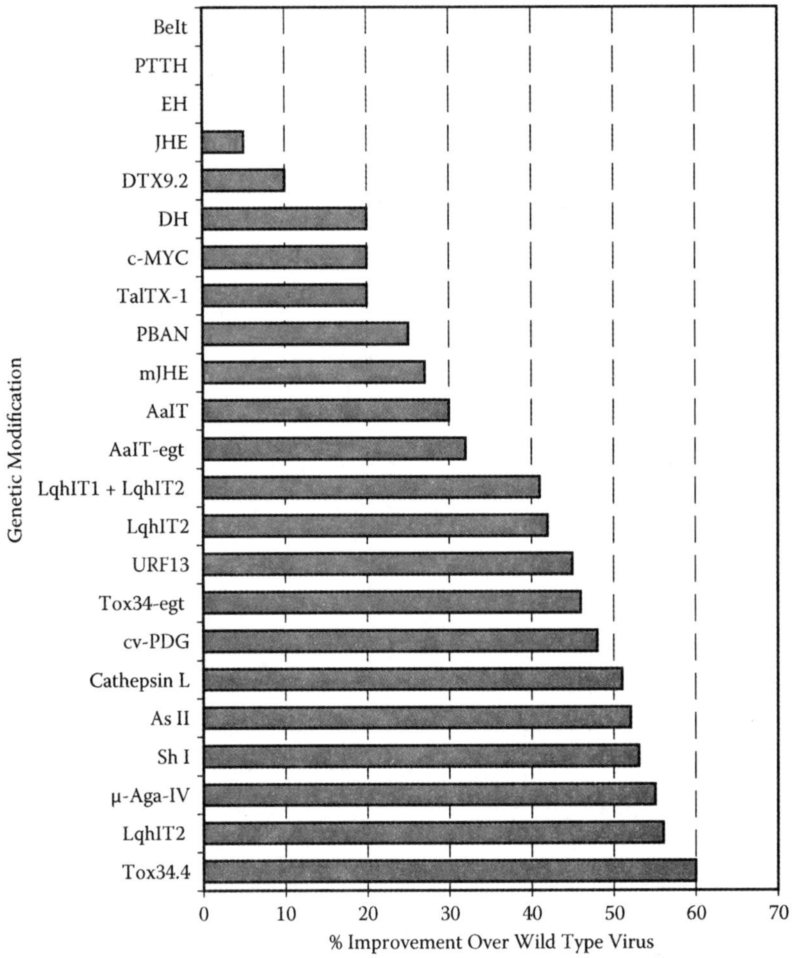

图 16.3　与野生型病毒相比较,表达体腔内毒素的重组昆虫杆状病毒杀虫效果的改进。与野生型病毒的杀虫速度相比较,重组昆虫杆状病毒的改进体现在把编码有体腔内效应物的基因插入到昆虫杆状病毒基因组中。因为使用的亲本病毒和方法的不同,把各种重组的病毒直接拿来比较是不恰当的。缩写和参考文献:Tox34.4,**螨毒素**(Carbonell et al. 1988; Burden et al. 2000);LqhIT2,iel **启动子下的蝎子毒素**(Harrison **和** Bonning 2000a);*μ*-Aga-IV,**蜘蛛毒素**(Prikhod'ko et al. 1996);Sh I,**海葵毒素**(Prikhod'ko et al. 1996);As II,**海葵毒素**(Prikhod'ko et al. 1996);Cathepsin L,**基底膜降解组织蛋白酶** L(Harrison **和** Bonning 2001);cv-PDG,**糖基化酶**(Petrik et al. 2003);Tox34-egt,**早期** DA26 **启动子在** egt **位点表达** Tox34(Popham,Li **和** Miller 1997);URF13,**玉米微孔形成蛋白质**(Korth **和** Levings 1993);LqhIT2,**蝎子** *Leiurus quinquestriatus* **昆虫毒素** 2(Froy et al. 2000);LqhIT1 + LqhIT2,**采自以色列杀人蝎** *L. quinquestriatus* **的** LqhIT1 **和** LqhIT2 **毒素的共同表达**(Regev et al. 2003);AaIT-egt,AaIT **在** egt **位点的插入**(Chen et al. 2000);AaIT,**采自黄尾肥蝎** *Androctonus australis* **的** AaIT **毒素**(McCutChen et al. 1991; Stewart et al. 1991);mJHE,**改良的** JHE(Bonning et al. 1997);PBAN,**信息素生物合成激活神经肽**(Ma et al. 1998);TalTX-1,**蜘蛛毒素**(Hughes et al. 1997);*c-MYC*,**转录因子** *c-myc* **基因的反义链表达**(Lee et al. 1997);DH,**利尿激素**(Maeda 1989);DTX9.2,**蜘蛛毒素**(Hughes et al. 1997);JHE,**保幼激素酯酶**(Hammock et al. 1990);EH,**羽化激素**(Eldridge,O'Reilly **和** Miller 1992);PTTH,**促前胸腺激素激素**(O'Reilly et al. 1995);Belt,**螨昆虫毒素**-1(Carbonell et al. 1988)。**继** Kamita **等之后**(2005)。

病毒杀虫剂是表达一种 L 样的组织蛋白酶，这种蛋白酶是以基底膜为靶标（Harrison 和 Bonning 2001）。这种来自麻蝇（食肉蝇，*Sarcophaga peregrina*）的蛋白酶能选择性地分解两种基底膜上的底物蛋白（Homma et al. 1994；Homma 和 Natori 1996；Fujii-Taira et al. 2000）。此重组病毒被用来探讨昆虫宿主体内被破坏的基底膜是否会促进病毒的传播进而加快杀虫的速度（Harrison 和 Bonning 2001）。然而，与其说加快病毒传播，还不如说昆虫杆状病毒表达的蛋白酶给基底膜及以下的组织带来的破坏足以杀死宿主昆虫（Li et al. 2007；Philip et al. 2007；Tang et al. 2007）。

2.2 神经毒素

昆虫选择性神经毒素已从胡蜂、蜘蛛、蝎子、海葵、螨等有毒生物中分离出来（De Lima et al. 2007；Gurevitz et al. 2007；King 2007）。以上神经毒素是分离昆虫特异性毒素的主要来源，比如蜘蛛的毒腺中就有上千种多肽毒素（Escoubas 2006；Escoubas，Sollod 和 King 2006）。这些毒素主要在钠、钾、钙、氯等离子通道中起作用，导致昆虫快速麻痹。这些神经毒素在昆虫杆状病毒中表达后，被感染幼虫取食植物所造成的危害明显比那些感染野生型病毒的幼虫小很多(图 16.3)。这些麻痹神经毒素还可使幼虫从植物上掉下来，减小了对植物进一步损害的可能性，同时也减少了对叶子的污染（Cory et al. 1994；Hoover et al. 1995）。北非蝎（*Androctonus australis*）和以色列金蝎（*Leiurus spp.*）的毒素都已经研究得比较透彻，北非蝎能产生起兴奋作用的神经毒素 AaIT，以色列金蝎能产生既起兴奋作用又起抑制作用的毒素（Zlotkin et al. 1991；Zlotkin，Fishman 和 Elazar 2000）。美国、中国和英国的研究人员都对表达这些毒素的重组昆虫杆状病毒进行了田间实验（Maeda et al. 1991；McCutchen et al. 1991；Stewart et al. 1991；Zlotkin et al. 1995；Black et al. 1997；Chen et al. 2000；Harrison 和 Bonning 2000a；Treacy，Rensner 和 All 2000；Zlotkin，Fishman 和 Elazar 2000；Sun et al. 2002；Sun et al. 2004）。

与直接注射毒素相比，通过昆虫杆状病毒在鳞翅目昆虫体内表达 AaIT 起到了出乎意料的效果。这是因为杆状病毒在气管上皮中的复制使其产生的毒素能绕过保护鳞翅目昆虫神经元的神经层。这种杆状病毒传递系统能让毒素直接进入到中枢神经系统，通过注射进入虫体的神经毒素却不能穿过这层障碍（Elazar，Levi 和 Zlotkin 2001）。研究者尝试在同一昆虫杆状病毒中表达多种神经毒素，并发现毒素间在增加毒性上有协同效应（Herrmann et al. 1995；Prikhod'ko et al. 1998；Regev et al. 2003）。

2.3 其他基因产物

除了毒素、昆虫激素和酶以外，研究者还将几种其他的体腔内效应因子基因插入到昆虫杆状病毒基因组中，加快了杀虫速度。

URF13 是一种来自玉米的线粒体编码的蛋白质，该蛋白能在线粒体内膜上形成微孔（Korth 和 Levings 1993）。与野生型病毒相比较，昆虫杆状病毒表达 URF13 的杀虫速度明显提高（提高了 45%）。其作用机理不涉及微孔的形成，但和细胞功能的破坏有关（Korth 和 Levings 1993）。

构建表达了反义 *c-myc* 基因的重组昆虫杆状病毒（Lee et al. 1997）。c-MYC 蛋白是一种参与细胞生长、增殖以及凋亡等多种生理过程的转录因子。反义 *c-myc* 基因的表达缩短了感染幼虫存活的时间，减少了感染幼虫对植物造成的取食危害。正如众多研究者所指出的一样，使用反义技术可以避免涉及翻译后过程和重组蛋白分泌的问题，因其没有外源蛋白产生，同时也降低了昆虫产生抗性的可能性。

为了使杆状病毒在紫外光下能存活，2003 年 Petrik 等研制了一种表达海藻病毒的嘧啶二聚体特异的糖基化酶（cv-PDG）的重组杆状病毒，这种糖基化酶与 DNA 的修复有关。出芽病毒比多角体病毒对紫外光灭活更具有抵抗性。出乎意料的是感染这种病毒的草地贪夜蛾的幼虫存活时间只相当于那些感染野生型病毒的 48%（Petrik et al. 2003）。

3. 真菌介导的体腔内毒素的传递

Prl 是一种由昆虫致病性真菌绿僵菌（*Metarhizium anisopliae*）所产生的蛋白酶，它能帮助真菌穿透宿主的表皮。研究者已经利用转基因绿僵菌过度表达降解宿主表皮的蛋白酶（Leger 1995）。Prl 在血淋巴中的过度表达激活了昆虫免疫系统中的前酚氧化酶系统，加速了杀灭真菌的速度，同时也减少了烟草天蛾（*Manduca sexta*）幼虫摄食对植物的危害（Leger et al. 1996；Hu 和 St. Leger 2002）。与此相反，内切壳多糖酶的过量表达能使它更容易穿透宿主的表皮，但不能增强真菌的毒性（较野生型真菌而言），由此表明，壳多糖酶的活性不是表皮穿透的限制因素（Screen，Hu 和 St. Leger 2001）。

就像通过重组昆虫杆状病毒表达一样，研究人员已证实通过绿僵菌传递体腔内毒素的方法是可行的（Wang 和 Leger 2007）。通过真菌介导，利用 MCL1 启动子在血淋巴中快速、高水平地表达昆虫选择性神经毒素 AaIT，能使真菌对 5 龄的烟草天蛾幼虫的感染力增加 22 倍。此外，为了减少灭杀烟草天蛾幼虫所需分子孢子的数量，AaIT 表达使其幼虫的存活时间比那些感染野生型绿僵菌的幼虫缩短了 28%。研究人员还观察到了由 AaIT 引起的收缩麻痹这一典型性状。这些经过遗传改良之后的绿僵菌将会为防治那些对 Bt 毒素不敏感的昆虫提供一个更有效可行的方法（Wang 和 Leger 2007）。

4. 来源于转基因植物的体腔内毒素的传递

根据体腔内毒素的定义，它的作用靶点在昆虫体内，而不是在昆虫肠内，由此，研究人员认为这些病毒没有口服活性。也许就因为这种假设，很少有关于体腔内毒素在鳞翅目昆虫中口服毒性的测试报告。但也有例外，来自蝎子的神经毒素 SFI1 和 AaIT 不具有口服活性（Zlotkin，Fishman 和 Shapiro 1992； Fitches et al. 2004），而来自蜘蛛的 atracotoxin ω-ACTX-Hvla 具有口服活性（Khan et al. 2006）。没有这些神经毒素食下的测试报告原因可能有二：一是因为很难从毒液中提纯到足够多可做活性鉴定的材料；二是因为很难生产有活性的重组毒素（Taniai，Inceoglu 和 Hammock 2002）。

ω-ACTX-Hvla 等具有口服活性的毒素必须具有足够的稳定性以耐受住昆虫中肠的恶劣环境。这些经由食下的毒素必须具有抵抗消化蛋白酶降解的特性，然后足够量的毒素能穿过中肠的上皮细胞到体腔中发挥它们的毒性。目前这些毒素到达鳞翅目昆虫体腔的机制还不是很清楚。

虽然已经证实来自蝎子毒液的非常小剂量(0.3%)的神经毒素 AaIT 可以经食下进入到绿头苍蝇(*Sarcophaga falculata*)的体腔,但 AaIT 在烟草天蛾(*M. sexta*)或棉铃虫(*Helicoverpa armigera*)却没有食下活性(Zlotkin,Fishman 和 Shapiro 1992)。Zlotkin 等运用放射性碘标记的毒素证实鳞翅目昆虫幼虫血淋巴中只存在已被降解的毒素,并没有发现完好的毒素(Zlotkin,Fishman 和 Shapiro 1992)。与这些结果形成鲜明对比的是,已有了好几例关于表达抗害虫毒素 AaIT 的具有昆虫抗性的转基因植物成功的报告:抗舞毒蛾(*Lymantria dispar*)的杨树(Wu et al. 2000)和抗棉铃虫的玉米(*Helicoverpa armigera*)(Yao et al. 1996; Wu et al. 2008)。然而,这些报告中没有足够的信息证实杀虫作用是由 AaIT 起的,他们并没有检测到这些植物表达的 AaIT 毒性,也没有提供 AaIT 能使幼虫麻痹的证据。由于注射鳞翅目昆虫幼虫体内的 AaIT 只有相对较低的毒性,要达到杀虫效果就需要大量的毒素(Elazar,Levi 和 Zlotkin 2001)。由于 AaIT 显示出对两种鳞翅目昆虫缺少食下毒性(Zlotkin,Fishman 和 Shapiro 1992),就应该更慎重对待转基因植物表达 AaIT 的数据。

此外,还有两例关于转基因植物表达毒素的报告。一例是表达蜘蛛毒液基因的转基因水稻(Huang et al. 2001);另一例是表达烟草天蛾幼虫壳多糖酶基因和来自满洲蝎(*Buthus martensii*)的蝎毒基因 *Bmk*(Wang et al. 2005)的转基因油菜。结果显示它们分别对钻杆虫(*Chino suppressalis*)幼虫和小菜蛾(*Plutella xylostella*(L.) syn. *P. maculipennis*)(Curtis)具有抗性。但是这些报告同样未能提供神经毒素起作用的证据。

4.1 食下活性毒素

近年来,研究人员把取自蜘蛛、蝎子和胡蜂的具有杀虫活性的昆虫毒素的基因重组到了植物基因组中,以防治鳞翅目昆虫中的植食性昆虫。研究者在转基因烟草表达了具有食下活性的来自澳大利亚漏斗蜘蛛(*Hadronyche versuta*)的毒素 atracotoxin ω-ACTX-Hvla。这些转基因烟草在 72h 内对第 2 龄期的棉铃虫(*Helicoverpa armigera*)和斜纹夜蛾(*Spodoptera littoralis*)的杀虫性达到 100%(Khan et al. 2006)。ω-ACTX-1 肽家族的毒素有 36～37 个氨基酸残基,拥有稳定的半胱氨酸结构,因其为小分子,所以它们能穿过血脑障碍到达位于中枢神经系统中的作用靶位点(Fletcher et al. 1997)。有趣的是,即使只是局部使用 ω-ACTX-Hvla,幼虫同样也会被杀死(Khan et al. 2006)。

4.2 TSP14

膜翅目寄生虫会在其卵孵化的时候将畸形细胞释放到鳞翅目昆虫幼虫的体腔中(关于鳞翅目寄生虫在第十七章介绍)。这些畸形细胞产生抑制鳞翅目宿主蛋白合成、生长和发育的畸形细胞分泌蛋白 TSP14,以维持体内寄生虫的生长(Dahlman et al. 2003)。研究人员已把编码 TSP14 的基因重组到了烟草基因组中(Maiti et al. 2003)。当烟草天蛾(*Manduca sexta*)和烟芽夜蛾(*Heliothis virescens*)取食这些转基因的烟草时,其生长速率降低并且死亡率上升。虽然烟草天蛾和烟芽夜蛾的死亡是由于 TSP14 抑制了宿主体内的蛋白质合成(Maiti et al. 2003),但至于 TSP14 起的杀虫作用是在幼虫的肠中,还是像人们所期望的一样在体腔内,从这个研究中还得不到答案。

4.3 凝集素介导的体腔内毒素的传递

凝集素是结合碳水化合物的蛋白质,能抵抗植食性动物肠道内的蛋白酶水解。在植物体内,凝集素是一种针对鳞翅目昆虫幼虫等植食性动物的防护性化合物。有资料显示,通过转基因植物表达的凝集素对甘蓝夜蛾(*Lacanobia oleracea*)等鳞翅目昆虫有副作用(Gatehouse et al. 1999)。凝集素在肠道内结合聚糖受体,从而产生毒性。凝集素拥有广泛的结合特异性,一些植物凝集素,比如甘露糖特异的雪莲凝集素(GNA),可以(可能是以胞吞的方式)通过中肠壁进入昆虫的体腔(Fitches et al. 2001)。这个特性使得我们可以用凝集素将毒素运送到昆虫体腔。昆虫吸收凝集素后,可在其肠道、马氏管和血淋巴中检测到雪莲凝集素(GNA)(Fitches et al. 2001)。

GNA 对昆虫具有高毒性,但对哺乳动物等其他非靶标物种却显示出低毒性,这种特性对病虫害的治理极为有利(Hilder et al. 1995; Down et al. 1996; Gatehouse et al. 1996; Sauvion et al. 1996)。虽然 GNA 对鳞翅目昆虫也显示出低毒性(Gatehouse et al. 1995; Fitches,Gatehouse 和 Gatehouse 1997; Gatehouse et al. 1997; Rao et al. 1998; Stoger et al. 1998),但是它和烟草天蛾的咽侧体抑制素结合后却能抑制第五龄期的甘蓝夜蛾(*Lacanobia oleracea*)的摄食和生长发育。而且,在血淋巴中能检测到咽侧体抑制素,表明 GNA 能穿过肠的上皮细胞将咽侧体抑制素运送到血淋巴中(Fitches et al. 2002)。GNA 同时也被用以将来自蜘蛛毒液的 *segestria florentina* 1 毒素(SFI1)运送到血淋巴中。通过口服法单独使用 SFI1 毒素或 GNA 对甘蓝夜蛾(*Lacanobia oleracea*)没有杀虫效果,但将二者的融合产物添加到食物中后,6 天内甘蓝夜蛾的死亡率能达到 100%(Fitches et al. 2004)。研究者还用免疫印迹实验证实了体腔中 GNA-SFI1 融合蛋白的存在。和单独使用毒素或 GNA 相比,把来自红蝎(*Mesobuthus tamulus*)毒素 ButalT 和 GNA 融合在一起使用能降低甘蓝夜蛾(*Lacanobia oleracea*)幼虫的生长率,提高死亡率。同样,可以在昆虫血淋巴中检测到完整的 ButalT-GNA 融合蛋白(Trung, Fitches 和 Gatehouse 2006)。这些研究表明,通过在植物体内表达凝集素一体腔内毒素融合体在鳞翅目害虫防治上具有很好的发展潜力。

5. 展望

(1)毒液仍是将来用以防治鳞翅目害虫毒素的优质来源。为防治鳞翅目害虫,研究人员同样在关注来源于昆虫致病性线虫的细菌(ffrench-Constant,Dowling 和 Waterfield 2007)。

(2)鉴于较其他体腔内效应物,昆虫选择性神经毒素具有更好的特异性和持久性,这些毒素在鳞翅目害虫防治上将会起到更大的作用。表达 TSP14 或 atracotoxin 毒素的抗鳞翅目害虫转基因植物很可能会得到商业化。

(3)在中国,经遗传改良的昆虫杆状病毒杀虫剂在防治棉铃虫(*Helicoverpa armigera*)方面有特别好的应用前景。在防治那些对 Bt 毒素不易感的害虫方面经遗传改良的昆虫致病性真菌也将会在杀虫剂市场中占有一席之地。

(4)ω-ACTX-Hvla(实际上是凝集素)等食下活性神经毒素进入到鳞翅目昆虫体腔的机理到现在为止还不清楚,但对其的研究将为鳞翅目害虫的防治提供新的策略。

找到 Bt 毒素转基因植物的替代品,为抗鳞翅目害虫转基因植物另辟蹊径越来越显重要,因此利用体腔内毒素防治鳞翅目害虫的研究扩展了该领域的研究空间。

6. 致谢

本研究部分由美国农业部(USDA NRI 2003-35302-13558)、哈奇法案和衣阿华州立基金资助。NRS 由 Henry 和 Sylvia Richardson 研究激励基金资助。

参考文献

[1]Adams,J. R. ,and J. T. McClintock. 1991. Baculoviridae. Nuclear polyhedrosis viruses,part 1: Nuclear polyhedrosis viruses of insects. In *Atlas of Invertebrate Viruses* ,ed. J. R. Adams and J. R. Bonami,87—204. Boca Raton,FL: CRC Press.

[2]Bates,S. L. ,J. Z. Zhao,R. T. Roush,and A. M. Shelton. 2005. Insect resistance management in GM crops: Past,present and future. *Nat. Biotechnol.* 23:57—62.

[3]Black,B. C. ,L. A. Brennan,P. M. Dierks,and I. E. Gard. 1997. Commercialization of baculoviral insecticides. In *The Baculoviruses* ,ed. L. K. Miller,341—88. New York: Plenum Press.

[4]Bonning,B. C. ,V. K. Ward,M. van Meer,T. F. Booth,and B. D. Hammock. 1997. Disruption of lysosomal targeting is associated with insecticidal potency of juvenile hormone esterase. *Proc. Natl. Acad. Sci. U. S. A.* 94:6007—12.

[5]Burden,J. P. ,R. S. Hails,J. D. Windass,M. -M. Suner,and J. S. Cory. 2000. Infectivity,speed of kill,and productivity of a baculovirus expressing the itch mite toxin Txp-1 in second and fourth instar larvae of *Trichoplusia ni*. *J. Invertebr. Pathol.* 75:226—36.

[6]Cao,J. ,J. -Z. Zhao,J. D. Tang,A. M. Shelton,and E. D. Earle. 2002. Broccoli plants with pyramided *cry*1Ac and *cry*1C Bt genes control diamondback moths resistant to Cry1A and Cry1C proteins. *Theor. Appl. Genet.* 105:258—64.

[7]Carbonell,L. F. ,M. R. Hodge,M. D. Tomalski,and L. K. Miller. 1988. Synthesis of a gene coding for an insect-specific scorpion neurotoxin and attempts to express it using baculovirus vectors. *Gene* 73:409—18.

[8]Chen,X. ,X. Sun,Z. Hu,M. Li,et al. 2000. Genetic engineering of *Helicoverpa armigera* single-nucleopoly-hedrovirus as an improved pesticide. *J. Invertebr. Pathol.* 76:140—46.

[9]Cory,J. S. ,M. L. Hirst,T. Williams,et al. 1994. Field trial of a genetically improved baculovirus insecticide. *Nature* 370:138—40.

[10]Dahlman,D. L. ,R. L. Rana,E. J. Schepers,T. Schepers,F. A. DiLuna,and B. A. Webb. 2003. A teratocyte gene from a parasitic wasp that is associated with inhibition of insect growth and development inhibits host protein synthesis. *Insect Mol. Biol.* 12:527—34.

[11]De Lima,M. E. ,S. G. Figueiredo,A. M. Pimenta,et al. 2007. Peptides of arachnid venoms with insecticidal activity targeting sodium channels. *Comp. Biochem. Physiol. C Toxicol. Pharmacol.* 146:264—79.

[12]Down, R. E., A. M. R. Gatehouse, G. M. Davison, et al. 1996. Snowdrop lectin inhibits development and decreases fecundity of the glasshouse potato aphid (*Aulacorthum solani*) when administered in vitro and via transgenic plants both in laboratory and glasshouse trials. *J. Insect Physiol.* 42:1035—45.

[13]Elazar, M., R. Levi, and E. Zlotkin. 2001. Targeting of an expressed neurotoxin by its recombinant baculovirus. *J. Exp. Biol.* 204:2637—45.

[14]Eldridge, R., F. M. Horodyski, D. B. Morton, et al. 1991. Expression of an eclosion hormone gene in insect cells using baculovirus vectors. *Insect Biochem.* 21:341—51.

[15]Eldridge, R., D. R. O'Reilly, and L. K. Miller. 1992. Efficacy of a baculovirus pesticide expressing an eclosion hormone gene. *Biol. Control* 2:104—10.

[16]Escoubas, P. 2006. Molecular diversification in spider venoms: A web of combinatorial peptide libraries. *Mol. Divers.* 10:545—54.

[17]Escoubas, P., B. Sollod, and G. F. King. 2006. Venom landscapes: Mining the complexity of spider venoms via a combined cDNA and mass spectrometric approach. *Toxicon* 47:650—63.

[18]Federici, B. A. 1999. Naturally occurring baculoviruses for insect pest control. Methods Biotechnol. 5:301—20. ffrench-Constant, R. H., A. Dowling, and N. R. Waterfield. 2007. Insecticidal toxins from *Photorhabdus* bacteria and their potential use in agriculture. *Toxicon* 49:436—51.

[19]Fitches, E., N. Audsley, J. A. Gatehouse, and J. P. Edwards. 2002. Fusion proteins containing neuropeptides as novel insect control agents: Snowdrop lectin delivers fused allatostatin to insect haemolymph following oral ingestion. *Insect Biochem. Molec. Biol.* 32:1653—61.

[20]Fitches, E., M. G. Edwards, C. Mee, et al. 2004. Fusion proteins containing insect-specific toxins as pest control agents: Snowdrop lectin delivers fused insecticidal spider venom toxin to insect haemolymph following oral ingestion. *J. Insect Physiol.* 50:61—71.

[21]Fitches, E., A. M. R. Gatehouse, and J. A. Gatehouse. 1997. Effects of snowdrop lectin (GNA) delivered via artificial diet and transgenic plants on the development of the tomato moth (Lacanobia oleracea) larvae in the laboratory and glasshouse trials. *J. Insect Physiol.* 43:727—39.

[22]Fitches, E., S. D. Woodhouse, J. P. Edwards, and J. A. Gatehouse. 2001. In vitro and in vivo binding of snowdrop (*Galanthus nivalis* agglutinin; GNA) and jackbean (*Canavalia ensiformis*; Con A) lectins within tomato moth (*Lacanobia oleracea*) larvae; mechanisms of insecticidal action. *J. Insect Physiol.* 47:777—87.

[23]Fletcher, J. I., R. Smith, S. I. O'Donoghue, et al. 1997. The structure of a novel insecticidal neurotoxin, omega-atracotoxin-HV1, from the venom of an Australian funnel web spider. *Nat. Struct. Biol.* 4:559—66.

[24]Froy, O., N. Zilberberg, N. Chejanovsky, J. Anglister, and E. Loret. 2000. Scorpion neurotoxins: Structure/function relationships and application in agriculture. *Pest Manag. Sci.* 56:472—74.

[25]Fujii-Taira, I., Y. Tanaka, K. J. Homma, and S. Natori. 2000. Hydrolysis and synthesis of substrate proteins for cathepsin L in the brain basement membranes of *Sarcophaga* during metamorphosis. *J. Biochem.* 128:539—42.

[26]Gatehouse, A. M. R., G. M. Davison, C. A. Newell, et al. 1997. Transgenic potato plants with enhanced resistance to the tomato moth (*Lacanobia oleracea*) larvae; mechanisms of insecticidal action. *Mol. Breeding* 3:49—63.

[27]Gatehouse, A. M. R., G. M. Davidson, J. N. Stewart, et al. 1999. Concanavalin A inhibits development of tomato moth (*Lacanobia oleracea*) and peach-potato aphid (*Myzus persicae*) when expressed in

transgenic potato. *Mol. Breeding* 5:153—65.

[28]Gatehouse, A. M. R. , R. E. Down, K. S. Powell, et al. 1996. Transgenic potato plants with enhanced resistance to the peach-potato aphid *Myzus persicae*. *Ent. Exp. Appl.* 79:295—307.

[29]Gatehouse, A. M. R. , K. S. Powell, E. J. M. V. Damme, and J. A. Gatehouse. 1995. Insecticidal properties of plant lectins. In *Lectins, biomedical perspectives*, ed. A. Pusztai and S. Bardocz, 35—57. London: Taylor and Francis.

[30]Gurevitz, M. , I. Karbat, L. Cohen, et al. 2007. The insecticidal potential of scorpion beta-toxins. *Toxicon* 49:473—89.

[31]Hammock, B. D. , B. C. Bonning, R. D. Possee, T. N. Hanzlik, and S. Maeda. 1990. Expression and effects of the juvenile hormone esterase in a baculovirus vector. *Nature* 344:458—61.

[32]Harrison, R. L. , and B. C. Bonning. 2000a. Use of scorpion neurotoxins to improve the insecticidal activity of *Rachiplusia ou* multicapsid nucleopolyhedrovirus. *Biol. Control* 17:191—201.

[33]Harrison, R. L. , and B. C. Bonning. 2000b. Genetic engineering of biocontrol agents for insects. In *Lectins, biomedical perspectives*, ed. A. Pusztai and S. Bardocz, 35—57. London: Taylor and Francis.

[34]Harrison, R. L. , and B. C. Bonning. 2001. Use of proteases to improve the insecticidal activity of baculoviruses. *Biol. Control* 20:199—209.

[35]Heckel, D. G. , L. J. Gahan, S. W. Baxter, et al. 2007. The diversity of Bt resistance genes in species of Lepidoptera. *J. Invertebr. Pathol.* 95:192—97.

[36]Herrmann, R. , H. Moskowitz, E. Zlotkin, and B. D. Hammock. 1995. Positive cooperativity among insecticidal scorpion neurotoxins. *Toxicon* 33:1099—1102.

[37]Hilder, V. A. , K. S. Powell, A. M. R. Gatehouse, et al. 1995. Expression of snowdrop lectin in transgenic tobacco plants results in added protection against aphids. *Transgenic Res.* 4:18—25.

[38]Homma, K. , S. Jurata, and S. Natori. 1994. Purification, characterization, and cDNA cloning of procathepsin L from the culture medium of NIH-Sape-4, an embryonic cell line of *Sarcophaga peregrina* (flesh fly), and its involvement in the differentiation of imaginal discs. *J Biol. Chem.* 269:15258—64.

[39]Homma, K. , and S. Natori. 1996. Identification of substrate proteins for cathepsin L that are selectively hydrolyzed during the differentiation of imaginal discs of *Sarcophaga peregrina*. *Eur. J. Biochem.* 240:443—47.

[40]Hoover, K. , C. M. Schultz, S. S. Lane, B. C. Bonning, S. S. Duffey, and B. D. Hammock. 1995. Reduction in damage to cotton plants by a recombinant baculovirus that causes moribund larvae of *Heliothis virescens* to fall off the plant. *Biol. Control* 5:419—26.

[41]Hu, G. , and R. J. St. Leger. 2002. Field studies using a recombinant mycoinsecticide (Metarhizium anisopliae) reveal that it is rhizosphere competent. *Appl. Environ. Microbiol.* 68:6383—87.

[42]Huang, J. , S. Rozelle, C. Pray, and Q. Wang. 2002. Plant biotechnology in China. *Science* 295:674—76.

[43]Huang, J. Q. , Z. M. Wel, H. L. An, and Y. X. Zhu. 2001. *Agrobacterium tumefaciens*-mediated transformation of rice with the spider insecticidal gene conferring resistance to leaffolder and striped stem borer. *Cell Res.* 11:149—55.

[44]Hughes, P. R. , H. A. Wood, J. P. Breen, S. F. Simpson, A. J. Duggan, and J. A. Dybas. 1997. Enhanced bioactivity of recombinant baculoviruses expressing insect-specific spider toxins in lepidopteran crop pests. *J. Invertebr. Pathol.* 69:112—18.

[45]Inceoglu, A. B. , S. G. Kamita, and B. D. Hammock. 2006. Genetically modified baculoviruses: A

historical overview and future outlook. *Adv. Virus Res.* 68:323—60.

[46]Kamita, S. G. , K. —D. Kang, B. D. Hammock, and A. B. Inceoglu. 2005. Genetically modified baculoviruses for pest insect control. In *Comprehensive molecular insect science*, ed. L. I. Gilbert et al. , Vol. 6,271—322. Oxford: Elsevier.

[47]Khan, S. A. , Y. Zafar, R. W. Briddon, K. A. Malik, and Z. Mukhtar. 2006. Spider venom toxin protects plants from insect attack. *Transgenic Res.* 15:349—57.

[48]King, G. F. 2007. Modulation of insect Ca(v) channels by peptidic spider toxins. *Toxicon* 49:513—30.

[49]Korth, K. L. , and C. S. Levings. 1993. Baculovirus expression of the maize mitochondrial protein URF13 confers insecticidal activity in cell cultures and larvae. *Proc. Natl. Acad. Sci. U. S. A.* 90:3388—92.

[50]Lawrence, S. 2005. Agbio keeps on growing. *Nat. Biotechnol.* 23:281.

[51]Lee, S. —Y. , X. Qu, W. Chen, et al. 1997. Insecticidal activity of a recombinant baculovirus containing an antisense *c-myc* fragment. *J. Gen. Virol.* 78:273—81.

[52]Leger, R. J. S. 1995. The role of cuticle-degrading proteases in fungal pathogenesis of insects. *Can. J. Bot.* (Suppl. 1) 73:S1119—25.

[53]Leger, R. J. S. , L. Joshi, M. J. Bidochka, and D. W. Roberts. 1996. Construction of an improved mycoinsecticide overexpressing a toxic protease. *Proc. Natl. Acad. Sci. U. S. A.* 93:6349—54.

[54]Li, H. , H. Tang, R. L. Harrison, and B. C. Bonning. 2007. Impact of a basement membrane-degrading protease on dissemination and secondary infection of *Autographa californica* multiple nucleopolyhedrovirus in *Heliothis virescens* (Fabricus). *J. Gen. Virol.* 88:1109—19.

[55]Ma, P. W. K. , T. R. Davis, H. A. Wood, D. C. Knipple, and W. L. Roelofs. 1998. Baculovirus expression of an insect gene that encodes multiple neuropeptides. *Insect Biochem. Mol. Biol.* 28:239—49.

[56]Maeda, S. 1989. Increased insecticidal effect by a recombinant baculovirus carrying a synthetic diuretic hormone gene. *Biochem. Biophys. Res. Commun.* 165:1177—83.

[57]Maeda, S. , S. L. Volrath, T. N. Hanzlik, et al. 1991. Insecticidal effects of an insect-specific neurotoxin expressed by a recombinant baculovirus. *Virology* 184:777—80.

[58]Maiti, I. B. , N. Oey, D. L. Dahlman, and B. A. Webb. 2003. Antibiosis-type resistance in transgenic plants expressing a teratocyte secretory peptide (TSP) gene from a hymenopteran endoparasite (Microplitis croceipes). *Plant Biotech. J.* 1:209—19.

[59]McCutchen, B. F. , P. V. Choudary, R. Crenshaw, et al. 1991. Development of a recombinant baculovirus expressing an insect-selective neurotoxin: Potential for pest control. *Bio/Technol.* 9:848—52.

[60]Moscardi, F. 1999. Assessment of the application of baculoviruses for control of Lepidoptera. *Annu. Rev. Entomol.* 44:257—89.

[61]Oerke, E. —C. 1994. Estimated crop losses due to pathogens, animal pests and weeds. In *Crop production and crop protection: Estimated losses in major food and cash crops*, ed. E. —C. Oerke et al. , 72—78. Amsterdam: Elsevier.

[62]O'Reilly, D. R. , T. J. Kelly, E. P. Masler, et al. 1995. Overexpression of *Bombyx mori* prothoracicotropic hormone using baculovirus vectors. *Insect Biochem. Mol. Biol.* 25:475—85.

[63]Petrik, D. T. , A. Iseli, B. A. Montelone, J. L. Van Etten, and R. J. Clem. 2003. Improving baculovirus resistance to UV inactivation: Increased virulence resulting from expression of a DNA repair enzyme. *J. Invertebr. Pathol.* 82:50—56.

[64]Philip, J. M. D., E. Fitches, R. L. Harrison, B. C. Bonning, and J. A. Gatehouse. 2007. Characterisation of functional and insecticidal properties of a recombinant cathepsin L-like proteinase from flesh fly (*Sarcophaga peregrina*), which plays a role in differentiation of imaginal discs. *Insect Biochem. Molec. Biol.* 37:589—600.

[65]Popham, H. J. R., Y. Li, and L. K. Miller. 1997. Genetic improvement of *Helicoverpa zea* nuclear polyhedrosis virus as a biopesticide. *Biol. Control* 10:83—91.

[66]Pray, C. E., J. Huang, R. Hu, and S. Rozelle. 2002. Five years of Bt cotton in China—the beneits continue. *Plant J.* 31:423—30.

[67]Prikhod'ko, G. G., H. J. R. Popham, T. J. Felcetto, et al. 1998. Effects of simultaneous expression of two sodium channel toxin genes on the properties of baculoviruses as biopesticides. *Biol. Control* 12:66—78.

[68]Prikhod'ko, G. G., M. Robson, J. W. Warmke, et al. 1996. Properties of three baculovirus-expressing genes that encode insect-selective toxins: μ-Aga-IV, As II, and Sh I. *Biol. Control* 7:236—44.

[69]Rao, K. V., K. S. Rathore, T. K. Hodges, et al. 1998. Expression of snowdrop lectin (GNA) in the phloem of transgenic rice plants confers resistance to rice brown planthopper. *Plant J.* 15:469—77.

[70]Regev, A., H. Rivkin, B. Inceoglu, et al. 2003. Further enhancement of baculovirus insecticidal efficacy with scorpion toxins that interact cooperatively. *FEBS Lett.* 537:106—10.

[71]Rosi-Marshall, E. J., J. L. Tank, T. V. Royer, et al. 2007. Toxins in transgenic crop byproducts may affect headwater stream ecosystems. *Proc. Natl. Acad. Sci. U. S. A.* 104:16204—08.

[72]Sauvion, N., Y. Rahbe, W. J. Peumans, E. J. V. Damme, J. A. Gatehouse, and A. M. R. Gatehouse. 1996. Effects of GNA and other binding lectins on development and fecundity of the peach-potato aphid *Myzus persicae*. *Ent. Exp. Appl.* 79:285—93.

[73]Screen, S. E., G. Hu, and R. J. St. Leger. 2001. Transformants of *Metarhizium anisopliae sf. anisopliae* overexpressing chitinase from *Metarhizium anisopliae sf.* acridum show early induction of native chitinase but are not altered in pathogenicity to *Manduca sexta*. *J. Invertebr. Pathol.* 78:260—66.

[74]Sisterson, M. S., Y. Carriere, T. J. Dennehy, and B. E. Tabashnik. 2007. Nontarget effects of transgenic insecti-cidal crops: Implications of source-sink population dynamics. *Environ. Entomol.* 36:121—27.

[75]Stewart, L. M. D., M. Hirst, M. L. Ferber, A. T. Merryweather, P. J. Cayley, and R. D. Possee. 1991. Construction of an improved baculovirus insecticide containing an insect-specific toxin gene. *Nature* 352:85—88.

[76]Stoger, E., S. Williams, P. Christou, R. E. Down, and J. A. Gatehouse. 1998. Expression of the insecticidal lectin from snowdrop (*Galanthus nivalis aggluitin*; GNA) in transgenic wheat plants: Effects on predation by the grain aphid *Sitobion avenae*. *Mol. Breeding* 5:65—73.

[77]Sun, X., X. Chen, Z. Zhang, et al. 2002. Bollworm responses to release of genetically modified *Helicoverpa armigera* nucleopolyhedroviruses in cotton. *J. Invertebr. Pathol.* 81:63—69.

[78]Sun, X., H. Wang, X. Sun, et al. 2004. Biological activity and field efficacy of a genetically modified *Helicoverpa armigera* single-nucleocapsid nucleopolyhedrovirus expressing an insect-selective toxin from a chimeric promoter. *Biol. Control* 29:124—37.

[79]Tang, H., H. Li, S. M. Lei, R. L. Harrison, and B. C. Bonning. 2007. Tissue specificity of a baculovirus-expressed, basement membrane-degrading protease in larvae of *Heliothis virescens*. *Tissue Cell* 39:431—43.

[80]Taniai,K. ,A. B. Inceoglu,and B. D. Hammock. 2002. Expression efficiency of a scorpion neurotoxin,AaHIT,using baculovirus in insect cells. *Appl. Entomol. Zool.* 37:225—32.

[81]Toenniessen,G. H. ,J. C. O'Toole,and J. DeVries. 2003. Advances in plant biotechnology and its adoption in developing countries. Curr. Opin. *Plant Biol.* 6:191—98.

[82]Treacy,M. F. ,P. E. Rensner,and J. N. All. 2000. Comparative insecticidal properties of two nucleopolyhedrovirus vectors encoding a similar toxin gene chimer. *J. Econ. Entomol.* 93:1096—1104.

[83]Trung,N. P. ,E. Fitches,and J. A. Gatehouse. 2006. A fusion protein containing a lepidopteran-specific toxin from the South Indian red scorpion (*Mesobuthus tamulus*) and snowdrop lectin shows oral toxicity to target insects. *BMC Biotechnol.* 6:18—30.

[84]Wang,J. X. ,Z. L. Chen,J. Z. Du,Y. Sun,and A. H. Liang. 2005. Novel insect resistance in Brassica napus developed by transformation of chitinase and scorpion toxin genes. *Plant Cell Rep.* 24:549—55.

[85]Wang,C. ,and R. J. S. Leger. 2007. Expressing a scorpion neurotoxin makes a fungus hyperinfectious to insects. *Nat. Biotechnol.* 25:1455—56.

[86]Wu,J. ,X. Luo,Z. Wang,Y. Tian,A. Liang,and Y. Sun. 2008. Transgenic cotton expressing synthesized scorpion insect toxin AaHIT gene confers enhanced resistance to cotton bollworm (*Heliothis armigera*) larvae. *Biotechnol Lett.* 30:547—54.

[87]Wu,N. F. ,Q. Sun,B. Yao,et al. 2000. Insect-resistant transgenic poplar expressing AaIT gene. *Sheng Wu Gong Cheng Xue Bao* 16:129—33.

[88]Yao,B. ,Y. Fan,Q. Zheng,and R. Zhao. 1996. Insect-resistant tobacco plants expressing insect-specific neurotoxin AaIT. *Chin. J. Biotechnol.* 12:67—72.

[89]Zlotkin,E. ,M. Eitan,V. P. Bindokas,et al. 1991. Functional duality and structural uniqueness of depressant insect-selective neurotoxins. *Biochemistry* 30:4814—20.

[90]Zlotkin,E. ,Y. Fishman,and M. Elazar. 2000. AaIT: From neurotoxin to insecticide. *Biochimie* 82:869—81.

[91]Zlotkin,E. ,L. Fishman,and J. P. Shapiro. 1992. Oral toxicity to flesh flies of a neurotoxic polypeptide. *Arch. Insect Biochem. Physiol.* 21:41—52.

[92]Zlotkin,E. ,H. Moskowitz,R. Herrmann,M. Pelhate,and D. Gordon. 1995. Insect sodium channel as the target for insect-selective neurotoxins from scorpion venom. *ACS Symp. Ser.* 591:56—85.

第十七章 携带多分DNA病毒的寄生物与鳞翅目昆虫宿主的相互影响

Michael R. Strand

1. 前言

正如本书其他章节所讨论的(见第一章),鳞翅目昆虫属完全变态昆虫,全世界已知约16万种,包括很多影响农业、林业生产的害虫(Powell 2003)。绝大多数蛾类和蝶类的幼虫危害各类植物,鳞翅目昆虫在整个进化谱系中是最大的植食群体。而鳞翅目昆虫的天敌也和它们一样数量巨大。这些天敌中最多的是寄生蜂(膜翅目昆虫)。几乎所有的鳞翅目昆虫都被一种或多种寄生蜂袭击,而这些数量巨大的寄生虫专门寄生在一种或几种宿主的特定发育时期(Whitfield 1998; Pennacchio 和 Strand 2006)。

根据这些寄生虫的发育和与宿主的相互影响,可以把它们分为抑性寄生性和容性寄生昆虫两大类(Askew 和 Shaw 1986; Pennacchio 和 Strand 2006)。抑性寄生性昆虫(idiobiont)又分为体表寄生和体内寄生两种。体表寄生者(ectoparasitoid)麻痹宿主;体内寄生者(endoparasitoid)袭击宿主的卵或蛹。这类寄生虫寄生后,宿主都会停止生长。多数容性寄生昆虫(koinobiont)是寄生在幼虫阶段的体内寄生者,并随着寄生者后代的成长,宿主也在发育。毫不足奇,面对寄生者的袭击,鳞翅目昆虫的幼虫有着令人惊叹的防御本领,同样,这些体内寄生者针对这些防御也演变出了多种应对策略以在宿主体内发育生长。其中一种策略就是利用共生体,如黄蜂在产卵时将一些共生的微生物一并注射到宿主体内(Pennacchio 和 Strand 2006)。这些共生生物的活动使得宿主的生理发生改变,让寄生者的后代在宿主体内能正常地生长发育。茧蜂、姬蜂以及多分DNA病毒中寄生者一微生物的关系是种类最大,也是研究最多的。多篇综述都已讨论过多分DNA病毒(PDV)基因组的结构和组成(Kroemer 和 Webb 2004a; Webb 和 Strand 2005; Dupuy, Huguet 和 Drezen 2006)以及多分DNA病毒(PDV)在寄生中起的作用(Schmidt, Theopold 和 Strand 2001; Webb 和 Strand 2005; Pennacchio 和 Strand 2006; Gill et al. 2006)。在本章中更新了相关知识,将从鳞翅目昆虫生物学角度关注多分DNA病毒(PDV)的功能活性以及多分DNA病毒(PDV)基因产物在虫害治理与生物技术方面的潜在用途。

2. 多分DNA病毒的生活周期

最近,PDV被分为两类,一是与茧蜂相伴的,叫茧蜂病毒(BVs);另外一种是与姬蜂相伴的叫姬蜂病毒(IVs)。茧蜂的7个类群(折脉茧蜂亚科 Cardiochilinae,甲腹茧蜂亚科 Cheloninae,Dirrhoponae,Mendesellinae,Khoikhoiinae,Miricinae,小腹茧蜂亚科 Microgastrinae,共17000种)都携带茧蜂病毒(BVs)。姬蜂的2个类群(高缝姬蜂亚科、栉姬蜂,共13000种)的所有种群都携带姬蜂病毒(IVs)(Stoltz 和 Vinson 1979; Whitfield 和 Asgari 2003; Webb 和 Strand 2005)。最近的一些研究明显揭示出BVs源自一种裸病毒(nudivirus),而IVs的来源却不得而知(Bezier et al. 2009)。比较基因组分析显示虽然每一种特定的寄生蜂所携带的PDV病毒在基因上都具有唯一性,但在进化上都有相似性(Whitfield 2002; Espagne et al. 2004; Webb et al. 2006; Tanaka et al. 2007; Murphy et al. 2008)。BVs和IVs也具有相似的生活周期(如图17.1所示)。在寄生蜂中,BVs和

IVs以前病毒的形式存在，这种前病毒稳定整合到雌雄寄生蜂包括生殖细胞在内的所有细胞的基因组中。这些前病毒通过精子和卵子的垂直传播，代代相传。然而，病毒的复制场所就仅限于雌蜂体内形成输卵管萼的细胞中。病毒的复制最先发现于蛹的晚期，且一般会持续到雌蜂发育成成年蜂后(参见 Webb 和 Strand 2005)。BVs病毒体只由一个单一的细胞膜组成，膜内包有一个或多个长度不一的柱状核衣壳，这些圆柱状核衣壳是在胞溶作用下由输卵管萼细胞释放出来的(Stoltz，Vinson 和 MacKinnon 1976)。姬蜂病毒(IVs)由两个细胞膜组成，膜内包有一个大小一样的纺锭状的核衣壳，这些纺锭状的核衣壳是由输卵管萼细胞通过出芽释放出来的(Volkoff et al. 1995)。到目前为止，还只从一种栉姬蜂中检测到病毒体，它同样由两个细胞膜组成，但包含有多个纺锭状的核衣壳(Lapointe et al. 2007)。随着病毒不断复制和从输卵管萼细胞中释放出来，病毒体在侧输卵管腔中的浓度越来越大，最后形成输卵管萼液(calyx fluid)(如图 17.1 所示)。寄生蜂把卵产到宿主体内时，也把一些能感染宿主多种细胞和器官的病毒传到了宿主体内。PDV 病毒不在宿主体内复制，但这些病毒基因的表达使得宿主的免疫系统不能杀死寄生蜂的后代，并使宿主的发育发生多种变化，以利于寄生蜂的生长(参见 Webb 和 Strand 2005；Dupuy，Huguet 和 Drezen 2006；Pennacchio 和 Strand 2006 的总结)。因为病毒的传递依赖于寄生蜂，而寄生蜂的生存也依赖于由病毒编码的基因产物的感染和表达，所以 PDV 病毒和寄生蜂之间形成了一种真正的互利共生关系。

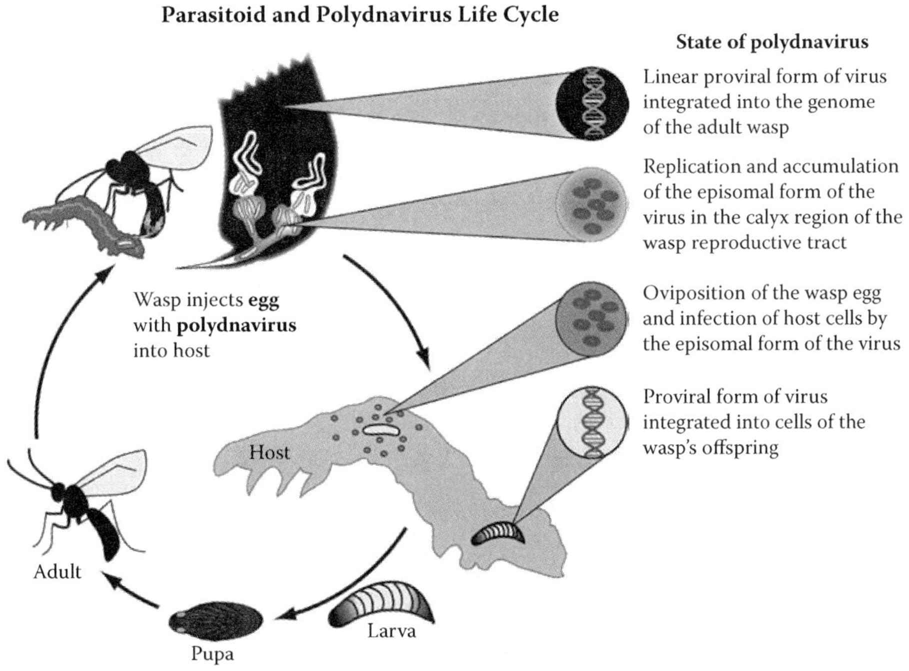

图 17.1　多分 DNA 病毒的生活周期。多分 DNA 病毒以前病毒的方式整合于寄生蜂的基因组中。寄生蜂通过把包在病毒附加体中的病毒体注射入鳞翅目昆虫宿主体内的方式来感染宿主。寄生蜂把一个或多个卵产到宿主体内，这些卵孵化成幼虫后以宿主的血淋巴或组织为食。完成发育后，寄生蜂幼虫就脱离宿主，然后化蛹，最后发育为成年寄生蜂。

3. 多分DNA病毒基因组及其在被寄生鳞翅目昆虫体内的表达活性

3.1 多分DNA病毒基因组的结构特点

正如前文所提到的，PDV基因组以两种状态存在：一种是在复制期间被包裹在核衣壳中的附加体；另一种是整合到寄生蜂基因组中的线性前病毒。BVs和IVs基因组的附加体包含有多个环状的双链DNA，这些DNA的大小从2000到30000多个碱基对不等(kb; Kroemer和Webb 2004a; Webb和Strand 2005)。PDV中DNA片段数量不等，有些BVs中仅有6个，而在有些栉姬蜂的病毒中却高达100个(Stoltz et al. 1995; Tanaka et al. 2007)。一般来讲，BVs中的DNA片段数量较少但体型较大，而IVs基因组中的DNA片段数量较多但体型较小。PDV总的基因组大小从180kb到500kb以上不等(Espagne et al. 2004; Webb et al. 2006; Lapointe et al. 2007; Tanaka et al. 2007)。众所周知，BVs和IVs的基因片段在输卵管萼液(calyx fluid)中不以等物质的量存在，其中一些片段的物质的量总比另一些要多。比如，在毁侧沟茧蜂病毒(MdBV)中，整个基因组由15个片段组成，但在输卵管萼液(calyx fluid)中，其中5个片段以及和它们相关联的核衣壳超过病毒DNA的60%(Beck, Inman和Strand 2007)。对BVs的研究表明每个核衣壳包含一个基因片段，衣壳的大小是由这个基因片段的大小决定的(Albrecht et al. 1994; Beck, Inman和Strand 2007)。与之相反，研究人员到现在为止还不知道IV的核衣壳是由一个还是多个基因片段组成的。与BV不同的是，IV基因片段的非等分产物同时和一种叫做片段嵌套的现象有关(参见Kroemer和Webb 2004a; Webb和Strand 2005)。事实上，IV中的有些基因片段是独一无二的，然而，通过分子内重组从寄生蜂基因组上切割下来后，又会产生一些附加的片段(Xu和Stoltz 1993; Cui和Webb 1997; Webb和Cui 1998)。

序列分析显示，在前病毒形态中，BVs的附加片段以一个或多个串联重复的排列方式整合到寄生蜂的基因组中，这些串联重复的排列方式形成巨大位点(macroloci)(Belle et al. 2002; Desjardins et al. 2007)。另一方面，IVs中的前病毒片段在姬蜂基因组中呈离散状(Fleming和Summers 1991)。对BVs的研究表明在切割成单个的基因片段之前，巨大位点已经得到扩增并被包进病毒体(Pasquier-Barre et al. 2002; Drezen et al. 2003; Marti et al. 2003)。对*Chelonus inanitus*茧蜂病毒(CiBV)的研究同样显示出附加片段越多，没有切割的前病毒中的拷贝数就越多(Annaheim和Lanzrein 2007)。与之相反的是，Webb(1998)表明，IVs可能以一种滚环型的机制进行复制，前病毒DNA切割后，以环状附加体的形式得以扩增。

到目前为止，来源于小腹茧蜂(*Cotesia congregata*茧蜂病毒[CcBV]和*Microplitis demolitor*毁侧沟茧蜂病毒[MdBV])的两种BVs的壳体化基因组和来源于高缝姬蜂亚科(*Campoletis sonorensis IV*齿唇姬蜂病毒[CsIV]，*Hyposoter fugitives IV*姬蜂病毒[HfIV]和*Tranosema rostrale IV*姬蜂病毒)的三种IVs病毒以及一种栉姬蜂(*Glypta fumiferanae*[GfV])病毒基因组都已完成或接近完成测序工作(Espagne et al. 2004;

Webb et al. 2006; Tanaka et al. 2007; Lapointe et al. 2007)。数据库检索进一步表明大约有另外 6 种寄生蜂的 BVs 或 IVs 已经部分测序。除了前文所提到的病毒基因组的共性外,这些结果也表明它们有几个共同的特征。这些特征包括:编码密度低,AT 碱基偏好性强,以及发现了 PDV 编码的大量预测基因。这些基因由相关的基因变体组成基因家族,其中多数在被寄生的宿主体内表达(见下文)。这些用壳体包裹的 PDV 基因组缺少编码为多聚酶基因或其他 DNA 复制所需的蛋白质,这也就解释了 PDV 病毒不在宿主体内复制的原因。来自单个寄生蜂的 PDV 同时也编码几个在其他 PDV 中没有发现的新的单拷贝基因或基因家族。

3.2 多分 DNA 病毒基因家族

到目前为止,通过比较研究只找到一个由 BVs 和 IVs 共同编码的基因家族—ankyrin 基因(细胞膜锚蛋白基因)(*ank*;图 17.2)。然而,系统发生研究显示 BVs 和 IVs 的 *ank* 基因形成了不同的基因群,这表明它们来自不同的祖先(Webb et al. 2006)。除了甲腹茧蜂,两个基因家族 *ank* 基因和蛋白酪氨酸磷酸酶基因(*ptp*),存在于已研究的小腹茧蜂和折脉茧蜂所携带的所有的 BVs 中;然而,来源于高缝姬蜂的 IVs 中有 6 个基因家族(*ank*,*innexin* 基因[*inex*],半胱氨酸基序基因[*cys-motif*],重复元件基因[*rep*],*N*-家族基因[*N-family*]和富含极性氨基酸基因[*pol-res*];见图 17.2)是共有的。迄今为止,唯一已测序的栉姬蜂病毒 GfV 表现出比已报道的来自高缝姬蜂的 IVs 病毒基因组的分割程度要大得多。GfV 基因组中的基因家族包括已预测的 *anks* 和 *ptps* 基因,但系统发生研究同样显示出 *ank* 基因并没有嵌入在高缝姬蜂中的 *ank* 基因中,而 *ptp* 基因家族的成员和那些 BVs 编码的 PTPs 并不一样。GfV 同时也编码了一个新的 NTPase-like 基因,这个基因不存在于其他 PDV 基因组内。引人注意的是,到目前为止,来自甲腹茧蜂的茧蜂病毒(比如 CiBV)(见上文)的基因组序列数据还远不能揭示它的基因和其他的 PDV 基因的相同性(Annaheim 和 Lanzrein 2007)。比较研究表明寄生相同或相近种类的寄生蜂的 PDV 编码类似的基因家族,而寄生亲缘关系远一些的寄生蜂的病毒在基因组上显示出更多的差异。比如,CcBV 病毒编码半胱氨酸蛋白酶抑制剂(*cys*),*crvl*-like 以及 C 型凝集素有关的基因,在来自绒茧蜂种类的其他寄生蜂中的 *BV* 病毒基因组上检测出了这些基因的同源序列(图 17.2)。与之相反,来自小腹茧蜂种类的寄生蜂的 BVs(比如 MdBV),缺少这些基因,但编码其他基因家族(*egf*),这些基因的同源序列存在于侧沟茧蜂种类的其他寄生蜂中的 BVs 病毒基因组上(图 17.2)。

PDV 基因家族似乎起源于基因复制以及基因间的片段重组(Espagne et al. 2004; Webb et al. 2006; Friedman 和 Hughes 2006)。不同种类之间的基因家族的大小变化相当大。例如,高缝姬蜂寄生蜂中 CsIV 病毒的半胱氨酸基序基因包含 10 个成员,然而在 HfIV 和 TrIV 病毒基因组中分别只有 5 个和 1 个半胱氨酸基序基因(Tanaka et al. 2007)。虽然基因家族的大小可能反映出相关寄生蜂的宿主范围或某一种特定病毒在寄生中的功能,但目前仍然不清楚为什么基因家族的大小在不同的种类中不同(见下文)。对表达的研究揭示出由 BVs 和 IVs 编码的不同基因家族成员在被病毒感染的宿主组织

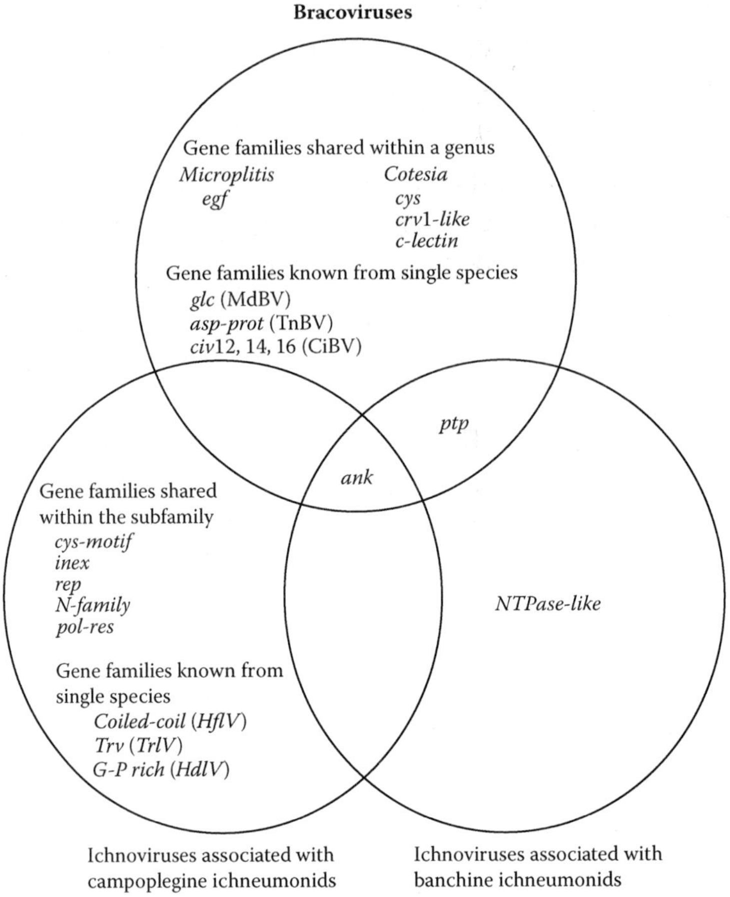

图 17.2 茧蜂病毒和姬蜂病毒中共有和独有基因家族的维恩图。迄今为止，在几乎所有研究过的 PDV 中都能检测到 *ank* 基因，但其他的基因家族仅限于：BVs 和来自栉姬蜂（*ptp*）的姬蜂病毒、高缝姬蜂的姬蜂病毒以及某一种属或某单一类别的寄生蜂的茧蜂病毒。

中的表达是不一样的（见 Kroemer 和 Webb 2004b；Pruijssers 和 Strand 2007）。宿主的血细胞、脂肪体和神经系统在其免疫和发育中非常重要，而这些都是多数病毒基因家族成员表达的主要场所。然而，PDV 在宿主的其他主要器官，比如肠和马氏管几乎没有表达活性。功能研究表明，在一些实例中 PDV 基因家族成员具有不同于其他基因家族成员的特异活性（见下文）。与某一特别类属的寄生蜂有关的病毒新基因和基因家族表现出受正向选择压力的驱使以获得作用于某一种宿主的特定功能。研究人员还不清楚 PDVs 编码的致病基因是否来源于祖先病毒，与之相关的寄生蜂或宿主，或是有其他的来源（Webb 和 Strand 2005）。例如，对 CcBV 的序列分析表明，许多基因高度分化，它们和相关的膜翅目昆虫以及其他类似昆虫基因的关系并不比它们和人类以及其他哺乳动物的基因的关系更密切（Bezier et al. 2008）。

BVs 病毒相互之间比 BVs 与 IVs 间有更多的相似之处，反之亦然，这完全和作为前病毒的相关寄生蜂的系统发生（种系发生）相一致，系统发生分析是根据孟德尔遗传理论来完成的。然而，形态上的差异结合序列数据也表明，来自高缝姬蜂的 BVs 和 IVs，与来自梢姬蜂的 IVs 有着不同的起源。如果真是这样，相似的生活史和它们相似的基因组结构体现了这些病毒在寄生现象中扮演角色的相似性（Webb 和 Strand 2005；Pennacchio 和 Strand 2006）。

4. 携带多分 DNA 病毒的寄生蜂与鳞翅目昆虫在免疫方面的相互影响

PDVs 最重要的功能是在宿主的免疫系统中保护发育中的寄生蜂。和其他昆虫一样，鳞翅目昆虫的先天免疫系统由体液免疫和细胞免疫反应组成（第十四章中详细讨论了烟草天蛾的免疫系统）。体液免疫反应是指一些有防御活性的可溶性分子，如抗菌肽（AMPs）、补体样蛋白，以及黑色素之类的酚氧化酶（PO）级联产物（Cerenius 和 Soderhall 2004；Theopold et al. 2004；Imler 和 Bulet 2005；Kanost 和 Gorman 2008）。细胞免疫反应是指直接由昆虫免疫细胞（血细胞）所介导的防御，如吞噬作用、包囊作用和凝聚作用（Gillespie，Kanost 和 Trenczek 1997；Strand 2008a）。鳞翅目昆虫体内循环的血细胞包括 4 类，分别叫做粒细胞（granular cells）、浆细胞、小球细胞和类绛色细胞（Lavine 和 Strand 2002；Strand 2008a）。粒细胞是体液循环中最丰富的血细胞类型，专门起吞噬作用。浆血细胞通常大于粒细胞，是形成囊状物的主要血细胞。非黏附血细胞包括含有酚氧化酶的类绛色细胞和小球细胞，这种细胞是表皮成分的潜在来源。被称之为前血细胞的祖细胞存在于造血器官中，循环较慢（Gardiner 和 Strand 2000；Nakahara et al. 2003；Nardi et al. 2003，2006；Ling et al. 2005）。综合而言，鳞翅目昆虫具有发达的针对寄生蜂和其他多种寄生虫和病原体的先天性免疫反应。

4.1 鳞翅目昆虫对寄生物的防御

鳞翅目昆虫防御寄生蜂攻击的最初方式就是包囊作用，宿主一旦认出外来的蜂卵或其幼虫，包囊作用就开始了（第一阶段）（Strand 2008a）。对寄生蜂和其他入侵者的识别涉及血球细胞表面的受体和体液中通过与外来物表面结合来达到识别效果的模式识别分子（即调理作用）。起吞噬或包囊作用的细胞表面受体包括清除受体（Ramet et al. 2001；Kocks et al. 2005；Philips，Rubin 和 Perrimon 2005），唐氏综合征细胞黏附蛋白跨膜形式（Dscam；Ramet et al. 2002；Moita et al. 2005；Dong，Taylor 和 Dimopoulos 2006）以及结合素（Lavine 和 Strand 2003；Irving et al. 2005；Levin et al. 2005；Wertheim et al. 2005；Moita et al. 2005）。体液识别模式受体包括类免疫球蛋白，脂多糖结合蛋白，革兰氏阴性菌结合受体（GNBPs），可溶性肽聚糖受体 PGRPs（PGRP-SA 和 PGRP-SD），GRPs，互补样含硫酯蛋白质（TEPs）以及免疫凝集素（Levashina et al. 2001；Irving et al. 2005；Moita et al. 2005；Dong，Taylor 和 Dimopoulos 2006；Ling 和 Yu 2006；Terenius et al. 2007）。识别的结果是一小部分血细胞，通常是粒性细胞结合在寄生蜂的表面，同时也刺激另外的血细胞的增生和分化（第 2 阶段；Pech 和 Strand 1996；Irving et al. 2005）。结合的粒细胞释放出细胞因子，这些细胞因子招募其他的血细胞，主要是浆细胞包裹寄生蜂，形成囊状物。最近，研究人员发现在烟草天蛾体内，四旋蛋白作为结合素蛋白的配基，介导了血细胞在囊状物中的结合(Zhuang et al. 2007)。粒细胞活性的刺激以及浆细胞的结合都需要免疫球蛋白超家族成员神经胶质蛋白的参与(Nardi et al. 2006)。在第 3 阶段囊状物通常会因为酚氧化酶级联系统的活化而黑化。窒息和/或因产生的黑色素能杀死寄生蜂。通常，囊状物在寄生蜂寄生鳞翅目昆虫 2～6h 后开始形成，在标准条件下 48h 内能完全形成。

4.2 多分 DNA 病毒对鳞翅目昆虫免疫系统的攻击

多数研究表明，PDV 为了防止与它相关的寄生蜂被识别和封装进囊状物中，它在宿主体内表达起免疫抑制作用的基因产物(Davies，Strand 和 Vinson 1987；Lavine 和 Beckage 1995；Strand 和 Pech 1995；Schmidt，Theopold 和 Strand 2001；Webb 和 Strand 2005；Strand 2008b)。在一些寄生蜂一宿主系统中，病毒感染的作用似乎只有抑制寄生蜂被封装进囊状物中(Doucet 和 Cusson 1996)，然而，另外一些感染的免疫抑制效果更强，使得一些外来目标不被封装进囊状物中，同时还可攻破其他细胞和体液的免疫防御(见 Webb 和 Strand 2005；Strand 2008b 的总结)。

仅有少数 PDV 基因通过实验证明与免疫抑制相联系，对宿主而言，与 PDV 基因相互作用的分子和信号通道的了解更是微乎其微。在茧蜂病毒(BVs)中，小腹茧蜂亚科毁侧沟茧蜂病毒所携带的 MdBV 通过抑制血细胞与外来物相结合，阻碍血细胞对其吞噬，诱导粒细胞凋亡，以及抑制可诱导的体液免疫来广泛实施对宿主的免疫抑制(Strand 2008b)。所有这些免疫抑制效果在宿主感染 MdBV 病毒 12h 内表现出来，该免疫抑制过程中所涉及的物质包括在血细胞表达并起作用的基因产物和那些在血细胞和脂肪体之类的组织中表达，然后分泌到血淋巴中的基因产物。功能研究暗示了 4 个基因家族的成员导致了这些变化。血细胞对寄生蜂的囊化和吞噬作用的抑制主要是由一种表面黏液素 Glcl. 8 和两种 PTPs(PTP-H2 和 PTP-H3)的相互作用所介导的，这种抑制使中心黏结综合体(focal adhesion complexes)出现在感染的免疫细胞中(Beck 和 Stand 2003，2005；Pruijssers 和 Strand 2007)。近期研究也表明 PTP-H2 是一种细胞凋亡的诱导因子(Suderman，Pruijssers 和 Strand 2008)。在被寄生物或其他有机体感染后，Toll 和 imd 信号通路调控多种免疫基因进行表达。两种通道都需要特定的 NF-κB 转录因子，这些转录因子通常由内生的抑制物 κB(IκB)蛋白调控。MdBV 病毒拥有与昆虫 IκBs 具有明显同源性的 12 个 *ank* 基因成员，以及另两个家族成员 ANK-H4 和 ANK-H5 蛋白，它们通过与昆虫 NF-κBs 结合来抑制 Toll 和 imd 信号通路(Thoetkiattikul，Beck 和 Strand 2005)。正如前文所提到的，黑色素的形成由 PO 级联系统调控，这个过程包含多个丝氨酸蛋白酶，直至最后的前酚氧化酶酶原(Kanost 和 Gorman 2008)。虽然研究人员还没搞清楚任何一种昆虫体内 PO 级联中蛋白水解的步骤，但是前酚氧化酶 (proPo)能被丝氨酸蛋白酶 proPAPs 激活，proPAPs 能被多条上游路径的蛋白酶激活。作为 MdBV *egf* 基因家族中的一员，EGF1. 0 能抑制黑化的发生，这当中，EGF1. 0 扮演的是一个双重活性抑制者的角色，它既阻碍 proPAP 的活化加工又抑制 PAPs 加工后的酶活性(Beck 和 Strand 2007；Lu et al. 2008)。相反，其他 EGF 蛋白基因家族成员(比如 EGF0. 4)也可能为丝氨酸蛋白酶抑制剂，但它们未被鉴定的靶标酶有着与 PO 级联无关的功能（Beck 和 Strand 2007）。

大多数其他关于 BV 介导的免疫抑制的研究都涉及来自绒茧蜂属寄生蜂的病毒。近期对菜蛾绒茧蜂病毒（CpBV）的研究表明，病毒编码的 PTP、凝集素、组蛋白和一个 EP1 样基因破坏了血细胞在异物表面的黏附以及血细胞的增殖（Gad 和 Kim 2008；Ibrahim 和 Kim 2008；Kwon 和 Kim 2008；Lee，Nalini 和 Kim 2008）。用微红盘绒茧蜂（*C. rubecula*）中的 CrBV 病毒感染菜粉蝶（*Pieris rapae*）造成血细胞中细胞支架暂时变形，也抑制了黑化。已经证实了 *CrVl* 基因致使血细胞中细胞支架的暂时变形，而黑化则涉及存在于微红盘绒茧蜂的毒液腺体中一种非病毒的丝氨酸蛋白酶同源物（Asgari，Schmidt 和 Theopold 1997；Zhang et al. 2004）。*C. congregata* 的 CcBV 对寄生蜂被封装进囊状物中这个过程的破坏与宿主血细胞支架的变形有关，使其形成聚合物，而不是失去黏着力。CcBV 和其他的绒茧蜂病毒能降低黑化作用。*Toxoneuron nigriceps* BV（TnBV）编码了一个基因（*TnBV*1），这个基因在没有细胞凋亡时能激活半胱氨酸蛋白酶（Lapointe et al. 2005），如同与昆虫 NF-κBs 相互作用的 *ank* 基因（Falabella et al. 2007）。不像其他携带 BV 的茧蜂，甲腹茧蜂属的寄生蜂全都是卵和幼虫寄生蜂。人们对甲腹茧蜂的 BVs 在免疫抑制中的作用还知之甚少，但对 CiBV 的研究表明病毒侵染宿主会选择性地保护 *C. inanitus* 幼虫不被封装进囊状物中（Lanzrein et al. 1998）。

与免疫抑制有关的 IV 基因包括半胱氨酸基序蛋白质，来自齿唇姬蜂的 CsIV 中的 VHv1.1 和 VHv1.4，它们在宿主的脂肪体内表达并分泌到血淋巴中，两种蛋白质都黏附在血细胞的表面。重组杆状病毒表达 VHv1.1 感染宿主时，VHv1.1 的表达也能降低宿主体内齿唇姬蜂的卵被封装进囊状物中的危险（Li 和 Webb 1994；Cui，Soldevila 和 Webb 1997）。因为与先前描述的 BV 基因的同源性或是因为它们在主要的免疫细胞（血细胞或脂肪体）中表达，其他的 IV 基因与免疫抑制有着间接的关系。例如，表达 IV 的 *ank* 基因的作用可能是抑制 NF-κB，但在结构上却与 BV 基因不同。而近来的表达研究显示它们可能有其他的功能（Kroemer 和 Webb 2004b；Webb et al. 2006；见下文）。*CsIV innex* 基因的产物形成功能性的间隙连接，这些连接与细胞间通讯有着不同的功能（Turnbull et al. 2005）。根据观察，间隙连接都是在很多囊状物里的血细胞中形成的，因此 IV innex 基因的产物有可能改变了免疫细胞的功能。虽然研究人员预测 HdIV 的 P30 开放阅读框同样也编码了一个黏蛋白样蛋白质，这与 MdBV 中的 Glcl.8 序列仅有一点相似，但还不清楚它在免疫中的功能（Galibert et al. 2003）。

虽然病毒的基因表达通常用于破坏宿主针对寄生蜂的宿主免疫系统，但是一些茧蜂病毒和姬蜂病毒却似乎只是被动地保护寄生蜂的新生卵，它们只是把卵的表面覆盖住，让那些宿主不把这些卵认作为外来物，阻碍宿主血细胞在它上面的固着。例如，黏虫盘绒茧蜂的卵外包裹着 CkBV 病毒体和卵巢细胞分泌出的其他蛋白质。研究人员已检测出两种卵巢免疫规避蛋白质（IEP），IEP1 和 IEP2，它们不存在于 CkBV 基因组中，但却在 CkBV 病毒体的表面被检测到。IEP 使得卵不被封在囊状物中，保护 CkBV 不被血细胞消除掉(Hayakawa 和 Yazaki 1997)。IEP 卵巢蛋白质同样也能保护绒茧蜂和 *Toxoneuron* 属中其他寄生蜂的卵(Davies 和 Vinson 1986; Asgari et al. 1998)。姬蜂中的仓蛾姬蜂(*Venturia canescens*)产生一种像病毒的颗粒(VLPs),这种颗粒外形上像其他高缝姬峰亚科成员所产生的姬蜂病毒,但其中却似乎没有核酸。仓蛾姬蜂所寄生的宿主仍然能够针对众多的外来靶标物形成囊状物,但由于 VLPs 和包在卵外的体液黏蛋白的存在,它却不能把仓蛾姬蜂封在里面(Theopold et al. 1996)。TrIV 也同样把 *Tranosema rosele* 的卵包裹起来,使其得到保护,不被封装形成囊状物(Cusson et al. 1998)。

4.3 鳞翅目昆虫与携带多分 DNA 病毒的寄生物在昆虫发育方面的相互影响

PDV 基因强烈表达造成鳞翅目昆虫生理上的第二大改变是宿主生长和发育的改变。所有携带 PDV 的姬蜂病毒都是单体的(每个宿主产生一个后代)。而携带 PDV 的茧蜂既有单体也有聚生的(每个宿主多个后代)。甲腹属茧蜂(甲腹茧蜂、革腹茧蜂等)寄生于鳞翅目的卵,但几乎所有其他携带 PDV 的寄生蜂则寄生于多龄期的幼虫。携带 PDV 拟寄生物的后代有的在宿主的最后龄期完成它们不成熟的发育,有的迅速发育并在更早的龄期完成发育(Pennacchio 和 Strand 2006)。被单体物种寄生的鳞翅目昆虫体重显著降低,蜕皮延长,不能化蛹。被聚生物种寄生的宿主化蛹也受到抑制,但体重没有急剧变化。甲腹属茧蜂与这些趋势不同,它引起宿主提前 1 龄期早熟变态,并且影响宿主化蛹(Lanzrein,Pfister-Wilhelm 和 von Niederhäusern 2001)。

4.4 感染或被寄生多分 DNA 病毒的鳞翅目昆虫体内内分泌系统和代谢系统的变化

上述改变与内分泌和代谢生理学的变化有关(Thompson 和 Dahlman 1998; Beckage 和 Gelman 2004; Pennacchio 和 Strand 2006)。内分泌改变包括血淋巴中保幼激素(JH)滴度的增加和蜕皮甾醇滴度在幼虫－幼虫或幼虫－蛹期蜕皮期间低于非寄生宿主体内正常发生的水平。大部分研究表明 JH 滴度的增加与 JH 酯酶等宿主代谢酶活性的减少(Balgopal et al. 1996; Dong,Zang 和 Dahlman 1996; Schafellner,Marktl 和 Schopf 2007)或拟寄生幼虫 JH 的分泌(Cole et al. 2002)有关。仅有几个研究报道宿主咽侧体 JH 的分泌和合成下降(Cole et al. 2002; Li et al. 2003)。相比之下,内分泌改变参与蜕皮甾体滴度抑制的一系列变化包括 PTTH 合成和释放的减少(Tanaka,Agui 和 Hiruma 1987; Hayakawa 1995),前胸腺对 PTTH 刺激的不敏感(Kelly et al. 1998),前胸腺生物合成活性降低(Tanaka,Agui 和 Hiruma 1987; Pennacchio,Digilio 和 Tremblay 1995)和前胸腺细胞成熟前的死亡(Dover,Tanaka 和 Vinson 1995)。不同的甲腹茧蜂品种实验表明因为咽侧体失活,保幼激素酯酶活性的增加和蜕皮甾醇滴度的改变使 JH 滴度比正常下降得更早,从而发生早熟变态(Reed 和 Brown 1998; Lanzrein,Pfister-Wilhelm 和 von Niederhäusern 2001)。代谢改变包括血淋巴蛋白质、游离氨基酸和碳水化合物丰度的改变(Thompson 1993; Thompson 和 Dahlman 1998)。蛋白质丰度的明显改变跟蛋白质合成的选择性改变有关。例如,齿唇姬蜂(*C. sonorensis*)的寄生和 CsIV 的感染大大地减少了血淋巴主要储藏蛋白中的芳基贮存蛋白 (arylphorin)的数量,但不会影响其他主要的血淋巴蛋白(如低密度脂蛋白和转铁蛋白)的含量(Shelby 和 Webb 1994,1997)。被齿唇姬蜂(*C. sonorensis*)寄生的宿主脂肪体中的三酰甘油和肝糖原的沉积也大大地减少,而血淋巴中游离糖的水平增加(Vinson 1990)。在被茧蜂寄生的宿主中也有相似的改变(参看 Thompson 和 Dahlman 1998; Pennacchio 和 Strand 2006)。

4.5 *PDV* 基因产物在改变宿主生长发育中的作用

根据昆虫内分泌系统在调控营养储藏和转运中的作用,以上所述的内分泌和代谢生理的改变是相关联的(Vinson 1990; Britton et al. 2002; Strand 和 Casas 2007)。提高宿主的 JH 滴度和阻碍 20E 水平的升高在一定程度上与鳞翅目幼虫发育晚期上调的特定储藏蛋白的滴度变化有关,而昆虫的碳水化合物和脂类代谢通过胰岛素信号途径调控(Brown et al. 2008)。Pennacchio 和 Strand(2006)强调在拟寄生物生活史中重要的主题是使变态和(或)繁殖,以及营养物质转运这些代谢不能正常运转,将宿主组织的能量资源转变为寄生物后代的营养储存物质。因此,改变宿主内分泌生理和繁殖的首要意义更多的是代谢方面的,而与此相关的诸如发育受阻、化蛹抑制的现象是宿主营养资源重新分配的间接后果。

人们对引起这些改变的 PDV 基因产物的特性还不太清楚。在宿主前胸腺(Falabella et al. 2006)TnBV 的 PTP 基因的表达已被证明是通过改变通路成员的磷酸化来扰乱 PTTH 信号传导途径和(或)蜕皮甾醇的生物合成的。几个来自毁侧沟茧蜂病毒(MdBV)编码的 PTP 基因在感染宿主的中枢系统和前胸腺中差异性地表达(Pruijssers 和 Strand 2007),而两个被 CiBV 编码的新基因明显参与了调控发育受阻和紧随的早熟变态的起始(Bonvin et al. 2005)。由 CsIV 编码的半胱氨酸结构域基因家族的两个成员 VHv1.1 和 VHv1.4 使一些宿主组织(脂肪体、血球细胞、精巢)以及来自粉纹夜蛾(*Trichoplusia ni*)的鳞翅目昆虫细胞系(TN368 细胞系)中的蛋白质翻译受阻(Shelby 和 Webb 1997; Kim 2005)。与对照组幼虫相比,注射或添食来自 CsIV 的重组 VHv1.1 和 VHv1.4 的宿主幼虫增重更慢,不正常化蛹数增加(Fath-Goodin et al. 2006)。有趣的是,来自茧蜂 *M. croceipes* 畸形细胞分泌的基因产物(TSP14)与 IV 半胱氨酸基序基因有一些结构上的相似性,在生物活性测定中与 VH1.1 和 VHv1.4 有相似的活性(Dahlman et al. 2003)。

5. *PDV* 基因在害虫防治上的应用前景

根据 PDVs 在改变宿主昆虫免疫、发育中的重要作用,科学家正着手探索 PDV 基因产物对鳞翅目和其他害虫控制上,以及作为生物技术应用工具上的潜在用途。需要再次强调的是,除了目前已经从不到 10 个物种中鉴定到的 PDV 基因和基因家族外,估计有 3 万个寄生物种携带这些病毒,而且这些寄生蜂大多都可以攻击大量重要的农业害虫。由于每种 PDV 寄生群在遗传上是特定的,因此这些病毒共同代表了大量与环境相容的新一代杀虫剂或具有多种用途的药物毒性分子。

目前，利用 PDV 基因的策略有三种。首先是利用具有免疫抑制活性的 PDV 基因产物来增强杆状病毒等其他杀虫剂的功效。抑制宿主免疫防护能增强致病性和（或）其他控制剂的宿主范围。Washburn（2000）等就采用了这种策略，他们用核型多角体病毒（AcMNPV）和茧蜂病毒 CcBV 来共同感染烟草天蛾幼虫。同时，Rivkin 等也用 AcMNPV 和姬蜂病毒 CsIV 感染另外的鳞翅目昆虫来进行研究（Rivkin et al. 2006）。在两个例子中，PDVs 的感染与杆状病毒的感染有协同效应。携带有 PDV 的拟寄生物的寄生或 PDVs 的感染能增加鳞翅目昆虫幼虫对细菌、其他的病毒和寄生物等天敌的易感性（Shelby 和 Webb 1999；Kadash，Harvey 和 Strand 2003；Stoltz 和 Makkay 2003）。以前的研究也已经证实在 PDV 基因表达模式和宿主范围间的关系（Cui，Soldevila 和 Webb 2000），也证实除在相关寄生物正常宿主范围之外的鳞翅目和非鳞翅目昆虫中，几种免疫抑制 PDV 基因有生物学活性（Thoetkiattikul，Beck 和 Strand 2005；Beck 和 Strand 2007）。总之，这些结果为 PDV 免疫抑制基因应用于抑制害虫免疫系统，并为提高几种潜在生物防治剂的功效提供了依据。

第二种利用 PDV 基因的策略主要是直接利用其生长抑制活性，通过在转基因作物的表达或其他常规的方法，直接作为候选的杀虫剂。正如前文所讨论的，尽管半胱氨酸结构域相关的基因产物在寄生阶段没有被天然地运送到宿主昆虫的肠道，但添食来自 CsIV 的重组 VHv1.1 和 1.4 蛋白，来自于 M. croceipes 的畸形细胞蛋白 TSP14 的结果表明，其可以减弱宿主的生长活性（Maiti et al. 2003；Fath-Goodin et al. 2006），这与寄生时所发生的情形一样。当这些蛋白质组成性地在转基因烟草中表达时，宿主生长活性的降低，对植物的取食危害也相应地减少（Gill et al. 2006），表明至少一些 PDV 基因有助于减少鳞翅目害虫对作物的损害。

正在考虑的第三种策略涉及利用源自于 PDV 的蛋白作为生物技术应用的试剂。最近已经有一例应用得到商业开发，即利用 PDV 基因增强杆状病毒表达系统（BEVs）的生产率，该表达系统被广泛地用于重组蛋白的生产。在研究由 CsIV 编码的 *ank* 基因家族成员的表达模式和细胞定位时，Kroemer 和 Webb（2004b）发现与对照或表达另外家族成员的重组病毒相比较，表达两个 *ank* 基因家族成员的重组杆状病毒寿命得以延长。昆虫细胞的寿命是表达蛋白质的关键限速因子，提高宿主细胞存活率能增加重组蛋白的产出。尽管其作用模式仍不清楚，但稳定表达 ANK 蛋白的细胞系或共表达 *ank* 基因和靶标新基因的二元 BEVS 系统使重组蛋白的产量增加了 4～15 倍。

参考文献

[1]Albrecht，U.，T. Wyler，R. Pfister-Wilhelm，et al. 1994. PDV of the parasitic wasp *Chelonus inanitus* (Braconidae)：Characterization，genome organization and time point of replication. *J. Gen. Virol*. 75：3353－63.

[2]Annaheim，M.，and B. Lanzrein. 2007. Genome organization of the *Chelonus inanitus* polydnavirus：Excision sites，spacers，and abundance of proviral and excised segments. *J. Gen. Virol*. 88：450－57.

[3]Asgari，S.，O. Schmidt，and U. Theopold. 1997. A polydnavirus-encoded protein of an endoparasi-

toid wasp is an immune suppressor. *J. Gen. Virol.* 78:3061—70.

[4]Asgari,S. ,U. Theopold,C. Wellby,and O. Schmidt. 1998. A protein with protective properties against the cellular defense reactions in insects. *Proc. Natl. Acad. Sci. U. S. A.* 95:3690—95.

[5]Askew,R. R. ,and M. R. Shaw. 1986. Parasitoid communities: Their size,structure and development. In Insect *parasitoids*,ed. J. K. Waage and D. Greathead,225—63. London: Academic Press.

[6]Balgopal,M. M. ,B. A. Dover,W. G. Goodman,and M. R. Strand. 1996. Parasitism by *Microplitis demolitor* induces alterations in the juvenile hormone titers and juvenile hormone esterase activity of its host,*Pseudoplusia includens*. *J. Insect Physiol.* 42:337—45.

[7]Beck,M. H. ,R. B. Inman,and M. R. Strand. 2007. *Microplitis demolitor* bracovirus genome segments vary in abundance and are individually packaged in virions. *Virology* 359:179—89.

[8]Beck, M. , and M. R. Strand. 2003. RNA interference silences Microplitis demolitor bracovirus genes and implicates glc1. 8 in disruption of adhesion in infected host cells. *Virology* 314:521—35.

[9]Beck,M. ,and M. R. Strand. 2005. Glc1. 8 from *Microplitis demolitor* bracovirus induces a loss of adhesion and phagocytosis in insect high five and S2 cells. *J. Virol.* 79:1861—70.

[10]Beck,M. ,and M. R. Strand. 2007. A novel polydnavirus protein inhibits the insect prophenoloxidase activation pathway. *Proc. Natl. Acad. Sci. U. S. A.* 104:19267—72.

[11]Beckage,N. E. ,and D. B. Gelman. 2004. Wasp parasitoid disruption of host development: Implications for new biologically based strategies for insect control. *Annu. Rev. Entomol.* 49:299—330.

[12]Belle,E. ,N. E. Beckage,J. Rousselet,M. Poirie,F. Lemeunier,and J. —M. Drezen. 2002. Visualization of polydnavirus sequences in a parasitoid wasp chromosome. *J. Virol.* 76:5793—96.

[13]Bézier, A. , M. Annaheim, J. Herbinière, et al. 2009. Polydnaviruses of braconid wasps derive from an ancestral nudivirus. *Science* 323:926—30.

[14]Bézier,A. ,J. Herbinière,C. Serbielle,et al. 2008. Bracovirus gene products are highly divergent from insect proteins. *Arch. Insect Biochem. Physiol.* 17:172—87.

[15]Bonvin,M. ,D. Marti,S. Wyder,D. Kojic,M. Annaheim,and B. Lanzrein. 2005. Cloning,characterization and analysis by RNA interference of various genes of the *Chelonus inantius* polydnavirus. *J. Gen. Virol.* 86:973—83.

[16]Britton,J. S. ,W. K. Lockwood,L. Li,S. M. Cohen,and B. A. Edgar. 2002. *Drosophila's* insulin/PI3-kinase pathway coordinates cellular metabolism with nutritional conditions. *Dev. Cell* 2:239—49.

[17]Brown,M. R. ,K. D. Clark,M. Gulia,et al. 2008. An insulin-like peptide regulates egg maturation and metabolism in the mosquito *Aedes aegypti*. *Proc. Natl. Acad. Sci. U. S. A.* 105:5716—21.

[18]Cerenius,L. ,and K. Soderhall. 2004. The prophenoloxidse-activating system in invertebrates. *Immunol. Rev.* 198:116—26.

[19]Cole,T. J. ,N. E. Beckage,F. F. Tan,A. Srinivasan,and S. B. Ramaswamy. 2002. Parasitoid-host endocrine relations: Self-reliance or co-optation? *Insect Biochem. Mol. Biol.* 32:1673—79.

[20]Cui,L. ,A. I. Soldevila,and B. A. Webb. 1997. Expression and hemocyte-targeting of a *Campoletis sonorensis* polydnavirus cysteine-rich gene in *Heliothis virescens* larvae. *Arch. Insect Biochem. Physiol.* 36:251—71.

[21]Cui,L. ,A. I. Soldevila,and B. A. Webb. 2000. Relationships between polydnavirus gene expression and host range of the parasitoid wasp *Campoletis sonorensis*. *J. Insect Physiol.* 46:1397—1407.

[22]Cui,L. ,and B. A. Webb. 1997. Homologous sequences in the *Campolitis sonorensis* polydnavirus genome are implicated in replication and nesting of the W segment family. *J. Virol.* 71:8504—13.

[23]Cusson, M. , C. Lucarotti, D. Stoltz, P. Krell, and D. Doucet. 1998. A polydnavirus from the spruce budworm parasitoid *Tranosema rostrale* (Ichneumonidae). *J. Invertebr. Pathol.* 72:50—56.

[24]Dahlman,D. L. ,E. L. Scheppers,T. Scheppers,R. L. Rana,F. A. Diluna,and B. A. Webb. 2003. A gene from a parasitic wasp expressed in teratocytes inhibits host protein synthesis,growth and development. *Insect Mol. Biol.* 12:527—34.

[25]Davies,D. H. ,M. R. Strand,and S. B. Vinson. 1987. Changes in differential haemocyte count and in vitro behaviour of plasmatocytes from host *Heliothis virescens* caused by *Campoletis sonorensis* PDV. *J. Insect Physiol.* 33:143—53.

[26]Davies,D. H. ,and S. B. Vinson. 1986. Passive evasion by eggs of braconid parasitoid *Cardiochiles nigriceps* of encapsulation in vitro by haemocytes of host *Heliothis virescens*. Possible role for fibrous layer in immunity. *J. Insect Physiol.* 32:1003—10.

[27]Desjardins,C. A. ,D. E. Gundersen-Rindal,J. B. Hostetler,et al. 2007. Structure and evolution of a proviral locus of *Glyptapanteles indiensis* bracovirus. *BMC Microbiol.* 7:61.

[28]Dong,Y. M. ,H. E. Taylor,and G. Dimopoulos. 2006. AgDscam,a hypervariable immunoglobulin domain-containing receptor of the *Anopheles gambiae* immune system. *PLoS Biol.* 4:1137—46.

[29]Dong,K. ,D. Zang,and D. L. Dahlman. 1996. Down-regulation of juvenile hormone esterase and arylphorin production in *Heliothis virescens* larvae parasitized by *Microplitis croceipes*. *Arch. Insect Biochem. Physiol.* 32:237—48.

[30]Doucet,D. ,and M. Cusson. 1996. Role of calyx luid in alterations of immunity in *Choristoneura fumiferana* larvae parasitized by *Tranosema rostrale*. Entomol. *Exp. Appl.* 81:21—30.

[31]Dover,B. A. ,T. Tanaka,and S. B. Vinson. 1995. Stadium-specific degeneration of host prothoracic glands by *Campoletis sonorensis* caly luid and its association with host ecdysteroid titers. *J. Insect Physiol.* 41:947—55.

[32]Drezen,J. —M. ,B. Provost,E. Espagne,et al. 2003. Polydnavirus genome: Integrated vs. free virus. *J. Insect Physiol.* 49:407—17.

[33]Dupuy,C. ,E. Huguet,and J. —M. Drezen. 2006. Unfolding the evolutionary history of polydnaviruses. *Virus Res.* 117:81—89.

[34]Espagne,E. ,C. Dupuy,E. Huguet,et al. 2004. Genome sequence of a polydnavirus: Insights into symbiotic virus evolution. *Science* 306:286—89.

[35]Falabella,P. ,P. Cacciaupi,P. Varricchio,C. Malva,and P. Pennacchio. 2006. Protein tyrosine phosphatases of *Toxoneuron nigriceps* bracovirus as potential disrupters of host prothoracic gland function. *Arch. Insect Biochem. Physiol.* 61:157—69.

[36]Falabella,P. ,P. Varricchio,B. Provost,et al. 2007. Characterization of the IkappaB-like gene family in polydnaviruses associated with wasps belonging to different braconid subfamilies. *J. Gen. Virol.* 88:92—104.

[37]Fath-Goodin,A. ,T. A. Gill,S. B. Martin,and B. A. Webb. 2006. Effect of *Campoletis sonorensis* *cys*-motif proteins on *Heliothis virescens* development. *J. Insect Physiol.* 52:576—85.

[38]Fleming,J. G. ,and M. D. Summers. 1991. PDV DNA is integrated in the DNA of its parasitoid

wasp host. *Proc. Natl. Acad. Sci. U. S. A.* 88:9770—74.

[39]Friedman,R. ,and A. L. Hughes. 2006. Pattern of gene duplication in the *Cotesia congregata* bracovirus. *Infect. Genet. Evol.* 6:315—22.

[40]Gad,W. ,and Y. Kim. 2008. A viral histone H4 encoded by *Cotesia plutellae* bracovirus inhibits haemocyte spreading behaviour of the diamondback moth,*Plutellae xylostella*. *J. Gen. Virol.* 89:931—38.

[41]Galibert,L. ,J. Rocher,M. Ravallec,M. Duonor-Cerutti,B. A. Webb,and A. N. Volkoff. 2003. Two *Hyposoter didymator* ichnovirus genes expressed in the lepidopteran host encode secreted or membrane-associated serine and threonine rich proteins in segments that may be nested. *J. Insect Physiol.* 49: 441—51.

[42]Gardiner,E. M. M. ,and M. R. Strand. 2000. Hematopoiesis in larval *Pseudoplusia includens* and *Spodoptera frugiperda*. *Arch. Insect Bich. Physiol.* 43:147—64.

[43]Gill,T. A. ,A. Fath-Goodin,I. I. Maiti,and B. A. Webb. 2006. Potential uses of Cys-motif and other polydnavirus genes in biotechnology. *Adv. Virus Res.* 68:393—425.

[44]Gillespie,J. P. ,M. R. Kanost,and T. Trenczek. 1997. Biological mediators of insect immunity. *Annu. Rev. Entomol.* 42:611—43.

[45]Hayakawa,Y. 1995. Growth-blocking peptide: An insect biogenic peptide that prevents the onset of metamorphosis. *J. Insect Physiol.* 41:1—6.

[46]Hayakawa,Y. ,and K. Yazaki. 1997. Envelope proteins of parasitic wasp symbiont virus,protects the wasp eggs from the cellular immune reactions by the host insect. *Eur. J. Biochem.* 246:820—26.

[47]Ibrahim,A. M. ,and Y. Kim. 2008. Transient expression of protein tyrosine phosphatases encoded by *Cotesia plutellae* bracovirus inhibits insect cellular immune responses. *Naturwissenschaften* 95:25—32.

[48]Imler,J. —L. ,and P. Bulet. 2005. Antimicrobial peptides in *Drosophila*,structures,activities and gene regulation. In *Mechanisms of epithelial defense*,ed. D. Kabelitz and J. M. Schroder,Vol. 86,1—21. Basil: Karger.

[49]Irving,P. ,J. Ubeda,D. Doucet,et al. 2005. New insights into *Drosophila* larval haemocyte functions through genome-wide analysis. *Cell Microbiol.* 7:335—50.

[50]Kadash,K. ,J. A. Harvey,and M. R. Strand 2003. Cross-protection experiments with parasitoids in the genus *Microplitis* (Hymenoptera: Braconidae) suggest a high level of *specfiicity in* their associated bracoviruses. *J. Insect Physiol.* 49:473—82.

[51]Kanost,M. R. ,and M. J. Gorman. 2008. Phenoloxidases in insect immunity. In *Insect immunity*, ed. N. E. Beckage,69—96. San Diego: Academic Press.

[52]Kelly,T. J. ,D. B. Gelman,D. A. Reed,and N. E. Beckage. 1998. Effects of parasitization by *Cotesia congregata* on the brain-prothoracic gland axis of its host,*Manduca sexta*. *J. Insect Physiol.* 44:323—32.

[53]Kim,Y. 2005. Identification of host translation inhibitory factor of *Campoletis sonorensis* ichnovirus on the tobacco budworm,*Heliothis virescens*. *Arch. Insect Biochem. Physiol.* 59:230—44.

[54]Kocks,C. ,J. —H. Cho,N. Nehme,et al. 2005. Eater,a transmembrane protein mediating phagocytosis of bacterial pathogens in *Drosophila*. *Cell* 123:335—46.

[55]Kroemer,J. A. ,and B. A. Webb. 2004a. Polydnavirus genes and genomes: Emerging gene families and new insights into polydnavirus replication. *Annu. Rev. Entomol.* 49:431—56.

[56]Kroemer, J. A. , and B. A. Webb. 2004b. Divergences in protein activity and cellular localization within the *Campoletis sonorensis* ichnovirus vankyrin family. *J. Virol*. 80:12219—28.

[57]Kwon, B. , and Y. Kim. 2008. Transient expression of an EP1-like gene encoded in *Cotesia plutellae* bracovirus suppresses the hemocyte population in the diamondback moth, *Plutella xylostella*. *Dev. Comp. Immunol*. 32:932—42.

[58]Lanzrein, B. , R. Pfister-Wilhelm, and F. von Niederhausern. 2001. Effects of an egg-larval parasitoid and its polydnavirus on development and the endocrine system of the host. In *Endocrine interactions of insect parasites and pathogens*, ed. J. P. Edwards and R. J. Weaver, 95—109. Oxford: BIOS Sci. Publishers.

[59]Lanzrein, B. , R. Pfister-Wilhelm, T. Wyler, T. Trenczek, and P. Stettler. 1998. Overview of parasitism associated effects on host haemocytes in larval parasitoids and comparison with effects of the egg-larval parasitoid *Chelonus inanitus on its host Spodoptera littoralis*. *J. Insect Physiol*. 44:817—31.

[60]Lapointe, R. , K. Tanaka, W. E. Barney, et al. 2007. Genomic and morphological features of a banchine poly-dnavirus: Comparison with bracoviruses and ichnoviruses. *J. Virol*. 81:6491—6501.

[61]Lapointe, R. , R. Wilson, L. Vilaplana, et al. 2005. Expression of a *Toxoneuron nigriceps* polydnavirus (TnBV) encoded protein, causes apoptosis-like programmed cell death in lepidopteran insect cells. *J. Gen. Virol*. 86:963—71.

[62]Lavine, M. D. , and N. B. Beckage. 1995. Polydnaviruses: Potent mediators of host insect immune dysfunction. *Parasitol. Today* 11:368—78.

[63]Lavine, M. D. , and M. R. Strand. 2002. Insect hemocytes and their role in cellular immune responses. *Insect Biochem. Mol. Biol*. 32:1237—42.

[64]Lavine, M. D. , and M. R. Strand. 2003. Hemocytes from *Pseudoplusia includens* express multiple alpha and beta integrin subunits. *Insect Mol. Biol*. 12:441—52.

[65]Lee, S. , M. Nalini, and Y. Kim. 2008. A viral lectin encoded in *Cotesia plutellae* bracovirus and its immunosuppressive effect on host hemocytes. *Comp. Biochem. Physiol. A* 149:351—61.

[66]Levashina, E. A. , L. F. Moita, S. Blandin, G. Vriend, M. Lagueux, and F. C. Kafatos. 2001. Conserved role of a complement-like protein in phagocytosis revealed by dsRNA knockout in cultured cells of the mosquito, *Anopheles gambiae*. Cell 104:709—18.

[67]Levin, D. M. , L. N. Breuer, S. F. Zhuang, S. A. Anderson, J. B. Nardi, and M. R. Kanost. 2005. A hemocyte specific integrin requered for hemocytic encapsulation in the tobacco hornworm, *Manduca sexta*. *Insect Biochem. Mol. Biol*. 35:369—80.

[68]Li, S. , P. Falabella, I. Kuriachan, et al. 2003. Juvenile hormone synthesis, metabolism, and resulting haemolymph titre in *Heliothis virescens* larvae parasitized by *Toxoneuron nigriceps*. *J. Insect Physiol*. 49:1023—30.

[69]Li, X. , and B. A. Webb. 1994. Apparent functional role for a cysteine-rich polydnavirus protein in suppression of the insect cellular immune response. *J. Virol*. 68:7482—89.

[70]Ling, E. , K. Shirai, R. Kanekatsu, and K. Kiguchi. 2005. Hemocyte differentiation in the hematopoietic organs of the silkworm, *Bombyx mori*: Prohemocytes have the function of phagocytosis. *Cell Tiss. Res*. 320:535—43.

[71]Ling, E. J. , and X. Q. Yu. 2006. Cellular encapsulation and melanization are enhanced by im-

mulectins, pattern recognition receptors from the tobacco hornworm *Manduca sexta*. *Dev. Comp. Immunol.* 30:289—99.

[72]Lu, Z., M. H. Beck, H. Jiang, Y. Wang, and M. R. Strand. 2008. The viral protein Egf1.0 is a dual activity inhibitor of prophenoloxidase activating proteinases 1 and 3 from *Manduca sexta*. *J. Biol. Chem.* 283: 21325—233.

[73]Maiti, I. B., N. Dey, D. L. Dahlman, and B. A. Webb. 2003. Antibiosis-type resistance in transgenic plants expressing teratocyte secretory peptide (TSP) gene from a hymenopteran endoparasite (*Microplitis croceipes*). *Plant Biotechnol.* 1:209—19.

[74]Marti, D., C. Grossniklaus-Burgin, S. Wyder, T. Wyler, and B. Lanzrein. 2003. Ovary development and polydnavirus morphogenesis in the parasitic wasp *Chelonus inanitus*. I. Ovary morphogenesis, ampliication of viral DNA and ecdysteroid titres. *J. Gen. Virol.* 84:1141—50.

[75]Moita, L. F., R. Wang-Sattler, K. Michel, et al. 2005. In vivo identification of novel regulators and conserved pathways of phagocytosis in *A.* gambiae. *Immunity* 23:65—73.

[76]Murphy, N., J. C. Banks, J. B. Whitfield, and A. D. Austin. 2008. Phylogeny of the parasitic microgastroid subfamilies (Hymenoptera: Braconidae) based on sequence data from seven genes, with an improved time estimate of the origin of the lineage. *Mol. Phylogenet. Evol.* 47:378—95.

[77]Nakahara, Y., Y. Kanamori, M. Kiuchi, and M. Kamimura. 2003. In vitro studies of hematopoiesis in the silkworm, cell proliferation in and hemocyte discharge from the hematopoietic organ. *J. Insect Physiol.* 49:907—16.

[78]Nardi, J. B., B. Pilas, C. M. Bee, S. Zhuang, K. Garsha, and M. R. Kanost. 2006. Neuroglian-positive plasmatocytes of *Manduca sexta* and the initiation of hemocyte attachment to foreign surfaces. *Dev. Comp. Immunol.* 30:447—62.

[79]Nardi, J. B., B. Pilas, E. Ujhelyi, K. Garsha, and M. R. Kanost. 2003. Hematopoietic organs of *Manduca sexta* and hemocyte lineages. *Dev. Genes Evol.* 213:477—91.

[80]Pasquier-Barre, F., C. Dupuy, E. Huguet, et al. 2002. Polydnavirus replication: The EP1 segment of the parasitoid wasp *Cotesia congregata* is amplified within a larger precursor molecule. *J. Gen. Virol.* 83:2035—45.

[81]Pech, L. L., and M. R. Strand. 1996. Granular cells are required for encapsulation of foreign targets by insect haemocytes. *J. Cell Sci.* 109:2053—60.

[82]Pennacchio, F., M. C. Digilio, and E. Tremblay. 1995. Biochemical and metabolic alterations in *Acyrthosiphon pisum* parasitized by *Aphidius ervi*. *Arch. Insect Biochem. Physiol.* 30:351—67.

[83]Pennacchio, F., and M. R. Strand. 2006. Evolution of developmental strategies in parasitic Hymenoptera. *Annu. Rev. Entomol.* 51:233—58.

[84]Philips, J. A., E. J. Rubin, and N. Perrimon. 2005. *Drosophila* RNAi screen reveals C35 family member required for mycobacterial infection. *Science* 309:1248—51.

[85]Powell, J. A. 2003. Lepidoptera (moths, butterflies). In *Encyclopedia of insects*, ed. V. H. Resh and R. T. Carde, 631—64. San Diego: Academic Press.

[86]Pruijssers, A. J., and M. R. Strand. 2007. PTP-H2 and PTP-H3 from *Microplitis demolitor* bracovirus localize to focal adhesions and are antiphagocytic in insect immune cells. *J. Virol.* 81:1209—19.

[87]Ramet, M., P. Manfruelli, A. Pearson, B. Mathey-Prevot, and R. A. B. Ezekowitz. 2002. Func-

tional genomic analysis and identification of a *Drosophila* receptor for *E. coli*. *Nature* 416:644—48.

[88]Ramet, M., A. Pearson, P. Manfruelli, et al. 2001. *Drosophila* scavenger receptor CI is a pattern recognition receptor for bacteria. *Immunity* 15:1027—38.

[89]Reed, D. A., and J. J. Brown. 1998. Host/parasitoid interactions: Critical timing of parasitoid-derived products. *J. Insect Physiol.* 44:721—32.

[90]Rivkin, H., J. A. Kroemer, A. Bronshtein, E. Belausov, B. A. Webb, and N. Chejanovsky. 2006. Response of immunocompetent and immunocompromised *Spodoptera littoralis* larvae to baculovirus infection. *J. Gen. Virol.* 87:2217—25.

[91]Schafellner, C., R. C. Marktl, and A. Schopf. 2007. Inhibition of juvenile hormone esterase in *Lymantria dispar* (Lepidoptera: Lymantriidae) larvae parasitized by *Glyptapanteles liparidis* (Hymenoptera: Braconidae). *J. Insect Physiol.* 53:858—68.

[92]Shelby, K. S., and B. A. Webb. 1994. Polydnavirus infection inhibits synthesis of an insect plasma protein, arylphorin. *J. Gen. Virol.* 75:2285—94.

[93]Shelby, K. S., and B. A. Webb. 1997. Polydnavirus infection inhibits translation of specific growth-associated host proteins. *Insect Biochem. Mol. Biol.* 27:263—70.

[94]Shelby, K. S., and B. A. Webb. 1999. Polydnavirus-mediated suppression of insect immunity. *J. Insect Physiol.* 45:507—14.

[95]Schmidt, O., U. Theopold, and M. Strand. 2001. Innate immunity and its evasion and suppression by hymenopteran endoparasitoids. *Bioessays* 23:344—51.

[96]Stoltz, D. B., N. E. Beckage, G. W. Blissard, et al. 1995. Polydnaviridae. In *Virus taxonomy*, ed. F. A. Murphy, C. M. Fauquet, D. H. L. Bishop, et al., 143—47. New York: Springer-Verlag.

[97]Stoltz, D., and A. Makkay. 2003. Overt viral diseases induced from apparent latency following parasitization by the ichneumonid wasp, *Hyposoter exiguae*. *J. Insect Physiol.* 49:483—90.

[98]Stoltz, D. B., and S. B. Vinson. 1979. Viruses and parasitism in insects. *Adv. Virus Res.* 24:125—71.

[99]Stoltz, D. B., S. B. Vinson, and E. A. MacKinnon. 1976. Baculovirus-like particles in the reproductive tracts of female parasitoid wasps. *Can. J. Microbiol.* 22:1013—23.

[100]Strand, M. R. 2008a. Insect hemocytes and their role in immunity. In *Insect immunity*, ed. N. E. Beckage, 25—47. San Diego: Academic Press.

[101]Strand, M. R. 2008b. Polydnavirus abrogation of the insect immune system. In *Encyclopedia of virology*, ed. B. W. J. Mahy and M. H. V. van Regenmortel, Vol. 4, 250—256. London: Elsevier.

[102]Strand, M. R., and J. Casas. 2007. Parasitoid and host nutritional physiology in behavioral ecology. In *Parasitoid behavioral ecology*, ed. E. Wajnberg, J. van Alphen, and C. Bernstein, 113—28. Oxford: Blackwell Press.

[103]Strand, M. R., and L. L. Pech. 1995. Immunological basis for compatibility in parasitoid-host relationships. *Annu. Rev. Entomol.* 40:31—56.

[104]Suderman, R. J., A. J. Pruijssers, and M. R. Strand. 2008. Protein tyrosine phosphatase-H2 from a polydnavirus induces apoptosis of insect cells. *J. Gen. Virol.* 89:1411—20.

[105]Tanaka, T., N. Agui, and K. Hiruma. 1987. The parasitoid *Apanteles kariyai* inhibits pupation of its host, Pseudaletia separata, via disruption of prothoracicotropic hormone release. Gen. Comp. Endocri-

nol. 67:364—74.

[106]Tanaka,K. ,R. Lapointe,W. E. Barney,et al. 2007. Shared and species-specific features among ichnovirus genomes. *Virology* 363:26—35.

[107]Terenius,O. ,R. Bettencourt,S. Y. Lee,W. Li,K. Soderhall,and I. Faye. 2007. RNA interference of hemolin causes depletion of phenoloxidase activity in *Hyalophora cecropia*. *Dev. Comp. Immunol.* 31:571—75.

[108]Theopold,U. ,C. Samakovlis,H. Erdjumentbromage,et al. 1996. *Helix pomatia* lectin,an inducer of *Drosophila* immune response,binds to hemomucin,a novel surface mucin. *J. Biol. Chem.* 271:12708—15.

[109]Theopold,U. ,O. Schmidt,K. Soderhall,and M. S. Dushay. 2004. Coagulation in arthropods,defence,wound closure and healing. *Trends Immunol.* 25:289—94.

[110]Thoetkiattikul,H. ,M. H. Beck,and M. R. Strand. 2005. Inhibitor kappaB-like proteins from a polydnavirus inhibit NF-kappaB activation and suppress the insect immune response. *Proc. Natl. Acad. Sci. U. S. A.* 102:11426—31.

[111]Thompson,S. N. 1993. Redirection of host metabolism and effects on parasite nutrition. In *Parasites and pathogens of insects*, ed. N. E. Beckage,S. N. Thompson,and B. A. Federici,Vol. 1,125—44, New York: Academic Press.

[112]Thompson,S. N. ,and D. L. Dahlman. 1998. Aberrant nutritional regulation of carbohydrate synthesis by parasitized *Manduca sexta*. *J. Insect Physiol.* 44:745—54.

[113]Turnbull,M. W. ,A. N. Volkoff,B. A. Webb,and P. Phelan. 2005. Functional gap junction genes are encoded by insect viruses. *Curr. Biol.* 15:R491—92.

[114]Vinson,S. B. 1990. Physiological interactions between the host genus *Heliothis* and its guild of parasitoids. *Arch. Insect Biochem. Physiol.* 13:63—81.

[115]Volkoff,A. —N. ,M. Ravallec,J. Bossy,et al. 1995. The replication of *Hyposoter didymator* PDV: Cytopathology of the calyx cells in the parasitoid. *Biol. Cell* 83:1—13.

[116]Washburn,J. O. ,E. J. Haas-Stapleton,F. F. Tan,N. E. Beckage,and L. E. Volkman. 2000. Co-infection of *Manduca sexta* larvae with PDV from *Cotesia congregata* increases susceptibility to fatal infection by *Autographa californica* M Nucleopolyhedrovirus. *J. Insect Physiol.* 46:179—90.

[117]Webb,B. A. 1998. Polydnavirus biology,genome structure,and evolution. In *The insect viruses*, ed. L. K. Miller and L. A. Balls,105—39. New York: Plenum Press.

[118]Webb,B. A. ,and L. Cui. 1998. Relationships between polydnavirus genomes and viral gene expression. *J. Insect Physiol.* 44:785—93.

[119]Webb,B. A. ,and M. R. Strand. 2005. The biology and genomics of polydnaviruses. In *Comprehensive molecular insect science*, Vol. 6,ed. L. I. Gilbert,K. Iatrou,and S. S. Gill,323—60. San Diego: Elsevier.

[120]Webb,B. A. ,M. R. Strand,S. E. Dickey,et al. 2006. Polydnavirus genomes relect their dual roles as mutualists and pathogens. *Virology* 347:160—74.

[121]Wertheim,B. ,A. R. Kraaijeveld,E. Schuster,et al. 2005. Genome wide expression in response to parasitoid attack in *Drosophila*. *Genome Biol.* 6: R94.

[122]Whitfield,J. B. 1998. Phylogeny and evolution of host-parasitoid interactions in Hymenoptera.

Annu. Rev. Entomol. 43:129—51.

[123]Whitfield, J. B. 2002. Estimating the age of the polydnavirus/braconid wasp symbiosis. *Proc. Natl. Acad. Sci. U. S. A.* 99:7508—13.

[124]Whitfield, J. B., and S. Asgari. 2003. Virus or not? Phylogenetics of polydnaviruses and their wasp carriers. *J. Insect Physiol.* 49:397—405.

[125]Xu, D., and D. Stoltz. 1993. Polydnavirus genome segment families in the ichneumonid parasitoid *Hyposoter fugitivus*. *J. Virol.* 67:1340—49.

[126]Zhang, G., Z. —Q. Lu, H. Jiang, and S. Asgari. 2004. Negative regulation of prophenoloxidase (proPO) activation by a clip-domain serine proteinase homolog (SPH) from endoparasitoid venom. *Insect Biochem. Mol. Biol.* 34:477—83.

[127]Zhuang, S., L. Kelo, J. B. Nardi, and M. R. Kanost. 2007. An integrin-tetraspanin interaction required for cellular innate immune responses of an inset, *Manduca sexta*. *J. Biol. Chem.* 282:22563—72.

第十八章 家蚕浓核病毒的抗性

Keiko Kadono-Okuda

1. 家蚕浓核病毒

家蚕浓核病毒(*Bm*DNV)在家蚕幼虫饲养期对蚕农造成了毁灭性的灾难,从而长久以来一直被视为影响养蚕业发展的一个非常严重的问题。感染这种病毒的家蚕中肠柱状细胞(病毒繁殖的地方)含有过度生长的细胞核。因此,感染细胞的核质能被 DNA-易染的甲基绿或 Feulgen 试剂浓染(Watanabe et al. 1976)。由于中肠组织被破坏,感染病毒的幼虫变得软弱、易腹泻(这被认为是蚕软腐病的症状),最终死亡。

*Bm*DNV 属于细小病毒科。这种细小病毒科被划分为两个亚家族,细小病毒病和浓核病毒。细小病毒病的宿主是脊椎动物,而浓核病毒的宿主是节肢动物。这些病毒含有单链 DNA,基因组大小为 4～6kb,是所有病毒中最小的病毒。

浓核病毒亚家族最早是在大蜡螟中 *Galleria mellonella* 被发现和鉴定的(Meynadier et al. 1964; Kurstak 和 Cote 1969)。目前它被划分为四种基因型:(1)浓核病毒,其宿主为鳞翅目昆虫,如鹿眼蛱蝶 *Junonia coenia* 和大蜡螟 *G. mellonella*,以及双翅目宿主如金腹巨蚊 *Toxorhynchites splendens*; (2) 重复病毒,其宿主为鳞翅目昆虫,如家蚕,*Casphalia extranea* 和 *Sibine fusca*; (3) 环星黑烟浓稠病毒,已知来自于黑胸大蠊 *Periplaneta fuliginosa* 浓核病毒;(4)短浓稠病毒,其拥有最小的基因组,发现于伊蚊和亚洲虎蚊 *A. albopictus*(双翅目)中。浓核病毒已经在 6 个目的昆虫中被发现,包括直翅目、半翅目和蜻蜓目,也在甲壳类,虾和螃蟹等节肢动物中有记载。然而,它们的属性分类还没有被证实(见 8th ICTV 报道,http://www.ncbi.nlm.nih.gov/ICTVdb/Ictv/index.htm)。最近有研究用蚊子的浓核病毒来控制如疟疾、登革热和西尼罗热等病原体的传播。这些研究利用了宿主的特异性和自然稳定性(Carlson, Suchman 和 Buchatsky 2006)。另一方面,有报道表明浓核病毒会慢性地感染蚊子培养细胞系,表明浓核病毒在自然界蚊子中是广泛分布的(Jousset et al. 1993; Chen et al. 2004)。

*Bm*DNV,即所谓的 Picorna 病毒,于 1968 年在日本一个养蚕农场中感染了软腐病的病蚕体内首次被发现,当时被认为是蚕软腐病毒(IFN, *Iflavirus*),但其特征不同于蚕软腐病毒 *Bm*IFV(Shimizu 1975)。相同的病毒从 *Bm*IFV 的 Sakashiro 品系的感染溶液中被分离出来,随后被命名为家蚕Ⅰ型浓核病毒(*Bm*DNV-1, *Iteravirus*)。后来,分别在日本和中国发现两种新型的 *Bm*DNV。来自日本的病毒被命名为家蚕Ⅱ型浓核病毒(*Bm*DNV-2),从中国分离到的被命名为 *Bm*DNV-3 或 Z(Kawase 和 Kurstak 1991)。*Bm*DNV-2 和-3/Z 表现出非常相似的症状(慢性感染,而 *Bm*DNV-1 为急性感染)和血清学的同源性。此外研究发现 *Bm*DNV-2 和 *Bm*DNV-3/Z 具有二重基因组(Bando et al. 1995),包括一个编码 DNA 多聚酶的基因(Hayakawa et al. 2000; Wang et al. 2007),这也说明它们具有密切的相关性。*Bm*DNV-1 或另外的细小病毒科没有这些特征;因此 Tijssen 和 Bergoin(1995)建议它们应该被重新划分为双浓核病毒的新属。然而,通过那些有差别的特征,最近 ICTV 第八次报导从细小病毒科排除了 *Bm*DNV-2 和-3/Z(Bergoin 和 Tijssen 2000; Fauquet et al. 2005),却未确定对它们更高的分类位置。现在它们暂时被认为是细小病毒样的病毒,但被单独列为一群。

在家蚕中,不同的基因控制着对 *Bm*DNV-1, *Bm*DNV-2 的抗性。因此,这些基因被

视为 *Bm*DNV-1,*Bm*DNV-2 抗性基因。这篇综述将简要地描述这些抗性基因以及它们本身的基因型特征,并将重点描述 *Bm*DNV-2 抗性基因 *nsd*-2 的定位克隆。

2. 家蚕中浓核病毒的抗性基因

有趣的是,有些家蚕品系对 *Bm*DNV 有完全的抵抗力(非易感)。这些品系不管被接种多少 *Bm*DNV 病毒都不会被感染。抗性和易感性家蚕的杂交表明抗性被显性和隐性基因共同控制。到目前为止,已经报道了四种抗性基因。它们是抗 *Bm*DNV-1 病毒的 *Nid*-1 和 *nsd*-1 基因及抗 *Bm*DNV-2 病毒的 *nsd*-2 和 *nsd*-*Z* 基因。以下是对当前研究中每个基因的详细描述。

2.1 针对 *Bm*DNV-1 的抗性基因

Nid-1(对 DNV-1 不感染)是控制对 *Bm*DNV-1 易感/非易感两个已知基因中的一个,如同另外的浓核病毒抗性基因(*nsd*-1,*nsd*-2 和 *nsd*-*Z*),它主要作为一个单一的主基因起作用。此外,它是唯一的由显性等位基因来负责的 *Bm*NDV 抗性基因(Eguchi,Furuta 和 Ninaki 1986)。*Nid*-1 在最近被检测到。迄今为止,仅有 5 个品系被证实为 *Nid*-1 携带者,而在日本却有许多 *nsd*-1 和 *nsd*-2 携带品系(Furuta 1995;K. Kadono-Okuda, unpubl. obs.)。据推测这个突变品系可能发生于日本的一个非商业品系(K. Kadono-Okuda,未发表数据),这能解释为什么它没有像 *nsd*-1 和 *nsd*-2 抗性突变品系一样广泛传播。

在抗性和易感品系之间的单配对回交(BC_1)品系表明 *Nid*-1 基因是与 17 连锁群上的两个突变:*Bm*(黑蛾)及 *bts*(褐头尾斑)相连锁。它的位置是通过三元显性($+^{bts}$,*Nid*-1,$+^{ow}$)和三元隐性(*bts*,$+^{Nid\text{-}1}$,*ow*)突变品种的一个三点杂交方法确定在 31.1cM 处,*ow* 是多种油蚕突变体中的一种,由于尿酸代谢的反常,这些油蚕突变体幼虫的表皮是半透明的(Eguchi et al. 2007)。Hara et al. (2008)通过使用表达序列标签作为探针,利用限制性片段长度多态性方法进行了 *Nid*-1 的连锁及定位分析。利用这个显性抗性基因,Eguchi 等用含有 *Nid*-1 的日系近等位基因系杂交野生型的中国种(1998),成功地得到了 *Bm*DNV-1 抗性的杂交品系。根据他们的结果,我们继续实施 *Nid*-1 的精确定位,且成功地缩小到了大约 34kb 的一个 *Nid*-1 基因候选区域(K. Kadono-Okuda,未发表数据)。将来的工作旨在利用已有的家蚕基因组资源来分离和鉴定候选基因(详见第 2 章关于家蚕基因组的综述)。

与 *Nid*-1 相比较,*nsd*-1(DNV-1 非易感)对 *Bm*DNV-1 的抗性是一个隐性突变,并且只在纯合状态下有表达。它的位点被 Eguchi,Ninaki 和 Hara(1991) 确定在第 21 连锁群上的 8.3cM 处,该位点随后被 5 个相连锁的 DNA 随机多态性扩增标记(RAPD)所证实(Abe et al. 1998)。培养的胚胎一经感染,在中肠形成的胚胎反转期之后表现出抗性或易感的表现型,除此以外,人们对 *nsd*-1 的作用模式知之甚少(Furuta 1983)。

有趣的是在$+^{nsd\text{-}1}$和 *Nid*-1 的杂交后代中,*nsd*-1 基因的显性易感位点与 *Nid*-1 的显性抗性位点相比较,后者是上位的(Abe,Watanabe 和 Eguchi 1987)。即使家蚕在 21 连锁群上拥有$+^{nsd\text{-}1}/+^{nsd\text{-}1}$基因型,在 17 连锁群的一个杂合 *Nid*-1 基因存在的情况下,它仍具有抗性。这就表明 *nsd*-1 和 *Nid*-1 是相互独立的基因，以不同方式参与病毒入侵和增殖过程。

2.2 针对 *Bm*DNV-2 的抗性基因

nsd-2(DNV-2 非易感)基因是一个对 *Bm*DNV-2 有抗性的隐性突变基因。Abe 等(2000)报道了两个与$+^{nsd\text{-}2}$相连锁的 RAPD 标记，这是一个在 *nsd*-2 位点的野生型易感等位基因。随后，Ogoyi 等(2003)在 17 连锁群上发现了跟 *nsd*-2 紧密连锁，交换率为零的 3 个 RFLP 标记。根据这些研究，我们研究小组进行了 *nsd*-2 的定位克隆，并成功地鉴定了一个候选基因。此外，我们通过导入 1 个$+^{nsd\text{-}2}$基因，功能性地恢复了突变(Ito et al. 2008)。这个研究的细节将在下一部分描述。

Qin 和 Yi(1996)对代表每个连锁群突变品系的连锁分析表明 *nsd-Z* 基因(抗 DNV-3，也被称为 DNV-Z，属于 DNV 的中国镇江株)与位于 15 连锁群上的 *Se*(腹白卵)标记连锁。此外，单序列重复(SSR)图谱检测到 7 个标记包围着 *nsd-Z*(Li et al. 2006)；然而，连锁图谱并没有被直接证实。Nakagaki 等(1999)在四龄期接种之后，通过定量 PCR 比较了在易感和耐受品系之间产生的 *Bm*DNV-Z DNA 的量。在易感幼虫中病毒 DNAs 呈对数形式的增加，并且感染后的 48h 达到最大值，但抗性幼虫中病毒 DNAs 仅少量地增加，直到感染后的 12h，然后病毒 DNA 迅速下降，因此作者推测在抗性品系的中肠里病毒 DNAs 复制后再被降解。Han 等人(2007)继续研究了 *nsd-Z* 抗性表达机制，他比较了中国种易感和抗性品系 5 龄期幼虫 *Bm*DNV-3 病毒的复制水平的差异。发现与前面研究相似的生长模式(Nakagaki et al. 1999)：抗性品系病毒的复制在感染后的 2h 从每个细胞 6～10 个拷贝增加到在感染后 96h 的每个细胞 150～200 个拷贝。该作者认为在一些家蚕品系中对 *Bm*DNV-3 的抗性是一种由宿主携带的不引起蚕软腐病的慢性病毒复制。研究人员曾试图育成 *nsd-Z* 近等位基因系，用包括 *nsd-Z* 连锁标记的 SSR 标记来观察连续回交过程。比如，属于 11 连锁群的一个标记(不与 *nsd-Z* 连锁)在回交第 4 代时，表现了迅速的替代和纯合性，而另一个跟 *nsd-Z* 紧密连锁的标记没有被替换(Li et al. 2007)。Chen 等(2007)利用荧光差显从 *nsd-Z* 近等位基因系检测到了一个在抗性品系高表达而不在易感家蚕中表达的蛋白激酶 C 抑制子基因。对于这个基因如何与浓核病毒抗性相关，以及是否这种现象在其他 *nsd-Z* 携带的抗性品系中仍然存在目前还不清楚，对潜在机制的证实需要进一步的分析。

2.3 抗性基因可能的表达位点

利用遗传嵌合体突变 *mo* 产生的嵌合体家蚕，这种蚕双受精所产生的细胞有杂合抗性(*Nid*-1/+)或纯合易感的(+/+)两种，从中发现中肠细胞在感染性上相互之间是不同的(Abe，Kobayashi 和 Watanabe 1990)。这就表明由 *Nid*-1 编码的显性抗性因子没有分泌到中肠腔起作用，而是在每个细胞或中肠细胞表面起作用。另外在 *nsd*-2 嵌合体家蚕品系中也有相似的结果，其单个细胞是纯合抗性(*nsd-2*/*nsd-2*)或杂合易感的(*nsd-2*/+)：每一个中肠细胞与同 *Nid*-1 相同的方式分别地参与了对病毒的抗性和易感性(Abe et al. 1993)。在不同的连锁群上，相互独立的显性和隐性突变都自发地作用于细胞，这些研究结果表明控制抗性的两个基因在病毒感染的不同阶段起作用。

3. *Nsd*-2 的分离和分析

家蚕基因组研究最新的进展使基于 SSRs,RFLPs,RAPDs 和单核酸多态性(SNPs)的连锁图谱的许多工具都可用于基因定位克隆(Miao et al. 2005; Nguu et al. 2005;Yasukochi et al. 2006;Yamamoto et al. 2006,2008)。此外,已报道了大量的位点信息,包括 EST 克隆,细菌人工染色体(BAC)克隆末段序列和在扩展的物理-遗传图谱和框架图上被预测的表达基因(Yamamoto et al. 2008; 家蚕遗传和基因组学的当前状况详见第二章)。对定位克隆的要求是:(1)标记以一定距离规则地位于每个连锁图谱;(2)在 F_2 代或 BC_1(回交)代群体的分离中能清楚地辨认亲代品系的靶标表型。即使对要鉴定的性状所在的连锁群是不清楚的,早期的连锁分析通过跟踪小量的 F_1 雌性的 BC_1 后代(大约 15 个个体,依赖该性状遗传的复杂性)中每个连锁群一个或两个代表性标记的遗传,很容易地完成了对家蚕的连锁分析。这利用了雌完全连锁的不交换特征(鳞翅目雌性减数分裂重组的缺乏在第三章阐述)。在建立了靶标基因的连锁群之后,一个粗略的连锁图谱能使用在亲本雌性(含有纯合隐性性状的亲本)和 F_1 雄性之间相互回交的大约 50 个 BC_1 分离群体来完成,在其中,减数分裂过程中发生了交换,使用的标记分散在靶标连锁群上(图 18.1A;Ogoyi et al. 2003;作图策略的例子见第 6 章)。

如果在连锁图谱上的靶标基因位置是已知的,相应于该区域的标记被优先使用。当该"候选区域"包括在最近的上游和下游标记中,但一些分离群体仍然没能与靶标特征连锁时,就可用远端侧面的任一标记表明重组片段的内部标记去缩小范围。当靶标区域减小到一定程度,所有的个体跟形状相关时,靶标区域的进一步缩小可通过用当前最近的上下游标记筛查新的 BC_1 分离群体。对于没有多态标记的区域,从家蚕基因组数据库中(http://sgp.dna.affrc.go.jp/KAIKObase/)获取靶标周围的序列来设计 PCR 引物。在一些情况下,用几百头家蚕就可把靶标区域缩减到足够小的范围,但在另外的一些情况下,成千的片段需要限定一个足够小的区域来预测一个候选基因。有些时候这个区域还未充分地被缩减到能定义一个单一基因之前,一些候选基因就能够得到证实。

最近,这个统一的物理-遗传连锁图谱(Yamamoto et al. 2008)以及中国和日本的基因组计划整合在一起的数据产生了来自 WGS 序列组装的特别长的框架(国际家蚕基因组协会 2008)。此外,由于大量连锁群序列的涌现,获得靶标区域的序列信息会相对更容易。早期的定位克隆实验采用候选区域的两端最近的侧面标记连续数轮对 BAC 文库进行筛选(Koike et al. 2003),通过 DNA 指纹图谱构建重叠群,然后用在重叠群两末端上的 BAC 末端序列设计的引物筛查重组的 BC_1 分离群体,以缩减在染色体上的候选区域。重复这些过程,BAC 重叠群被扩展到了靶标基因。现在,不再需要如此麻烦的程序;比如,仅仅通过连锁分析检测指定区域基因组序列的预测,就可能证实该基因。接下来简要描述如何鉴定浓核病毒 2 型抗性基因(*nsd-2*),以及通过转基因技术恢复浓核病毒的易感性来证实其功能。

3.1 通过染色体步移法和连锁分析缩减 *Nsd*-2 候选区域

首先,实施染色体步移寻找标记来缩减 *nsd*-2 的候选区域。在 17 连锁群上的 3 个

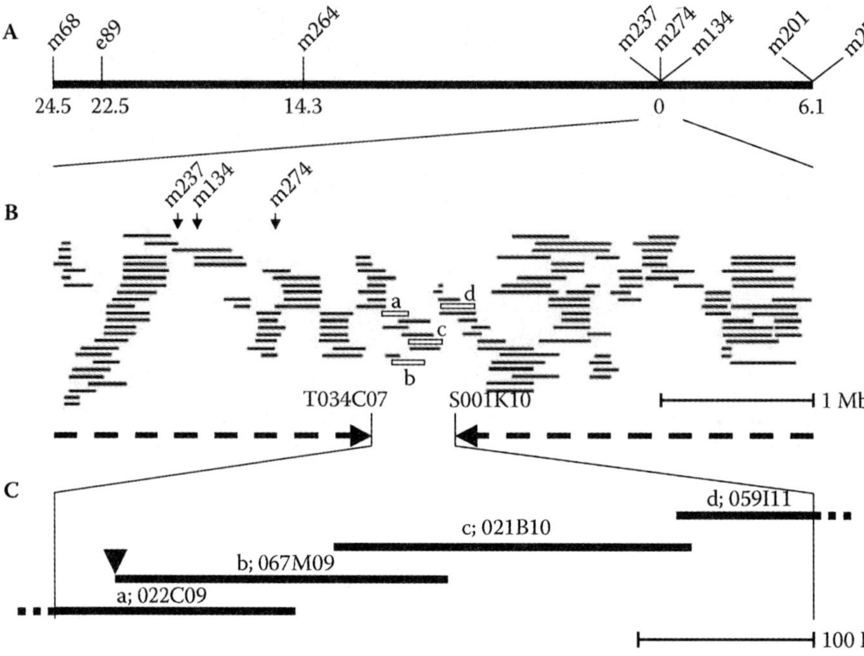

图 18.1　*nsd*-2 定位克隆策略。(A) *nsd*-2 的 RFLP 图谱，在以前的报道(Ogoyi et al. 2003)基础上修改。在图上标明了 EST 标记；在标记和 *nsd*-2 之间的距离用图下的 cM 单位表示。在最初的涉及 49 个雄个体回交后代(BC_1)的研究中表明 EST 标记 m237，m274 和 m134 与 *nsd*-2 没有重组交换，这作为了步行的起始点(Ogoyi et al. 2003)。(B)覆盖了 3 个 EST 标记的 BAC 重叠群。每条线表明了一个 BAC 克隆。有方框线的 BACs(a～d)在 C 图中阐述。虚线表明用 206BC_1 后代病毒处理存活调查以缩小跟 *nsd*-2 有关区域连锁分析的结果。(C)与 *nsd*-2 紧密连锁区域的最小 BAC 覆盖模式(tiling path)。箭头表明在抗性品系 J150 中发现的缺失位点(源自 Ito et al.，2008)。

EST 标记，ml34，m237 和 m274 是与 Ogoyi 等(2003)证实的 *nsd*-2 是紧密连锁的，它们被用作开始步移时的探针。通过用这些探针筛查家蚕 BAC 文库的高密度复制滤膜，然后再用末端序列重复筛查，构建了大约 5Mb 长的 BAC 重叠群，该重叠群覆盖了三个标记起始点的上游和下游区域(图 18.1C)。之后，基于建立的物理图谱，用 BC_1 后代病毒感染后存活率进行连锁分析，把与 nsd-2 连锁的区域缩减到了大约 400kb(图 18.1C)。根据该区域紧密间隔的基因座位设计的 PCR 引物被用于扩增 BC_1 分离群体的亲代品系的基因组 DNA。伴随 34bp 序列插入片段的 1 个 6.3kb 序列缺失在抗性(J150)品系，而不在易感品系(No.908)中被找到。使用这个序列信息，设计特定引物进行 PCR，比较 3 个抗性品系和 7 个易感品系。序列缺失和 34bp 的插入仅是抗性品系中的普遍特征(Ito et al. 2008)。

3.2 *Nsd*-2 候选基因的全长 cDNA 序列

使用 KAIKOGAAS 自动注释系统(http://kaikogaas.dna.affrc.go.jp/)，对 6kb 缺失序列进行研究，预测出一个候选基因，用 RACE 法在抗性和易感性品系中得到了候选基因的全长 cDNA 序列(cDNA 末端的迅速扩增)。易感类型由 14 个外显子构成，而抗性类型缺乏 5～13 外显子。此外，序列缺失引起了阅读框的改变，外显子 14 的起始部位出现了新的终止密码(图 18.2)。为了验证这个候选基因的产物，根据 5'UTR 和 3'UTR 区域设计引物，用 RT-PCR 对抗性品系和易感品系幼虫中肠来源的 RNA 进行扩增。发现所有抗性型候选基因产物的大小比易感性的基因产物更小，反映了来自基因组 DNA 扩增的 PCR 产物的大小差异(Ito et al. 2008)。

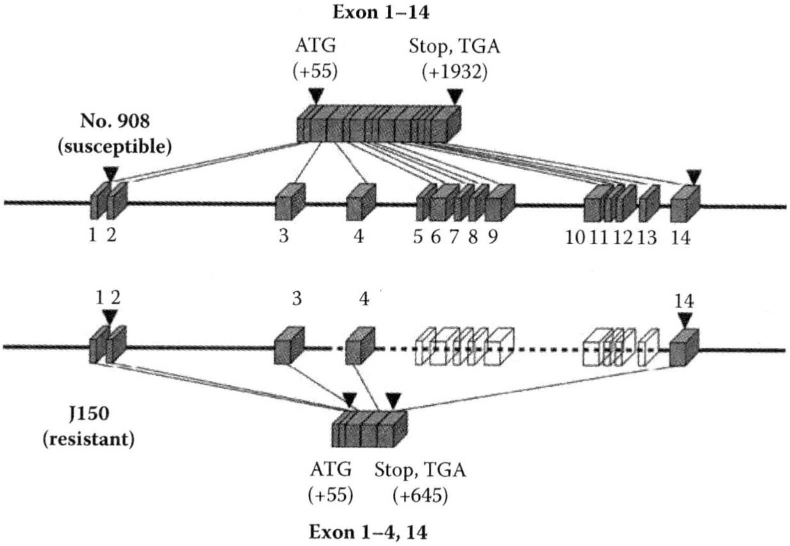

图 18.2　*nsd*-2 和＋nsd-2 候选基因的概要图。表明了在易感品系 No. 908(上)和抗性品系 J150(下)中外显子和内含子的相对位置和大小。虚线表明 J150 中的缺失。箭头表明起始和终止密码。

3.3 *Nsd-2* 候选基因的表达分析

从易感和抗性品系四龄 1 天幼虫的丝腺、前肠、中肠、后肠、马氏管、脂肪体、精巢和卵巢及中枢神经系统提取 RNA 进行 RT-PCR。RT-PCR 产物仅在中肠被检测到(Ito et al. 2008)。这与 Seki 和 Iwashita(1983)发现浓核病毒的感染是中肠特异性的研究结果相一致。这个基因的发育表达模式在 No. 908 和 J150 品系从卵发育 1、4、10 天及幼虫、蛹和成虫中进行检测；发现候选基因 RNA 仅出现在胚胎发育晚期和幼虫的取食阶段。重要的是在胚胎发育早中期，每一个龄期的蜕皮阶段，蛹或成年蚕蛾没有检测到候选基因的表达(Ito et al. 2008)。这些结果表明这个基因仅在中肠进行食物的消化和吸收时表达。

3.4 *Nsd*-2 候选基因的结构分析

No. 908 的 *nsd*-2 基因翻译后氨基酸序列与烟草天蛾氨基酸转运子有 63％的高度同源性(Castagna et al. 1998; Feldman, Harvey 和 Stevens 2000)。氨基酸转运子是位于上皮细胞内腔面的膜蛋白，这与通过 SOSUI(http://bp. nuap. nagoya-u. ac. jp/sosui/)，(一个从氨基酸组成的疏水性预测蛋白跨膜区域结构的软件)预测得到的结果一致，SOSUI

预测估计 No. 908 的 *nsd*-2 基因翻译后氨基酸序列的二级结构中有 12 个跨膜蛋白。反之，因为大量的缺失，抗性品系 J150 仅有 3 个蛋白前端部分的跨膜结构(图 18.3)。由于这个蛋白质在烟草天蛾主要作为一个单体转运特定的氨基酸(Castagna et al. 1998; Feldman, Harvey 和 Stevens 2000)，因此推测家蚕 *nsd*-2 候选基因编码了一个参与氨基酸转运的功能蛋白。另一方面，因为抗性品系这个编码基因缺乏内部外显子却能正常生长，因此认为它对家蚕是非必需的。然而在 *nsd*-2 的分离过程中，我们发现在那个基因的附近有一个与 *nsd*-2 同源性特别高(69%)的另一个基因。当前，这个基因也许代替 *nsd*-2 作为氨基酸转运载子的候选基因而被详细分析(K. Kadono-Okuda，未发表数据)。另一方面，没有功能的 *nsd*-2 就不能感染病毒，表明该基因的产物是病毒必需的因子。

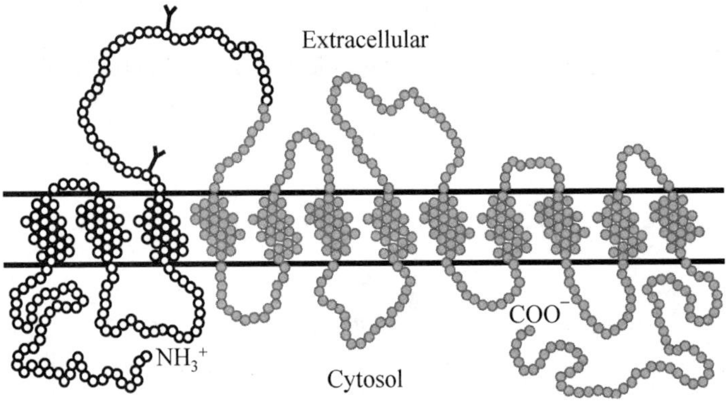

图 18.3 NSD-2 的预测二级结构。在跨膜结构域 3 和 4 之间表明的两个假定的 N 端糖基化位点(Y 样标记)。灰色区域表明了在 *NSD*-2 中的缺失。二级结构是基于拓扑学的预测方法，SOSUI(http://bp.uap.nagoya-u.ac.jp/sosui)；这个简图参考了烟草天蛾中报道的 12 个跨膜结构的膜蛋白(Feldman, Harvey 和 Stevens 2000)(源自 Ito et al., 2008)。

3.5 *Nsd*-2 候选基因的功能证据

互补测试是鉴定所分离候选基因功能的最直接和必需的证据。当前，没有体外系统或培养的细胞系来研究这种病毒的感染。此外，成功地在家蚕中用 RNAi 敲掉野生型基因功能的例子很少(Ohnishi, Hull 和 Matsumoto 2006)，RNAi 在家蚕中效率较低(Tabunoki et al. 2004)，这使得该方法不能很好地用于鉴定 *nsd*-2 候选基因。因此，为了鉴定是否被分离的候选基因是负责病毒抗性的，利用 GAL4-UAS 转基因系统把 $+^{nsd\text{-}2}$ 候选序列转化抗性家蚕(Tamura et al. 2000)。选用在中肠高表达 GAL4 蛋白的 GAL4 品系。一旦用这个病毒感染，仅运载和表达野生型(易感)候选基因的改造 GAL4/UAS 家蚕表现出明显的易感表型(图 18.4; Ito et al. 2008)。根据这些结果，可以推断被分离的候选基因 *nsd*-2，就是 *Bm*DNV-2 抗性基因，病毒感染需要的是 $+^{nsd\text{-}2}$ 表达的野生型膜蛋白。

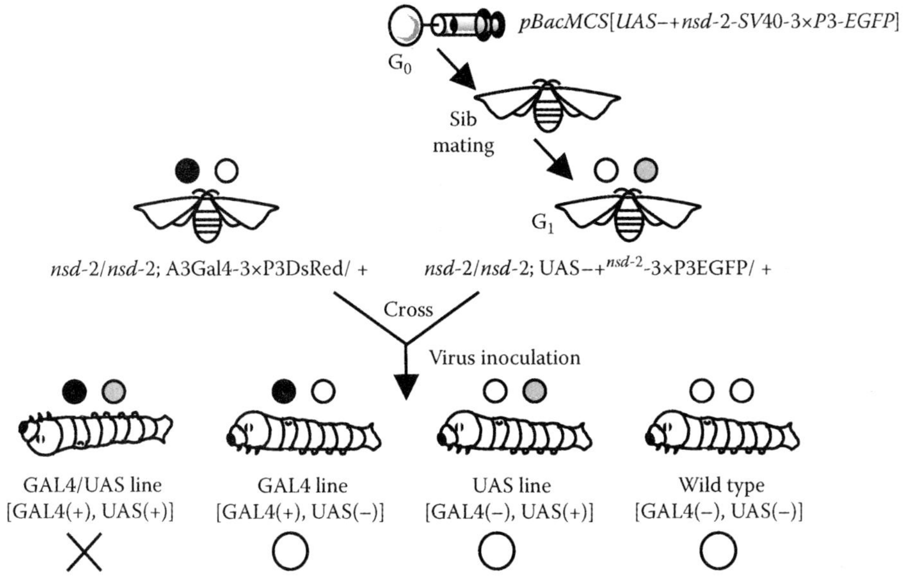

图 18.4　转基因家蚕后代群体对病毒的易感性。通过带有 DsRed 标记的 GAL4 系(●○)与带有靶标基因和 GFP 标记的 UAS 系(○◎)交配产生 4 种后代群体(●◎/●○/○◎/○○)。转基因($+^{nsd\text{-}2}$)只在 GAL4/UAS 系(●◎)中表达,GAL4 蛋白识别 UAS 序列,激活插入基因的转录。只有 GAL4/UAS 系表现出明显的易感表现型(X);其他的表现出对病毒的抗性(○)。

4. *Nsd*-2 基因分离的应用前景

*Bm*DNV-2 抗性基因 *nsd*-2 的获得是在家蚕中用图谱/定位克隆的方法分离突变基因的第一个成功的例子。它的意义不仅在于阐述了宿主昆虫针对致病性微生物抗性机制的关键方面,而且也揭示了病毒完全抗性形成的原因。

昆虫病毒中宿主或组织特异性的感染机制在 NPV 中已经得到研究(Bonning 2005);然而,研究集中在病毒基因,而不在宿主的反应机制。因此,*Bm*DNV-2 和中肠膜蛋白之间的相互作用将非常有助于从宿主昆虫的角度理解病毒感染机制。对感染人、犬或猫的细小病毒的感染机制进行比较研究是重要的。在细小病毒亚科,众所周知,人的细小病毒 B19 仅感染人、犬和猫细小病毒仅分别感染狗和猫家族成员。这表明在每一种病毒和它的受体之间的关系是高度特异的。Brown,Anderson 和 Young(1993)在首先阐明细小病毒的感染机制中发现 B19 病毒的受体是成红血细胞表面的 P 抗原。另一方面,犬和猫细小病毒与宿主特定的肠道细胞表面的转铁蛋白受体结合,通过胞吞作用进入细胞,然后通过细胞质进入到细胞核。每个病毒特定地感染宿主犬或猫与病毒表面结构和每种转铁蛋白受体不同的氨基酸序列有关(Truyen,Agbandje 和 Parrish 1994; Truyen et al. 1996; Parker 和 Parrish 2000; Palermo,Hueffer 和 Parrish 2003)。脊椎和无脊椎动物细小病毒是基因组 DNA 为 4～6kb 大小的小 DNA 病毒。令人惊讶的是,在如此小的基因组结构上的差异能决定宿主和组织的特异性。研究家蚕和 *Bm*DNV 之间的这些现象将提供关于发生在细小病毒(包括浓核病毒)上的感染机制的新信息。即使

*Bm*DNV-2 是一个不属于细小病毒的新病毒,它也能用于研究家蚕针对新形式主要昆虫病原体的一种完全抗性基因行为。

通过将病毒抗性基因引入到有病毒易感等位基因的品系,比如通过替代抗性等位基因或破坏它,将可能为蚕丝产业提供更为精确的标记辅助筛选或育种,其中这个基因本身可直接用作标记。通过基因分型选择含有隐性基因的抗性型后代比通过病毒感染进行选择要迅速和精确得多,因为额外的侧交和病毒感染测试在从相同易感表型的野生型纯合子区别隐性突变杂合子的每一步都是必需的(参看近等位基因系或同基因型品系的产生;Abe et al. 1995,2000;Chen et al. 2007)。此外,如果通过转入易感基因,如$+^{nsd\text{-}2}$,建立一个高度易感的细胞系,有望使目前由于没有被这个病毒感染的易感培养细胞系而远远滞后的病毒分子和细胞生物学研究得以迅速发展。

我们运用经典的定位克隆来分离 *nsd*-2 的方法已被证明是极其有效的。在家蚕中,至少另外 3 个以上的 *Bm*DNV 形式的抗性基因已经被研究,它们是与 *Bm*DNV-1 相对应的显性的 *Nid*-1 和隐性的 *nsd*-1 突变体,以及与 *Bm*DNV-Z/3 相对应的隐性 *nsd*-Z 突变体。这些基因和 *nsd*-2 之间的关系是令人感兴趣的,它们可能会为养蚕业提供新的保护工具。尽管通过定位克隆对显性突变分离的前提是必须在大量 BC_1 或 F_2 种群中能明确地区分靶标特征,这对这些病毒抗性基因是可行的。此外,伴随家蚕基因组数据库的日渐改善,正迅速地提高定位克隆效率(Xia et al. 2004; International Silkworm Genome Consortium 2008;Yamamoto et al. 2008; 参看第二章家蚕基因组细节),不再需要像图 18.1 所示的劳力密集型的染色体步移法。当前,正在分离 *Bm*DNV 抗性基因和家蚕另外的许多突变基因。

5. DNV 抗性基因起源的进化注释

我们能通过检验在日本找到的三个 DNV 抗性基因追踪家蚕的进化吗? 野桑蚕 *B. mandarina* 与家蚕有相同的祖先,是家蚕最近的亲缘生物(Arunkumar, Metta 和 Nagaraju 2006),其载有$+^{nsd\text{-}2}$(易感)基因(Ito et al. 2008)。通常,印度和中国的地方种对 *Bm*DNV-2 是易感的,而现代的日本种和改善种则趋向于抗性(Furuta 1994,1995)。这意味着家蚕是在中国驯化的,然后被带到日本,是否在品种改良的过程中出现了 *nsd*-2 的缺失突变? 事实上,大部分日本祖先品种如 Seihaku, Dainyorai, 或 Matamukashi 仍然对 *Bm*DNV-2 是易感的, *Bm*DNV-2 有可能是家蚕和野桑蚕 *B. mandarina* 共同祖先的一种致病性病毒。因此,了解存在于中国本地变异种中的几种抗性品系是否含有相同的 *nsd*-2 突变是非常有趣的。反之,品种表现出的对 *Bm*DNV-1 的抗性/易感同 *Bm*DNV-2 有着镜像关系(Furuta 1994, 1995),如日本的野桑蚕 *B. mandarina* 以隐性遗传的方式抗 *Bm*DNV-1(K. Kadono-Okuda,未发表数据)。因此在中国野桑蚕 *B. mandarina* 中调查对 *Bm*DNV-1 的抗性是非常有趣的。如果它也有抗性,我们能推测 *Bm*DNV-1 不是家蚕天然的病毒,而是在日本品种改良的过程中出现了易感的突变体。

Watanabe, Kurihara 和 Wang(1988)报道,在桑园中几乎一半的桑树螟蛾(*Glyphodes pyloalis*)幼虫潜在地含有与针对 *Bm*DNV-1, *Bm*DNV-2 和感染蚕软腐病病毒三种抗体反应的抗原。此外,那些桑树螟蛾的匀浆液经口服能感染家蚕。如果它们与

*Bm*DNVs 相同，不明显地或亚临床地感染的桑树螟蛾也许是最早的宿主。与血清学诊断是检测病毒唯一方法的时代相比，我们已经能够解决这个问题。至于 *Nid*-1，到目前为止，3 个独立的研究组织发现，仅有 6 个品系对 *Bm*DNV-1 是显性抗性的（R. Eguchi 和 H. Abe，未发表数据）。因此，我们假设 *Nid*-1 是最近产生的突变。在当前已经建立了大部分的商业品种，这个基因很难通过育种扩展到其他的品种中。如果有这 6 个品种系谱的记录，就能确定突变在什么时候，在哪个品种中产生的。

参考文献

[1]Abe, H. , T. Harada, M. Kanehara, T. Shimada, F. Ohbayashi, and T. Oshiki. 1998. Genetic mapping of RAPD markers linked to the densonucleosis refractoriness gene, *nsd*-1, in the silkworm, *Bombyx mori*. *Genes Genet. Syst.* 73:237－42.

[2]Abe, H. , M. Kobayashi, and H. Watanabe. 1990. Mosaic infection with a densonucleosis virus in the midgut epithelium of the silkworm, *Bombyx mori*. *J. Invertebr. Pathol.* 55:112－17.

[3]Abe, H. , K. Kobayashi, T. Shimada, et al. 1993. Infection of a susceptible/nonsusceptible mosaic silkworm, *Bombyx mori*, with densonucleosis virus type-2 is not lethal. *J. Seric. Sci. Jpn.* 62:367－75.

[4]Abe, H. , T. Shimada, G. Tsuji, T. Yokoyama, T. Oshiki, and M. Kobayashi. 1995. Identification of random amplified polymorphic DNA linked to the densonucleosis virus type-1 susceptibility gene of the silkworm, *Bombyx mori*. *J. Seric. Sci. Jpn.* 64:262－64.

[5]Abe, H. , T. Sugasaki, M. Kanehara, et al. 2000. Identification and genetic mapping of RAPD markers linked to the densonucleosis refractoriness gene, *nsd*-2, in the silkworm, *Bombyx mori*. *Genes Genet. Syst.* 75:93－96.

[6]Abe, H. , H. Watanabe, and R. Eguchi. 1987. Genetical relationship between nonsusceptibilities of the silkworm, *Bombyx mori*, to two densonucleosis viruses. *J. Seric. Sci. Jpn.* 56:443－44.

[7]Arunkumar, K. P. , M. Metta, and J. Nagaraju. 2006. Molecular phylogeny of silkmoths reveals the origin of domesticated silkmoth, *Bombyx mori* from Chinese *Bombyx mandarina* and paternal inheritance of Antheraea proylei mitochondrial DNA. *Mol. Phylogenet. Evol.* 40:419－27.

[8]Bando, H. , T. Hayakawa, S. Asano, K. Sahara, M. Nakagaki, and T. Iizuka. 1995. Analysis of the genetic information of a DNA segment of a new virus from silkworm. *Arch Virol.* 140:1147－55.

[9]Bergoin, M. , and P. Tijssen. 2000. Molecular biology of Densovirinae. *Contrib. Microbiol.* 4:12－32.

[10]Bonning, B. C. 2005. Baculoviruses: Biology, biochemistry, molecular biology. In *Comprehensive molecular insect science*, Vol. 6: *Control*, ed. S. Gill, K. Iatrou, and L. Gilbert, 233－70. Oxford: Elsevier.

[11]Brown, K. E. , S. M. Anderson, and N. S. Young. 1993. Erythrocyte P antigen: Cellular receptor for B19 parvovirus. *Science* 262:114－17.

[12]Carlson, J. , E. Suchman, and L. Buchatsky. 2006. Densoviruses for control and genetic manipulation of mosquitoes. *Adv. Virus. Res.* 68:361－92.

[13]Castagna, M. , C. Shayakul, D. Trotti, V. F. Sacchi, W. R. Harvey, and M. A. Hediger. 1998. Cloning and characterization of a potassium-coupled amino acid transporter. *Proc. Natl. Acad. Sci. U. S. A.* 95:5395－400.

[14]Chen, K. －P. , H. －Q. Chen, X. －D. Tang, Q. Yao, L. －L. Wang, and X. Han. 2007. *bmpkci* is highly expressed in a resistant strain of silkworm (Lepidoptera: Bombycidae): Implication of its role in re-

sistance to *Bm*DNV—Z. *Eur. J. Entomol.* 104:369—76.

[15]Chen, S., L. Cheng, Q. Zhang, et al. 2004. Genetic, biochemical, and structural characterization of a new densovirus isolated from a chronically infected *Aedes albopictus* C6/36 cell line. Virology 318:123—33.

[16]Eguchi, R., Y. Furuta, and O. Ninaki. 1986. Dominant nonsusceptibility to densonucleosis virus in the silkworm, *Bombyx mori*. *J. Seric. Sci. Jpn.* 55:177—78.

[17]Eguchi, R., W. Hara, A. Shimazaki, et al. 1998. Breeding of the silkworm race "Taisei" non-susceptible to a densonucleosis virus type 1. *J. Seric. Sci. Jpn.* 67:361—66.

[18]Eguchi, R., K. Nagayasu, O. Ninagi, and W. Hara. 2007. Genetic analysis on the dominant non-susceptibility to densonucleosis virus type 1 in the silkworm, *Bombyx mori*. *Sanshi-Konchu Biotec.* 76:159—63.

[19]Eguchi, R., O. Ninaki, and W. Hara. 1991. Genetical analysis on the nonsusceptibility to densonucleosis virus in the silkworm, *Bombyx mori*. *J. Seric. Sci. Jpn.* 60:384—89.

[20]Fauquet, C. M., M. A. Mayo, J. Maniloff, U. Desselberger, and L. A. Ball, eds. 2005. *Virus taxonomy. VIIIth report of the international committee on taxonomy of viruses*. London: Elsevier/Academic Press.

[21]Feldman, D. H., W. R. Harvey, and B. R. Stevens. 2000. A novel electrogenic amino acid transporter is activated by K^+ or Na^+, is alkaline pH-dependent, and is Cl^-- independent. *J. Biol. Chem.* 275: 24518—26.

[22]Furuta, Y. 1983. Multiplication of the infectious flacherie virus and the densonucleosis virus in cultured silkworm embryos. *J. Seric. Sci. Jpn.* 52:245—46.

[23]Furuta, Y. 1994. Susceptibility of Indian races of the silkworm, *Bombyx mori*, to the nuclear polyhedrosis virus and densonucleosis viruses. *Acta Seric. Entomol.* 8:29—36.

[24]Furuta, Y. 1995. Susceptibility of the races of the silkworm, *Bombyx mori*, preserved in NISES to the nuclear polyhedrosis virus and densonucleosis viruses. *Bull. Natl. Inst. Seric. Entomol. Sci.* 15:119—45.

[25]Han, X., Q. Yao, L. Gao, Y. J. Wang, F. Bao, and K. P. Chen. 2007. Replication of *Bombyx mori* Densonucleosis Virus (Zhenjiang isolate) in different silkworm strains. *Sheng Wu Gong Cheng Xue Bao.* 23:145—51.

[26]Hara, W., Y. An, R. Eguchi, et al. 2008. Mapping of a novel virus resistant gene, *Nid*-1, in the silkworm, *Bombyx mori*, based on the restriction fragment length polymorphism (RFLP). *J. Insect Biotech. Sericol.* 77:59—66.

[27]Hayakawa, T., K. Kojima, K. Nonaka, et al. 2000. Analysis of proteins encoded in the bipartite genome of a new type of parvo-like virus isolated from silkworm—structural protein with DNA polymerase motif. *Virus Res.* 66:101—08.

[28]International Silkworm Genome Consortium. 2008. The genome of a lepidopteran model insect, the silkworm *Bombyx mori*. *Insect Biochem. Mol. Biol.* 38:1036—45.

[29]Ito, K., K. Kidokoro, H. Sezutsu, et al. 2008. Deletion of a gene encoding an amino acid transporter in the midgut membrane causes resistance to a *Bombyx* parvo-like virus. *Proc. Natl. Acad. Sci. U. S. A.* 105:7523—27.

[30]Jousset, F. X., C. Barreau, Y. Boublik, and M. Cornet. 1993. A parvo-like virus persistently infecting a C6/36 clone of *Aedes albopictus* mosquito cell line and pathogenic for *Aedes aegypti* larvae. *Virus*

Res. 29:99—114.

[31]Kawase,S. ,and E. Kurstak. 1991. Parvoviridae of invertebrates: Densonucleosis viruses. In *Viruses of invertebrates*,ed. E. Kurstak,315—43. New York: Marcel Dekker.

[32]Koike,Y. ,K. Mita,M. G. Suzuki,et al. 2003. Genomic sequence of a 320—kb segment of the Z chromosome of *Bombyx mori* containing a *kettin* ortholog. *Mol. Genet. Genomics* 269:137—49.

[33]Kurstak,E. ,and J. —R. Cote. 1969. Proposition de classiication du virus de la densonucleose (VDV) basee sur l'etude de la structure moleculaire et des proprietes physicochimiques. *C. R. Acad. Sci. Paris*. 268:616—19.

[34]Li,M. ,Q. Guo,C. Hou,et al. 2006. Linkage and mapping analyses of the densonucleosis non-susceptible gene nsd-Z in the silkworm *Bombyx mori* using SSR markers. *Genome* 49:397—402.

[35]Li,M. ,C. Hou,Y. Zhao,A. Xu,X. Guo,and Y. Huang. 2007. Detection of homozygosity in near isogenic lines of non-susceptible to Zhenjiang strain of densonucleosis virus in silkworm. *Afr. J. Biotechnol*. 6:1629—33.

[36]Meynadier,G. ,C. Vago,G. Plantevin,and P. Atger. 1964. Virose de un type inhabituel chez le Lépidoptère. "*Galleria. mellonella.*" *Rev. Zool. Agric. Appl*. 63:207—08.

[37]Miao,X. X. ,S. J. Xub,M. H. Li,et al. 2005. Simple sequence repeat-based consensus linkage map of *Bombyx mori*. *Proc. Natl. Acad. Sci. U. S. A.* 102:16303—08.

[38]Nakagaki,M. ,T. Morinaga,C. Zhow,Z. Kajiura,and R. Takei. 1999. Increasing curves of two virus DNAs in the midgut epithelium of silkworm infected with *Bombyx mori* densonucleosis virus type 2 (*Bm*DNV—2). *J. Seric. Sci. Jpn*. 68:173—80.

[39]Nguu,E. K. ,K. Kadono-Okuda,K. Mase,E. Kosegawa,and W. Hara. 2005. Molecular linkage map for the silkworm, *Bombyx mori*,based on restriction fragment length polymorphism of cDNA clones. *J. Insect Biotechnol. Seriocol*. 74:5—13.

[40]Ogoyi,D. O. ,K. Kadono-Okuda,R. Eguchi,et al. 2003. Linkage and mapping analysis of a non-susceptibility gene to densovirus (*nsd*-2) in the silkworm, *Bombyx mori*. *Insect Mol. Biol*. 12:117—24.

[41]Ohnishi,A. ,J. J. Hull,and S. Matsumoto. 2006. Targeted disruption of genes in the *Bombyx mori* sex pheromone biosynthetic pathway. *Proc. Natl. Acad. Sci. U. S. A.* 103:4398—403.

[42]Palermo,L. M. ,K. Hueffer,and C. R. Parrish. 2003. Residues in the apical domain of the feline and canine transferrin receptors control host-specific binding and cell infection of canine and feline parvoviruses. *J. Virol*. 77:8915—23.

[43]Parker,J. S. ,and C. R. Parrish. 2000 Cellular uptake and infection by canine parvovirus involves rapid dynamin-regulated clathrin-mediated endocytosis,followed by slower intracellular traficking. *J. Virol*. 74:1919—30.

[44]Qin,J. ,and W. Z. Yi. 1996. Genetic linkage analysis of *nsd-Z*,the nonsusceptibility gene of *Bombyx mori* to the Zhenjiang (China) strain densonucleosis virus. *Sericologia* 36:241—44.

[45]Seki,H. ,and Y. Iwashita. 1983. Histopathological features and pathogenicity of a densonucleosis virus of the silkworm, *Bombyx mori*,isolated from sericultural farms in Yamanashi prefecture. *J. Seric. Sci. Jpn*. 52:400—05.

[46]Shimizu,T. 1975. Pathogenicity of an infection flacherie virus of the silkworm *Bombyx mori*,obtained from sericultural farms in the suburbs of Ina city. *J. Seric. Sci. Japan*. 44:45—48.

[47]Tabunoki,H. ,S. Higurashi,O. Ninagi,et al. 2004. A carotenoid-binding protein (CBP) plays a crucial role in cocoon pigmentation of silkworm (*Bombyx mori*) larvae. *FEBS Lett*. 567:175—78.

[48]Tamura T. ,C. Thibert,C. Royer,et al. 2000. Germline transformation of the silkworm *Bombyx mori* L. *using a piggy*Bac transposon-derived vector. *Nat. Biotechnol.* 18:81—84.

[49]Tijssen,P. ,and M. Bergoin. 1995. Densonucleosis viruses constitute an increasingly diversified subfamily among the parvoviruses. *Semin. Virol*. 6:347—55.

[50]Truyen,U. ,M. Agbandje,and C. R. Parrish. 1994. Characterization of the feline host range and a specific epitope of feline panleukopenia virus. *Virology* 200:494—503.

[51]Truyen,U. ,J. F. Evermann,E. Vieler,and C. R. Parrish. 1996. Evolution of canine parvovirus involved loss and gain of feline host range. *Virology* 215:186—89.

[52]Wang,Y. J. ,Q. Yao,K. P. Chen,Y. Wang,J. Lu,and X. Han. 2007. Characterization of the genome structure of *Bombyx mori* densovirus (China isolate). *Virus Genes* 35:103—08.

[53]Watanabe,H. ,Y. Kurihara,and Y. -X. Wang. 1988. Mulberry pyralid,*Glyphodes pyloalis*: Habitual host of nonoccluded viruses pathogenic to the silkworm, *Bombyx mori*. *J. Invertebr. Pathol.* 52:401—08.

[54]Watanabe,H. ,S. Maeda,M. Matsui,and T. Shimizu. 1976. Histopathology of the midgut epithelium of the silkworm, *Bombyx mori*,infected with a newly-isolated virus from the flacherie-diseased larvae. *J. Seric. Sci. Jpn.* 45:29—34.

[55]Xia,Q. ,Z. Zhou,C. Lu,et al. 2004. A draft sequence for the genome of the domesticated silkworm (*Bombyx mori*). *Science* 306:1937—40.

[56]Yamamoto,K. ,J. Narukawa,K. Kadono-Okuda,et al. 2006. Construction of a single nucleotide polymorphism linkage map for the silkworm, *Bombyx mori*,based on bacterial artificial chromosome end sequences. *Genetics* 173:151—61.

[57]Yamamoto,K. ,J. Nohata,K. Kadono-Okuda,et al. 2008. A BAC-based integrated linkage map of the silkworm *Bombyx mori*. *Genome Biol*. 9:R21. 1—14.

[58]Yasukochi,Y. ,L. A. Ashakumary,K. Baba,A. Yoshido,and K. Sahara. 2006. A second-generation integrated map of the silkworm reveals synteny and conserved gene order between lepidopteran insects. *Genetics* 173:1319—28. Erratum 2008 178:1837.

后记

美国罗德岛大学的玛丽安·戈德史密斯教授是中国蚕学界的老朋友，也与我是忘年之交。1995年，她与威尔金斯共同编著了《鳞翅目昆虫分子模式系统》一书。该书侧重于鳞翅目昆虫与黑腹果蝇的比较，给读者展现了鳞翅目昆虫作为模式系统在分子生物学研究中做出的贡献。时隔十五年，随着分子生物学，特别是基因组科学的快速发展，该领域的研究成果源源不断地增长，又到了需要对最前沿的研究成果进行总结的时候。这次，她与弗兰蒂塞克·马莱克携手编著了本书。这本书第一次对家蚕基因组研究作了全面介绍，详细描述了蝴蝶翅模式以及眼斑进化和发育遗传学的最新研究进展，对鳞翅目昆虫的化学受体系统、宿主范围、杀虫剂抗性的分子机制作了详细的探讨，所著内容也涵盖了鳞翅目昆虫的免疫反应、昆虫毒素和病毒的相关知识，提出了现阶段害虫防治最新的策略。其翔实的研究内容和富于创见性的描述给我以深刻启迪，希望读者能从中获取自己感兴趣的知识。

作为译者，我首先要向恩师向仲怀院士致以敬意，没有他的指导和鼓励，此书是不可能出版的。

同研究所的查幸福、童晓玲、王菲老师以及甘玲、梁九波、徐云敏、付强、刘碧朗、亓希武和王长春参与了初译稿的形成。对他们的付出表示衷心的感谢。

对支持和帮助过这本书出版的陈力老师、冯丽春老师表示诚恳的谢意。

感谢西南师范大学出版社社长周安平、总编辑李远毅对本书的关心和支持。同时感谢编辑杜珍辉、任志林、伯古娟、卢旭等所付出的辛勤劳动。

本书的翻译出版历时一年，经过多次的校对修改，不足与欠妥之处尚存，请读者指正。

何宁佳

2011年初春于西南大学

1.1 含有模式系统的总科的代表。A：蚕蛾总科，*Anthela oressarcha* (A.Zwick)；B：蚕蛾总，*Antheraea larissa* (A. Kawahara)；C：夜蛾总科：粉纹夜蛾（*Trichoplusia ni*）(M. Dreiling)；夜蛾总科，朱砂夜蛾（*Tyria jacobaeae*）(D. Dictchburn)；E：凤蝶总科，热带蝴蝶（*Bicyclus* *nana*）(A. Monteiro and W. Piel)； F：凤蝶总科，邮差蝴蝶（*Heliconius erato*）(K. Garwood)；螟蛾总科，玉米螟（*Ostrinia nubilalis*）(S. Nanz)；H：卷蛾总科，苹果蠹蛾（*Cydia* *onella*）(N. Schneider)。

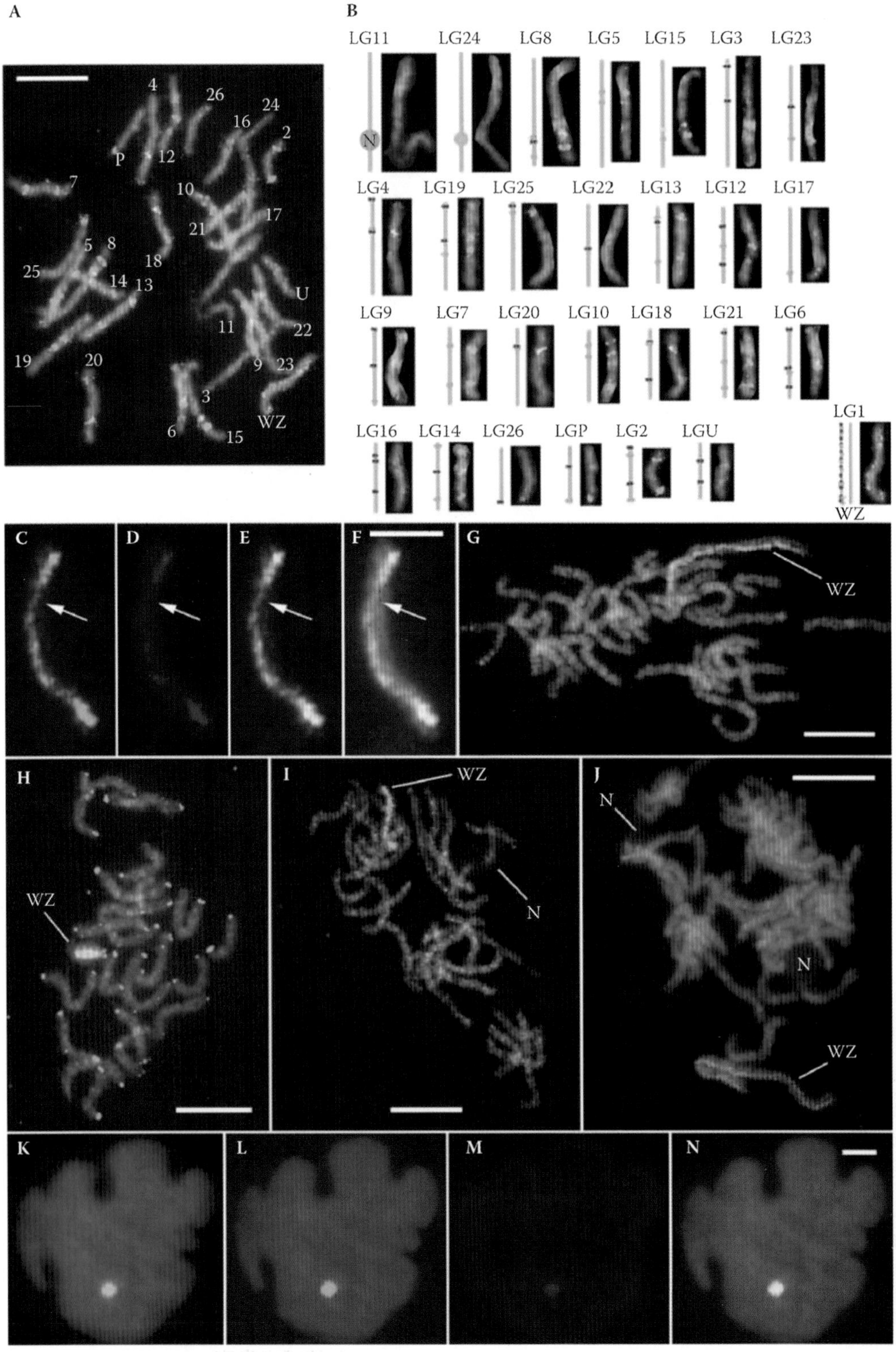

图3.1 FISH杂交图片。来自鳞翅目昆虫雌性的粗线期（A～J）和多倍体化的分裂间期的核（K～N）的FISH杂交图片。DAPI（蓝色）复染色的染色体和多倍体核。

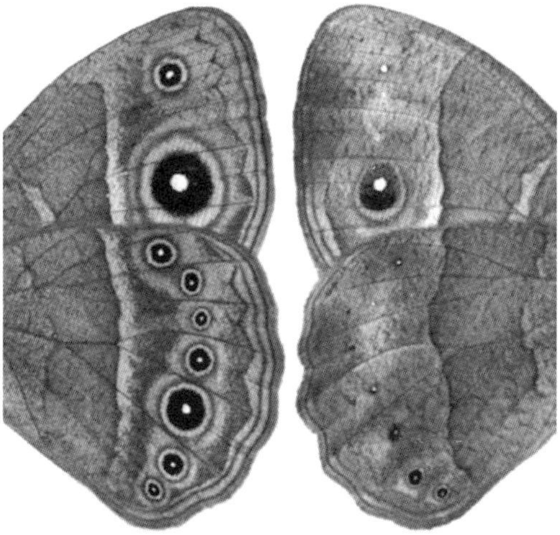

图5.1 热带蝴蝶（*Bicyclus anynana*） 翅模式的表型。在野外发生的两种季节型斑纹模式亦可在实验室内模拟，饲育条件为幼虫期27 ℃或20 ℃，成虫将分别呈现类似于自然界的潮湿季节（左）和干燥季节（右）的两种表型。图片展示的是前后翅的腹面，其背面由于饲育温度并未显示出差异。文章中将详细讨论这种交替选择表型的适应性意义和潜在的生理学基础。

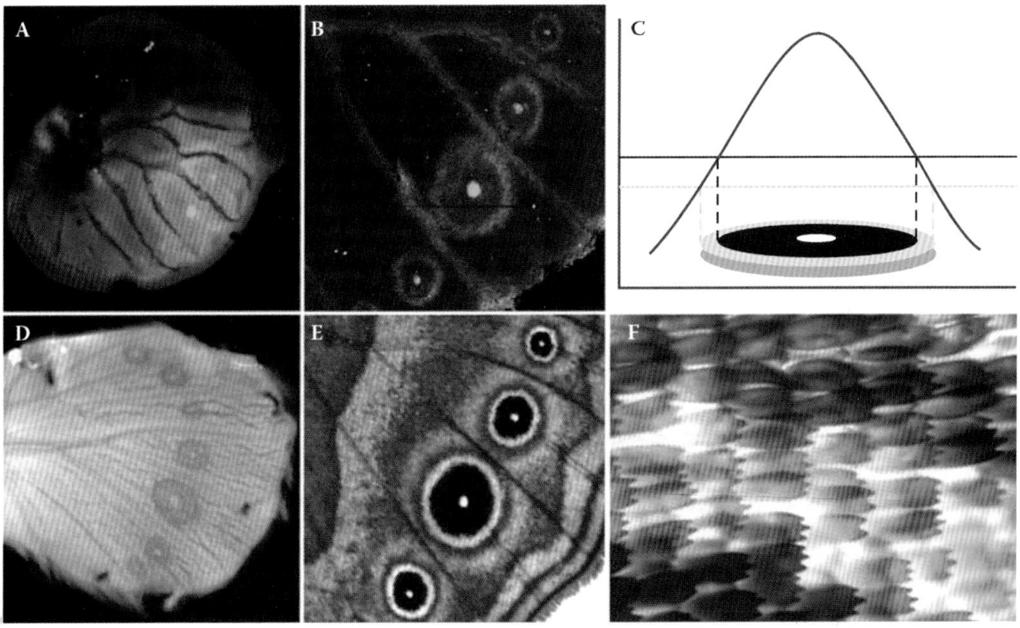

图5.2 热带蝴蝶（*Bicyclus anynana*）的眼斑发育。

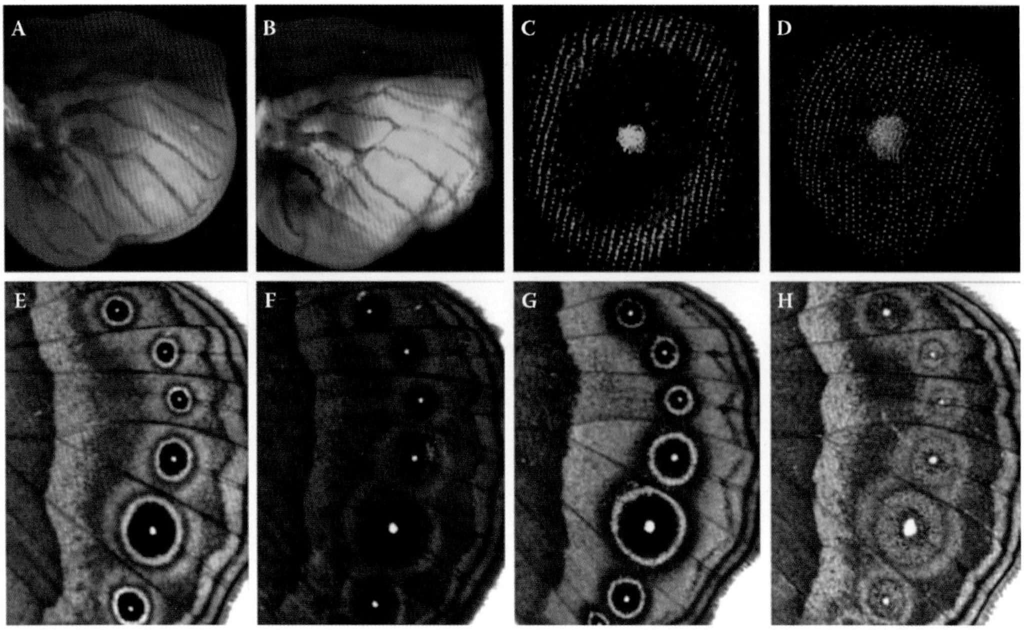

图5.4 热带蝴蝶（*Bicyclus anynana*）翅模式的突变表型。

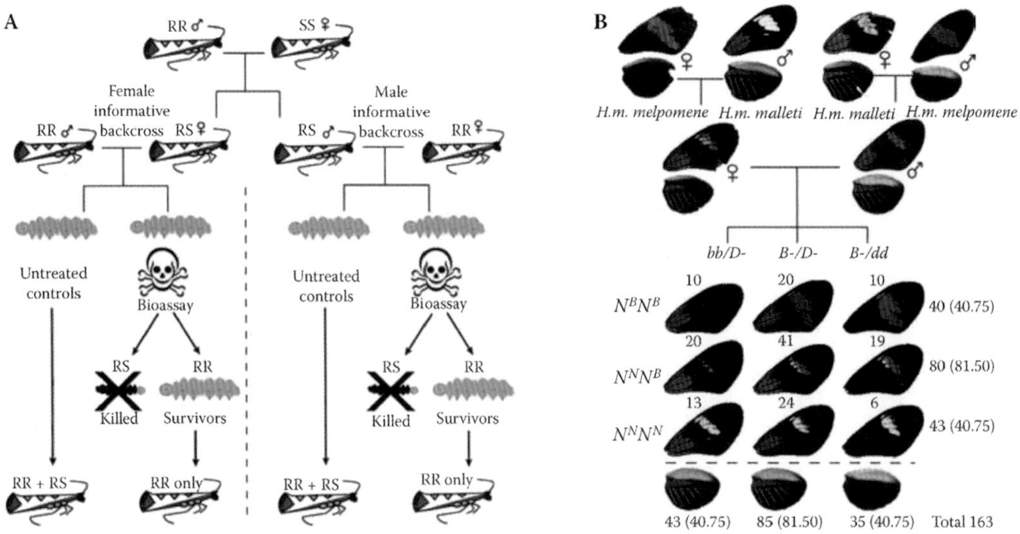

图6.1 遗传定位的杂交组合。

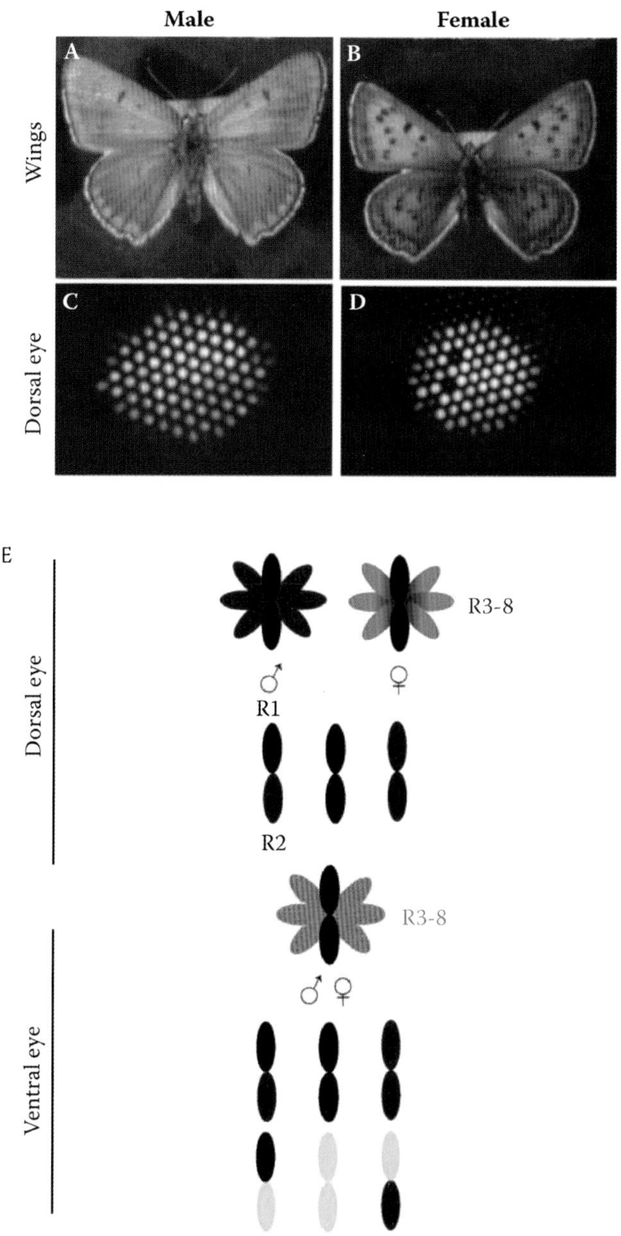

7.5 灰蝶*Lycaena rubidus*中翅膀颜色模式，点睛和视蛋白表达模式的性别差异。(A)在雄性低翅和后翅边缘反射UV的鳞毛(闪光紫色)。(B)在雌性翅膀上不反射UV的鳞毛。雄性(C)和雌性)背侧眼上的点睛展现出很强的着色性别二态性。(E)概括了*L. rubidus*眼中视蛋白表达模式的示图。深蓝色代表*BRh1*视蛋白mRNA的表达。橙色代表*LWRh*视蛋白mRNA的表达。深蓝色和橙代表*BRh1*和*LWRh*视蛋白mRNAs的共表达。黑色代表*UVRh*视蛋白mRNA的表达。淡蓝色代表*h2*视蛋白mRNA的表达。修改于Sison-Mangus等(2006)。

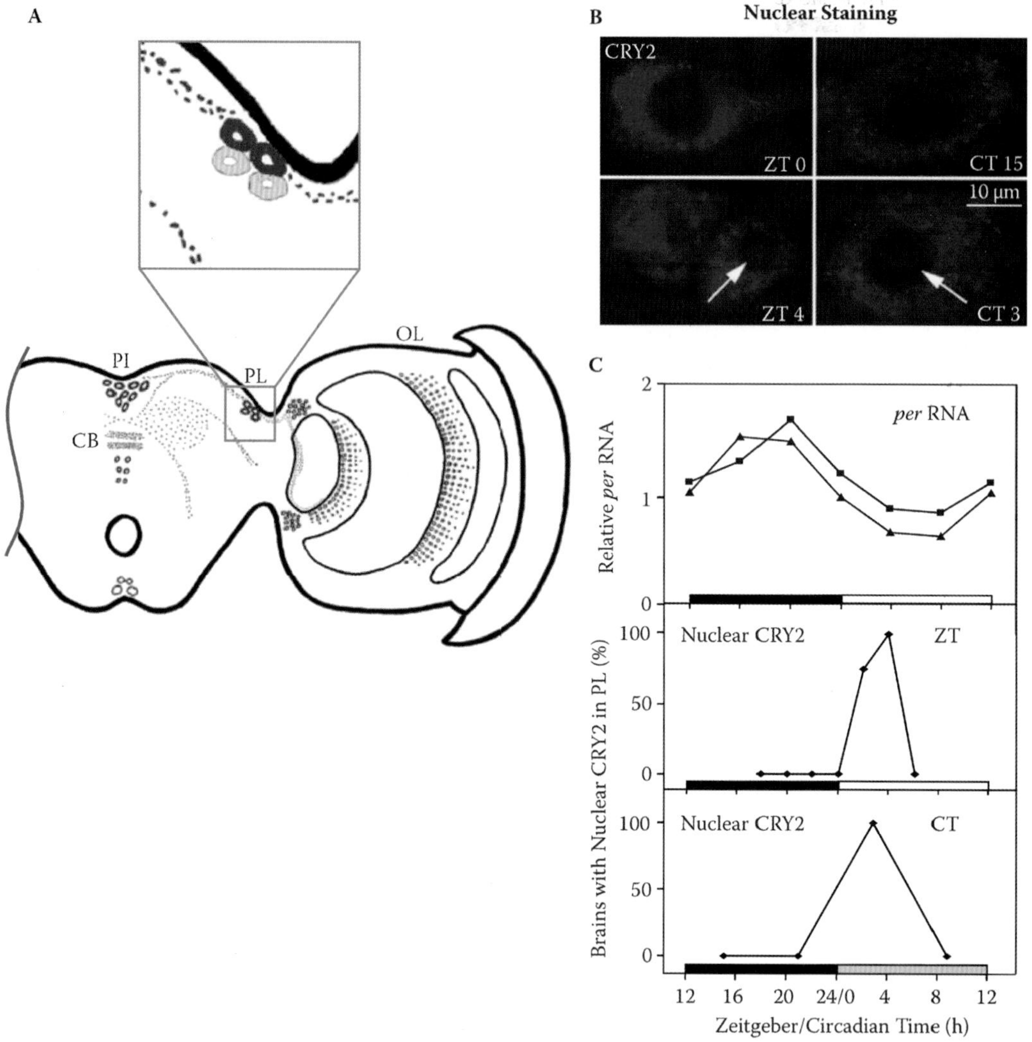

图8.3 帝王蝴蝶脑中生物钟细胞的细胞定位。(A)通过帝王蝴蝶特异的抗体，前脑切片的局部示意图展示了CRY2-阳性的细胞和神经投影的图像。(OL)光叶；(PL)外侧部；(PI)脑间部；(CB)中心体。上面放大的PL区域显示了对4个生物钟蛋白显阳性的细胞；两个红色细胞共表达PER、TIM, CRY1和CRY2，而两个粉红色的细胞共表达TIM和CRY2。修改自Zhu等(2008)。(B)PL细胞中对CRY2的细胞核染色。左上，同步时间[ZT]0；左下，同步时间4；右上，昼夜节律时间[CT]15；右下，昼夜节律时间3。在ZT0或CT15时细胞核中的CRY2染色没有被发现，但是在ZT4和CT3时发现在PL中的CRY2被染色(箭头所指)。来自Zhu等 (2008)。(C)脑中*per* RNA的表达水平与PL中CRY2细胞核染色的时间表达模式的比较。在一天中以4个小时为间隔，记录了组缺乏光受体的脑中*per* RNA的表达水平(上部)。在ZT7(中部)和CT(下部)时对细胞核中的CRY进行了半定量的免疫染色，图以被检测的脑的百分数绘制(每个时间点使用了4～5个脑)。来自Zhu等 (2008)。

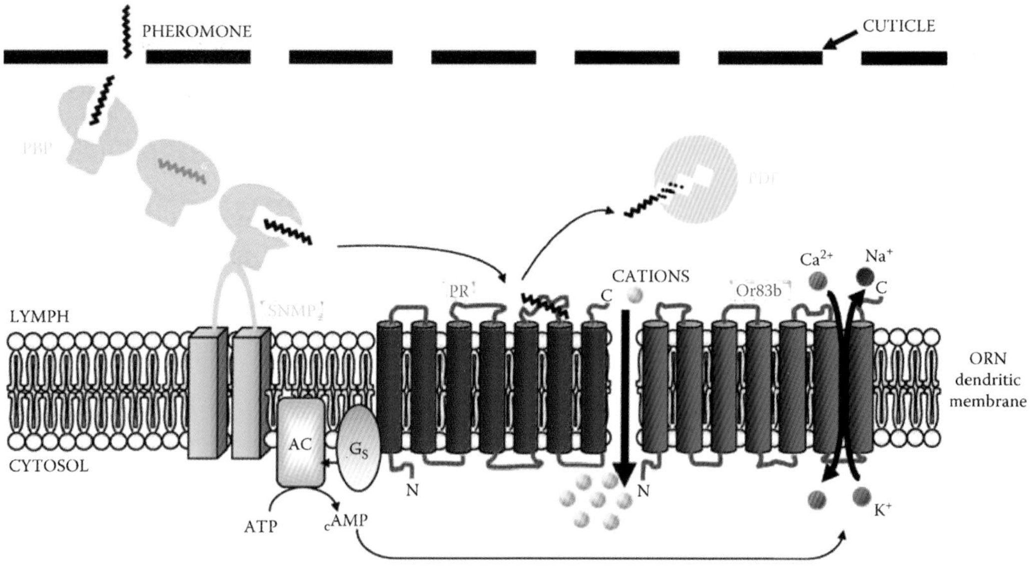

图10.3 蛾类信息素受体反应的示意图（修改自Rützler和Zwiebel，2005）。

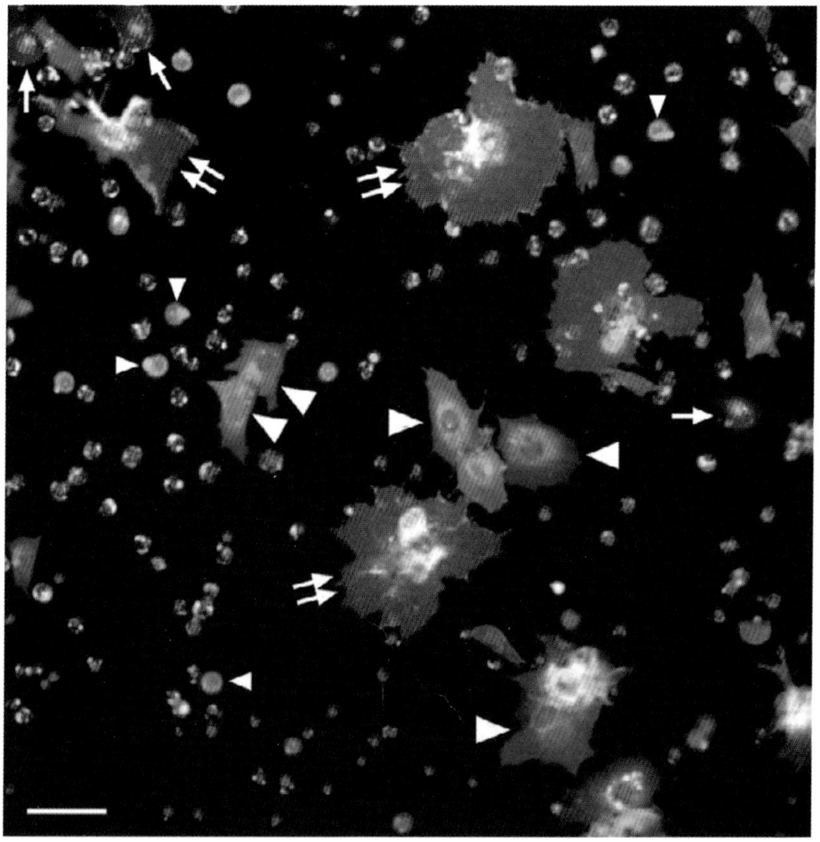

图14.2 在Grace's 培养基中培养60 min后的烟草天蛾血细胞。

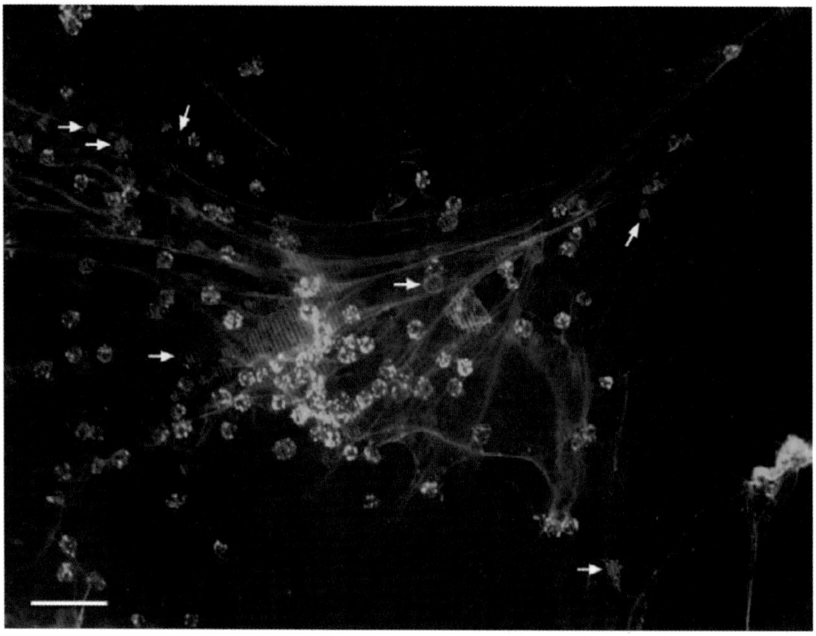

图14.4 在凝结的烟草天蛾血细胞中标记黏附分子。

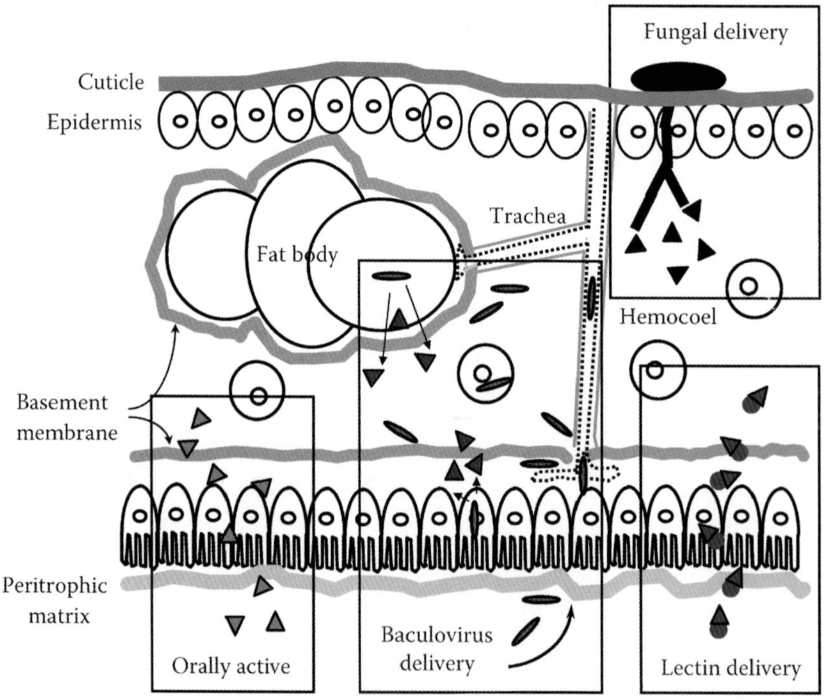

图16.1 体腔内毒素的传递与鳞翅目害虫防治。